Book 1

Plumbing

SECOND EDITION

LEVEL 3 APPRENTICESHIP (9189)
LEVEL 2 TECHNICAL CERTIFICATE (8202)
LEVEL 2 DIPLOMA (6035)
T LEVEL OCCUPATIONAL SPECIALISMS (8710)

Stephen Lane
Peter Tanner

The City & Guilds textbook

HODDER EDUCATION
AN HACHETTE UK COMPANY

Orders: please contact Hachette UK Distribution, Hely Hutchinson Centre, Milton Road, Didcot, Oxfordshire, OX11 7HH. Telephone: +44 (0)1235 827827. Email education@hachette.co.uk
Lines are open from 9 a.m. to 5 p.m., Monday to Friday. You can also order through our website: www.hoddereducation.co.uk

ISBN: 978 1 3983 6161 4

First published in 2019
This edition published in 2022 by
Hodder Education,
An Hachette UK Company
Carmelite House
50 Victoria Embankment
London EC4Y 0DZ

www.hoddereducation.co.uk

Impression number 10 9 8 7 6 5

Year 2026 2025 2024

Typeset by Integra Software Services Pvt. Ltd., Pondicherry, India

Printed and bound in Great Britain by Bell & Bain Ltd, Glasgow

A catalogue record for this title is available from the British Library.

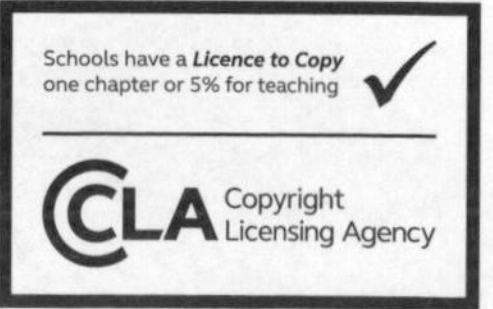

Contents

About your qualification v
Acknowledgements vii
Picture credits viii
How to use this book x

1 **Health and safety practices and systems** 1
Health, safety and welfare legislation and regulation 1
Recognising and responding to hazardous situations at work 16
Personal protection 26
Responding to accidents 33
Electrical safety in the workplace 40
Working safely with heat-producing equipment 47
Working safely with access equipment 54
Working safely in excavations and confined spaces 63

2 **Common processes and techniques** 69
Using hand and power tools 69
Types of pipework, bending and jointing techniques 80
Preparation techniques 103
Using pipe clips and pipe brackets 109
Pipework installation techniques 114

3 **Scientific principles** 127
Units of measurement used in the plumbing and heating industry: the SI system 127
The properties of materials 129
The relationship between energy, heat and power 144
The principles of force and pressure, and their application in the plumbing and heating industry 151
The mechanical principles in the plumbing and heating industry 155
The principles of electricity in the plumbing and heating industry 160

4 **Planning and supervision** 164
The role of the construction team within the plumbing and heating industry 164
Information sources in the building services industry 172
Communicating with others 177
The responsibilities of relevant people in the building services industry 181
Work programmes in the plumbing and heating industry 184
Risk assessments and method statements for the plumbing and heating industry 189

5 **Cold water systems** 201
Sources and properties of water 201
The types of water supply to dwellings 206
The water treatment process and distribution of water 208
Sources of information relating to cold water systems 212
The water service pipework to dwellings 214
Selecting cold water systems 217
The system layout features of cold water systems fed from private water supplies 223
The components used in boosted (pumped) cold water supply systems from private sources for single-occupancy dwellings 225
The Water Supply (Water Fittings) Regulations 1999 240
Backflow protection 240

Installing cold water systems and components 256
Replacing or repairing defective components: planned and unplanned maintenance 273
Decommissioning of systems 280

6 **Hot water systems** **285**
Sources of information relating to work on hot water systems 285
Hot water systems and components 294
System safety and efficiency 337
Prepare for the installation of systems and components 340
Install and test systems and components 341
Decommission systems and components 354
Replace defective components 355

7 **Central heating systems** **361**
Understand central heating systems and their layouts 361
Install central heating systems and components 424
Understand the decommissioning requirements of central heating systems and their components 431

8 **Rainwater systems** **436**
Understand layouts of gravity rainwater systems 436
Installation of gravity rainwater systems 446
The maintenance and service requirements of gravity rainwater systems 453
Decommission rainwater and gutter systems and components 456
Perform a soundness test, and commission rainwater and gutter systems and components 457

9 **Sanitation systems** **460**
Sanitary pipework and appliances used in dwellings 460
Install sanitary appliances and connecting pipework systems 498
Service and maintenance requirements for sanitary appliances and connecting pipework systems 516
The principles of grey water recycling 518

10 **Domestic fuel systems** **524**
Identify the types of fuels used in appliances 524
Factors that affect the selection of fuels 530
Sources of information for fuel supply installation 531
Regulatory bodies that govern the installation of fuel systems 531
Storage requirements for fuels 532
Considerations that could affect the storage requirements of fuels 536

11 **Working with electricity** **542**
Electrical principles 542
Conductors and insulators 544
Resistances in series and parallel 546
Protection against electric shock 550
Electrical supply systems 554
Protective devices 556
Working on electrical systems 558
Installing wiring systems 561

Glossary 569
Index 574
Answers can be found online at: hoddereducation.co.uk/construction

About your qualification

INTRODUCTION TO THE PLUMBING QUALIFICATIONS

You are completing one of the following qualifications:

- Level 2 Technical Certificate in Plumbing (8202-25)
- Level 2 Diploma in Plumbing Studies (6035-02)
- Level 3 Diploma in Plumbing and Domestic Heating (9189)
- T Level Technical Qualification in Building Services Engineering for Construction (8710).

The Level 2 Technical Certificate and Level 2 Diploma are for learners who are interested in developing the specific technical and professional skills that can support development towards becoming a plumber, or progression to Level 3 qualifications.

The Level 3 Diploma is the on-programme qualification for the Plumbing and Heating Technician Apprenticeship and is designed to provide the apprentice with the opportunity to develop the knowledge, skills and core behaviours that are expected of a competent Plumbing and Domestic Heating Technician operating in a number of regulated areas.

T Levels are new Level 3 vocational qualifications available following the completion of GCSEs. They are the same size as three A Levels and you will sit them across two years. They offer a mixture of classroom, workshop and on the job experience through industrial placements. The Plumbing Engineering (356) and Heating Engineering (355) occupational specialisms will offer knowledge and experience needed to open the door to skilled employment or further study.

HOW TO BECOME A PLUMBING AND HEATING TECHNICIAN

To become a fully recognised plumber, you must complete the following:

- Plumbing and Heating Technician Apprenticeship (9189).

The 8202 Technical Certificate and 6035 Level 2 Diploma provide the knowledge and practical skills to prepare you for an apprenticeship.

The apprenticeship and 9189 Level 3 Diploma will give you an understanding of suitable on-site skills and further knowledge required to work in the plumbing industry. Once qualified, there are many specialist qualifications available, such as environmental technology systems and designing and planning complex water systems.

How to achieve your qualification

The requirements for successfully obtaining your qualification depend on which programme you are enrolled on.

6035

The 6035 diploma is assessed by a range of multiple choice exams, assignments and practical tests. You will be assessed, by one of these methods, at the end of each unit.

For details on which assessments will follow which units, you should consult the City & Guilds qualification handbook. For details on when you will complete your assessments, consult your tutor.

8202

Level 2 is assessed using one multiple choice examination and one practical synoptic assignment.

For the synoptic assignment, a typical brief might be to install a cold water supply and hot water distribution pipework connected to all sanitary appliances. You will need to draw on skills and understanding developed across the qualification content in order to consider the specific requirements of the particular system and related plumbing principles, and carry out the brief. This includes the ability to plan tasks, such as plant, materials and equipment for an installation, and apply the appropriate practical and hand skills to carry them out using appropriate tools and equipment.

You will also demonstrate that you are following health and safety regulations at all times by drawing upon your knowledge of legislation and regulations.

The exam draws from across the content of the qualification, using multiple choice questions to:

- confirm breadth of knowledge and understanding
- test applied knowledge and understanding – giving the opportunity to demonstrate higher-level integrated understanding through application, analysis and evaluation.

9189

Level 3 is assessed using multiple choice tests and practical assignments. These will happen at the end of each phase of learning, with there being four phases in total. Learners will also be expected to keep a work log for the duration of the programme.

The apprenticeship is assessed separately to the on-programme qualification and is assessed by an end-point assessment (EPA). In order to progress through the end-test gateway to end-point assessment, you must complete the following:

- Level 3 Diploma in Plumbing and Domestic Heating qualification (9189)
- Level 2 Maths
- Level 2 English.

The graded EPA will be comprised of the following assessment methods:

- multiple choice test
- design project
- practical installation test
- practical application test
- professional discussion.

T Level (8710)

This Level 3 course, which runs alongside the apprenticeship programme, offers the opportunity for learners to gain essential skills that will enable them to enter employment within the plumbing and heating sector.

The course is a two-year programme. All learners studying a Building Services Engineering for Construction T Level will complete the core component (350), which introduces the foundational industry principles. This component is assessed by two written exams and an employer-set project. This core component is covered in another Hodder Education textbook: *Building Services Engineering for Construction T Level: Core*.

You will also choose one or two occupational specialisms. These include:

- 355 Heating engineering
- 356 Plumbing engineering

Although these specialisms will involve practical work, which you will cover with your tutor in the workshop, and will be assessed by observation of practical tasks, the key underpinning plumbing and heating content needed for these specialisms is covered across this book and *The City & Guilds Textbook: Plumbing Book 2* (also Hodder Education).

Acknowledgements

Michael Maskrey was the author of the previous edition of this book and we are indebted to him for his work and expertise.

This book draws on several earlier books that were published by City & Guilds, and we acknowledge and thank the writers of those books:

- Michael Maskrey
- Neville Atkinson
- Andrew Hay-Ellis
- Trevor Pickard
- Eamon Wilson.

We would also like to thank everyone who has contributed to City & Guilds photoshoots. In particular, thanks to: Jules Selmes and Adam Giles; Martin Biron and the staff at the College of North West London and the following models: Vivian Chioma, Jennifer Close, Peko Gayle-Reveault, Adam Giles, Michael Maskrey, Nahom Sirane, Zhaojie Yu; Michael Maskrey and the staff at Stockport College and the following models: Michael Maskrey, Jordan Taylor; Jocelynne Rowan, Steve Owen and Mick Gibbons/Baxi Training Centre; Jamie Purser, Graham Fleming, John Pierce and Sabir Ahmed/Hackney Community College; Rob Wellman/National Skills Academy; models Anup Chudasama, Michaela Opara and Sami Simela.

Picture credits

Every effort has been made to trace and acknowledge ownership of copyright. The publishers will be glad to make suitable arrangements with any copyright holders whom it has not been possible to contact.

Fig.1.1 © markus_marb/stock.adobe.com; Fig.1.2 © auremar – Fotolia; Fig.1.3 courtesy of Facelift Access Hire; Fig.1.4 © Алина Бузунова/stock.adobe.com; Fig. 1.5 © Health and Safety Executive; Fig.1.6 © Lucaz80/stock.adobe.com; Fig.1.7 Michael Maskrey; Table 1.3 1st © ambassador806 – Fotolia, 2nd © nazar12/stock.adobe.com, 3rd © Ricochet64/stock.adobe.com, 4th/5th © Distraction Arts/stock.adobe.com; Fig.1.8 © markobe/stock.adobe.com; Fig.1.9 © jusep/stock.adobe.com; Fig.1.10 Image & lead work by Paul Dooley, Plannet Plumbing Services Ltd; Fig.1.11 City & Guilds; Fig.1.12 © Health and Safety Executive; Fig.1.13 © Andrei Rybachuk/stock.adobe.com; Fig.1.14 courtesy of Snickers Workwear; Figs.1.15 & 1.16 © JSP Ltd; Figs.1.17 & 1.18 © Jack Sealey Ltd; Fig.1.19 © JSP Ltd; Figs.1.20 & 1.21 City & Guilds; Fig.1.22 used with permissions from Machine Mart; Fig.1.23 © SPLAV/stock.adobe.com; Fig.1.24 City & Guilds; Fig.1.25 © Alex White/stock.adobe.com; Figs.1.26–1.28 City & Guilds; Fig.1.30 courtesy of Martindale Electric; Fig.1.31 © Reece Safety Products Ltd; Fig.1.33 courtesy Lincoln Electric, Inc. Unauthorized use not permitted; Fig.1.34 © Calor Gas Ltd; Fig.1.35 © Monument Tools Ltd; Fig.1.37 1st © Alan Stockdale/stock.adobe.com, 2nd & 3rd © Jenny Thompson/stock.adobe.com, 4th © Hartphotography /stock.adobe.com; Fig.1.40 © Ladders-direct.com; Fig.1.41 © Werner UK Operations Ltd; Figs.1.42 & 1.45 City & Guilds; Fig.1.51 courtesy of Facelift Access Hire; Fig.1.52 © www.vpgroundforce.com/gb; p.66 © markus_marb/stock.adobe.com; Table 2.2 1st © paketesama/stock.adobe.com, 2nd © Revenaif/Shutterstock.com; Table 2.3 Images courtesy of Draper Tools Ltd (www.drapertools.com); Table 2.4 1st © vvoe/stock.adobe.com, 2nd © dp3010/stock.adobe.com, 3rd © aldorado/stock.adobe.com, 4th/5th Images courtesy of Draper Tools Ltd (www.drapertools.com); Table 2.5 1st © vj/stock.adobe.com, 2nd © remedia/stock.adobe.com, 3rd © Screwfix Direct Limited, 4th © artburger/stock.adobe.com, 5th Image courtesy of Draper Tools Ltd (www.drapertools.com); Table 2.6 1st modustollens/stock.adobe.com, 2nd/4th © Screwfix Direct Limited, 3rd © Vladimir Liverts/stock.adobe.com, 5th Image courtesy of Draper Tools Ltd (www.drapertools.com); Table 2.7 1st © Alexstar/stock.adobe.com, 2nd © maxximmm/stock.adobe.com, 3rd © Sergey Sosnitsky/stock.adobe.com, 4th © cristi180884/stock.adobe.com; Table 2.8 1st © lunglee/stock.adobe.com, 2nd © Molnia/stock.adobe.com; Table 2.9 1st Image courtesy of Draper Tools Ltd (www.drapertools.com), 2nd © michaklootwijk/stock.adobe.com, 3rd © Dmitriy Syechin/stock.adobe.com; Fig.2.1 © David J. Green/Alamy Stock Photo; Fig.2.3 © Rapheephat/stock.adobe.com; Fig.2.4 Photograph by kind permission of ROTHENBERGER UK Ltd; Fig.2.5 Image courtesy of Draper Tools Ltd (www.drapertools.com); Table 2.10 1st Image courtesy of Draper Tools Ltd (www.drapertools.com), 2nd © kasinv/stock.adobe.com, 3rd/4th © Screwfix Direct Limited; Table 2.11 1st © Metabo, 2nd/3rd © Screwfix Direct Limited; Fig.2.6 © stoleg/stock.adobe.com; Fig.2.7 © Roman Milert/stock.adobe.com; Fig.2.8 © Eugene Shatilo/stock.adobe.com; Table 2.12 1st Image courtesy of RIDGID®. RIDGID® is the registered trademark of RIDGID, Inc., 2nd Photograph by kind permission of ROTHENBERGER UK Ltd; Table 2.13 Image courtesy of RIDGID®. RIDGID® is the registered trademark of RIDGID, Inc.; Table 2.14 1st © bradcalkins/stock.adobe.com, 2nd © Anton/stock.adobe.com, 3rd © Vladimir Zubkov/stock.adobe.com, 4th/5th © Screwfix Direct Limited, 6th © Luckylight/stock.adobe.com; Fig.2.19 © Pegler Yorkshire Group; Fig.2.20 © Toolstation Ltd; Figs.2.21–2.24 City & Guilds; Fig.2.25 Image courtesy of RIDGID®. RIDGID® is the registered trademark of RIDGID, Inc.; Fig.2.26 City & Guilds; Tables 2.17–2.21 & p.92 © Pegler Yorkshire Group; Table 2.24 top row 1st © arbalest/stock.adobe.com, 2nd © Dionisvera/stock.adobe.com, 3rd © amnach/stock.adobe.com, 4th © Unkas Photo/stock.adobe.com, bottom row 1st © cegli/stock.adobe.com, 2nd © amnach/stock.adobe.com, 3rd © sompob wongnuksue/123RF; Fig.2.30 © Hawle Armaturenwerke GmbH; Fig.2.34 © John Guest; Table 2.27 1st © Wavin Limited, 2nd © John Guest, 3rd © Trading Depot; Figs.2.35 & 2.36 City & Guilds; Table 2.28 top row © Toolstation Ltd, middle row 1st © LisAnn/stock.adobe.com, 2nd/3rd © Toolstation Ltd, bottom row 1st/3rd Images courtesy of drainageonline.co.uk, 2nd © Toolstation Ltd; Table 2.29 top row 1st/2nd © Images supplied by Polypipe Building Products, 3rd ©MTG/stock.adobe.com, bottom row 1st/3rd © Images supplied by Polypipe Building Products, 2nd © Toolstation Ltd; Table 2.30 © Toolstation Ltd; Table 2.31 1st © Pegler Yorkshire Group, 2nd © Philmac; Fig.2.43 © Trading Depot; Figs.2.44–2.46 © Toolstation Ltd; Fig.2.47 © Screwfix Direct Limited; Fig.2.48 © Toolstation Ltd; Fig.2.49 © remus20/stock.adobe.com; Fig.2.50 © Screwfix Direct Limited; Fig.2.51 © Toolstation Ltd; Figs.2.52 & 2.53 © Screwfix Direct Limited; Table 2.37 © Screwfix Direct Limited; Fig.2.54 © cvetanovski/stock.adobe.com; Figs.2.55 & 2.56 © Screwfix Direct Limited; Fig.2.59 © Regin Products Ltd; Fig.2.60 © Astroflame Fireseals Ltd; p.124 Image courtesy of www.cromwell.co.uk; Table 3.7 © Jo Edkins except 3rd © Scott Horvath, USGS. Public domain; Fig.3.5 © http://corrosion-doctors.org; Figs.3.7 & 3.8 © Phillip Munn, Midland Corrosion Services Ltd; Figs.3.12–3.14 © S. Brannan & Sons Ltd.; Fig.3.31 © Tony Zaccarini/Shutterstock.com; Fig.4.2 © Sebastiano Fancellu/stock.adobe.com; Fig.4.3 1st © Max Tactic/stock.adobe.com, 2nd © Lisa F. Young – Fotolia, 3rd © Kadmy/stock.adobe.com; Fig.4.5 © Phovoir/Shutterstock.com; Fig.4.6 © kemaltaner/stock.adobe.com; Fig.4.7 © Pimlico Plumbers; Fig.4.8 left 1st © Stephen Coburn/stock.adobe.com, 2nd © adiruch na chiangmai/stock.adobe.com, 3rd © fotofabrika/stock.adobe.com, right 1st © didesign/stock.adobe.com, 2nd © Africa Studio/stock.adobe.com; Fig.4.9 © Phovoir/Shutterstock.com; Fig.4.10 © eric/stock.adobe.com; Fig.4.13 © Mile Atanasov/Shutterstock.com; Fig.4.14 © jusep/stock.adobe.com; Fig.4.15 ©

Andrei Rybachuk/stock.adobe.com; Fig.4.16 © Image courtesy of Snickers Workwear; Fig.4.17 © Vladimir Breytberg/ Shutterstock.com; Fig.4.20 © Commissioned by Kingspan Environmental; Figs.5.17 & 5.18 © Tycowaterworks (now part of Talis UK); Fig.5.21 © Image provided by Groundbreaker Systems; Fig.5.31 © AquaTech Pressmain; Fig.5.36 © KSB; Fig.5.37 © Whisper Pumps Ltd, www.whisperpumps.com; Fig.5.40 © Rosemount Measurement Ltd. A business unit of Emerson; Fig.5.42 Pressure Switch DS 200 / © by BD|SENSORS GmbH / www.bdsensors.com; Fig.5.43 © AquaTech Pressmain; Fig.5.47 © www.stevensonplumbing.co.uk/part-2-ballcock-with-float.html; Fig.5.56 © Screwfix Direct Limited; Fig.5.65 © Paxton WT75 Drinking Trough; Figs.5.68 & 5.77 © Arrow Valves Ltd; Fig.5.79 © Hans Sasserath GmbH & Co. KG; Fig.5.82 © Arrow Valves Ltd; Figs.5.91 & 5.93 © Pegler Yorkshire Group; Fig.5.98 © www.stevensonplumbing.co.uk/part-2-ballcock-with-float.html; Fig.5.99 © Pegler Yorkshire Group; Fig.5.100 © silvergull/ stock.adobe.com; Fig.5.101 © ILYA AKINSHIN/stock.adobe.com; Fig.5.102 © photographyfirm/stock.adobe.com; Fig.5.103 © ChiccoDodiFC/stock.adobe.com; Fig.5.106 © gavran333/stock.adobe.com; Fig.5.107 City & Guilds; Fig.5.109 © Image supplied by GROHE; Fig.5.110 © Aqualisa; Figs.5.111 & 5.112 © Salamander Pumps; Figs.5.113 & 5.114 © Scalemaster Limited; Fig.5.116 © Dyna Pipe; Fig.5.118 © editonepankaj16/stock.adobe.com; Figs.5.121 & 5.122 © Photograph by kind permission of ROTHENBERGER UK Ltd; Figs.5.123 & 5.124 City & Guilds; p.282 1st © www.stevensonplumbing.co.uk/part-2-ballcock-with-float.html, 2nd © RZ/stock.adobe.com; Fig.6.8 © Coballes, Spain; Fig.6.12 © Kingspan Environmental Ltd; Fig.6.14 © Maksymiv Iurii/stock.adobe.com; Fig.6.21 © Crown Water Heaters Ltd; Fig.6.33 © Dimplex; Fig.6.38 © Toolstation Ltd; Fig.6.41 top left © anton_antonov/stock.adobe.com; Fig.6.44 © Zilmet UK; Fig.6.47 © anton_antonov/stock.adobe.com; Fig.6.62 City & Guilds; Fig.6.63 © ddukang/stock.adobe.com; Fig.6.64 © Michael Maskrey; Fig.7.16 left © Toolstation Ltd, right © Air22 image Tower TFC Group LLP www.tfc-group.co.uk; Fig.7.21 1st/2nd © Pegler Yorkshire Group, 3rd © John Guest; Fig.7.23 © Photograph of Grant back boiler by permission of Grant Engineering Ireland; Fig.7.25 © Alex Tihonov/stock.adobe.com; Fig.7.26 © Graham Thornhill Range Cookers; Fig.7.41 © flashpics/stock.adobe.com; Fig.7.43 © imagesab/stock.adobe.com; Fig.7.44 © fabioberti.it/stock.adobe.com; Fig.7.45 © Warm Rooms LTD; Fig.7.49 © Zilmet UK; Fig.7.50 © anton_antonov/ stock.adobe.com; Fig.7.51 © Toolstation Ltd; Fig.7.56 © ADEY Innovation Limited; Fig.7.57 © Joe Gough/stock.adobe.com; Figs.7.64 & 7.65 © Pump images supplied by kind permission of Grundfos Pumps Ltd; Fig.7.68 © Queen soft/Shutterstock.com; Fig.7.69 © Rigamondis/Shutterstock.com; Fig.7.73 © Installation of underfloor heating by Geotek Heating Ltd.; 7.75 © Benjamin Haas/Shutterstock.com; 7.77 © Peter Gudella/Shutterstock.com; 7.78 © Jiri Hera/stock.adobe.com; p.434 © Pump image supplied by kind permission of Grundfos Pumps Ltd; Fig.8.1 © www.hunterplastics.co.uk; Fig.8.2 © Image reproduced by kind permission of Rainclear Systems Ltd. UK; Fig.8.3 © Andy Dean/ stock.adobe.com; Fig.8.4 © British Crown copyright, Met Office; Fig.8.7 © www.hunterplastics.co.uk; Fig.8.11 © Toolstation Ltd; Fig.8.12 © Burton Wire & Tube Co. Ltd; Figs.8.15–8.17 © Brett Martin; Fig.8.19 © Yeoman Shield; Fig.8.20 © Image reproduced by kind permission of Rainclear Systems Ltd. UK; Fig.8.22 © Brett Martin; Fig.8.23 © www.hunterplastics.co.uk; Fig.8.24 © Images courtesy of drainageonline.co.uk; Fig.8.26 © Kara/stock.adobe.com; 458 left © Marley Alutec, right © Ed Buziak/Alamy Stock Photo; Fig.9.11 © Wavin Limited; Fig.9.18 © Screwfix Direct Limited; Fig.9.20 © Avalon/Photoshot License/Alamy Stock Photo; Fig.9.21 © steheap/stock.adobe.com; Figs.9.24 & 9.25 © Screwfix Direct Limited; Fig.9.26 © Fluidmaster 3" Adjustable Flapper; Fig.9.28 © Günter Menzl/stock.adobe.com; Fig.9.29 © Stephen VanHorn/stock.adobe.com; Fig.9.30 © bennnn/stock.adobe.com; Fig.9.31 © bouybin/stock.adobe.com; Fig.9.32 © Jrstock/stock.adobe.com; Fig.9.33 © worldwide_ stock/stock.adobe.com; Fig.9.34 © Katarzyna Białasiewicz/123RF; Table 9.3 1st © Andrew Twort/Alamy Stock Photo, 2nd © Francesco/stock.adobe.com, 3rd © Tang Wai Chung/Truphotos.com/Alamy Stock Photo, 4th © yampi/stock.adobe.com; Fig.9.35 © Screwfix Direct Limited; Fig.9.36 © Vado; Fig.9.39 © sharafmaksumov/stock.adobe.com; Fig.9.40 © yampi/stock.adobe.com; Figs.9.41 & 9.43 © Photographee.eu/ stock.adobe.com; Fig.9.44 © siraphol/stock.adobe.com; Fig.9.45 © Image courtesy of Highlife Bathrooms by Alliance; Fig.9.46 © mitev/stock.adobe.com; Figs.9.47 & 9.48 © Ultra Finishing Ltd; Fig.9.49 © Macdee Wirquin; Fig.9.50 © Pegasus Whirlpool Baths Ltd; Fig.9.51 © Aleks Kend/Shutterstock.com; Figs.9.52 & 9.53 © Copyright Mira Showers. All rights reserved.; Fig.9.54 1st © Molotok289/stock.adobe.com, 2nd/3rd © Pro3DArtt/ Shutterstock.com; Fig.9.55 © Armitage Shanks, Alder Sink; Fig.9.57 © Richards of Hull Ltd; Fig.9.58 © Shaws of Darwen 2018; Fig.9.61 © Aleksandr Kurganov/stock.adobe.com; Fig 9.62 © Image reproduced with permission from Cistermiser Ltd; Fig.9.63 © Saniflo; Figs.9.70 & 9.71 © Images supplied by Polypipe Building Products; Fig.9.72 © Toolstation Ltd; Fig.9.73 © Image supplied by Polypipe Building Products; Fig.9.74 © Screwfix Direct Limited; Figs.9.75–9.77 © Images supplied by Polypipe Building Products; Fig.9.78 © Wavin Limited; Fig.9.93 © www.hunterplastics.co.uk; Figs.9.98 & 9.99 © Images owned and supplied by Brett Martin Ltd; Figs.9.100 & 9.101 © Trading Depot; p.503 1st © Image supplied by Polypipe Building Products, 2nd © Wavin Limited, 3rd © Lal Chand Anand And Sons, 4th © Image reproduced by kind permission of Rainclear Systems Ltd. UK; Figs.9.102–9.108 City & Guilds; Fig.9.114 © Hugh Threlfall/Alamy Stock Photo; Fig.9.115 © Trade Counter Direct; Figs.9.116 & 9.117 © Toolstation Ltd; p.521 © Screwfix Direct Limited; p.522 Image supplied by Polypipe Building Products; p.523 © Wavin Limited; Fig.10.2 © Jimj0will/ Shutterstock.com; Fig.10.3 © Andrii/stock.adobe.com; Fig.10.5 © Jiri Hera/stock.adobe.com; Fig.10.8 © OFTEC; Fig.10.9 © HETAS; Fig.10.13 © Lenscap/Alamy Stock Photo; Fig.10.15 © Biomass Silo Systems; Fig.10.16 © Calor Gas Ltd; Figs.11.5 & 11.7 © Peter Tanner; Fig.11.10 © Microgen/stock.adobe.com Fig.11.12 © Andrew Hay-Ellis; Fig.11.14 © Toolstation Ltd; Figs.11.18–11.20 City & Guilds; Figs.11.21 & 11.22 © test-meter.co.uk; Figs.11.23, 11.25–11.28 & 11.31–11.34 City & Guilds.

How to use this book

Throughout this book you will see the following features:

Industry tips and **Key points** are particularly useful pieces of advice that can assist you in your workplace or help you remember something important.

KEY POINT

It is vital that fuels are kept dry and that they are delivered in good condition for optimum combustion efficiency to occur.

INDUSTRY TIP

While many companies have their own style of working, others employ plumbers for specific tasks, i.e. those operatives that work on-site and those that work in private houses.

Key terms in bold purple in the text are explained in the margin to aid your understanding. (They are also explained in the Glossary at the back of the book.)

KEY TERM

Corrosion: any process involving the deterioration or degradation of metal components, where the metal's molecular structure breaks down irreparably.

Health and safety boxes flag important points to keep yourself, colleagues and clients safe in the workplace. They also link to sections in the health and safety chapter for you to recap learning.

HEALTH AND SAFETY

A fire extinguisher should always be available when using any form of soldering equipment.

Activities help to test your understanding and learn from your colleagues' experiences.

ACTIVITY

What would motivate you to improve your work? Make a note and discuss with your team to see what motivates them.

Values and behaviours boxes provide hints and tips on good workplace practice, particularly when liaising with customers.

VALUES AND BEHAVIOURS

It is good practice to keep customers informed of any inconveniences that could be caused by the work that may affect their day-to-day routine.

Improve your maths items provide opportunities to practise or improve your maths skills.

Improve your English items provide opportunities to practise or improve your English skills.

At the end of each chapter there are some **Test your knowledge questions** and **Practical tasks**. These are designed to identify any areas where you might need further training or revision.

Apprenticeship only flagging identifies content that is relevant to apprenticeship learners only.

CHAPTER 1

HEALTH AND SAFETY PRACTICES AND SYSTEMS

INTRODUCTION

Plumbers that work on construction sites are at risk from hazards and accidents every day. Construction is one of the UK's largest industries and arguably the most dangerous. In the past 25 years, nearly 3000 people have been killed on construction sites or as a direct result of construction work. Recent years have seen a fall in the fatality figures, yet accidents continue to be a cause for concern within the industry. While total elimination of accidents is an impossibility, we can ensure that, by proper health and safety management, this figure is reduced still further.

The overriding factor that you need to remember is that health and safety is everyone's responsibility.

In this first chapter we will look at the health, safety and welfare of the people that work on construction sites, and the protection from hazards and harm of the general public. We will investigate the health and safety legislation that helps to keep us safe, as well as look at the methods we should employ for safe working at height, and in excavations and confined spaces. We will investigate how we should deal with toxic and dangerous substances, such as lead and asbestos, solvents, flammable materials and gases, and discuss how we can keep ourselves from harm by the correct use of personal protective equipment (PPE).

By the end of this chapter, you will have knowledge and understanding of the following areas of health, safety and welfare in the construction and building services industries:

- health, safety and welfare legislation and regulation
- recognising and responding to hazardous situations
- personal protection methods and equipment
- responding to accidents and incidents
- electrical safety in the workplace and the home
- safe working practices with heat-producing equipment
- safe working practices at height
- safe working practices in excavations and confined spaces.

1 HEALTH, SAFETY AND WELFARE LEGISLATION AND REGULATION

Hazards encountered by plumbers in particular include asbestos, strained muscles, broken bones, falls, slips, trips and noise. Diseases they risk include dermatitis, asbestosis and emphysema.

In many instances, when the work is subcontracted on a construction project, there is confusion as to who is responsible for safety. However, **legislation** is very clear that everyone has duties and responsibilities regarding health and safety, from the worker to each contractor, to the architect up to the client and the owner of the structure that is being built.

KEY TERMS

Hazard: a danger; something that can cause harm.

Legislation: a law or group of laws that have come into force; health and safety legislation for the plumbing industry includes the Health & Safety at Work Act and the Electricity at Work Regulations.

In this the first section of this chapter we will look at some of the many pieces of legislation surrounding health and safety in the construction industry.

INDUSTRY TIP

- You can access the Health and Safety at Work etc. Act 1974 (HASAWA) at: www.legislation.gov.uk/ukpga/1974/37
- You can access the Control of Substances Hazardous to Health (COSHH) Regulations 2002 at: www.legislation.gov.uk/uksi/2002/2677/regulation/7/made
- You can access the Personal Protective Equipment (PPE) at Work Regulations 1992 at: www.legislation.gov.uk/uksi/1992/2966/contents/made

Protecting the workforce and the general public

General health and safety legislation

The Health and Safety at Work etc. Act 1974

The Health and Safety at Work etc. Act 1974 (HASAWA) is the principal piece of legislation covering occupational health and safety in the UK.

The Act lays down the principles for the management of health and safety at work, enabling the creation of more specifically targeted legislation and codes of practice, such as the Control of Substances Hazardous to Health (COSHH) Regulations 2002 and the Personal Protective Equipment (PPE) at Work Regulations 1992. In other words, all other health and safety legislation has been written as an addition to and because of the HASAWA 1974.

The Act covers all people at work (except domestic servants in private employment) whether they are employers, employees or the self-employed. It is specifically aimed at people and their activities at work rather than premises or processes. It includes provisions for both the protection of people at work and members of the general public who may be at risk as a consequence of the workplace activities.

The main objectives of the HASAWA 1974 are:

- to secure the health, safety and welfare of all people at work
- to protect others from the risks arising from work activities
- to control the obtaining, keeping and use of explosives and highly flammable substances
- to control emissions into the atmosphere of noxious or offensive substances.

Sections 2, 3, 7 and 8 of the HASAWA 1974 cover more general duties that relate directly to you, your employer and the general public.

The general duties of the HASAWA 1974 – Section 2

Section 2 of the HASAWA deals specifically with the general duties of the employer towards its employees. It states that:

> 'It is the duty of every employer, so far as is reasonably practicable, to ensure the health, safety and welfare at work of their employees.'

More specifically, this applies to ensuring that:

- plant and systems are safe and without risk to health
- there is no risk to health in connection with the use, handling, storage and transport of articles and substances
- information, instruction and supervision with regard to the health and safety at work of employees is available
- the working environment for employees is safe, without risk to health, and adequate with regards to facilities and arrangements for their welfare at work
- the place of work is maintained in a safe condition and without risk to health, and the means of access to it and egress from it are safe and without risk.

This legislation also states that employers must have a health and safety policy and, if the company has five or more employees, that policy must be written down. It must be revised as necessary at regular intervals and all employees must have access to and be informed of any changes made to the policy.

Every employer must consult with health and safety representatives appointed by their employees with a view to making and maintaining arrangements that will enable co-operation between employer and employees in promoting and developing health and safety measures and checking their effectiveness.

> **HEALTH AND SAFETY**
>
> Every employer must consult with health and safety representatives. These people are appointed by employees of an organisation to act on their behalf. Their role is to make and maintain arrangements that will enable the employer and employees to promote and develop health and safety measures, and to check their effectiveness.

The general duties of employers and the self-employed to people other than their employees – Section 3

Every employer must ensure, so far as is reasonably practicable, that people not in their employment who may be affected by their work are not exposed to risks to their health and safety. These duties also apply to the self-employed.

Every employer and self-employed person must give to those people who are not in their employment information on the way that aspects of their work might affect the health and safety of others.

Additional employer responsibilities

In addition, the HASAWA 1974 tells us that any employer must:

- carry out risk assessments of all the company's work activities
- identify and implement adequate control measures
- inform all employees of the risk assessments and associated control measures
- review the risk assessments at regular intervals
- make a record of the risk assessments if five or more operatives are employed.

The general duties of employees at work – Section 7

It is the duty of every employee while at work to take reasonable care for the health and safety of themselves and others who may be affected by their acts or omissions at work, and to co-operate with their employer so far as is necessary to enable any duty or requirement to be performed or **complied** with.

> **KEY TERM**
>
> **Comply:** act in accordance with; meet the standards of.

Duty not to interfere with or misuse anything provided – Section 8

Section 8 is often referred to as the 'horseplay section'. According to the HASAWA:

> 'Employees must not intentionally or recklessly interfere with, or misuse, anything provided in the interests of health, safety or welfare, for example, the fooling with and the misuse of a fire extinguisher.'

> **INDUSTRY TIP**
>
> - You can access the PUWER Regulations at: www.legislation.gov.uk/uksi/1998/2306/contents/made
> - You can access the Electricity at Work Regulations 1989 at: www.legislation.gov.uk/uksi/1989/635/contents/made

The Provision and Use of Work Equipment Regulations (PUWER) 2009

These Regulations lay down the minimum standards for the use of all work-related tools and equipment, and are usually used in conjunction with other more specific regulations, such as the Electricity at Work Regulations or similar. The requirements contained within the Regulations are aimed specifically at employers, who must:

- take notice of working conditions and hazards on-site and at work when selecting equipment
- provide work equipment that is fit for purpose and conforms to relevant safety standards
- ensure that the work equipment is used only for its intended purpose
- maintain all equipment in good working order
- ensure that appropriate safety devices are available
- issue operatives with appropriate instructions, training and supervision to enable them to use the work equipment safely
- make sure that all equipment is inspected regularly and at least after installation or assembly at a new location.

The Personal Protective Equipment at Work Regulations 1992

Employers have basic duties concerning the provision and use of personal protective equipment (PPE) at work wherever there are risks to health and safety that cannot be adequately controlled in other ways.

PPE is defined in the Regulations as all equipment that is intended to be worn or held by a person at work and that protects them against one or more risks to their health or safety. Examples of this would be safety helmets, gloves, eye protection, high-visibility clothing, safety footwear and safety harnesses.

Hearing protection and respiratory (breathing) protective equipment (RPE) provided for most work situations are not covered by the PPE Regulations because other regulations are in force that deal specifically with these areas. However, these items need to be compatible with any other PPE provided.

The Regulations require that PPE is:

- properly assessed before use to ensure it is suitable
- maintained and stored correctly
- provided with instructions on how to use it safely
- used correctly by employees.

All employers must provide PPE free of charge whether the PPE is returnable or not (this also applies to agency workers not in the employer's full employment). There are no exemptions from using or wearing PPE. PPE must also be provided to members of the public who are at risk – for example, site visitors. If PPE is provided it must be used.

▲ Figure 1.1 Mandatory helmet sign

INDUSTRY TIP

You can access the Control of Substances Hazardous to Health Regulations 2002 at: www.legislation.gov.uk/uksi/2002/2677/pdfs/uksi_20022677_en.pdf

The Control of Substances Hazardous to Health (COSHH) Regulations 2002

The Control of Substances Hazardous to Health Regulations, known as COSHH, are intended to protect people from illness caused by exposure to hazardous substances. The Regulations require employers to:

- assess the risks to health and safety
- decide what precautions are needed to prevent ill health
- prevent or control exposure
- make sure that the control measures are used and maintained
- monitor exposure and carry out health checks if needed
- make sure that all employees are properly informed, trained and supervised.

To comply with COSHH, eight steps should be followed (Table 1.1).

▼ Table 1.1 The eight steps needed to comply with COSHH

1 Assess the risks	Your employer should assess the risks to health from hazardous substances used in or created by your workplace activities.
2 Decide what precautions are needed	Your employer must not carry out work that could expose you to hazardous substances without first considering the risks and the necessary precautions.
3 Prevent or adequately control exposure	Your employer must prevent you being exposed to hazardous substances. Where preventing exposure is not reasonably practicable, then your employer must adequately control it.
4 Ensure that control measures are used and maintained	Your employer must ensure that control measures are used and maintained properly, and that safety procedures are followed.
5 Monitor the exposure	Your employer should monitor the exposure of employees to hazardous substances, if necessary.

6 Carry out health surveillance	Your employer must carry out appropriate health surveillance where the risk assessment has shown that this is necessary or where COSHH sets specific requirements.
7 Prepare plans and procedures to deal with accidents, incidents and emergencies	Your employer must prepare plans and procedures to deal with incidents and emergencies involving hazardous substances, where necessary.
8 Ensure employees are properly informed, trained and supervised	Your employer should provide you with suitable and sufficient information, instruction and training.

Source: Health and Safety Executive (2005) *COSHH: A brief guide to the Regulations*

COSHH data sheets

There are many forms of hazardous substance for which manufacturers and suppliers produce COSHH data sheets. These are an invaluable source of safety information, designed to make you aware of the known hazards associated with a material or substance, advise you of safe handling procedures, and recommend the most effective response to accidents.

KEY POINT

There are many forms of hazardous substance, for which manufacturers and suppliers produce COSHH data sheets. The data sheet is an invaluable source of safety information and is designed to make you aware of the known hazards associated with a material or substance, advise you of safe handling procedures and recommend the most effective response to accidents.

Under the COSHH Regulations, hazardous substances include:

- chemicals – classified under 'Chemicals Regulations' and identifiable by red and white diamond-shaped warning symbols on the container; care should be taken with unmarked containers
- any substance that has been assigned a workplace exposure limit
- dusts in concentrations in air greater than 10 mg/m^3 for inhaled dust or 4 mg/m^3 of respirable dust
- biological agents such as bacteria, viruses, fungi and parasites
- asphyxiants such as carbon dioxide and nitrogen
- carcinogens such as radon gas or tobacco smoke.

Routes of entry into the body include:

- breathing in vapours, gases, dusts and fumes
- eating or drinking substances or foods contaminated by hazardous substances
- contact with the skin or absorption into the body through the skin, causing harm to internal organs, or via cuts or wounds, causing harm to internal organs
- contact with the eyes by fumes, vapours, liquids and dusts.

INDUSTRY TIP

You can access the Reporting of Injuries, Diseases and Dangerous Occurrences Regulations 2013 at: www.legislation.gov.uk/uksi/2013/1471/contents

The Reporting of Injuries, Diseases and Dangerous Occurrences Regulations 2013

The Reporting of Injuries, Diseases and Dangerous Occurrences Regulations (RIDDOR) 2013 apply to all work activities. They place a legal duty on your employer, the self-employed and people in control of work premises to report some work-related accidents, diseases and dangerous occurrences by the fastest means possible, usually first by telephone and then in writing. RIDDOR applies to all work activities but not all incidents are reportable. Those that must be reported are:

- deaths
- over-three-day injuries – where an employee or self-employed person is away from work or unable to perform their normal work duties for more than three consecutive days. This must be reported within 15 days
- injuries to members of the public or people not at work where they are taken from the scene of an accident to hospital
- certain work-related diseases, for example illnesses such as cancers which can be linked to hazards that a person may have been exposed to in their work
- dangerous occurrences – where something happens that does not result in an injury, but could have done.

Gas Safe-registered gas fitters must also report dangerous gas fittings they find, and gas conveyors/suppliers must report some flammable gas incidents.

HEALTH AND SAFETY

How to report an incident

Call: 0845 300 9923

Email: riddor@connaught.plc.uk

Report online at: www.hse.gov.uk/riddor/report.htm

Write to: Incident Contact Centre, Caerphilly Business Park, Caerphilly CF83 3GG

INDUSTRY TIP

You can access the Electricity at Work Regulations 1989 at: www.legislation.gov.uk/uksi/1989/635/contents/made

The Electricity at Work Regulations 1989

The Electricity at Work (EAW) Regulations place legal responsibilities on employers and employees to ensure that fixed electrical equipment and portable appliances are tested (**PAT test**) and maintained, and regular inspections carried out to ensure they are safe to use. Verifiable evidence is required in the form of:

- documented inspection and testing records, such as portable appliance test (PAT) records and test certificates
- evidence that training has been carried out
- electrical authorisations
- the control of work activities
- **competent** persons.

KEY TERM

Portable appliance test (PAT test): the process of checking electrical appliances and equipment to ensure they are safe to use.

KEY TERM 

Competent: having the necessary ability, knowledge or skill (trained, tested and received a certificate).

The Regulations ensure precautions are taken to avoid death or personal injury from electricity during work activities. The main requirements are to:

- make sure that all persons working on or near electrical equipment are competent
- maintain electrical systems in safe condition
- carry out electrical work safely
- ensure equipment is suitable and safe to use in terms of:
  - strength and capability
  - use in adverse or hazardous environments – for example, weather, dirt, dust, gases, mechanical hazards and flammable atmospheres
- ensure effective insulation of conductors in a system
- ensure effective earthing of the system
- ensure that if work is carried out to the earthing system that involves breaking the flow of current, other precautions are taken to maintain the earth continuity
- ensure all components of the electrical system are suitable and safe for use
- protect against system overload
- provide suitable means for cutting off the supply of electrical current to any electrical equipment and effective isolation of electrical equipment
- ensure that work is not carried out on or near a live conductor unless absolutely essential and suitable precautions are taken to prevent injury
- ensure adequate working space, access and lighting to all electrical equipment where work is undertaken.

INDUSTRY TIP

You can access the Work at Height Regulations 2005 at: www.legislation.gov.uk/uksi/2005/735/contents/made

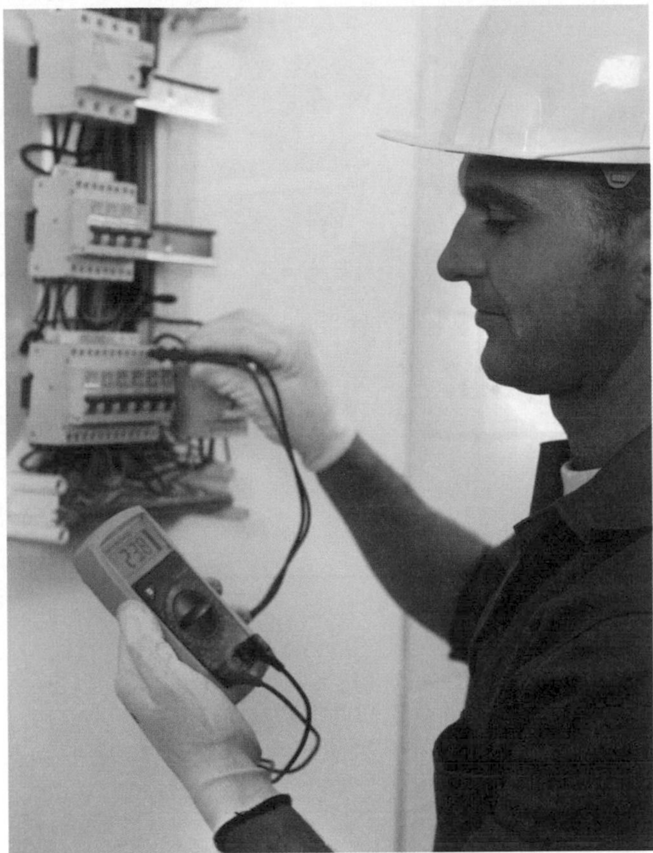

▲ Figure 1.2 Electrical testing

The Work at Height Regulations 2005

The Work at Height Regulations 2005 apply to all work at height where there is a risk of a fall that may cause personal injury. They place duties on employers, the self-employed and any person that controls the work of others, such as managers, supervisors or building owners who may use contractors to work at height. As part of the Regulations, **duty holders** must ensure that:

- all work at height is properly planned and organised
- those people working at height are competent
- the risks from working at height are assessed and the correct work equipment is selected and used
- equipment for working at height is regularly inspected and properly maintained.

KEY TERM

Duty holder: a person who controls, reduces or eliminates health and safety risks that may arise during the construction of a building or during future maintenance. They must also provide information for the health and safety file.

Duty holders must also:

- ensure working at height is avoided where possible
- use work equipment or other measures to prevent falls where working at height is unavoidable
- where they cannot eliminate the risk of a fall, use work equipment or other measures to reduce the distance of the fall.

The Regulations also include requirements for existing places of work and means of access for working at height, collective fall prevention equipment such as guardrails and working platforms, collective fall arresters such as nets and airbags, personal fall protection such as harnesses and work restraints, and ladders.

▲ Figure 1.3 Working at height

The Manual Handling Operations Regulations 1992

The Manual Handling Operations Regulations apply to a wide range of manual handling activities, including lifting, lowering, pushing, pulling and carrying. In the Regulations, loads are described as being either inanimate – for example, a gas boiler – or animate, such as a person or animal.

The Regulations require employers to:

- avoid hazardous manual handling operations so far as is reasonably practicable
- assess any hazardous manual handling operations that cannot be avoided
- reduce the risk of injury so far as is reasonably practicable, including automating or mechanising the lifting process as much as possible.

Employees have a duty to make full and proper use of any system of work provided for employees by their employer, to reduce risks of manual handling injuries. A useful resource is 'Manual handling at work: a brief guide' on the HSE website.

INDUSTRY TIP

You can access the Manual Handling Operations Regulations 1992 at: www.legislation.gov.uk/uksi/1992/2793/contents/made

The Safety Signs and Signals Regulations 1996

The Safety Signs and Signals Regulations require employers to provide specific safety signs whenever and wherever there is a risk that has not been avoided or controlled in other ways, including the use of road traffic signs within workplaces to control road traffic movements. They also place a duty on employers to keep the safety signs in good condition and explain unfamiliar signs to their employees, giving instructions on what they need to do when they see a safety sign.

The Regulations apply to all places of work and cover other methods of conveying health and safety information, including the use of illuminated signs, hand and audible signals such as fire alarms, fire safety signs and the marking of pipework containing dangerous substances. These are in addition to the traditional safety signs such as prohibition and warning signs.

▲ Figure 1.4 Good manual handling at work

KEY POINT

The Safety Signs and Signals Regulations apply to all places of work, but do not include signs and labels used in connection with the supply of substances, products and equipment or the transport of dangerous goods.

INDUSTRY TIP

You can access the Safety Signs and Signals Regulations 1996 at: www.legislation.gov.uk/uksi/1996/341/made

The Control of Lead at Work Regulations 2002

The Control of Lead at Work Regulations apply to all work that exposes any person to lead in any form whereby the lead may be ingested, inhaled or absorbed into the body. This is relevant to plumbers as the lead may be absorbed through the skin when it is being handled or the fumes breathed in when they lead weld.

An Approved Code of Practice (ACOP), 'Control of Lead at Work', is available and should be used in conjunction with the Regulations.

The Regulations state that the employer must assess the nature and extent of the exposure to lead so that the measures of control will be adequate based on that assessment. Where there is 'significant' exposure to lead all the Regulations will apply, but below this level only some of the Regulations will apply.

The basic measure to protect employees from absorbing lead is the prevention of the escape of lead dust, fume or vapour into the workplace. Personal hygiene is important in controlling lead absorption, and the provision and use of adequate washing facilities and PPE is a basic requirement. Food and drink should not be consumed in any place that may be contaminated by lead and the employer should provide alternative arrangements.

Employees should be given sufficient information and training regarding hazards, precautions and duties under the Regulations.

Working with lead and the symptoms of lead poisoning will be covered in detail later in this chapter (see page 22).

INDUSTRY TIP

You can access the Control of Lead at Work Regulations 2002 at: www.legislation.gov.uk/uksi/2002/2676/contents/made

The Control of Asbestos Regulations 2006

The Control of Asbestos Regulations 2006 **prohibit** the importing, supplying and use of all forms of asbestos. They continue the ban introduced in 1985 for blue and brown asbestos and, in 1999, for white asbestos. The ban on the second-hand use of asbestos products, such as asbestos cement sheets and asbestos boards and tiles, also remains in place.

KEY TERM

Prohibit: prevent or forbid by law.

The ban applies to new use of asbestos. If existing asbestos-containing materials are in good condition, they may be left in place provided that their condition is monitored and managed to ensure they are not disturbed.

Asbestos will be covered later in this chapter (see page 24).

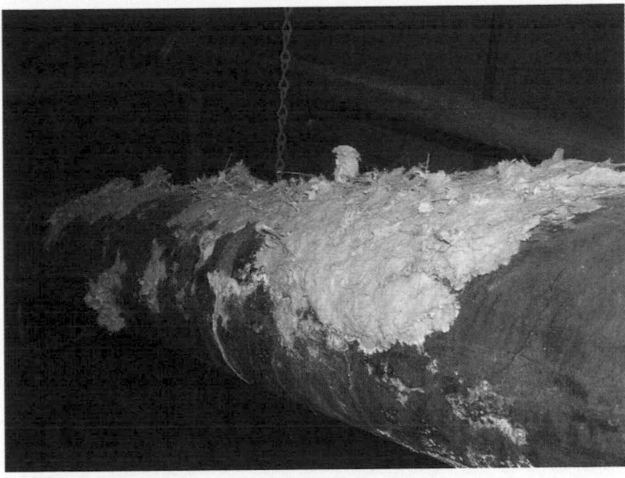

▲ Figure 1.5 Asbestos in poor condition

INDUSTRY TIP

You can access the Control of Asbestos Regulations 2006 at: www.legislation.gov.uk/uksi/2006/2739/contents/made

Health and Safety (First Aid) Regulations 1981 (with 2013 amendment)

These Regulations set out what employers need to do to address the issue of first aid provision in the workplace:

- Managing the provision, i.e. first aid kit, equipment, room etc.
- The requirement for training first aiders
- The requirement for appointed persons
- Making employees aware of these provisions
- First aid for the self employed
- Examples of where the Regulations do not apply.

Confined Spaces Regulations 1997

Under the HASAWA 1974, employees are responsible where the work carries a risk when working in confined spaces. This responsibility is outlined in the Confined Spaces Regulations 1997. The key duties are:

- to avoid working in confined spaces wherever possible by completing the work from the outside
- to follow a safe system of work, if confined space working cannot be avoided
- to put in place adequate emergency arrangements BEFORE work starts.

INDUSTRY TIP

You can access the Construction (Design and Management) Regulations 2015 at: www.legislation.gov.uk/uksi/2015/51/contents/made

INDUSTRY TIP

A summary of the duties of each party and how they are applied is given in Table 1.2. This is taken from the **Health and Safety Executive (HSE)** publication L153 *Managing health and safety in construction,* (published 2015), available on the HSE's website at: www.hse.gov.uk/pubns/priced/l153.pdf

KEY TERM

Health and Safety Executive (HSE): the government body in the UK responsible for the encouragement, regulation and enforcement of workplace health, safety and welfare regulations and government legislation.

Construction-specific legislation

The Construction (Design and Management) Regulations 2015

The Construction (Design and Management) (CDM) Regulations 2015 are the principal piece of health and safety legislation specifically written for the construction industry. They came into force on 6 April 2015, replacing and updating previous regulations.

The main aim of the CDM Regulations 2015 is to combine health and safety into the management of large construction projects and to encourage everyone involved to work together to:

- improve the planning and management of projects from the very start
- identify hazards early on, so they can be eliminated or reduced at the design planning stage and the remaining risks can be properly managed
- target effort where it can do the most good in terms of health and safety, and discourage unnecessary red tape.

The aim is for health and safety considerations to be treated as an essential part of a project's development and not as an afterthought or added extra. This ensures that the responsibility lies firmly with all individuals, from management at the highest level, the client, the designer (architect) and the main contractor, down to the subcontractors, tradespersons and apprentices on-site.

The CDM Regulations require the appointment of a principal designer whose job it is to advise the client on health and safety issues during the design and planning phases of construction work. They should:

- help the client prepare the pre-construction information, and ensure that this is received by the designers and principal contractor in good time
- ensure that the designers fulfil their roles
- plan, manage and monitor pre-construction phase, co-ordinating any matters relating to health and safety during this phase to ensure that the project is without health and safety issues
- eliminate and control any risks throughout the design work
- ensure that there is co-operation and co-ordination between all duty holders
- liaise with the principal contractor to share information relevant to the planning, management and monitoring of the construction phase, and co-ordinate any health and safety issues during construction
- prepare the health and safety file.

▼ Table 1.2 CDM roles and duties

CDM duty holders: who are they?	Summary of role/main duties
Clients are organisations or individuals for whom a construction project is carried out.	Make suitable arrangements for managing a project. This includes making sure: ● other duty holders are appointed ● sufficient time and resources are allocated. Make sure: ● relevant information is prepared and provided to other duty holders ● the principal designer and principal contractor carry out their duties ● welfare facilities are provided.
Domestic clients are people who have construction work carried out on their own home, or the home of a family member that is not done as part of a business, whether for profit or not.	Domestic clients are in scope of CDM 2015, but their duties as a client are normally transferred to: ● the contractor, on a single contractor project, or ● the principal contractor, on a project involving more than one contractor. However, the domestic client can choose to have a written agreement with the principal designer to carry out the client duties.
Designers are those who, as part of a business, prepare or modify designs for a building, product or system relating to construction work.	When preparing or modifying designs, to: ● eliminate, reduce or control foreseeable risks that may arise during construction, and the maintenance and use of a building once it is built ● provide information to other members of the project team, to help them fulfil their duties.
Principal designers are designers appointed by the client in projects involving more than one contractor. They can be an organisation or an individual with sufficient knowledge, experience and ability to carry out the role.	Plan, manage, monitor and co-ordinate health and safety in the pre-construction phase of a project. This includes: ● identifying, eliminating or controlling foreseeable risks ● ensuring designers carry out their duties ● preparing and providing relevant information to other duty holders. Provide relevant information to the principal contractor to help them plan, manage, monitor and co-ordinate health and safety in the construction phase.
Principal contractors are contractors appointed by the client to co-ordinate the construction phase of a project where it involves more than one contractor.	Plan, manage, monitor and co-ordinate health and safety in the construction phase of a project. This includes: ● liaising with the client and principal designer ● preparing the construction phase plan ● organising co-operation between contractors and co-ordinating their work ● ensuring suitable site inductions are provided ● taking reasonable steps to prevent unauthorised access ● consulting workers and engaging in securing their health and safety ● making sure welfare facilities are provided.
Contractors are those who do the actual construction work and can be either an individual or a company.	Plan, manage and monitor construction work under their control so that it is carried out without risks to health and safety. For projects involving more than one contractor, co-ordinate their activities with others in the project team; in particular, comply with directions given to them by the principal designer or principal contractor. For single-contractor projects, prepare a construction phase plan.
Workers are the people who work for or under the control of contractors on a construction site.	They must: ● be consulted about matters that affect their health, safety and welfare ● take care of their own health and safety, and that of others who may be affected by their actions ● report anything they see that is likely to endanger either their own or others' health and safety ● co-operate with their employer, fellow workers, contractors and other duty holders.

Source: Health and Safety Executive (2015) *Managing health and safety in construction*

INDUSTRY TIP

You can access the Building (Amendment) Regulations 2013 at: www.legislation.gov.uk/uksi/2013/1105/contents/made

The Building (Amendment) Regulations 2013

The Building Regulations in England and Wales come under the Building Act 1984. They set the standards for the design and construction of buildings to ensure the safety, health and welfare of the people who live and work in buildings, including provision for those people with a physical disability.

The Building Regulations are set out in a series of Approved Documents titled from A to R; these describe the technical detail.

Those documents listed below have specific implications for plumbers, heating engineers and building services operatives:

- **Approved Document A: Structure**
  Where the components of a system affect the loading placed on the structure of a building or excavations are close to the building. Has to be followed when joists are notched or drilled.
- **Approved Document B: Fire safety**
  Where holes in walls have to be made which could reduce the fire resistance of the building between areas.
- **Approved Document C: Site preparation and resistance to contaminates and moisture**
  Where holes are made in walls for pipes and fixings which may reduce the moisture resistance or allow moisture to ingress the building.
- **Approved Document D: Toxic substances**
  Prevents toxic substances and fumes from entering a property.
- **Approved Document E: Resistance to sound**
  Where holes in the building fabric may reduce the soundproof integrity of the building or systems may cause a noise nuisance to nearby buildings.
- **Approved Document F: Ventilation**
  Building ventilation and guidance on air quality within the building preventing the build up of condensation.
- **Approved Document G: Sanitation, hot water safety and water efficiency**
  Outlines cold water supply, flow and efficiency use within a property. States the daily usage per person. States the requirement for safe working temperatures and controls for hot water. Outlines suitable sanitary appliances within a property.
- **Approved Document H: Drainage and waste disposal**
  Outlines the requirements for soil stack and guttering installation and design.
- **Approved Document J: Combustion appliances and fuel storage systems**
  Outlines the safe installation and usage of heat producing appliances, including boilers, chimneys and flues, and offers advice on safe fuel storage installations, including solid fuel, liquid oil fuels and gas-fired heating.
- **Approved Document K: Protection from falling, collision and impact**
  Outlines the requirements for stairs, ramps and loading bays. The protection against impacts with glazing and use of safety windows and prevention from being trapped by doors.
- **Approved Document L: Conservation of fuel and power**
  Outlines the energy efficiency standards for properties including boilers, controls and insulation.
- **Approved Document M: Access to and use of buildings**
  Outlines the ease of access to and use of buildings, including facilities for disabled visitors or occupants, and the ability to move through a building easily including to toilets and bathrooms.
- **Approved Document P: Electrical safety**
  Outlines electrical safety in dwellings, including detailed information about what procedures need to be in place and who may carry these out, such as when a professional electrician must be hired. Details electrical safety to avoid injuries and fires caused by electrical installations, including the design, installation, inspection and testing of any electrical works made within a dwelling.

- **Approved Document Q: Security in dwellings**
  Outlines security in new dwellings, including measures taken to avoid any unauthorised entrance to dwellings and flats within a building.
- **Approved Document R: High-speed electronic communications networks**
  Outlines high-speed electronic communications networks, for example the use of physical infrastructures within a building to ensure it may be connected to a broadband network.
- **Approved Document 7: Material and workmanship**
  Outlines materials and workmanship, for example the use of the appropriate materials for a construction and how those who are working on the building must behave in a workmanlike manner.

INDUSTRY TIP

All Approved Documents can be accessed from the index at: www.gov.uk/government/collections/approved-documents

Building services-specific legislation

The term 'building services' is used to describe those activities not connected with the construction of the building but related to the services that are installed within the building as it is constructed. The services in a building are:

- water
- gas
- electricity
- heating and ventilation
- telecommunications.

The building services industry has specific legislation to ensure the health and safety of the general public.

The Water Supply (Water Fittings) Regulations 1999

These relate to the supply of safe, clean, wholesome drinking water to properties and dwellings, specifically targeting the prevention of contamination, waste, undue consumption, misuse and erroneous metering.

INDUSTRY TIP

- You can access the Water Supply (Water Fittings) Regulations 1999 at: www.legislation.gov.uk/uksi/1999/1148/contents/made
- You can access the Gas Safety (Installation and Use) Regulations 1998 at: www.legislation.gov.uk/uksi/1998/2451/contents/made

The Gas Safety (Installation and Use) Regulations 1998

These cover the safe installation, maintenance and use of gas and gas appliances in private dwellings and business premises, aimed at preventing carbon monoxide (CO) poisoning, fires and explosions. The Regulations state that all gas engineers must be registered on the Gas Safe Register to prove their competency, and it is the responsibility of landlords to ensure that their tenants' pipework and appliances are checked annually and certified safe to use. Homeowners and other gas consumers are also recommended by the HSE to have their appliances serviced and checked annually by a registered Gas Safe installer.

The 18th Edition IET Wiring Regulations (BS 7671)

These are the national standard to which all wiring – industrial or domestic – should now conform. All wiring must be designed to the specifications laid down in the Regulations, and any person involved in the design, installation, inspection and testing of electrical installations must have a sound knowledge of this document.

KEY POINT

The IET Regulations are produced by the Institute of Engineering and Technology (IET), the industry body that covers electrical installation. The 18th edition contains many major changes that align it with other similar European documents. To find out more, visit: http://electrical.theiet.org/bs-7671/

Health and safety responsibilities

According to the CDM Regulations, each member of the construction team has certain responsibilities towards health, safety and welfare during the planning stage, the construction stage and after the building is completed. The main document to be produced as a result of the CDM Regulations is the health and safety file, which must stay with the building until its demolition. The main responsibilities are as follows.

The employer

The responsibilities of the employer are vast and are detailed in law to comply with government legislation and regulation. The main duty is to ensure health, safety and welfare by providing a safe working environment for all employees. This includes providing safe systems of work, safe handling, storage, training and supervision. Employers are obligated to provide an up-to-date health and safety policy that is accessible for all employees and any site visitors.

The employee

Under the HASAWA, employees must act with due care for themselves and anyone else who may be affected by their acts or omissions. They must co-operate with the employer in respect of health and safety matters, and must not recklessly interfere with or misuse equipment that is provided for health and safety.

The client

The client must demonstrate an acceptable standard of health and safety by appointing a principal designer (see below) to monitor and advise on all health and safety matters. They must also make suitable arrangements for managing a project. This includes making sure that:

- other duty holders are appointed
- sufficient time and resources are allocated
- the relevant information is prepared and provided to other duty holders
- the principal designer and principal contractor carry out their duties
- welfare facilities are provided.

The principal designer

These are designers appointed by the client in projects involving multiple contractors. They can be an organisation or an individual with the knowledge, experience and ability to carry out the role, which includes planning, managing, monitoring and co-ordinating health and safety in the construction phase of a project. This covers:

- liaising with the client and principal contractor
- preparing the construction phase plan
- organising co-operation between contractors and co-ordinating their work
- ensuring that:
  - suitable site inductions are provided
  - reasonable steps are taken to prevent unauthorised access
  - workers are consulted and engaged in securing their health and safety
  - welfare facilities are provided.

Principal contractors

These are appointed by the client to co-ordinate the construction phase of a project where it involves more than one contractor. Their responsibilities include planning, managing, monitoring and co-ordinating health and safety in the construction phase of the project, which covers:

- liaising with the client and principal designer
- preparing the construction phase plan
- organising co-operation between contractors and co-ordinating their work
- ensuring that suitable site inductions are provided
- taking steps to prevent unauthorised site access
- consulting workers, and engaging them in securing their health and safety
- ensuring that welfare facilities are provided.

Contractors

Contractors are those who do the actual construction work and can be either an individual or a company. Their role is to:

- plan, manage and monitor the construction work under their control so that it is carried out without risks to health and safety
- for projects involving more than one contractor, to co-ordinate their activities with others in the project team, and comply with directions given to them by the principal designer or principal contractor
- for single-contractor projects, prepare a construction phase plan.

Subcontractors

Subcontractors do not have direct contact with health and safety issues. However, they must abide by the law in respect to health and safety and be provided with relevant safety information and PPE. They must also complete an initial site induction before they are allowed on-site alone.

The legal status of health and safety guidance

Health and safety guidance can be divided into two distinct groups: **mandatory** and **advisory**.

KEY TERMS

Mandatory: required by law; compulsory.

Advisory: recommended but not enforced.

Those that are mandatory (the law) are:

- **Acts of Parliament:** these create a new law or change an existing one. Their implementation is the responsibility of a specific government department; in the case of health and safety acts, this is the Health and Safety Committee.
- **Regulations:** rules, procedures and administrative codes set by authorities or governmental agencies to achieve an objective. They are legally enforceable and must be followed to avoid prosecution.

Those that give guidance and advice are:

- **Approved codes of practice (ACOPs):** documents that give practical guidance on complying with regulations. Although it is not an offence not to comply with an ACOP, in the case of health and safety ACOPs, proof that their advice has been ignored could be seen as evidence of guilt if an employer or employee faces criminal prosecution under health and safety law. Following an ACOP is considered good practice.
- **Guidance notes:** these are produced by the HSE to help people interpret and understand what is required by a law, and to comply with it. They also give technical advice. Courses of action set out in guidance notes are not compulsory, but if the guidance is followed it is usually enough to comply with the law.

▲ Figure 1.6 Managing and working with asbestos

Who enforces health and safety regulations?

Health and safety law is enforced by the HSE and local authority working in partnership under the Health and Safety Executive/Local Authorities Enforcement Liaison Committee (HELA). Both employ health and safety inspectors whose job it is to ensure that the law is complied with.

The role of the health and safety inspectors

Inspectors have the legal right to enter your workplace without giving notice, although notice may be given where the inspector considers it appropriate. On a normal inspection visit, the inspector would look at your place of work, work activities, management of health and safety, and check that your employer is complying with health and safety law. The inspector may offer guidance and advice or talk to employees, take photographs and samples, serve improvement notices or take action if a risk to health and safety is perceived.

If a breach of health and safety law is found, the inspector will decide what action to take. The action will depend on the severity of the breach. The inspector should provide employees or their representatives with information relating to the breach and any necessary action.

There are several ways in which an inspector may take enforcement action to deal with a breach of the regulations. These are as follows.

- **Informal action:** where the breach of the law is comparatively small, the inspector will advise the duty holder what action to take in order to conform with the requirements of the law. If requested, this can be given in writing.
- **Improvement notice:** more severe breaches will receive a direct order to take specific action to comply with the law. The inspector will discuss with the duty holder the improvement notice and resolve points of difference before serving it. The notice will say what has to be done, why and by when. The time period to take the corrective action will be a minimum of 21 days, to allow the duty holder time to appeal to an industrial tribunal.
- **Prohibition notice:** where an activity involves a risk of serious personal injury, the inspector may issue a prohibition notice forbidding the activity either immediately or after a specified time period. This notice will not be lifted and work will not be allowed to resume until corrective action has been taken.
- **Prosecution:** in some cases, prosecution may be deemed necessary. Failure to comply with an improvement or prohibition notice, or a court remedy order, carries a fine of up to £20,000 or six months' imprisonment, or both. Unlimited fines and in some cases imprisonment may be given by higher courts.

2 RECOGNISING AND RESPONDING TO HAZARDOUS SITUATIONS AT WORK

We will now look at construction site safety from a general and personal point of view. We will examine general site hazards, and how we can either help or hinder our own health and safety and that of those around us.

Preventing accidents at work

Accidents do not just happen; they are caused. The first step towards preventing accidents is finding out what the causes are. Accident prevention is something that everyone needs to practise. It means being able to recognise and take steps to remove danger, and is the responsibility of everyone working, in any way, on a construction site.

KEY TERM

Accident: an unexpected or unplanned event that could result in personal injury, damage and, occasionally, death. When an accident occurs, there are always reasons for it and if there's a reason, then there is usually blame.

HEALTH AND SAFETY

Next time you are tempted to take a risk, STOP and THINK safety!

Learning to spot a dangerous situation is not difficult because accidents follow a regular pattern. Every day, the same set of dangerous conditions build up and the same unsafe acts take place. Consequently, the same kinds of accident happen over and over again.

ACTIVITY

Do any of the things you normally see and do at work add up to a source of danger? Write a list of potential accidents and, against each one, note down an action you could take to reduce the risk of it happening.

Identifying hazards at work

Hazards on-site can be divided into three specific groups:

1. general site and work area cleanliness, which can lead to trips, slips and falls
2. using equipment and PPE that is inadequate for the job, non-existent (in the case of PPE) or defective

3. personal conduct such as:
   - incorrect manual handling methods
   - incorrect methods of working at heights, in trenches and on excavations
   - not taking enough care and attention in dangerous environments
   - using equipment or carrying out activities without appropriate training
   - taking risks.

Here are a few examples of things that can lead to accidents in the workplace:

- excessive haste or taking shortcuts in order to get the job done
- lack of preparation, and failure to comply with instructions and rules of safety
- lack of concentration due to distraction or lack of interest in the job
- PPE or clothing not used or worn
- inadequate training and supervision
- inadequate lighting, heating or noise
- poor storage of materials
- unsafe methods of handling and lifting
- defective tools and equipment
- poor weather conditions
- electrical faults
- failure to use guards provided
- working under the influence of drugs and/or alcohol.

In many cases, these can be prevented by following safe working practices, including the use of risk assessments, method statements and permits to work.

Risk assessments

A risk assessment is a detailed examination of any factor that could cause injury, so that you or your employer can assess whether sufficient steps have been taken to prevent harm. Other workers and the general public have a right under health and safety law to be protected from any harm that may be caused by the failure to take reasonable control measures. Your employer is legally required to assess the risks in the workplace and implement measures to control those risks. The law does not expect you to eliminate all risks, but you are expected to take steps to ensure health and safety as far as is reasonably practicable.

> **KEY POINT**
>
> Remember that a **hazard** is anything that may cause harm, such as chemicals, electricity, gas, working from ladders, etc. The **risk** is the chance, no matter how high or low, that somebody could be harmed by these and other hazards, together with an indication of how serious the harm could be.

Risk assessment step by step

1. Identify the hazards.
   - Work out how people could be harmed by:
     - walking around the site
     - asking employees what they think
     - visiting the HSE website for practical guidance
     - contacting trade associations for advice
     - checking manufacturers' instructions and COSHH data sheets.
2. Decide who might be harmed and how.
   - Identify the groups of people at risk.
3. Evaluate the risks and decide on precautions.
   - Consider:
     - whether you can get rid of the hazard altogether
     - if not, how can you control the risks so that harm is unlikely?
4. Record your findings and implement them.
   - Ensure a proper check was made by:
     - asking who might be affected
     - dealing with all the significant hazards, taking into account the number of people who could be involved
     - making sure all precautions are reasonable, and the remaining risk is low
     - involving your staff or their representatives in the process.
5. Review your assessment and update it if necessary.
   - Ensure you review risk assessments every year taking into account whether:
     - more employees have joined the company
     - new machinery and/or equipment has been installed
     - any fellow workers have spotted any problems
     - anything has been learned from accidents or near misses.

RISK ASSESSMENT RA01: ALL OPERATIONS – CORE ASSESSMENT

Operation/Task:	All operations – core assessment	Employees at risk:	All site personnel
Location/Area:	Any	Other persons at risk:	General public
Assessor:		Key responsible personnel:	Contract managers and supervisors

Activity	Hazard	Risks	Pre-control risk ratings			Control measures	Post-control risk ratings			Comments
			1*	2**	1x2		1*	2**	1x2	
Construction plant operation	Plant and vehicle movement	Contact by Major injury	5	6	30	Controlled operations with use of banksman as necessary. Competent operators to CITB CTA, EPIC or CSCS standards where applicable. Clear and reasonable access/egress for plant. Well-maintained construction plant.	2	5	10	This is a general assessment only. See also assessments relating to specific items of construction plant.
Signing and guarding of works	Public, traffic, site traffic	Contact by Major injury	7	7	49	Signing and guarding to Chapter 8 of Traffic Signs Manual. Installation procedures consistent with those recommended by Chapter 8 and the Traffic Management Contractors' Association (TMCA). Competent operators carrying out the installation. Additional training for those carrying out signing etc. on high-speed roads.	2	6	12	This is a high-risk activity but methods of working and control measures keep risk to a minimum. Consequences can still be serious, especially on high-speed roads. High level of discipline required. See other signing specific Risk Assessments.
Driving around the site	Plant and vehicle movements, obstruction	Contact by Major injury Contact with	5	6	30	Site speed limit usually 15 m.p.h. High level of personal awareness. Rotating flashing amber beacon displayed prominently on top of vehicle.	2	2	4	See also task specific Risk Assessments.
Pedestrian activity	Plant and vehicle movement	Contact by	5	6	30	Personal awareness of operations on site. Use of pedestrian only routes where available/possible. Wearing high viz clothing to EN471.	2	5	10	See also task specific Risk Assessments.
All tasks	Incompetence	Various	4	6	24	All site personnel to be competent to perform the tasks they are asked to do. Compliance with the Site Managers' rules. Skills/competencies as per Company Health & Safety Policy.	1	5	5	High level of awareness and most operations of a 'pass-by' nature thereby minimising the Laeq values.
	Noise from operations	Hearing damage	3	4	12	All construction plant and vehicles constructed to national standards and industry norm that includes noise attenuation. Noise levels identified on machine when possible. Noise Risk Assessment carried out for various items of plant and hearing.	1	2	2	

▲ Figure 1.7 Example of a risk assessment form

Method statements

A method statement, sometimes called a safe system of work, is usually completed after the risk assessment. It is a document that details the work task or process, outlines the hazards involved and includes a step-by-step guide on how the work should be completed safely. The method statement must also detail which control measures have been initiated to ensure the safety of anyone affected by the task or process. Method statements are frequently requested as part of the tendering process as this allows the client to gain an insight into the company and the way it operates.

Permits to work

When work has been identified as high risk, strict health and safety controls are required. In this instance the work must be carried out against a pre-agreed permit to work. A permit to work is a document put together by those authorising the work and those carrying it out, which gives authorisation for named persons to carry out specific work within a nominated time frame. It lists the precautions that are required to complete the work safely based on a written risk assessment. It describes the work and how it will be carried out (more detail is given in the method statement). On completion of the work, and before equipment or machinery is reinstated, it will require a written declaration from the permit originator that normal practice may be returned to.

Work affecting the public and their health and safety

It is not only construction workers that suffer accidents as a result of construction work. Members of the public can also be killed and injured. Accidents can often occur when people are walking close to where buildings are being constructed, refurbished or demolished. It must be remembered that work near to where the general public have access needs must be planned and executed correctly, taking into account people with pushchairs, people with disabilities and the elderly.

The best way of protecting the public from the dangers of construction sites is to restrict access – in other words, keep them out! Here are just a few pointers to remember:

- Erect a 2 m high perimeter fence. If parts of it need to be taken down for access, make sure these are put back at the end of the day.
- Lock the site gates and any windows and doors at night.
- If work is being done in an occupied property, clear responsibilities need to be established with the occupier for maintaining the fencing.
- If the work is near a school or residential area, enlist the help of the head teacher or the residents' association to discourage children and young people from entering the site.
- Young children should be protected from the dangers of building sites. Steps taken should include:
  - cover trenches, excavations and scaffolds, removing all ladders
  - store materials so there is no risk of them toppling over
  - lock away hazardous substances
  - initiate other security methods such as security guards.
- Protect passers-by from falling objects from scaffolds by the use of toe boards, brick guards and netting.
- Use plastic sheeting to retain dust, drips and splashes.
- Tie down or remove loose materials from scaffolds.
- Ensure that warning and danger signs are posted on and around the scaffold.

INDUSTRY TIP

These are just a few of the precautions you can take but there are many others. For more information, see the HSE website at: www.hse.gov.uk

Safety signs

Safety signs are used on construction sites where risks have not been avoided by other means. Employers are required to provide and maintain safety signs, and workers need to be trained in the recognition of safety signs and symbols so that they understand their meaning. To ensure that the correct number and type of safety signs have been used, an employer must carry out a number of simple tasks. They must:

- conduct a risk assessment
- ensure fire equipment and emergency exits are clearly indicated
- use signs to prohibit entry into dangerous areas
- make sure that mandatory requirements, such as wearing PPE, are clearly shown
- clearly indicate all first-aid areas and equipment
- use signs to show prohibited behaviour, such as 'no smoking'.

The signs used must communicate their message clearly and effectively, and many have to conform to strict legal and accessibility obligations. Safety signs must comply with the European Council's Safety Signs Directive (92/58/EEC), the purpose of which is to encourage the standardisation of safety signs throughout the European Union so that safety signs have the same meaning. Safety signs are divided into five separate groups as shown in Table 1.3.

INDUSTRY TIP

You can access more information on the European Council's Safety Signs Directive (92/58/EEC) via: www.unece.org/trans/danger/publi/ghs/pictograms.html

▼ Table 1.3 Six groups of safety signs

Category	Description	Example
Prohibition – 'Stop!'	**Colour:** A red circular band with a diagonal cross bar on a white background; the symbol within the circle to be black. **Purpose:** To indicate that a certain behaviour is prohibited. **Meaning:** Stop / Do not / You must not	
Warning – 'Danger'	**Colour:** A yellow triangle with a black border and black symbol. **Purpose:** To warn of any type of hazard. **Meaning:** Danger / Hazard / Caution / Beware / Careful	
Mandatory – 'Obey'	**Colour:** A blue circle with a white symbol. **Purpose:** Indicates that a specific course of action must be taken. **Meaning:** Obey / You must / Carry out instructions shown / Do	
Fire equipment – 'Fire'	**Colour:** A red rectangle or square with a white symbol. **Purpose:** To describe the location of fire-fighting equipment. **Meaning:** Location of fire-fighting equipment	
Safe condition – 'Safety'	**Colour:** A green rectangle or square with a white symbol or text. **Purpose:** To provide information about safe conditions. **Meaning:** The safe way / Where to go in emergencies / First aid	

Occasionally, a sign may be seen that is a mixture of many different types of signs on one signboard. These are known as combination signs; an example is shown in Figure 1.8.

▲ Figure 1.8 Example of a combination safety sign board

Identifying hazardous substances

Section 7 of the HASAWA states that:

> 'Every employer must ensure that the exposure of his/her employees to substances hazardous to health is either prevented or adequately controlled.'

In most cases, hazardous substances can be divided into six main categories, as presented in Table 1.4.

▼ Table 1.4 Classifications of hazardous substances

Category	Description	Example
Toxic	Cyanide, asbestos, lead	Poisons and dangerous substances that have the ability to cause death if ingested, inhaled or absorbed into the body.
Harmful	Fluxes, solvents, cleaning fluids, chemicals, dust	Harmful substances could be in any form, liquid, solid (dust particles) or gas.
Corrosive	Hydrochloric acid, sulphuric acid, caustic soda	Such substances have the ability to cause severe burns to exposed parts of the body.
Irritant	Fibreglass roof insulation, some paints, solvents and sealants	Can cause irritation of the skin, eyes, nose and throat.
Oxidising	Oxygen from welding bottles	Materials are induced to burn fiercely by adding oxygen to a fire.
Extremely flammable	Petrol, LPG, acetylene gas, solvent weld adhesives and cleaning agents	These have the potential to burn fiercely if the substance is either exposed to a source of ignition or subjected to temperatures close to its flashpoint, so that it spontaneously combusts.

Labels on packaging use the Globally Harmonised System (GHS) on the classification and labelling of hazardous substances and mixtures. This system was developed by the United Nations to ensure the labelling of hazardous substances is consistent around the world, so that they can be easily identified. It uses standard symbols or pictograms (see Figure 1.8a) alongside signal words (for example, 'danger' or 'warning') and a hazard statement (a standard phrase to describe the hazard, for example 'wear eye protection'). The system categorises hazards into classes: physical hazards (for example, explosives, flammable gases, oxidising liquids, corrosive to metals); health hazards (for example, acute toxicity, skin corrosion/irritation, eye damage/irritation, respiratory/skin sensitisation); and environmental hazards (for example, hazardous to the aquatic environment).

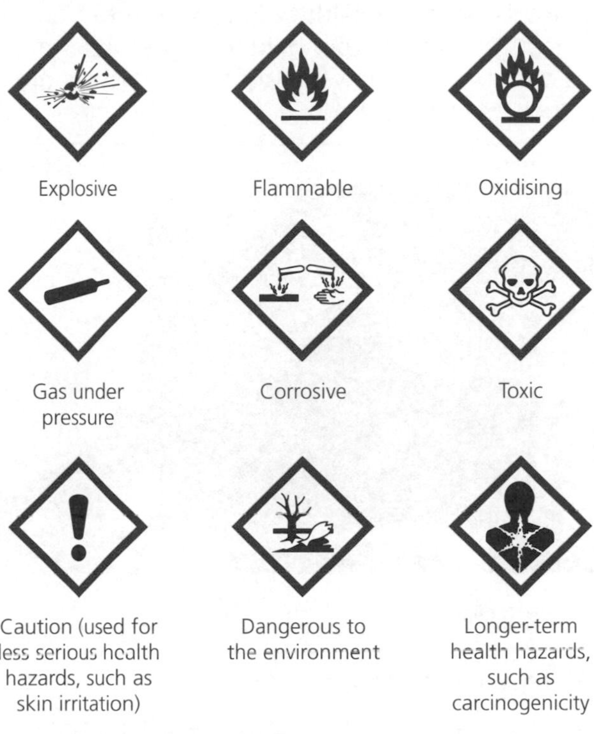

▲ Figure 1.9 Hazard pictograms

INDUSTRY TIP

Use the HSE website to identify the GHS pictograms: www.hse.gov.uk/pubns/indg136.pdf

Chemicals

There are many chemicals that may be found on construction sites, from new build and refurbishment to demolition:

- lead
- fluxes
- solvents (these have many uses on construction sites, such as cleaning agents)
- asbestos
- cadmium (found in plastics like PVCu)
- carbon monoxide (from use of blowtorches, welding, generators, gas heaters, etc.)
- welding fumes (from welding metals like steel pipes)
- spray paints

- cutting oil mists (cutting and threading mild steel tubes)
- jointing compounds.

KEY POINT

The effects on your health from exposure to chemicals can range from mild to very severe. In some cases, it may be years before the effects are felt, such as with asbestos.

▲ Figure 1.10 Lead work

Working with lead

As part of your job as a plumber, you may be asked to work with lead, whether it is replacing a lead pipe or installing lead sheet weatherings (sheet lead shaped and positioned on roofs and chimneys to prevent the rain getting in) and roof work. Lead is a highly toxic metal that can enter the body through:

- absorption – touching and handling lead without the use of barrier cream
- ingestion – not observing personal hygiene by not washing your hands before eating and drinking after handling lead
- inhalation – by breathing lead fumes when lead welding or soldering with leaded solder.

Lead is a very powerful neurotoxin that damages the central nervous system and leads to brain and blood disorders. Lead oxide in the form of a white powder from the corrosion of lead is particularly dangerous. The symptoms of lead poisoning are:

- headaches
- tiredness
- irritability
- constipation
- nausea
- stomach pains
- anaemia (lack of healthy red blood cells)
- loss of weight.

Continued uncontrolled exposure could cause more serious symptoms, such as:

- kidney damage
- nerve and brain damage
- infertility.

Employer and employee health and safety responsibilities when working with lead

If you are exposed to lead or lead compounds, such as lead oxide, dust, fume or vapour from lead welding or smelting, while you are at work, your employer must:

- assess the risk to your health, to decide whether or not your exposure is 'significant' and what precautions are needed to protect you
- put in place systems of work, such as fume and dust extraction, to prevent or control your exposure to lead and keep equipment in good working order
- provide washing and changing facilities, and places free from lead contamination where you can eat and drink
- inform you about the risks to your health from working with lead, and the precautions you should take
- train you to use any control measures and protective equipment correctly
- provide you with protective clothing and arrange for that clothing to be laundered
- measure the amount of lead in the air that you are exposed to and tell you the results; if your exposure to lead cannot be kept below a certain level then your employer must issue you with respiratory protective equipment
- arrange to measure the level of lead in your body; this is done by a simple blood test administered by a doctor at your place of work; you must be told the results of your tests.

There are ways you can help yourself too, which include the following.

- Make sure you have all the information and training you need to work safely with lead, including knowing what to do in an emergency such as the sudden uncontrolled release of lead dust or fume into the atmosphere.

- Use all the equipment provided by your employer and follow its instructions for use.
- Follow good work practices, keeping your immediate work area as clean and tidy as possible and taking care not to take home any PPE such as overalls or protective footwear.
- Wear any necessary PPE clothing and respiratory protection.
- Report any damaged or defective equipment to your employer.
- Eat and drink only in designated areas that are free from lead contamination.
- Practise a high standard of personal hygiene, washing your hands, face and nails regularly and showering before leaving the site when necessary.
- Do not miss medical appointments with the doctor at your place of work.

Working with fluxes

Flux is a paste compound that helps solder to adhere to copper tubes and copper-based fittings. There are two basic types of flux used today in the plumbing industry:

1. **Traditional fluxes:** these grease-based fluxes often contain a chemical called **rosin** (also known as colophony) or **zinc chloride**. Caution should be exercised when using this kind of flux. It is recommended that you check COSHH data sheets for further information regarding these products.
2. **Self-cleaning fluxes:** this type of flux is also known as 'active' flux because of its aggressive nature. Most are based on zinc chloride or hydrochloric acid, both of which can cause burns and severe skin irritation, so careful handling and use is very important. Other self-cleaning fluxes may use natural enzymes as cleaning agents but these are also known to irritate the skin.

All flux should be handled with care. Use a brush to apply the paste and always wash your hands thoroughly after use.

The process when the flux has been added, the joint is heated and the solder is added (if required) is referred to as **sweating** the joint.

KEY TERMS

Rosin: a natural solid, resin-type material obtained from pine trees, which, when heated, forms acidic particles that can irritate the breathing. This could lead to occupational asthma.

Zinc chloride: a corrosive substance that can cause skin irritation, burns and eye damage if it gets in the eye.

Sweating: the term sweating a joint refers to the process of fluxing, heating and soldering the joint to create a watertight connection.

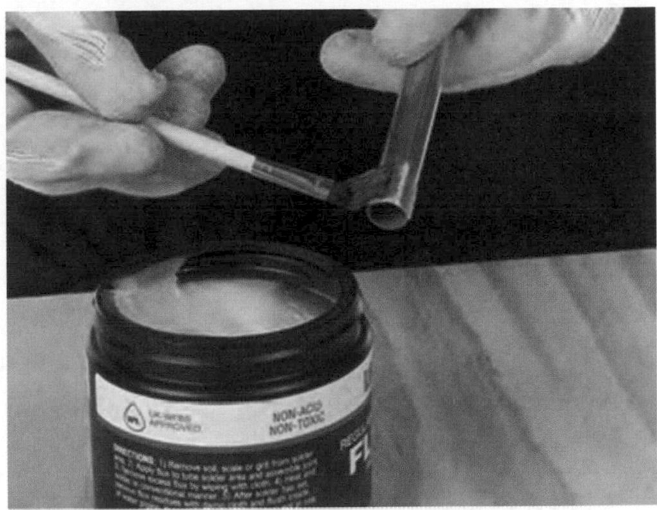

▲ Figure 1.11 Typical self-cleaning flux

INDUSTRY TIP

Again, it is recommended that you check COSHH data sheets for further specific health and safety information regarding these products.

Working with solvents

A variety of solvents with differing degrees of toxicity are used in construction. They are in paints, adhesives, epoxy resins and other products.

Generally, exposure to excessive amounts of solvent vapour is greater when solvents are handled in enclosed or confined spaces. Care should be taken when using solvent adhesives to solvent weld PVCu pipes and fittings in confined spaces. Solvents can:

- irritate your eyes, nose or throat
- make you dizzy, sleepy, give you a headache or cause you to pass out

- affect your judgement or co-ordination
- cause internal damage to your body
- dry out or irritate your skin.

When working with solvents, follow these basic instructions:

- avoid contact with the skin
- avoid contact with the eyes
- use only in an open, well-ventilated space
- keep away from naked flames as solvents are flammable
- store in a well-ventilated, secure area.

Identifying the types of asbestos

Asbestos is a naturally occurring fibrous material that can cause major illnesses. It has been used as a building material since the end of the 1940s and, because asbestos is often mixed with other materials such as cement, it is hard to know if you're working with it or not. The problem is that if you work in a building built before the year 2000, it is likely that asbestos has been used during its construction in one form or another.

KEY TERM

Asbestos: a fibrous silicate material highly resistant to heat.

There are three main types of asbestos:

1 Chrysotile (white asbestos): a white curly fibre, chrysotile accounts for 90 per cent of asbestos in products and is a member of the serpentine group. It is a magnesium silicate.
2 Amosite (brown or grey asbestos): straight amosite fibres belong in the amphibole group, and contain iron and magnesium.
3 Crocidolite (blue asbestos): a member of the amphibole group, crocidolite takes the form of blue, straight fibres. It is a sodium iron magnesium silicate.

Other forms of asbestos include:

- anthophyllite
- tremolite
- actinolite.

▲ Figure 1.12 Domestic uses of asbestos

Materials that may contain asbestos

Asbestos is one of the most dangerous materials that you will come across during your work as a plumber. Many people die each year from asbestos-related diseases.

In the past, it was used extensively for the following plumbing-specific applications:

- flue pipes
- gutters and rainwater pipes
- soil and vent pipes
- pipe insulation (both sprayed on and applied as a paste and wrapped in linen)
- boiler gaskets and fireproof ropes
- cold water cisterns.

It may also be found in:

- Artex
- roof and ceiling tiles
- soffit boards
- plaster coatings
- floor tiles and coverings
- asbestos sheeting and corrugated roofing.

Safe working practices when dealing with asbestos

The presence of asbestos alone does not necessarily mean there is a health risk. Provided that the fibres are intact and are not disturbed the risk is relatively low. However, once the fibres are loose and enter the atmosphere the risk increases dramatically; the asbestos is inhaled into the lungs, which causes certain types of lung disease.

Lung disease from exposure to asbestos can be divided into three main types:

1. Asbestosis: a process of widespread scarring of the lungs.
2. Disease of the lining of the lungs, called the pleura. This has a variety of signs and symptoms and is the result of inflammation and the hardening (calcification) and/or thickening of the lining tissue.
3. Mesothelioma: a rare form of lung cancer.

ACTIVITY

Visit the HSE's page on the risks of asbestos at: www.hse.gov.uk/asbestos/dangerous.htm

All of the commonly available commercial forms of asbestos have been linked to cancerous and non-cancerous lung disease. Although asbestos is not used in any new builds, continuing sources of exposure are asbestos removal and general construction industries. The delay between exposure to asbestos and the development of cancer is generally 20 years or more.

Asbestos-containing materials should have been identified before work begins, but there is always the risk that some may be hidden on-site and is not found until work has started. If you think you have found asbestos, STOP WORK AT ONCE and alert people that asbestos may be present. Asbestos is a difficult substance to identify, so it is better to assume a material contains asbestos until proven otherwise. Do not return to the site until it has been deemed safe to do so.

Do not start work if:

- you are not sure if there is asbestos where you are working
- the asbestos materials are sprayed coatings, board or insulation and lagging on pipes and boilers; only licensed contractors should work on these
- you have not been trained on non-licensed asbestos work – basic awareness is not enough.

You should continue only if:

- the work has been properly planned, the right precautions are in place and you have the correct equipment
- the materials are asbestos cement, textured coatings and certain other materials that do not need a licence; these are listed in HSE's 'Asbestos essentials' (see www.hse.gov.uk/asbestos/essentials/)
- you have had training in asbestos work and know how to work with it safely.

ACTIVITY

Watch the HSE video about the risks of asbestos at: www.hse.gov.uk/asbestos/videos/index.htm

If you work with asbestos:

- use hand tools and not power tools
- keep materials damp, not too wet

- wear a properly fitted, suitable mask (e.g. disposable FFP3 type); an ordinary dust mask will not be effective
- don't smoke, eat or drink in the work area
- double-bag asbestos waste and label the bags properly
- clean up as you go and use a special (class H) vacuum cleaner, not a brush
- after work, wipe down your overalls with a damp cloth, or wear disposable overalls (type 5)
- always remove overalls before removing your mask
- do not take overalls home to wash
- wear boots without laces, or disposable boot covers
- put disposable clothing items in asbestos waste bags and dispose of them properly
- do not carry asbestos into your car or home.

KEY POINT

Remember: do not take chances with asbestos! If you are in any doubt, seek expert advice.

Licensed asbestos removal

Asbestos removal requires a licence for all asbestos contamination situations where the risk of airborne asbestos particles is high. The Health and Safety Executive Asbestos Licensing Unit issues the appropriate documentation. To be granted a licence, a company must demonstrate the necessary skills, competency, expertise, knowledge and experience of work with asbestos, together with excellent health and safety management systems.

Licences, which act as a permit to work, are issued for a fixed time period, after which they have to be renewed. At this time, the recorded performance of the company through the HSE and local authority inspectors will be taken into account.

Waste management will also be covered in Chapter 10, Domestic fuel systems.

Asbestos disposal

In order of safety, there are three ways to dispose of asbestos and asbestos-containing materials (ACM). These are described in Table 1.5.

▼ Table 1.5 Safe ways to dispose of asbestos

1	Hire a specialist asbestos removal company.	Recommended
2	Dismantle the asbestos material yourself, taking the correct precautions with regard to health and safety, and hire a licensed asbestos waste company to dispose of the waste.	Not recommended
3	Transport it yourself to a site licensed by the Environment Agency. The asbestos will require double-wrapping in strong plastic bags and must be clearly marked as asbestos waste. (The site will usually make a charge for this service.) Before you arrive at the site you will need to telephone ahead to advise them of the type, quantity and intended time of arrival of the asbestos you wish to dispose of.	

Most licensed sites will accept only certain types and quantity of ACM. Usually these are:

- asbestos produced by the householder from domestic properties
- cement-bonded asbestos sheeting, pipes, gutters or flues in pieces of 150 mm or less
- asbestos sheeting that is in pieces of 150 mm or less
- a maximum of six small bags.

3 PERSONAL PROTECTION

The purpose and use of personal protective equipment

Personal protective equipment (PPE) is designed to create a barrier against workplace hazards. Health and safety law states that:

- your employer must try to make the wearing of PPE unnecessary if at all possible; if not, then:
  - your employer must provide you with PPE
  - your employer must train you in how to use PPE
- you must use the PPE provided by your employer
- you must take care of your PPE
- you must report lost or damaged PPE to your supervisor.

Depending on the type of workshop or site situation, the wearing of correct safety clothing and safe working practices are the best methods of avoiding accidents or injury.

All construction operatives have a responsibility to safeguard themselves and others. Making provision to protect yourself often means wearing the correct protective clothing and safety equipment. Your employer is obliged by law to provide:

- suitable protective clothing for working in the rain, snow, sleet, etc.
- eye protection or eye shields for dust, sparks or flying objects
- respirators, to avoid breathing dangerous dust and fumes
- shelter accommodation for use when sheltering from bad weather
- storage accommodation for protective clothing and equipment when not in use
- ear defenders where noise levels cannot be reduced below 80 dB(A) 8 hour
- adequate protective clothing when exposed to high levels of lead, lead dust or fumes, or paint
- safety helmets for protection against falls of materials or protruding objects
- industrial gloves for handling rough abrasives, sharp and coarse materials, e.g. rough cast concrete or when using toxic or corrosive materials.

Safety helmets

While on-site there is always a danger of materials or objects falling into excavations or from scaffolds, and there is also a danger that you will hit your head on protruding objects.

▲ Figure 1.13 Safety helmet

Always wear your personal safety helmet, which you will have to adjust to fit your head snugly. Do not add paint or stickers to your helmet, as these may reduce its effectiveness.

Safety helmets, approved to **BS EN 397**, are designed to:

- protect the head of a wearer against falling objects by resisting penetration and reducing the shock absorption by the head and body
- be used in temperatures as low as –30°C and as high as +150°C
- have electrical resistance up to 440 V
- be resistant against molten metal, marked as MM
- be resistant against side squeeze, marked LD, for lateral deformation.

Hard hats should be replaced once a year or if they have been struck by an object.

Safety footwear

You need to protect your feet against various hazards, including damp, cold, sharp objects, uneven ground and crushing. Flimsy footwear and ordinary trainers will not give the protection required. A good pair of boots with steel toecaps, **EN 20345**, and steel midsole for underneath protection is a mandatory requirement on construction sites.

INDUSTRY TIP

Look up the following webpage to see how safe your safety boots are: www.rockfall.com/conformity/standards-and-testing/

Overalls and work wear

There are numerous types of clothing produced to wear over your normal clothes for protection from dust, dirt and grime. Some have protective kneepad provision, which is especially useful for plumbers, and are designed to last longer. Plumbers should always consider flame-retardant work wear where possible.

▲ Figure 1.14 Plumbers' trousers

High-visibility jackets and vests are now a mandatory requirement for all construction site workers. The usual colours are fluorescent yellow or orange.

Eye protection

There are, on average, 1000 injuries to people's eyes every working day. Some injuries are so severe that they may cause partial or even total blindness. Your eyes are very vulnerable and an accident or injury can completely change your way of life.

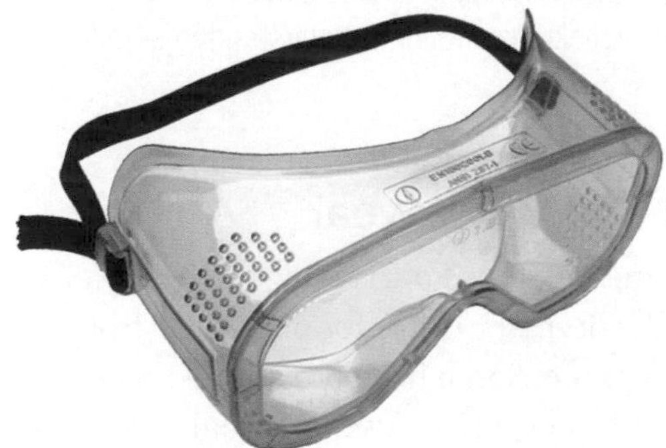

▲ Figure 1.15 Impact-resistant goggles

The majority of eye injuries would have been prevented if the correct eye protection had been worn. The protection that should be provided includes goggles, visors, spectacles, face screens and fixed shields.

As well as providing protection, suitable signs must be displayed where there is a chance of anyone sustaining an eye injury.

Types of hazard that can cause eye injuries

Some of the hazards and risks that might be encountered in the workplace are:

- using hammers and chisels
- handling or coming in contact with corrosive or irritant substances such as acids and alkalis
- the use of gas or vapour under pressure
- molten metals
- instruments that emit light or lasers
- abrasive wheels
- chipped or broken tools
- work involving welding or soldering
- threading steel pipe.

All eye protection should be CE approved to the relevant European standards, including **EN 166** and **EN 172**. Eye protection is a requirement by law under Regulation 4 of the Personal Protective Equipment at Work Regulations 1992 when working in a hazardous area.

In the event of an eye injury:

- no medication is to be applied to the eye
- the eye involved should be washed with clean, cold water if needed, and covered with clean, dry material (if possible, cover the unaffected eye as well, to reduce eye movement)
- immediate medical attention should be sought
- a thorough ophthalmic examination should be carried out within 24 hours.

Respirators (respiratory protective equipment)

Dust and fumes are a known hazard to health, especially when inhaled for long periods.

The greatest problem on-site and in the workshop is the dust from common substances such as wood, cement, stone, silica and plastics. Cutting and grinding of these materials can often produce large amounts of dust, which can cause breathing problems such as asthma and emphysema. In general, the dust is too fine to be seen with the naked eye, but problems and symptoms can appear in later years.

Fumes from solvents, paints and adhesives can also cause serious health problems, especially if used in confined or unventilated spaces.

The Personal Protective Equipment at Work Regulations 1992 make provision for the protection of employees at work from dust and fumes, and also persons not employed who may be at risk.

As well as providing respiratory protective equipment (RPE), suitable signs must be displayed where there is a chance of anyone coming into contact with dust and fumes from hazardous substances.

It is the responsibility of the employer to carry out a risk assessment to determine when RPE is required and what type is appropriate to control the exposure to the hazardous material.

Selecting the correct respirator

The selection of the correct RPE must be carried out by a competent person. The choice will depend upon:

- the nature of the hazard and material
- the amount of dust present
- the period of exposure
- the weather conditions, if working outdoors
- whether the respirator is suitable for the user, field of vision, communication, etc.

There are many types of RPE available, including:

- disposable face masks
- half dust respirators
- high-efficiency dust respirators
- ventilator visor or helmet respirators
- compressed air line breathing apparatus
- self-contained breathing apparatus.

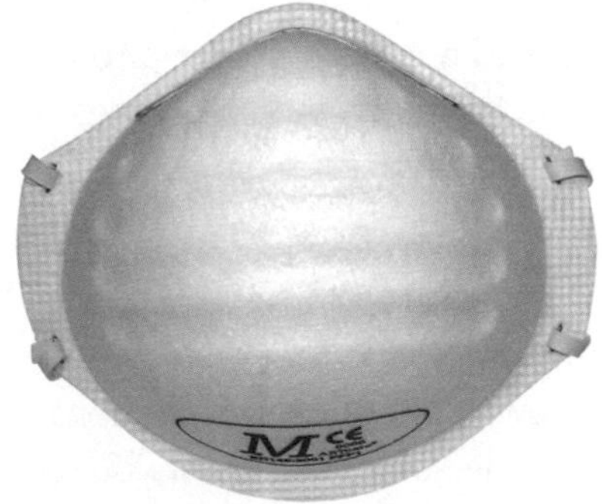

▲ Figure 1.16 Disposable dust mask

Gloves

There are many instances in construction where the correct hand protection is a necessity. The type of glove required depends on the type of work. Your hands are vulnerable to a wide range of hazards, such as cuts, blows, chemical attack and temperature extremes, making it vital you choose the right gloves.

- **EN 388** is the classification for gloves designed to protect the hands against mechanical risks associated with the handling of rough or sharp objects, which could cut or graze. A mechanical hazard does not mean moving machinery.
- **EN 407** – Protective gloves against thermal hazards: heat can be convected, conducted or radiated, or it may be a naked flame. Cold can be anything from cold water to freezing pipe gases.
- **EN 374** – Protective gloves against chemicals and micro-organisms: any substance that would irritate, inflame or burn the skin is classed as a chemical hazard. Some substances can cause the skin to become sensitive over a period of time, while others have an immediate, painful effect. This type of glove gives protection against chemical splashes and micro-organism hazards. They are often recommended specifically by the COSHH Regulations 2002.

▲ Figure 1.17 Gloves for mechanical risk

▲ Figure 1.18 Gloves for chemical risk

- **EN 12477** is the standard for protective gauntlets for welders.
- **EN 421** – Protective gloves against ionising radiation and radioactive contamination.

Hearing protection

The noise level at which employers must make a risk assessment and provide information and training is 80 decibels. There is also an upper noise limit of 87 decibels (taking into account hearing protection), above which workers should not be exposed. The British Standards for ear protection are:

- ear defenders **BS EN 352–1:2002**
- earplugs **BS EN 352–2:2002**
- ear defenders on safety helmet **BS EN 352–3:2002**
- level dependent ear defenders **BS EN 352–4:2001**
- active noise reduction ear defenders **BS EN 352–5:2002**
- ear defenders with electrical audio input **BS EN 352–6:2002**
- level dependent earplugs **BS EN 352–7:2002**.

▲ Figure 1.19 Ear defenders

The type of hearing protection you use will depend on the work you are doing. For very noisy situations or long-duration work, ear defenders would be the best solution as they offer greater protection than earplugs.

Manual handling

Manual handling operations are an important part of the construction industry. They are probably the biggest cause of back problems and time off work. Here, we will look at the following safe manual handling techniques:

- how to avoid manual handling injuries by using correct lifting methods
- how to assess your own lifting capability
- how to decide whether a manual handling activity is safe
- how to safely lift a load, transport it and put it down
- ways of reducing the load
- ways of avoiding manual handling.

Often, manual handling and lifting can cause immediate pain and injury; this type of injury is called an **acute** injury. Sometimes the result of an injury can take weeks, months or even years to develop. These types of injuries are called **chronic** injuries.

KEY TERMS

Acute injury: occurs when manual handling or lifting causes immediate pain and injury.

Chronic injury: type of injury that can take weeks, months or even years to develop.

As already mentioned, the Manual Handling Operations Regulations 1992 control manual handling and lifting, and require employers to reduce the risks from manual handling and employees to adopt the safe working practices set down by the employer.

Here are some points for you to consider before attempting any lifting or handling operation.

- Be aware of your own strength and limitations.
- Decide if it is a one-person operation or you require help.

- Always use mechanical equipment or aids if available and ensure you are trained in their use.
- Be sure of the weight of the item before lifting.
- Wear gloves to protect your hands.
- Wear safety boots to protect your feet.
- Make sure the area around is clear and safe to carry out lifting and movement.

KEY POINT

Manual handling

For further information and advice on manual handling, download the HSE's information leaflet at: www.hse.gov.uk/pubns/indg383.pdf

KEY POINT

Remember: even a light weight can cause injury, especially if it's too big and you can't see where you are going!

Lifting and handling techniques

INDUSTRY TIP

The basic principle of good manual handling when lifting is: 'knees bent – back straight'. This is known as the Kinetic Lifting Technique.

The one-person lift

To avoid injury, the principles listed in Table 1.6 should be followed.

▼ Table 1.6 Kinetic lifting

Step 1	Think before lifting/handling. Plan the lift. Can handling aids be used? Where are you moving the load to? Will you need help with the load? Remove obstructions in your way. For a long lift, consider resting the load midway on a table or bench to change your grip.
Step 2	Adopt a stable position. The feet should be apart, with one foot slightly forward to maintain balance (at the side of the load, if it is on the ground). You should be prepared to move your feet during the lift to maintain your stability.
Step 3	Get a good hold. The load should be hugged as close to your body as possible. This may be better than gripping it tightly with hands only.
Step 4	Start with a good posture. At the start of the lift, slight bending of the back, hips and knees is preferable.
Step 5	Don't flex the back any further while lifting. When you lift the load, your legs and the load should move together to avoid flexing the back.
Step 6	Avoid twisting the back or leaning sideways. This puts excessive strain on the back muscles, especially while the back is bent. Keep shoulders level and facing in the same direction as the hips. Turn by moving your feet rather than twisting your body.
Step 7	Keep the load close to your waist, and close to the body for as long as possible while lifting. Keep the heaviest side of the load next to your body.
Step 8	Keep your head up when handling. Look ahead, not down at the load, once it has been held securely.
Step 9	Move smoothly. Do not jerk or snatch the load as this can make it harder to keep control and can increase the risk of injury.
Step 10	Don't lift or handle more than can easily be lifted. If the load is too heavy, seek advice or get help.

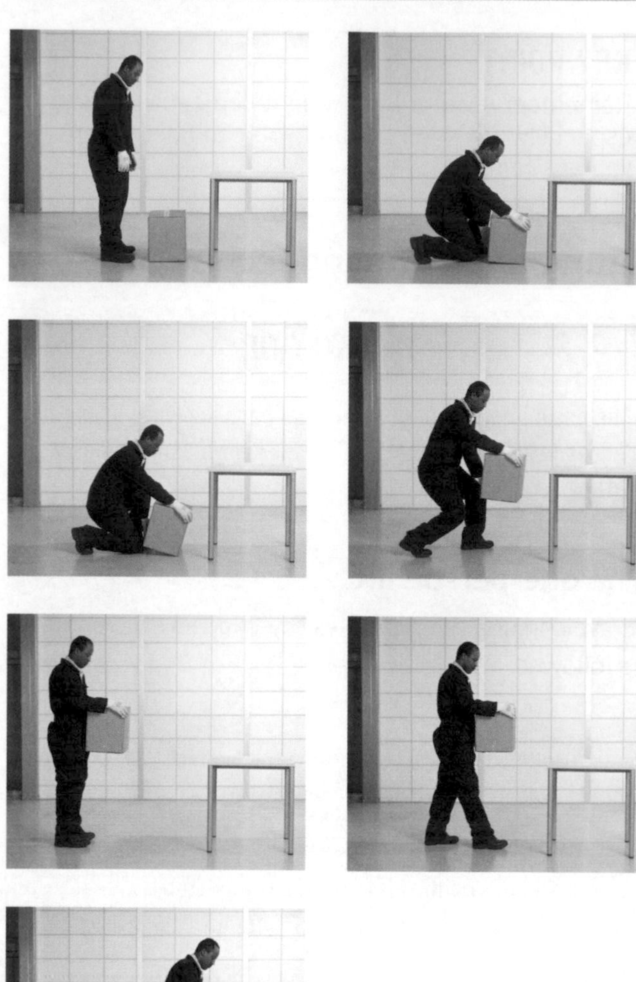

▲ Figure 1.20 Stages of a one-person lift

The two-person lift

Awkward shaped and very heavy objects should be moved or carried only with the help of other work mates. Appoint a team leader and obey his or her instructions.

When an object has been assessed as being too heavy or awkward to lift on your own, team lifting may be employed. The same rules of lifting should be applied. Try to pick someone of the same height and size so that the effort of each person is the same.

▲ Figure 1.21 Two-person lift

Mechanical lifting aids

There are numerous items of small lifting equipment available to assist with handling materials on-site and in the workshop. Use these only if you are qualified to do so. These range from small brick lifts, slings, barrows and dumpers through to mechanical forklift trucks.

- A pallet truck (Figure 1.21) can be used on hard areas for moving heavy loads.
- Barrows are the most common form of equipment for moving materials on-site.
- A sack truck (Figure 1.22) can be used for moving bagged materials, heavy boilers and other heavy pieces of plumbing materials.

▲ Figure 1.22 A pallet truck (top) and a sack truck

Most large construction sites will have a hired crane of some description, whether it is a fixed crane or a mobile crane. These are sometimes the only method of getting heavy equipment and appliances to where they are needed. They are operated by trained personnel only.

For very large sites, the crane operator will be guided by a 'banksman', who uses hand signals to the crane operator to guide the load to its destination.

▲ Figure 1.23 Mobile crane

Care should be taken if cranes are on-site, and you should be aware of where the jib (the lifting hook) is when you are walking to and from different areas of the site; the area it covers should be off-limits to all non-essential personnel.

4 RESPONDING TO ACCIDENTS

First-aid provision in the workplace

People at work can suffer injuries or fall ill at any time. The most important thing is that they receive immediate and appropriate attention. First aid covers the arrangements that should be made to ensure this happens. It can prevent minor injuries from becoming major incidents, and can often save lives.

What the law requires

Health and safety regulations require the provision of adequate and appropriate equipment, facilities and personnel to enable first aid to be given if an employee suffers an accident or injury or falls ill at work. While different working environments have different needs, the minimum first-aid provision in any workplace or on any construction site should include:

- a suitably stocked and maintained first-aid box; HSE advice suggests this should include at least:

  - 24 wrapped sterile adhesive dressings in assorted sizes
  - two sterile eye pads
  - four individually wrapped triangular bandages
  - six safety pins
  - six medium-sized and two large individually wrapped sterile unmedicated wound dressings
  - a pair of disposable gloves
- an appointed person to take charge of first-aid arrangements
- around-the-clock fast access to first-aid equipment
- a trained first-aider present at all times during working hours.

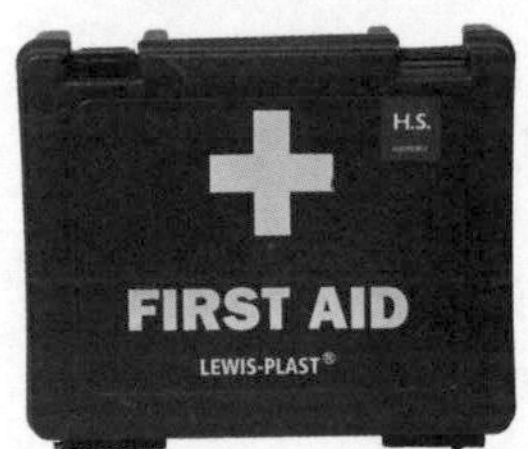

▲ Figure 1.24 A first-aid kit

What is an appointed person?

An appointed person is someone your employer chooses to:

- take charge when someone is injured or falls ill, including calling an ambulance if required
- keep stock of the first-aid box and replenish supplies
- be available at all times that people are working on-site.

What is a first-aider?

A first-aider is someone who has undergone a recognised first-aid training course such as a course given by the Association of First Aiders (AoFA) and recognised by the HSE. The first-aider must hold a current First Aid at Work certificate.

What your employer should consider when assessing first-aid needs

- Your employer is required by law to make an assessment of significant risks in your workplace, and to assess the risks of potential injury and ill health. If a significant number of risks exist, more than one first-aider may be needed.
- Your employer needs to assess whether there are any specific risks, such as working with hazardous substances, dangerous tools or machinery, etc., that could necessitate specific training for first-aiders or extra first-aid equipment.
- If there are different parts of your workplace that present different degrees of risk, your employer will need to make sure each area has the relevant provisions.
- Your employer may need to review the accident record book to find out about types of injury and how often they are occurring. This may influence the number of first-aid boxes and their exact location.
- If your workplace or site is spread out over different floors and buildings, adequate provision must be made for all locations.
- For shift work or out-of-hours working, your employer needs to ensure there are enough first aiders to cover all hours of operation.
- If any employees travel or work alone, the employer should issue a personal first-aid kit to them and provide training on how to use it.
- There are no legal responsibilities for guests and site visitors, but it is good practice to include them in first-aid provision.

Your employer must inform all employees of the first aid arrangements by putting up notices telling staff who the first-aiders are and where they can be found, as well as where the nearest first-aid box is kept. It is also good practice to make provision here for people who have reading difficulties or whose first language is not English.

▲ Figure 1.25 First-aid sign

Responding to an accident or an emergency

In an emergency, time is of the essence. The faster the emergency services arrive at the scene, the greater the chance that lives will be saved. The ambulance service advises that there is around an eight-minute response time to incidents. Their help and assistance is vital but we also need to help them get to the emergency. If calling for help:

1. Dial 999 and ask for the service that you require: police, fire or ambulance.
2. Once you are connected, speak clearly and logically to the operator. Tell them the nature of the incident, the location and possible entry points to your workplace or site.
3. Send work colleagues to wait at all the entrances for their arrival and to assist the emergency services when they arrive at the scene so that they can be directed straight to the incident. If necessary, have a chain of people to direct them to where they are needed if the site is large.
4. On no account leave the injured person. Stay with them and let the emergency services come to you.
5. Stay at the scene until you are not needed. Ask if the injured person should be accompanied to hospital and, if necessary, go with them.
6. Ask someone to advise the injured person's next of kin, wife/husband, etc., without alarming them unduly.

Dealing with minor injuries at work

Even the most cautious person will suffer from minor injuries from time to time. As a plumber you will experience minor cuts and burns. Here, we will look at the following minor injuries:

- minor cuts
- minor burns
- objects in the eyes
- exposure to fumes.

Tending minor cuts and burns

Cuts

Minor cuts need treatment to prevent dirt getting into the wound, causing infection. Some minor cuts will bleed quite a lot, depending on where the cut is and how deep it is. The area around the wound should be cleaned thoroughly with soap and warm water. If it is still bleeding, apply direct pressure to stem the flow of blood. It is a good idea to wear protective gloves when dealing with cuts that are bleeding.

The edges of a cut can be held together using butterfly bandages, and applying an antiseptic cream will help reduce the chance of infection. The wound can then be covered by a bandage or a sticking plaster. Care should be taken when using plasters as some people can suffer reactions to the adhesive, which can cause a rash.

Burns

Burns need to be treated immediately. First, cool the area with cold running water. Alternatively, the burn can be cooled by submersing the affected area in a clean bucket of clean, cold water. Keep the burn in the water for at least 10 minutes as this is the single most effective way of stopping the pain. Remove anything that could cause constriction (e.g. watches, jewellery) before the area starts to swell.

Once the burn has cooled sufficiently, it should be washed gently with clean water and covered with a sterile burns sheet or other suitable non-fluffy material. If no other materials are available, cling film or a clean plastic bag could be used. Do not apply any antiseptic cream or ointments as these have the effect of sealing the heat inside the burn, resulting in a more intense pain. Do not pierce or pop any blisters that develop as this could result in the burn becoming infected.

Depending on the severity of the burn, the person should be accompanied to the nearest hospital accident and emergency department or a doctor.

KEY POINT

Remember: these tips are for minor injuries only. You should seek expert medical attention if you think the wound is more serious or the following circumstances are present:

- the wound will not stop bleeding
- the injury is to the eye or ear
- the wound was caused by a rusty or dirty object
- the cut is deep or wide
- the person's last tetanus injection was more than 10 years ago
- the burn is larger than the palm of your hand or is situated on the neck, face, groin, foot or back of the hand
- signs of infection, such as redness of the skin or fever, are present
- the person has lost consciousness.

Objects in the eye

Objects in the eye can be painful and irritating. Loose objects like an eyelash or a speck of dirt, and even a contact lens, can float on the white of the eye. Usually, these can be rinsed off easily but you must never touch anything that penetrates the eyeball or rests on the coloured part of the eye (the pupil and iris) because this may permanently damage the eye. Faced with this situation, the casualty should seek immediate medical attention.

KEY POINT

Remember:

- do not touch anything that is embedded in the eye
- place a pad over the eye
- take the person to the nearest accident and emergency department.

The signs to look for are whether the person may be suffering from:

- blurred vision
- pain or discomfort
- redness or watering of the eye
- eyelids screwed up in a spasm.

The aim of any treatment you give is to avoid permanent damage, so:

- sit the casualty down, facing the light
- stand behind the casualty and very gently part the eyelids with a finger and thumb.

Make sure you examine every part of the eye by getting the person to:

- look up, then
- look down
- look to the left, and then
- look to the right.

If you spot an object on the white of the eye:

- wash it out with clean, cold water from a glass or fresh running water from the tap
- tilt the person's head towards the injured eye and place a towel or pad on the shoulder
- pour water from the bridge of the nose so that the water runs across the eye to flush the object out
- if this does not work, then lift the object off with a damp corner of a clean tissue or swab
- if this still does not work, seek medical advice.

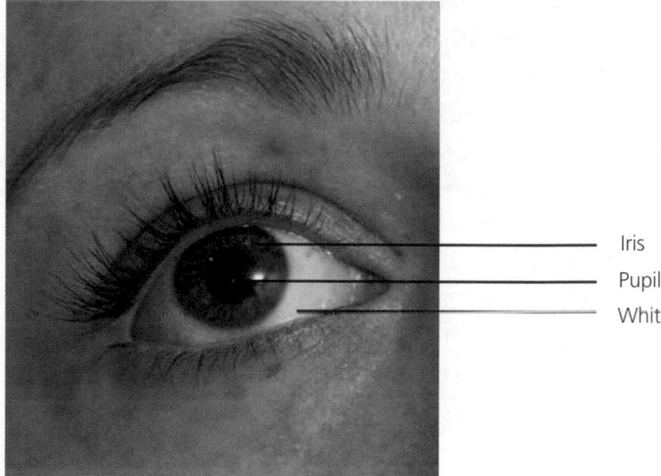

▲ Figure 1.26 Parts of the eye

Exposure to fumes

Dealing with a person who is suffering from exposure to fumes is a difficult area. The one thing you do not want to happen is that you become overcome with the fumes yourself. You will have to consider the following points.

- The nature of the fumes:
  - What are they?
  - Where have they come from?
  - Can they be stopped?
- Can the area be sufficiently ventilated?
- Can I get the person out without falling victim to the same fumes?

If the person is unconscious, then getting them out of the area and into fresh air is absolutely vital. The following should only be carried out if you can minimise your own risk.

- Immediately carry or drag the person to fresh air.
- Minimise your exposure to the fumes.
- If the person is not breathing, start cardiopulmonary resuscitation (CPR) immediately if you are trained to do so, and continue it until the person is breathing or help arrives.
- Send someone to fetch help as quickly as possible.

Dealing with serious injury at work

In this section we will examine the best way of dealing with those injuries that are more serious, such as:

- fractures
- unconscious co-workers
- electric shock.

Fractures and breaks

A fracture is a break or crack in the bone. There are two types of fracture:

1. a simple fracture, where the skin is intact and there is no wound present; there may be a swelling around the area of the fracture
2. a compound fracture, where the bone causes a wound or the breaking of the skin; the bone may or may not be visible with this kind of injury.

You cannot always tell if the bone is broken, but if you are in any doubt always assume that it is. There are some signs to help you and a few rules to observe to ensure that the injured person is kept comfortable until the emergency services arrive:

- Check for deformity of the limb by comparing it with the opposite side of the body, i.e. left arm, right arm.
- Look for an open wound, which may indicate a hidden fracture.
- Check for pain. The injured person will be able to tell you where the pain is, if they have any. Check by gently feeling along the area. The person will almost certainly complain of discomfort.
- Check for swelling.

In a few cases there may be no pain associated with the fracture and the person may be able to move the injured limb. In most cases, however, the person will be in great pain and any movement will cause severe pain.

Talk to the person. Ask them questions. They might have heard the bone snap at the time of the injury. Overall, the best approach for limited treatment is as follows.

1. It is recommended that you check and monitor the person's airway, breathing and circulation.
2. Treat the person for shock, if necessary.
3. Ask questions to try to find out how the accident happened. This will be vital information when the emergency services arrive.
4. Examine around the area for wounds and cuts. Feel along the area carefully for tenderness, swelling and deformities.
5. Check the injured limb for a pulse. No pulse indicates a more serious problem that could require immediate surgery. If this is the case, seek emergency help IMMEDIATELY.
6. Lightly squeeze the person's fingers or toes. A lack of sensation may indicate a spinal injury or nerve damage. Again, if this is the case seek emergency help IMMEDIATELY. Stay with the person at all times until the emergency services arrive at the scene, and try to reassure them.

If you have not completed a first-aid or CPR course, you MUST seek immediate medical attention for the person to ensure that no more damage is inadvertently done.

Dealing with unconscious people

A person can faint or fall unconscious for many reasons:

- after strenuous work or exercise
- shock or emotional upset
- excessive heat
- the side effects of drugs or medication
- a blow to the head (concussion)
- a fit or seizure.

Fainting involves loss of blood to the brain, leading to dizziness, nausea, cold sweats and a partial or complete loss of consciousness, which usually is brief and the person makes a full recovery in a matter of minutes. The real danger here is not the period of unconsciousness, but the damage that can arise from the resulting fall.

More serious unconsciousness comes from a blow to the head (called concussion), a fit or a seizure. In these cases, recovery can take much longer and have underlying health implications later. The watchwords here are 'always seek immediate medical advice'.

What should you do?

1. Try to break the victim's fall.
2. Loosen any items of clothing that might restrict the flow of blood, such as neck ties and shirt buttons.
3. Lie the patient on their back and raise their legs to encourage blood to flow to the brain.
4. Make sure they are breathing and that their airway is clear. If they are not breathing, start artificial respiration STRAIGHT AWAY!
5. If they are breathing, place them in the recovery position (see page 39).

ATTEMPT THE FOLLOWING ONLY IF YOU HAVE BEEN TRAINED TO DO SO.

1. Opening and maintaining the airway is your first priority, to ensure there are no obstructions, like the tongue, that could prevent normal breathing. To do this, place one hand on the casualty's forehead and gently tilt their head backwards, then lift their chin using only two fingers.
2. Look, listen and feel for the victim's breathing for no more than 10 seconds. Is the chest rising and falling? Can you feel their breath against your cheek? If breathing is normal, you can place the victim in the recovery position. If breathing is not present, you will need to start cardiopulmonary resuscitation (CPR) immediately (see below).

Dealing with electric shock

Electricity is one of the most dangerous elements that we have to deal with. It is indiscriminate, you can't see it or smell it, but if you touch it, it could kill you. The measure of shock's intensity lies in the amount of current (measured in amperes) that is forced through the body, and not the voltage. Any electrical appliance used on a house wiring circuit of 230 V can transmit a fatal current.

While any amount of current over 10 **milliamps** (0.01 **amp**) is capable of producing painful to severe shock, currents between 100 and 200 mA (0.1 to 0.2 amp) are lethal.

KEY TERM

Amp (and milliamp): unit of electrical current, the measurement of ampere.

It is vital to know how to deal with a person who has had direct contact with a live electricity power source, how to isolate them from the power supply and administer CPR.

If you see someone who is in direct contact with electrical current, they need immediate help. The victim may be unable to move because of muscle spasms, or they may be unconscious. Helping such a person is very dangerous. If you touch them, you may get caught by the current yourself and become a second victim.

First, you should try to turn off and unplug the appliance or, better still, turn off the power at the electrical consumer unit (fuse box). If you cannot turn off the power, try to get a long piece of dry wood (a broom handle will do) or any non-conducting material, and try to break the contact between the victim and the electricity.

Do not move the victim if there is any suspicion of neck or spinal injuries, unless there is an immediate danger. Keep them lying down and check for a pulse and their breathing.

If the victim is not breathing, apply mouth-to-mouth resuscitation. If the victim has no pulse, begin CPR if you are trained to do so.

Once a pulse and breathing have been established, cover the victim with a blanket to maintain body heat, keep their head low and get medical attention. Stay with the victim until help arrives.

Cardiopulmonary resuscitation

There are many instances where a person may need cardiopulmonary resuscitation (CPR), from exposure to fumes to a blow on the head and electrocution, but what is CPR and how is it administered?

CPR is a manual method of maintaining a heartbeat and air supply to a person who has collapsed, is unconscious, is not breathing and has no pulse. The idea of CPR is to keep blood pumping around the body to maintain a supply of oxygen to the brain so that brain damage does not occur, until the person can breathe by themselves.

How do I perform CPR?

If someone is with you, you must send them to telephone for an ambulance immediately. If you are alone, telephone for an ambulance and then quickly return to the victim.

1. Check for a pulse or other signs of circulation.
2. Carefully place the victim on their back on a firm surface.
3. Kneel next to the victim's chest.
4. Remove or open the clothes around the victim's chest area so that the rib cage and sternum are visible.
5. Place the heel of one hand directly above the sternum, close to the point where the lower ribs meet.
6. Place the other hand on top of the first hand and interlock the fingers. Keep the fingers off the chest so that only the heel of the hand is touching, otherwise you risk further injury to the victim.
7. Move forward until you are directly above the sternum, straighten your arms and lock the elbows. You must push down about 4–5 cm for an adult on every chest compression, but you must release after every downward movement.
8. Start the 30 chest compressions at a speed of 100–120 times a minute, followed by two rescue breaths.
9. After the two rescue breaths, return to the chest and repeat the cycle.
10. Continue the 30:2 ratio until emergency help or a defibrillator arrives.

Continue CPR until:

- breathing, coughing or movement is seen
- the ambulance service paramedics arrive and you are asked to stop, or
- you are too exhausted to carry on.

The recovery position

Why use the recovery position?

Placing someone in the recovery position will ensure their airway remains open and clear. It also enables any vomit or other fluids to flow away from the casualty's airway so that they do not choke.

Recovery position for adults

This is the best position for a casualty who is unconscious but still breathing.

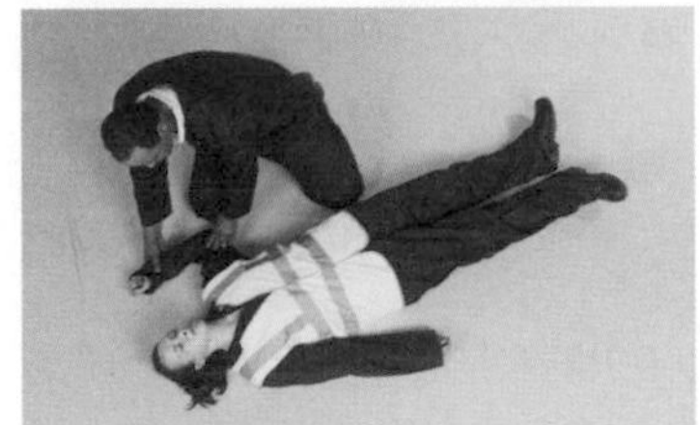

STEP 1 Place the arm that is nearest to you at a 90° right angle.

STEP 2 Move the other arm and rest the back of their hand against their cheek. Then take hold of the knee furthest away from you and pull it up until the foot is flat against the floor.

STEP 3 Pull the knee towards you, keeping the person's hand pressed against their cheek, and position the leg at a 90° angle.

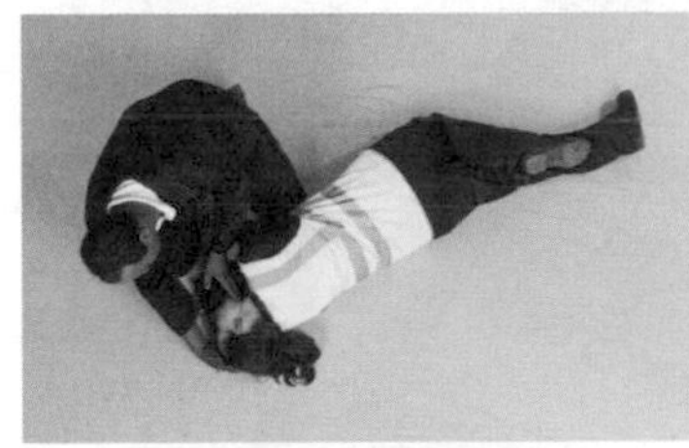

STEP 4 Make sure that the airway stays open by tilting the head back and lifting the chin. Check the casualty's breathing once more.

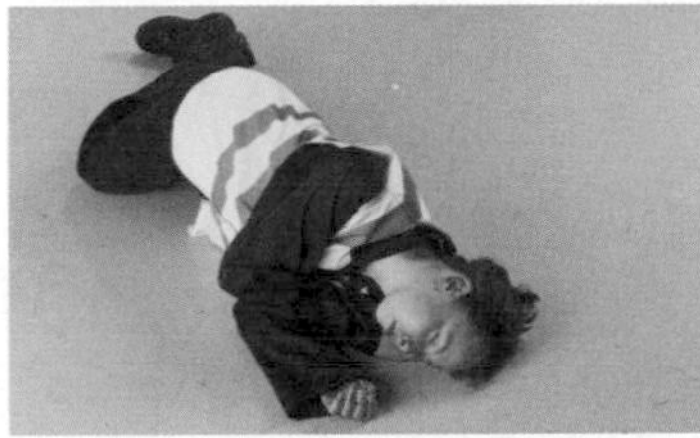

STEP 5 Monitor the casualty's condition and try to reassure them until help arrives. You must not leave them unattended for more than three minutes.

▲ Figure 1.27 The recovery position

Recording and reporting accidents and near misses at work

All accidents and incidents on-site should be properly recorded in the company's accident book and, if necessary, reported.

It is important that you have a basic understanding of accident reporting in terms of what the law says and what you should do.

Every accident should be reported – an accident report book should be on every site or place of work, usually with the site manager or whoever is in charge of the site or workshop. Make sure that you report any accident that you are involved in as soon as possible. Serious injuries are reportable under RIDDOR (see page 5).

Obviously, some accidents are more serious than others. Any accident that results in death, major injury or more than three days' absence from work is called a 'reported accident'. Any accident of this type should be reported to the HSE as soon as possible. Accidents where persons require hospital treatment must be recorded at the place of work, even if no treatment was given there.

▲ Figure 1.28 Accident report book

There is no set place to keep an accident book, but it needs to be kept in a place that is accessible and often is kept where first aid is available. Employers must make employees aware of where the accident book is kept.

All accidents MUST be entered in the accident book and the following information must be recorded:

- name, address and occupation of the injured person
- signature of the person making the entry, address, occupation and date
- when and where the accident happened
- brief description of the accident, its cause and what injury occurred
- whether the accident is of such a nature that it has to be reported to the HSE.

All accidents that cause death or major injury to an employee or member of the public must be reported to the HSE or your Local Authority Administrator for Health and Safety. A major injury is specified as certain fractures, amputations, loss of sight or anything that requires hospital treatment for more than 24 hours.

All accidents, whether fatal or otherwise, are investigated. Those involved in this investigation may include:

- the employer
- an investigator from an insurance company, acting on behalf of the employer or employee
- a safety representative, usually from a trade union
- a health and safety inspector from the local authority or HSE.

5 ELECTRICAL SAFETY IN THE WORKPLACE

During your work in the building services industry you will encounter many types of specialist equipment, some of it directly related to your job and some of it not. This part of the chapter covers how to work safely with or around the main types of equipment you will find on-site.

Electrical hazards on construction sites and in the home

Electrical hazards occur through:

- faulty installations
- lack of maintenance
- faulty or misused electrical equipment

- trailing cables
- buried or hidden cables
- inadequate fuse and over-current protection
- cables too close to pipework
- overloading electrical sockets and outlets
- using electrical equipment in wet or damp situations.

Electric shock is a major hazard; the severity of the shock will depend on the level of current and the duration of the contact.

- At low levels of current (about 1 milliamp) the effect may be only an unpleasant tingle but enough to cause loss of balance or a fall.
- At medium levels of current (about 10 milliamp) the shock can cause muscular tension or cramp so that anything grasped is hard to release.
- High levels of current (about 50 milliamps and above) for a period of one second can cause fibrillation of the heart, which can be lethal.
- Electric shock also causes burning of the skin at the points of contact.

Electric shocks are caused by a contact between a live conductor and earth. An electric current will always attempt to earth itself, therefore if anything comes between the flow of current and earth, the current will pass through it depending upon its resistance to the flow of current. The human body, because it contains 70 per cent water, is a very good conductor of electricity that offers very little resistance to the flow of electric current.

Some materials are poor conductors and will therefore offer greater resistance to the flow of electric current. Some of these materials, like PVC, are used to shield the electricity and are called insulators.

Electric cables consist of a copper wire (an excellent conductor) and a PVC outer cover or sheath (an excellent insulator). The result is a safe electric cable that can be used as an electrical supply for tools and equipment.

Electric shock is not the only problem because electricity can produce great amounts of heat, depending upon the size of the current and, if the current passes through a flammable material, it can ignite the material, causing a fire or explosion.

Methods of safe electrical supply on construction sites

To comply with the EAW Regulations, employers are required to maintain their electrical systems at work in a safe condition. According to the HSE, periodic inspections and testing should be completed as part of this maintenance. More than 1000 electrical accidents and incidents at work are reported to the HSE every year and around 30 people, across all sectors of industry, die from their injuries. The HSE reports that many deaths and injuries arise from:

- the use of poorly maintained electrical equipment
- work near overhead power lines
- contact with underground power cables during excavation work
- work on or near 230 V domestic electricity supplies, and
- fires started by poor electrical installations and faulty electrical appliances.

Electricity supply

The supply of electricity to homes and construction sites will normally be provided by either:

- a public supply from a local electricity company
- a site generator, where the use of the public supply is not practicable or is uneconomic.

The supply of electricity to a construction site

To maintain site safety, the supply of electricity to a construction site or workshop should always be distributed by means of a reduced voltage system. This system ensures that the correct voltage is supplied to where it is required:

- woodworking machines in a workshop require a 400 V 3 phase supply (Industrial/commercial voltage)
- site office lighting requires a voltage of 230 V 1 phase supply (domestic voltage)
- site portable power tools and site lighting require a 110 V 1 phase supply (safe site voltage).

Each site voltage has its own colour coding, as shown in Table 1.7.

▼ Table 1.7 Colour coding for site voltage

AC operating voltage	Voltage colour coding	Use
25 V	Violet	Lighting in damp conditions
50 V	White	Lighting in damp conditions
110 V	Yellow	General site voltage
230 V	Blue	Domestic and site offices
400 V	Red	Fixed machinery

The reduced voltage system must comply with the EAW Regulations 1989, and the distribution units, sockets and plug adapters should comply with **BS 4363:1998** (Specification for distribution assemblies for reduced low voltage electricity supplies for construction and building sites).

To avoid plugs designed for one voltage being connected to sockets of another voltage, there are different positions for the connecting pins in the plugs and sockets.

The voltage used on construction sites for site lighting and portable power tools is 110 V, colour-coded yellow. A 110 V 1 phase supply is much safer than 230 V and so the risk of serious injury from an electric shock is much reduced; 1 phase simply refers to the fact that there is only one live conductor, phase meaning live.

However, 230 V (colour-coded blue) for general site use is not allowed unless it is through a residual current device (RCD), which disconnects the supply immediately in the event of a fault or shock condition occurring.

Electrical installations in the workplace and domestic properties

All electrical installations should comply with **BS 7671** and be maintained to prevent danger. The HSE recommends that this includes an appropriate system of visual inspection and, where necessary, periodic testing. Electrical risks can be controlled by a simple system of looking for visible signs of damage or faults. This will need to be reinforced by thorough testing of the system as necessary.

It is recommended that fixed installations (the wiring to sockets, lights and fixed equipment) are inspected and tested periodically by a competent electrician.

KEY POINT

BS 7671 is the British Standard for the requirements for electrical installations. This is the national standard in the UK for low-voltage electrical installations. It is also used as a national standard by Mauritius, St Lucia and several other countries that base their wiring regulations on **BS 7671**.

Formal visual inspections and tests

During formal electrical inspections, the system will be checked and tested to ensure that:

- the polarity (live and neutral) of the system is correct
- all the fuses, miniature circuit breakers (MCBs) and RCDs are correct and working
- all the cables and cores are effectively terminated
- the equipment is suitable for its environment.

Working in domestic properties

When you are working in domestic properties, there are things that you can do to help prevent electrical hazards and accidents from occurring:

- Be aware of any concealed cables in solid and stud walls. Use a cable finder and check the wall before using drills and chisels.
- Do not install pipework too close to electrical cables. Heating pipework can cause the cable to overheat and faulty cables can arc across to the pipe causing a potential electric shock hazard. Pipework must be a minimum of 25 mm away from electrical cables and 150 mm from electrical apparatus.
- Take care when lifting or replacing floorboards. There may be cables underneath.
- Do not overload sockets and outlets with too many appliance connections as this can cause the system to overheat, sometimes with disastrous consequences. As a general rule, one socket = one plug unless a recognised, independently fused multi-socket is used.

- Look out for damaged cables, sockets and fittings. Report any problems to the customer or your supervisor.

Portable power tool safety

All portable power tools, such as drills, jig saws, circular saws and angle grinders, should be of the double insulated type, which simply means that the power tool has two levels of protective insulation built in to the appliance. The symbol for double insulated tools is shown in Figure 1.29.

▲ Figure 1.29 Double insulated symbol

Power tools must be subjected to safety tests; these are as follows.

- User checks – should be performed before use.
- A formal visual inspection – to be scheduled in accordance with your maintenance schedule and Health and Safety Policy.
- Combined inspection and test – to be carried out by a competent person, usually an external contractor (portable appliance testing).

Portable appliance testing

The 'Inspection and Testing of In-Service Electrical Equipment' (portable appliance testing, or PAT) was introduced to enable companies and organisations to comply with the EAW Regulations. To meet these regulations it is necessary to have in place a programme of inspection and electrical safety testing of portable appliances.

Records should be kept of all inspections and tests made and these should be up to date at all times. PAT testing helps to ensure:

- earlier recognition of potentially serious equipment faults, such as poor earthing
- discovery of inappropriate electrical supply
- discovery of incorrect fuses being used
- the misuse of portable equipment can be monitored
- an increased awareness of hazards linked to electricity.

ACTIVITY

Check the electrical power tools you use to see when they were last tested.

Type of equipment	Formal visual inspection	Combined inspection and testing
Construction sites 110 V equipment		
Stationary equipment	Monthly	3 months
IT equipment	Monthly	3 months
Movable equipment	Monthly	3 months
Portable equipment	Monthly	3 months
Hand-held equipment	Monthly	3 months

Visual inspections of power tools

Before using any portable electrical appliance, you should always carry out safety checks:

- Is there a recent PAT label attached to the equipment?
- Are there overheating or burn marks on the plug, cable, sockets or the equipment?
- Are any bare wires or conductors visible?
- Is the cable covering undamaged and free from cuts and abrasions?
- Is the cable too long, or too short?
- Is the cable a trip hazard?
- Is the plug in good condition (not cracked and the pins are not bent)?
- Are there any taped or other non-standard joints in the cable?
- Is the outer covering of the cable where it should be (i.e. no coloured wires are visible)?
- Is the outer casing of the equipment damaged or loose?
- Are 'trip-out' devices working effectively (i.e. RCD adapters)?

When using portable electrical power tools, you should always:

- wear or use PPE or clothing that is appropriate for the work you are doing
- switch off the tools before connecting them to a power supply
- if a power cord feels too warm or if a tool is sparking, have it checked by an electrician or other qualified person
- disconnect the power supply before making adjustments or changing accessories such as blades and drill bits
- remove adjusting tools before turning on the tool
- inspect the cord for fraying or other damage before each use
- tag defective tools clearly with an 'out of service' tag and replace immediately with a tool in good running order
- use clamps, a vice or other device to hold and support the piece being worked on, when practical to do so; this will allow you to use both hands for better control of the tool and will help prevent injuries if a tool jams
- use only approved extension cords that have the proper size flex for the length of cord and power requirements of the electric tool that you are using; this will prevent the cord from overheating
- fully unwind any extension cable being used; a coiled extension cable is likely to overheat, which could cause a fire
- for outdoor work, use outdoor extension cords marked 'w-a' or 'w'
- suspend power cords over aisles or work areas to eliminate tripping hazards
- pull the plug, not the cord, when unplugging a tool; pulling the cord causes wear, and may adversely affect the wiring to the plug and cause electrical shock to the operator
- keep the work area free of clutter and debris that could be a tripping or slipping hazard
- keep power cords away from heat, water, oil, sharp edges and moving parts; these can damage the insulation and cause a shock
- ensure that cutting tools such as drill bits and blades are kept sharp, clean and well maintained
- store tools in a dry, secure location when they are not being used.

Battery-powered cordless tools

In recent years, the use of battery-powered cordless tools such as drills and jig saws has become widespread both on construction sites and in domestic use. Voltages tend to be from 9 V to 36 V. Cordless tools offer many benefits over their mains-powered cousins:

- often the tools are smaller and lighter, giving greater flexibility of use
- no extension cables to cause trip hazards
- much less risk from electric shock.

On the downside:

- the power packs tend to wear out quickly and are costly to replace
- most are not as powerful as their mains counterparts
- power packs require constant recharging
- there is still an electric shock risk from the battery charger.

Cordless tools are still subject to health and safety inspection and testing with regard to:

- PAT testing of the battery charger
- disposal of spent battery packs in line with local authority guidelines as they contain nickel-cadmium and should not be disposed of in domestic waste
- leaking batteries, which contain acid; do not allow a leaking battery to come into contact with your skin
- not burning spent battery packs as they are liable to explode
- storage restrictions, as set storage temperatures exist for most cordless power tools.

Procedures for portable electrical tools that fail inspections

From time to time, electrical tools fail inspections and tests. If this should happen, on no account must the tool be used. The tool should be labelled as faulty and taken to the stores or for repair. A record of the fault must be logged on the maintenance and inspection documents.

A checklist like the one shown below can help you identify the hazards related to maintenance of portable tools. This will help you to take the necessary preventive measures. Depending on the power source, different checklists may be necessary:

General questions	Yes	No
Is there a maintenance plan?		
Are portable tools periodically tested and labelled with the date of test?		
Are instructions and operating manuals available?		
Are damaged tools labelled 'Do not use'?		
Are maintenance records kept of all tools that are used on the site?		
Are all tools used at the workplace in good condition and clean?		
Are all tools properly lubricated?		
Are blades, bits and other cutting parts sharp and well fixed, and not worn, cracked or loose?		
Are tools stored in a dry and safe place?		
Are blades removed when tools are being transported, stored or not in use?		
Are maintenance workers trained in safe working procedures?		
Electric power-operated tools	**Yes**	**No**
Are tools disconnected from the power source?		
Are the cables or plugs damaged?		
Have the electrical tools been subjected to unsuitable conditions (wet or dusty)?		
Are flexible extension cables in safe condition?		
Are there signs of overheating?		
Do all tools have safety guards on their blades, bits, rollers, chains, gears, sprockets and other dangerous moving parts?		

Safe isolation procedures for electrical supplies

Plumbers often need to work on electrical supplies for repair or replacement of equipment such as electric showers and immersion heaters. You should bear in mind, however, that to work on electrical installations you must have proved your competency and have gained Part P (**BS 7671**) certification.

The correct isolation of electrical supplies and systems is vital if accidents are going to be avoided.

In domestic properties, the type of electricity supply is 230 V, single phase. The EAW Regulations require that LIVE WORK IS NOT ATTEMPTED unless it is impracticable to work on the circuit when it is dead.

All electrical circuits must be properly switched off, isolated and, whenever possible, locked in the off position.

You must then prove that the circuit is dead by the use of an approved voltage indicator. Volt sticks and neon screwdrivers are NOT suitable for this purpose.

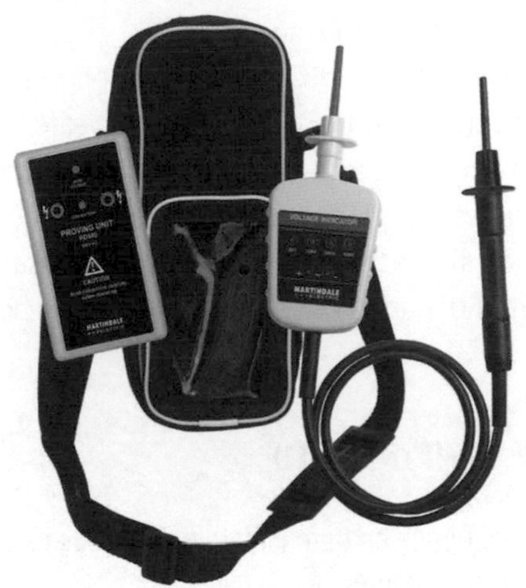

▲ Figure 1.30 Voltage proving meter and unit

▲ Figure 1.31 MCB safety lock

To safely isolate an electrical supply, you must do the following.

1. Identify the circuit or equipment to be isolated.
2. Make sure it is convenient to isolate the supply.
3. Select approved voltage indicator and test on proving unit.
4. Using the approved voltage indicator, verify that the circuit or equipment is live.
5. Isolate supply at consumer unit by switching off the miniature circuit breaker (MCB) or spur by removing the fuse.

6. Using an approved voltage indicator:
   a. First, check the indicator is working on a known live supply by testing live/neutral, live/earth, neutral/earth.
   b. Then use the indicator to check that the circuit you wish to work on is dead.
   c. Then re-check that the indicator is still working on the known live supply.
7. Lock off the isolator (RCD, MCB) using an approved lock or keep the fuse you have removed in a safe place. To be absolutely sure that no one can put the fuse back in, the safest place is in your pocket!
8. Place a notice or sign at the consumer unit advising that the circuit is off and must not be turned back on.

INDUSTRY TIP

Read the following webpage on safe isolation. The page can be downloaded for future reference.

www.electricalsafetyfirst.org.uk/media/1201/best-practice-guide-2-issue-3.pdf

Temporary continuity bonding

Temporary continuity bonding involves the use of two crocodile clips joined by 10 mm^2 earth cable. This is called a temporary continuity bonding clip.

All gas, water and central heating copper pipework should be bonded to the main electrical **equipotential bonding** system. In other words, copper pipework must be earthed. When we cut into a copper pipe to make a tee connection, for instance, we in effect disconnect all pipework after the cut from the earth system and, if a fault to earth already exists, then all the pipework after the cut could become live.

KEY TERM

Equipotential bonding: a system where all metal fixtures in a domestic property, such as hot and cold water pipes, central heating pipes, gas pipes, radiators, stainless steel sinks, pressed steel enamelled washbasins and steel and cast iron baths, are connected together through earth bonding so that they are at the same potential voltage everywhere.

The reason for temporary continuity bonding before removing or replacing metal pipework is to provide a continuous earth for the pipework, to prevent an electric shock in the event of any electrical fault. Installing temporary continuity bonding clips before we make the cut will ensure that, once the pipe has been severed, any bonding applied to the pipework will not be interrupted and the pipework will not become live. Once the connection has been made, the clips can be removed safely.

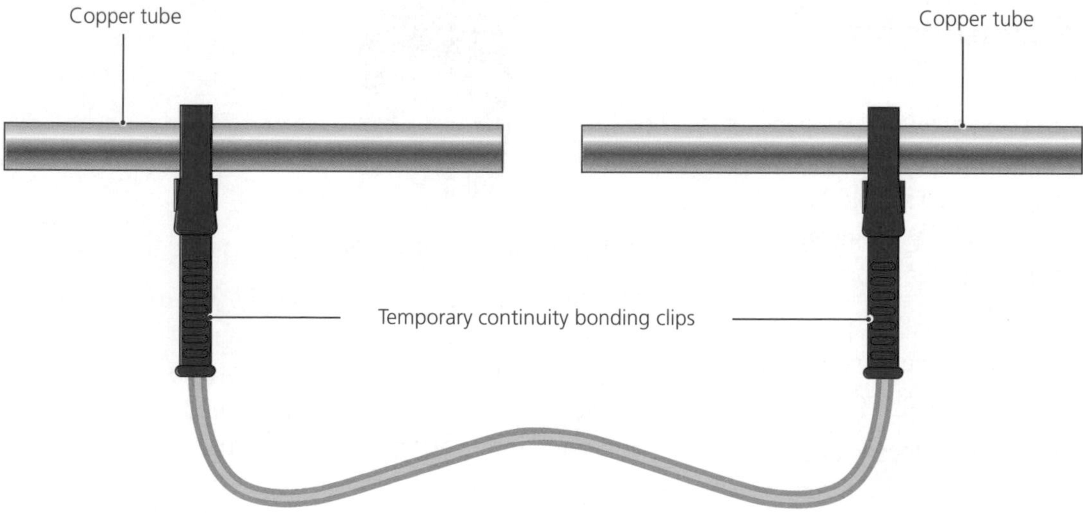

▲ Figure 1.32 Use of temporary continuity clips

6 WORKING SAFELY WITH HEAT-PRODUCING EQUIPMENT

Part of a plumber's work involves the use of heat-producing tools, such as blowtorches and possibly welding and brazing torches. Invariably these will use bottled gases, both flammable and non-flammable types. Using bottled gas of any kind can be dangerous and requires special consideration.

Identifying bottled gases

The types of gases you may come across are described below.

- Propane (C_3H_8) is a highly flammable, liquid petroleum gas (LPG) that is heavier than air, which makes it especially dangerous when working in trenches and confined spaces as any leaks would collect at low ground. Propane has a distinctive smell like rotten eggs. It is used for soldering processes.
- MAPP (methylacetylene-propadiene propane) gas is also used for soldering processes but has a much hotter flame than propane. Usually only supplied in small cylinders for plumbing work, MAPP gas has a distinctive garlic smell.
- Acetylene (C_2H_2) is used in conjunction with oxygen when undertaking welding and brazing processes. Plumbers usually use oxyacetylene sets only when lead welding. Acetylene is a colourless, odourless gas. When contaminated with impurities it has a garlic-like odour. Acetylene burns with a sooty flame that produces lots of carbon when used without oxygen. It is lighter than air.
- Oxygen (O_2) in the form of bottled liquid oxygen is a very powerful oxidising agent and organic materials will burn rapidly in the presence of oxygen. Used in conjunction with acetylene, oxygen hardens the flame, increasing the temperature. Although oxygen itself is not flammable, it can induce other materials to combust fiercely. NEVER use oxyacetylene near jointing compounds or grease as oxygen reacts violently in their presence and can spontaneously combust.

▼ Table 1.8 The four main types of gases you may come across in your work, along with cylinder colour and thread direction

Bottled gas	Cylinder colour	Thread direction
Propane	SIGNAL RED	Left-hand thread
MAPP	YELLOW	Left-hand thread
Acetylene	MAROON	Left-hand thread
Oxygen	BLACK	Right-hand thread

Many companies operate a written permit to work system when using fuel gases. This is known as hot work. The permit details the type of work to be done, how and when it is to be carried out and the precautions to be taken. Anyone carrying out hot work must have public liability insurance.

Training

Oxy/fuel gas equipment should not be used unless you have received adequate training in:

- the safe use of the equipment
- the precautions to be taken
- the use of the correct type of fire extinguishers
- the means of escape, raising the fire alarm and calling the fire brigade.

Safe storage, transportation and handling of bottled gases

- Oxygen cylinders should be stored at least 3 m away from those containing acetylene or LPG, or separated by a wall.
- Gas cylinders should preferably be kept on a hard surface (not soft ground) in a secure, open-air compound. The enclosures must be properly labelled.
- If stored in a storeroom, oxygen cylinders must not be kept in the same storeroom as LPG or acetylene cylinders.
- Acetylene and LPG cylinders should always be kept upright, even if they are empty.
- Oxygen cylinders can be stacked horizontally a maximum of four cylinders high, and wedged to prevent rolling.

- Vertically stacked cylinders should be secured against falling.
- Always keep full cylinders separate from empty ones.
- Cylinders should be shielded from direct sunlight or other heat sources to avoid excessive internal pressure build-up as this could lead to a gas leakage or, in extreme cases, bursting of the cylinder.
- Gas cylinders must be treated with care and not subjected to shocks or falls.
- NEVER lift oxyacetylene or LPG bottles by their control valves.
- When they are transported around a site, cylinders should be secured upright to avoid any violent contact that could weaken the cylinder walls.
- When they are unloaded from a vehicle they should not be dropped to the ground.
- Acetylene cylinders must always be transported and used in the vertical position. If they have been left in the horizontal position they must be stood upright for approximately 12 hours to allow them to settle before they are used.
- Cylinders should be transported only on purpose-designed trolleys of the correct size. Three-wheeled trolleys are safer than two-wheeled.
- Trolleys for transporting cylinders should be manufactured to **BS 2718**.

Equipment used with oxy/fuel gases

As well as the bottles themselves, there are several other pieces of equipment that we need before we can start using our oxyacetylene bottle set. The main components of oxy/fuel gas equipment are shown in Figure 1.33.

A flashback arrester to protect cylinders from **flashbacks** and backfires. Flashback arresters (also called flame traps) must be fitted into both oxygen and acetylene gas lines to prevent a flashback flame from reaching the regulators.
A pressure regulator fitted to the outlet valve of the gas cylinder, used to reduce and control gas pressure. Most modern regulators work with a two-stage system: the initial stage dispenses the gas at a set rate from the storage cylinder; the second stage handles the pressure reduction. On a two-stage system, the device has two pressure gauges. One gauge tells how much gas is remaining in the cylinder, and the other tells the pressure of the gas being released.
A control valve to shut off or isolate the gas supply, usually situated at the top of the cylinder. It has a square key to open and close the valve. As a general rule, when using oxyacetylene, both the oxygen and acetylene bottles should have their own key, which should be left on the bottle during the welding process so that the bottle can be isolated quickly in an emergency.
Cylinders of oxygen and fuel gas (propane or acetylene).
Non-return valves to prevent oxygen reverse flow into the fuel line and fuel flow into the oxygen line. The valves can be used to prevent conditions leading to flashback, but should always be used in conjunction with flashback arresters.
Flexible hoses to convey the gases from the cylinders to the blow pipe. Hoses between the torch and the gas regulators should be colour-coded: red for acetylene and blue for oxygen. Fittings on the oxygen hose have right-hand threads (non-flammable gas), while those on the acetylene hose have left-hand threads (flammable gas).
A blow pipe or other burner device where the fuel gas is mixed with oxygen and ignited.

▲ Figure 1.33 Oxyacetylene set

KEY TERM

Flashback: where the flame burns in the torch body, accompanied by a high-pitched whistling sound. It will occur when flame speed exceeds gas flow rate so that the flame can pass back through the mixing chamber into the hoses. Most likely causes are incorrect gas pressures giving too low a gas velocity, hose leaks or loose connections.

Oxyacetylene equipment safety checks

Before using welding equipment, it is wise to check its condition and operation. As well as normal equipment and workplace safety checks, there are specific procedures for oxyacetylene.

You should check that:

- flashback arresters are present in both oxygen and acetylene lines
- the hoses are the correct colour, with no sign of wear or damage, as short as possible and not taped together
- the regulators are the correct type for the gas being used
- a bottle key is in each bottle
- the bottles are securely fastened by chains to the bottle trolley and the trolley is in good condition.

It is recommended that oxyacetylene equipment be checked at least annually. Regulators should be taken out of service after five years. Flashback arresters should be checked regularly in line with the manufacturer's instructions; it may be necessary to replace some types if flashback has occurred.

Assembling and purging the oxyacetylene equipment

Before assembling the equipment, you must:

- check that all cylinders have been handled properly
- check that there are no physical signs of damage to the cylinder
- check the valve assembly on each cylinder for damage
- inspect the chains or other device used to secure the acetylene and oxygen cylinders.

Assembly of the gauges, hoses and blow pipe

- Make sure that each regulator is the correct type for the cylinder it is to be attached to and that the regulator is designed for the pressure of the cylinder.
- Open the oxygen valve assembly briefly before attaching the oxygen regulator. This is to eliminate the potential for a dust explosion.
- NEVER open the acetylene control valve to 'blow-out' as this could cause a fire.
- Inspect the regulator and cylinder valve for the presence of any oils or grease. If present, DO NOT USE.
- Make sure the adjusting screw on the regulator has not been damaged.
- Wipe the connection seats with a clean cloth.
- Connect the gauges to the cylinders. Typically, oxygen hoses are blue and acetylene hoses are red. Remember that the acetylene hose will have left-hand threads and the fitting will be a male type, to prevent accidental switching of the two hoses/gauges. Tighten them with the correct size of open-ended spanner. Take care not to damage the brass threads and do not over-tighten or the brass thread could snap.
- Make sure that the flashback arresters and non-return valves are in good condition and fitted to the gauges.
- Inspect the torch. Check that the inlet connection is in good condition for a tight connection. Check for obvious physical damage to the torch.
- Make sure the acetylene regulator is turned off by turning the regulator handle anti-clockwise out a few turns, then turn on the gas valve on top of the cylinder. Only turn the control valve half a turn. This allows the bottle to be turned off quickly in an emergency. Never allow acetylene gas pressure to exceed 15 PSI. At higher pressures acetylene becomes unstable and may ignite spontaneously or explode.
    1. After turning on the acetylene cylinder control valve, open the regulator valve by turning the handle clockwise. This should be done very slowly, while watching the low pressure gauge. Open only until the pressure indicated is between 5 and 8 PSI.

2. Open the gas valve on the blow pipe handle until you hear gas escaping. This is to purge the air from the acetylene hose. Then observe the low pressure gauge to see if the pressure remains steady during flow, to ensure you have the regulator set correctly.
3. Close the acetylene valve on the torch.
4. Check for leaks by using suitable non-greasy leak detection fluid.

- Turn the oxygen regulator pressure off by turning the regulator handle a few turns anti-clockwise then proceed with the following steps to adjust the oxygen pressure.
  1. Open the oxygen cylinder control valve all the way.
  2. Open the regulator valve slowly, watching the low pressure gauge as you do so, until the pressure reads between 25 and 40 PSI.
  3. Open the oxygen valve on the blow pipe to allow the air to vent out of the hose until the hose is purged, about three to five seconds for an 8 m hose.
  4. Close the blow pipe valve.
  5. Check for leaks by using suitable non-greasy leak detection fluid.
- Make sure that you purge both acetylene and oxygen lines (hoses) prior to igniting the torch. Failure to do this can cause serious injury to personnel and damage to the equipment.

Safe lighting and extinguishing procedures for oxyacetylene equipment

To light

1. Open the acetylene blow pipe valve a quarter turn and light the acetylene with a friction-type lighter.
2. NEVER LIGHT THE OXYACETYLENE TORCH WITH MIXED GAS.
3. Adjust the acetylene flame to the desired velocity.
4. For welding mild steel, open the oxygen blow pipe valve and adjust to neutral flame (equal amounts of acetylene and oxygen).
5. For brazing or bronze welding, open the oxygen blow pipe valve and adjust to a slightly oxidising flame (slightly more oxygen than acetylene).

> **KEY POINT**
>
> Make sure that you purge both acetylene and oxygen lines (hoses) prior to igniting the torch. Failure to do this can cause serious injury to personnel and damage to the equipment.

To extinguish

1. Close the acetylene blow pipe valve first, then close the oxygen blow pipe valve.
2. Turn off both acetylene and oxygen control valves on the cylinders.
3. Turn the acetylene regulator handle anti-clockwise until it is loose.
4. Open the acetylene blow pipe valve to release the pressure off the regulator.
5. Close the acetylene blow pipe valve.
6. Turn the oxygen regulator handle anti-clockwise until it is loose.
7. Open the oxygen blow pipe valve to release the pressure off the regulator.
8. Close the oxygen blow pipe valve.

What to do in the event of leakage

If you smell acetylene:

- NEVER use an open flame to check for leaks
- use commercial leak detector solution that is compatible with oxygen and acetylene to check all equipment connections before starting work
- NEVER use a leaking cylinder.

If the cylinder leaks:

- close the cylinder valve
- label the bottle as 'leaking'
- remove the cylinder to an outdoor location and post 'no smoking' and 'keep clear signs'
- call the gas supplier to collect the cylinder as soon as possible.

Safe use of liquid petroleum gas

Liquid petroleum gas (LPG) is the generic name for the family of carbon-based flammable gases that are found in coal and oil deposits deep below the surface of the earth. They include:

- methane
- ethane
- butane
- propane.

Of these, generally only two – butane and propane – are commercially available as bottled LPG. Plumbers regularly use propane when soldering copper tubes and fittings.

▲ Figure 1.34 Commercial propane cylinder

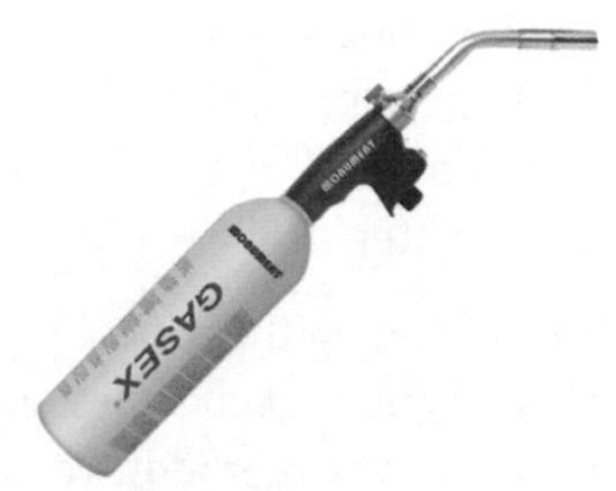

▲ Figure 1.35 A modern plumber's blowtorch

Commercially available propane

Propane has many uses, from camping to industrial processes, and is available in a variety of cylinder types and sizes. The colour of propane bottles is signal red. Propane turns from its liquid state to a gas, in other words it boils, at −42°C, whereas butane boils at −4°C. This means that propane can be used when the outside temperature is much colder – a distinct advantage when working on a construction site in the winter.

LPG regulators, hoses and blowtorches

Most blowtorches that we use today require a regulator to control the amount of gas that flows from the cylinder and a hose that connects from the regulator to the blowtorch. The regulator should have an adjustable pressure setting control. High-pressure hoses are usually coloured orange and are manufactured to **BS 3212**.

There are many different types of blowtorch available. Most have a range of interchangeable aeration nozzles of differing sizes so that the correct nozzle can be chosen for the type of work. Some blowtorches connect straight onto a small propane or MAPP 400 g gas cylinder.

Precautions with LPG

The main dangers caused by LPG are fire/explosion, carbon monoxide poisoning, asphyxiation and extreme cold, but there are others, as described below.

- LPG (propane or butane) is a colourless liquid, which easily evaporates into a gas when exposed to the outside air. One litre of liquid propane creates 250 litres of gas.
- It has no smell. Its distinctive odour is added to help detect leaks.
- It can burn or explode when it is mixed with air in the correct ratios and if it comes into contact with a source of ignition.
- It is heavier than air, so tends to sink towards the ground. It can flow for long distances along the ground, and can collect in drains, gullies, cellars and trenches.
- LPG is supplied in pressurised cylinders to keep it liquefied. The cylinders are strong and not easily damaged, but the control valve at the top can be vulnerable to damage if knocked.
- Leaks can occur from valves and pipe connections, mostly as a gas.
- If the gas is drawn from the cylinder too quickly, the control valve is likely to freeze.
- LPG liquid can cause cold burns if it comes into contact with the skin.
- LPG equipment should be used in a well-ventilated space to prevent the build-up of carbon dioxide (CO_2). Take particular care when using in a confined dry space, such as a loft.

The basic rules of LPG storage

- LPG cylinders should be used and stored in an upright position.
- They should be stored in well-ventilated places, away from sources of heat, sources of ignition and combustible materials.
- Cylinders must not be stored or used below ground level or in high-rise flats where LPG gas is prohibited by law.
- Cylinders should preferably be stored in a lockable cage outdoors, away from entry and exit points into buildings, and away from inspection chambers and drains.
- Do not keep LPG cylinders near to any corrosive, toxic or oxidant material.

- Propane cylinders may be used indoors in commercial and industrial premises only on a temporary basis, i.e. blowtorch use etc., but the cylinders must always be stored outdoors.
- When connecting hoses and blowtorches, always check for leaks with a suitable leak-detection fluid.
- Always turn the cylinder off at the control valve when it is not in use.

Fire safety

An important part of learning and understanding fuel gases such as propane and acetylene is awareness of what they produce as an end result: fire.

Fire is one of the most destructive elements known to man and it is something that plumbers risk on an almost daily basis when we solder, braze and weld. But what is combustion? What are the circumstances that are needed before combustion and the resulting fire takes place? How can we control it and reduce the risk of it occurring? And what do we do if a fire breaks out?

Combustion

Combustion is a chemical reaction in which a substance (the fuel) reacts violently with oxygen to produce heat and light. The fuel can be a solid such as wood, a liquid such as petrol, or a gas such as propane. Oxygen is known as an oxidiser or an oxidising agent. To create combustion or fire, we need a third element in the form of heat or an ignition source. These three elements – fuel, oxygen and heat – combine into what is known as the fire triangle.

▲ Figure 1.36 The fire triangle

All three need to be in place for combustion to happen. Take any of the three away and combustion will not take place.

If we remove the fuel then combustion will not occur simply because there is nothing for the fire to consume. The fuel can be removed naturally as the resulting fire consumes it, mechanically by removing the fuel or chemically by rendering the fuel incombustible.

Similarly, if we remove the oxygen, the fire will extinguish itself because the fuel has nothing to react with. There are several ways that we can 'suffocate' a fire: using foam, powder or CO_2.

Without a source of heat or ignition, fire can neither start nor continue. Take away the heat and a fire will die. If we douse a wood fire with water, the water turns to steam, which effectively removes the heat from the fire as the heat is transferred from the wood to the water.

Understanding these simple processes is the basis for all fire-fighting techniques and the fire extinguishers we use, as the methods we employ to effectively fight fires involve removing the heat, removing the fuel or suffocating the flames.

Fire safety in the plumbing industry

The use of soldering and welding equipment presents plumbers with the potential to cause fires in homes, factories and commercial properties. You should take precautions to eliminate as much as possible the fire risk from your everyday work. You can do this by:

- always carrying a dry powder or CO_2 fire extinguisher with you when soldering or welding
- always using heatproof mats when soldering next to wall coverings and skirting boards
- moving furniture and carpets away from the soldering area
- never pointing your blowtorch directly at combustible materials
- when soldering joints under a suspended floor, before you solder checking to make sure there is nothing that could catch fire
- never replacing floorboards etc. after soldering activities until you are sure there is nothing smouldering underneath the floor; wait at least an hour
- when lead welding on a flat roof, damping off the substrate before welding begins.

Classes of fire and fire extinguisher

There are six classes of fire (as shown in Table 1.9), each involving a different source of fuel. Because of this, each class of fire requires a different class of fire extinguisher, although some extinguishers can be used on more than one class of fire.

▼ Table 1.9 The six classes of fire

Class A	SOLIDS such as paper, wood, plastic
Class B	FLAMMABLE LIQUIDS such as paraffin, petrol, oil
Class C	FLAMMABLE GASES such as propane, butane, methane
Class D	METALS such as aluminium, magnesium, titanium
Class E	Fires involving ELECTRICAL APPARATUS
Class F	Cooking OIL and FAT, etc.

▲ Figure 1.37 Types of fire extinguisher

There are four classes of fire extinguisher. Each fire extinguisher is coloured red but has a different coloured panel on it to show its content (see Figure 1.37).

Table 1.10 explains where each of these can be used.

▼ Table 1.10 The uses of different classes of fire extinguisher

Class A	Class B	Class C	Class D	Class E	Class F
Water					Special wet chemical fire extinguisher
Foam	Foam				
Powder	Powder	Powder	Powder	Powder	
	CO_2	CO_2	CO_2	CO_2	

Fighting small localised fires

The following steps describe the correct procedure for dealing with small fires.

There is a simple way to remember the steps to take when using a portable fire extinguisher. Start by standing at least 3 m back from the fire, then follow the acronym P.A.S.S.

- **P**ull the pin on the extinguisher. The pin is there as a safeguard and locks the handle. Pulling it out enables it for use.
- **A**im low. The hose or nozzle should be pointed at the base of the fire to be effective.
- **S**queeze the lever above the handle. This will shoot the extinguishing substance from the hose or nozzle. Keep in mind that most small extinguishers will run out of their extinguishing agent in 10 to 25 seconds.
- **S**weep from side to side. Move slowly towards the fire, keeping the hose or nozzle aimed at the base of the fire. If the flames appear to be out, release the handle and watch closely. If the fire reignites, repeat the process.

Keep in mind

- Before you use an extinguisher to fight small fires, make sure everyone has left the area and that the fire service has been called by dialling 999.
- Always have an exit route behind you. Never let the fire get between you and your escape route.
- Call the fire service to inspect the fire area, even if you are sure you have extinguished the fire.

7 WORKING SAFELY WITH ACCESS EQUIPMENT

Most work in the construction industry is carried out above ground level. Some of this work can be done at a normal working height of up to 1.5 m without the assistance of steps and ladders, but there will be occasions when you will be required to work at heights above normal working level.

There are various types of equipment that can assist you when required to work at height. Each of these types of equipment is designed for a specific purpose and use and should not be used outside of its limitations. These are:

- stepladders
- ladders
- roof ladders
- trestle scaffolds
- tower scaffolds (mobile and fixed)
- tubular scaffolds (fixed)
- mobile elevated working platforms (MEWPs) and mobile mini tower scaffolds.

Stepladders

These are used for internal work but can be used outside if there is a firm base to stand them on. They are manufactured from either timber, aluminium or glass-reinforced plastic (GRP) in various sizes and heights, and consist of a set of stiles supporting flat steps spaced at approximately 250 mm intervals. A back supporting frame is hinged to the top and secured at the bottom with a cord or a metal locking bar. When the steps are extended the locking bar ensures the correct working angle and this prevents the steps from collapsing.

Timber stepladders are susceptible to damage, warping and twisting, whereas the aluminium and GRP-type stepladders are much lighter and stronger, as well as rot-proof. Stepladders must not be used if they are broken, damaged, have been repaired or have missing parts. The points to remember are:

- timber stepladders must not be painted as this may hide defects
- aluminium ladders must not be used near overhead electric power lines
- on finding defects, a 'Do not use' notice must be displayed and the defects reported as soon as possible to your supervisor.

Aluminium and GRP stepladders have several advantages over their wooden counterparts.

- They are lighter than timber steps, very strong, rot-proof, and will not twist, warp or bend.
- The treads are not less than 76 mm deep and are non-slip and horizontal when the steps are open.
- The working top is not less than 100 mm deep.
- The back is attached with a single hinge extending the full width of the back of the stepladder. This ensures stability when the steps are open.
- The locking bar clips in place when fully opened.
- The feet are made of a non-slip material.

When working with stepladders, the following precautions should be observed.

- Never use a stepladder that has been painted or repaired.
- Never stand on the top platform of a stepladder. You are at a safe working height when your knees touch the top platform.
- Never stand side-on to work. Always face the job.
- Use stepladders only on firm, level ground.
- Use only the right size ladder for the job. Never be tempted to place the steps on, say, a pile of bricks, to gain extra height.
- Always use in accordance with the manufacturer's instructions.

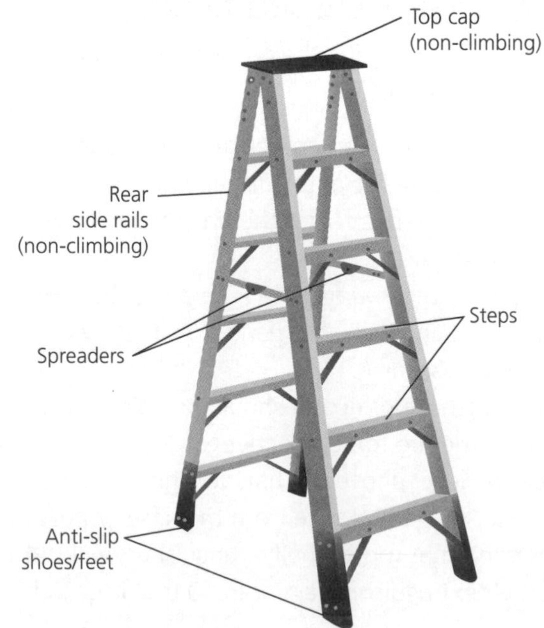

▲ Figure 1.38 A stepladder

Ladders

Ladders are used to gain access to scaffolds or light work at high levels. All ladders that are manufactured and supplied in the UK and the European Union should be constructed to identical standards and be classified correctly. These requirements apply to all portable ladder types, including stepladders, platform steps and extension ladders. Ladders are manufactured to **BS EN 131**.

> **INDUSTRY TIP**
>
> Look at the following website to get more details about the individual types of ladders: https://ladderassociation.org.uk/standards

All ladders are now made to **BS EN 131** and all have a 150 kg minimum capacity. Each class indicates the safe working load that a ladder is designed to support, which includes the weight of a single person plus their equipment. This is referred to as the 'maximum static load'.

The standards detail factors such as dimensions, markings, and testing requirements such as deflection, torsion, rigidity, straightness, loading and performance.

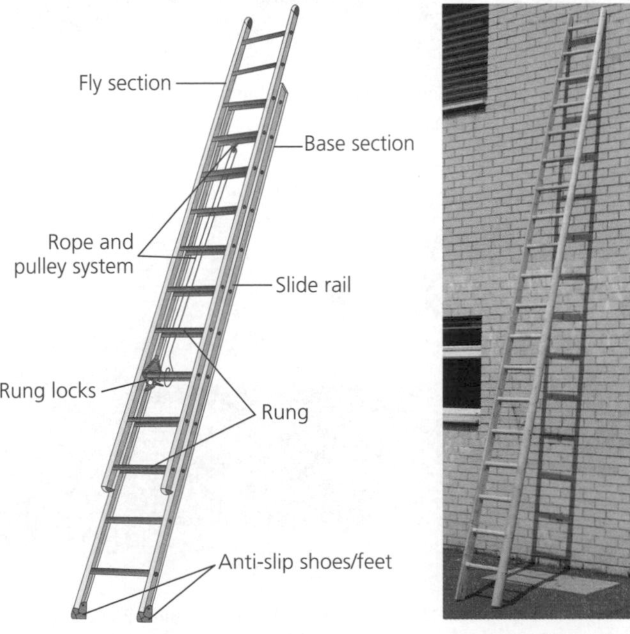

▲ Figure 1.39 An extension ladder

▲ Figure 1.40 A timber pole ladder

Types of ladder

There are several types of ladder available.

Pole ladders are generally made of timber. The stiles are cut from one tree trunk sliced down the middle. This ensures strength and durability. Pole ladders are used on fixed ladder installations for access to scaffolds and can be up to 12 m in length. Some pole ladders have wire reinforcement to provide extra strength. They have a Class 1 rating and will safely support a maximum load of 175 kg. They are made to **BS EN 131**.

Single-section ladders are usually made to Class 1 standard from lightweight aluminium. These ladders are often called standing ladders.

Multi-section ladders are often called extension ladders. They consist of two or three sections that can be slid apart to give the required height. They are available as two or three extensions and in various 'closed' lengths of 2.5 to 3.5 m. A double extension ladder can give a length of up to about 8 m and should be suitable for most two-storey properties. Three-section ladders can give lengths up to about 10 m.

On smaller ladders, the ladder may be extended by hand and secured with stay locks, which rest on a selected rung. On larger ladders, the sections are extended by means of a rope loop and pulley system running down the side of the ladder.

▲ Figure 1.41 A roof ladder

Multi-section ladders can be made of timber, aluminium and GRP.

Roof ladders: this type of ladder should always be used when working on a pitched roof. It should ALWAYS be accessed from a scaffold, not a ladder. The roof ladder has two wheels at the upper end, which allow it to be pushed up the roof without damaging the slates or tiles. On the other side to the wheels, the ladder is formed into a hook, which fits over the top ridge of the roof and stops the ladder from slipping down the roof.

Crawlboards are used for working on fragile roofs. They help to spread the weight across the roof to lower the risk of the roof giving way. They are used for access only and are not intended for carrying tools or materials. They should be used with extreme care.

Ladder safety check before use

Ladders, roof ladders and crawlboards should NOT be used if they have any of the following defects:

- broken, missing or makeshift rungs
- broken, weakened or repaired stiles
- broken or defective ropes, guide brackets, latching hooks or pulley wheels
- they are painted; paint can hide defects on wooden ladders
- they have missing safety feet.

Raising and lowering ladders

Ladders should be raised with the sections closed. Extension ladders with long sections should be raised one section at a time and slotted into position before the ladders are used. TWO people are required to raise and lower heavier-type ladders. The following is a step-by-step procedure for raising heavier-type ladders.

STEP 1 Lay the ladder flat.

STEP 2 One person stands on the bottom rung of the ladder and holds the stiles to steady the ladder as it is lifted.

STEP 3 The second person stands at the other end of the ladder and lifts the ladder over their head, moving hand over hand, walking towards the foot, raising the ladder as they go.

STEP 4 This is continued until the ladder is upright. When erected, the correct safety angle is 75° or a ratio of 4 up to 1 out.

▲ Figure 1.42 Raising heavier-type ladders

To lower the ladder from the upright position, the above process is reversed.

Lighter ladders can be raised by one person, but the bottom must be placed against a firm stop before lifting is commenced.

KEY POINT

Remember: ladders must extend at least 1 m above the working platform.

Tying ladders

Ladders must have a firm and level base on which to stand and, if more than 3 m long, they must be fixed at the top or, if this is not possible, at the bottom. If neither way is possible, a person must **foot the ladder**. They must hold both stiles and pay attention all the time. This prevents the base from slipping outwards and the ladder from falling sideways.

KEY TERM

Foot a ladder: stand with one foot on the bottom rung, the other firmly on the ground.

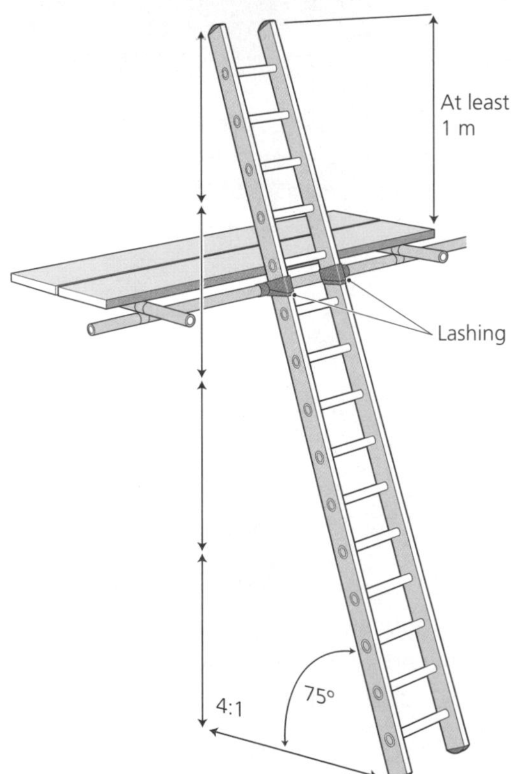

▲ Figure 1.43 A correctly erected ladder

▲ Figure 1.44 Tying and lashing ladders

Lifting and carrying ladders

To lift and carry ladders over short distances, rest them on the shoulder before lifting vertically by grasping the rung just below normal reach. The correct balance and angle must be found before moving.

When moving ladders more than a few metres, they should be lowered and carried on the shoulders by two people, one on either end.

▲ Figure 1.45 Carrying a ladder

Storing ladders

- ALWAYS store ladders in a covered, well-ventilated area, protected from the weather and away from too much dampness or heat.
- NEVER leave ladders leaning against a wall or building. Ladders can fall if stored vertically, so take

particular care. If possible, secure the top – with a bracket, for instance.

- NEVER hang a ladder vertically from a rung.
- DON'T store a ladder anywhere a child might be able to climb it.
- STORE the ladder horizontally; a rack or wall bracket is ideal. Always support on the lower stile and every 1 m (maximum).
- KEEP wooden ladders clear of the ground to avoid contact with damp.

Safe working with ladders

- Ladders MUST extend at least 1 m above the working platform.
- NEVER stand ladders on uneven, soft or loose ground.
- NEVER support a ladder on a fragile surface.
- NEVER stand a ladder on a box, drum or other unsteady base.
- NEVER use a makeshift ladder.
- NEVER use a ladder that is too short.
- NEVER climb ladders with slippery, icy or greasy rungs.
- NEVER wear soft-soled shoes or footwear with slippery soles.
- NEVER try to carry too much equipment up a ladder.
- NEVER over-reach when working on a ladder.
- TAKE CARE when raising to avoid overhead obstructions such as electric cables.
- ALWAYS ensure the ladder is at the correct angle of 75° (1 in 4), projecting above the working platform by at least 1 m and securely fixed.
- OVERNIGHT SAFETY: ladders should be lowered and stored in a safe place overnight. If this is not possible, a scaffold board at least 2 m long should be firmly lashed to the rungs to prevent access.

Scaffolds

Scaffolds are a much safer way of working at height but extreme care must still be taken. You need to be aware of your surroundings at all times and take care with tools and equipment.

Tower scaffolds

There are two types of tower scaffold:

1. Static – stationary, remains in one place and cannot be moved
2. Mobile – can be moved to a different location when needed.

Both kinds of tower scaffold can be either tailor-made for a particular job and constructed in situ (in place), or made by a manufacturer with standard sections that fit together (proprietary). You must be certificated and registered to erect all types of fixed scaffolding.

Static tower scaffolds

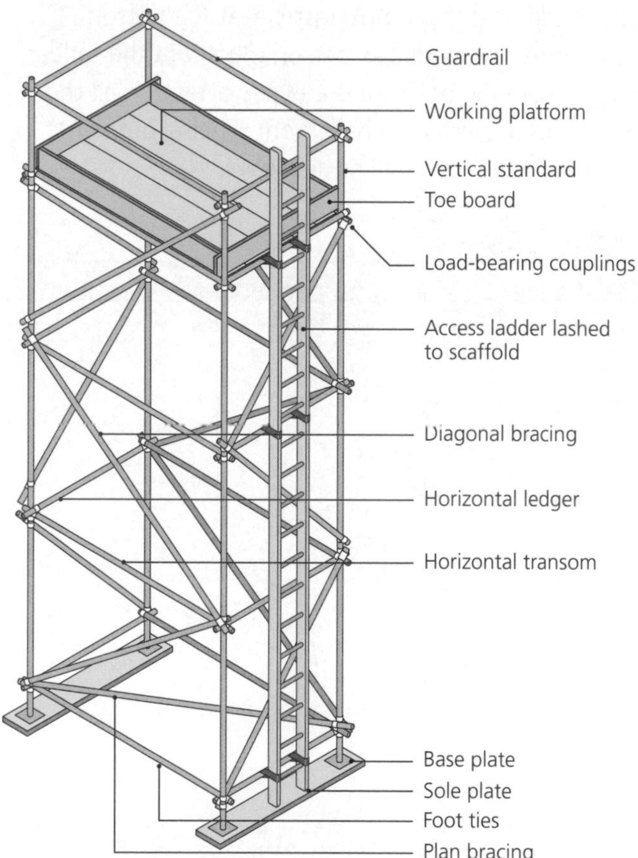

▲ Figure 1.46 A correctly erected static tower scaffold

This type of scaffold is constructed from individual tubular scaffold components using standard scaffolding clips, and strengthened by braces and ledgers (horizontal scaffold tubes that the boards rest on).

Access to the tower via a ladder may be from the inside or outside of the tower, but care must be taken to ensure that the tower is not destabilised by leaning ladders on the outside.

Scaffold towers should be designed to carry a load of 150 kg/m² spread over the whole working platform, in addition to their own weight. A special design will be required for any tower that may be subject to extra loadings from wind or materials on the working platform.

Proprietary static tower scaffolds

These towers are constructed with steel interlocking sections that simply slot together. The sections are pre-formed H-shaped welded units to give strength. These scaffolds are easy to erect but are often expensive and not as versatile as other static towers.

Static tower scaffold requirements

- Tubes must be straight.
- Tubes and fittings must be undamaged and free from corrosion.
- The ground that the tower is to be erected on must be firm and even.
- Base plates must be used. Adjustable base plates can be used on ground that has different levels.
- Sole plates must be used to provide even weight distribution if the tower is to be used on soft ground.
- Any couplers must be load bearing.
- Horizontal members must be fixed to uprights, with the exception of transoms (diagonal strengthening tubes) under the working platform.
- Towers must have diagonal and plan bracing built in.
- Foot ties or plan bracing must be fitted as low as possible.
- Working platforms must be close boarded.
- Overhang boards must not exceed four times the thickness of the boards and must not be less than 50 mm past the support.
- Working platforms above 2 m high must have toe boards fitted at least 150 mm high.
- Working platforms above 2 m high must have guardrails between 920 mm and 1150 mm high.
- Access ladders must be lashed vertically, preferably on the inside of the tower.
- The minimum base measurement for any tower is 1.25 m.
- When erected indoors, the height of any tower must not be greater than three and a half times its smallest base measurement.
- When erected outdoors, the height of the tower must not be greater than three times its smallest base measurement.
- Towers higher than 6.4 m must be tied to the building or have adequate outriggers fitted.
- ALWAYS use towers on firm, even ground – NEVER on sloping ground.
- NEVER place steps on the working platform.
- NEVER '**sheet out**' a tower.

KEY TERM

Sheeting out: sheeting out a tower scaffold means covering the outside of the scaffold with tarpaulins. This can be extremely dangerous as the tarpaulins act like the sails on a ship and could easily blow the scaffold over.

Mobile tower scaffolds and proprietary mobile scaffolds

These scaffolds are useful for light work of a short duration, such as installing boilers and flues. They are usually manufactured from aluminium for lightness. This scaffold should be used only where the ground is sufficiently firm, level and smooth to maintain stability. All wheels on mobile types must be the lockable type and kept locked when the scaffold is in use. The scaffold should only be moved by exerting force on the bottom of the tower and never pulled from the top.

Tower scaffolds should never be moved with persons still on them.

Outriggers

▲ Figure 1.47 A correctly erected mobile tower scaffold

If extra working height is required, then the base measurement can be increased by the use of outriggers. These are tubes or special units that connect to the bottom of the tower at the corners, giving a greater overall base measurement. Outriggers also help to stabilise a scaffold tower, as well as giving extra working height.

IMPROVE YOUR MATHS

Working out the height of a tower scaffold

If the base of the scaffold measures, say, 3 m × 2 m, take the shortest measurement and multiply by 3. So:

$$2\,\text{m} \times 3 = 6\,\text{m}$$

This means the tower scaffold can be erected safely to 6 m high.

ACTIVITY

If the base of a scaffold measures, say, 2 m × 1.5 m, calculate the maximum height of the tower scaffold.

Mobile tower scaffold requirements

The requirements for mobile tower scaffolds are the same as for static tower scaffolds, but with the following additions:

- castors must only be fitted to standards (vertical uprights)
- castors must be of a swivel type fitted with a locking device
- castors must fitted by a method that prevents the wheel from falling out if not in contact with ground.

Tubular scaffolds

Independent scaffold

This is a scaffold that, apart from the necessary building ties, stands completely free of the building. The main applications for this scaffold are:

- access for stonework on masonry buildings
- access to solid or reinforced concrete structures
- maintenance and repair work.

The independent scaffold consists of two rows of standards joined together by ledgers, which in turn are joined by transoms. The standards must always be upright and slightly towards the building. The inner row must be as near to the building as possible, but never more than 375 mm away from the wall. It is essential to include triangulation by cross-bracing at every lift to ensure rigidity.

Only load-bearing couplers should be used between the standards and ledgers, and on through ties. The ground should be firm and level, and base plates should be used under every standard. If this scaffold does need to be placed on soft ground, wooden sole plates 225 mm × 38 mm thick should be used. While it is called an independent scaffold, it does need tying to the building to prevent movement and for stability. This scaffold would usually be tied through door and window openings.

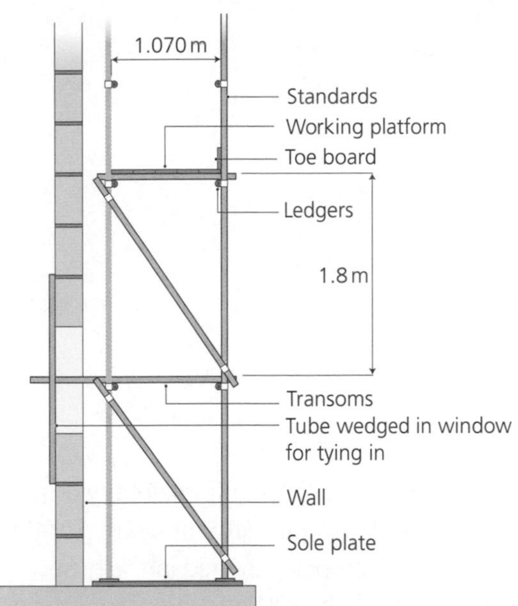

▲ Figure 1.48 Independent scaffold, side view

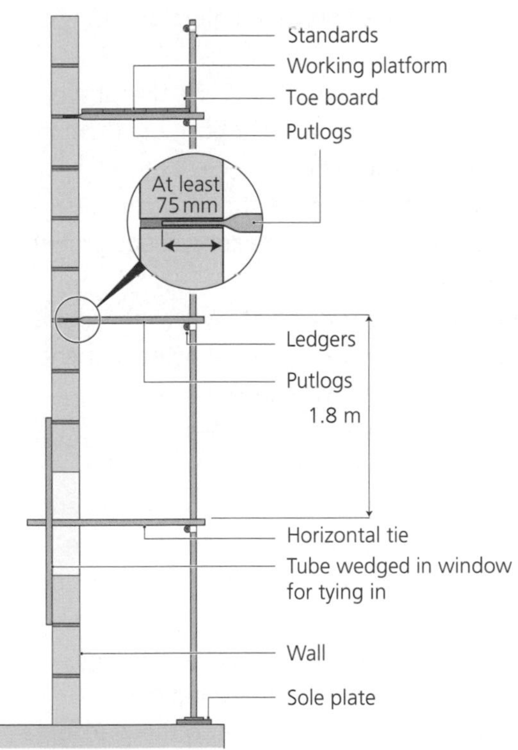

▲ Figure 1.49 Putlog scaffold, side view

Putlog scaffold

This is also known as a dependent, or bricklayers', scaffold and is similar to the independent scaffold but has only one row of standards, with the inner row replaced by the brickwork. This means that the inside ledgers and ledger bracing are not required.

The remaining scaffold functions in the same way as the independent scaffold.

The scaffold can be erected to existing brickwork, but is usually erected along with new building work. The working platform is supported by putlogs and not transoms. The putlog, in turn, is supported by the new brickwork by allowing the spade (flat) end of the putlog to rest flat on the brickwork. Putlogs should never be removed or the scaffold will be in danger of collapse. Putlog scaffolds should be tied to the building at least every 4 m vertically and 6 m horizontally.

Access to tubular scaffolds

It is usual to access a scaffold from a ladder, and provision must be made so that this can be done easily and safely. A suitable gap should be left in the handrail and toe board to allow operatives to access the scaffold. The ladder should be secured both at the top and bottom, and extend at least 1 m (approx. five rungs) above the platform.

The final rung of the ladder from which the operative steps onto the platform should ideally be just above the surface of the platform. The gap left between the ladder and the guardrail should not be more than 500 mm.

Guardrails and toe boards

The risk of falling materials causing injury should be minimised by keeping platforms clear of loose materials. Access platforms more than 2 m high must, therefore, have guardrails, brick guards and toe boards. They provide a method of preventing materials or other objects from rolling or being kicked off the edges of working platforms, so must be fixed at all open edges of the working platform.

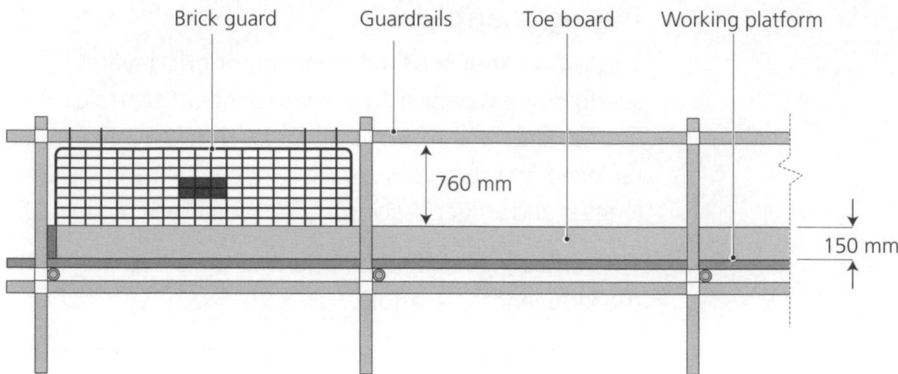

▲ Figure 1.50 Guardrails and toe boards

Working platforms

This is the level at which the work will be carried out. The following points should be noted.

- The loading of materials must be spread as evenly as possible.
- Working platforms must be kept free from ice, snow, grease and other hazards.
- Gaps between boards should be kept as small as possible.
- Boards must be at least:
  - 150 mm wide if they are 50 mm thick
  - 200 mm wide if they are between 32 mm and 38 mm thick.
- Boards must rest evenly on their supports.
- No board must project more than four times its thickness beyond its end support, and no less than 50 mm.
- All board ends should be bound with a steel strap to prevent splitting.
- Split or damaged boards must not be used.

Mobile elevated working platforms

Mobile elevated work platforms (MEWPs) include cherry pickers, scissor lifts and vehicle-mounted booms. MEWPs can provide a safe way of working at height. They:

- allow the worker to reach the task quickly and easily
- have guardrails and toe boards, which prevent a person falling
- can be used indoors or out.

Your employer should:

- choose the right MEWP for the job
- identify and manage the risks involved with working from MEWPs.

However, if used improperly or poorly maintained, these devices can cause serious injury or death. For this reason, you need to be properly trained in their use. Safety checks that should be carried out before use are as follows.

- The mid and top chains, railing and gate enclosures must be in the closed position before elevating the device.
- To prevent movement of the MEWP after it has been moved into the final work position, the control panel must be turned off.
- The maximum operating weight capacity must not be exceeded (including personnel, equipment, supplies and tools).
- Personnel must not work on MEWPs when exposed to high winds or storms, or when the equipment or materials are covered with ice or snow.
- MEWPs must not be used as cranes.
- No MEWP must travel with personnel in the basket while it is elevated, unless the equipment is designed for this activity and operations are conducted in accordance with the regulations and standards.
- MEWPs must be operated on stable, flat and structurally sound flooring or ground only.
- Where moving vehicles are present, the work area must be marked with warnings such as flags or roped-off areas, or other effective means of traffic control must be provided.
- Unstable objects, such as barrels, boxes, loose brick, tools and debris, must not be allowed to gather on the floor of the MEWP.

▲ Figure 1.51 MEWP in use

Using safety lines, harnesses and nets

Equipment that prevents falls, such as safety harnesses, restraints and safety nets, should be used only as a last resort when the risk cannot be eliminated by other means. They should be used only by operatives trained in their use, and must be inspected and tested at regular intervals.

8 WORKING SAFELY IN EXCAVATIONS AND CONFINED SPACES

Every year construction workers are killed and injured when the excavations and trenches they are working in collapse suddenly. Deaths have occurred in both shallow and deep excavations, so it is important that any excavation work is properly planned, managed, supervised and carried out to prevent accidents.

Many types of ground are, to some degree, self-supporting but this should not be relied upon when working in a trench. It is vital that precautions are taken to ensure that excavations are adequately supported.

Working in trenches and excavations

Plumbers are fortunate in that the need for working in trenches and excavations is limited. Most of the outside work below ground is done by others on construction sites, such as groundworkers and the services providers such as the gas company or the water authority. The services – i.e. water, gas, building drainage – that plumbers may have to deal with are relatively shallow, but there is still a need to understand the planning and working practices of working in trenches and excavations.

Trenches and excavations

The maximum depth that a trench or excavation can be dug without support is 1.2 m. On the average person, 1.2 m would be around waist height. The significance here is that the chest would be above ground level and so breathing would not be restricted in the event of trench collapse.

After this depth, the trench sides should be either:

- battered – a method by which the sides of the trench are sloped away from the trench bottom; the angle of the slope would be decided by the type of ground but usually 45° is considered adequate
- benched – benching or stepping back simply means that the sides of the trench are cut into steps away from the trench side
- shored up – supported using a proprietary trench support system.

The general requirements for safe trench and excavation design are as follows.

- The ladder used to gain access should be secured in position to the trench supports, and in long trenches access should be spaced at regular intervals.
- The spoil from the trench should be at least 1 m away from the edge of the trench to prevent trench collapse; 1 m^3 of earth can weigh as much as a tonne and the added weight against the weak edge of the trench could cause collapse or earth slide.

- The edge of the trench must have a 2 m-high barrier placed around it, at least 1 m away from the edge, to stop people from falling into the trench. It must also have a toe board to stop tools and materials from being accidentally kicked in.
- Vehicle stops must be used to prevent vehicles and plant getting too near the edge and to stop a build-up in the trench of poisonous carbon monoxide fumes.
- The use of propane gas is prohibited as the gas is heavier than air and any leak could gather at the trench floor.
- Trenches and excavations must have a secure ladder (or several if the trench is long) for fast emergency evacuation.
- Warning notices and signs should be placed at regular intervals along the trench length.

Proprietary trench support systems

In order to support the walls of an excavation and prevent trench collapse, a preliminary trench is dug and its walls shored up by means of a trench box or trench shield placed inside the trench. A series of piles are driven into the soil below the trench box or trench shield as the excavation is made deeper.

▲ Figure 1.52 Trench support sheets and braces

Trench safety

Safety when working in trenches and excavations is crucial. There are many things you have to be aware of to maintain your own personal safety.

- Always wear the correct PPE. Arguably the most important piece of PPE is the high-visibility (hi-viz) jacket or vest, followed by your hard hat.
- Never work in an unsupported trench deeper than 1.2 m and never work ahead of the trench supports.
- Be aware of where the access points and ladders are. This could be vital in an emergency situation.
- Be aware of plant and vehicles approaching the trench.

Working in confined spaces

A confined space is a place that is considerably enclosed, where there is a risk of death or serious injury from hazardous substances or dangerous conditions, such as lack of oxygen or being overcome by fumes.

During plumbing installations and maintenance, you may be required to work in:

- tanks and cisterns
- trenches
- sewers
- drains
- flues
- ductwork
- unventilated or poorly ventilated rooms
- under floors and in small roof spaces
- drainage systems
- plant rooms
- cylinders.

All of these constitute a confined space and precautions need to be put in place to ensure your health and safety.

What are the risks of working in confined spaces?

Every year, a number of people are killed or seriously injured working in confined spaces in the construction industry, from those involving complex plant to unventilated or poorly ventilated rooms.

Those killed include not only people working in the confined space but those who try to help them without the proper training and equipment. Dangers occur because of:

- lack of oxygen
- poisonous gas, fumes or vapour
- liquids and solids suddenly filling the space
- fire and explosions
- residues left behind, which may give off fumes, vapour or gas
- poor lighting conditions
- hot working conditions.

Legal duties and obligations relating to working in confined spaces

The Management of Health and Safety at Work Regulations 1999 require that a suitable assessment of the risks for all work activities is carried out so that decisions can be made as to what measures are necessary for safety.

For work in confined spaces, this means identifying the hazards present, assessing the risks and determining what precautions to take. In most cases the assessment will include consideration of:

- the task
- the working environment
- tools and materials
- the suitability of those carrying out the task, including pre-existing medical conditions
- the risk of lone working (if relevant)
- arrangements for emergency rescue.

SUMMARY

It is no coincidence that this chapter is the longest in the book, such is the importance of health and safety in the modern construction industry. We, as plumbers and apprentices, have a duty of care towards ourselves, those we work with and those we come into contact with. The ultimate responsibility of how we behave, how we work and how we respond to accidents and incidents rests with us. By taking notice of health and safety and following the rules that are in place to safeguard us, we too can reduce the likelihood of accidents and, ultimately, save lives. The key message that we must always remember is: health and safety is everyone's responsibility.

Test your knowledge

1 As a plumber, who should provide your PPE?
   a The client
   b The HSE
   c Your employer
   d Yourself

2 Which of the following regulations specifically enforces the use of power tools in the workplace?
   a COSHH
   b RIDDOR
   c IPAF
   d PUWER

3 When replacing damaged guttering on a two-storey dwelling, what method of access would be the safest?
   a Pole ladder
   b Mobile tower scaffold
   c Stepladder
   d Trestles and boards

4 What is the purpose of a health and safety site induction?
   a To ensure that all operatives are aware of the safety procedures on-site
   b To provide details of the pay and working conditions for employees
   c To inform staff of the type of plumbing work that they will do on the job
   d To encourage better relations between the HSE and construction workers

5 What type of safety sign is shown below?

   a Mandatory
   b Hazard
   c Prohibition
   d Fire and first aid

6 Fires containing flammable liquids fall into which class?
   a Class A
   b Class B
   c Class C
   d Class D

7 What colour are the lead and plug of a 110 V power tool?
   a White
   b Yellow
   c Blue
   d Red

8 What is the maximum depth a trench should be dug before supports are used?
   a 1.0 m
   b 1.2 m
   c 1.6 m
   d 2.0 m

9 Toe boards are required to be fitted on a mobile tower when the working platform exceeds what height?
   a Required on all mobile towers
   b When exceeding 1.5 m
   c When exceeding 2.0 m
   d When exceeding 3.0 m

10 Which of the following defines a major injury?
   a An injury on-site that results in first-aid treatment
   b An injury requiring five or more stitches
   c An injury that results in hospital treatment
   d An injury that results in more than 24 hours' hospitalisation

11 Which of the following is NOT a colour of an LPG cylinder?
   a Maroon
   b Black
   c Red
   d Blue

12 When would be the correct time to display a warning sign on-site?

a In the hours of darkness

b Only times when the public have access

c All the time

d Only when there are more than five workers on-site

13 What is the maximum recommended weight an adult male can lift from ground level?

a 25 kg

b 15 kg

c 10 kg

d 5 kg

14 Which of the following installations would require the need for an excavation to take place?

a Installing the mains service pipe to a property

b Installing pipework under a suspended floor

c Installing underfloor heating pipework in a new build property

d Installing pipework below a screed

15 Which of the following tools will offer the LEAST hazards in a customer's property?

a Battery

b 240v

c 110v

d 50v

16 What is a method statement meant to promote?

a List the appropriate welfare facilities for site

b Inform the workforce about health and safety legislation

c Request employees to suggest appropriate work methods

d Inform employees of the approved way of working on-site

17 Under the Manual Handling Regulations, what should be completed before a lift is undertaken?

a Handling assessment

b Risk assessment

c Weight assessment

d Lift assessment

18 Which one of the following is NOT a confined space?

a Under a suspended floor

b Outside toilet

c Loft area

d Duct work

19 When referring to a fire and combustion, which of the following statements is correct?

a Fire square

b Fire circle

c Fire line

d Fire triangle

20 Which item would be used to gain access across a fragile roof area?

a Crawling boards

b Extension ladder

c Mobile scaffold tower

d Trestle

21 On inspection, the body of a power tool has a crack in it. What would be the correct course of action to take?

a Replace the tool immediately

b Replace the tool after the job is finished

c Report the situation at the next toolbox talk

d Replace the tool at the end of the day

22 Which of the following items is most likely to contain asbestos?

a A washbasin

b A hot water cylinder

c An old boiler

d Electrical wiring centre

23 Under the COSHH regulations, what should happen?

a Employees need to carefully record exposure to hazardous substances

b Employees to limit their exposure to hazardous substances

c Employers to control employees' exposure to hazardous substances

d Customers to warn engineers about hazardous substances in the property

24 Which of the following is important when choosing your PPE?
   a It must be cleaned before use
   b It must be made in Britain
   c It must be BBA approved
   d It must be suitable for its purpose

25 What would be the greatest danger when soldering in a loft area?
   a Rodents
   b Fire
   c Dust from insulation
   d Impact to the head

26 What is the overriding duty of Section 2 of the HASAWA?

27 What is the purpose of a permit to work?

28 What is the most dangerous type of asbestos?

29 Give six examples of locations in which asbestos may be found within a building.

30 What is the correct term for safe manual lifting?

31 You are required to use an oxyacetylene welding kit for a specialised installation. Write out the correct method to assemble the cylinders to the blowtorch.

32 You find a fellow worker on the ground who needs to be put into the recovery position quickly. Outline the procedure to position the patient into the recovery position.

33 You are asked to drill a series of holes in a masonry wall. State what PPE you will be required to wear and what it will be protecting you from.

34 You need to inspect some lead weathering around a chimney. Why would it be important to use a roof ladder?

Answers can be found online at www.hoddereducation.co.uk/construction.

Practical task

Complete a suitable risk assessment for the following task.

Maintenance needs to be carried out to clear blocked gutters at your training centre/place of work. The gutters are 6 m from the ground and an extension ladder has been provided. Consideration needs to be given to your safety and that of others who may be present.

CHAPTER 2

COMMON PROCESSES AND TECHNIQUES

INTRODUCTION

A plumber's job is to install the systems of hot and cold water, central heating, sanitation and gas in a professional, efficient and organised manner, using materials safely, economically and correctly. This involves planning and setting out the work, and using installation techniques that not only satisfy the requirements of the customer and protect their property, but that also comply with the relevant regulations, British Standards and codes of good practice.

In this chapter, we will explore the wide variety of tools we use, the range of materials available, creating watertight joints and the correct methods of working we need to install them.

By the end of this chapter, you will have knowledge and understanding of the following:

- how to use hand tools
- types of pipework, bending and jointing techniques
- preparation techniques
- how to use pipe clips and pipe brackets
- pipework installation techniques.

1 USING HAND AND POWER TOOLS

A wide range of tools are used in plumbing for the different tasks and installations customers require. You must know how to use them correctly and safely, as well as how to keep them in good working order to ensure a long working life.

Hand tools

Screwdrivers

There are many different types of screwdrivers, some with specialist applications and uses, such as insulated electrical screwdrivers and long-bladed types. A plumber should have a wide selection of screwdrivers available. The common head types are shown in Table 2.1.

▼ Table 2.1 Common types of screwdriver head

Flat blade For use with slotted screws. Care should be taken to ensure the correct blade size for the screw slot.	
Phillips head Originally designed in the 1930s to intentionally 'ride out' of the screw head, to prevent over-tightening.	
Pozidriv head Similar to the Phillips head but has an eight-pointed star shape for better grip. Not compatible with Phillips screws.	
Hexagon head (Allen key) Mainly used in the gas industry for appliance servicing and installation.	
Star head Not often used except in specialist installations and appliances. Also known as Torx screwdrivers.	

Each screwdriver has a particular use and, when used correctly, should give long-lasting service.

HEALTH AND SAFETY

Problems often occur and accidents happen if screwdrivers are mistreated or used improperly.

- A screwdriver is not a chisel and should not be used as such.
- Use the correct screwdriver for the screw – for example, a Pozidriv screw needs a Pozidriv screwdriver, not a Phillips screwdriver.
- Never over-tighten the screw as this can damage the screw head, making it difficult to withdraw the screw in future.
- Choose the right-sized blade for slotted screws; using too small a blade will result in the screwdriver slipping out of the head, causing damage.
- Keep fingers behind the blade.
- Use an insulated screwdriver when working with electricity.

Hammers

There are two primary types of hammer used by plumbers. These are described in Table 2.2.

▼ Table 2.2 The primary types of hammer used by plumbers

Claw hammer Used for driving nails into, and extracting nails from, wood. The head is made from forged steel and the handle is made from wood, fibreglass or steel. The claw splits down the middle, forming a 'V' shape that, when used in conjunction with the handle, gives leverage for taking out nails.	
Club/lump hammer Used for heavy hammering work, mainly with cold chisels and bolster chisels. May also be used in light demolition work.	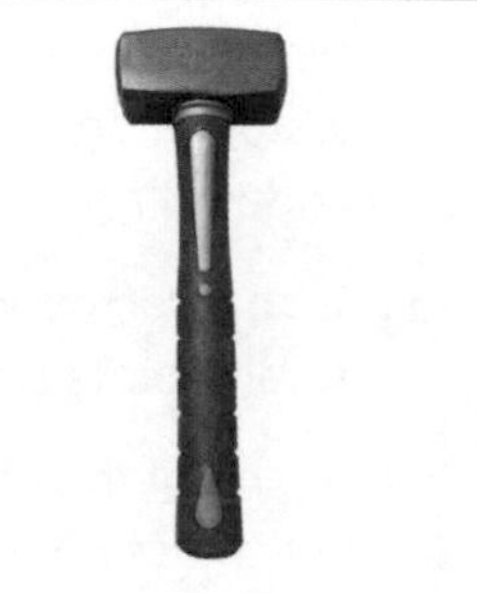

Chisels

Again, there are two types of chisel a plumber will use (see Table 2.3). Both have very different uses.

1. Cold chisels are used for breaking and cutting masonry and concrete. These include:
   - bolster chisels
   - plugging chisels
   - flat chisels.
2. The second type are wood chisels.

▼ Table 2.3 Types of chisel used for plumbing

Cold chisels	
Bolster chisel A bolster chisel is used for cutting masonry, brick and concrete. It can also be used when lifting floorboards, e.g. for cutting out the tongue from tongue and groove floorboards. Often called a floorboard chisel.	
Plugging chisel Mostly used for cutting out and removing the mortar joints in brickwork and masonry.	
Flat chisel A general-purpose tool for cutting, breaking and cutting brickwork, masonry, stone and concrete.	

Wood chisels	
There are many different types of wood chisel, including flat-bladed and bevelled-edge chisels. Mainly used by plumbers for notching joists. Care should be taken as the blades can be extremely sharp.	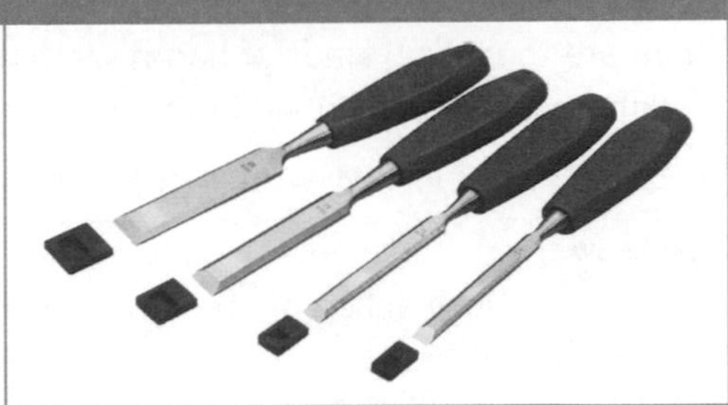

HEALTH AND SAFETY

Here are some important points to remember about the safe use of chisels.

- Eye protection must be worn when using any type of chisel because of the risk of flying debris.
- Always keep the cutting blade sharp and well ground.
- Watch out for 'mushrooming' on the heads of cold chisels. This is where the metal begins to fold over and split due to being repeatedly hit with a hammer. Mushrooming should be removed by grinding on a grinding wheel.
- Always wear gloves when using cold chisels. They help to protect the hands from cuts and abrasions.
- Always keep fingers away from the cutting blade, especially when using very sharp wood chisels.

Spanners

A plumber's toolbox should contain a variety of spanners. Different types will be needed depending on the type of work. Types of spanners include those listed in Table 2.4.

INDUSTRY TIP

Always adjust the spanner to fit correctly across the flats. This will avoid 'rounding off' the nut head and slipping off the nut and hitting your hand.

INDUSTRY TIP

Immersion heater spanners come in three main designs; box (as pictured), flat and cranked.

▼ Table 2.4 Types of spanner used for plumbing

Adjustable spanner A general-purpose spanner, used for tightening compression joints, radiator valve unions, nuts and bolts. Three pairs of spanners of varying sizes are the optimum number for a plumber's toolbox. This tool should be kept oiled and clean.	
Open-jawed spanner These are mostly used for boiler and appliance servicing. A small set is recommended for the toolbox.	
Ring spanner Mostly used for boiler and appliance servicing.	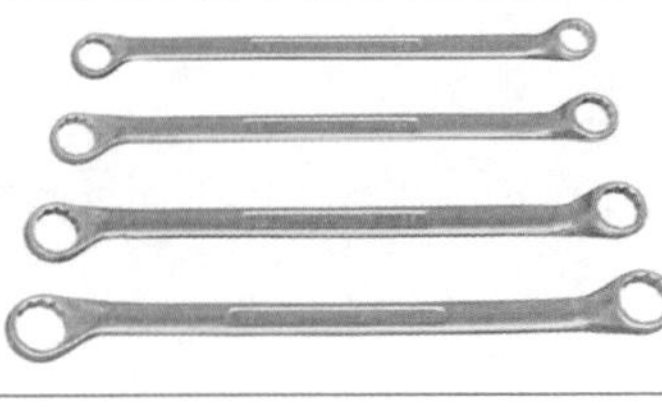
Box spanner The main tool for fixing taps to sanitary ware in sizes 13 mm for monobloc mixers, ½-inch for sink and washbasin taps and ¾-inch for baths. These are used to tighten the back nut of the tap.	
Immersion heater spanner A specialist tool for installing and removing immersion heaters from hot water storage cylinders and vessels.	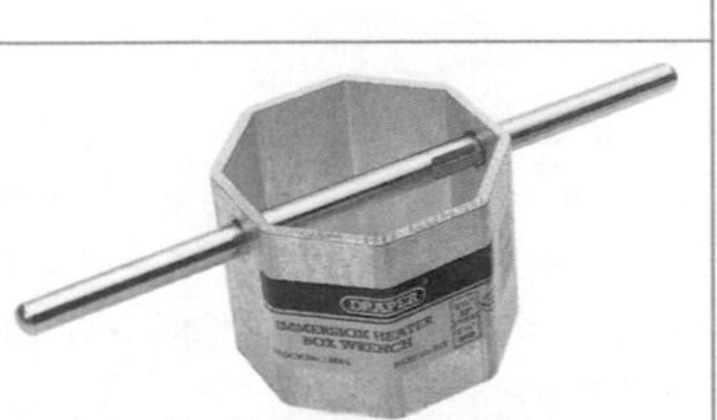

Handsaws

There are four main kinds of handsaw that a plumber would find a use for and that should be included in their toolkit. These are described in Table 2.5.

▼ Table 2.5 Types of handsaw used for plumbing

Hacksaw Used to cut copper tubes, plastic waste pipes, gutters, soil pipes and low carbon steel pipes. Not suitable for cutting wood. Always ensure the correct type of blade is fitted, that the teeth are facing forwards and the tension of the blade is not loose.	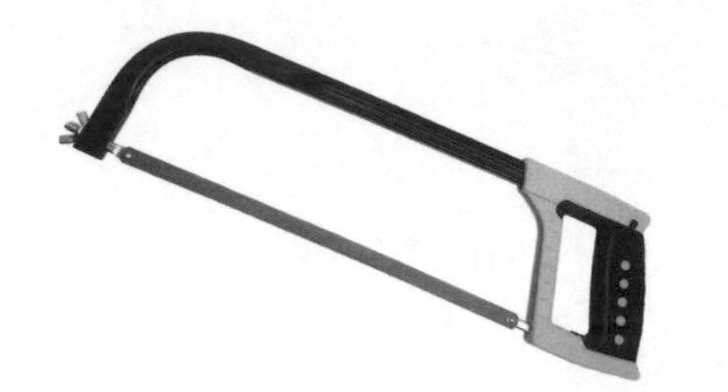
Junior hacksaw An essential saw for the toolkit. This small saw is used to cut small copper tubes and plastic pipes. Excellent for cutting tubes in position in tight situations where access is difficult. When replacing the blade, always ensure the teeth of the blade face forwards.	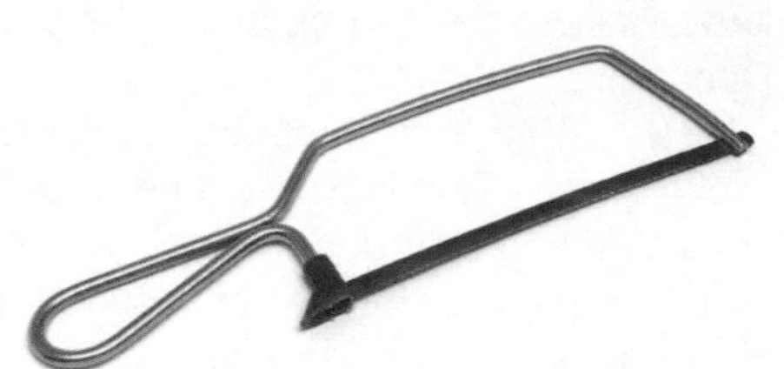
Universal hard point saw A general-purpose wood saw that is used for lifting and cutting floorboards, building platforms and stillages. May be used for cutting plastic pipes. The teeth have tungsten steel hard points so that they remain sharp for longer.	
Floorboard saw A saw made specifically for cutting and lifting floorboards in position. It has teeth on the end of the saw for cutting through floorboards while they are still in position.	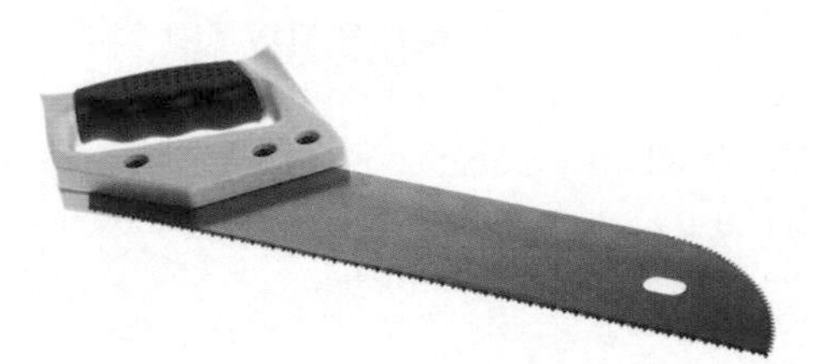
Pad saw Often called a 'keyhole saw' or 'drywall saw'. A long, narrow saw used for cutting small, awkward holes and shapes in building materials, such as wood and plasterboard. There are two types of pad saw: the fixed-blade type and the retractable-blade type.	

Grips and wrenches

Grips and wrenches are tools that are used almost every day by plumbers. They are essential tools for tightening and gripping. There several different types, as described in Table 2.6.

▼ Table 2.6 Types of grips and wrenches used for plumbing

Water pump pliers These are a plumber's general-purpose grips. Three pairs should be available in the toolbox: 175 mm, 250 mm and 300 mm.	
Footprints Another general-purpose grip used by plumbers for tightening fittings and unions in low carbon steel pipework. Care should be taken when using these as they can easily trap fingers if used incorrectly.	
Stillsons (Pipe wrench) Stillsons are used when installing low carbon steel pipe. They are available in many sizes, ranging from 10-inch to 36-inch.	
Basin wrench This is used for tightening and loosening tap connections in hard-to-reach areas such as behind wash hand basins, baths and kitchen sinks. It has a spring loaded/interchangeable jaw and can have a telescopic handle.	
Mole grips Mole grips are a locking type of pliers. They give a high clamping force and can be locked to allow hands-free gripping.	

Pliers

Pliers are two-handled, two-jawed hand tools used mainly for gripping, twisting and turning. The jaws meet at the tip, which means they can grip with precision. Some types are also made for cutting cable and wire. Some of the main types of pliers used by plumbers are described in Table 2.7.

▼ Table 2.7 Types of pliers used for plumbing

General-purpose pliers A useful addition to the toolbox, these general-purpose pliers are used to grip and tighten small nuts and bolts. They can also be used to cut thin wire and electrical flex.	
Long nose pliers Long nose (or needle nose) pliers are useful both for gripping small items and for reaching into small, deep spaces. They are used to tighten small nuts and bend wire. They often include a wire cutter.	
Circlip pliers Circlip pliers have a specific use for removing the circlips from tap spindles and shower valves. Mainly used in maintenance and repair operations.	
Wire cutters Used for cutting electrical cables and flex. These are a useful addition to a plumber's toolbox.	

Spirit levels

Spirit levels are used to ensure that appliances and pipework are installed **level** and **plumb**. They use a bubble positioned between two markers. Electronic and laser spirit levels are also available. The most common ones used by plumbers are described in Table 2.8.

KEY TERMS

Level: when pipework is perfectly horizontal.

Plumb: when pipework is perfectly vertical.

▼ Table 2.8 Types of spirit level used for plumbing

Torpedo level (Boat level) A small, 300 mm level with a magnetic strip on the bottom, which makes it easier to level appliances such as boilers.	
Spirit level Two sizes, 600 mm and 1200 mm, are advisable for levelling large appliances such as baths and wash hand basins.	

Plumbing-specific tools

So far we have looked at the more common hand tools. As well as these, there are many 'plumbing-specific' hand tools that plumbers must have in their toolkits, such as:

- pipe-cutting tools:
  - pipe slices
  - adjustable copper pipe cutters
  - plastic pipe cutters
- pipe-bending tools:
  - scissor benders
  - tripod benders
  - bending springs
    - internal
    - external
- pipe-soldering equipment:
  - blowtorch, hose and gas governor
  - brazing torch
- socket-crimping tools
- manual pipe-threading equipment.

We will look at each of these in turn.

Pipe-cutting tools

▼ Table 2.9 Types of pipe-cutting tool used for plumbing

Pipe slices An essential tool for cutting copper tube. The pipe slice can be used in tight situations where junior hacksaws and adjustable pipe cutters cannot. Non-adjustable sizes are available to suit copper pipe: 15 mm, 22 mm and 28 mm. There are also sizes available to cut plastic waste pipe: 32 mm and 40 mm. Always ensure the cutting wheel, wheel spring and rollers are lubricated and free from dirt.	
Adjustable pipe cutters An essential tool that can be adjusted to cut many sizes of copper tube. Periodic maintenance of this tool is recommended, such as changing the cutting wheel and regular oiling.	
Plastic pipe cutters This tool can be used to cut all forms of plastic pipe. It gives a clean cut, which is essential when jointing push-fit pressure plastic pipe and waste pipe.	

Pipe-bending tools

Types of pipe-bending tool used for plumbing include:

- **Scissor-type benders:** these bending machines, also known as handi-benders, are excellent for precision bending of copper tube.

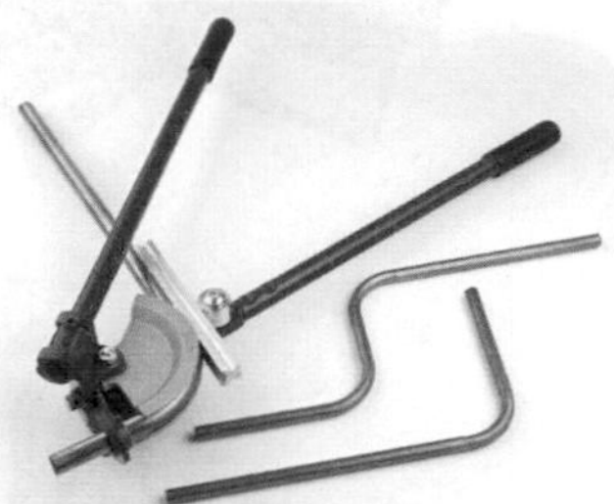

▲ Figure 2.1 Scissor-type benders

They are light in weight and portable. For bending copper tube in sizes 15 mm and 22 mm.

- **Tri-pod bending machines:** these are static bending machines for bending copper tubes from sizes 15 mm to the larger sizes up to 42 mm. Particular attention should be paid to the bending roller to prevent excessive **rippling** of the tube, which can occur when the roller is not tight against the bending guide.

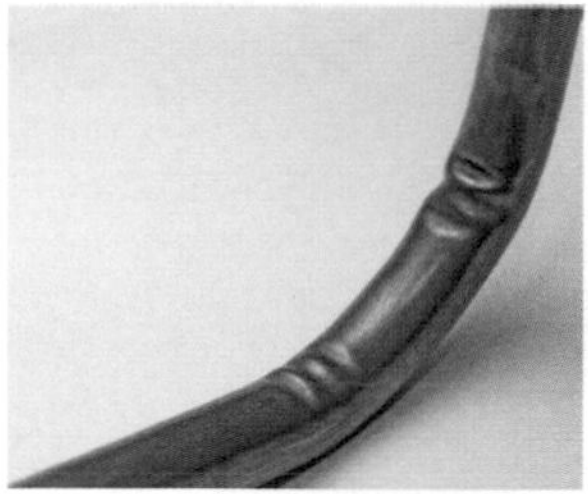

▲ Figure 2.2 Example of rippling in copper pipe

If the roller is too tight, then **throating** of the copper tube will occur.

- **Internal bending springs:** not used as much since the development of the scissor bender, the internal bending spring can be used to bend half-hard copper grade R250 or soft copper grade R220. These can be used to adjust existing pipework.
  It is recommended that the tube be **annealed** first before bending to prevent excessive rippling.
- **External bending springs:** used in the same way as internal springs but the spring is placed on the outside of the tube.
  Usually used with microbore soft copper grade R220 of sizes 8 mm and 10 mm.
- **Microbore scissor bender:** a small version of the scissor bender, for microbore tubing of sizes 6 mm, 8 mm and 10 mm.

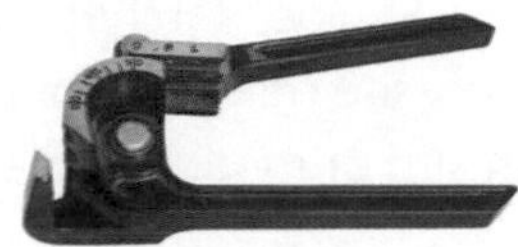

▲ Figure 2.3 Microbore scissor bender

Pipe-soldering equipment

Types of pipe-soldering equipment used for plumbing include:

- **Blowtorch with separate governor, hose and LPG bottle:** the traditional plumber's blowtorch.
  The governor can be pre-set or adjustable, and the nozzles on the blowtorch are interchangeable with varying sizes for different tube sizes.
  These are not as controllable as other torches.
- **Soldering and brazing torch:** this type of blowtorch is much more portable and gives a hotter flame that is far more controllable.
  It can be used with propane and MAPP gas, but gas usage tends to be high.

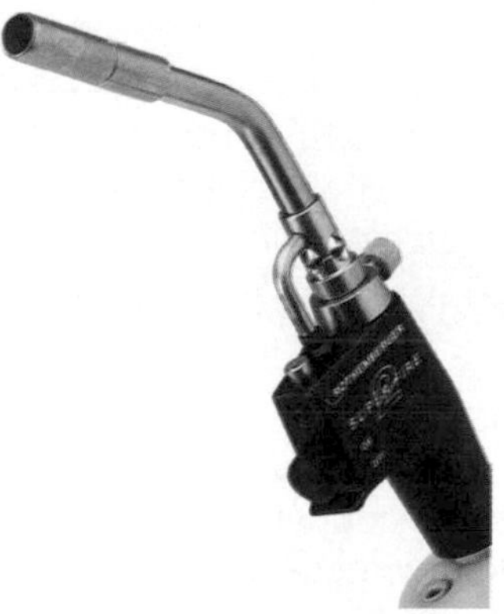

▲ Figure 2.4 Soldering and brazing torch

▲ Figure 2.5 Ratchet stocks and dies

Manual pipe-threading equipment

Although not strictly plumbing tools, manual pipe-threading equipment may be used occasionally when installing low carbon steel (LCS) pipe. Ratchet stocks and dies are tools used for on-site threading of BSP mild steel pipes, whether in situ or mounted in a pipe vice.

INDUSTRY TIP

When using threading tools, plenty of oil should be applied as this helps to lubricate and cool the cutting heads. Threading tools have a reversible action. This allows the cutting head to be removed from the pipe and also cleans the newly cut threads of all cut steel and excess oil (known as 'swarf').

British Standard Pipe (**BSP**) and British Standard Pipe Threads (**BSPT**) relate to the type of thread used on screwed low carbon steel pipes and fittings. Although the pipe is measured in mm, it is universally referred to in imperial measurements, e.g. 'x-inch BSPT' (meaning x-inch British Standard Pipe Thread). See the following BSPT page for further information: www.bspt.co.uk

KEY TERMS

Rippling: an unwanted, wavy pattern made on the inside face of a machine bend when the bending arm roller is not tight enough.

Throating: a slight indentation that the bending machine makes when the bend is formed.

Annealing: a process that involves heating the copper to a cherry-red colour and then quenching it in water. This softens the copper tube so that the copper can be worked without fracturing, rippling or deforming.

BSP: British Standard Pipe.

BSPT: British Standard Pipe Thread; the type of thread used on screwed low carbon steel pipes and fittings.

Other hand tools

As well as the tools we have already looked at, there are others a plumber may need. These are general tools that are useful additions to the toolkit and include those listed in Table 2.10.

▼ Table 2.10 Other hand tools used for plumbing

Files and rasps Essential for filing the ends of tubes to remove internal and external burrs. Three types should be included in the toolkit: **1** flat files **2** half-round files **3** rat-tail files.	
Allen keys These small hexagonal keys are used mainly in maintenance tasks, e.g. for repairing and servicing shower valves.	
Tap reseating tool A widely used plumber's tool for repairing the seats of taps by grinding the seat to a smooth surface. This ensures that the tap washer sits properly on the tap seat, preventing taps from dripping.	
Radiator spanner A specialised spanner for inserting radiator valve tails into radiators and convectors.	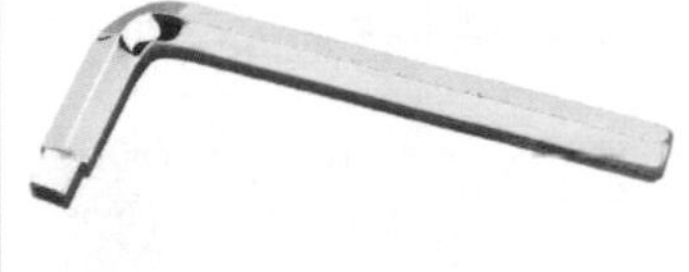

Hand tool safety and maintenance

A large number of accidents occur every year in the construction industry because of the unsafe use of manual and power hand tools, such as using a screwdriver as a chisel or a lever.

Most accidents involving tools result from:

- using the wrong tool for the job
- using the tool incorrectly
- not wearing personal protective equipment (PPE)
- not following approved safety guidelines
- poor maintenance.

The most common tools involved in accidents are:

- chisels
- saws
- screwdrivers
- files

- snips
- hammers
- wrenches, grips and pliers.

> **HEALTH AND SAFETY**
> - Hand tools are an important part of your job. They must be treated, cared for and used in a professional manner. By following these safety rules, many hand tool injuries can be avoided.
> - Don't forget that, under the Health and Safety at Work etc. Act 1974, you have a duty of care to yourself, your employer and others who may be affected by your acts or omissions. You can access this act at: www.legislation.gov.uk/ukpga/1974/37/section/7

The safety rules to follow when using hand tools are as follows.

- Know the purpose of each tool in your toolbox, and use it for the task it was designed for.
- Never use any tool unless you are trained to do so.
- Inspect tools before each use and replace or repair if they are worn or damaged.
- Always clean your tools when you have finished using them.
- Always keep the cutting edges of chisels and saws sharp.
- Always keep any moving parts free from dirt and make sure they are well oiled.
- Select the right size tool for the job.
- When working on ladders or scaffolding, be sure that you and your tools are secure. Falling tools could injure people working or passing below.
- Do not put sharp or pointed tools in your pockets. Use a sheath or holster instead.
- Do not throw tools, as they are easily damaged.
- Do not use a tool if the handle is missing or has splinters, burrs or cracks, or if the head of the tool is loose.
- Do not use cold chisels that have mushroomed heads.
- When using tools such as jig saws, chisels and drills, always wear PPE such as safety goggles, face masks or gloves.
- Worn, damaged or defective tools should be taken out of service and not used until they have been repaired or replaced.

Power tools

> **HEALTH AND SAFETY**
> Power tools for use on-site should be 110 V, which is colour-coded yellow for easy identification, or they should be battery operated. You should not use 230 V; 110 V is a safer voltage.

Apart from the hand tools we have looked at, a plumber needs certain power tools to help with installation processes. Here, we will take a brief look at the essential power tools and accessories, including:

- power drill
- circular saw
- jig saw
- reciprocating saw
- portable pipe threading machine
- hydraulic machine bender
- hydraulic crimping kit
- portable pipe freezing kit.

Power drills

Power drills come in three basic types, as described in Table 2.11.

▼ Table 2.11 Types of power drill

Rotary hammer drill This type of drill has a standard chuck so accessories such as metal drills and hole saws can be used.	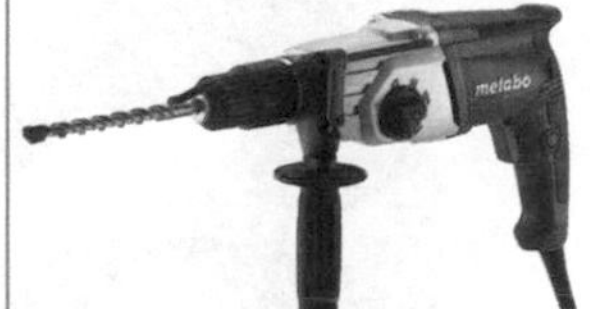
SDS hammer drill This type of drill uses the secure drill system (SDS) bayonet-type fixing to secure the drill bits into the chuck. This type of drill is necessary when using core bits.	
Cordless drill Typical voltages are from 14.4 V to 36 V. These drills are available in many forms, from screwdriver-type drills to large-voltage SDS types.	

INDUSTRY TIP

When purchasing power tools it can be advantageous to purchase the same make, as in many instances the same battery will fit many tools.

INDUSTRY TIP

The chuck on a rotary hammer drill should be kept well oiled to prevent breakdown.

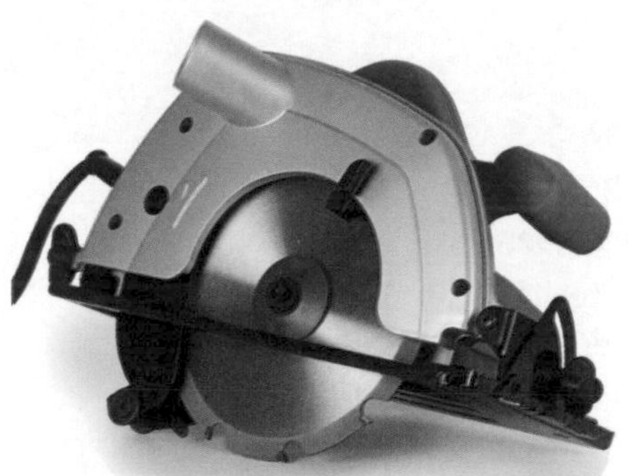

▲ Figure 2.6 Circular saw

▲ Figure 2.7 Jig saw

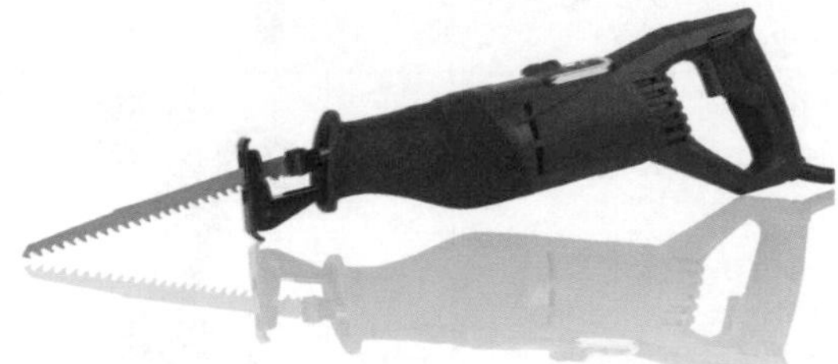

▲ Figure 2.8 Reciprocating saw

Circular saw

Circular saws are very useful tools for lifting floorboards and notching joists. Care should be taken to ensure the blade guard is in place and that the blade is securely fastened.

Jig saw

Jig saws are used for cutting out sinks and wash hand basins in worktops in kitchens and bathrooms. Always ensure the blade guard and blade are securely in place.

Reciprocating saw

Reciprocating saws are a useful addition to the toolkit; the reciprocating saw should not be used where accuracy is required. Different blades can be used to cut different materials, such as wood, plastic, metal, tile and stone.

Hydraulic low carbon steel bending machines

Hydraulic low carbon steel bending machines use pressure from hydraulic oil to bend steel pipe.

INDUSTRY TIP

The oil level of hydraulic low carbon steel bending machines should be checked periodically and topped up as necessary.

Pipe threading machines

▼ Table 2.12 Types of pipe threading machine

Hand-held electric pipe threader This is an easy to use hand-held electric pipe threading tool for threading mild steel BSP pipes, in situ with a pipe clamp or mounted in a pipe vice. Threads ½-inch to 2-inch BSP pipe.	
Pipe threading machine Used on-site, these electric floor-mounted tools will cut, de-burr and thread LCS pipe easily and quickly. They need regular maintenance.	

Copper pipe socket crimping tool

INDUSTRY TIP

This type of jointing is becoming more common due to the speed and the reduction of hazards in the process.

▼ Table 2.13 Copper pipe socket crimping tool

Copper pipe socket crimping tool A fairly new tool, used for **crimping** press fit-type fittings onto copper tubes.	

KEY TERM

Crimping: the process of pressing the fittings into a copper pipe using a 'press fit' or crimping tool.

Pipe freezing kits

HEALTH AND SAFETY

Gloves should always be worn when using pipe freezing kits because of the risk of frostbite.

Pipe freezing kits create a plug of ice to hold back water while maintenance and repair tasks are undertaken. There are generally two types available:

1. electric freezing kits
2. freezing kits using refrigerants.

Power tool safety and maintenance

HEALTH AND SAFETY

Remember: power tools have the potential to seriously risk the health and safety of those on-site if not properly checked and maintained.

As with hand tools, power tools need regular inspection and maintenance. There are certain points that should be followed.

- Power tools should be PAT tested every three months (see Chapter 1, Health and safety practices and systems, page 43).
- Always wear safety goggles or safety glasses when using power tools.
- Always check the tool, the cord and the plug before use for any signs of wear or damage.
- Always check to make sure the tool is the correct voltage for the power supply.
- Never drag the tool or the power cord across the floor.
- Never lift or lower a power tool by its cord.
- Never use a tool that is damaged or not working properly. Damaged tools should be taken out of use, tagged and sent for repair.
- Use a dust mask in dusty conditions and wear hearing protection if the tool is being used for an extended period of time. Remember: prolonged use of hammer-type power tools can cause vibration injury.
- Make sure the work area is clean and free of debris that might get in the way, and always make sure the work area has plenty of light.
- Make sure all appropriate safety guards are in place and never remove a safety guard.
- Always turn off and unplug the tool before any adjustments or change of blades takes place.
- Never use power tools in wet or damp conditions.
- Make sure extension cords are the correct type, and don't use cords designed for inside use outside.
- Make sure cutters or blades are clean, sharp and securely in place. Never use bent or broken blades or cutters.
- Never over-reach when using a power tool, and always take care when using power tools at height.
- When using hand-held power tools, always grip with both hands.
- Always unplug, clean and store the tool in a safe, dry place when the job is finished.

Drills bits, core drills and hole saws

There are many types of drill bits that should be included in a plumber's toolkit. Each one has a specific job, as detailed in Table 2.14.

▼ Table 2.14 Drill bits, core drills and hole saws used for plumbing

Masonry drill bits The tip of this drill bit is made from tungsten carbide steel to enable the bit to penetrate masonry, concrete and stonework.	
Wood drill bits Also known as a spur point or dowel bit, these have a central point and two raised spurs that help keep the bit drilling straight.	
Metal drill bits Also known as twist bits, these can be used on timber, metal and plastics. Most twist bits are made from high speed steel (HSS), which is suitable for drilling most types of material. When drilling metal, the HSS stands up to the high temperatures.	
Spade bits Also known as flat bits, these are for power drill use only. The centre point locates the bit and the flat steel on either side cuts away the timber. These bits are used to drill fairly large holes in floorboards and joists.	
Core drills These are diamond impregnated and are used for drilling very large holes through masonry, stone and concrete. Used in the installation of boiler flues and large pipes such as waste and soil pipes.	
Hole saws Hole saws are ideal for drilling holes in equipment and appliances such as cold water storage cisterns and acrylic baths, which have no tap holes. Some hole saws can also be used on metal, wood and plastic.	

2 TYPES OF PIPEWORK, BENDING AND JOINTING TECHNIQUES

In this part of the chapter, we will take a brief look at the pipe materials that plumbers use in their everyday installation work. We will see how the different methods of jointing, bending and installation practices dictate the methods of working we need to employ. We will look at:

- copper tubes and fittings
- low carbon steel pipes and fittings
- the various types of plastic pipes and fittings.

Copper tubes to **BS EN 1057**

Copper tube has been used in the UK since the 1940s and today still accounts for around 60 per cent of all new installations. The type of copper used in the manufacture of tubes is phosphorus de-oxidised copper, with a minimum copper content of 99.90 per cent.

De-oxidised copper tube can be safely soldered, welded or brazed. The density of copper is 8900 kg/m³. It has a melting point of 1083°C and its coefficient of linear expansion is 0.0000166 per °C (between 20°C and 100°C).

The standard for copper tubes for water, gas and sanitation installations is **BS EN 1057**, which is available in three **tempers**, as outlined in Table 2.15.

KEY TERM

Temper: the temper of a metal refers to how hard or soft it is.

▼ Table 2.15 Grades of copper tubes to **BS EN 1057**

Grade	Description
R220	This is softer copper tube, fully annealed and supplied in coils. It is thicker walled than other grades of copper tube. Used for underground water services (sizes 15, 22, 28 mm) and microbore central heating systems (sizes 6, 8 and 10 mm).
R250	This is the most widely used grade of copper tube for plumbing and heating applications. Commonly supplied in straight lengths of 3 or 6 m, in sizes 15, 22, 28, 35, 42 and 54 mm. It is known as half-hard tempered.
R290	This grade is hard tempered, thin walled and totally unsuitable for bending. Not normally used in the UK.

INDUSTRY TIP

R250 grade of copper, although commonly supplied in 3 or 6 m lengths, can be supplied in shorter lengths or 1 or 2 m lengths, but this will reflect in the price of the material.

KEY POINT

During your time as a plumber you will come across many materials, and each will have its own unique working properties, including different melting points and expansion rates. It is important that we recognise these properties so that we can choose the correct material for a given installation. You will come across other such properties as you work through this book.

Tubes supplied in half-hard (R250) and hard drawn (R290) condition are supplied in straight lengths of 3 or 6 m. Tubes in the soft, fully annealed (R220) condition, up to 28 mm outside diameter (OD), are supplied in coils. The length of the coils is between 10 and 50 m, depending on the diameter.

Copper tubes are generally used in buildings for the following services:

- domestic hot and cold water supplies under pressure, usually up to mains pressure (typically up to 4 bar but can be up to 10 bar in some parts of the UK) or head pressure from a storage cistern
- central heating systems (with radiators/convectors)
- underfloor heating systems
- natural gas installations for heating and cooking
- oil installations for heating
- sanitary pipe work in rare installations
- medical gases (when de-greased).

Copper tube is available chromium plated for situations where there are aesthetic considerations and plastic coated in various colours where protection from corrosion is necessary.

Bending copper tube

INDUSTRY TIP

Practice and master forming copper pipe using your pipe benders. This is an important skill for the professional plumber.

Bending copper tubes becomes easy with practice. The two methods used to correctly bend copper tubes are:

1. machine bending – the preferred method of bending copper tubes
2. spring bending – using a bending spring; not so widely used now since scissor benders have become available.

Here, we will look at each method and investigate its advantages and disadvantages.

Bending copper tube using a bending machine

Bending copper tubes using a bending machine is an economical method of installation, especially where lots of bends or changes of direction are required. There are many types of bending machine available for copper tubes up to 42 mm diameter, all of which are worked by hand. For larger diameters, ratchet-action machines are required. The most useful type of machine for 15 and 22 mm tube is the portable type or scissor bender (see page 75 of this chapter), which is light in weight and requires no adjustment before use.

The advantages of machine bending over spring bending are:

- bends can be formed quickly
- multiple bends can be formed easily
- bends can be formed close to the end of the tube
- bend radius, quality and accuracy are consistent.

Producing accurately positioned bends depends on determining the bending point and the position of the tube in the machine. Figures 2.9–2.14 demonstrate this.

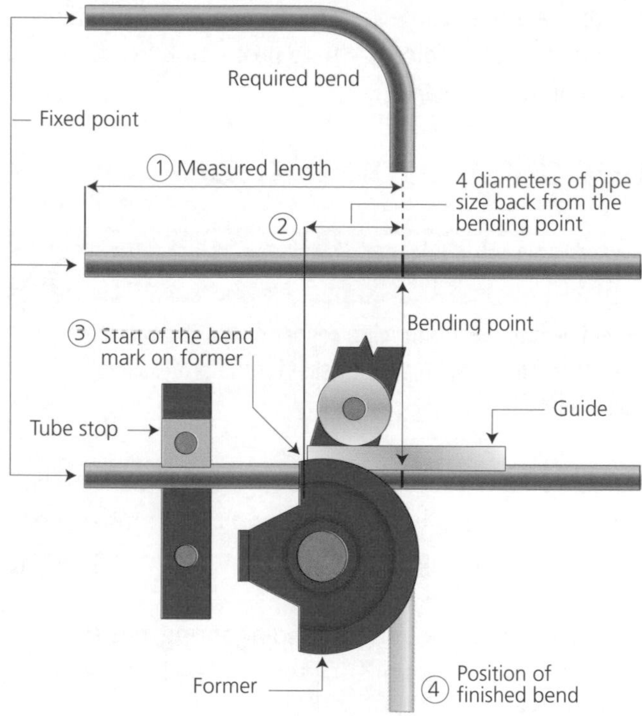

▲ Figure 2.9 90° bends: method 1

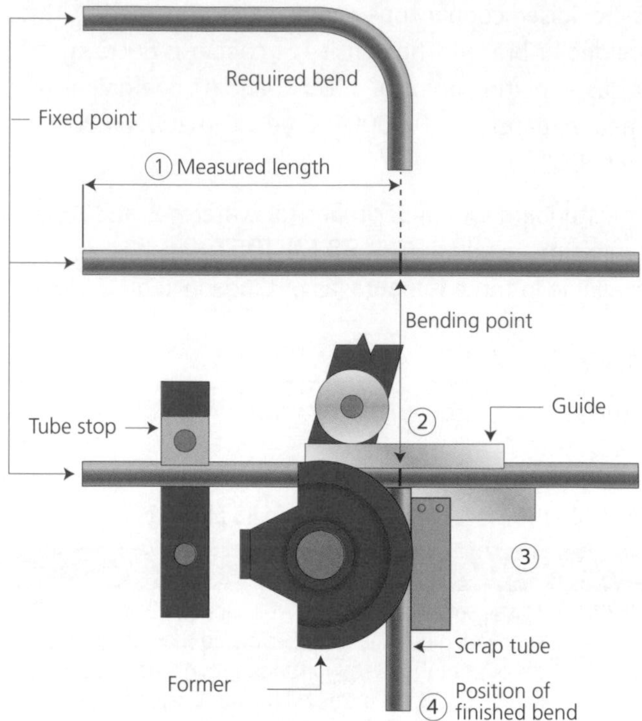

▲ Figure 2.10 90° bends: method 2

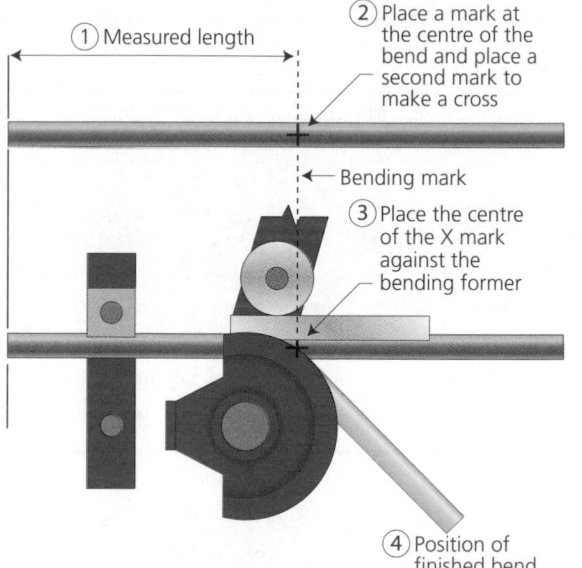

▲ Figure 2.11 45° sets

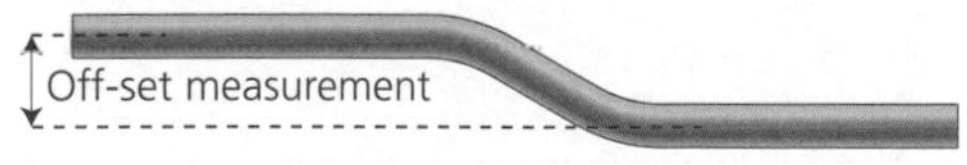

① To find the correct off-set angle, the size of the off-set should be deducted from the 600 mm and the 600 mm folding rule opened to the measurement, i.e. off-set 50 mm. 600 – 50 = 550

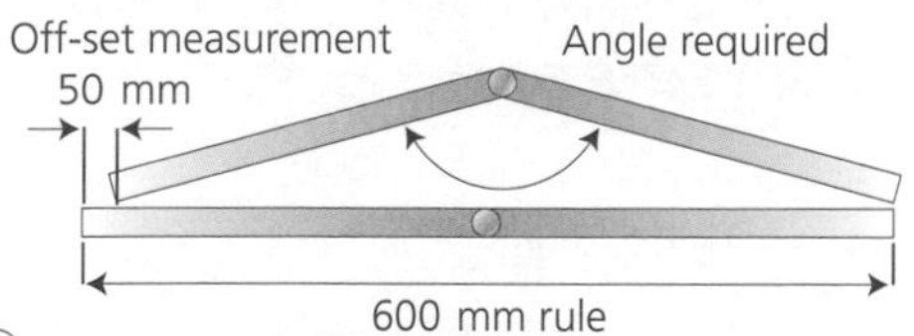

600 mm rule

② Bend tube to the angle set by the rule

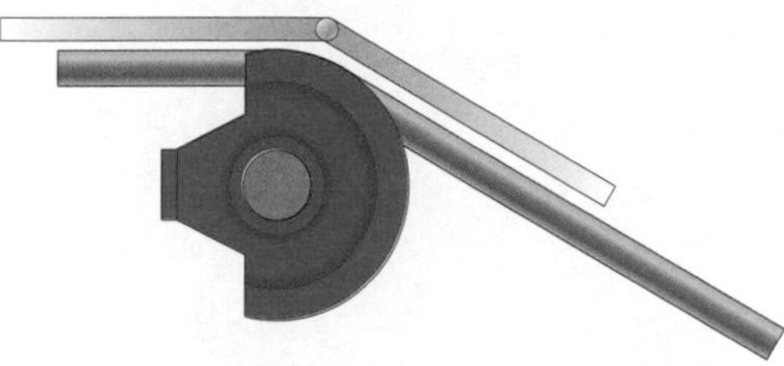

③ Remove tube from machine and mark for the second bend measuring from inside edge of tube using a straight edge

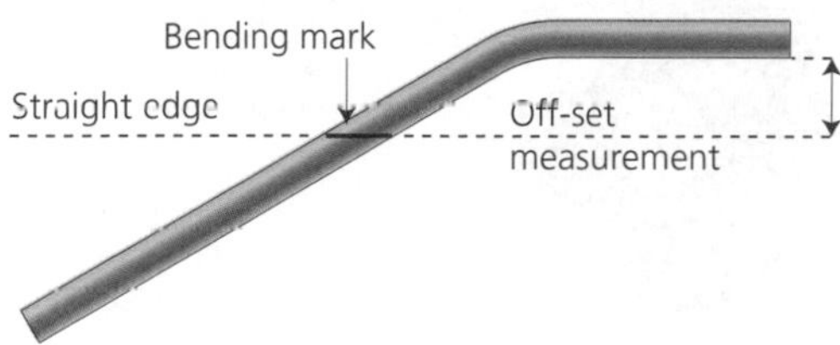

④ Re-position tube in the machine so that the mark forms a tangent to the former

⑤ Re-position the rule to give the correct angle for the second bend

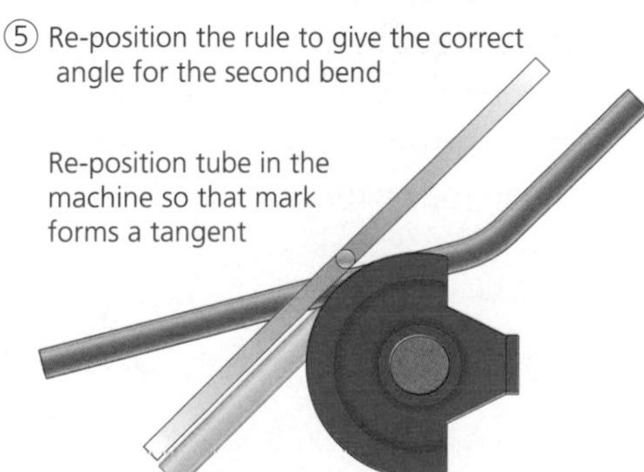

▲ Figure 2.12 Off-set bends

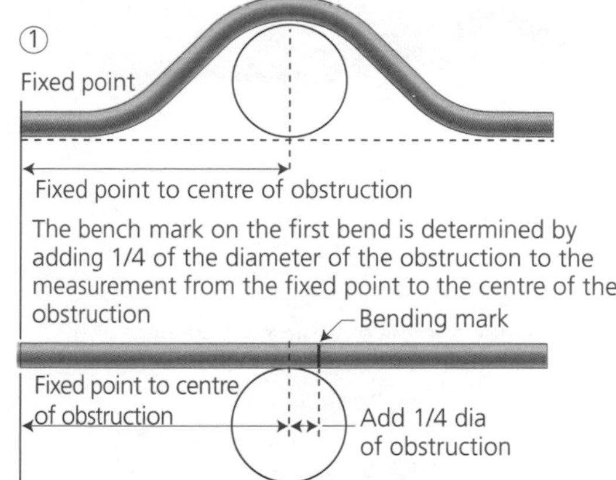

The bench mark on the first bend is determined by adding 1/4 of the diameter of the obstruction to the measurement from the fixed point to the centre of the obstruction

② To find the correct angle for the first bend, multiply the diameter of the obstacle by 3 and close the folding rule by this amount then position the tube in the machine so that the bending mark and the centre of the angle align

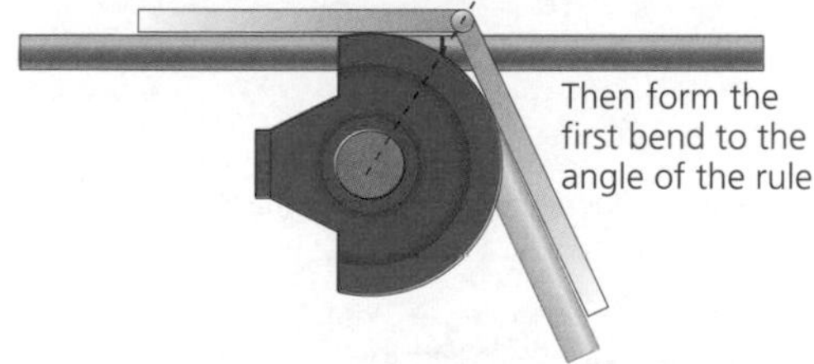

Then form the first bend to the angle of the rule

③ Making sure that the bend clears the obstruction, place a straight edge over the tube and mark the bending marks on both sides

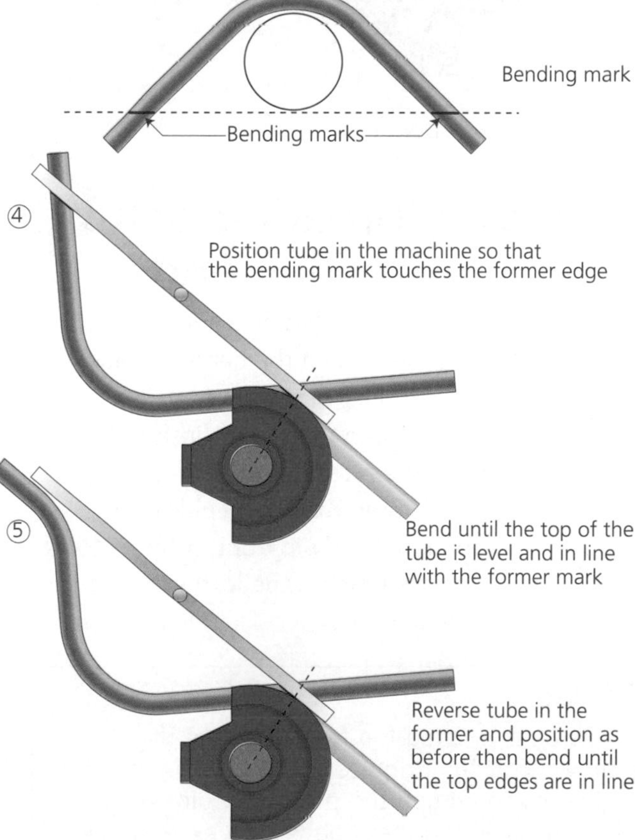

Bend until the top of the tube is level and in line with the former mark

Reverse tube in the former and position as before then bend until the top edges are in line

▲ Figure 2.13 Passover bends

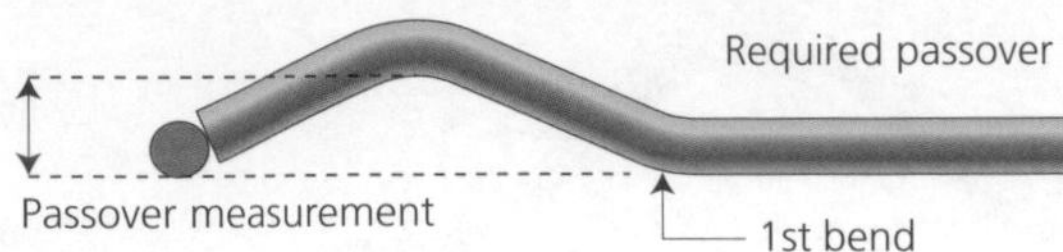

(1)
Close a folding rule down to the passover measurement to obtain the angle for the first bend

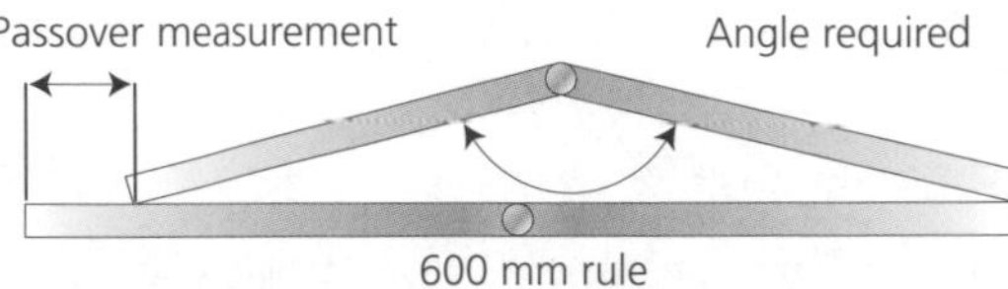

(2)
Bend tube to the angle required by the folding rule

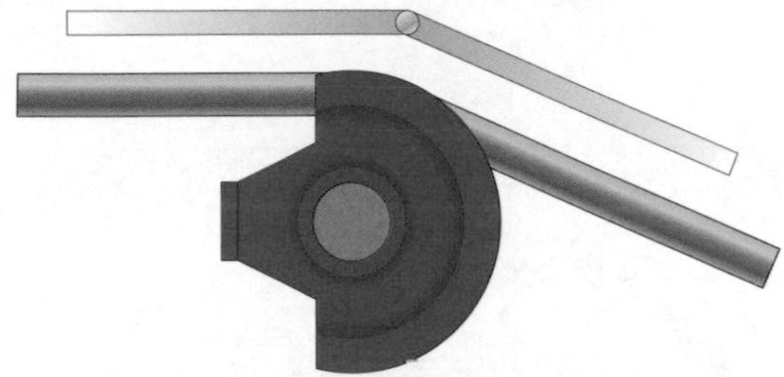

(3)
Close folding rule down to twice the passover measurement to obtain the angle for the second bend

2 × passover measurement
Angle for second bend

(4)
Mark for the second bend by measuring from the inside edge of tube

Bending mark
Passover measurement
Straight edge

(5)
Re-position tube in the machine so that mark forms a tangent to the former

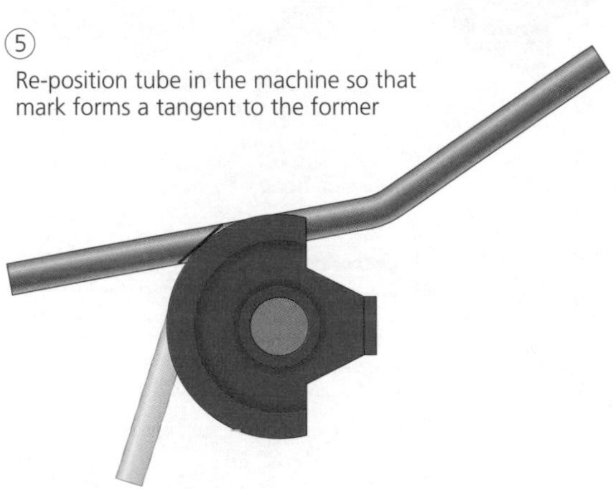

▲ Figure 2.14 Partial passover bends

Rippling or throating of tube in machine-made bends

Bending machines are designed to give a 4 diameter bend and so that the former and the bending guide supports the **throat** and sides of the tube against collapse. Ripples will occur in the throat (inside of the bend) of a bend if the pressure of the roller on the guide is insufficient or in the wrong place. The correct pressure point is slightly in front of the bending position, where the tube touches the former before the actual bending process occurs.

KEY POINT

If pressure is exerted too far forward of the bending point, then ripples will occur. If the roller is tightened too much the pressure point will be too far back and the tube will be excessively 'throated' or made oval in section.

KEY TERM

Throat: the inside face.

With scissor (handy) benders, rippling occurs with use. This is because the bending former, being made of aluminium, stretches over time and, because the pressure roller is fixed, it cannot be tightened or repositioned to give the correct bending pressure. If ripples appear when using fixed-position 'handy benders', the pressure point can be readjusted by inserting a thin piece of strip steel (the thickness of a hacksaw blade) between the guide and the roller to cure the problem.

IMPROVE YOUR MATHS

Read the key point on pipe gain, then work out how much gain a 15 mm and 22 mm copper pipe would make on a 90 degree bend.

15 mm =

22 mm =

KEY POINT

Pipe gain

When bending copper tube using a bending machine, the tube appears to gain length. This is called pipe gain or stretch and we have to take it into account when precision bending. Gain or stretch will naturally occur whenever copper pipe is formed into a radius around the former. The smaller the angle, the smaller the gain. The pipe gain on a 90° bend is 1.5 times the diameter of the pipe. Let's say we have to put a 90° bend on a piece of 22 mm copper tube so that the finished measurements are 150 mm end to centre and 250 mm end to centre. The length of pipe needed appears to be 400 mm but, because the bend cuts the corner, we can deduct a certain amount of pipe. If pipe gain is 1.5 times the diameter and the diameter is 22 mm, we can deduct 33 mm, so the actual pipe length needed for the bend is 367 mm. Pipe gain occurs only with 90° bends (Figure 2.15).

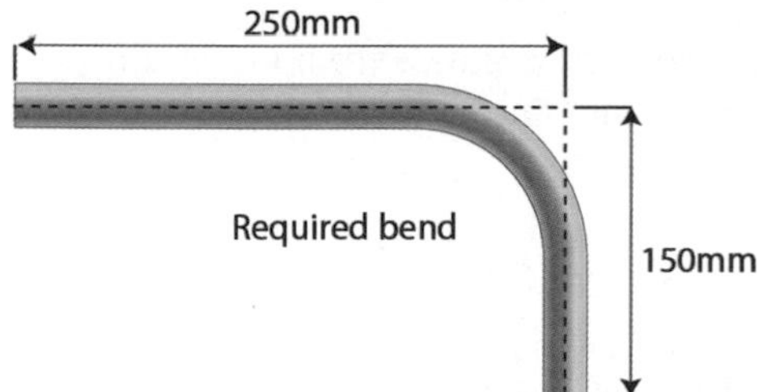

▲ Figure 2.15 Pipe gain occurs only with 90° bends

Bending copper tube using a bending spring

Bending springs are used to support the walls of the tubes against collapse while the bend is being formed. The British Standard for bending springs is **BS 5431:1976** and they are available for copper grades R220 and R250. It is important that the correct-sized spring is used or wrinkling and even snapping of the tube may occur.

INDUSTRY TIP

As a rule, the bend radius should be four to five diameters of the tube. When using a bending spring, the desired radius should be four to five times the diameter of the tube being formed.

For example, for 15 mm copper the bend radius should be between 60–75 mm.

One advantage that a spring bend has over a machine-made bend is that the bend radius can be varied because it is not fixed by a bending former. This allows the tube centres to be carried around bends. In other words, the radii of the bends can be enlarged so that the aesthetic appearance of the bends is enhanced and the gap between the tubes remains even (see Figures 2.16–2.18).

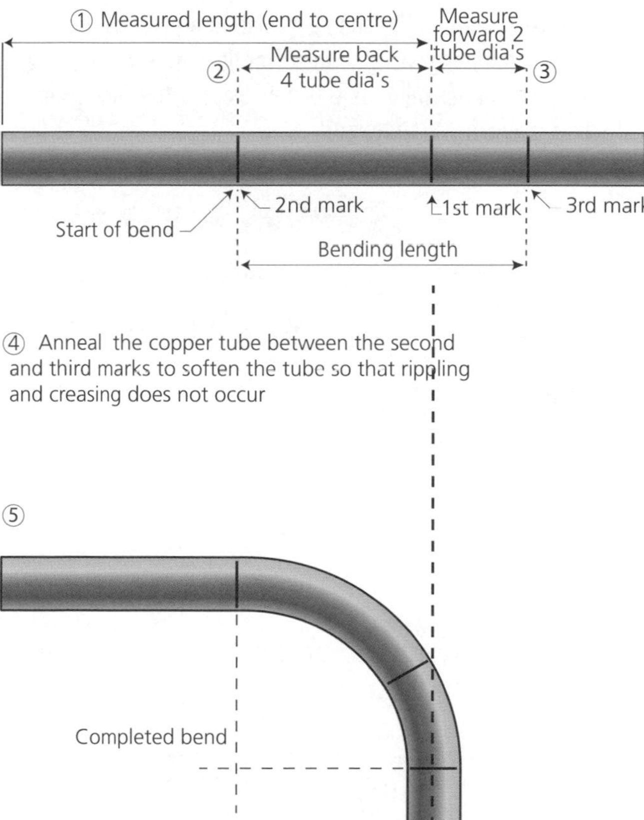

▲ Figure 2.16 Bending a 90° bend by spring

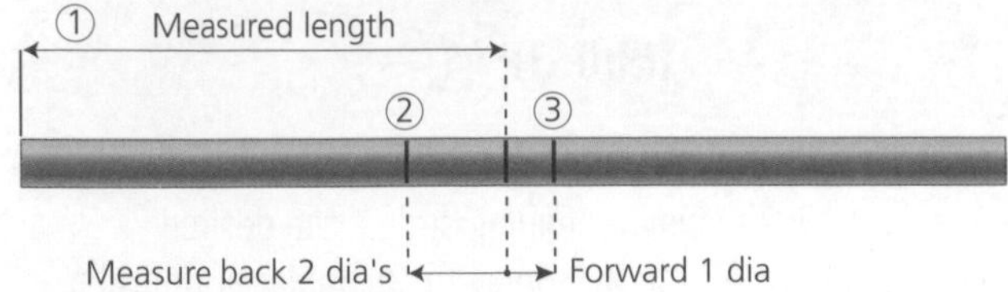

④ Anneal the copper tube between the second and third marks to soften the tube so that rippling and creasing does not occur

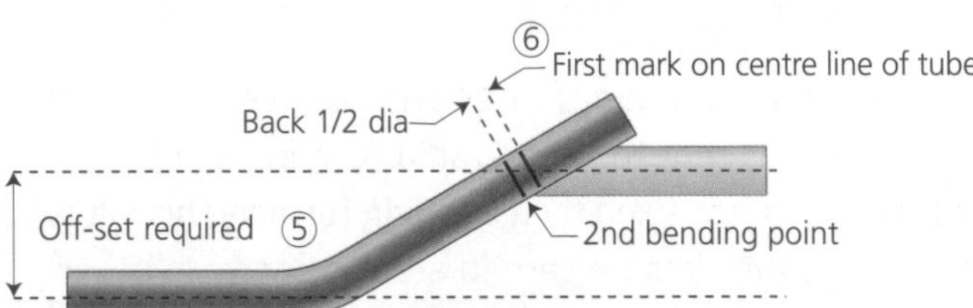

▲ Figure 2.17 Bending an off-set by spring

INDUSTRY TIP

Remember to anneal the copper, which is a process of heating and cooling the pipe to soften it, before attempting to bend the copper tube, as this will prevent the tube from rippling, creasing or snapping.

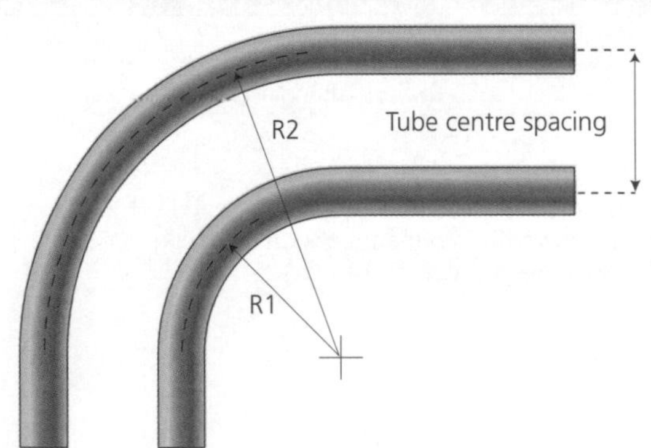

R1 = 4 times dia of pipe
R2 = R1 + tube centre spacing
For two 22 mm diameter tubes at 80 mm centres:
R1 = 4 × 22 = 88 mm
R2 = 88 + 80 = 168 mm

So, set out inner bend as before then for outer bend
Measure back distance for outer bend = 168 mm
Measure forward distance = 84 mm (both from first mark)

▲ Figure 2.18 Concentric spring bends

Setting out spring bends

Spring bends should be limited to copper tube R250 up to 22 mm diameter as bending tubes by hand over this diameter, although possible, is very difficult because of the amount and thickness of the tube. Remember, when setting out:

- for a spring 90° bend, there is gain of the tube in the same way as when the bend is formed with a machine
- allowances have to be made for the 'gain in material'
- the bend must first be pulled in the correct position in relation to the fixed point.

IMPROVE YOUR MATHS

Setting out, step by step

1 Decide on bend radius, which is usually taken as four times the diameter of the pipe (4D), although Yorkshire Copper Tube recommend 5D. It could be any radius determined by a drawing.

The length of the pipe taken up by a 90° bend can be calculated using the formula:

$$\frac{\text{Radius (R)} \times 2 \times \pi\ (3.142)}{4}$$

2 Assuming that a 15 mm pipe is to be bent to a radius of 4D and we need to find out how much pipe will be taken up by the bend:

Radius of bend is 4D = 4 × 15 = 60 mm

Using the formula:

$$\frac{60 \times 2 \times 3.142}{4}$$

Length of bend = 94.26 mm (95 mm)

3 Mark off the required distance from the end of the tube to the centre line of the bend (the end-to-centre measurement).

4 Then divide the calculated length of pipe by three (95 mm / 3 = 31.6 or 32 mm).

5 From the original measurement, mark 32 mm forward and 64 mm back.

6 The bend can then be pulled, ensuring that it is kept within the three 32 mm measurements; this will keep the centre of the bend the correct distance from the fixed point.

Jointing copper tube

There are generally four methods of jointing for copper tubes:

1 capillary fittings:
   a integral solder ring
   b end feed

2 compression fittings:
   a type A – non-manipulative
   b type B – manipulative
3 push-fit fittings
4 press-fit fittings.

Capillary fittings to **BS EN 1254:1998**

Capillary fittings use the principle of capillary action to draw solder into the fitting when they are heated by a blowtorch. There are two different types:

1. **Integral solder ring:** this type of fitting has a band of lead-free solder housed inside a raised ring on the fitting socket, so extra solder is not needed.
2. **End feed:** this type of fitting needs solder to be fed at the end or the mouth of the fitting. It does not have solder in the fitting.

▲ Figure 2.19 Integral solder ring

▲ Figure 2.20 End feed

How to complete a soldered fitting, step by step

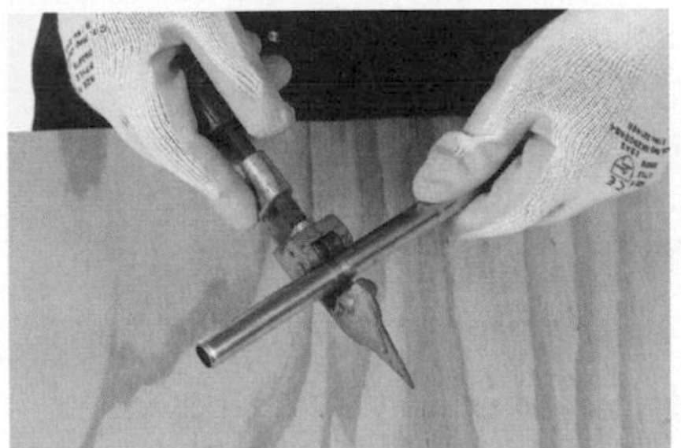

STEP 1 Cut and de-burr the tube.

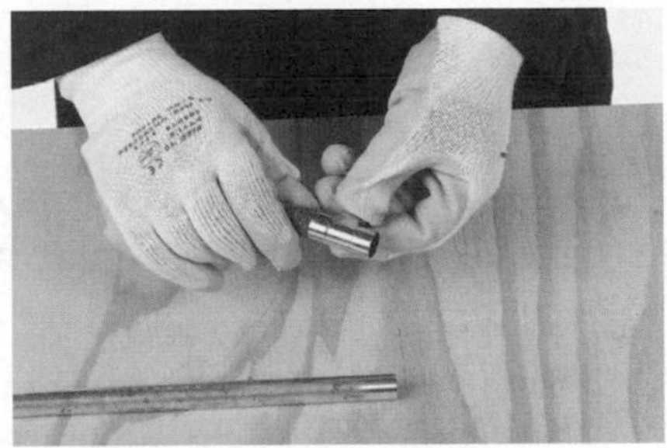

STEP 2 Clean the end of the tube and the inside of the fitting with either wire wool or emery cloth.

STEP 3 Apply flux to the end of the tube only. Do not apply the flux to the inside of the fitting. Insert the tube into the fitting. Twist the tube slightly when inserting it. This ensures an even spread of flux on the tube and fitting.

STEP 4 Apply heat to the fitting and wait 10 seconds. If the fitting is an integral soldered ring type, then solder will appear at the mouth of the fitting. If the fitting is an end feed type, then apply solder to the mouth of the fitting, ensuring that the solder flows all around the socket. Do not use too much heat or the fitting and flux will turn black and the fitting will not solder.

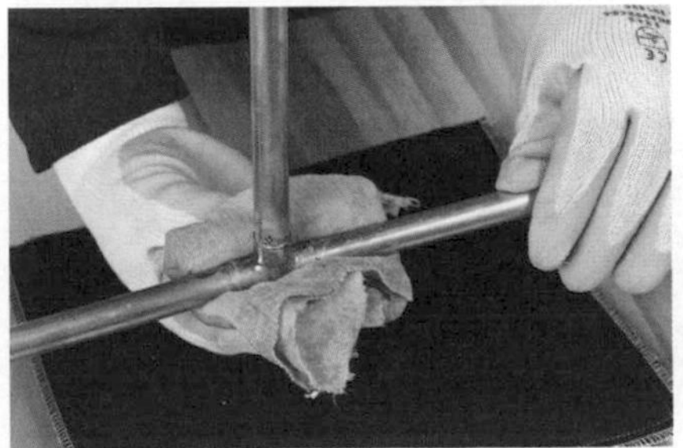

STEP 5 While the fitting is still hot, use a clean cloth to wipe any excess solder off the fitting. Try not to disturb the fitting as you may cause a leak. When the fitting has cooled down a little, clean off any excess flux with a damp cloth.

▲ Figure 2.21 Completing a soldered fitting

Fluxes and solders used with capillary fittings

As we have already seen, integral solder ring fittings have a bead of lead-free solder inside the fitting and so solder is not required for this type of joint. End feed fittings, however, require that solder be added during the soldering process to the mouth of the fitting. For hot and cold water pipework installations this solder MUST be lead-free to comply with the Water Supply (Water Fittings) Regulations 1999. There are several types of lead-free solder available, the most popular being a mixture of tin and copper to BS EN ISO 9453:2020 (known as number 23 tin-based solder), which has a melting point of 230°C to 240°C and is suitable for making end feed capillary joints on all domestic plumbing, heating and gas systems.

The use of leaded solder is permitted for use on gas and central heating installations, but there is always a risk that this solder will be used on the wrong system and, if this occurs, the plumber risks a hefty fine and a criminal record if prosecuted.

Fluxes are used to clean oxides from the surface of the copper and to help with the flow of solder into the fitting. There are two basic forms of flux available:

1. **Active fluxes:** otherwise known as 'self-cleaning' flux because it cleans the copper tube and the fitting during the soldering process. Cleaning of the tube and fittings beforehand is not necessary. Some types of active flux are known to contain hydrochloric acid, which can be harmful if not used correctly and can promote corrosion in copper tubes if excess flux is not removed after soldering has been completed. They are, however, suitable for use in potable water systems, because they dissolve in contact with water and are flushed out when initial flushing of the system takes place.
2. **Traditional flux paste:** usually made from zinc chloride and/or zinc ammonium chlorides. Some fluxes contain other active ingredients such as amines. Cleaning of the tube and fitting is required with this type of flux and will not dissolve in water. It will remain in the pipe after the soldering process has been completed and will not flush out during commissioning, so it should be used sparingly.

Compression fittings to **BS EN 1254:1998**

Compression-type fittings are mechanical fittings that require tightening with a spanner to make a watertight joint. There are two different types:

1. type A – non-manipulative compression fittings
2. type B – manipulative compression fittings.

Fitting type A: non-manipulative compression fittings, step by step

This type of fitting consists of three main parts: the fitting body, a metal 'O' ring called an olive, and the back nut. It is called 'non-manipulative' simply because neither the tube nor the fitting need working, or 'manipulating', to make the joint. When the nut is tightened, the olive is slightly compressed onto the copper tube. To make a type A fitting, follow the steps shown in Figure 2.22.

STEP 1 Cut and de-burr the tube.

STEP 2 Take apart the fitting, and slip the nut and olive over the tube.

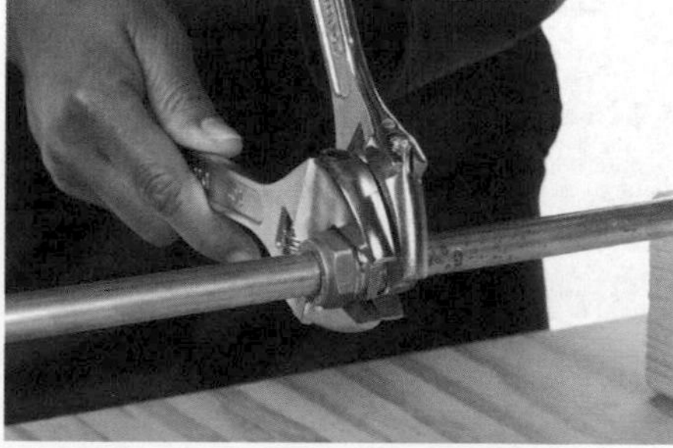

STEP 3 Assemble the fitting and tighten by hand. Then, using an adjustable spanner, turn the nut clockwise 1.5 to 2 turns to fully tighten the joint.

▲ Figure 2.22 Making a type A fitting

Do not over-tighten the joint as this will crush the olive onto the tube too much and may cause the fitting to leak.

This joint does not require any jointing paste or PTFE tape to make the joint. This should be used only if the joint shows signs of leakage.

Fitting type B: manipulative compression fittings, step by step

Unlike type A fittings, type B fittings require that the end of the tube is worked, or more specifically flared, with a special tool called a swaging tool, before a successful joint can be made. This type of fitting is made for jointing soft copper tube (type R220) for below-ground water services. The parts of the fitting are the fitting body, the compression nut, the compensating ring and the adapter piece. To complete a type B compression joint, follow the steps shown in Figure 2.23.

Push-fit fittings for copper tube

Push-fit fittings for copper tube are made from either copper or DZR brass, and are available in sizes 10 mm to 54 mm. They can be used on hot and cold water services above ground, and central heating systems.

Push-fit joints rely on a stainless steel grab ring and a sealing ring to make a watertight joint. There are a number of different makes available and all use a similar method of jointing. When a piece of copper tube is pushed into the joint it passes through a release collar and then through a stainless steel grip ring. This has a number of teeth that grip on to the tube, securing it in place. It can only be released using a de-mounting tool. When the tube is pushed further into the joint it passes through a support sleeve, which helps to align the tube and compresses a pre-lubricated EPDM 'O' ring between the wall of the fitting and the tube. When the tube has passed through the 'O' ring and has reached the tube stop, a secure joint is made.

The pressures and temperatures that apply to push-fit fittings are listed in Table 2.16.

▼ Table 2.16 The pressures and temperatures that apply to push-fit fittings

Temperature not exceeding	Max. working pressure
30°C	16 bar
65°C	10 bar
90°C	6 bar

To complete a push-fit joint, follow the steps shown in Figure 2.24.

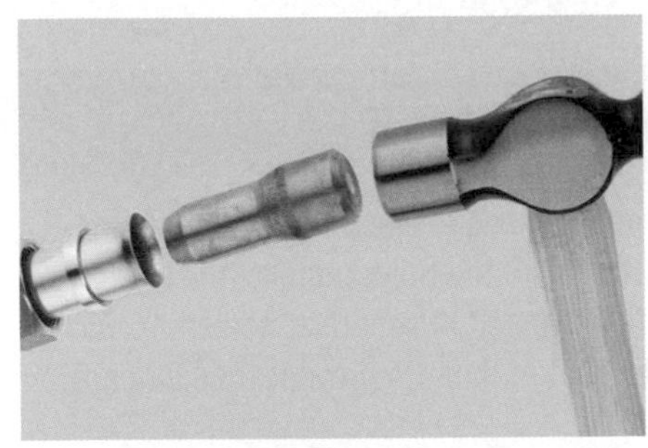

STEP 1 Cut and de-burr the copper tube. Slip the compression nut and the compensating ring over the end of the tube and swage open the end using the special type 1882 swaging tool.

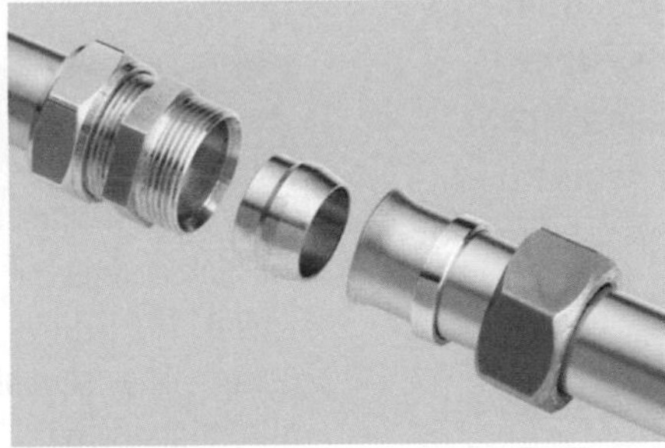

STEP 2 Insert the plain end adapter into the socket.

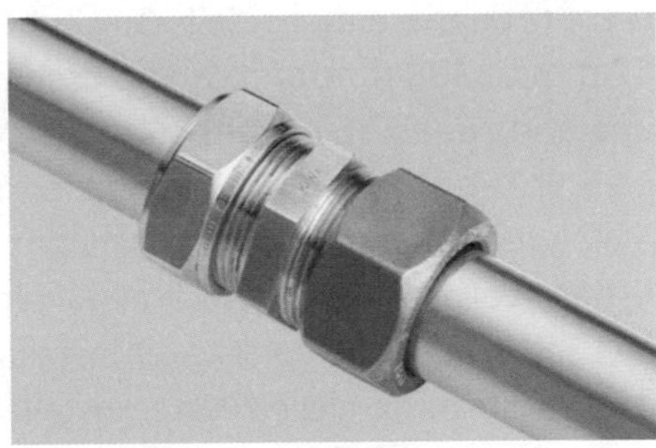

STEP 3 Locate the flared end of the copper tube onto the tapered face of the adapter piece, screw the compression nut on the fitting body and tighten with a spanner.

▲ Figure 2.23 Completing a type B compression joint

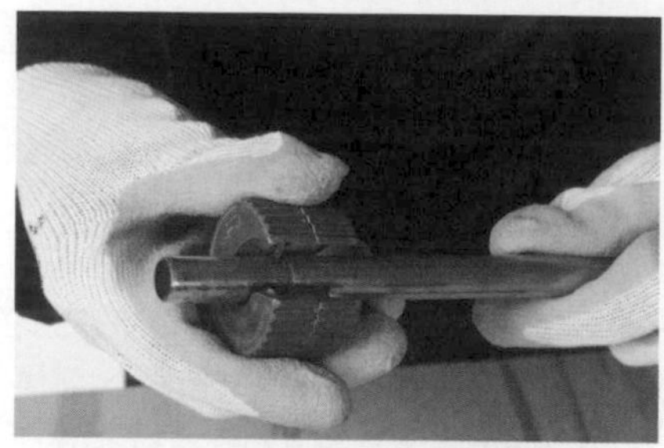

STEP 1 Cut the tube using a tube cutter (not a hacksaw). The tube needs to be round and free from damage.

STEP 2 De-burr the end of the tube so that it is free from any burrs or sharp edges.

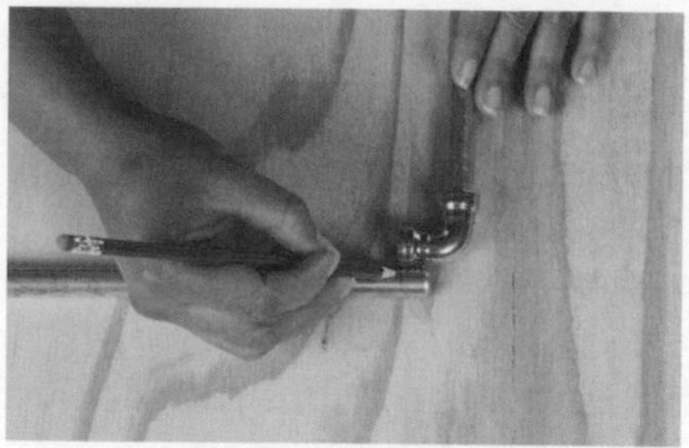

STEP 3 Mark the socket insertion depth to provide a visual marker that the tube has been pushed fully into the socket. Bear in mind that the X dimension will have been taken off the overall length of the pipe (see page 117).

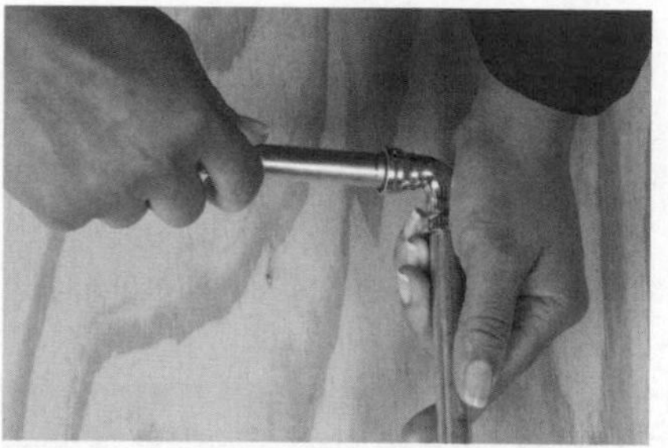

STEP 4 Keep the fitting and tube in line. Push the tube through the release collar to rest against the grab ring.

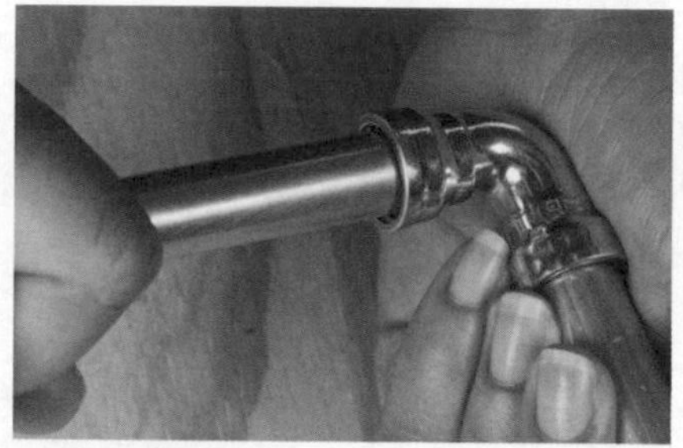

STEP 5 Push the tube firmly, with a slight twisting action, until it reaches the tube stop with a 'click'.

▲ Figure 2.24 Completing a push-fit joint

Press-fit fittings for copper tube

Press fittings are available to suit tube sizes from 12 mm to 108 mm, and can be used for systems operating up to 16 bar pressure at 20°C and 6 bar pressure at 110°C. They are ideal for use where using a blowtorch is not possible. There are several different fitting types available, which allow press-fit fittings to be used on hot and cold water installations, central heating systems, chilled water installations, solar hot water systems and gas installations (using a special yellow rubber 'O' ring).

▲ Figure 2.25 Electrical press tool

A press-fit fitting consists of the fitting body, a rubber seal and a stainless steel grab ring.

INDUSTRY TIP

Press-fit fittings require a special electrical press tool, which crimps the fitting onto the tube to make a secure joint. The fittings are packaged in separate, sealed plastic bags. They should be kept in them to prevent the lubricant from drying out.

To complete a press-fit joint for sizes up to 35 mm, follow the steps shown in Figure 2.26.

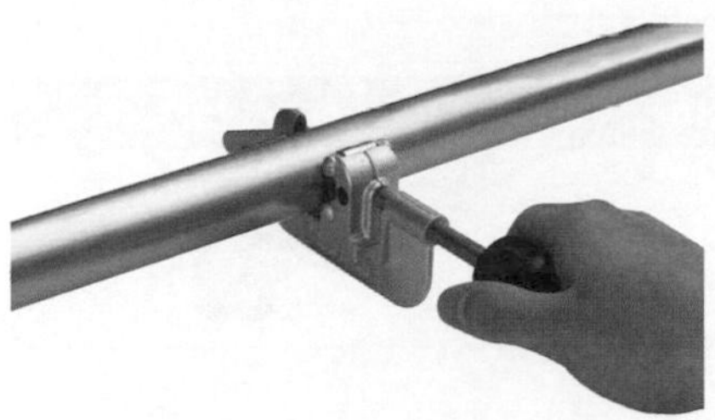

STEP 1 Cut the tube with, preferably, a tube cutter and de-burr the pipe. Care should be taken to ensure the tube is cut square.

STEP 2 The tube must be fully inserted into the socket. To ensure this, use a socket depth gauge to mark the depth of the socket onto the tube or, alternatively, measure and mark using a rule.

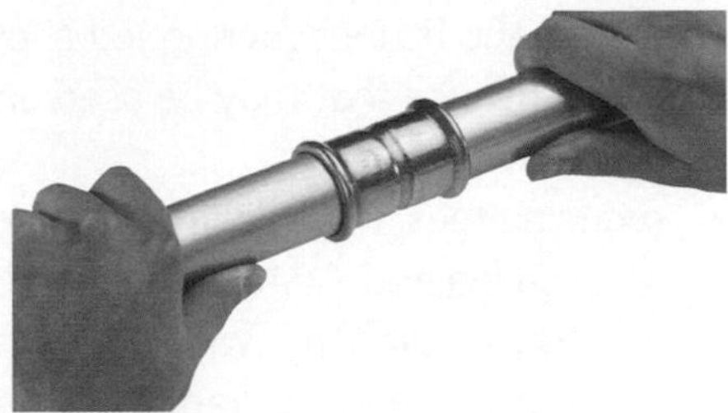

STEP 3 Insert the tube into the fitting all the way to the tube stop. The fitting depth mark previously made on the tube will help as a guide.

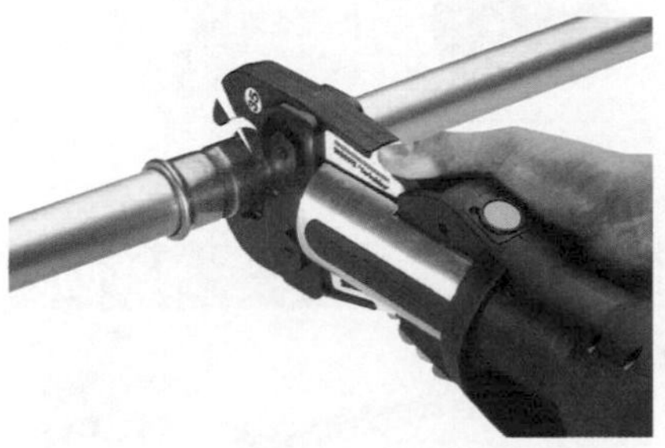

STEP 4 Place the jaws of the press-fit tool over the bead of the fitting, making sure the jaws of the tool are well lubricated.

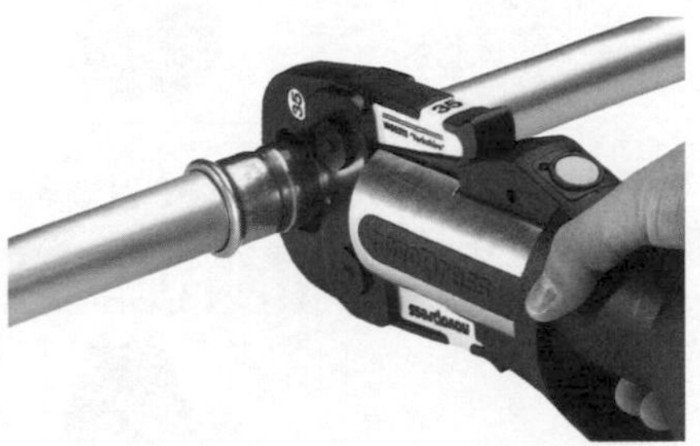

STEP 5 A 90° angle between the tool and the socket must be maintained when making the joints.

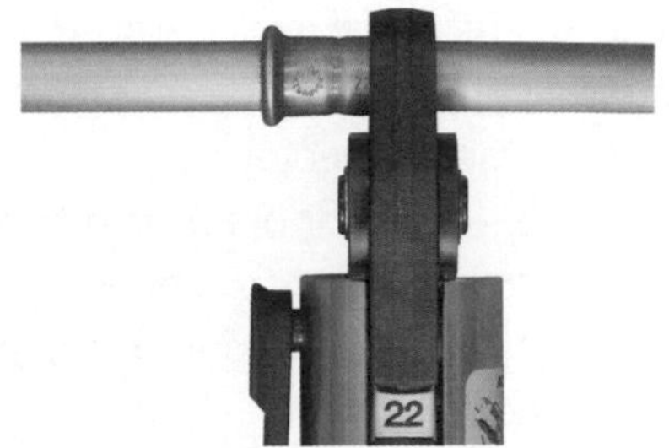

STEP 6 Press the trigger on the press-fit tool to start the jointing process, making sure that fingers are kept away from the jaws.

▲ Figure 2.26 Completing a press-fit joint

Fittings recognition

Fittings recognition is a part of a plumber's job. Choosing the right fitting for the right application is a key element of a successful installation. There are four fittings that are used more than all others. These are couplings, equal tees, elbows and reducers. Table 2.17 shows these four fittings in each of the jointing types.

▼ Table 2.17 Copper tube fittings recognition

	Couplings	Equal tees (all three connections equal size)	Elbows	Reducers
End feed				
Integral solder ring				
Compression				
Push fit				
Press fit				

As well as the fittings mentioned above, there are other common fittings that may be used on a regular basis. These are:

- reducing tees, which come in three different forms:
    1. reduced end – where one end is reduced
    2. reduced branch – where the branch is reduced
    3. reduced end and branch – where one end and the branch are reduced – and two reduced ends (sometimes called pendant tees), where both ends are reduced
- tap connectors – used for connecting to taps and float-operated valves:
    - straight tap connector
    - bent tap connector
- cap ends – used for blanking off the ends of the tube; also known as stop ends
- tank connectors – used for making connections to tanks and cisterns
- flexible connectors – often used instead of tap connectors on sanitary ware
- central heating manifolds – a specialist fitting used in microbore central heating systems.

INDUSTRY TIP

When ordering tees with a mixture of end and branch sizes, care should be taken to ensure that the correct configuration is quoted. The method to use when ordering tees is to quote the largest end, then the smallest end, and the branch last.

In the photograph above, the tee would be ordered as:

22 mm × 22 mm × 15 mm

▼ Table 2.18 Reducing tees

Reduced end	Reduced branch	Reduced end and branch	Two reduced ends

▼ Table 2.19 Tap connectors

Straight tap connector	Bent tap connector

▼ Table 2.20 Cap ends

Compression cap end	Push-fit cap end	End feed capillary cap end

▼ Table 2.21 Connectors and manifolds

Tank connector	Flexible connector	Manifold

Low carbon steel pipes to **BS EN 10255:2004** (formerly **BS 1387:1985**)

Low carbon steel pipe is used occasionally in domestic installations but its use should be restricted to wet central heating systems, gas installations and oil lines. It must not be used to supply hot or cold water supplies for domestic purposes because of the risk of rusty water being drawn from the taps. Often referred to as mild steel pipe, low carbon steel pipes are usually supplied painted red or black and it can also be galvanised coated. Its carbon content is low.

It is available in three grades, each grade being identified by a colour code, as shown in Table 2.22.

▼ Table 2.22 Carbon steel pipe grades and colour codes

Grade	Colour code
Light	Brown
Medium	Blue
Heavy	Red

The grades of low carbon steel have identical external diameters but the pipe wall thickness will vary according to the grade – heavy grade having the thickest pipe wall and light grade the thinnest.

Medium-grade pipe is the most common grade used in plumbing installations but heavy grade may be used where a long system life is expected. Heavy-grade pipe can also be used below ground. Light-grade pipe is seldom used, except in some dry sprinkler installations for fire prevention.

Low carbon steel pipe is available in 3 m and 6 m lengths, may be supplied with threaded ends or plain ends, and is referred to by imperial pipe sizes, which are specified as nominal bore. The common pipe sizes for domestic purposes are shown in Table 2.23.

▼ Table 2.23 Common low carbon steel pipe sizes

Thread size/fitting size	1/8"	1/4"	3/8"	1/2"	3/4"	1"	1 1/4"	1 1/2"	2"
Nominal diameter mm	6	8	10	15	20	25	32	40	50

Bending low carbon steel pipe

There are two methods of bending low carbon steel pipe:

1 **By hydraulic bending machine:** this method uses a hydraulic bending machine (see page 78 of this chapter) to bend the pipe. It uses an oil to exert hydraulic pressure. The oil, being incompressible, exerts great force on the pipe through the bending former to bend the pipe when the handle of the machine is pumped. Steel is very tough to bend and tends to 'spring' back once the bend is formed. Because of this, bends should be over-bent about 5° to allow for the bend springing back slightly. This is the method used in domestic installations.

2 **By heat:** mainly used on industrial installations. This involves the use of oxyacetylene torches to heat the steel almost to white hot to soften the pipe. This allows the steel pipe to be bent easily by hand.

Here, we will look at how to bend a 90° bend and an off-set bend using a hydraulic bending machine.

Bending a 90° bend, step by step

1 Mark a line on the pipe at the required distance from the fixed point to the centre line of the required bend (Figure 2.27).

2 From this measurement, measure back towards the fixed point 1 nominal bore (the internal diameter) of the pipe to point A.

3 Place point A at the centre of the correct size bending former on the bending machine.

4 Pump the handle of the bending machine until an angle of 90° + 5° (allowance for springing back) has been achieved. Make sure you are standing to the side of the machine. NEVER stand in front of it while bending is taking place.

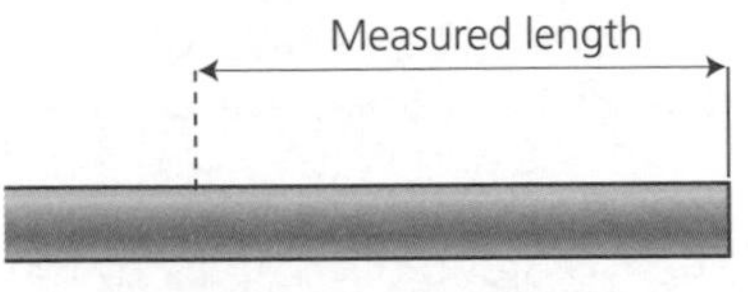

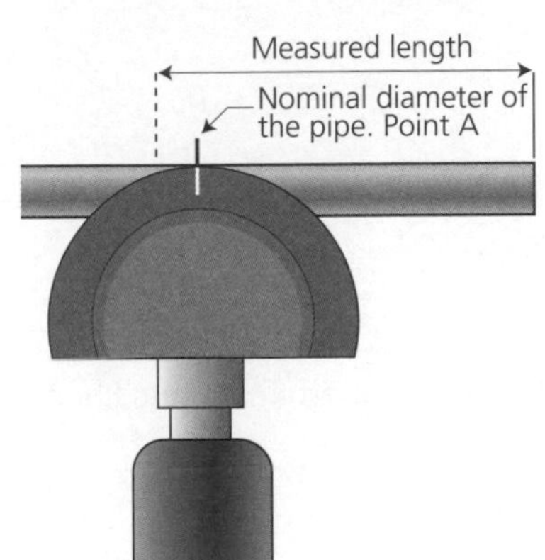

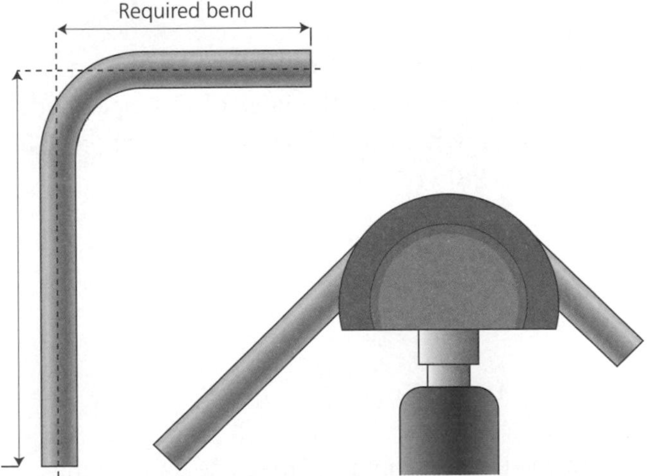

▲ Figure 2.27 Bending a 90° bend

INDUSTRY TIP

You may find it easier to judge the angle of the bend by making a template from a welding rod bent to 90° or by the use of a steel set square. The template can be placed on the bending machine so that you can see where to stop the bend.

Bending an off-set bend, step by step

1 Mark a line at the required measurement for the first bend onto the pipe.

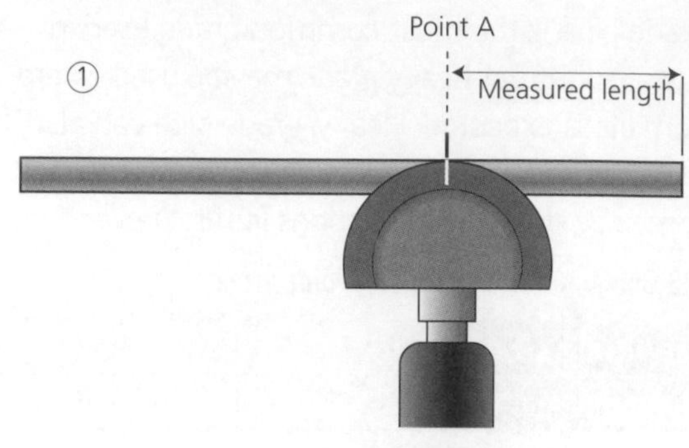

2 Place the pipe in the machine but this time do not make any deduction. The mark goes directly on the centre of the former. The measurement A (Figure 2.28) is from the fixed end of the pipe to the centre of the set.

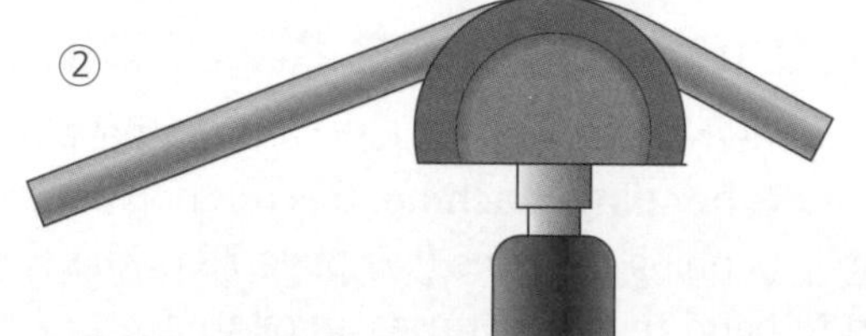

3 Make the first bend to the required angle and check the angle using the template.

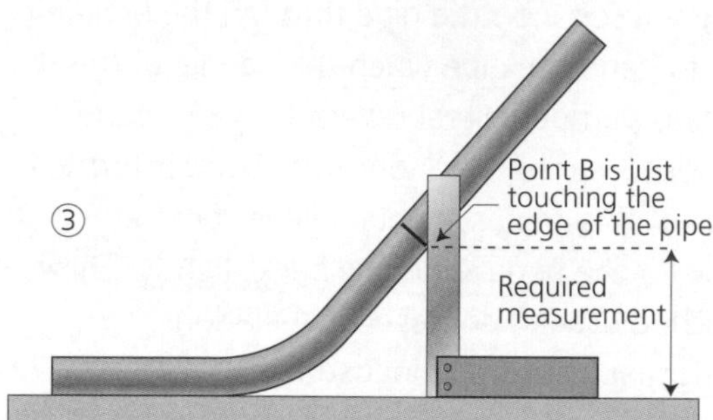

4 Take the pipe from the machine and place a straight edge against the back of the pipe. Mark the measurement of the second bend at point B (Figure 2.28).

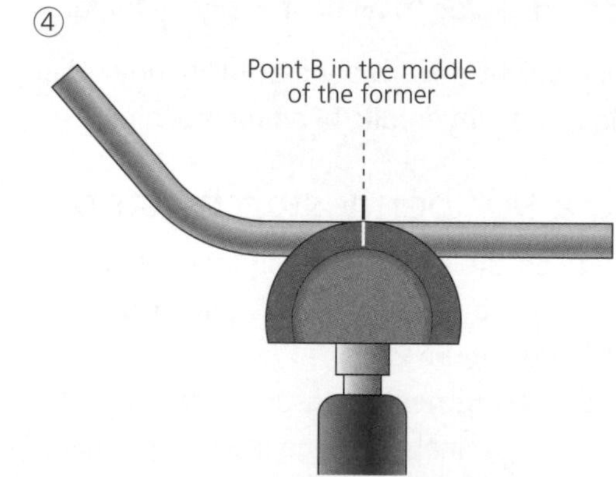

5 Put the pipe back into the machine and line the mark up with the centre of the former.
6 Bend the second bend and check with the template.

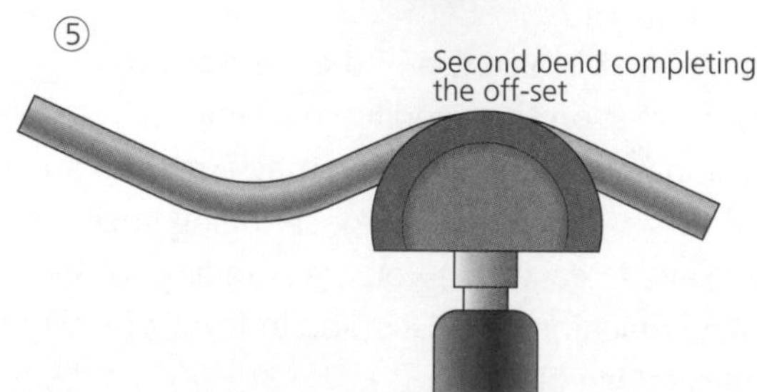

▲ Figure 2.28 Bending an off-set bend

INDUSTRY TIP

With the off-set bend, you may find it easier to make a welding rod template bent to the required angle (say 45°). This can be used for both the first and second bends. Remember to over-bend both bends by 5° to allow for spring back.

Jointing low carbon steel pipe

There are three ways to joint low carbon steel pipe. These are:

1 threaded joints
2 compression joints
3 welded joints.

We will look at the first two only, as welded joints are generally used only on larger pipes in industrial applications and installations.

Threaded joints

Low carbon steel pipes can be jointed using threads to **BS 21**, which are cut into the end of the pipes using either manual stocks and dies or electric threading machines (see page 78 of this chapter). There are two kinds of thread, as follows.

1 **Tapered threads:** a standard thread cut onto the ends of pipes and blackheart malleable, male fittings to ensure a watertight, gas-tight or steam-tight joint. The tube tightens the further it is screwed into the fitting.
2 **Parallel threads:** a screw thread of uniform diameter used on fittings such as sockets.

KEY POINT

Male threads are external threads; female threads are internal threads.

There are two types of fittings that use threads. Fittings for low carbon steel pipe are made from steel and malleable iron to **BS EN 10242** (formerly **BS 143**) and **BS 1256**. Steel fittings, although stronger than malleable iron, tend to be more expensive. Malleable iron fittings fall into two groups:

1 **Blackheart fittings** with tapered female threads are identified by a square-edged bead around the mouth of the fitting. These fittings are quite brittle and susceptible to splitting if over-tightened.
2 **Whiteheart fittings** with parallel female threads are identified by a rounded bead around the mouth of the fittings. These fittings are slightly softer, and therefore more flexible, and tend to stretch if over-tightened.

INDUSTRY TIP

All threads are BSPT, which stands for British Standard Pipe Threads.

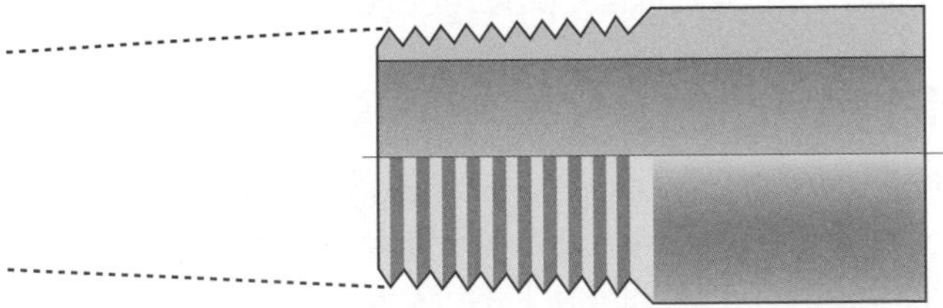

Threads taper towards the end of the tube

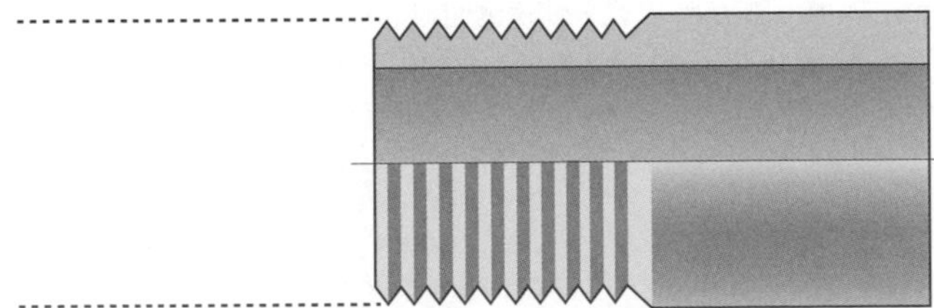

Threads remain parallel throughout the length of the tube

▲ Figure 2.29 Tapered thread (left) and parallel thread (right)

▼ Table 2.24

Couplings	Equal tees	Elbows	M/F elbows
Unions	**Nipples**	**Bushes**	

When cutting a thread onto a length of pipe, the length of the thread should be such that, once the joint is made, one and a half to two threads should be visible when the joint is completed.

There are a variety of jointing compounds that can be used with threaded joints. Jointing compounds are used to make leak-free joints. Each one has a specific use, although some are universal and can be used on a number of different installations. Jointing compounds include those listed in Table 2.25.

▼ Table 2.25 Jointing compounds

Linseed oil-based compounds (boss white, hawk white and templars paste)	Can be used in conjunction with hemp on wet central heating systems and compressed air lines. Must not be used on natural gas installations.
Unsintered polytetrafluoroethylene (PTFE tape or string)	A thin, white (or yellow if used on gas) tape that can be used on most installations, including hot and cold water, central heating and gas installations.
PTFE-based jointing compounds (boss green)	A compound specially made for use on hot and cold water supplies. Not suitable for natural gas installations.
Hematite paste	A truly universal compound that can be used on many installation types, such as oil, gas, hot and cold water, central heating, compressed air lines and vacuum lines.
Manganese paste	These are specialist compounds for use with high temperature hot water and steam installations.
Graphite paste	
Gas seal paste	A specialist compound for use with natural and liquid petroleum gas installations.

Compression joints

There are a number of different manufacturers of compression joints for low carbon steel pipes. They incorporate a rubber compression ring to ensure a leak-free joint. They tend to be rather expensive but can save time on installation costs. They are often referred to as transition fittings.

Low carbon steel compression fittings can be used on new installations, pipe repair and pipework extensions on the following installations:

- water (hot and cold water, central heating systems)
- gas (natural gas, LPG)
- oil
- compressed air.

They have several advantages to screwed fittings:

- very versatile connection suitable for connecting LCS pipe to different pipe materials, such as copper and lead
- quick and easy to make joints
- no special tools necessary
- no threads on steel pipe required.

Plastic pipe

Plumbers should understand the properties of the types of plastics they use to prevent mistakes being made during their installation. Plastics have revolutionised modern plumbing systems but it is all too easy to use plastics for what they are not designed. There are two main types of plastics used in plumbing:

1 those plastics that can be used for hot and cold water supply and central heating services (plastic pressure pipe), such as:
   - polyethylene (PE)
   - polybutylene (PB-1)
2 those plastics that can be used for sanitation, drainage and rainwater systems, such as:
   - polyvinyl chloride
   - acrylonitrile butadiene styrene (ABS)
   - polypropylene.

Plastic pressure pipe: polyethylene (PE)

PE is used extensively in the plumbing industry for mains cold water pipes. Two grades are used below ground on cold water services:

1 medium-density polyethylene (MDPE), manufactured in accordance with the requirements of **BS EN 12201-2**; it is blue in colour
2 high-density polyethylene (HDPE); this was used until the mid-1980s for mains cold water pipes until superseded by MDPE; it is still manufactured but is not used as extensively as MDPE; coloured black, it is available in grades A, B, C and D.

MDPE is a hard-wearing plastic for water pipes, gas pipes and fittings. It is available in a variety of colours. It is resistant to shock (and subsequent fractures) and has good performance in freezing weather conditions. It is, however, susceptible to ultraviolet (UV) and direct sunlight, and it is recommended that a maximum of 150 mm of pipe is showing when it enters the building. It must not be used above ground except for temporary installations.

MDPE piping and pipe fittings are available in sizes of 20 mm to 63 mm; 25 mm is the most common pipe size used for cold water services for domestic properties. It is supplied in coils of 25 m to 150 m.

Jointing medium-density polyethylene (MDPE) pipe

MDPE pipe can be jointed in a variety of ways. The most common types of fitting are:

- compression fittings made from brass
- compression fittings made from plastic
- push-fit fittings made from plastic
- fusion welded.

Compression fittings made from brass

These require a pipe insert, which can either be made of copper or nylon. The insert is placed inside the pipe to strengthen the wall of the pipe so that the fitting does not blow off under mains pressure.

▲ Figure 2.30 Pipe inserts must be used when using brass compression fittings on plastic pressure pipe

To make a compression joint on MDPE pipe using brass compression fittings, follow the steps below.

1 Measure and cut the pipe to the required length, ensuring that the cut is square. A plastic pipe cutter should be used to do this.
2 De-burr the pipe inside and out.
3 Slip the compression nut and the olive over the pipe.
4 Put the pipe insert inside the pipe. Make sure that the nut and olive are in place before you do this as placing the insert inside the pipe makes slipping the olive over the pipe difficult.
5 Put the pipe inside the fitting body and hand tighten the nut.
6 Now, using a suitable spanner, fully tighten the fitting 1 to 1.5 turns.

Compression fittings made from plastic

These are known as 'Philmac' fittings, and give the ability to connect MDPE to MDPE or MDPE to most forms of pressure pipe, including copper tube and lead pipe.

> **KEY POINT**
> For more information on Philmac fittings, visit the website at: www.philmac.co.uk

Push-fit fittings made from plastic

These fittings are the simplest form of jointing for MDPE. They simply push onto the pipe to make a secure, watertight joint. No tightening is needed. The fitting contains a stainless steel grab ring to grab and hold the pipe, and a neoprene rubber seal. A pipe insert made of nylon is required inside the pipe.

Fusion welded

Large underground water mains use fusion-welded fittings, where the fitting and the pipe are welded together by heat created by electricity. A special fitting is used that has an electrical element inside the fitting body, which when subjected to electricity, generates heat, which melts the fitting and the pipe together.

Polybutylene (PB-1)

Polybutylene is the latest plastic to be manufactured into pipe for pressurised plumbing systems. Polybutylene is very flexible, allowing it to be cabled easily and quickly through timber joists during the installation process. It has a high temperature and pressure resistance, low noise transmission, low thermal expansion and low thermal transmission. Its internal bore is very smooth, giving it good flow rate characteristics and it does not suffer from corrosion or scaling. It is, however, micro-porous, allowing air to be leeched through the walls of the pipe.

INDUSTRY TIP

When PB-1 pipe was first introduced in the late 1980s, central heating systems suffered failure due to increased black oxide sludge created by excess air in the system. This has since been cured with the introduction of barrier pipe, which has an impermeable barrier placed in the walls of the pipe. Barrier pipe is not needed for hot and cold water installations.

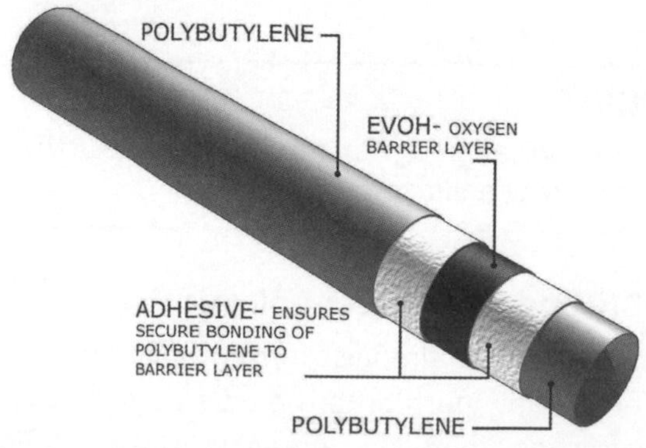

▲ Figure 2.31 Polybutylene (PB-1) pipe

Polybutylene pipe is usually coloured white or grey, but older PB pipe (known as Acorn) is usually coloured brown. It can be used on hot and cold water installations, wet central heating systems and underfloor heating. It is available in sizes 10 mm, 15 mm, 22 mm and 28 mm, in straight lengths of 3 m, and coils of 25 m, 50 m and 100 m lengths. The pipe sizes are compatible with copper tubes to **BS 7291-2:2010**.

The benefits of using polybutylene pipe

In recent years, polybutylene pipe has become very popular with both installers and architects for new-build installations. There are many reasons for this:

- It does not affect the taste or colour of the water.
- There is minimal internal resistance, thereby increasing flow rates.
- It is very flexible, even at very low temperatures.
- It is highly resistant to stress.
- It is non-corrosive.
- It involves safe installation processes; no flame is needed or chemicals such as flux required during installation, and it therefore presents no risk to installers.
- It has high resistance to frost damage.
- It is not affected by water hardness or softness.
- It is not affected by chemical central heating inhibitors or anti-freezes.
- It is unaffected by micro-biological growth.
- It has high impact strength.
- It is suitable for heating and cooling applications.
- There are a multitude of coil lengths for economical installation with minimal waste.
- It has a low environmental impact in terms of soil, water and air pollution.

Bending polybutylene pipe

Polybutylene pipe can be bent by hand without the use of a bending machine. However, the use of cold forming bend fixtures is recommended. These are pre-formed metal braces, which hold the pipe in a 90° position.

Alternatively, it is possible to brace the bend using pipe clips, ensuring the radius of the bend is not less than those shown in Table 2.26.

▼ Table 2.26 Bend radii for polybutylene pipes

Diameter of pipe – mm	10	15	22	28
Radius dimension A – mm	80*	120*	160*	224*
*Depending on the pipe manufacturer				

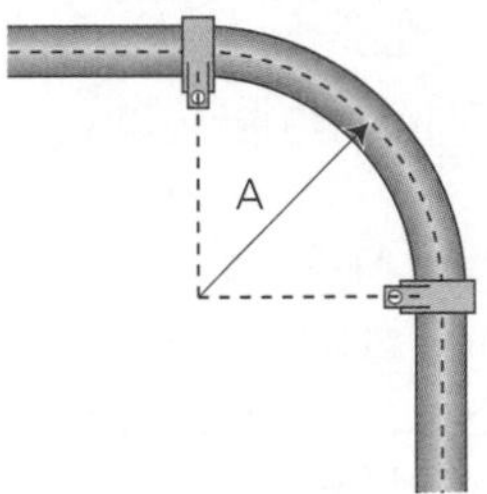

▲ Figure 2.32 Bending polybutylene pipe

Jointing polybutylene pipe

Polybutylene pipe can be joined in two ways:

1 push-fit fittings
2 standard type A non-manipulative type compression fittings to **BS EN 1254:1998**.

Push-fit fittings

These have a stainless steel grab wedge to hold the pipe firm, and a neoprene rubber 'O' ring to make a watertight joint.

A pipe insert usually made from either plastic or stainless steel (depending on the pipe manufacturer) must be placed inside the pipe before the joint is made. The procedure for making a push-fit joint on polybutylene pipe is as follows.

1 Cut the pipe using a scissor-type plastic pipe cutter. This ensures a clean cut to the pipe end. Do not use a hacksaw.
2 Push the pipe insert into the pipe. Most pipe manufacturers put marks on the pipe at fitting depth distance. This helps to visually ensure that the pipe is pushed fully into the joint.
3 Lubricate the end of the pipe with silicone spray lubricant.
4 Push the pipe fully into the fitting until the fitting stop is felt.

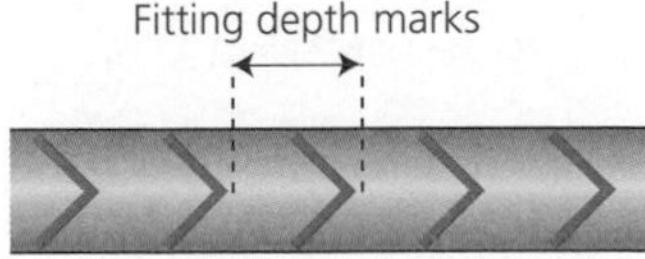

▲ Figure 2.33 Pipe showing fitting depth marks

Standard type A non-manipulative type compression fittings to **BS EN 1254:1998**

As polybutylene is manufactured to the same pipe sizes as copper tubes, type A compression fittings can be used. Again, if using a compression fitting, a pipe insert must be pushed inside the tube. This is because the polybutylene pipe is too soft to support the olive being crushed onto it. The pipe insert (or liner) supports the pipe wall.

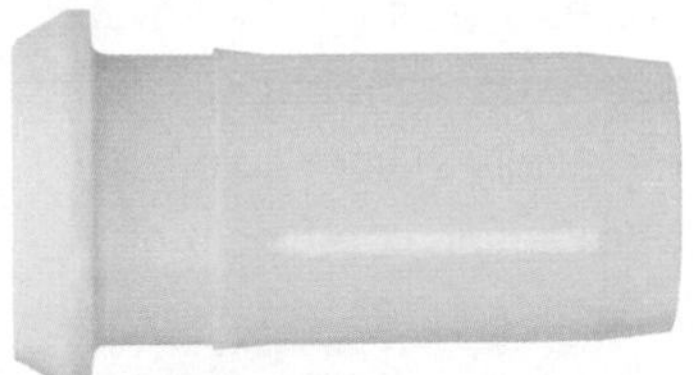

▲ Figure 2.34 Pipe insert

The different types of push-fit fittings for polybutylene pipe

There are many different manufacturers of polybutylene pipe and each one has its own type of push-fit fitting. The general arrangement is almost always the same. Each one has:

- a fitting body
- a rubber 'O' ring to make the joint
- a stainless steel grab ring or grab wedge to hold and lock the pipe into the fitting body
- a spacer washer between the 'O' ring and the grab ring.

Most of the fittings are de-mountable, which means they can be taken off the pipe and reused.

▼ Table 2.27 Common styles of push-fit fitting

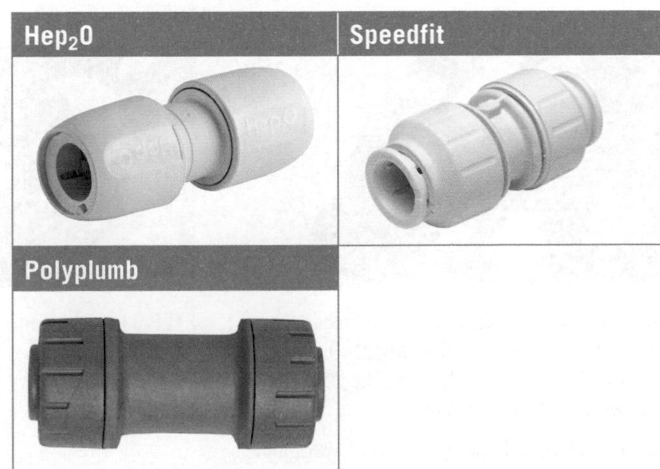

Plastic pipe for sanitation, drainage and rainwater

Polyvinyl chloride (PVC) and acrylonitrile butadiene styrene (ABS)

Polyvinyl chloride is available in four different types:

1. **Unplasticised polyvinyl chloride (PVCu) to BS 4514:2001; BS EN 1401-1:2019; BS EN 1329-1:2020:** used mainly for push-fit and solvent weld soil and vent pipes, below-ground drainage, solvent weld waste and overflow pipes, gutters and rainwater pipes. It has good resistance to UV light but can suffer from photodegradation, especially in light colours such as white and grey. It has a high coefficient of linear expansion. Sizes available are 110 mm, 50 mm, 40 mm, 32 mm and 21.5 mm for soil/vent pipes and waste and overflow pipes. Gutters and rainwater pipes are available in a variety of sizes and styles, which are discussed in Chapter 8, Rainwater systems, page 429.
2. **Modified unplasticised polyvinyl chloride (MuPVC) to BS 4514:2001; BS EN 1401-1:2019; BS EN 1329-1:2020:** used for solvent weld waste and overflow pipes. It is more durable than PVCu and performs better than other plastics, especially at higher temperatures. Sizes available are 50 mm, 40 mm, 32 mm and 21.5 mm.
3. **Chlorinated unplasticised polyvinyl chloride (CuPVC):** used for solvent weld cold water service pipes in the late 1970s. Fittings are still available for repairs, but pipe is increasingly difficult to find. It has a tendency to snap, especially at low temperatures and if mishandled.
4. **Acrylonitrile butadiene styrene (ABS) to BS 1455-1:2000:** these pipes and fittings are usually used for soil and waste pipes and, because of their toughness, can also be used for mains cold water pipes. ABS degrades quickly when exposed to UV light. It possesses extremely good impact strength and high mechanical strength, which makes it suitable for plumbing pipework and installations. The jointing methods used, pipe sizes and clipping distances are the same as for PVCu.

Jointing methods for PVCu, MuPVC and ABS

PVCu can be jointed using:

- solvent weld – used on soil/vent pipes, waste pipes and overflow pipes
- push-fit – used on soil and vent pipes.

Making a solvent weld joint on PVCu, MuPVC and ABS soil/vent pipes, step by step

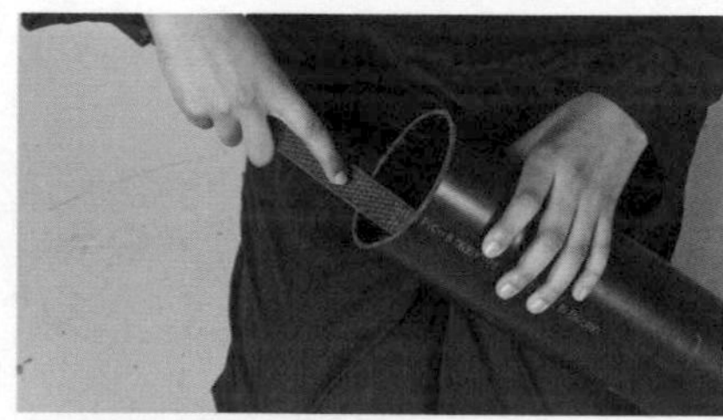

STEP 1 Cut the pipe square using a hacksaw.

STEP 2 Wipe the pipe to remove excess dirt and swarf.

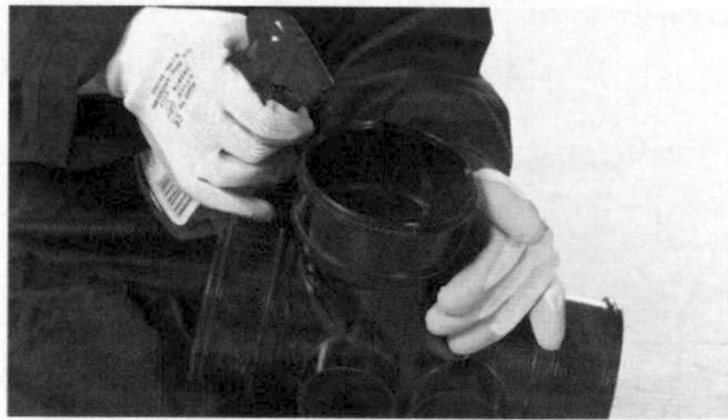

STEP 3 Clean inside the socket and the pipe spigot with solvent cleaner.

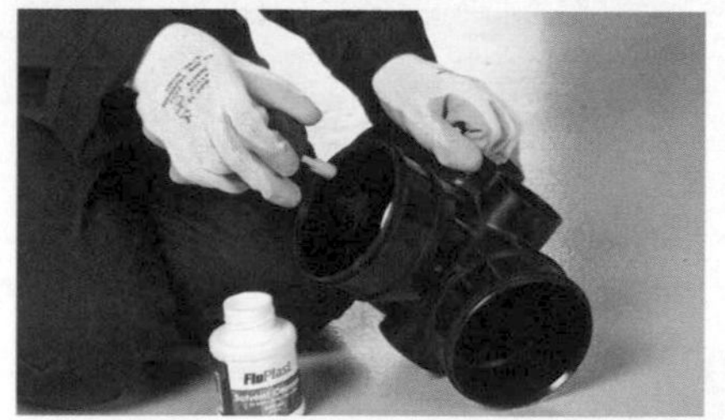

STEP 4 Apply solvent weld cement inside the socket first and then to the **spigot**. This will allow a little more time to make the joint before the cement begins to dry out.

STEP 5 Insert the pipe into the socket and twist fully into the socket.

STEP 6 Wipe off excess cement using a dry cloth.

▲ Figure 2.35 Making a solvent weld joint

Making a push-fit joint on PVCu, MuPVC and ABS soil/vent pipes, step by step

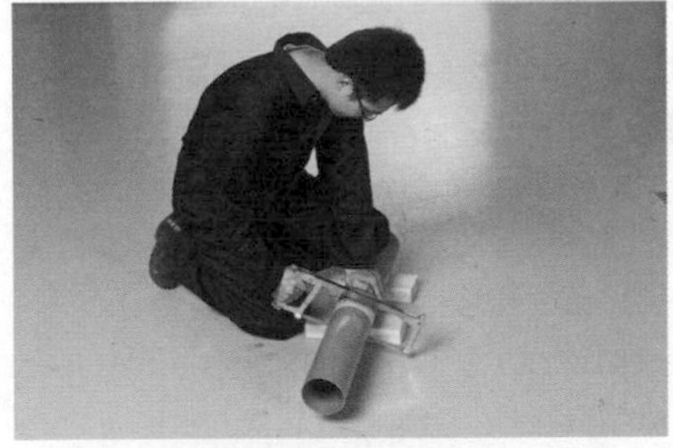

STEP 1 Cut the pipe square using a hacksaw.

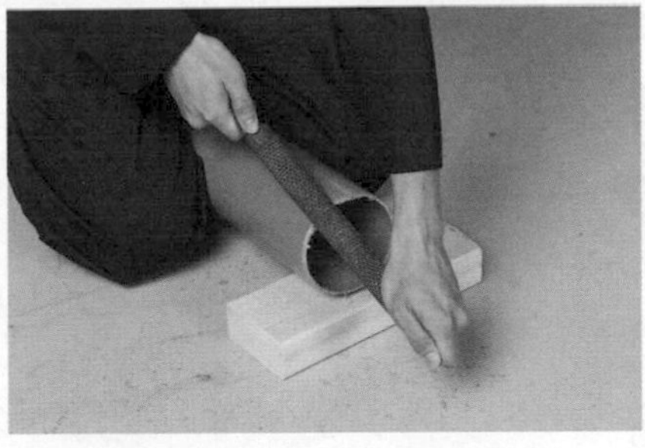

STEP 2 **Chamfer** the pipe using a file or a rasp.

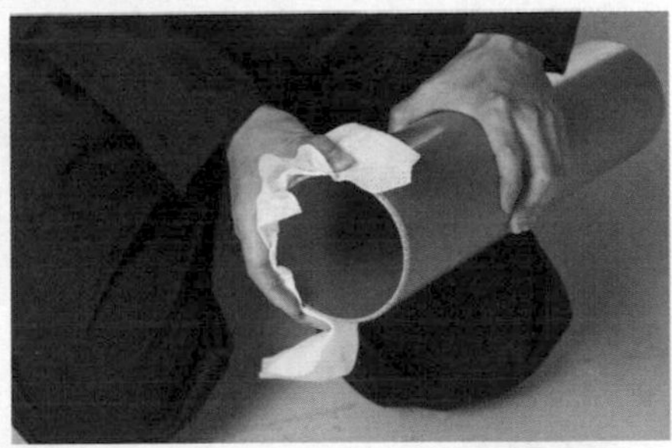

STEP 3 Wipe the pipe to remove excess dirt and swarf.

STEP 4 Lubricate the end of the pipe using silicone grease. Do not use liquid soap as this can adversely affect the rubber seal.

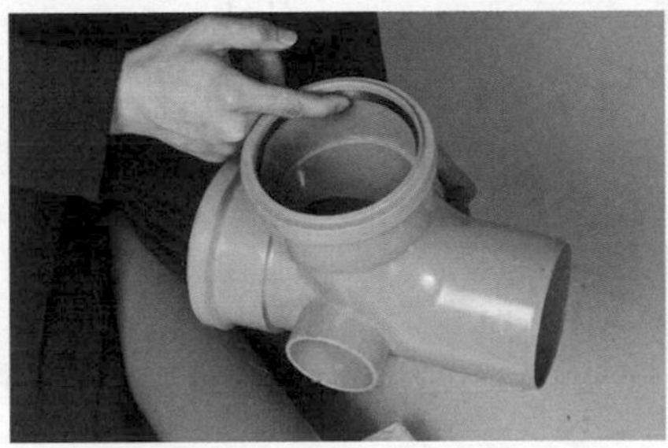

STEP 5 Check that the seal is in the correct position in the fitting.

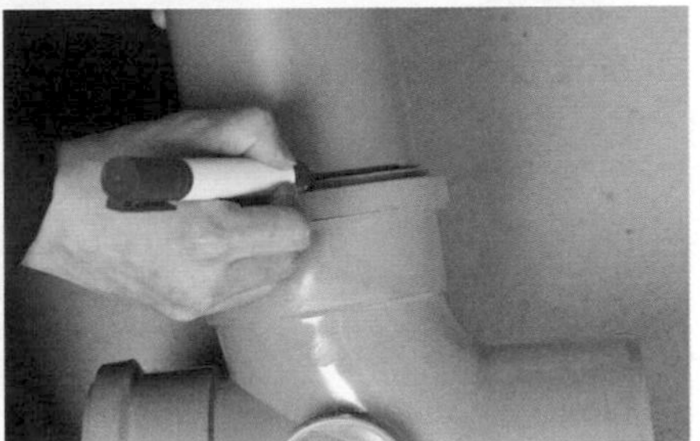

STEP 6 Push the pipe all the way into the fitting and mark the pipe at the end of the fitting using a pencil.

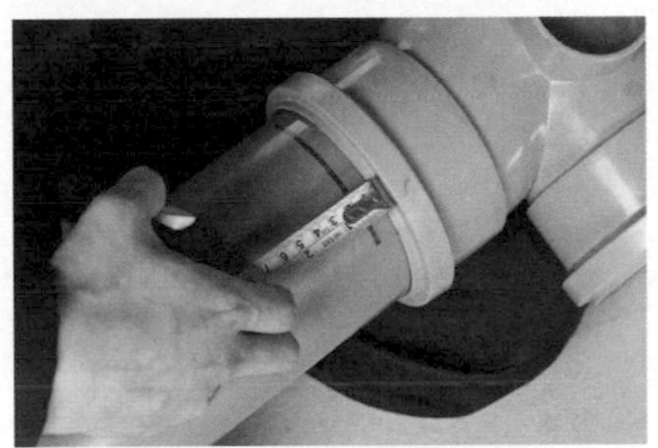

STEP 7 Withdraw the pipe 10 mm from the fitting. This is to allow for expansion of the pipe. Fittings **must** be supported by a pipe bracket to prevent the fitting from slipping.

▲ Figure 2.36 Making a push-fit joint

KEY TERMS

Spigot: another name for the plain end of a pipe. If the fitting we buy has a plain pipe end, we call this a spigot end.

Chamfer: to take off a sharp edge at an angle. If we chamfer a pipe end, we are taking the sharp, square edge off the pipe.

Fittings for PVCu soil/vent and waste pipe installations

Fittings for PVCu and MuPVC soil/vent and waste pipes are of the same size. This means that the two systems are interchangeable. Table 2.28 lists some of the more common types of soil pipe fitting.

▼ Table 2.28 Common soil pipe fittings

90° bends	45° bends	Junctions	Sockets
Boss pipes	Strap boss	Access pipes	
Boss pipe adapters	Pipe clips	Waste pipe manifolds	

PVCu 82 mm, 110 mm and 160 mm soil pipe is available in 2.5 m, 3 m and 4 m lengths in a variety of colours. The pipe and fittings are manufactured to **BS EN 1329–1:2000**. The pipe has a socket on one end and a chamfered spigot on the other.

Solvent weld waste pipe fittings

▼ Table 2.29 Solvent weld waste pipe fittings

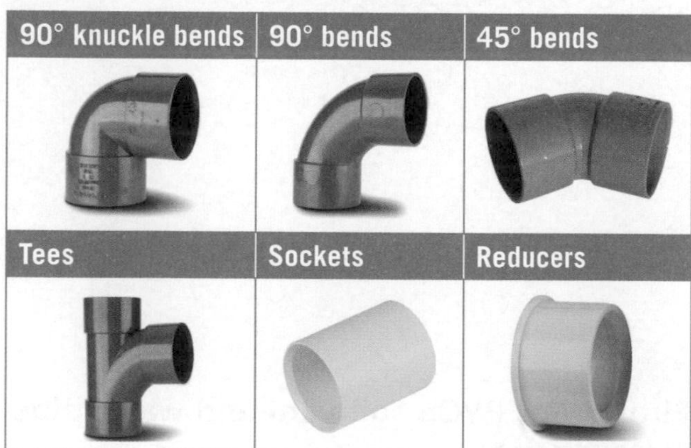

90° knuckle bends	90° bends	45° bends
Tees	Sockets	Reducers

PVCu waste pipe is manufactured to **BS EN 1455** and is available in 3 m lengths in sizes 21.5 mm (overflow pipe), 32 mm, 40 mm and 50 mm.

Polypropylene (PP)

Polypropylene is a common plastic in plumbing systems. It is used to manufacture cold water cisterns, WC siphons and push-fit waste and overflow pipe. It is the waste pipe we will look at here.

Polypropylene waste pipe manufactured to **BS EN 1451–1:2017** and **BS 5254** is flexible, tough and resistant to most acids and alkalis. It melts at a relatively low temperature of 160°C and starts to soften at 100°C. For this reason, its use as a waste pipe is limited to waste water below 100°C. It is also adversely affected by direct sunlight and cannot be solvent welded. It is jointed by the use of push-fit fittings, which have a rubber sealing ring inside the fitting.

Polypropylene push-fit waste pipe fittings

Polypropylene pipe is supplied in 3 m lengths and in various colours, including white, black, grey and brown. Some of the most common fittings are shown in Table 2.30.

▼ Table 2.30 Polypropylene push-fit waste pipe fittings

90° bends	90° swivel bends	45° bends
Swept tees	Sockets	Reducers

Proprietary fittings

Proprietary fittings are those that will connect tubes and pipes of different materials such as copper and lead or lead and medium-density polyethylene (MDPE). There are several different types of proprietary fittings, including leadlocks and Philmac. These are described in Table 2.31.

▼ Table 2.31 Proprietary fittings: leadlocks and Philmac

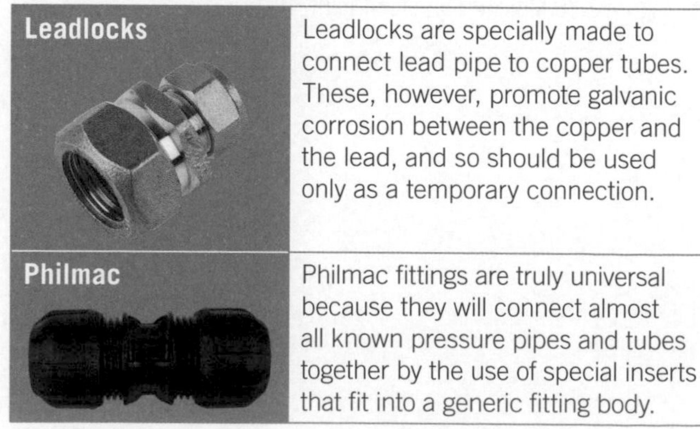

Leadlocks	Leadlocks are specially made to connect lead pipe to copper tubes. These, however, promote galvanic corrosion between the copper and the lead, and so should be used only as a temporary connection.
Philmac	Philmac fittings are truly universal because they will connect almost all known pressure pipes and tubes together by the use of special inserts that fit into a generic fitting body.

3 PREPARATION TECHNIQUES

The successful installation of a domestic hot and cold water system or a domestic central heating system is the result of a series of processes. These involve design, planning, installation, commissioning and maintenance activities, all of which you will be involved in during your career as a plumber.

Installation processes, however, are not just about successful installations. Much of a plumber's work involves other tasks, including repair, maintenance, removal, replacement and decommissioning of existing installations.

Much of the work we do involves the need for skills other than the bending and jointing of tubes and fittings – for example, the installation of a central heating system may involve taking up floorboards and the making good of any holes made in brickwork.

VALUES AND BEHAVIOURS

Remember: you have a responsibility to also protect your customers' property and possessions, such as caring for furniture, fixtures and decorations. Following this code of practice will help you build a good reputation and a successful career as a plumber.

In this section of the chapter, we will look at the processes that are involved when working on new and existing installations. We will also investigate the associated skills we need for some of those jobs that are outside a plumber's skill base, and look at how we can care for and protect customers' valuables and possessions.

INDUSTRY TIP

While many companies have their own style of working, others employ plumbers for specific tasks, i.e. those operatives that work on-site and those that work in private houses.

Pre-installation activities on new and existing installations

Working on-site requires two completely different styles of working, depending on whether you are in a new-build house or an occupied dwelling. While many of the working practices we use on-site can be used in an occupied dwelling, care and attention to detail is absolutely crucial when you are in someone's home.

There are three concerns when working in an occupied dwelling:

1 protecting the customer's property
2 protecting the building fabric
3 installing in accordance with the customer's wishes while maintaining the quality of the installation against the regulations in place.

Many instances have occurred in the past where a good installation has been marred by carelessness by the plumber and a failure to **liaise** with the customer. This often results in disputes, withholding of money owed and, occasionally, court action.

KEY TERM

Liaise: establish a co-operative working relationship.

Working in private houses

Many customers complain about the lack of information given to them. In many cases, this is down to poor customer liaison. So, before an installation takes place, ensure you have covered the following points.

- The customer knows what day and time you will be arriving, or agree a start time with the customer and stick to it.
- Walk around the house with the customer, pointing out any existing damage to furniture, fixtures, carpets and wall coverings. This will prevent any misunderstandings regarding damage and marks already in place.
- Point out which carpets and pieces of furniture will need to be removed before you begin work, and ask the customer to remove them.

- Let the customer know when any of the services, i.e. water, gas or electricity, or appliances such as the WC, are going to be turned off or taken out of service, and ensure that they have collected enough water for the period of temporary decommission; or, if working on a central heating system, ensure they have access to other forms of heat, especially during cold weather.

INDUSTRY TIP

If you are going to be working outside, politely ask the customer to move their car before you begin work so that it does not get damaged.

- Cover with dust sheets all furniture, carpets and fixtures that cannot be removed in the area where you are going to work.
- Before work begins, agree with the customer the position of radiators, boilers and all visible pipework. When fitting sanitary ware, make sure you are fitting the appliances in the position that the customer wants.
- Keep the customer informed about any problems that arise that may need them to make a decision.

VALUES AND BEHAVIOURS

Early-morning arrivals are not always welcome.

It is good practice to keep customers informed of any inconveniences that could be caused by the work that may affect their day-to-day routine.

IMPROVE YOUR ENGLISH

Clear and open communication with your customer will always be appreciated. Make sure you explain to them any unforeseen problems that arise, and allow them to discuss possible solutions so they feel well informed about any changes that may incur extra costs or impact the schedule.

Preparation of the work area: lifting floorboards, notching and drilling joists, and chasing walls

Much of the work in occupied and existing dwellings involves installing pipework under floors, in walls and through walls. In this part of the chapter, we will look at the procedure for lifting floorboards, notching and drilling joists, and chasing walls to allow the installation of pipework.

Lifting floorboards using power tools, step by step

1 Decide on the boards to be lifted, and mark them with a pencil.
2 Locate the position of the joists. This can be done by searching for the row of nails holding the board to the joist.
3 Mark the centre of the joist where the board is to be cut. If this is not possible, a cut can be made inside the joists and a supporting noggin (or cleat) fitted before the board is replaced. Number the boards as this makes replacement easier.

▲ Figure 2.37 Step 3

4 Using a nail punch, punch the floorboard nails below the surface of the board.
5 Set the depth on the circular saw just less than the depth of the board. This is to ensure that any cables or services already installed are not damaged.

The blade of the circular saw just above the thickness of the floorboard

▲ Figure 2.38 Step 5

6 Run down the length of the boards to be lifted with the circular saw to cut the tongue of the board.
7 Now, using the marks on the joist previously made, carefully cut across the board at the joist using the circular saw.
8 The board can now be lifted using a bolster chisel to prise it up.

Lifting floorboards using hand tools, step by step

Follow points 1 to 4 of the previous method.

5 Break the tongue of the board. This can be done by either carefully driving the bolster chisel through the tongue with a claw hammer or cutting down it with a hand floorboard saw.

6 Now, using the marks on the joist previously made, carefully cut across the board at the joists using a hand floorboard saw.

7 The board can now be lifted using a bolster chisel to prise it up.

Lifting chipboard flooring, step by step

Chipboard flooring is quite different to timber floorboards. It is laid in large sheets measuring 2 m × 0.6 m and glued at the tongue and groove joint. The boards break very easily if mishandled. When part of a board is lifted, unlike timber boards they require support at every edge, including the edges where there are no joists. Lifting chipboard flooring is best done with a circular saw.

1 Decide on the boards to be lifted, and mark them with a pencil.

2 There is no need to mark the joists with chipboard as the long joints indicate where the joists are. All that is needed is to mark the area of the board that needs lifting.

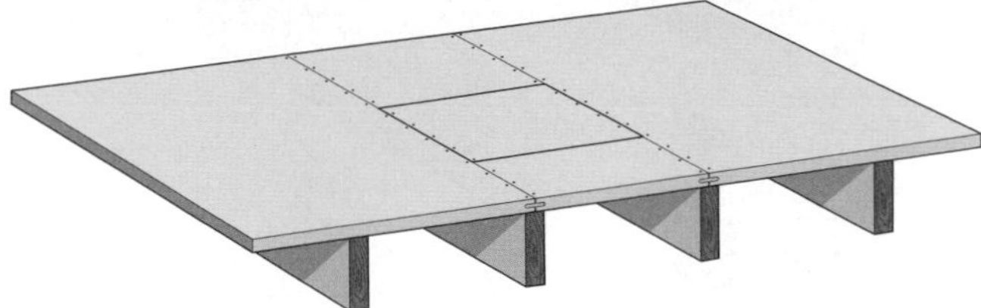

▲ Figure 2.39 Step 2

3 Using a nail punch, punch the nails below the surface of the board.

4 Set the depth on the circular saw just less than the depth of the board. This is to ensure that any cables or services already installed are not damaged.

5 Run down the length of the boards to be lifted with the circular saw to cut the tongue of the board.

6 Now, using the marks previously made, carefully cut across the board using the circular saw.

7 The board can now be lifted using a bolster chisel to prise it up.

8 When replacing the board, the edges need to be supported by a wooden noggin (or cleat). This can be done as shown in Figure 2.40.

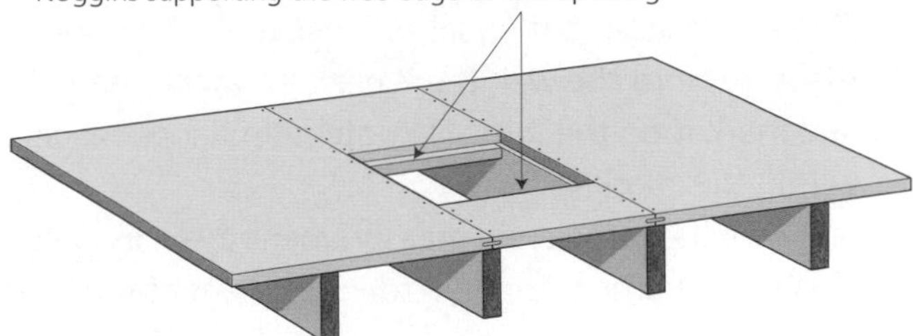

▲ Figure 2.40 Step 8

Notching and drilling joists

Many installations require the notching and drilling of timber joists to accommodate tubes and fittings under the floor. If these operations are not carried out correctly, it could result in a weakening of the joist and, in some extreme cases, structural damage to the property.

Holes or notches that are made too close together, holes drilled too near the end of a joist, and holes or notches incorrectly positioned too near to the centre of the joist span can weaken joists to the point where they become useless as structural supports.

> **KEY POINT**
>
> The strength and the stiffness of the joist must not be compromised.

Notches must be made as shown in Figure 2.41.

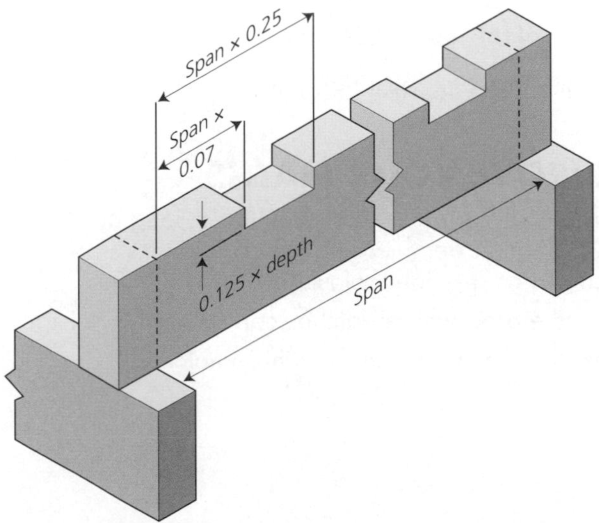

▲ Figure 2.41 Notching measurements

Finding out where notches can be made in a joist, step by step

1 Measure the span of the joist from wall to wall.
2 Multiply the span measurement by 0.07. This will give a measurement equal to 7 per cent of the span.
3 Measure from the wall the 7 per cent measurement and mark it on the joist. No notches must be made within this mark.
4 Now, multiply the span measurement again by 0.25. This measurement is equal to 25 per cent of the joist's span.
5 Measure from the end of the joist again, find the 25 per cent distance and mark it on the joist.
6 All notches must be within the 7 per cent and 25 per cent marks.

To put this into practice we must look more closely at the calculation:

$$\text{Length of span of the joist} = 4 \text{ m}$$

$$7\% \text{ of the span} = 4 \times 0.07 = 0.28 = 280 \text{ mm}$$

$$25\% \text{ of the span} = 4 \times 0.25 = 1 = 1 \text{ m}$$

Therefore, notches in the joist must start 280 mm from the end of the joist and must finish 1 m from the end of the joist. All notches required must be made within a distance of 720 mm. This can be done from both ends of the joist, so two sets of notches can be made.

The depth of the notch must not exceed 12.5 per cent (or $^1/_8$) of the depth of the joist. So, if the above joist measured 250 mm in depth, then the depth of any notches must not exceed:

$$\text{Depth of the joist} = 250 \text{ mm}$$

$$12.5\% \text{ of the depth} = 250 \times 0.125 = 31.25 \text{ mm}$$

IMPROVE YOUR MATHS

1 The span of a joist measures 4.5 m long and 200 mm in depth. Using the calculations shown above as a guide, calculate:
- **a** the area where notches can be made
- **b** the maximum depth of those notches.

2 The span of a joist measures 3 m long and 250 mm in depth. Using the calculations shown above as a guide, calculate:
- **a** the area where notches can be made
- **b** the maximum depth of those notches.

3 The span of a joist measures 3.6 m long and 300 mm in depth. Using the calculations shown above as a guide, calculate:
- **a** the area where notches can be made
- **b** the maximum depth of those notches.

Holes drilled or cut into joists follow a similar procedure. A hole must not begin within 25 per cent of measurement of the span measured from the end of the joist and must stop at a point equal to 40 per cent of the span, again measured from the end. The size of the hole must not exceed a measurement equal to 25 per cent of the depth of the joist when measured from the centre line. This is illustrated in Figure 2.42.

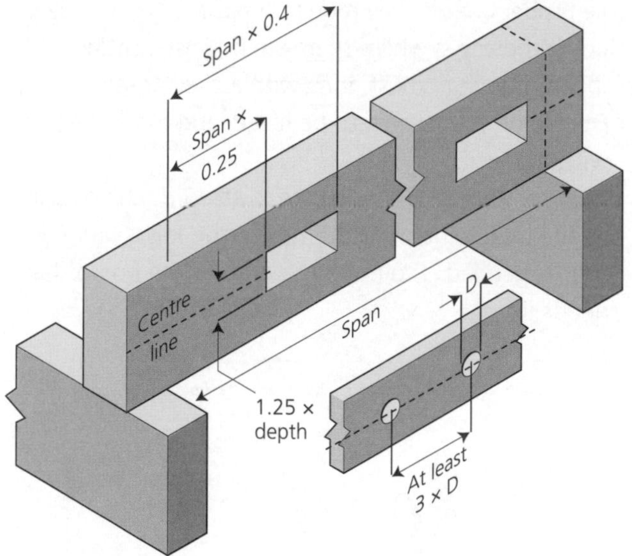

▲ Figure 2.42 The positioning of cut or drilled holes in joists

IMPROVE YOUR MATHS

Again, to understand this fully we must look at the calculation. Let us take the above joist measurement once more:

$$\text{Length of span of the joist} = 4\ \text{m}$$

$$25\%\ \text{of the span} = 4 \times 0.25 = 1\ \text{m}$$

$$40\%\ \text{of the span} = 4 \times 0.4 = 1.6\ \text{m}$$

Therefore, holes drilled or cut in the joist must start 1 m from the end of the joist and must finish 1.6 m from the end of the joist. All holes required must be made within a distance of 600 mm. Again, this can be done from both ends of the joist, so two sets of holes can be made.

To calculate the size of the holes:

$$\text{Depth of the joist} = 250\ \text{mm}$$

$$25\%\ \text{of the depth} = 250 \times 0.25 = 62.5\ \text{mm}$$

This measurement must be measured equally either side of the centre line of the joist. No holes can be drilled in a joist within 100 mm of a notch and circular holes must be at least three diameters of the hole size apart measured **centre to centre**.

ACTIVITY

The span of a joist measures 4.5 m long and 200 mm in depth. Using the calculations shown above as a guide, calculate:

a the area where holes can be made

b the maximum size of those holes.

KEY TERM

Centre to centre: measuring from the centre line of one pipe to the centre line of another, so that all the tube centres are uniform. This ensures that the pipework will look perfectly parallel because all of the tubes will be at equal distance from one another.

Cutting chases in walls

Occasionally, it may be necessary to cut a chase in a wall to conceal pipework; for example, burying pipes for a downstairs radiator. This will involve the use of an angle grinder to, first, cut the outline of the chase onto the wall and then carefully removing the unwanted masonry from between the cuts. Caution should be exercised.

Chases cut in walls must be cut to no more than the following depths:

- horizontal chases must not be deeper than one-sixth of the wall thickness
- vertical chases must not be deeper than one-third of the wall thickness.

VALUES AND BEHAVIOURS

It is advisable not to cut chases in walls in a room containing carpets and furniture but, if this is unavoidable, ensure that all furniture and carpets are either placed to the far side of the room or covered with dust sheets, and that all doors out of the room are closed. Using an angle grinder on masonry, concrete and stone produces excessive amounts of dust and this must, wherever possible, be prevented from escaping from the room you are working in. If possible, open a window to allow some of the dust out.

INDUSTRY TIP

Be wary of installing hot and cold water pipes in a wall where they are going to be concealed or tiled over. It is an offence under the Water Supply (Water Fittings) Regulations to bury pipework carrying hot and cold water in any wall, floor or ceiling where that pipework will, eventually, be inaccessible. All valves should be easily accessible, but ultimately even surfaces like tiles could be removed if deemed necessary to repair pipework and fittings.

HEALTH AND SAFETY

- Always wear the correct PPE. Cutting chases in walls will require the use of safety goggles (not glasses), gloves, overalls and a very good dust mask of the correct type to stop the plumber from breathing in the dust.
- Always check the angle grinder beforehand to ensure that:
  - it is in good condition and carries an in-date PAT test certificate
  - the correct masonry cutting wheel is installed
  - the wheel is secure and the wheel guard is in place.

Associated trade skills: making good the building fabric

During the installation process, there will be many occasions where the building fabric will need to be worked on. Holes will need to be drilled or broken through with a hammer and chisel, chases will need to be made to accommodate pipework, and floorboards will need to be lifted and replaced. Unless it is specified in the contract that these will be repaired by other tradespersons on-site, they will have to be repaired by you, the plumber.

Making good involves having a few basic skills of another associated trade such as a bricklayer, plasterer and joiner. We have already seen the methods of lifting and replacing floorboards (see page 104 of this chapter); here, we will look at making good the holes we have made in walls.

By far the easiest holes to repair are those made by drills and masonry bits. These will require pointing with a 4:1 (four parts sand to one part cement) mortar mixed to fairly stiff consistency. A pointing trowel should be used for this. Larger holes may need the replacement of broken or half bricks. Any new bricks used should match the existing wall bricks. The finished wall should be pointed with a pointing trowel and cleaned with a soft brush.

Patching plaster can be a tedious task. The type of plaster used will depend on the wall surface. Sand and cement rendering will need a smooth-finish plaster and plasterboard will need a plasterboard-finish plaster. The two are very different: board finish dries much faster and so is harder to 'skim' to a smooth finish.

VALUES AND BEHAVIOURS

Making good a customer's home so it is as you found it always leaves a lasting impression. You will be remembered as a professional. At the end of the making-good procedures, make sure the area is cleared of all waste materials and cleaned.

Sources of information

Information regarding the planning of installations, installation of materials and components, and maintaining the integrity of structural components of the building can be found in the following documents and these should be consulted wherever possible:

- statutory regulations – Building Regulations, Water Supply (Water Fittings) Regulations, IET Regulations, Gas Safety (Installation and Use) Regulations, etc.
- industry standards – British and European Standards
- manufacturer technical instructions
- building plans – architects' plans, schematic drawings and sketches, etc.
- specifications.

Storing tools and materials

The storage of tools and materials is an important aspect of any job. For the plumber on-site, the safe storage of tools becomes a major concern, since theft of tools and materials costs the construction industry millions of pounds per year. Here are some points to remember regarding delivery and storage of tools and materials.

- When working on a large housing site, make sure that all tools and materials are locked away in a secure lock-up when not in use. Materials that are left in uncompleted houses uninstalled are not covered by theft insurance. Materials should not be left in the open and all unused materials should be returned to the store.
- Ensure that materials such as sanitary ware, boilers and radiators are stacked to a safe height and are covered to prevent damage.
- Have a materials requisition system in place so that materials can be booked out of the stores for use and any unused materials can be booked back in. This ensures that a close check can be kept on the stock of pipes, tubes and fittings, which can help to prevent theft and over-ordering.
- Keep large pieces of equipment and tools in a separate part of the store. This can help to prevent accidental damage to fragile materials such as washbasins and WC pans.
- Keep a file of all delivery and advice notes so that a check can be made against the stock of materials delivered and the materials used.

- When undertaking work in a private dwelling, the delivery and storage of materials should be agreed with the customer so that they can be delivered at a convenient time and stored in a place that will cause as little disruption as possible to the day-to-day activities of the household.
- Partially installed items, such as baths, washbasins and WCs should be protected from damage. Any protective tape or plastic coverings on sanitary ware should be removed before installation so that they can be visually checked for any damage that may have occurred in transit.

4 USING PIPE CLIPS AND PIPE BRACKETS

Much thought should be given to the positioning of pipework because not all of the pipes we install can be hidden. The golden rule is that visible pipework needs to be as neat as possible. A pipe that is not plumb or level looks unsightly and the eye is drawn to it immediately. Most people believe that surface-mounted pipework is an eyesore and customers will invariably ask the plumber to hide pipes wherever possible. There are occasions, however, because of the constraints of regulations and approved good practice, where this cannot be done and surface-mounted pipes are the only solution. In these cases, the correct positioning, marking and installation of pipework is essential.

Positioning of pipework

The routes taken by surface-mounted pipework should be well planned to take the shortest practicable route but not be intrusive, and there should be as little marking out as possible so as not to deface the customer's decorations. The area must be well protected by dust sheets and coverings.

Select an appropriate pipe clip. Large, sturdy pipe clips in a domestic dwelling would look obtrusive and plastic pipe clips used on large commercial/industrial installations would not stand up to every knock. If a number of pipes are to be installed in one place, say, in an airing cupboard, they can be arranged in banks, so that all the pipe clips are in a neat line.

The use of machine-made bends over elbows should be considered wherever possible, as these not only provide a visually attractive installation but also aid better flow rates. The finished pipework should be as **aesthetically pleasing** as possible, with even spaces between the pipe clips and supports, and even gaps between different lines. The tube should be installed plumb and level, or installed with the correct fall where this is needed.

KEY TERM 

Aesthetically pleasing: beautiful in appearance, good-looking, in keeping with the rest of the surroundings.

Finally, make sure that when the pipework is in position it is wiped down with a damp cloth. This will seem a menial task but it will ensure that any flux that has run down the pipe during soldering operations is removed. Where possible, remove any setting-out marks and fingerprints from the wall with a damp, soapy cloth.

Types of pipe clip available for copper tubes

As we have already seen, copper is relatively easy to joint and bend, and can produce an installation that not only looks good but is also economical in terms of tube usage and installation costs. By adopting a systematic approach to copper tube installation, fabrication and planning, savings can be made on labour costs and material usage. A big part of installing copper tubes is the planning of pipework routes, ensuring that surface-mounted pipework, once installed, looks neat, and is well clipped, unobtrusive and performs to the design criteria.

The correct clipping of copper tube is essential. It prevents excessive noise and fittings failure from vibration, movement and water hammer, and can assist in preventing accidental or intentional damage of the pipework. There are many different types of pipe clip available for copper tubes and each one has a specific use.

For most domestic installations, plastic stand-off pipe clips are preferred, the most common type being the interlocking clip lock type where several banks of pipes of different sizes can simply be clipped together. This ensures a uniformity that is often hard to accomplish with single, individually fixed pipe clips because, once assembled, all of the pipe clips have exactly the same tube centres and, provided the first clips are installed correctly, the others will be perfectly aligned. Single plastic pipe clips are also available for single runs of tube, and double pipe clips are a good idea when installing hot and cold pipework or flow and returns for radiator installations and central heating systems, as these also ensure uniform tube centres.

▲ Figure 2.43 Interlocking clip lock

▲ Figure 2.44 Nail-on clip

Nail-on clips are also available, but should be used with caution with copper tubes as the expansion of copper can loosen the clips making the copper tube vulnerable, especially in places where the pipework is hidden, such as under a suspended timber floor.

▲ Figure 2.45 Copper saddle clips

When fixing copper tubes to a skirting board, the use of copper saddle clips is recommended as the copper is fixed close to the skirting, which makes the tube a little less noticeable. Again caution should be exercised with saddle clips as they are not suitable for fixing to masonry or plastered walls. This can create corrosion of the copper due to the reaction between the copper tube and the wall surface, and can also encourage condensation on the tube surface.

▲ Figure 2.46 Strip brass school board pipe clip

For installations that require a more rigid fixing, such as light commercial/industrial installations, strip brass school board clips or cast brass school board clips should be used. These types of tube brackets give more resistance to tube movement and subsequent damage.

▲ Figure 2.47 Brass munsen ring

Industrial installations require a very secure type of fixing. Brass munsen rings fastened with 10 mm tapped rod and back plates are the strongest types of bracket available for copper tube installations. As well as being screwed to the building fabric, munsen rings can also be hung from the ceiling in banks of pipes using a special metal slotted channel.

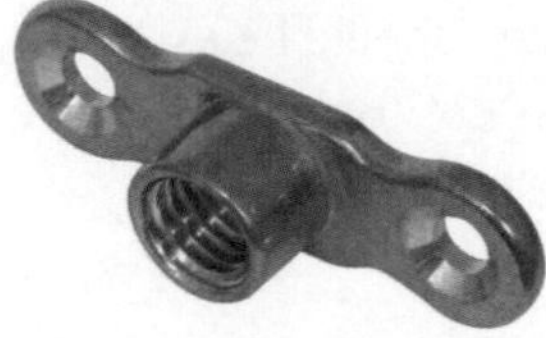

▲ Figure 2.48 Brass back plate

Table 2.32 shows the clipping distances of the common sizes of copper tube.

▼ Table 2.32 Clipping distances for copper tubes

Tube size	Horizontal distance between the clips	Vertical distance between the clips
10 mm	0.8 m	1.2 m
15 mm	1.2 m	1.8 m
22 mm	1.8 m	2.4 m
28 mm	1.8 m	2.4 m
35 mm	2.4 m	3.0 m
42 mm	2.4 m	3.0 m
54 mm	2.7 m	3.0 m

Types of pipe clip available for low carbon steel tubes

Low carbon steel pipe is a very rigid material, and is heavier than most types of pipes and tubes. The clips and fastenings need to be capable of carrying the weight of the material. Because of this, the clips available tend to be very robust. The types of clips and fastenings for low carbon steel pipe are limited. For fixing to walls, cast steel school board clips are recommended.

Since most low carbon steel is used in industrial installations, the use of munsen rings and tapped rod is recommended; these can be used with backing plates or, if being hung from a ceiling, with anchor bolts.

▲ Figure 2.49 Anchor bolt

The clipping spacings for low carbon steel are listed in Table 2.33.

▼ Table 2.33 Clipping distances for low carbon steel

Pipe size	Horizontal	Vertical
½"	1.8 m	2.4 m
¾"	2.4 m	3 m
1"	2.4 m	3 m
1¼"	2.7 m	3 m
1½"	3 m	3.6 m
2"	3 m	3.6 m

Clipping distances for PVCu pipes

Clipping distances for PVCu soil and waste pipes are listed in Table 2.34.

▼ Table 2.34 Clipping distances for PVCu pipes

	Maximum support distance		Maximum distance between expansion joints
	Vertical	Horizontal	
Pipe size – soil			
82 mm	2 m	0.9 m	4 m
110 mm	2 m	1 m	4 m
160 mm	2 m	1 m	4 m
Pipe size – waste			
32 mm	1.2 m	0.5 m	2 m
40 mm	1.2 m	0.5 m	2 m
50 mm	1.2 m	0.9 m	2 m

Clipping distances for polypropylene push-fit waste pipes

Clipping distances for polypropylene push-fit waste pipes are listed in Table 2.35.

▼ Table 2.35 Clipping distances for polypropylene pipes

	Maximum support distance		Maximum distance between expansion joints
	Vertical	Horizontal	
Pipe size – polypropylene waste pipe			
32 mm	1.2 m	0.5 m	2 m
40 mm	1.2 m	0.5 m	2 m
50 mm	1.2 m	0.6 m	2 m

Clipping and supports for polybutylene pipe

Unlike copper tubes and low carbon steel pipe, polybutylene is very flexible. It can sag if not clipped correctly and, if the pipework is visible, this can look unsightly. Because of its flexible qualities, polybutylene pipe should be clipped at the distances shown in Table 2.36.

▼ Table 2.36 Clipping distances for polybutylene pipes

Pipe diameter	Horizontal spacing	Vertical spacing
10 mm	0.3 m	0.5 m
15 mm	0.3 m	0.5 m
22 mm	0.5 m	0.8 m
28 mm	0.8 m	1.0 m

If the pipework is adequately supported or is run in concealed spaces, such as through joists on a suspended timber floor, pipe clips need not be fitted, provided that:

- the pipe is not part of an open vent connected to a heat source or an appliance, such as a boiler or hot water storage cylinder, where the pipework is liable to become hot
- the pipe is not part of a distribution pipe or circuit where poor pipe alignment may affect the venting of air
- no hot water or heating pipe will come into contact with a cold water supply pipe
- there is no risk of the pipe coming into contact with sharp or abrasive edges.

Fixings for masonry, timber and plasterboard

In this section of the chapter, we will take a brief look at the various fixings for brickwork, concrete, stone, timber and plasterboard that we use during our working life.

Fixings can be classified into four distinct types:

1. nails – for both masonry and timber
2. screws
3. heavy-duty fixings
4. plasterboard and lightweight fixings.

Nails

There are many different types of nails that are used for a variety of jobs. It is not important that we know every type of nail but it would be beneficial for us to become familiar with some types, such as floor brads and oval nails.

Nails are usually described by their head shape and their dimensions in mm, e.g. 150 × 4 is 150 mm long and 4 mm in diameter. Some of the different nail types you may use from time to time include:

- **Masonry nails:** used for making fixings to masonry. Normally made of hardened zinc.
- **Copper nails:** used by plumbers to fix sheet lead. They are made of copper to prevent corrosion between the lead and the nail and, because they do not rust, they have a long life.
- **Floor brads:** used to fasten floorboards. Generally, these are 50 mm long.
- **Galvanised clout nails:** used for fixing slates and roof tiles.
- **Round bright wire nails:** used generally for rough joinery work where strength is more important than appearance.
- **Oval bright wire nails:** suitable for joinery work where appearance is important. The head is lost when driven into the timber.

Screws

There are many types of screws available for different applications:

- brass wood screws
- turn-threaded wood screws
- steel countersunk screws
- chipboard screws
- mirror screws
- self-tapping screws.

Screws can be made from steel, stainless steel and brass, and come with a range of screw head types (see the section on screwdrivers on page 69). They can be coated with corrosion protection such as bright zinc and black japanned coatings. Screws are specified by their length in mm or inches, and gauge. The most common lengths used in plumbing range from 15 mm for fixing copper saddle clips to skirting boards, to 50 mm × 10 mm for fixing radiator brackets.

▲ Figure 2.50 Masonry nail

▲ Figure 2.51 Copper nail

▲ Figure 2.52 Round bright wire nail

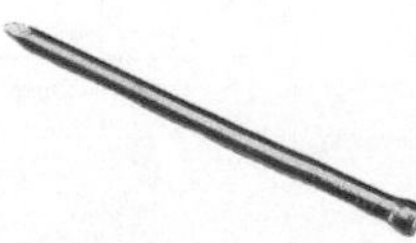

▲ Figure 2.53 Oval bright wire nail

Screw length and gauge

▼ Table 2.37 Screw types

Countersunk screw: used for general work. The head sinks flush, or a little below the wood surface.	
Crosshead/Pozidriv screw (countersunk): used for general work but, unlike the countersunk screw, needs a crosshead screwdriver, which does not slip out of the screw head. Ideal for pipe clips.	
Raised countersunk screw: used for fixing decorative fittings with countersunk holes. The head is designed to be visible.	
Round head screw: used for fixing copper saddle clips.	
Mirror screw: used for fixing mirrors and bathroom fittings such as bath panels. The chromed cap threads into the screw head to hide the screw.	
Coach screws: these usually come with purpose-made wall plugs. They are used for fixing heavy constructions such as boilers. Can be tightened with a spanner but some have Pozidriv screw heads.	
Chipboard screw: used for securing chipboard and medium-density fibreboard (MDF). Various types of head are available.	

Heavy-duty fixings

There are a number of heavy-duty fixings that plumbers use occasionally. These are:

- Coach bolts – these are not usually used by plumbers but can be useful for building structures and platforms for carrying heavy loads like cold water cisterns and hot water cylinders. They are usually made from galvanised steel.
- Rawlbolts – these are also known as heavy-duty expansion anchors. They are easy to use, with good load-carrying capacity, and can be used in concrete, brickwork and stone for fixing heavy appliances and large-diameter pipework.

▲ Figure 2.54 Rawlbolt

Plasterboard and light structure fixings

These are used where the wall is lightweight, such as a plasterboard stud wall. Plasterboard is extremely difficult to fix to. Generally, if a fixing is required, it is better to ask a joiner to put a wood **noggin** in the wall before it is plasterboarded and skimmed with the plaster top coat. When working in existing properties, this is not always possible without damaging the wall's surface and decoration. In this situation, plasterboard fixings are the only option.

> **KEY TERM**
>
> **Noggin:** a term often used on-site to describe a piece of wood that supports or braces timber joists or timber-studded walls. They are particularly common in timber floors as a way of keeping the joists rigid and at specific centres, but they can also be used as supports for appliances such as wash hand basins and radiators that are being fixed to plasterboard.

There are several different types, as described below:

- **Collapsing cavity fixings:** these are probably the strongest plasterboard fixing. They can be used to hang sanitary ware, radiators and many other types of appliance.
  First, a hole is made large enough to pass the fixing through. Then, the fixing is tightened, collapsing the fixing on to the plasterboard.
- **Self-drill plasterboard fixings:** these are used to hang small appliances and radiators. The body of the fixing is self-drilling and is simply screwed into the wall using a screwdriver. The hanging screw is then screwed into the fixing body.
- **Rubber nut fixings:** because of their lack of strength, rubber nut fixings can be used only as lightweight fixings. As the fixing is tightened, the rubber compresses onto the plasterboard.
- **Spring loaded toggle bolts:** also known as butterfly bolts, the spring toggle is an excellent plasterboard fixing that can be used to hang radiators and other small appliances.
  First, a hole large enough to pass the toggle through is made in the wall. Then, as the bolt is pushed through the hole, the spring opens the toggle, allowing it to be drawn up against the wall and creating the fixing.

▲ Figure 2.55 Collapsing cavity fixing

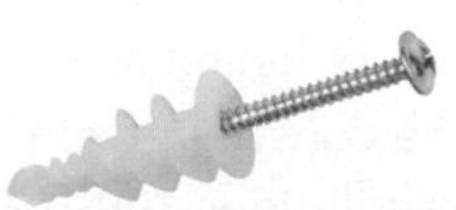

▲ Figure 2.56 Self-drill plasterboard fixing

Plastic wall plugs

Plastic wall plugs are used in conjunction with screws to fasten appliances, sanitary ware and many other pieces of equipment to masonry, concrete and stone walls. They are available in different sizes to match screw gauge and are colour coded for easy identification. The wall must be drilled with a masonry drill bit of a specific size for the colour of the plug (see Table 2.38).

▼ Table 2.38 Rawlplug size to hole diameter

Colour	Screw gauge/ size	Plug length mm	Hole diameter mm
Yellow	No. 4–10	25	5
Red	No. 6–12	35	6
Brown	No. 10–14	45	7
Orange	No. 8–12	25	6.5
Grey	No. 8–10	35	5.5
White	No. 8–12	35	6.5
Blue	No. 10–14	35	8

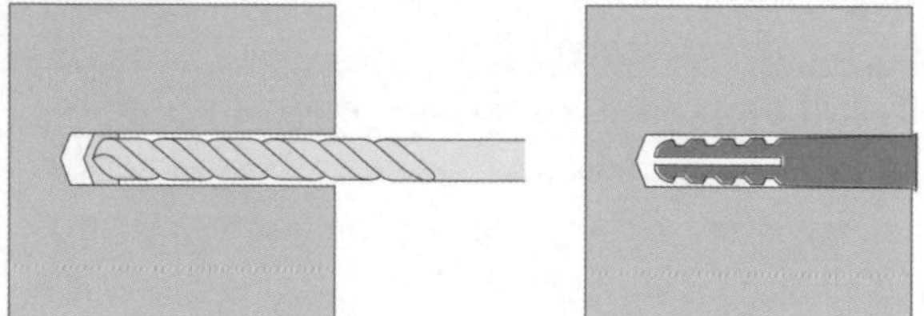

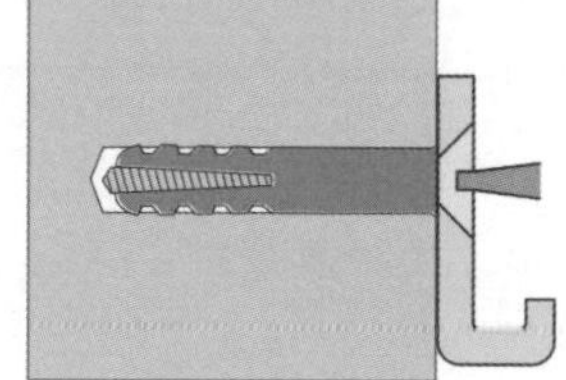

▲ Figure 2.57 Rawlplugs: hanging a radiator

INDUSTRY TIP

If you are using a plastic wall plug in a tiled surface, make sure the wall plug is inserted beyond the tile and into the masonry. Otherwise there is a high chance the tile will crack when the screw is tightened in position.

Installing rawlplugs (to hang a radiator bracket), step by step

1. Mark and drill the hole using the correct size drill to the correct depth for the plug.
2. Insert the plug into the hole and push it slightly below the wall's surface.
3. Using the correct size screw, fasten the radiator bracket to the wall using a screwdriver.

5 PIPEWORK INSTALLATION TECHNIQUES

Installation activities on new and existing installations: marking out, positioning and installation of pipework

As already stated in Section 4, the positioning of surface-mounted pipe clips requires careful consideration. Pipework that is neither level nor plumb is an eyesore. Many customers specifically ask for the pipework to be hidden. However, there are situations where hiding pipework is not possible because of the constraints of the Water Supply (Water Fittings) Regulations 1999. Where this becomes an issue, careful positioning and consultation with the customer will most likely resolve the problem.

Installation fixes

The installation of plumbing, heating, sanitation and gas systems can be broken down into a series of stages, known as fixes. These are described below.

First fix

The first fix involves everything plumbing related in the first phase of construction. It is now that joists are notched and holes drilled ready for the installation of the pipework. Prefabrication of pipework can help here, especially when repetitious work, such as the same house type or the same bathroom layout is expected.

First fixing is basically installing the pipework for the hot and cold water supply, heating system, sanitation system and gas installation that would otherwise not be seen, such as the pipework under floors or in walls. This stage requires very careful planning to avoid issues later on in the build. All too often plumbers have had to return to their work because pipes positioned during the first fix are in the wrong place for the appliance they are supplying.

Whether you are installing copper or plastic pipework, it is a good idea to install some extra fittings that will help to make maintenance much easier once the system has been commissioned and is in use:

- single and double check valves – these protect against contamination and backflow of water
- service valves – these are mandatory prior to a float-operated valve, but are not required elsewhere in a system. It is considered good practice to fit them to aid maintenance of systems.

At the end of the first fix stage, the systems should be pressure tested in accordance with the Water Supply Regulations to make sure there are no leaks. This will be discussed on page 119.

Above-ground sanitation systems are quite easy to install and most of the work can be completed at first fix stage using either push-fit or solvent cement pipework.

Second fix

This is the process of installing the sanitary appliances, boiler, radiators, etc., and connecting them up to the first-fix pipework. It is at this stage that shower valves are fitted, and any shower doors and the like installed.

Final fix

The final fix is where tap heads are fitted, and WC seats fitted and adjusted. Sealing in sanitary appliances with waterproof sealant often takes place at this stage. Not all plumbers perform a separate final fix, preferring instead to complete this task during second fix.

Snagging

When the systems are completed and commissioned, then the clerk of works will inspect the work to ensure that it complies with the customer's requirements. Any problems found, such as loose tap heads and incorrectly adjusted showers, are noted down and the list of snags passed over to the plumber for **rectification**.

KEY TERM

Rectification: putting something right, correcting.

Installing plastic pressure pipe: polybutylene

Building Regulations Document A and **BS 8558**
A allows for joists to be notched or drilled for the installation of pipes and cables. On new buildings, one of the major benefits to plumbers offered by polybutylene pipe is during the installation process, as its flexibility allows the pipe to be installed through holes drilled in the centre of the joists rather than placed in notches. This is known as cabling and has several advantages for the building structure:

- The integrity of the joist is maintained with little or no loss of strength.
- Because the pipe is supplied in coils, longer runs of pipe without joints are possible, which means less likelihood of damaging leaks.
- It allows the floorboards to be fitted before installation takes place, giving the building added strength and rigidity.
- Pipes are less likely to be damaged by nails when the floor is laid.

INDUSTRY TIP

You can access the Building Regulations 2010 Approved Document A: Structure at: www.gov.uk/government/uploads/system/uploads/attachment_data/file/429060/BR_PDF_AD_A_2013.pdf

The benefits to the installer are:

- faster installation leads to savings on installation costs
- push-fit joints ensure there is no fire risk
- the use of a bending machine is not required as the pipe is flexible enough to be bent without pipe wall collapse, with a minimum radius of eight diameters of pipe
- testing can begin immediately after the installation is completed.

Positioning of pipework

The routes taken by surface-mounted pipework should be well planned to take the shortest practicable route without being intrusive. There should be as little marking out as possible so as not to deface the customer's decorations. The area must be well protected by dust sheets and coverings.

Select an appropriate pipe clip. Large, sturdy pipe clips in a domestic dwelling would look too noticeable and plastic pipe clips used on large commercial/industrial installations would not stand up to the knocks. If a number of pipes are to be installed in one place, for example in an airing cupboard, they can be arranged in banks, so that all the pipe clips are in a neat line.

The use of machine-made bends over elbows should be considered wherever possible, as these not only provide a visually attractive installation but also aid better flow rates. The finished pipework should be as aesthetically pleasing as possible, with even spaces between the pipe clips and supports, and even gaps between different lines. The tube should be installed plumb and level, or with the correct fall where this is needed.

Finally, make sure that when the pipework is in position it is wiped down with a damp cloth. This might seem a menial task but it will ensure that any flux that has run down the pipe during soldering operations is removed. Where possible, remove any setting-out marks and fingerprints from the wall with a damp, soapy cloth.

Prefabrication of pipework

Prefabrication of pipework often takes place on large housing contracts where many houses will be built of the same type and style. Pre-forming pipework can often save time in this situation as the pipes can be bent beforehand to fit a particular part of the job, saving time and installation costs, and can be of benefit where hot working, i.e. the use of blowtorches, is forbidden. It can also be used where making joints in the fitted position may be difficult.

Prefabrication of pipework involves precise marking, cutting and forming, with measurements taken either on-site or from a drawing and then fabricated in a workshop and delivered to site ready for use. In this way, many units can be made at once and stored on-site ready for installation.

On-the-job working will also involve some prefabrication of pipework. Precise measurements, cutting and bending are essential if the pipework is to look good. Consider the drawing in Figure 2.58.

INDUSTRY TIP

The X dimension is the measurement from the insert point of the pipe to the centre line of the component. The X dimension is taken off to give you the cut length of the material. It may be taken off both ends of the pipe.

The pipework is to be fabricated on-site from one piece of tube from elbow 1 to elbow 2 using measurements taken on-site.

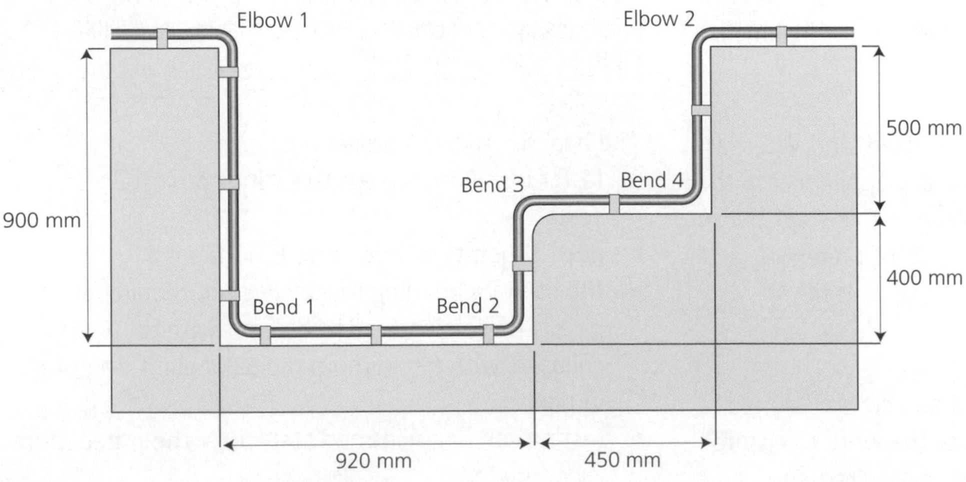

▲ Figure 2.58 Pipework layout drawing

IMPROVE YOUR MATHS

Method

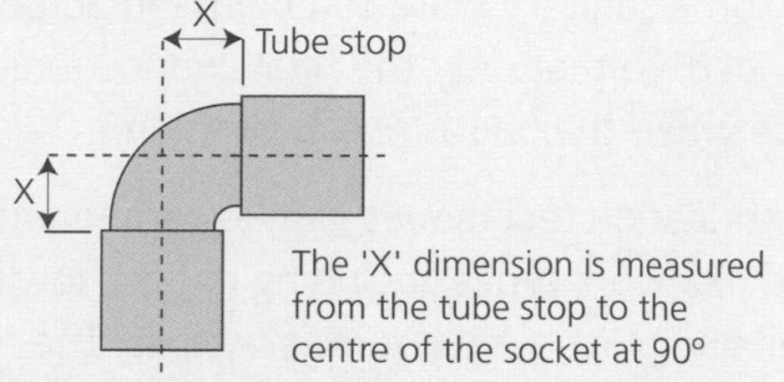

The 'X' dimension is measured from the tube stop to the centre of the socket at 90°

When calculating and marking out tube for one-piece bending, there are several pieces of information we require:

- the 'X' dimension of any fittings
- the distance to the centre of the clip
- the measurements of the space where the tube is going to be installed
- the pipe gain of any machine bends.

Look again at the drawing in Figure 2.58. We can see that the tube has to fit in an alcove. Elbows will be required at elbow 1 and elbow 2 simply because the wall has sharp corners at those points. All other changes of direction can be achieved using machine-made bends. For this example, we will assume that:

The 'X' dimension of a 15 mm elbow = 12 mm

Distance to the centre of the clip = 15 mm

Total up the amount of tube required for the one-piece bend, as follows.

From elbow 1 to bend 1

The distance is 900 mm and because pipe clips are present at elbow 1 and bend 1, the distance is the same. However, because we need to make an end feed elbow joint, we have to deduct the 'X' dimension of the elbow:

$900 - 12 = 888 \text{ mm}$

So, measurement 1 = 888 mm. Therefore, bend 1 can be marked and bent at this distance.

From bend 1 to bend 2

The distance here is 920 mm, but the bends are fixed between clips either side, so deduct the distance to the centre of the clip each side:

$920 - (15 + 15) = 890 \text{ mm}$

Therefore, bend 2 can be marked and bent at this distance.

From bend 2 to bend 3

The distance here is 400 mm and because there are clips at both bends, the distance between the bends does not change.

Bend 3 can be marked and bent at the distance of 400 mm.

From bend 3 to bend 4

The distance here is 450 mm and because there are clips at both bends, the distance between the bends does not change.

Bend 4 can be marked and bent at the distance of 450 mm.

From bend 4 to elbow 2

The distance is 500 mm and because there is an elbow at the end, 12 mm should be deducted for the 'X' dimension:

$500 - 12 = 488 \text{ mm}$

Therefore, length of pipe:

$888 + 890 + 400 + 450 + 488 = 3116 \text{ mm or } 3.116 \text{ m}$

There are four machine bends on the 15 mm pipe and, as we have seen, these have a pipe gain of 21.5 mm each. Therefore:

$21.5 \times 4 = 86$ mm. This can be deducted from the total length:

$3116 - 86 =$ actual tube length $= 3030$ mm or **3.030 m**

Many of the appliances we fit arrive on-site prefabricated. Boilers, hot water storage cylinders and some sanitary ware can be manufactured 'pre-plumbed' so that only the final connections have to be made when the unit is put in position.

Prefabrication techniques can be carried out on most fixed pipework types, including copper, low carbon steel, and plastic soil and waste pipes. The techniques will differ depending on the material used.

Sleeving of pipework through walls

Pipes passing through masonry, stone and concrete should be sleeved by a piece of tube one size larger than the pipe being installed, to allow for expansion and pipe movement, and to prevent damage to the pipe by building movement. The sleeve should then be sealed with an approved sealant to prevent the ingress of rain, insects and vermin. Where gas pipes are sleeved, the sleeve should be sealed only on the inside of the wall; the outside part of the sleeve should be sealed only to the building fabric and NOT the pipe. The pipe and sleeve should be left open.

Working on existing installations: in situ working

Working on existing installations is challenging. There is always a risk of disturbing joints and causing further problems. Situations often occur where it is necessary to cut into existing pipework and it should be treated with care. Problems can occur when connecting to old imperial-sized pipework when the pipe sizes differ from new metric fittings and tubes. **In situ** installation operations include:

- cutting in fittings, such as isolation valves and tees into an existing hot or cold water pipe
- capping off existing pipework
- removing existing bath, washbasin and sink taps
- changing WC pans and cisterns, and other bathroom equipment
- boiler swaps on existing central heating installations.

KEY TERM

In situ: in situ, in plumbing terms, simply means pipework or appliances that are already in place. They are already 'in situation', hence the term 'in situ'.

Protection of the building fabric and its surroundings

HEALTH AND SAFETY

Soldering is a fire risk on any job. Ensure you take the proper precautions, such as using a heatproof soldering mat or heat-dissipating spray gel.

INDUSTRY TIP

A solder mat 'deflects' the heat from a flame. Hold the blowtorch at an angle so that the heat is deflected.

▲ Figure 2.59 Soldering mat

So far in this chapter we have seen how a customer's personal belongings should be protected from dust and damage from the installation process, but there are other ways that we can protect the building and its surroundings.

- When soldering is taking place in the building, the risk of fire is ever present. To protect the building fabric, a heatproof soldering mat should be used. It should be remembered, however, that these will not protect if the flame is directly on the mat. A shallow angle should be applied to the blowtorch, if possible, to deflect the heat away from the wall/floor/ceiling/skirting board. There are three different types of mat available that will resist temperatures of 600°C, 1000°C and 1300°C.
- One other way we can protect against heat is to use heat-dissipating spray gel. This offers protection against the scorching of wallpaper and paint, and loosening of existing joints, and it also reduces the risk of fire by protecting surfaces and dissipating heat.
- When drilling walls, to prevent **blowing** the surface of the backside of the wall you are drilling, first, drill a small pilot hole and drill from both sides. This will ensure that the wall surface around the hole is not damaged.

- Before drilling a wall, check it first with a cable/pipe detector to ensure that there are no services already in the wall.

INDUSTRY TIP

When removing old equipment, where there is a risk of spillage of dirty water, such as removing old radiators, to prevent the spillage turn the radiator upside down so that the valve tails are at the top.

VALUES AND BEHAVIOURS

When working outside the building, protect the customer's garden by the use of walk boards across flowerbeds and protective sheeting across grass lawns. Do not dig ladders into lawns.

Always remember to protect those carpets that cannot be removed during simple maintenance operations such as these.

Fire stopping

Where pipes (including soil and vent pipes) pass between floors, the holes around the pipe must have a **fire stop** installed to prevent the spread of smoke and fire. This can be done in two ways:

1. by the use of an intumescent collar – this is a collar that is placed around the pipe that expands in the presence of heat to stop the spread of fire
2. by the use of intumescent sealant – this is sealant that acts in the same way as an intumescent collar.

Correctly used, these techniques will help to contain fire in the room where it started, reducing damage.

▲ Figure 2.60 Intumescent collar

KEY TERMS

Blowing: this describes the outside surface of a brick being damaged when drilling a hole from the inside.

Fire stop: a barrier is placed where the pipe passes through a floor, ceiling or wall to prevent the spread of fire and smoke.

Testing and commissioning procedures

Testing of installations is the first time we see whether the installation is watertight. For pressure systems and sanitary systems, testing procedures are set out in the relevant British Standards and Regulations.

Pre-testing checks

Before commissioning takes place:

- walk around the installation; check that you are happy that the installation is correct and meets installations standards
- check that all open ends are capped off and all valves isolated
- check that all capillary joints are soldered and that all compression joints are fully tightened
- check that sufficient pipe clips, supports and brackets are installed, and that all pipework is secure.

Testing procedures

Testing procedures differ depending on the type of pipework installed. The process involves filling the system with water to a specific pressure, letting it stand for a period of time to temperature stabilise and then checking it for pressure loss. Here, we will look at those different testing procedures.

Hot and cold water systems testing is detailed in **BS EN 806** and The Water Regulations, but is commonly followed in **BS 6700**; central heating systems testing is detailed in **BS 5449**; above-ground sanitation systems should be tested in accordance with Document H of the Building Regulations.

- **Copper tubes and low carbon steel pipes:** systems installed in copper tube and low carbon steel pipes should be tested to 1.5 times normal operating pressure. They should be left for a period of 30 minutes to allow for temperature stabilisation and then left for a period of one hour with no visible pressure loss.
- **Plastic (polybutylene) pressure pipe systems:** these are tested rather differently to rigid pipes. There are two tests that can be carried out. These are known as test type A and test type B and are detailed in **BS EN 806** and The Water Regulations, but are commonly followed in **BS 6700**:
  - **Test type A:** slowly fill the system with water and raise the pressure to 1 bar (100 kPa). Check and re-pump the pressure to 1 bar if the pressure drops during this period, provided there are no leaks. Check for leaks. After 45 minutes, increase the pressure to 1.5 times normal operating pressure and let the system stand for 15 minutes. Now release the pressure in the system to one-third of the previous pressure and let it stand for a further 45 minutes. The test is successful if there are no leaks.

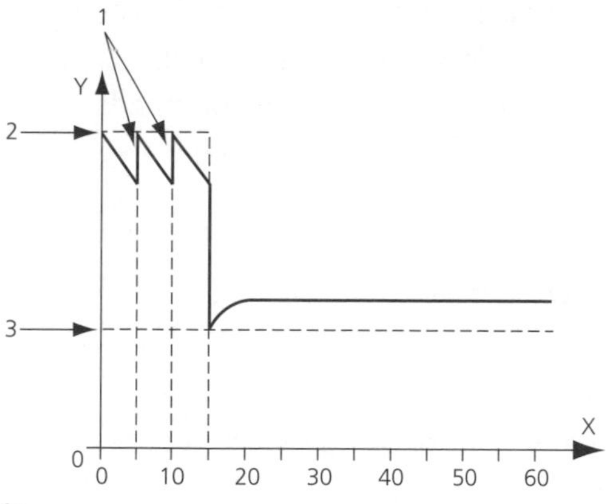

Key
1 Pumping
2 Test pressure 1.5 times maximum working pressure
3 0.5 times maximum working pressure
X Time (minutes)
Y Pressure

▲ Figure 2.61 Pressure test A chart

INDUSTRY TIP

You can access the Building Regulations 2010 Approved Document H: Drainage and waste disposal at: www.gov.uk/government/uploads/system/uploads/attachment_data/file/442889/BR_PDF_AD_H_2015.pdf

  - **Test type B:** slowly fill the system with water, pump the system up to the required pressure and maintain the pressure for a period of 30 minutes. Note the pressure after this time. The test should continue without further pumping. Check the pressure after a further 30 minutes. If the pressure loss is less than 60 kPa (or 0.6 bar), the system has no visible leakage. Visually check for leakage for a further 120 minutes. The test is successful if the pressure loss is less than 20 kPa (0.2 bar).

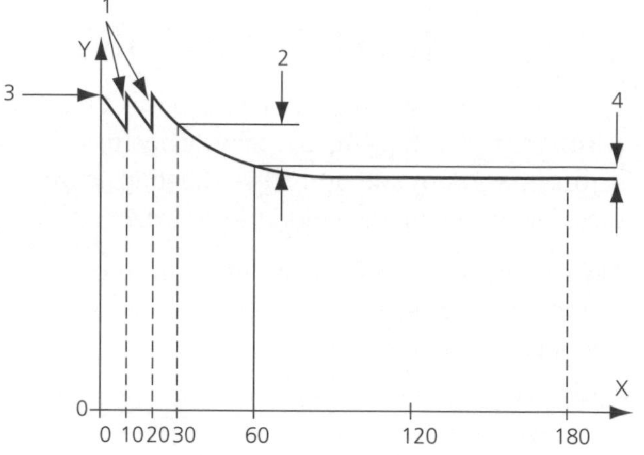

Key
1 Pumping
2 Pressure drop < 60 kPa (0.6bar)
3 Test pressure
4 Pressure drop < 20 kPa (0.2bar)
X Time (minutes)
Y Pressure

▲ Figure 2.62 Pressure test B chart

- **Above-ground sanitation systems:** these should be tested in accordance with Document H of the Building Regulations. They should be tested to a pressure of 38 mm water gauge (w/g) for a period of three minutes with no pressure loss.

Commissioning

Commissioning is the part of the installation where the system is filled and run for the first time. It is when we see if it works as designed. The first task is to fill the system and check for leaks at the appliances. This is best carried out in stages so that sections of the installation, i.e. cold water, hot water, central heating, can be filled and tested separately. At each stage of the filling process, the system should be checked for leaks before moving on to the next section.

INDUSTRY TIP

If the word 'commissioning' is used on-site, it is referring to a system or appliance being tested to make sure it is working correctly and according to the manufacturer's instructions. Common tests are temperature, flow and pressure.

Once the systems have been filled they should be drained down and flushed through with clean water, then refilled. The water levels in WC cisterns, cold water storage cisterns, and feed and expansion cisterns (if fitted) should be checked for compliance with the relevant regulations.

Gas installations should be checked for tightness, and central heating systems should be run up to full operating temperature before being drained down while they are still hot. Refill the system and add inhibitor before running the system again.

Check the flow rates at all taps to see if they deliver the flow rates demanded by the manufacturer's literature, and check the operation of all controls, including thermostats and motorised valves. Set the temperature of any cylinder thermostats and let the water reach full temperature. Using a thermometer, check the temperature of all radiators and the temperature of the hot water.

Benchmarking

At this stage of the installation, it is time to **benchmark** the system. Here, the boiler and any hot water cylinder installed are checked for compliance with the manufacturer's instructions, including:

- hot water flow rates
- flow and return temperatures
- hot water temperature
- operation and types of control
- gas rates.

The benchmark certificate should then be signed by the commissioning engineer.

KEY TERM

Benchmarking: this is now a compulsory requirement to ensure that systems and appliances are installed in accordance with the regulations and the manufacturer's instructions. It also safeguards any guarantee against bad workmanship.

INDUSTRY TIP

Always complete the benchmark certificate and hand it over to the customer.

Building Regulations Compliance certificates

Since 1 April 2005, the Building Regulations have demanded that all new builds, retro-fit and refurbishment installations must be issued with a Building Regulations Compliance certificate. This is to ensure that all Building Regulations relevant to the installation have been followed and complied with. This includes:

- the heating installation
- the sanitation system
- the hot and cold water systems
- the gas installation
- any electrical controls.

Certificates are issued by Local Authority Building Control.

Handover to the customer

When the system has been tested, commissioned and benchmarked, it can then be handed over to the customer. The customer will require all documentation regarding the installation:

- all manufacturers' installation and servicing instructions for boilers, electrical controls, taps, sanitary ware and any other equipment fitted to the installation
- the benchmarking certificate
- the Building Regulations Compliance certificate.

The customer must be shown around the system and guided as to how to use any controls, thermostats and time clocks. Isolation points on the system for gas, water and electricity should be pointed out and a demonstration given of the correct isolation procedure in the event of an emergency. Explain to the customer how the systems work and ask if they have any questions. Finally, point out the need for regular servicing of the appliances and leave emergency contact numbers.

Decommissioning of systems

Decommissioning a system or an appliance simply means taking it out of service. This falls into two categories:

1. **Temporary decommissioning:** this is where a system or an appliance is taken out of service for a period of time for repairs, replacement or maintenance. The customer must be kept informed of when the system is being shut down, the expected length of time of the decommission and the expected reinstatement time. If the period of time is considerable, ensure that the customer has access to vital services, i.e. gas, water and electricity.
2. **Permanent decommissioning:** this is where a system or appliance is permanently disconnected and/or removed. This will involve disconnection and making safe of any services. Pipes should be cut back and capped and, if necessary, tested for soundness. All electrical disconnections should be made by a qualified operative or an electrician.

VALUES AND BEHAVIOURS

With temporary decommissioning, the key to good customer service is information: keep the customer informed and aware of any disruptions to services such as water and electricity.

Maintenance activities

Maintenance falls into two categories:

1. **Planned preventative maintenance:** on larger installations, it may be necessary to have a planned maintenance schedule so that systems and equipment can be serviced and checked at regular intervals to ensure optimum performance. Maintenance activities should be recorded in a logbook, together with the results of any tests performed. Planned preventative maintenance operations include:
   - checking and repairing float-operated valves and setting water levels in cisterns
   - cleaning out cold water cisterns of all sediment as required
   - routine boiler maintenance
   - checking and re-washering taps as required
   - routine testing of above-ground drainage systems
   - checking the operation of any safety valves
   - checking the operation of all external controls and isolation valves, including:
     - stop taps
     - gate valves
     - isolation valves
     - motorised valves
     - thermostats.
2. **Breakdowns, repairs and emergencies:** these are unplanned maintenance activities that can occur at any time and include:
   - burst pipes
   - boiler breakdowns
   - running overflows
   - blockages
   - dripping taps
   - WC cistern problems.

KEY TERM

Maintenance: preserving the working condition of appliances and services.

Drawing symbols of plumbing valves and appliances

Working drawings for plumbing and heating installations often contain symbols that represent pipes, valves and appliances. It is important that these symbols are recognised for systems to be installed properly. All symbols shown will be in accordance with **BS EN ISO 19650-1/2 2018** and **BS EN 806-1:2000**.

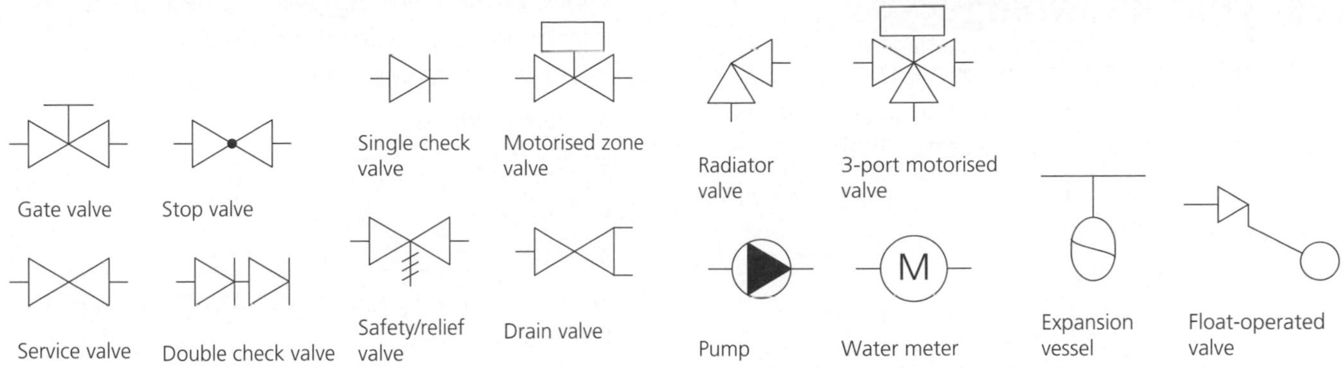

▲ Figure 2.63 Plumbing symbols drawings

Making good

As part of an installation, there will be times where plumbers need to 'make good'. This is the process of repairing or bringing something up to a finished standard or restoring it to its previous condition. This can include:

- filling in holes in masonry or wood
- painting
- minor brickwork
- minor plastering and minor carpentry.

It is always useful to pick up methods of work from other trades that may help. Making good, although seemingly a minor job, is a very important part of an installation and will lead to customer satisfaction and repeat work along with increased skills.

If a small amount of brickwork has been damaged or requires replacing externally, you will need to match the replacement bricks to the existing. The brick will need to be secured in position using a mixture of sand and cement mortar, commonly in a 4:1 ratio. If there is just a small hole in some masonry, filling it with mortar is quite acceptable. External surfaces need to be treated differently to internal surfaces as they are exposed to the elements and it is a requirement to ensure moisture cannot enter a building.

Inside the property, using off-the-shelf filler for small holes is acceptable. If plastering is damaged, being able to smoothly finish the surface again is an important skill. The surface will need to be gently rubbed down using abrasives, removing any loose parts, and then filled, left to dry and then re-rubbed down, leaving a smooth surface.

If plasterboard requires repairing, it should never be rubbed down using abrasive paper, but can be re-filled using a plaster mix and trowel.

Asking other trades on site to help or advise is regularly undertaken. Being able to finish jobs off to a high standard is an important part of any installation.

SUMMARY

During this chapter, we have explored the tools required, the materials we use and the installation practices we need to master to enable us to install good, working systems that not only meet the requirements of the regulations, but also satisfy the customer's needs and expectations. Good working practices at the start of your plumbing career will serve you well as you broaden your experience, gain knowledge and improve your skills.

Test your knowledge

1 Which of the following would be the most suitable masonry drill bit to use to make a hole in brickwork for a brown plastic plug?

a 5.5 mm HSS bit
b 7.0 mm HSS bit
c 5.5 mm SDS bit
d 7.0 mm SDS bit

2 What is the purpose of the tool shown below?

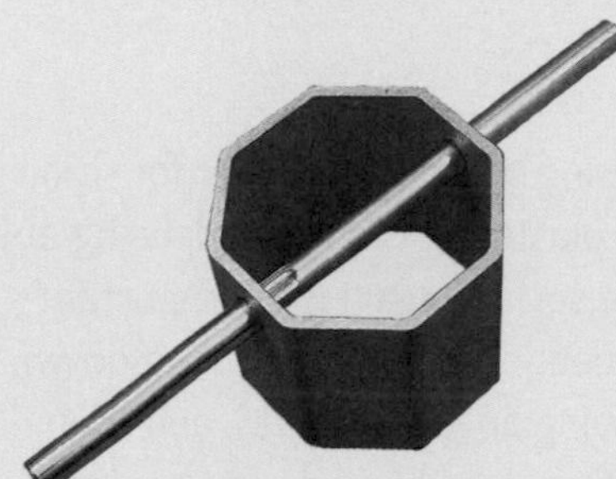

a To install a sacrificial anode
b To remove an immersion heater
c To remove a tap back-nut
d To tighten a compression nut

3 Which of the following is the British Standard for the manufacture of copper pipes used in the plumbing and heating industry?

a BS EN 806
b BS 1710
c BS 1212
d BS EN 1057

4 What is the minimum total length of pipe required to machine bend 15 mm copper pipe to 90°?

a 60 mm
b 95 mm
c 100 mm
d 115 mm

5 LCS pipe is given a colour band to indicate its grade. What grade is indicated by a blue band?

a Light
b Medium
c Heavy

6 Which of the following plastic pipe materials is commonly used for the distribution of underground mains cold water supplies?

a Polybutylene
b ABS
c PVCu
d MDPE

7 Which of the following is not a common size for PVCu soil pipes?

a 65 mm
b 82 mm
c 110 mm
d 160 mm

8 What is the maximum diameter of hole that can be drilled in a joist?

a 10% of the depth of the joist
b 20% of the depth of the joist
c 25% of the depth of the joist
d 30% of the depth of the joist

9 When installing 28 mm copper pipe in the horizontal plane, what is the recommended clipping distance?

a 1.8 m
b 2.4 m
c 2.7 m
d 3.0 m

10 Which of the following is the maximum recommended clip distance for 40 mm plastic waste pipe in the vertical position?

a 0.5 m
b 0.9 m
c 1.2 m
d 2 m

11 Which colour identifies wholesome water when installing MDPE pipework?

a Yellow
b Orange
c Green
d Blue

12 Which power tool is commonly used to remove the 'tongue' from a floorboard?

a Jig saw

b Circular saw

c SDS drill

d Reciprocating saw

13 Which Building Regulation outlines the criteria for cutting a notch in a joist?

a Approved document P

b Approved document H

c Approved document A

d Approved document J

14 Which type of joint uses an olive to create the watertight joint?

a Compression

b Press fit

c Push fit

d Solder

15 What type of screw is used to secure a floorboard back in position?

a Countersunk

b Round head

c Coach

d Mirror

16 What material would a cavity fixing be used on?

a Wood

b Masonry

c Plastic

d Plasterboard

17 When forming a right angle bend in low carbon steel, what angle would you most likely form the pipe to in the hydraulic bender?

a 45 degree

b 95 degree

c 30 degree

d 75 degree

18 Which one of the following methods is the most effective way of cutting a hole for a condensing boiler flue in a masonry wall?

a Pneumatic drill with chisel bit

b SDS drill with diamond core bit

c Hammer drill with masonry drill

d Rotary drill with a spade bit

19 What creates the watertight joint in a push fit waste pipe fitting?

a Olive

b PTFE tape

c Neoprene O ring

d Solder

20 What is the maximum size hole that can be drilled into a joist?

a 1/3rd of the joist width

b 28 mm

c 5% of the span of the joist

d 1/4 of the joist depth

21 What is the maximum clipping distance for a 15 mm vertical copper pipe?

a 1.2 m

b 1800 mm

c 1.7 m

d 2400 mm

22 The centre to centre distance of two tees is 300 mm. The X dimension of the fittings is 12 mm. What is the cut length?

a 300 mm

b 276 mm

c 324 mm

d 288 mm

23 Which of the following is important when replacing a hacksaw blade?

a The blade must have 24tpi

b The handle must be cleaned

c The teeth must face forward

d The butterfly nut must tighten anti-clockwise

24 Which tool is a slip or guide used with?

a Scissor type bender

b Adjustable spanner

c Blowtorch

d Water pump pliers

25 Where might a plumber use a 'mirror' type screw?

- **a** Pipe clip
- **b** Securing a toilet pan
- **c** Securing a bath panel
- **d** Replacing a floorboard

26 Complete the table below to indicate which gauge of screw is suitable for each plastic rawlplug type.

Yellow	
Red	
Brown	
Grey	
White	
Blue	

27 Calculate the maximum depth of notch if preparing to install copper pipe in a joist that is 300 mm deep.

28 What grade of copper tube is most commonly supplied in coils and used for microbore heating installations?

29 Give three statutory regulations relevant to the installation of a central heating system.

30 A joint made on copper pipe, which uses an electrically operated tool to compress a fitting incorporating a rubber seal onto the pipe, is known as what?

31 List the three grades of low carbon steel pipe and state their potential use.

32 You are asked to install a new WC in a customer's property. Your boss has asked you for a list of common tools that would be required for this installation. List the tools and what they would be used for.

33 State the two types of flux available to plumbers and describe any potential dangers or hazards.

34 What are the five common ways in which copper pipework can be connected together?

Answers can be found online at www.hoddereducation.co.uk/construction

Practical activity

Practise your copper pipe fabrication by producing the pipe bends shown in the diagram below, to the dimensions given.

400 mm centre to centre

100 mm end to centre

35 mm off-set at 30°

125 mm centre to start of bend

CHAPTER 3

SCIENTIFIC PRINCIPLES

INTRODUCTION

Plumbing contains a lot of science. The laws of physics and chemistry are involved in one form or another in almost everything that we do, from the installation of cold water systems and hot water systems to central heating and drainage. It is the science behind these laws that gives us the theory to enable us to design and install these systems correctly and efficiently. In this chapter, we will be investigating some of the laws of physics and chemistry that we use in our day-to-day activities.

By the end of this chapter, you will have knowledge and understanding of the following:

- units of measurement used in the plumbing and heating industry
- properties of materials
- the relationship between energy, heat and power
- the principles of force and pressure, and their application in the plumbing and heating industry
- the mechanical principles in the plumbing and heating industry
- the principles of electricity in the plumbing and heating industry.

Before we begin, it is important that we familiarise ourselves with the SI units of measurement so that we can use these as reference points during this chapter.

1 UNITS OF MEASUREMENT USED IN THE PLUMBING AND HEATING INDUSTRY: THE SI SYSTEM

The SI system of measurement is a universal, unified, self-consistent system of measurement units based on the m/k/s (metre/kilogram/second) system. We will use these measurement units as reference points throughout this chapter. The international system is commonly referred to throughout the world as SI after the initials of 'Systeme International Unite'. The units can be categorised into two main groups:

1 base units
2 **derived units**.

KEY TERM

Derived units: combinations of the seven base units by a system of multiplication and division calculations. There are 21 derived units of measurement, some of which have special names and symbols.

SI base units

▼ Table 3.1 SI base units

Measure of:	Base SI unit	Symbol
Length	metre	m
Mass	kilogram	kg
Time	second	s
Electric current	ampere	A
Thermodynamic temperature	kelvin	K

SI derived units

▼ Table 3.2 SI derived units

Measure of:	Unit	Symbol
Area (length × width)	square metre	m^2
Volume (length × width × height)	cubic metre	m^3
Volume of liquid (length × width × height × 1000)	litre	l
Velocity	metre per second	m/s
Acceleration	metre per second squared	m/s^2
Density	kilogram per cubic metre	kg/m^3
Specific volume	cubic metre per kilogram	m^3/kg
Force (mass (kg) × acceleration (m/s^2))	newton (kg m/s^2)	N
Pressure	pascal	Pa
Energy, work, quantity of heat	joule	J
Power	watt	W
Electric potential	volt	V
Electric resistance	ohm	Ω

Using unit conversion tables

Despite efforts to adopt the metric system in the 1970s, it is obvious that there are still many imperial units in use in the UK today. We still measure distances in miles rather than kilometres and often buy our food in pounds rather than kilograms. It is therefore helpful to know how to convert from one type of unit to another.

In plumbing, we may come across many different imperial units that are still in use. An example of an imperial/metric conversion that we still use today is shown in Table 3.3.

▼ Table 3.3 Copper pipe imperial and corresponding metric sizes

Imperial	Metric
½ inch	15 mm
¾ inch	22 mm
1 inch	28 mm
1¼ inch	35 mm
1½ inch	42 mm
2 inch	54 mm

Before 1973, copper pipe was manufactured in diameters by the inch and its subdivisions.

Table 3.4 gives some of the common conversion factors that are still in use in the UK.

▼ Table 3.4 Common conversions

Imperial	Actual measurement	Metric
1 inch [in]		2.54 cm
1 foot [ft]	12 in	0.3048 m
1 yard [yd]	3 ft	0.9144 m
1 mile	1760 yd	1.6093 km
1 int. nautical mile	2025.4 yd	1.853 km
1 sq inch [in^2]		6.4516 cm^2
1 sq foot [sq ft]	144 in^2	0.0929 m^2
1 sq yd [yd^2]	9 sq ft	0.8361 m^2
1 acre	4840 yd^2	4046.9 m^2
1 sq mile [$mile^2$]	640 acres	2.59 km^2
1 cu inch [in^3]		16.387 cm^3
1 cu foot [ft^3]	1728 in^3	0.0283 m^3
1 fluid ounce [fl oz]		28.413 ml
1 pint [pt]	20 fl oz	0.5683 l
1 gallon [gal]	8 pt	4.5461 l
1 ounce [oz]	437.5 grain	28.35 g
1 pound [lb]	16 oz	0.4536 kg
1 stone	14 lb	6.3503 kg
1 hundredweight [cwt]	112 lb	50.802 kg
1 long ton (UK)	20 cwt	1.016 t

ACTIVITY

There may be instances during our work when we have to convert from one unit to another. The following example shows how to use the conversions in Table 3.4.

A plumber has to travel 25 miles to work every day but claims 35p per kilometre in travelling expenses. How much does he claim?

Now, try these examples:

1 A plumber is asked to replace a cold water cistern in a roof space with a new like-for-like cistern. The capacity of the cistern is quoted on the existing cistern as a 25-gallon nominal capacity. What size cistern in litres is required?

2 A customer has requested that you quote for a new bathroom suite installation and sends you a plan of the existing bathroom. The measurements are in feet and inches.

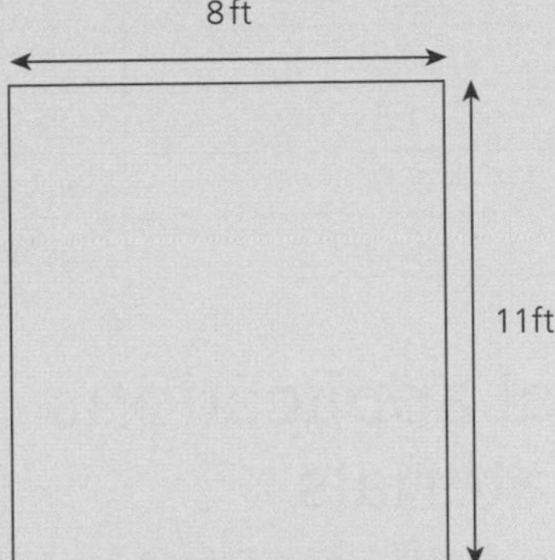

a Convert the dimensions into metres.

b What is the area of the room in square metres?

2 THE PROPERTIES OF MATERIALS

There are many materials that you, as a plumber, will deal with in your working life. Each one will have different characteristics, such as weight, melting point, density and strength. It is important that we know and understand the materials we work with to ensure that the correct material is used for a given application. Here, we will investigate some of the many different materials we use, together with their working properties and their uses.

Relative density of solids, liquids and gases

Relative density is the ratio of the density of a substance to the density of a standard substance under specific conditions. For liquids and solids, the standard substance is usually distilled water at 4°C. For gases, the standard is usually air at the same temperature and pressure as the substance being measured.

When we talk about a material's relative density, we are basically comparing the mass of that material against water or air (see Table 3.5). In both cases the water and air have a relative density (or specific gravity) of 1.

▼ Table 3.5 Relative densities of common substances used in the plumbing industry

Solids		
Substance	**Relative density**	**Mass/m^3**
Water (1 m^3 of water has a mass of 1000 kg at 4°C)	1	1000 kg
Copper	9	9000 kg
Steel	7.48–8.0 (depending on the grade)	7480–8000 kg
Lead (milled) Lead (cast)	11.34 11.30	11,340 kg 11,300 kg
Brass	8.4	8400
uPVC	1.35	1350 kg
Polypropylene	0.91	910 kg

INDUSTRY TIP

Another phrase for relative density is 'specific gravity' and this usually refers to gases.

▼ Table 3.6 Gases' specific gravity

Gas		
Gas	**Specific gravity**	
Air	1	
Natural gas	0.7	Lighter than air
Propane	1.5	Heavier than air
Butane	2.0	Heavier than air
Hydrogen	0.069	Lighter than air

Principal applications of solid materials

The solid materials used in the plumbing industry can be classified into three distinct groups:

1 those made from metals
2 those made from plastics
3 those made from ceramics and fireclays.

Metals

Metals are one of the main materials used in the plumbing industry. They can be found in the form of pipes, tubes and fittings, and in the manufacture of boilers, radiators and other heating appliances, as well as sundry items such as solder, screws and nails.

Metals can be subdivided into four specific groups, as described below.

1 **Pure metals:** these are the metals that are derived directly from the ore and contain very little in the way of impurities. Table 3.7 lists the more common metals and the ores from which they are extracted.
2 **Alloys:** an alloy is a mixture of two or more metals. This type of mixed metal is used extensively in the plumbing industry. Alloys used include brass (copper/zinc), bronze (copper/tin), gunmetal (copper/tin/zinc), lead-free solders (nickel/tin or copper/tin) and steel (iron/carbon).
3 **Ferrous metals:** those metals that contain iron, such as steel and cast iron. These corrode easily because of the formation of ferrous oxide, otherwise known as rust.
4 **Non-ferrous metals:** these metals do not contain iron and are known as pure metals. Non-ferrous metals include copper, lead, tin, zinc, aluminium and nickel. Non-ferrous metals do not rust but can **corrode** over time.

KEY TERM

Corrosion: any process involving the deterioration or degradation of metal components, where the metal's molecular structure breaks down irreparably.

▼ Table 3.7 Origin of common metals

Metal	Ore	Country	Type	
Iron	Pyrite Marcasite Haematite Magnetite	England Mexico Brazil Australia	Ferrous	
Copper	Copper Malachite Chalcopyrite Turquoise Azurite	North America Chile Cyprus Canada Germany	Non-ferrous	
Aluminium	Gibbsite Bauxite Cryolite	Brazil Jamaica India Australia Guinea	Non-ferrous	

Metal	Ore	Country	Type	
Lead	Galina Cerussite	England Germany Australia North America	Non-ferrous	
Zinc	Sphalerite Zincite	Australia Canada China Peru North America	Non-ferrous	
Tin	Cassiterite	Malaysia Thailand China Indonesia Bolivia Russia	Non-ferrous	

Plastics

Just as plumbers should know their metals, they should also know their plastics if mistakes during installation are to be avoided. There are many different plastics that plumbers use in their day-to-day work for installing hot and cold water supplies, central heating, guttering and rainwater pipes, and above-and below-ground drainage systems.

There are two basic types of plastics: thermoplastics and thermosetting.

1. **Thermoplastics:** a thermoplastic is a type of plastic made from polymer resins that becomes liquid-form when heated and hard when cooled. When frozen, however, a thermoplastic becomes brittle and subject to fracture. These characteristics are reversible and it can be reheated, reshaped and frozen repeatedly. This quality also makes thermoplastics recyclable. There are many different types of thermoplastics, some of which are used extensively in plumbing systems. Each type varies in melting temperature (crystalline organisations) and density. Table 3.8 lists the plastics commonly used in the plumbing industry and describes what they are used for.
2. **Thermosetting:** thermosetting plastics, such as polyester and epoxies, are rigid plastics, resistant to higher temperatures than thermoplastics. Once it has set, a thermosetting plastic cannot be remoulded. Its shape is permanent and it does not melt when heated.

▼ Table 3.8 Common plastics used in the plumbing industry

Type of plastic	Uses	Characteristics
uPVC **CuPVC**	Unplasticised polyvinyl chloride is used extensively for: cold water mains cold water installations (chlorinated unplasticised polyvinyl chloride) solvent-welded and push-fit soil and vent pipes solvent-welded waste and overflow pipes underground drainage pipes gutters and rainwater pipes.	Not suitable for hot water installations. Can be solvent welded.

Type of plastic	Uses	Characteristics
Polyethylene **MDPE** **HDPE**	MDPE (medium-density polyethylene) is used for: underground cold water mains (coloured blue) cold water storage cisterns underground gas pipes. HDPE (high-density polyethylene) is used for: underground cold water mains (coloured black).	Cannot be solvent welded. Degrades under direct sunlight.
Polypropylene	Used for: push-fit waste and overflow pipe cold water storage cisterns.	Cannot be solvent welded. Slightly greasy to the touch. Degrades under direct sunlight.
Polybutylene	Used for: push-fit hot and cold water installations central heating installations.	Cannot be solvent welded.
ABS	Acrylonitrile butadiene styrene. Used for: water supply – potable water for apartments, offices, commercial installations solvent-welded waste and overflow pipes.	Can be solvent welded. Degrades severely under direct sunlight.

Ceramics and fireclays

Ceramics and fireclays are used mainly for sanitary appliances and tiles. There are three varieties that plumbers may use widely in their work:

1 **Vitreous china:** this is a clay material with an enamelled surface used to manufacture bathroom appliances such as WCs and cisterns, wash hand basins and bidets, as well as soap dishes and other sundry bathroom items. It is made from very watery clay, known as 'slip', which is then spray enamelled and fired in a kiln at high temperature.
2 **Fireclay:** this is used primarily for heavy-duty appliances, such as Belfast sinks, London sinks, cleaners' and butler's sinks and shower trays, where there is greater risk of damage and a higher water temperature may be needed. Like other clays, this clay is highly malleable in its raw form. It can be moulded, extruded and shaped by hand. It is also used in the manufacture of building products such as chimney pots.
3 **Ceramic tiles:** these have many applications and are used extensively in bathrooms, kitchens, floors and swimming pools. The origin of the tile can be identified from looking at the reverse of the tile. This is known as the 'biscuit' of the tile. Tiles made in the UK usually have a white-coloured biscuit, Italian tiles usually have biscuit that is cream in colour, and Turkish and Spanish tiles have a dark red biscuit.

Principal properties of solid materials

Solid materials are made up of many molecules. How these molecules are arranged and how they behave under certain conditions will determine their properties. A solid material is assessed by its:

- strength – tensile, compressive and shear
- ductility
- malleability
- hardness
- conductivity – heat and electricity.

Tensile strength

Broadly speaking, the tensile strength of a material is a measure of how well or badly it reacts to being pulled or stretched until it breaks. Some materials, such as plastics, will stretch or elongate before breaking; others, such as metals, will also deform in a similar way but not by as much, and hard materials such as concrete and brick will not deform at all but will simply snap.

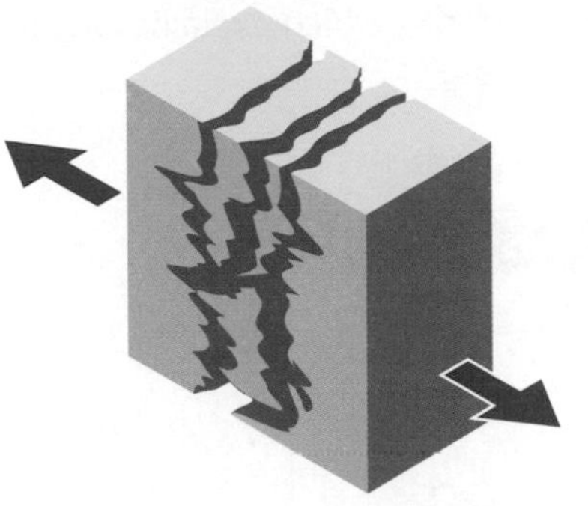

▲ Figure 3.1 Tensile strength

INDUSTRY TIP

Think of a munsen ring attached to a ceiling holding pipework in position which will give a downward pulling/stretching force.

A tensile strength test is also known as a tension test and is the most fundamental type of mechanical test that can be performed on a material. The tests are simple and relatively inexpensive. By simply pulling on a material under specific conditions, how the material will react to being pulled apart will quickly become apparent. The point at which the material fractures is its tensile strength.

IMPROVE YOUR MATHS

Tensile strength is measured in units of force per unit area. In the SI system, the unit is newton per square metre (N/m^2 or Pa – pascal).

Compressive strength

Compressive strength is the maximum stress a material can sustain when being crushed. Hard materials, such as concrete or cast iron, will shatter under compressive stress, while others, like plastics and some metals, may distort in shape. This is called barrelling.

Compressive strength is calculated by dividing the maximum load by the original cross-sectional area of a specimen in a compression test, and is measured in units of force per unit area.

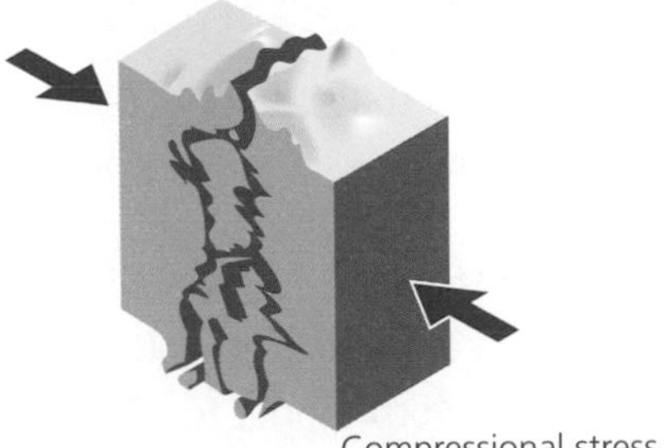

▲ Figure 3.2 Compressive strength

IMPROVE YOUR MATHS

In the SI system, compressive strength is measured using the unit newton per square metre (N/m^2 or Pa – pascal).

INDUSTRY TIP

Think of feet under a bath and the amount of force under each leg placed on the floor, or a cold water storage cistern resting on the joists in a loft.

Shear strength

Shear strength is the stress state caused by a pair of opposing forces acting along parallel lines of action through the material. In other words, the stress caused by sliding faces of the material relative to one another – for example, cutting paper with scissors or ripping a substance apart.

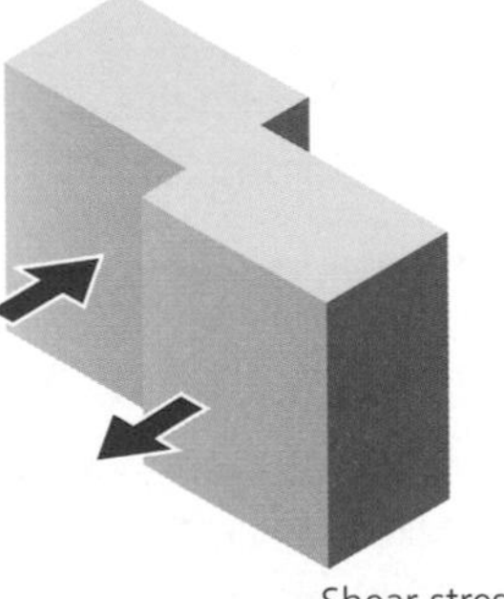

▲ Figure 3.3 Shear strength

INDUSTRY TIP

Think of a radiator or boiler attached to a wall. The wall is offering an upward force and gravity is pushing the appliance down. The two opposing forces are happening where the screws are holding the appliance to the wall.

Ductility of a material

Ductility is a mechanical property that describes by how much solid materials can be pulled, pushed, stretched and deformed without breaking. It is often described as the toughness of a material to withstand plastic deformation. In materials science, ductility specifically refers to a material's ability to deform under tensile stress. This is often characterised by the material's ability to be stretched into a wire. Copper is one of the most ductile materials a plumber will use because it is easily bent and softened into various shapes.

Malleability of a material

Malleability can be defined as the property of a material, usually a metal, to be deformed by compressive strength without fracturing. If a metal can be hammered, rolled or pressed into various shapes without cracking or breaking, or other detrimental effects, it is said to be malleable. This property is essential in sheet metals, such as lead, that need to be worked into different shapes.

INDUSTRY TIP

Think of a pressed steel radiator – the flutes have been pressed into place but the steel has not fractured.

Hardness

Hardness is the property of a material that enables it to resist bending, scratching, abrasion or cutting.

INDUSTRY TIP

Think of a core drill using impregnated diamonds to cut through brickwork. The brick is 'hard' but the diamond is harder.

Hardness of minerals can be assessed by reference to the Mohs scale, which ranks the ability of materials to resist scratching by another material. There is a good reason for grouping materials this way. If an unknown material is discovered, it is one way how to find out what it is by seeing how hard it is.

The Mohs hardness scale starts at 1 for the softest material and goes up to 10 for the hardest.

Diamond is the hardest material, which explains why it is used on many cutting edges.

▼ Table 3.9 The Mohs hardness scale

Material	Hardness scale
Talc	1
Gypsum	2
Calcite	3
Fluorite	4
Apatite	5
Feldspar	6
Quartz	7
Topaz	8
Corundum	9
Diamond	10

Conductivity

Conductivity is the property that enables a metal to carry heat (thermal conductivity) or electricity (electrical conductivity).

- **Thermal conductivity:** here, heat is transferred from molecule to molecule through the substance. How fast or how well the heat travels will determine the material's thermal conductivity. For example, metals, such as copper, transfer the heat quickly and are said to be good conductors of heat, whereas other materials, such as polyurethane, allow the passage of heat only very slowly and so are poor conductors of heat. The inability of polyurethane to allow the passage of heat makes it a very good insulator with the ability to keep heat in. Thermal conductivity is measured in watts per metre kelvin (W/mK).
- **Electrical conductivity:** this is the ability of a material to allow an electrical charge or current to pass through it. It is measured in ohms (Ω). Materials that allow an electrical current to flow freely, such as copper and gold, are known as good conductors, whereas those that do not allow the passage of an electrical current, such as wood, ceramics and PVC, are known as insulators.

IMPROVE YOUR MATHS

Thermal conductivity is measured in watts per metre kelvin (W/mK). Electrical conductivity is measured in ohms (Ω).

Oxidation, corrosion and degradation of solid materials

All solid materials will corrode or degrade over time. The amount that materials corrode or degrade will depend upon the material's resistance and the environment in which the material exists. In this section of the chapter, we will investigate these three processes and how they affect plumbing materials.

> **INDUSTRY TIP**
>
> Corrosion is why chemical additives/inhibitors are used in central heating systems.

Oxidation of metals

Metals are oxidised by the presence of oxygen in the air. This process is more commonly called corrosion. Electrons jump from the metal to the oxygen molecules (see the page on electron flow later in this chapter). The negative oxygen ions that are formed penetrate into the metal, causing the growth of an oxide on the metal's surface. As the oxide layer increases, the rate of electron transfer decreases. Eventually, the corrosion stops and the metal becomes **passive**. However, the oxidation process may possibly continue if the electrons succeed in entering the metal through cracks, pits or impurities in the metal, or if the oxide layer is dissolved.

> **KEY TERM**
>
> **Passive:** when a metal becomes passive, it means that an oxide film has formed that prevents further attack on the metal.

Corrosion

Corrosion is the main reason for metals deteriorating. Most metals will corrode on contact with water (and moisture in the air), acids, salts, oils, and other solid and liquid chemicals. Metals will also corrode when exposed to some gases, such as acid vapours, ammonia gas and any gas containing sulphur.

Corrosion specifically refers to any process involving the deterioration or degradation of metal components. The best-known case is that of the rusting of steel and iron where the formation of ferrous oxide occurs. The corrosion process is usually electrochemical.

When rusting occurs, the metal atoms are exposed to an environment containing water molecules. Here, they give up electrons and become positively charged ions.

> **KEY POINT**
>
> The electrochemical process involves the passage of a small electrical charge between two metals that are at opposite ends of the electromotive series of metals. The stronger, noble metal is called the cathode and the weaker metal is known as the anode. When these two dissimilar metals are placed in an electrolyte such as water, an electric charge is generated and the anodic metal is 'eaten' away by the cathodic metal. A by-product of this reaction is the generation of hydrogen gas. The process accelerates when heat is present.

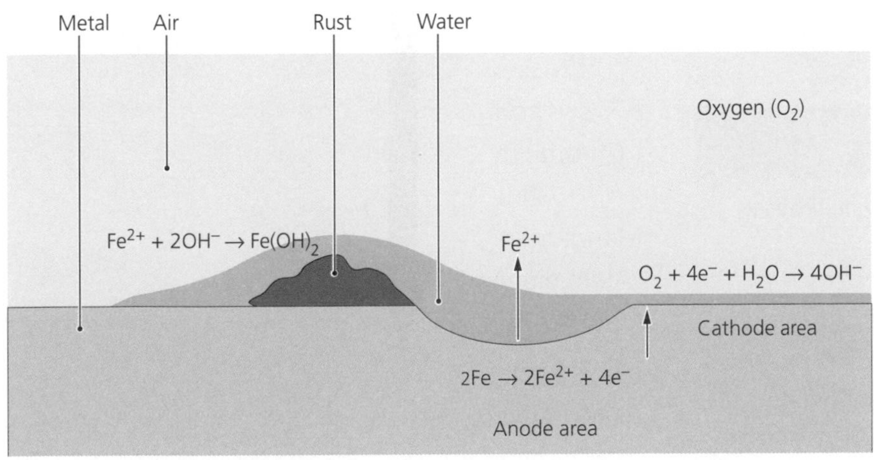

▲ Figure 3.4 How rust is formed

This effect can occur locally to form a pit or a crack, or it can extend across a wide area to produce general corrosion. In effect, iron (Fe) will rust if it is exposed to water and oxygen.

Other forms of metal corrosion that occur in plumbing and heating systems

There are many forms of metal corrosion that can occur within plumbing and heating systems, including:

- de-zincification
- galvanic corrosion
- erosion corrosion
- pitting corrosion.

De-zincification of brass

Brass is an alloy mixture of copper and zinc. De-zincification of brass is a form of selective corrosion (often referred to as de-alloying) that happens when zinc is leached out of the alloy, leaving a weakened brittle porous copper fitting. This commonly occurs in chlorinated tap water or in water that has high levels of oxygen. Signs of de-zincification are a white powdery zinc oxide coating the surface of the fitting, or if the yellow brass turns a shade of red. Selective corrosion can be a problem because it weakens a fitting, leaving it vulnerable to possible failure and eventual leaks.

▲ Figure 3.5 De-zincification and its effects

INDUSTRY TIP

Modern brass fittings do not suffer from de-zincification and are marked DZ or DZR.

Galvanic corrosion

Galvanic corrosion (also called galvanic action, 'dissimilar metal corrosion' and often wrongly termed 'electrolysis') occurs when two dissimilar metals are in contact with each other through the presence of an **electrolyte**. Metals are graded through the electromotive series (also known as the electrochemical series) of metals. The further the metals are apart in the series, the greater the chance of galvanic corrosion.

KEY TERM

Electrolyte: a fluid that allows the passage of electrical current, such as water. The more impurities (such as salts and minerals) there are in the fluid, the more effective it is as an electrolyte.

INDUSTRY TIP

A magnesium rod (sacrificial anode) is used in a hot water cylinder to prevent galvanic corrosion taking place. See Chapter 5 on hot water for more details.

For galvanic corrosion to occur, three conditions must be present:

1 electrochemically opposed metals must be present
2 these metals must be in electrical contact
3 the metals must be exposed to an electrolyte.

One of the metals is the most noble, cathodic metal and the other is the weaker, least noble anodic metal. When an electrolyte is introduced, such as water, a small electrical direct current (DC) is generated between the two metals. The stronger of the two metals will destroy the weaker metal, with hydrogen being produced as a by-product.

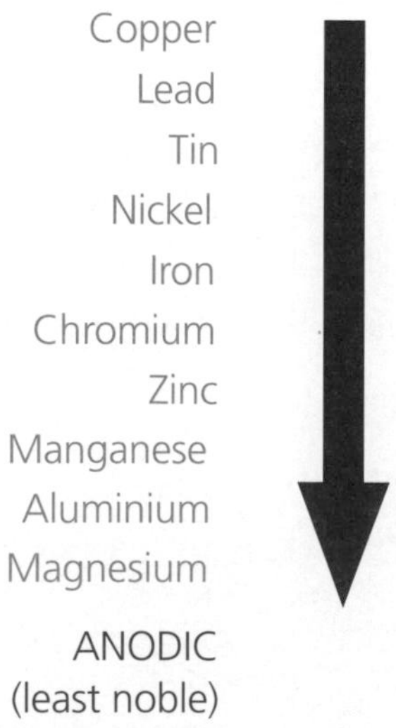

Copper
Lead
Tin
Nickel
Iron
Chromium
Zinc
Manganese
Aluminium
Magnesium

ANODIC
(least noble)

▲ Figure 3.6 Electromotive series of metals

Erosion corrosion

Erosion corrosion occurs in tubes and fittings because of the fast-flowing effects of fluids and gases. The increased turbulence caused by pitting on the internal surfaces of a tube can result in rapidly increasing erosion rates and eventually a leak. Erosion corrosion can also be encouraged by poor workmanship. For example, burrs left at cut tube ends can cause disruption to the smooth water flow, and this can cause localised turbulence and high flow velocities, resulting in erosion corrosion.

INDUSTRY TIP

Make sure all of the burrs are removed from the end of cut pipework before installing them to prevent erosion taking place.

▲ Figure 3.7 Erosion corrosion

Pitting corrosion

Pitting corrosion is the localised corrosion of a metal surface and is confined to a point or small area that takes the form of cavities and pits. Pitting is one of the most damaging forms of corrosion in plumbing, especially in central heating radiators, as it is not easily detected or prevented.

INDUSTRY TIP

A system needs to be correctly flushed out to prevent pitting, which can be caused by flux remaining on the inside of the pipework after soldering.

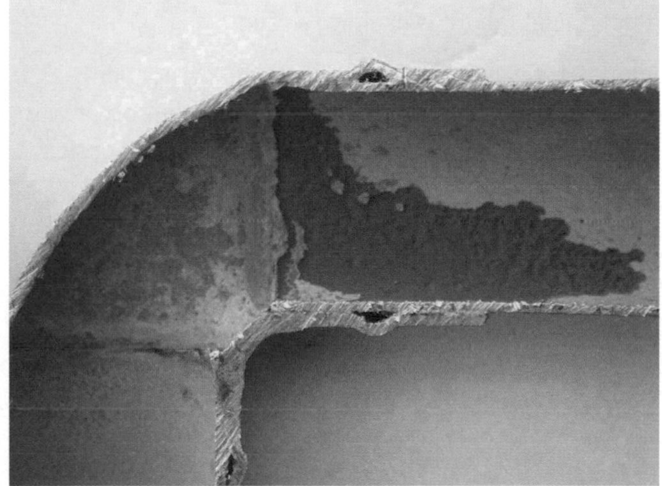

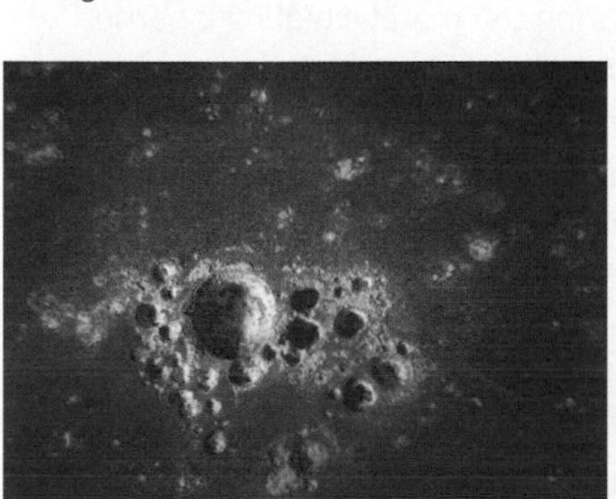

▲ Figure 3.8 Pitting corrosion

Degradation of plastics

The use of plastics is becoming common in the plumbing industry. Everything from hot and cold water services to central heating and drainage can now be installed in some form of plastic material. Problems, however, can occur with plastics under certain conditions. Degradation of plastics can occur from a variety of causes such as:

- heat
- light
- oxygen
- ultraviolet (UV) degradation.

Heat (thermal degradation)

One of the limiting factors when using plastics in high temperature applications is their tendency to not only soften but also to thermally degrade. In some instances, thermal degradation can occur at temperatures much lower than those at which mechanical failure is likely to occur.

All plastics experience some form of degradation during their life. The chemical reactions that occur with thermal degradation lead to both physical and optical changes, such as:

- reduced ductility and embrittlement
- chalking
- colour changes
- cracking.

Light (photodegradation)

This occurs due to the action of light, whether from natural sunlight or electrical fluorescent lighting, and generally causes a yellowing of the plastic material. It is usually more pronounced on light-coloured plastics but can affect all colours.

Oxygen (oxidative degradation)

This is decomposition of the plastic due to the presence of oxygen, which alters the plastic's properties. Colour change is often the first sign of oxidative degradation, coupled with a change in flow, mechanical and electrical properties of the plastic, even if the colour change is not noticeable. Polypropylene, polyethylene and ABS are the plastics most severely affected. PVC, however, is unaffected by oxidative degradation.

KEY POINT

Photodegradation takes place in direct light, even electric light, whether heat is present or not. UV degradation takes place in daylight, whether the Sun is present or not. Its effects occur even on cloudy days and as such it is generally down to the climate.

UV degradation

Most plastics are vulnerable to degradation by the effects of direct exposure to the UV part of the daylight spectrum. UV solar radiation is present even on cloudy days. When UV attack occurs, the colour of the plastic may change and its surface will become brittle and chalky. This can happen over a very short time period and will lead to cracking and eventual failure.

Polypropylene waste pipes and MDPE water pipes are adversely affected by UV degradation, with ABS pipework and fittings being severely compromised by prolonged exposure to the UV daylight spectrum.

Preventing corrosion

Corrosion is one of the most destructive processes to plumbing and heating systems, but there are methods we can employ to prevent and protect from corrosion:

- Galvanisation is one method of protecting steel from rusting by coating with a thin layer of zinc. Galvanising is a process by which the steel is dipped in a bath of molten zinc.
- Greasing and oiling are some commonly used methods to prevent rusting. The grease and oil prevent water and moisture penetration.
- Chrome plating and **anodising** prevent corrosion of metal by coating the metal, creating a barrier between it and the corrosive environment.
- Wet central heating systems can be protected from corrosion by the use of corrosion inhibitors mixed with the system water.
- Plastics can be protected from the effects of UV light by painting.
- Sacrificial anodes (magnesium rods) placed inside hot water storage cylinders protect the cylinder from electrolytic corrosion.
- Metals can be coated with enamel for protection. Enamel consists of a thin layer of glass heated to a high temperature which then fuses on to the surface of the metal.

> **KEY TERM**
>
> **Anodising:** coating one metal with another by electrolysis to form a protective barrier against corrosion.

The properties of liquids

The plumbing industry is primarily concerned with liquids in one form or another, with water being the most common fluid we deal with. Liquids you may come across in your working life include:

- water
- refrigerants
- glycols and anti-freeze
- fuel oils
- lubricants.

Here, we will investigate these liquids and their uses within the building services industry.

Water

Water is the most abundant compound on earth. It covers seven-tenths of the Earth's surface and is the key to life on Earth. Water has many uses, including hot and cold water supplies and wet central heating systems. Yet, what do we actually know about water?

The properties of water

▼ Table 3.10 The energy of sensible and latent heat of water from 0°C of water to 100°C of steam (see page 146 for more on sensible and latent heat)

Pressure		Boiling point of the water	
bar	kPa	°C	kJ/kg
0	0	100.00	419.06
1	100.0	120.42	505.6
2	200.0	133.69	562.2
3	300.0	143.75	605.3

- **Water is a colourless, odourless and tasteless liquid:** any taste it does have comes from the minerals that may be dissolved in it, and this can often explain why water tastes different in different parts of the country.
- **Water can exist in all three states of matter:** liquid (water), solid (ice) and gas (steam).
- **Water has a maximum density of 1000 kg per cubic meter (m^3) at 4°C:** at this temperature, water is at its densest. When the temperature of water is either raised or lowered from 4°C, water loses density. This peculiar behaviour is known as the 'anomalous expansion' of water. At 100°C, water has a density of 958 kg/m^3 and at 0°C, its density is 915 kg/m^3. This can be expressed as a percentage. When heated, water expands by 4 per cent; when cooled it expands by 10 per cent. When water is turned to steam, it expands by 1600 times, so 1 m^3 of water will transform into 1600 m^3 of steam!

> **KEY POINT**
>
> The effects of the changes in density of water can benefit water heating by creating heat circulation by convection. We will deal with heat transfer through water later in the chapter.

- **The boiling point of water at sea level is 100°C:** if the pressure is raised from this, the boiling point increases. At 1 bar pressure, the boiling point of water is 120°C. Similarly, if the pressure is lowered, then the boiling point decreases. At the top of Mount Everest, the boiling point of water is 69°C.
- **Water freezes at 0°C:** again, pressure can affect this. If the pressure increases then the freezing point is lower. Dissolved minerals can also affect the freezing point. Plumbers also add anti-freeze (glycol) to water in systems where pipework is vulnerable to freezing, for example in solar thermal systems.
- **The relative density of water is 1:** this is the measurement that all other solids and liquids are measured against.
- **The specific heat capacity of water is 4.187 kJ/kg°C:** the specific heat capacity of a substance is the amount of heat required to raise the temperature of 1 kg of the substance by 1°C (or by 1 K). In the case of water, it takes 4.187 kJ of heat to raise 1 kg of water by 1°C. Compared to other materials, water requires a lot of energy to heat up.
- **Water itself is a poor conductor of electricity:** it is the presence of dissolved minerals that makes water a good conductor of electricity. Sea water, for example, is a very good conductor of electricity because of the dissolved salts and minerals it contains.
- **Water is a poor conductor of heat:** compared to most metals, water is a poor conductor of heat. In fact, water is a better insulator of heat than it is conductor. That is why it takes so much energy to raise the temperature of water by 1°C (see specific heat capacity, above).

▼ Table 3.11 Classification of water

Type of water	pH value	Base	Notes
Neutral	7	N/A	Neutral water is neither soft nor hard.
Soft	Below 7	Acidic	Water is made soft by the presence of carbon dioxide (CO_2). It is particularly destructive to plumbing systems containing lead as it can dissolve the lead, making the water contaminated. Because of its lead-dissolving capability, soft water is known as 'plumbo-solvent'. Soft water lathers soap easily.
Temporary hard water	Above 7	Alkali	Temporary hard water contains calcium carbonate ($CaCO_3$), otherwise known as limestone. This kind of water can be softened by boiling but leaves behind limescale residues, which can block pipes and other plumbing fittings and appliances. When water reaches 65°C, the calcium in the water re-forms in a process known as precipitation, causing scaling within plumbing systems. Lathering of soap is difficult.
Permanently hard water	Above 7	Alkali	Permanently hard water contains magnesium and calcium chlorides, and sulphates in the solution. It cannot be softened by boiling.

- **Water is known as the 'universal solvent':** almost all substances dissolve in water to a certain extent. Because of this, it is almost impossible to get chemically pure water on Earth.
- **Water is classified as being hard or soft:** the hardness and softness of water affects its pH value (see Table 3.11).
- **Water goes through several stages to be turned into steam:** at atmospheric pressure, the boiling point of water is 100°C. To raise the temperature of the water from 0°C to 100°C takes 419 kJ/kg of energy. To turn the boiling water at 100°C to steam at 100°C takes a further 2257 kJ/kg of energy; this is when it expands by 1600 times. At this point, the steam is said to be saturated steam. In other words, it is saturated with heat. The total heat, therefore, to turn water at 0°C to steam at 100°C takes 2676 kJ/kg of heat energy. Any further heat added after this does not increase the temperature of the steam; it remains at 100°C and the steam is known as 'superheated' steam because of the extra heat energy. To increase the temperature of the steam, the initial pressure of the water will have to be increased.

INDUSTRY TIP

The pH scale allows engineers to test water for its acidity or alkalinity. This in turn will allow the corrective treatment to be offered to gain a pH level of 7 (natural water).

Capillary attraction

Capillary attraction is the process where water (or any fluid) can be drawn upwards through small gaps against the action of gravity. The wider the gap, the less capillary attraction takes place. It is of particular interest to plumbers as it has the ability to cause problems within some plumbing systems, such as:

- it can cause water to be drawn up underneath tiles and roof weatherings, resulting in water leaks inside the building
- it can initiate water trap seal loss in above-ground drainage systems; in this instance, there are two forces at work – capillary attraction and siphonic action.

Conversely, it is also the process we use to make soldered capillary joints on copper tubes and fittings.

Before capillary attraction can take place, two processes need to be present. These are adhesion and cohesion.

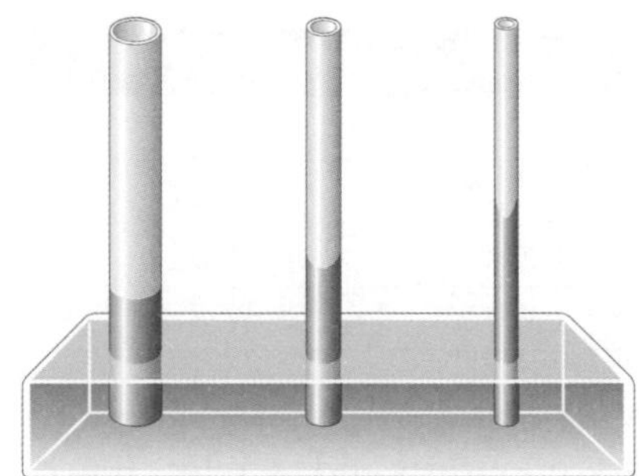

▲ Figure 3.9 Capillary attraction

INDUSTRY TIP

Capillary attraction is used to a plumber's advantage when solder is melted into a liquid and is drawn in between the pipe and fitting.

Adhesion and cohesion

Water is fluid because of **cohesion**. The cohesive quality gives water a slight film on its surface, which is known as the surface tension.

Water is also attracted to other materials, and so it tends to stick to whatever it comes into contact with. This is known as **adhesion**. When water is placed in a vessel or a glass, the adhesion qualities of the water give it a slightly curved appearance. This is known as the meniscus and can be convex (outward curve) or concave (inward curve).

> **KEY TERMS**
>
> **Cohesion:** the way in which the water molecules 'stick' to one another to form a mass rather than staying individual. This is because water molecules are attracted to other water molecules.
>
> **Adhesion:** the way in which water molecules 'stick' to other molecules they come into contact with.

Refrigerants

Refrigerants are fluorinated chemicals that are used in both liquid and gas states. They can, therefore, be classified as both liquid (when compressed) and gas (vapour). All refrigerants boil at extremely low temperatures, well below 0°C.

When a refrigerant gas is compressed, it changes its state to a liquid. During this process a lot of heat and pressure are generated. When the pressure is released quickly, it generates cold. Refrigerants' ability to change their state quickly with such wide temperature changes allows them to be used in refrigeration plants, air conditioning systems and heat pumps. The process is known as the vapour compression refrigeration cycle.

The refrigerant vapour enters the compressor, which compresses it, generating heat. The compressed vapour then enters the condenser, where the useful heat is removed and the vapour condenses to a liquid refrigerant. From here, the liquid refrigerant then passes into the expansion valve, where rapid expansion takes place, converting the warm liquid into a super-cold vapour/liquid mix, which creates the refrigeration effect. The vapour/liquid mix passes through the evaporator, where final expansion to a vapour takes place. This then enters the compressor for the cycle to begin again.

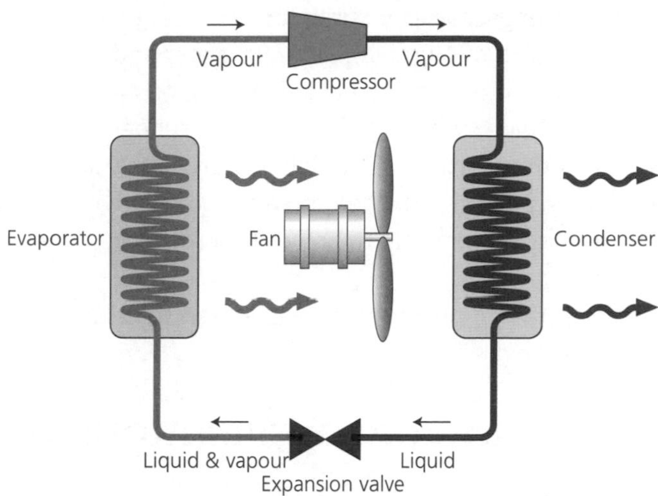

▲ Figure 3.10 The vapour compression refrigeration cycle

Glycol

Glycol is the name used for solar hot water system anti-freeze solution. It is used for protecting solar panels from freezing during the winter when mixed with water in the sealed solar panel circuit. It is available in two forms: propylene glycol and ethylene glycol. Propylene glycol is the preferred chemical for solar panels as ethylene glycol is highly toxic. The anti-freeze should be checked regularly as its anti-freezing capability diminishes with time and the solution can become corrosive with age.

Fuel oils (kerosene)

Kerosene is a fuel oil that is used with most domestic oil-fired boilers (see Chapter 7, Central heating systems, page 379). Kerosene is a thin, clear liquid formed from hydrocarbons, and has a density of 0.78–0.81 g/cm^3. It is made from the distillation of petroleum at temperatures between 150°C and 275°C. The flashpoint of kerosene is between 37°C and 65°C, and it will spontaneously combust at 220°C. The heat of combustion of kerosene is 43.1 MJ/kg, and its higher heating value is 46.2 MJ/kg.

Lubricants

A lubricant is a substance, often a liquid or grease, introduced between two moving surfaces to reduce friction, thus improving efficiency and reducing wear. There are many types of lubricant in use in the plumbing industry:

- **Silicone grease and spray:** used for general lubrication of plumbing parts for water and drainage systems. It is also used when jointing push-fit plastic pipe systems to lubricate the rubber seals.
- **Graphite paste:** used for lubrication of gas taps.
- **Cutting oils:** used when threading low carbon steel pipe. They help to prevent overheating of the cutting dies.
- **Penetrating oils:** used to help loosen tight and rusted joints.

The principal applications of gases

In this section of the chapter, we will look at the principal uses of gases in the building services industry, together with their properties and the scientific laws that apply to them.

Types of gases

The principal gases in the building services industry are listed below.

- **Air:** this has limited uses within the plumbing industry.
  - It can be used as a heating medium in warm-air heating systems. Here, the air is warmed by a warm-air heater, usually fired by gas. The warm air is distributed to the property by means of a fan.
  - It can be used as a pressure charge in expansion vessels. These are usually installed in sealed heating systems and some unvented hot water storage vessels.
  - Air at high pressure can be used to clear blocked drains.
- **Steam:** once the preferred method of heating, the use of steam has declined over recent years. However, because of new, more efficient system designs, steam is being used as a heating medium for:
  - new combined heat and power applications – steam can be used to generate electricity and warm properties in district heating systems
  - electricity generation
  - hot water production using large hot water calorifiers
  - heating systems – the steam is used instead of water in the heat emitters.
- **LPG:** liquid petroleum gas (LPG) can be used for heating appliances such as boilers, cookers and fires. It is also used with plumbers' blowtorches for soldering capillary fittings. There are two basic types:
  1. butane – used mainly as a camping gas
  2. propane – the most widely used LPG in the building services industry.
- **Natural gas:** the most widely used fuel in the UK, natural gas has many applications, both domestic and industrial. It is used as a fuel for:
  - gas fires
  - cookers
  - room heaters
  - condensing central heating boilers
  - water heaters
  - electricity generation
  - industrial heating and processes.
- **Carbon dioxide:** used as a freezing agent with pipe-freezing kits, and is also used in fire extinguishers.
- **Refrigerant gas:** see the section on refrigerants (page 141).

Gas laws

Gases behave very differently from the other two states of matter we have studied so far: solids and liquids. Gases, unlike solids and liquids, have neither a fixed volume nor a fixed shape. They are moulded completely by the container in which they are held. There are three variables by which we measure gases. These are as follows.

Pressure

This is the force that the gas exerts on the walls of its container; it is equal on all sides of the container. For example, when a balloon is inflated, the balloon expands because the pressure of air is greater on the inside of the balloon than the outside. The pressure is exerted on all surfaces of the balloon equally and so the balloon inflates evenly. If the balloon is released, the air will move from the area of high pressure (inside the balloon) to the area of low pressure (outside the balloon).

KEY POINT

Pressure is measured as force per unit area. The standard SI unit for pressure is the pascal (Pa). However, in plumbing it is more likely that pressure will be measured in bar pressure (1 bar = 100 kPa) or millibar (1 mbar = 100 Pa).

Volume

The volume of gas in a given container is affected by temperature and pressure. Pressure is constant if temperature is constant. If temperature is increased, then both the volume and pressure increase.

Temperature

An important property of any gas is its temperature. The temperature of a gas is a measure of the mean kinetic energy of the gas. The gas molecules are in constant random motion (kinetic energy). The higher the temperature, then the greater the kinetic energy and greater the motion. As the temperature falls, the kinetic energy decreases and the motion of the gas molecules diminishes.

Charles's law

Charles's law was discovered by Jacques Charles in 1802. It states that the volume of a quantity of gas, held at constant pressure, varies directly with the kelvin temperature. But what does that mean?

It relates to how gases expand when they are heated up and contract when they are cooled. In other words, as the temperature of a quantity of gas at constant pressure increases, the volume increases. As the temperature goes down, the volume decreases.

INDUSTRY TIP

Charles's law can be explained with the following analogy. If a sealed copper pipe were pressurised to 20 mb at room temperature and then placed in direct sunlight where the pipe could warm up, then the pressure inside the pipe would rise. The rise in pressure would be directly proportional to the rise in temperature. If the pipe were allowed to cool down to room temperature, then it would return to its original pressure.

IMPROVE YOUR MATHS

The mathematical expression for Charles's law is shown below:

$$V_1 \div T_1 = V_2 \div T_2$$

Where:

V = volume

T = temperature

Boyle's law

Boyle's law states that the volume of a sample of gas at a given temperature varies inversely with the applied pressure. In other words, if the pressure is doubled, the volume of the gas is halved. Table 3.12 illustrates this point.

INDUSTRY TIP 

The principle of Boyle's law applies to a child's balloon. If the balloon is inflated to a set pressure and then squeezed, the pressure inside increases as the space inside the balloon decreases. If the space inside the balloon were halved, then the pressure would double.

IMPROVE YOUR MATHS

Boyle's law can also be expressed as:

'Pressure multiplied by volume is constant for a given amount of gas at constant temperature.'

To put this in mathematical terms:

$P \times V$ = constant (for a given amount of gas at a fixed temperature)

Since $P \times V = K$, then:

$$P_i \times V_i = P_f \times V_f$$

Where:

V_i = initial volume

P_i = initial pressure

V_f = final volume

P_f = final pressure

K = constant

▼ Table 3.12 Sample of gas at constant temperature and varying pressure

Test	Pressure	Volume	Formula	Calculation
1	100 kPa	50 cm^3	$P \times V = K$	$100 \times 50 = 5000$
2	50 kPa	100 cm^3	$P \times V = K$	$50 \times 100 = 5000$
3	200 kPa	25 cm^3	$P \times V = K$	$200 \times 25 = 5000$
4	400 kPa	12.5 cm^3	$P \times V = K$	$400 \times 12.5 = 5000$
5	25 kPa	200 cm^3	$P \times V = K$	$25 \times 200 = 5000$

3 THE RELATIONSHIP BETWEEN ENERGY, HEAT AND POWER

The relationship between energy, heat and power is such that it is almost impossible to have one without the other two. Below is a list of units for energy, heat and power.

- **The unit of power:** the watt is the SI unit for power. It is equivalent to one joule per second (1 J/s) or, in electrical units, one volt ampere (1 V·A).
- **The unit of heat:** the joule is the unit of heat; 4.186 joule of heat energy (which equals one calorie) is required to raise the temperature of 1 g of water from 0°C to 1°C.
- **The unit of energy:** also the joule (see above).
- **Specific heat capacity:** the specific heat capacity of a substance is the amount of heat required to change a unit mass of that substance by one degree in temperature. It is measured in kilojoules per kilogram per degree Celsius or Kelvin (kJ/kg°C or kJ/kgK).

INDUSTRY TIP

Specific heat capacity is important to understand when working out how much energy is required to heat up a hot water cylinder.

Heat energy is transferred because of temperature difference – for example, heat passes from a warm body with high temperature to a cold body with low temperature. The transfer of energy as a result of the temperature difference alone is referred to as **heat flow**. The **watt**, which is the SI unit of power, can be defined as 1 joule per second (**J/s**) of heat flow.

In this part of the chapter, we will investigate the energy/heat/power/temperature relationship, and its implications for the building services industry.

Temperature

Temperature is simply the degree of hotness or coldness of a body or environment, and is expressed in terms of units or degrees designated on a standard scale, usually celsius (centigrade) (°C) or kelvin (K).

Celsius (°C)

This scale, using increments of 1 degree (1°), is the most widely used by the building services industry. In simple terms, it has a zero point (0°C), which corresponds to the temperature at which water will freeze. When this scale is used, the degree symbol (°) should accompany it, i.e. 21°C.

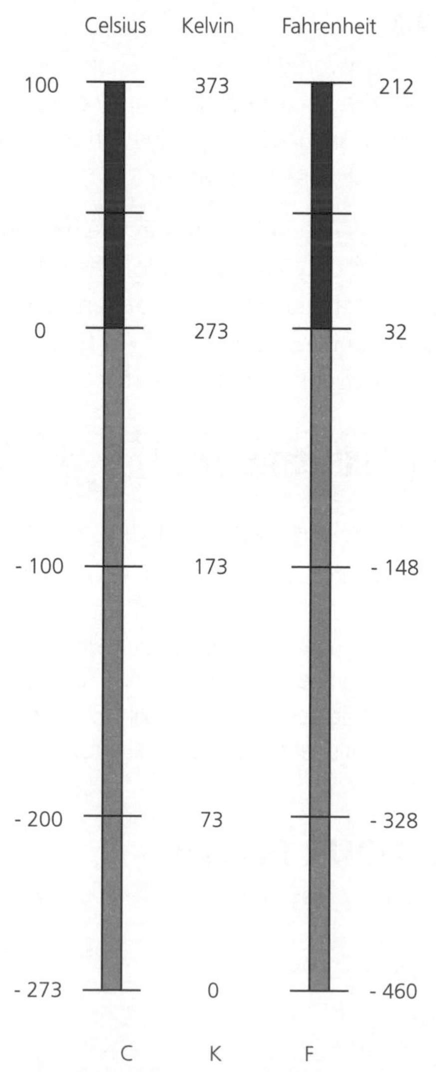

▲ Figure 3.11 The relationship between celsius, kelvin and fahrenheit

Kelvin (K)

This has the same increments as the Celsius scale, but has a minimum temperature that corresponds to the point at which all molecular motion will stop. This temperature is often called absolute zero and is equal to −273°C. Therefore:

- −273°C = 0K, or
- temperature K = temperature °C + 273.

The degree symbol (°) is not used when using the Kelvin scale, i.e. 21 K. The two scales (C and K) are, for the most part, interchangeable. The SI unit of temperature is the kelvin; however, when discussing temperature difference, celsius or kelvin may be used and, since both scales correspond with each other, temperature difference is uniform. In other words, a 1°C temperature difference is equal to a 1 K temperature difference.

INDUSTRY TIP

Celsius is named after the Swedish astronomer, Anders Celsius (1701 1744). The Kelvin scale is named after the Belfast-born engineer and physicist William Thomson, First Baron Kelvin (1824–1907).

Measuring temperature

Many methods have been developed for measuring temperature. Most rely on measuring some physical property of a working material that varies with temperature. Temperature measuring devices include the following.

- **Glass thermometer:** one of the most common devices for measuring temperature. This consists of a glass tube filled with mercury or some other liquid. Temperature increases cause the fluid to expand, so the temperature can be determined by measuring the volume of the fluid. These thermometers are usually calibrated so that the temperature can be read by observing the level of the fluid in the thermometer.
- **Gas thermometer:** this measures temperature by the variation in volume or pressure of a gas.

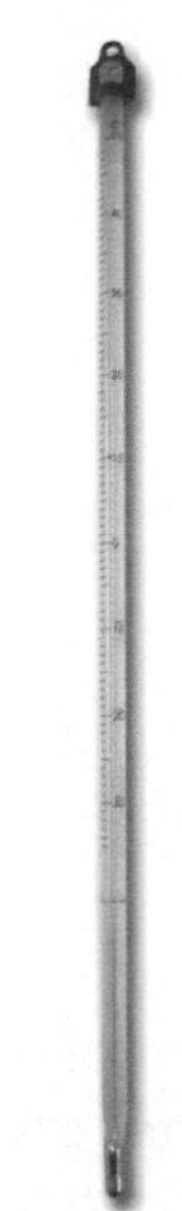

▲ Figure 3.12 Glass thermometer

- **Thermocouple:** this device is a connection between two different metals that produces an electrical voltage when subjected to heat. This senses a temperature difference. Thermocouples are a widely used type of temperature sensor for measurement and control when used with digital thermometers (see below). They can also be used to convert heat into electrical power.
- **Thermistor:** thermistors are resistors that vary with temperature. They are constructed of semiconductor material with a resistivity that is especially sensitive to temperature. When the temperature is measured, the resistance of the thermistor responds in a predictable way.
- **Infrared thermometers:** these use infrared energy to detect temperatures. They detect actual energy levels by the use of an infrared beam and so the thermometer does not need to actually touch the surface to take an accurate temperature measurement.

INDUSTRY TIPS

Thermistors are used in modern boilers to measure temperature on the return pipe.

Infrared thermometers are used to image a radiator and identify potential cold spots.

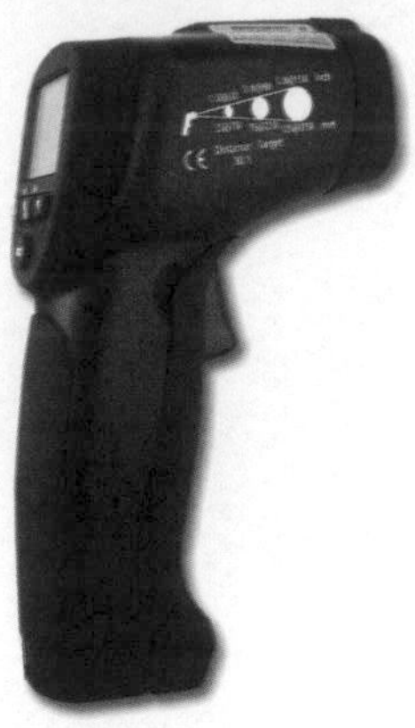

▲ Figure 3.13 Infrared thermometer

- **Digital thermometers:** these are probably the most common thermometer used in the plumbing industry. Dual digital thermometers can read two temperatures simultaneously, instantly giving the temperature difference between two points, which is essential when benchmarking central heating boilers for reading the temperature of both flow and return pipes.

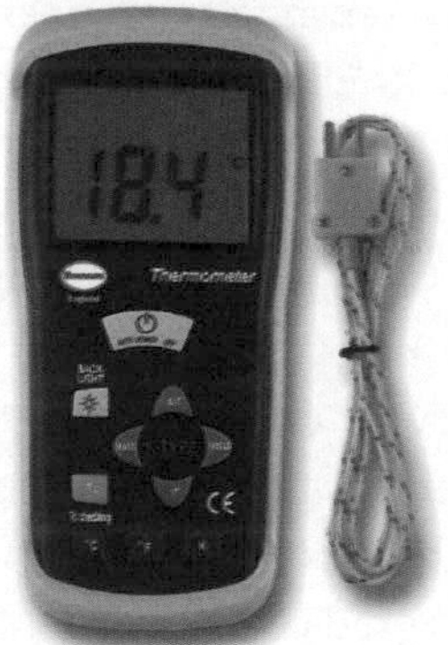

▲ Figure 3.14 Digital thermometer

INDUSTRY TIP

Digital thermometers are commonly used when commissioning a system as they offer a quick and easy way to measure water temperature.

States of matter

Everything around us is made up of matter, which can exist in three classic states: solid, liquid and gas. Each of the phase changes is associated with either an increase or decrease in temperature. For example, if heat energy is applied to ice, it melts to form water and, if more heat energy is applied to the water, it reaches its boiling point, where it vaporises, evaporating to steam. The process can also work in reverse. When the heat is given up by the steam, it condenses back to water. Each of these phase changes is given a name:

- ice (solid) to water (liquid) is called melting
- water (liquid) to steam (gas) is called evaporation/ vaporisation
- steam (gas) back to a water (liquid) is called condensation
- water (liquid) to ice (solid) is called freezing (solidification)
- ice (solid) to steam (gas) is known as sublimation
- steam (gas) to ice (solid) is known as deposition.

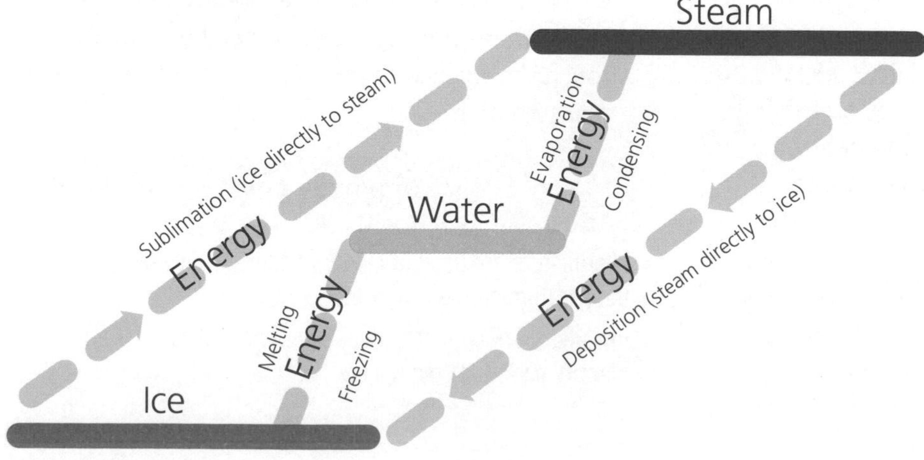

▲ Figure 3.15 States of matter

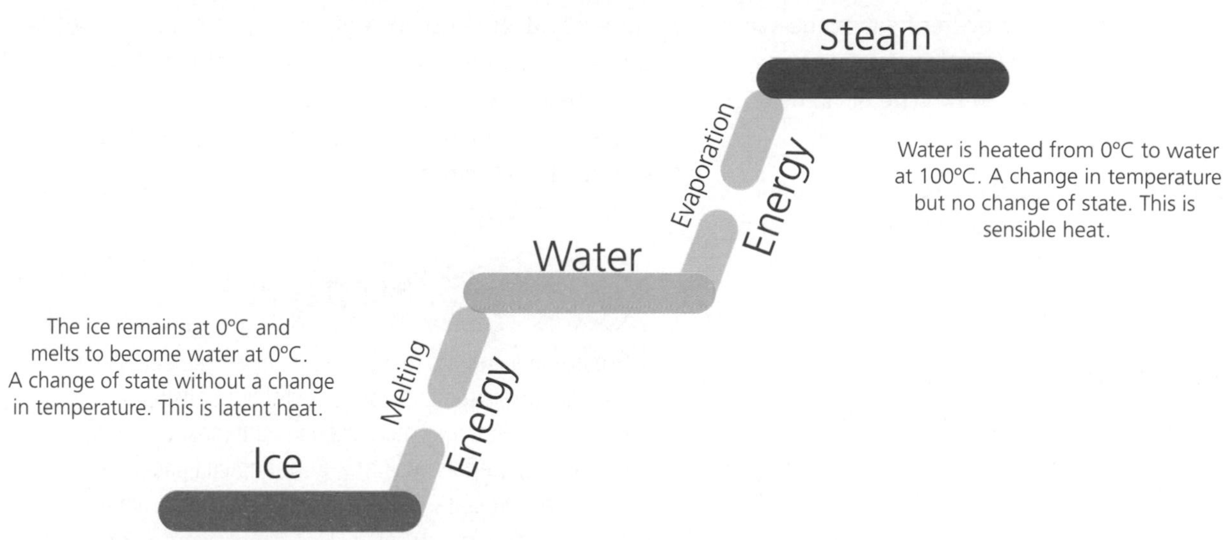
Steam
Evaporation
Energy
Water is heated from 0°C to water at 100°C. A change in temperature but no change of state. This is sensible heat.
Water
The ice remains at 0°C and melts to become water at 0°C. A change of state without a change in temperature. This is latent heat.
Melting
Energy
Ice

▲ Figure 3.16 How sensible and latent heat work together

Sensible and latent heat

Sensible heat of liquid and gases

When heat is applied to a liquid, its temperature will rise as heat is added without a change of state. The resulting increase in heat is known as sensible heat. This process can be reversed. When heat is removed from the liquid and its temperature decreases, the heat that is removed is also called sensible heat. Therefore, any heat that causes a change in temperature without a change of state is known as sensible heat.

Latent heat of liquid and gases

Changes of state, as we have already seen, are the result of a change in temperature. Solids can become liquids, liquids can become gases and each change of state is reversible. The heat that causes any change of state is known as latent heat. Latent heat, however, does not affect the substance's temperature. For example, water boils at 100°C. The heat required to raise the water to its boiling point of 100°C is called sensible heat. The heat required to keep it boiling at 100°C is latent heat.

Methods of heat transfer

So far we have investigated temperature and heat, and how these affect the different states of matter. Now, we will consider the methods of heat transfer.

There are three methods by which heat can be transferred through a substance or from one substance to another. These are:

1 conduction
2 convection
3 radiation.

We will look at each one in turn.

Conduction

Conduction happens when heat travels through a substance, with the heat being transferred from one molecule to another.

INDUSTRY TIP

To understand conduction, imagine holding a cold short piece of copper pipe. While still holding the pipe a blowtorch flame heats the opposite end up. The heat will travel down the pipe until the pipe is too hot to hold.

Consider a piece of copper tube. If heat is applied to one end, before long the heat will have travelled through the material so that the effects of the heat will be felt at the other end. This occurs because kinetic energy in the form of heat is being passed from one copper molecule to another very quickly. When the copper is cold, the atoms move very slowly. As heat is applied, these atoms gain speed and collide with the slower, cooler atoms. In this way, some of the kinetic energy is passed through the material, the slow atoms becoming faster and colliding with other slow atoms, and so on.

Not all substances, however, transfer heat at the same rate. Some materials, such as plastic or wood, are very poor at transferring heat, with little or no heat transference occurring at all.

Most metals are very good conductors of heat and, because of this, they are also very good at conducting electricity. Materials that do not transfer heat well, such as plastic, are known as insulators.

IMPROVE YOUR MATHS

The rate at which a material will transfer heat is known as the coefficient of thermal conductivity, which is measured in W/m/K. It can be found using the following equation:

$$\text{Thermal conductivity} = \frac{\text{heat} \times \text{distance}}{\text{area} \times \text{temperature difference}}$$

Table 3.13 lists some common substances, together with their coefficient of thermal conductivity.

▼ Table 3.13 Coefficient thermal conductivity of common substances

Material	Thermal conductivity W/m/K
Silver	406.0
Copper	385.0
Gold	310
Aluminium	205.0
Brass	109.0
Steel	50.2
Lead	34.7
Concrete	0.8
Polyethylene HD	0.5
Wood	0.12–0.04
Polystyrene expanded	0.03

From Table 3.13, it can be seen that silver is the best conductor of heat, with copper coming a close second.

The poorest conductor of heat is expanded polystyrene, which is an excellent insulator of heat.

The coefficient of linear expansion

Most materials expand when they are heated. When copper pipework expands it can often be heard as a ticking when the central heating is on. The copper expands in length by 0.000018 m/m/°C. This may not seem a lot, but when it is considered that this figure is for every degree rise in temperature, then the length of expansion can be significant. On larger installations, it may mean the use of expansion joints to accommodate the amount of expansion so that damage to the pipework is eliminated. PVCu expands by a greater amount of 0.0005 m/m/°C.

INDUSTRY TIP

Expansion joints are required in many systems to allow material to expand and contract without causing problems for the installation. Plastic expands the most out of all plumbing materials; look at a guttering component and note the expansion mark. For more information see Chapter 8.

IMPROVE YOUR MATHS

Let's see by how much copper expands.

20 m of 22 mm copper pipe contains water that rises from 4°C to 85°C. By how much does the copper expand?

There is 20 m of copper pipe, an 81°C temperature difference and a 0.000018 coefficient of expansion of copper, so:

$$20\text{ m} \times 81°\text{C} \times 0.000018\text{ m/m/°C} = 29.16\text{ mm}$$

Convection

Convection is heat transfer through a fluid substance, which can be water or air.

Convection occurs because heated fluids, due to their lower density, rise and cooled fluids fall.

As water or air is heated it expands, which makes it less dense and therefore lighter. If a cooler, denser material is above the warmer layer, the warmer material will rise through the cooler material. The lighter, rising material will release its heat into the surrounding environment, become denser (cooler), and will fall because of the effect of gravity, to start the process over again.

In a hot water system, this process is known as gravity circulation.

INDUSTRY TIP

Gravity circulation is no longer used to heat water but can still be found in a few older properties. See the old direct system in Chapter 6 and the C-plan plus system in Chapter 7.

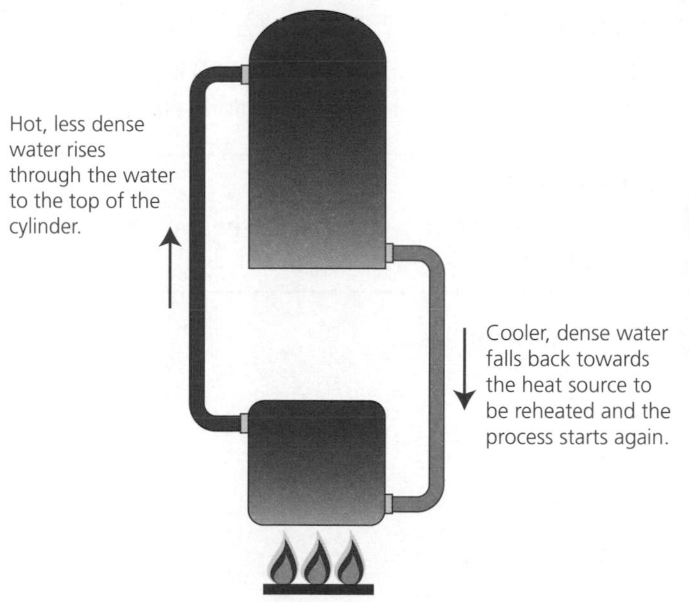
Hot, less dense water rises through the water to the top of the cylinder.
Cooler, dense water falls back towards the heat source to be reheated and the process starts again.

▲ Figure 3.17 Gravity circulation in a hot water system

Modern radiators in central heating systems use two methods of heat transfer, with convection being the main heat transfer method. The other is radiation.

Radiation

The third method of heat transfer is radiation. Radiation heat transfer is thermal radiation from infrared light, visible or not, which transfers heat from one body to another without heating the space in between. Like all forms of light, thermal radiation travels in straight lines.

> **INDUSTRY TIP**
>
> Radiation is used in solar thermal and solar PV panels which use dark coloured fins to absorb the energy from the Sun.

Consider the heat from the Sun, which travels millions of miles through the vacuum of space to heat the Earth. The heat can be felt from a distance because it travels in waves, which are emitted from the heat of the Sun. Radiation is the heat transfer method that makes solar hot water collectors in solar hot water systems so effective.

Radiation heat can also be felt from a hot radiator, even though there is no visible heat source or flame. This is because the heat is being radiated as thermal energy.

Radiated heat is better absorbed by some materials than others. The colour and texture of a surface can also affect the heat absorption. A dull matt surface will absorb heat more effectively than a shiny polished surface. This is the reason that solar thermal panels are dark and dull, to allow them to absorb the Sun's heat more effectively. This is also why a lot of cars in hot countries are coloured white, to reflect the heat.

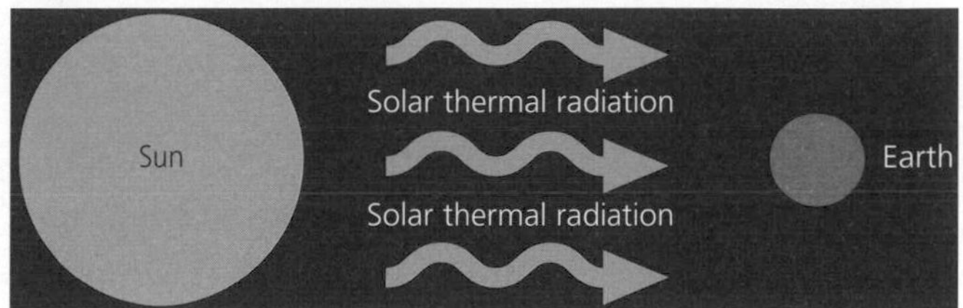
Solar thermal radiation
Sun
Earth
Solar thermal radiation

▲ Figure 3.18 Thermal radiation from the Sun

Energy, heat and power calculations

In this part of the chapter, we will look at simple energy, heat and power calculations using information we have previously discovered. To recap, the SI units of measurement of energy, heat and power are:

- energy – the joule (J)
- heat – the joule (J)
- power – the watt (W)
- specific heat capacity – kilojoules per kilogram per degree celsius (kJ/kg/°C).

Calculations using the specific heat capacity of water

> **KEY POINT**
>
> Remember: the specific heat capacity of water is 4.186 kJ/kg/°C.

Example 1

How many kilojoules would it take to heat 100 litres of water from 30°C to 80°C?

The formula for this is:

$$L \times \Delta t \times \text{SHC of water}$$

Where:

L = litres

Δt = temperature difference

SHC of water = 4.186

Therefore:

$$100 \times (80-30) \times 4.186 = 20930 \text{ kJ}$$

ACTIVITY

Using the formula shown in Example 1, calculate how many kilojoules it would take to heat 140 litres of water from 4°C to 65°C.

Example 2

We can develop this concept further to calculate how many kilowatts it would take to raise the temperature of the 100 litres of water by 50°C. To do this, we need to state a time frame. Let us assume that the 100 litres of water is required in one hour. The calculation would then become:

$$\frac{\text{L} \times \Delta\text{t} \times \text{SHC of water}}{\text{Time (in seconds)}}$$

Where:

L = litres

Δt = temperature difference

SHC of water = 4.186

1 hour in seconds = 3600

Therefore:

$$\frac{100 \times (80-30) \times 4.186}{3600} = 5.81 \text{ kW}$$

ACTIVITY

Using the formula shown in Example 2, calculate how many kilowatts it would take to raise the temperature of the 140 litres of water from 4°C to 65°C in two hours.

Example 3

How many seconds would it take for 20 kg of water to be heated by 15°C using a 3 kW heating element?

KEY POINT

Remember: water has a specific heat capacity of 4.186 kJ/kg/°C and that 1 W = 1 J/s.

The formula for this is:

$$\frac{\text{kg} \times \text{t} \times \text{SHC}}{\text{kW}}$$

Where:

kg = kilograms

t = temperature

kW = kilowatts

SHC = specific heat capacity

Therefore:

$$\frac{20 \times 15 \times 4.186}{3} = 418.6 \text{ s or } 6.976 \text{ minutes}$$

ACTIVITY

Using the formula shown in Example 3, calculate how many seconds it would take for 42 kg of water to be heated by 30°C using a 3 kW heating element.

4 THE PRINCIPLES OF FORCE AND PRESSURE, AND THEIR APPLICATION IN THE PLUMBING AND HEATING INDUSTRY

In this part of the chapter, we will look at the scientific principles of force and pressure, and investigate how they apply to the building services industry.

The SI units of force and pressure

▼ Table 3.14 SI units of force and pressure

Velocity	metres per second	m/s
Acceleration	metres per second squared	m/s^2
Flow rate	metres cubed per second	m^3/s
Force	newton (equal to kg m/s^2)	N
Pressure, stress	pascal (equal to N/m^2)	Pa

Velocity and acceleration

- **Velocity** is the measurement of the rate at which an object changes its position. In order to measure it, we need to know both the speed of the object and the direction in which it is travelling. It is measured in metres per second (m/s).
- **Acceleration** is a measure of the rate at which an amount of matter increases its velocity. It is measured in a change of velocity over a period of time and, as such, is directly proportional to force. It will increase and decrease linearly with an increase or decrease in force if the mass remains constant. It is measured in metres per second squared (m/s^2).
- **Acceleration due to gravity** is the rate of change of velocity of an object due to the gravitational pull of the Earth. If gravity is the only force acting on an object, then the object will accelerate at a rate of 9.81 m/s^2 downwards towards the ground.

Flow rate

In plumbing, flow rate is defined as an amount of fluid that flows through a pipe or tube over a given time. It is usually measured in metres cubed per second (m^3/s). However, in plumbing systems, flow rate is usually measured in litres per second (l/s) or in litres per minute (l/m) when using a flow/wier cup. When converting l/s into l/m you multiply by 60. When converting from l/m to l/s you divide by 60 (60 seconds per minute).

IMPROVE YOUR MATHS

To convert from m^3/s to l/s, multiply m^3/s by 1000.

To convert from l/s to m^3/s, multiply l/s by 0.001.

Flow rate can also be measured in kilograms per second (kg/s). Since 1 litre of water has a mass of 1 kilogram, then 1 litre per second (l/s) = 1 kilogram per second (kg/s).

Force

Force is an influence on an object at rest that, acting alone, will cause the motion of the object to change. If the object at rest is subjected to a force, it will start to move. For example, consider water in a pipe connected to a cistern at one end and a tap at the other. When the tap is closed, the water is not moving and so is said to be at rest. When the tap is opened, the force of gravity will move the water out of the tap, causing water to flow. It is measured in newtons (N).

IMPROVE YOUR MATHS

The unit of the force of gravity is the newton. It is the force required to accelerate a mass of 1 kg at 1 metre per second, every second. On Earth, that force of acceleration (known as gravitational pull) is 9.81 metres per second per second, or 9.81 m/s^2. Therefore, if we multiply the mass of an object (in kg) by 9.81, the result is measured in newtons (kgm/s^2).

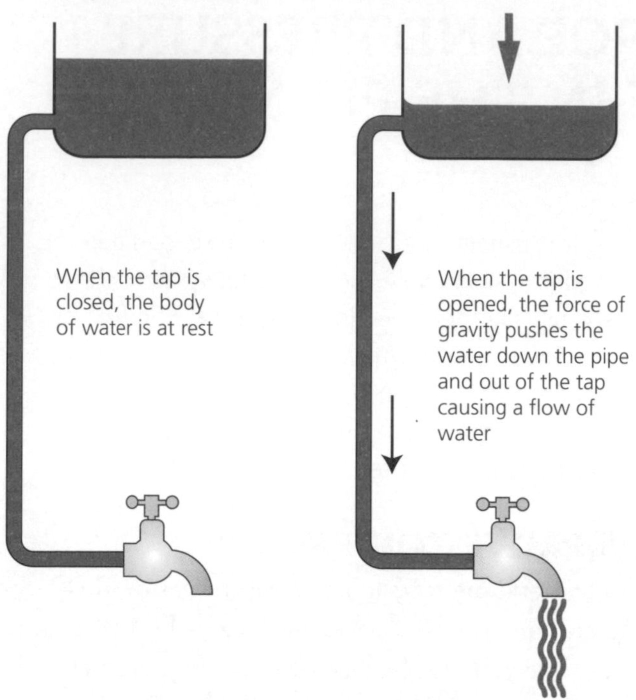
When the tap is closed, the body of water is at rest
When the tap is opened, the force of gravity pushes the water down the pipe and out of the tap causing a flow of water

▲ Figure 3.19 The force of gravity on a cold water system

ACTIVITY

Calculating force

Consider the cistern in Figure 3.19. If it contained a mass of water equal to 40 kg, then by multiplying the mass by the force of gravity, the force of the cistern acting downwards can be calculated:

$$40 \times 9.81 = 392.4 \text{ N}$$

If a cistern in a roof space contains a volume of 100 litres of water and 1 litre = 1 kg, what is the force acting on the platform it is standing on?

Pressure

In physics, pressure is defined as force per unit area. For an object sitting on a surface, the force pressing on the surface is the weight of the object measured in newtons per square metre (N/m^2). However, in different orientations it might have a different area in contact with the surface and will therefore exert a different pressure. For example, if a cistern measuring 1 m long × 0.5 m wide × 0.7 high was placed in a roof space, then what pressure would it exert if:

- it was placed on its bottom
- it was placed on its side
- it was placed on its end?

IMPROVE YOUR MATHS

Before we can attempt these calculations, we must first find the mass of the cistern in kg. The formula for this is:

$$\text{length} \times \text{width} \times \text{height} = \text{volume in m}^3$$

From earlier calculations, we know that to find the force of an object we use the formula:

$$\text{kg} \times \text{gravity} = \text{N}$$

Since 1 litre of water has a mass of 1 kg, a cistern measuring 1 m × 0.5 m × 0.7 m has a force of 3433.5 N.

The formula for finding pressure is:

$$\frac{\text{force}}{\text{area}} = \text{N/m}^2$$

From these calculations we can see that the greater the surface area for a given mass, the less force will be exerted by that mass. This is of particular importance when placing large cisterns in roof spaces since the greater the surface area we can rest the cistern on, the more we can spread the load of the cistern.

ACTIVITY

Pressure

What is the pressure exerted by a block of lead with a cross-sectional area of 4 m^2 and a mass of 4000 kg?

Static pressure of water (head)

The unit of water pressure is the pascal. The pressure exerted by water is due to its mass and is determined by the height of the column of water. For instance, if the pressure exerted by a water main is 300 kilopascals (kPa) it will balance a column of water about 30 m high. This pressure is equivalent to a head of water of 30 m. Therefore, 10 m of head = 100 kPa.

Water pressure in plumbing systems is usually measured in bar pressure.

Static head of water in plumbing systems is measured from the bottom of the water source, i.e. the cistern, to the outlet, as shown in Figure 3.20.

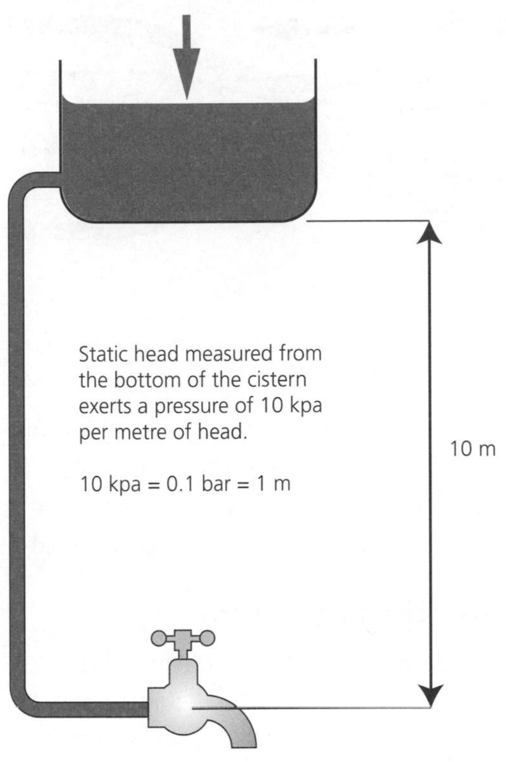
Static head measured from the bottom of the cistern exerts a pressure of 10 kpa per metre of head.
10 kpa = 0.1 bar = 1 m
10 m
Static head at the tap is 100 kpa = 1 bar = 10 m

▲ Figure 3.20 The head of pressure on a cold water system

ACTIVITY

Static pressure of water

If the vertical distance between the bottom of a cold water cistern and the tap is 16 m, what is the pressure at the tap in:

a kilopascals
b bar?

Table 3.15 shows the conversions for common units of head of pressure.

Dynamic pressure

Also called working pressure, dynamic pressure is the pressure of water while it is in motion. In other words, it is the pressure of flowing water. If the pressure of the water is increased, the velocity and flow rate will also increase.

Atmospheric pressure

Atmospheric pressure is the amount of force or pressure exerted by the atmosphere on the Earth and the objects located on it. The more pressure there is, the stronger that force will be; at sea level, the atmospheric pressure is 101.325 kPa. This is known as 1 atmosphere (atm). Atmospheric pressure decreases with height.

▼ Table 3.15 Conversions for common units of head of pressure

Kilopascals (kPa)	Bar	Metres head of water
10	0.1	1
20	0.2	2
30	0.3	3
40	0.4	4
50	0.5	5
100	1	10
150	1.5	15
200	2	20
250	2.5	25
300	3	30
350	3.5	35
400	4	40
450	4.5	45
500	5	50

The principle of a siphon (siphonic action) due to atmospheric pressure

The principle of a siphon is to discharge water from a high vessel to a lower vessel using atmospheric pressure and the cohesive properties of water.

The principle of a siphon can be understood with reference to the diagram (see Figure 3.21). The two beakers are both at atmospheric pressure, but they are at different levels. The pressure at beaker 'B' is greater because it is lower. The outlet from the hose at 'B' must be lower than the inlet of the hose at 'A' for flow to take place. When suction is applied to the end of the hose at 'B', the water will flow upwards over the top of beaker 'A', where the atmospheric pressure is slightly lower. Here, gravity and the cohesive nature of water will empty the contents of beaker 'A' into beaker 'B'.

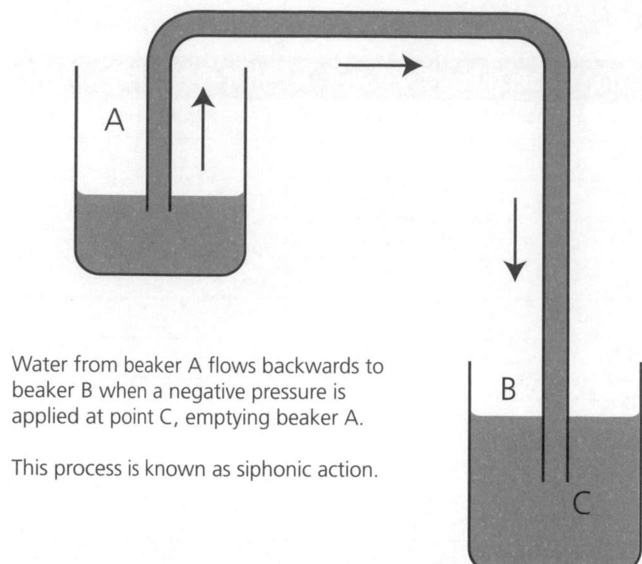
A
B
C
Water from beaker A flows backwards to beaker B when a negative pressure is applied at point C, emptying beaker A.
This process is known as siphonic action.

▲ Figure 3.21 Siphonic action

INDUSTRY TIP

Siphonic action is one way in which a toilet flushes. The water in the toilet cistern is transferred to the toilet pan via the flush pipe. See Chapter 9 for details of how a WC siphon works.

The relationship between velocity, pressure and flow rate in plumbing systems

As we have already discovered, if pressure is applied to a pipe full of water, the effect is to increase the velocity and therefore the flow rate of the water. The more pressure that is applied, the greater the velocity and flow rate becomes.

A similar effect occurs when a pipe is suddenly reduced in size; this can be seen in a hosepipe. If the end of a flowing hosepipe is suddenly reduced, then the speed increases and the water shoots further away, but the pressure and flow rate will be reduced. This is called the Bernoulli effect.

This describes the result of a reduction in pipe size, where the speed of fluid increases at the same time as the pressure or the fluid's potential energy decreases.

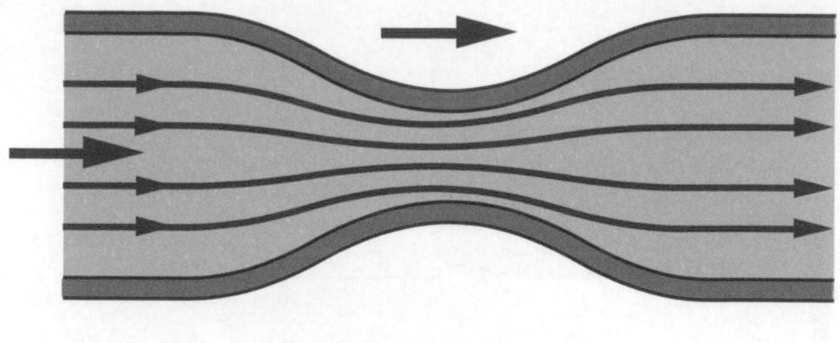
Increased fluid speed, decreased internal pressure

▲ Figure 3.22 The Bernoulli effect

Similarly, if the pipe suddenly increases in size, then the velocity of the water will decrease but the pressure will increase slightly. The flow rate remains constant. This can be seen in the different settings on a showerhead.

Factors affecting flow rate

As we have seen, flow rate is unaffected by sudden increases in pipe size but, as described below, there are elements in plumbing systems that can severely affect the flow rate.

- **Changes in direction:** any change in direction of a pipe will offer resistance to the flow of the water. That resistance will, in effect, be an increase in the overall length of the pipe. For example, an elbow installed in the run of copper pipe will offer resistance equivalent to 0.37 m of pipe. So, if 10 elbows are used, then the length of the pipe has, theoretically, increased by 3.7 m. Machine-made bends offer slightly less resistance at 0.26 m of pipe. This will also vary with the material of the pipe (see 'Frictional resistance of the internal bore of the pipe' below).

▼ Table 3.16 Resistances in the form of equivalent lengths of common fittings

	Nominal pipe size* (mm)					
	8	10	12	15	22	28
Type of fitting	Equivalent length (m)					
Capillary elbow	0.16	0.21	0.28	0.37	0.60	0.83
Compression elbow	0.24	0.33	0.42	0.60	1.00	1.30
Square tee piece	0.27	0.37	0.49	1.00	1.6	2
Swept tee piece	0.22	0.29	0.38	0.60	0.75	1
Manifold connection	0.60	1.00	1.20	n/a	n/a	n/a
Minimum radius (machine) bend	0.12	0.16	0.20	0.26	0.41	0.58

* Copper tubes to **BS EN 1057 R250**

- **Size of pipe:** the greatest factor in the flow rate of any system is the size of the pipe itself. The bigger the bore of the pipe, the better the flow rate will be.
- **Pressure:** pressure increases flow rate. The greater the pressure, the greater the flow rate.
- **Length of the pipe:** flow rate diminishes with length because of the frictional resistance of the wall of the pipe. Water flows faster down the centre of the pipe than it does at the pipe wall. The nearer the water is to the wall of the pipe, then the greater the frictional resistance and so the slower the water becomes. The frictional resistance of the pipe is slowing the flow rate constantly. The greater the length, the more frictional resistance, the greater the loss of flow rate. To counter this effect, the pipe size should be increased initially at the start of the pipe run and then reduced as length increases.
- **Frictional resistance of the internal bore of the pipe:** different materials offer different frictional resistance. Polybutylene pipe, for instance, has the smoothest bore of all common pipe materials and low carbon steel the roughest. Therefore, low carbon steel at like-for-like sizes will have a much lower flow rate than polybutylene pipe.
- **Constrictions such as valves and taps:** taps and valves offer a lot of resistance to the flow of water. Some stop taps can increase pipe length by up to 6 m per valve.

5 THE MECHANICAL PRINCIPLES IN THE PLUMBING AND HEATING INDUSTRY

Simple machines are those that aid with the lifting and moving of loads that are too heavy to lift or move on their own. There are four main types:

1 levers
2 wheel and axles
3 pulleys
4 screws.

IMPROVE YOUR MATHS

The calculation for finding out how a lever functions is:

$$\text{Mechanical advantage} = \frac{\text{Load}}{\text{Effort}}$$

These machines give a mechanical advantage (velocity ratio) to human effort, meaning they multiply the force that is put into them. There are two types of mechanical advantage:

1 **Ideal mechanical advantage (IMA):** purely theoretical, based upon an 'ideal machine', which does not exist.
2 **Actual mechanical advantage (AMA):** this is the mechanical advantage of a real machine such as a wheelbarrow (lever). AMA takes into consideration real-world factors such as energy lost because of friction.

Simple machines

Here, we will look at the machines themselves and their possible uses in everyday working life.

Levers

In physics, a lever is a rigid object that can be used with a pivot point or fulcrum to multiply the mechanical force that can be applied to another, heavier object. Levers are examples of mechanical advantage.

There are three classes of lever, as follows.

- **First class lever:** a simple see-saw arrangement where the long arm (force effort) is proportional to the short arm (load). Examples of this are:
  - the lever arm of a float-operated valve
  - claw hammer
  - water pump pliers (double lever).
- **Second class lever:** a variation on the first class lever. Examples of this are:
  - wheelbarrow
  - crowbar.
- **Third class lever:** examples of this are:
  - the human arm
  - tools, such as a hoe or scythe
  - spades and shovels.

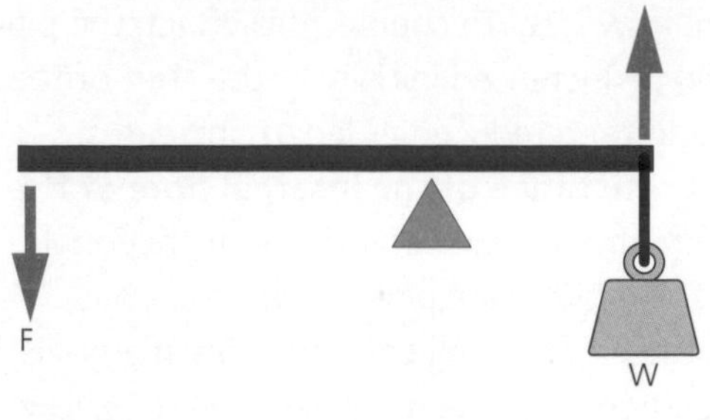

▲ Figure 3.23 First class lever

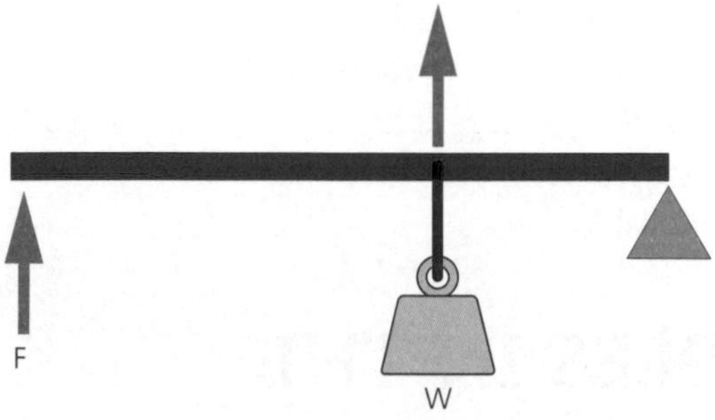

▲ Figure 3.24 Second class lever

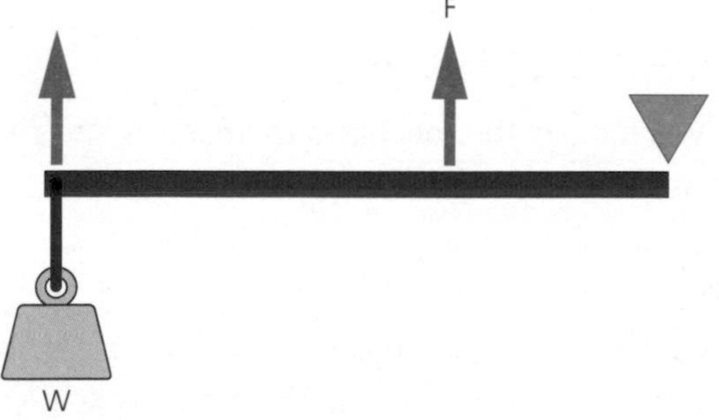

▲ Figure 3.25 Third class lever

Wheel and axles

The **wheel and axle** is composed of a wheel, which is larger than the diameter of the axle. Either of these can be used as the effort arm and the resistance arm, and this depends where the force is applied. The force is usually applied to the wheel rather than the axle to gain the maximum output. The point where the axle joins the wheel is known as the fulcrum and this acts as the point where the force from the larger wheel is transferred to the smaller axle.

KEY TERM

Wheel and axle: a mechanical device used to wind up weight; includes a grooved wheel, turned by a cord/chain, and a rigid axle.

The wheel and axle multiplies the 'torque' during the turning motion.

Both the wheel and the axle have ropes wound around them. The load is lifted by pulling on the rope around the wheel so that the wheel and axle is rotated once, therefore:

$$\text{Mechanical advantage} = \frac{\text{Radius of the wheel}}{\text{Radius of the axle}} = \frac{R}{r}$$

Spanners and screwdrivers use the principle of wheel and axle.

Pulleys

A pulley is a collection of one or more wheels over which a rope or chain is looped to aid lifting heavy objects. Pulleys are examples of simple machines. In other words, they multiply the lifting forces.

How do pulleys work?

A single pulley reverses the direction of the lifting force. When the rope is pulled down, the weight lifts up. If a lift of 100 kg is needed, an equal force of 100 kg must be exerted. A lift of 1 m high needs to be pulled downwards 1 m.

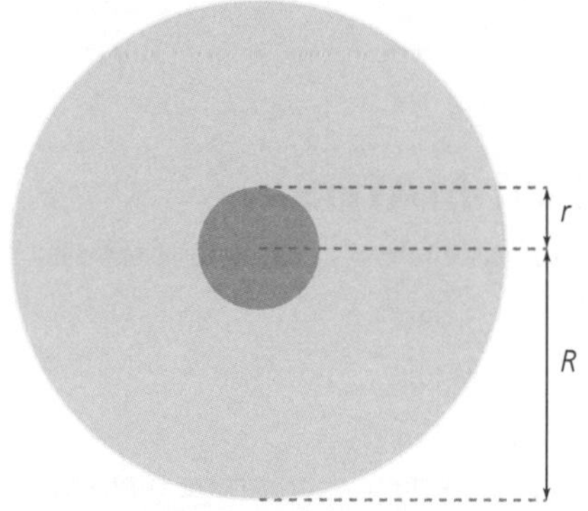

▲ Figure 3.26 The pulley wheel

If more ropes and wheels are added, the effort needed to lift the weight is reduced. The 100 kg weight is now supported by two ropes instead of one, so the lift effort is halved. This gives a positive mechanical advantage. The bigger the mechanical advantage, the less force is needed.

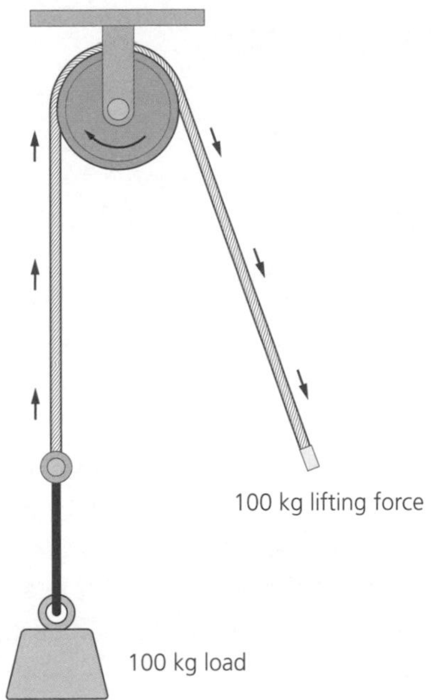

▲ Figure 3.27 Single pulley system

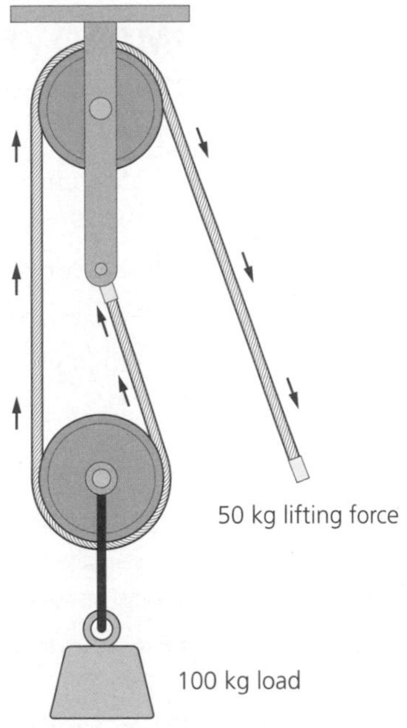

▲ Figure 3.28 Two pulley system

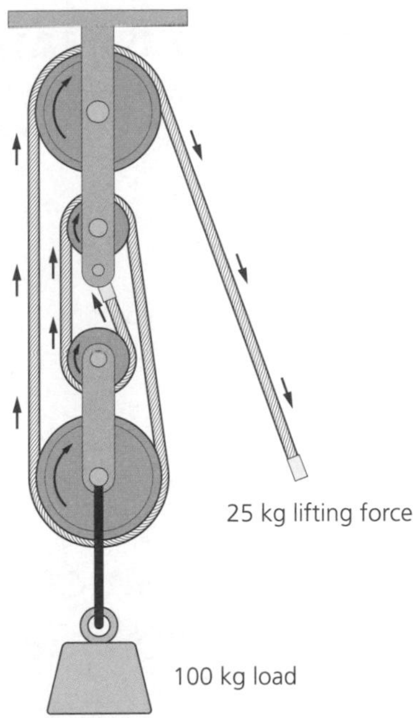

▲ Figure 3.29 Four pulley system

> **INDUSTRY TIP**
>
> In a pulley system, the load to be lifted can be divided by the number of wheels to get the lifting force, for example, a 100 kg load divided by four wheels gives a 25 kg lifting force.

If four wheels are used and held together by a long rope or chain that loops over them, the 100 kg weight is now supported by four ropes, which means that each rope is supporting a quarter of the total 100 kg weight, or 25 kg. This means that only a quarter of the force (25 kg) is needed to lift the weight (100 kg). This system is known as a block and tackle.

Screws

In terms of simple machines, a screw is a machine that converts rotation into a straight-line motion that can be placed vertically, horizontally or at an angle. It is basically a cylinder or wedge with an incline plane wrapped around it. It was originally designed as a simple water pump (the Archimedes screw), a task for which it is still used today. It can be found in many objects, such as screw fixings, bolts and threads on pipe. It can also be seen on drills and auger bits, and as a means of moving solid fuel, such as coal, towards a boiler by its rotary motion.

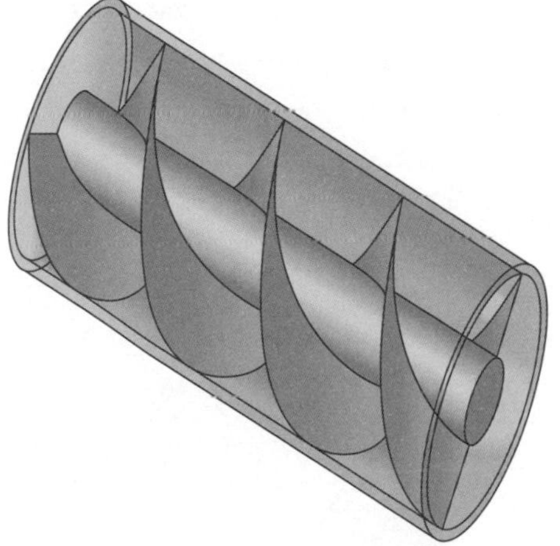
▲ Figure 3.30 The Archimedes screw

▲ Figure 3.31 The Archimedes screw in action as a water lifter

IMPROVE YOUR MATHS

The following formula is used to calculate the mechanical advantage of a screw:

$$MA = \frac{\pi \times D}{L}$$

Where:

MA = mechanical advantage

π = 3.142

D = diameter

L = length

Basic mechanics: moments of a force (torque)

In physics, the moment of a force is the measure of the turning effect (or torque) produced by a force acting on a body. It is equal to the applied force and the perpendicular distance from its line of action to the pivot, about which the body turns. The turning force around the pivot is called the moment. Its unit of measurement is the newton.

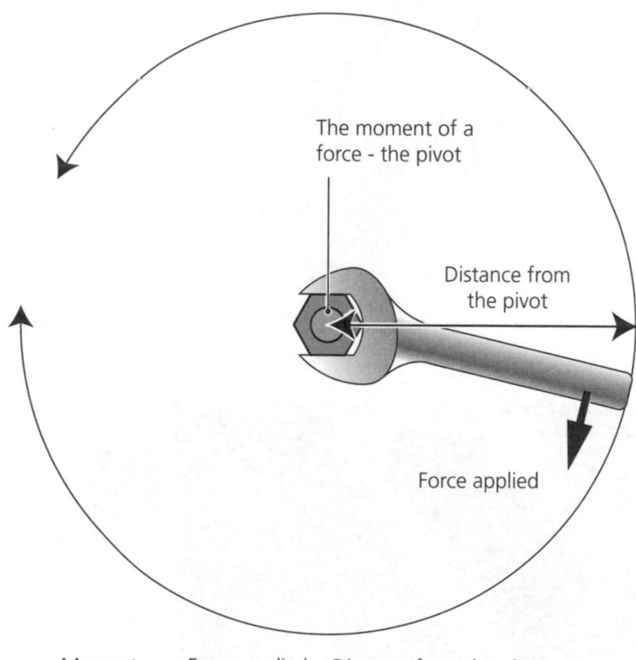

Moment = Force applied × Distance from the pivot
= Newtons

▲ Figure 3.32 The moment of force

IMPROVE YOUR MATHS

The moment of a force can be worked out using the formula:

Moment = force applied × perpendicular distance from the pivot

If the magnitude of the force is F and the perpendicular distance is d, then:

$$\text{Moment} = F \times d$$

An example of this would be a spanner turning a bolt. It is much easier to turn the bolt using a long spanner than it is using a short spanner. This is because more torque (turning force) can be applied at the bolt (pivot) for less effort. A long spanner is an example of a force multiplier.

Centre of gravity

In physics, the centre of gravity of an object is the imaginary point where all of the weight of the object is concentrated. This concept is especially important when designing large structures such as multi-storey buildings and bridges, or making a prediction of the gravitational effect on a moving object or body. Another term for it is the 'centre of mass'.

The centre of gravity will vary from object to object. In symmetrically shaped objects, it will coincide with the geometric centre.

In irregularly (asymmetrically) shaped objects, the centre of gravity may be some distance away from the centre of the object; in hollow objects, such as a ball, it may be in free space, away from the object's physical form.

KEY POINT

For many solid objects, the location of the geometric centre follows the object's symmetry. For example, the geometric centre of a cube is the point of intersection of the cube's diagonals.

Action and reaction: Newton's third law of motion

A push or a pull (action) on an object can often result in movement (reaction) when the pull or push is greater than the weight of the object. If both action and reaction are equal, then no movement takes place because the object is pushing or pulling against the action with equal force. This is known as contact force and is a result of contact interactions (normal, frictional, tensional, and applied forces are all examples of contact forces). Other forces are a result of 'actions-at-a-glance' interactions (gravitational pull, electrical and magnetic). These two types of force have one thing in common: for every force applied there is an equal opposing force and as such is subject to **action** and **reaction**.

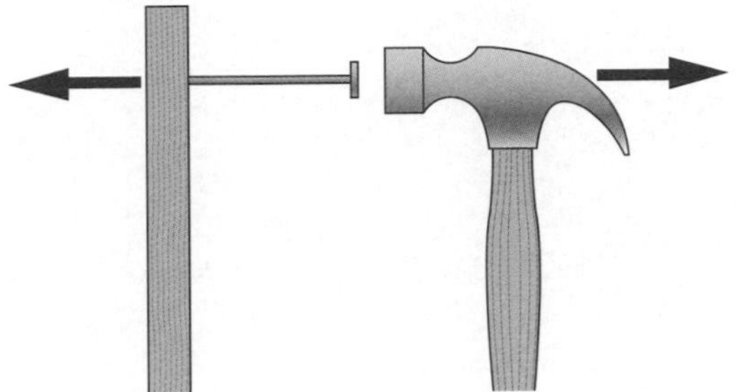

▲ Figure 3.33 Action and reaction

There are many ways in which this can be seen. For example, when a person sits on a chair (action), the downward force of the person provokes an upward force in the chair (reaction). The person and the chair have equal force and so equilibrium exists. If the person were too heavy for the chair, then the chair would collapse (reaction).

This is Newton's third law of motion, which states:

> Every action has an equal but opposite reaction.

This means that, for every force that an object is subjected to:

1 there is an opposing force from the object
2 both action and reaction forces are equal
3 forces always come in pairs (points 1 and 2).

Equilibrium

When all the forces acting on a stationary object are balanced, the object is said to be in a state of equilibrium. The forces are balanced when all forces (left, right, front, back, up and down) are the same. In Figure 3.34 (top), all forces are 50 N and are therefore equal forces in equilibrium.

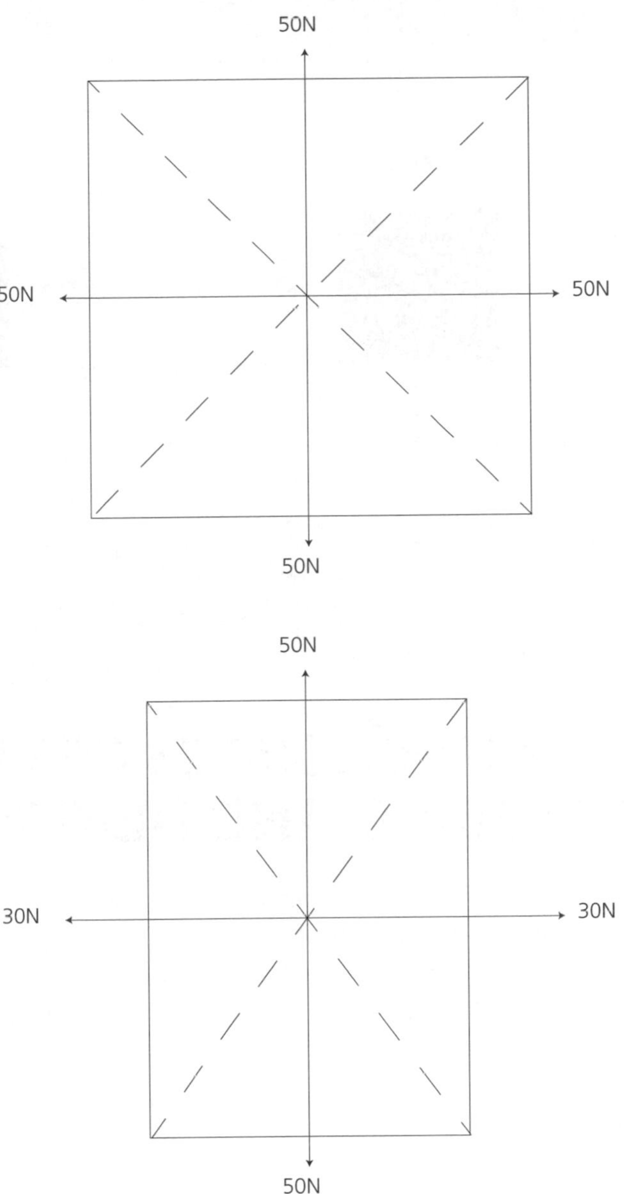

▲ Figure 3.34 Balanced forces in equilibrium (top) and unbalanced forces in equilibrium (bottom)

The same can apply for unequal forces. They, too, can be in a state of equilibrium provided left and right forces are equal but not necessarily the same as the equal up and down forces.

The key word here is balanced. All forces, whether equal or not, must be balanced. The forces cancel each other out and so add up to zero. In other words, for an object to be in equilibrium, the sum of the forces on each part of the system must be zero. Look at Figure 3.35.

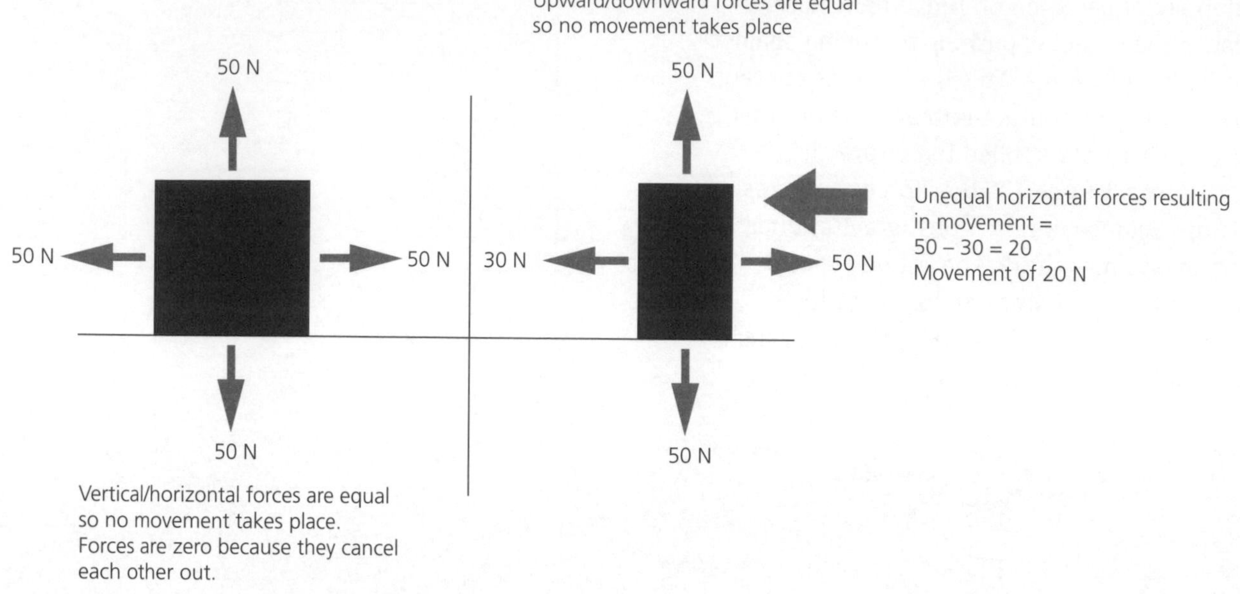

▲ Figure 3.35 Forces acting on an object

6 THE PRINCIPLES OF ELECTRICITY IN THE PLUMBING AND HEATING INDUSTRY

Electricity is a vital part of everyday life. It powers lighting, household appliances and heating systems, but its danger cannot be overstated. We cannot see it, hear it or smell it, yet if we touch it, it can kill. Because of the obvious dangers, it is necessary for us to have a better understanding of what electricity is and how it works.

You can find out about electricity, its scientific principles, basic circuitry and dangers, in Chapter 11.

SUMMARY

In this chapter, we have seen how even simple actions, such as hammering a nail or using a screwdriver, have a scientific explanation. The actions we perform and the materials we use employ the laws of physics and chemistry to useful effect that allow us to install systems of plumbing safely and professionally. We have also investigated the limitations of some materials and how we must always be aware of what we are using and how we use it, if problems of corrosion and poor workmanship are to be avoided. These are points that will become clearer as we move forward through the following chapters of this book.

Test your knowledge

1 A cold water storage cistern contains 250 litres when filled to its working capacity. If the base of the cistern measures 600 mm by 1200 mm, what pressure will be exerted by the cistern on the base beneath it?

a 1765 N/m^2

b 2542 N/m^2

c 3406 N/m^2

d 3740 N/m^2

2 Which of the following measurements is a base SI unit?

a kg

b m/s

c m^2

d l/s

3 Which of the following components is manufactured from a pure metal?

a 22 mm end feed elbow

b Double panel, single convector, welded seam radiator

c Lead-free solder

d 15 mm type A compression coupling

4 Which of the following statements describes the ductility of a material?

a Its ability to be stretched without breaking

b Its ability to conduct heat

c Its ability to resist atmospheric corrosion

d Its ability to return to its original shape once released from tension

5 If 0.150 m^3 of water were heated beyond its boiling point, what volume of steam would be produced?

a 0.150 m^3

b 150 m^3

c 0.240 m^3

d 240 m^3

6 What is the unit of heat?

a Joule

b Watt

c Kelvin

d Ampere

7 What size of earthing cable should be used to bond gas and water mains pipework?

a 4 mm^2

b 6 mm^2

c 8 mm^2

d 10 mm^2

8 When different metals are present together in a system of pipework, which of the following types of corrosion is likely to occur?

a Erosion

b Oxidic

c Atmospheric

d Electrolytic

9 In an old hot water system that does not use a pump, how is heat transferred from the boiler to the cylinder?

a Conduction

b Convection

c Impulse

d Radiation

10 How does water grip the side of a beaker or pipe?

a Adhesion

b Gravity

c Cohesion

d Convection

11 Which of the following materials is used because it is a good conductor of heat?

a Copper

b Steel

c Lead

d Plastic

12 What is the relative density of water?

a 1

b 9.81

c 10

d 3.14

13 At which of the following temperatures does water change state (latent heat)?

a 65 degree Celsius

b 100 degree Celsius

c 4 degree Celsius

d 10 degree Celsius

14 At what temperature is water at its most dense?

a 100 degree Celsius

b 0 degree Celsius

c 4 degree Celsius

d 65 degree Celsius

15

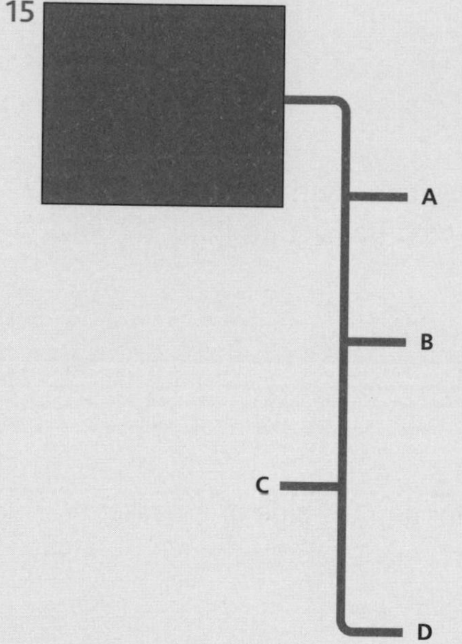

Which of the tappings will have the highest pressure?

a A

b B

c C

d D

16

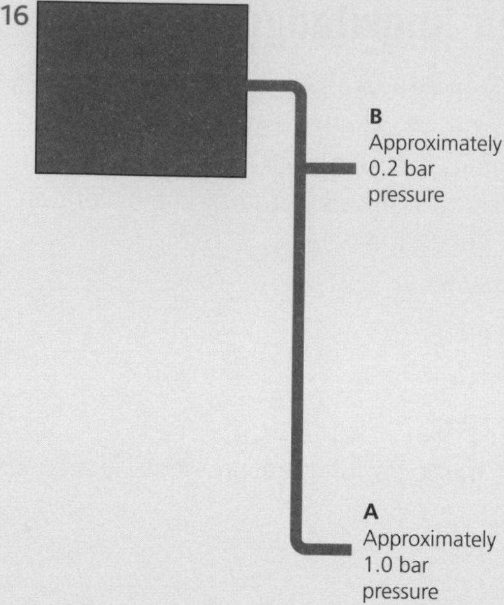

On the system shown in the diagram, the distance from the cold water storage cistern to outlet A is 10.0 m which gives approximately 1 bar pressure. Approximately how far below the cistern has outlet B been installed?

a 2 cm

b 20 m

c 2 m

d 200 mm

17 What property of water is identified by using the pH scale?

a Heat conductivity

b Acidity/alkalinity

c Electrical conductivity

d Specific heat capacity

18 How would you test a piece of material for its tensile strength?

a Squashing

b Hitting

c Pulling

d Twisting

19 What is convection the result of?

a Heat travelling through or along a solid

b Differences in the density of a solid

c Heat travelling through the air to a solid

d Difference in the density of liquids and gasses

20 What SI unit measures mass?

a Kilograms
b Kelvins
c Square centimetres
d Amperes

21 How can the freezing point of water be reduced?

a Adding insulation
b Adding glycol
c Adding inhibitor
d Adding alkalinity

22 A 150-litre cylinder is to be installed within an airing cupboard. The cylinder will require its temperature raising from 10°C to 60°C. Calculate the amount of heat energy required.

23 If a cylinder were installed at a height of 3.5 m above the level of the lowest hot water outlet, what head pressure will be provided at the outlet?

24 Copy and complete the table below.

Measure of:	SI unit	Symbol
Area		
Volume		
Velocity		
Density		

25 Calculate the thermal expansion in a 3 m length of copper pipe if its temperature were increased from 10°C to 80°C. Take the coefficient of linear of copper expansion as 0.000016 m/m/°C.

26 Give two examples of class 2 levers.

27 Explain by how much cold water will expand when heated up in a hot water cylinder, and where this expansion is taken up in various systems.

28 When a boiler is installed on a kitchen wall it may be secured in position by coach bolts. Explain what forces and material strengths are required to keep the boiler in position.

29 Outline what happens in the dezincification process.

30 Briefly state what two properties water has that can cause a resistance to the flow of water inside pipework.

31 State the relationship of Celsius and kelvin in temperature measurement.

Answers can be found online at www.hoddereducation.co.uk/construction.

CHAPTER 4

PLANNING AND SUPERVISION

INTRODUCTION

The plumbing services industry encompasses a vast number of roles, each associated with specific skills and responsibilities. It is important to understand each role in order to oversee building services work. You should also understand the reasons for risk assessments and method statements, and how to plan work programmes for work tasks in the building services industry.

By the end of this chapter, you will have knowledge and understanding of the following:

- the role of the construction team within the plumbing and heating industry
- information sources in the building services industry
- communicating with others
- the responsibilities of relevant people in the building services industry
- work programmes in the plumbing and heating industry
- risk assessments and method statements for the plumbing and heating industry.

1 THE ROLE OF THE CONSTRUCTION TEAM WITHIN THE PLUMBING AND HEATING INDUSTRY

The construction of any building is a complex process that requires a group of professionals, known as the construction team, working together to produce what the client has requested. In this first section of the chapter, we will take a closer look at the construction team. We will consider the role that each individual has in the overall construction project and their responsibilities within the management structure.

The structure of the site management team

Within each construction project, there is a site management team. This usually follows a recognised structure by which the team operates and communicates. This is illustrated in Figure 4.1.

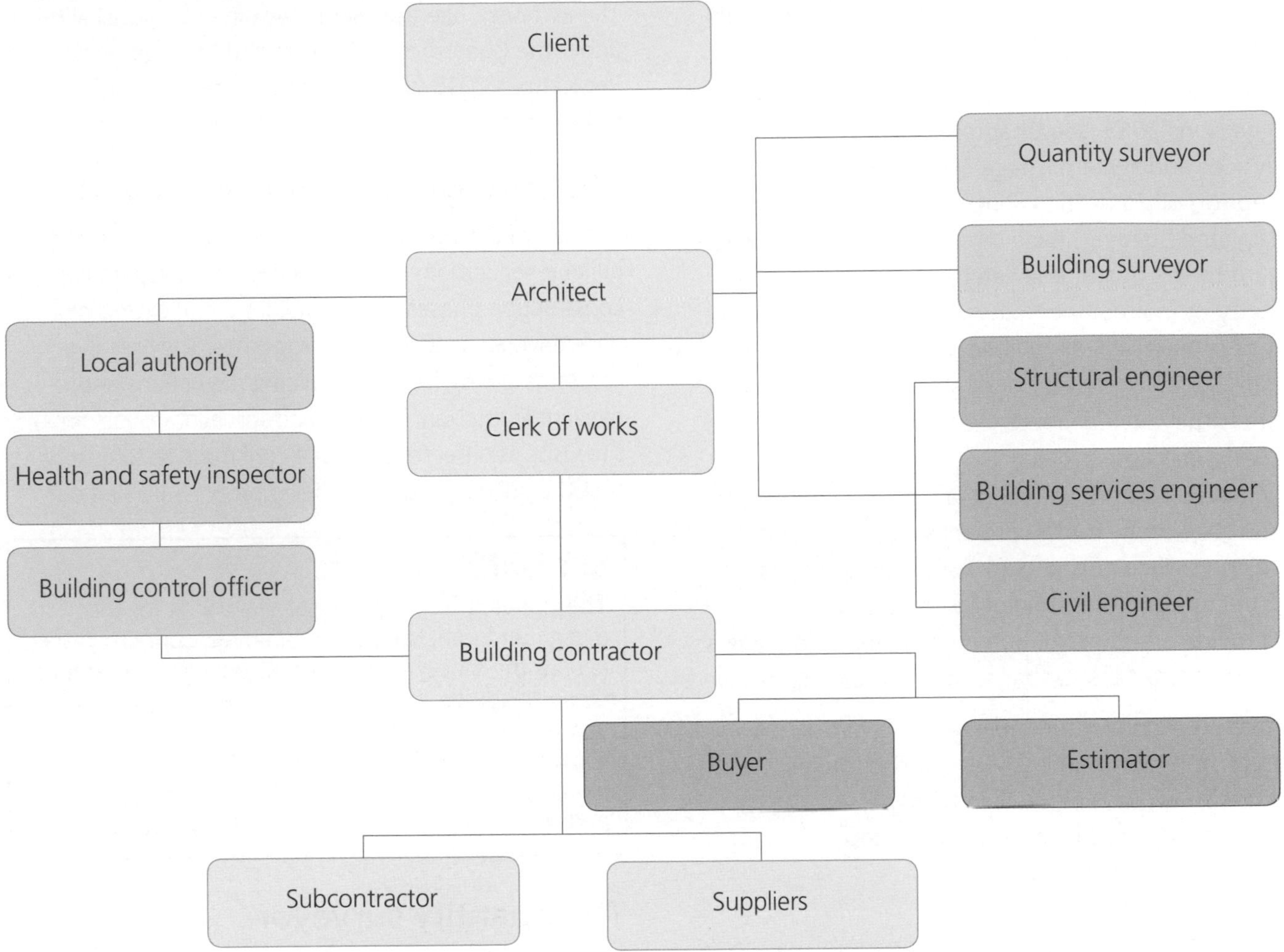

▲ Figure 4.1 The structure of the site management team

It is important that all members know their roles and responsibilities within the management structure to ensure the smooth running of the project and that any problems are dealt with as quickly as possible.

> **ACTIVITY**
>
> Do you know the management structure of the company you work for?
>
> Using the management structure diagram in Figure 4.1 as a guide, draw a chart of your own company.

The key roles of the site management team

The management of construction projects requires a good understanding of modern management systems as well as expert knowledge of the design and construction process. Construction projects have a specific set of objectives, which must be completed within a given time frame and on budget to a specific set of rules and regulations.

The management of any large construction site usually falls into two tiers:

1. those that visit the site only occasionally, usually senior management
2. those that are permanently site based.

In this section we will look at the first tier.

The client

The client is arguably the most important part of the project because they are the reason for the construction of the building. They, either directly or indirectly, employ everyone else who has a connection with the construction project because, without them, the work would not exist. They finance the whole project.

The client can vary from a single individual to a large consortium or organisation.

Under the Construction (Design and Management) Regulations 2015 (see Chapter 1, Health and safety practices and systems, page 10), clients (with the exception of domestic clients who intend to live in the completed building) have direct responsibilities with regard to the health and safety of all those people directly or indirectly employed as part of the project. On all projects clients will need to:

- ensure the competence of all team members, and that they are adequately resourced and appointed early on in the project
- ensure there are suitable management arrangements for the project's welfare facilities
- allow sufficient time and resources at every stage of the project from concept to completion
- provide pre-construction information to designers and contractors so that regulations can be followed.

Where projects are notifiable under Construction (Design and Management) Regulations (projects lasting more than 30 days or involving 500 person-days of construction work), clients must also:

- appoint a principal designer
- appoint a principal contractor
- make sure that construction work does not start unless a construction phase plan is in place, and that there are adequate welfare facilities on-site
- provide information relating to the health and safety file to the construction design and management (CDM) co-ordinator
- keep the health and safety file and provide access to it if required.

The architect

The architect (or designer) is considered to be the leader of the management team. It is their responsibility to transform the client's requirements into a building design and working drawings. Architects generally supervise all aspects of the construction work until handover to the client. They must be registered with the Architects Registration Board (ARB), whose duties and functions are defined by the Architects Act 1997. This was established to regulate the architect profession in the UK. Many architects are also members of the Royal Institution of British Architects (RIBA).

The architect, like the client, has direct responsibilities under the Construction (Design and Management) Regulations 2015 (discussed in Chapter 1, Health and safety practices and systems, page 10).

The surveyor (building surveyor)

The role of the building surveyor is to ensure that the building regulations are followed during the planning and construction phases of new buildings and extensions, and conversions to existing properties. They resolve problems arising from the building regulations and relevant legislation. The building surveyor will also make site visits at different stages of construction to ensure that the building process is being properly carried out.

KEY POINT

The Building Regulations set standards for the design and construction of buildings, primarily to ensure the safety and health of people in or around those buildings, but also for energy conservation and access to buildings. They are divided into 'Documents' or 'Parts' named after letters of the alphabet, such as Document L Conservation of Fuel and Power, and Document H Building Drainage.

The quantity surveyor

The quantity surveyor, or QS, is an accountant who advises as to how the building can be constructed within the client's finances. The QS also measures the amount of labour and materials needed to complete the building according to the architect's drawings. These details are then combined into a document called the Bill of Quantities, which is used by building contractors to produce an estimate.

As work progresses, the QS will produce measurements and variations of the work carried out to date so that the main contractor can receive interim payments. At the end of the contract, the QS will also prepare the final account to be presented to the client. In addition to these duties, the QS may also advise the architect on the cost of any variations to the original contract or any additional work completed.

Specialist engineers

These are hired as part of the architect's team to assist in the design of the building with regard to their specialist fields. There are three major engineering roles:

1 **Civil engineer:** the designer of the roads in to and out of the building, along with any bridges, tunnels etc. that may be required. May also be involved in the design of drainage and water requirements to the building or complex.
2 **Structural engineer:** works closely with the designer to find the most efficient method to construct the project. The engineer calculates the loads, taking into account wind, rain and the weight of the building itself. The frame and foundations can then be designed to support these loads.
3 **Building services engineer:** the designer of the internal services within the building, such as heating and ventilation, hot and cold water supplies, air conditioning and drainage. The building services engineer will produce calculations for heat loss through the building fabric, and take into account solar heat gain from windows and internal heat gains from plant, computers, lighting and people, so that accurate calculations can be made for thermal comfort within the structure. Most reputable engineers belong to the Chartered Institute of Building Services Engineers (CIBSE).

INDUSTRY TIP

Although their role increasingly demands a **multi-disciplinary approach**, building services engineers tend to specialise in one of the following areas:
- electrical engineering
- mechanical engineering
- public health.

KEY TERM

Multi-disciplinary approach: using skills from other professions or trades to overcome problems outside the normal scope of your skill set, trade or profession to reach satisfactory solutions, conclusions or outcomes.

The clerk of works

Appointed by the architect, the clerk of works (CoW) – also referred to as the project manager – is the architect's representative on-site. They ensure that the building is constructed in accordance with the drawings, while maintaining quality at all times. This includes checking the standard of the work and the quality of the materials. The CoW will make regular reports back to the architect as work progresses, and will also keep a diary in case of any disputes, make any necessary notes on the weather and note any stoppages.

On large sites, the CoW will be a resident member of the management team, while on smaller sites they will visit only periodically.

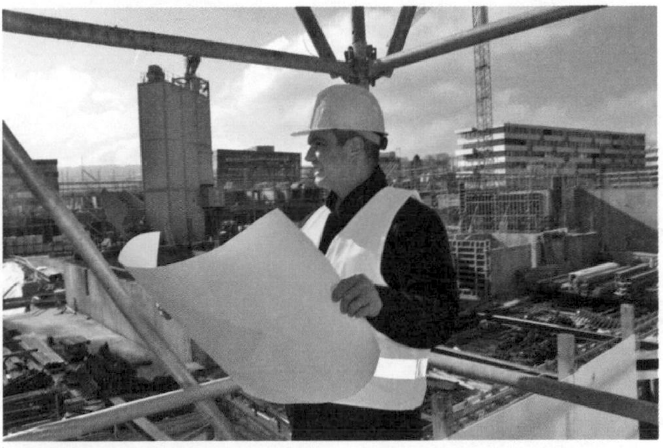

▲ Figure 4.2 The clerk of works

The local authority

The local authority has overall responsibility for ensuring that all works carried out conform to the requirements of the relevant planning and building regulations. They may also show interest in site health and safety in co-operation with the HSE. They employ the following people.

- Planning officer: they are responsible for processing planning applications, listed building consent applications, conservation area consent applications and advising on planning issues.
- Building control officer: responsible for ensuring that regulations connected with public health, safety, energy conservation and disabled access are met. They work to the Building Regulations. A building control officer's job involves:
  - checking plans and details of new constructions and alterations of existing buildings
  - regular inspections of work in progress to ensure that the construction work is in accordance with the Building Regulations
  - management of buildings and structures identified as being in a dangerous condition

- management of the demolition of derelict buildings
- management of improved access to buildings for people with disabilities
- guidance and advice on all types of buildings and constructional problems outside of Building Regulations control.

ACTIVITY

Who is the local authority in your area? Check out its website and see what services it offers to the construction industry.

The building contractor and their employees

In this section, we will examine the role of the building contractor and the members of the team directly employed by them.

The building contractor will enter into a contract with the client to carry out the work in accordance with the drawings, the Bill of Quantities and the specification. Every contractor develops their own methods of pricing and tendering for the work and, depending on the size of the job, this will determine the company's staff requirements.

The building contractor will employ specialists within the construction industry to undertake certain key roles. These include those listed in Table 4.1.

KEY TERM

Overheads: costs that include such things as site offices and staff salaries.

Members of the on-site team

So far we have looked at the roles and responsibilities of the site management team. Here, we consider the on-site workers who report to the site management.

Subcontractors

Subcontractors play an important role within the construction industry. Subcontractors will enter into a contract with the main building contractor for a specific or specialised part of the contract, such as plumbing, heating and ventilation, air conditioning, electrical installation, plastering, bricklaying and joinery/carpentry. The contract may be labour only, where the building contractor purchases the materials, or it may be on a supply-and-fix basis. The architect may specify a nominated subcontractor in the initial contract, who must then, with the client's permission, be used.

The site supervisor

Also known as the construction manager or project manager, they are the building contractor's main representative on-site, responsible for the general day-to-day running of the site. This can include preparing budgets, hiring team members, handling deliveries and overseeing construction duties.

▼ Table 4.1 Key specialist roles in the construction industry

The estimator	Breaks down the Bill of Quantities into labour, materials and plant, and applies a set payment rate for each one. This represents the amount it will cost the contractor to complete each stage of the project. Added to this will be a set percentage for **overheads** (site office costs and site/administration salaries) and profit.
The buyer	Responsible for sourcing and purchasing all the materials needed. They will obtain quotes for the materials in the quantities required, together with delivery times and quality assurances.
The planning engineer	Responsible for the pre-contract planning, and identifying the most economic and efficient way to use labour, plant and materials.
The plant manager	Responsible for all the items of mechanical plant used by the building contractor – from stock plant owned by the contractor or hire companies – to carry out a specific task. The plant manager is also responsible for maintenance and repair and the training of plant operators.
The safety officer	Accountable to the senior management for all health and safety aspects on-site (safety inspections, safety records, accident investigations, and safety training and inductions).
The contracts manager	Supervises the creation and management of planning and building operations contracts, liaising with head office staff and site agents as needed.

▲ Figure 4.3 On-site trades

The trade supervisor

Each of the different trades on-site will have its own supervisor. They will be responsible for the overall running of their company's contract on the site. Their tasks include:

- determining work requirements and the allocation of duties to the operatives under their direct control
- consulting with other managers to co-ordinate activities with other trades
- maintaining attendance records and rosters
- explaining and enforcing regulations
- overseeing the work of the workforce, and suggesting improvements and changes
- holding discussions with workers to resolve grievances
- perhaps performing the tasks of their trade.

The on-site trades

No construction site can function without the on-site trades. Working to the architect's drawings, it is the trades that build the architect's vision. The trades can be divided into two main groups:

1. craft operatives
2. building operatives.

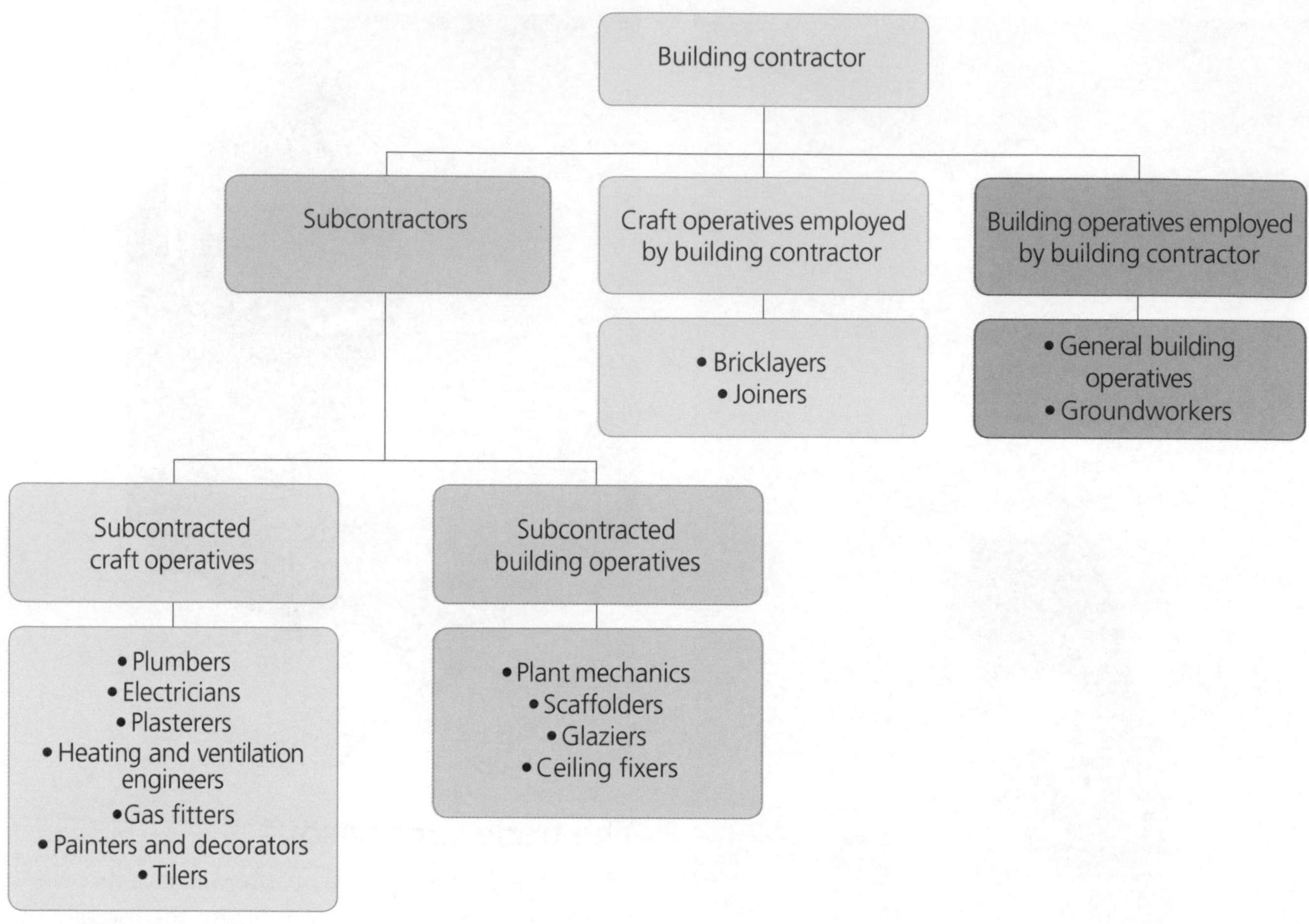

▲ Figure 4.4 Structure and roles of the building contractor and employees

Craft operatives

Craft operatives are skilled craftspeople who perform specialist tasks, such as those listed below.

- **Bricklayers:** construct the building to the architect's specifications using a range of building materials, including brick, block and stone.
- **Carpenters/joiners:** the wood trades provide a vital function on-site during the initial building phase, fitting door and window frames, floor joists and roof trusses. During the second phase they will fix internal doors, skirting boards, architraves, etc.
- **Plumbers:** on domestic construction sites, plumbers perform three key functions. They install:
  1. hot and cold water supplies
  2. central heating
  3. gas.

  On large construction sites, the plumber's work will be restricted to hot and cold water supplies only. In most cases, specialist companies will perform the gas and heating installations.
- **Electricians:** install and test all electrical installation work on-site, including power, lighting, fire and smoke alarms, and security systems, usually running the cables in trunking or conduits for neatness.
- **Heating and ventilation/air conditioning engineers:** this is a very specialist trade, especially where the installation of air conditioning is concerned. Their work mainly involves the installation of large diameter pipework for heating systems and air conditioning ductwork.
- **Gas fitters:** install natural gas lines in domestic properties and in commercial or industrial buildings. On some sites they may also install large appliances and pipelines.
- **Plasterers:** responsible for wall and ceiling finishing, dry lining and external rendering, if required, using a mixture of both modern and traditional techniques.
- **Painters and decorators:** responsible for wall and ceiling finishing, including painting skirting boards,

architraves and any specialist decorating such as murals, frescos, etc.

- **Tilers:** responsible for internal and external tiling of walls and floors, and specialist tiling such as swimming pools and wet rooms.

Building operatives

Building operatives are labourers who carry out practical tasks, such as those listed below.

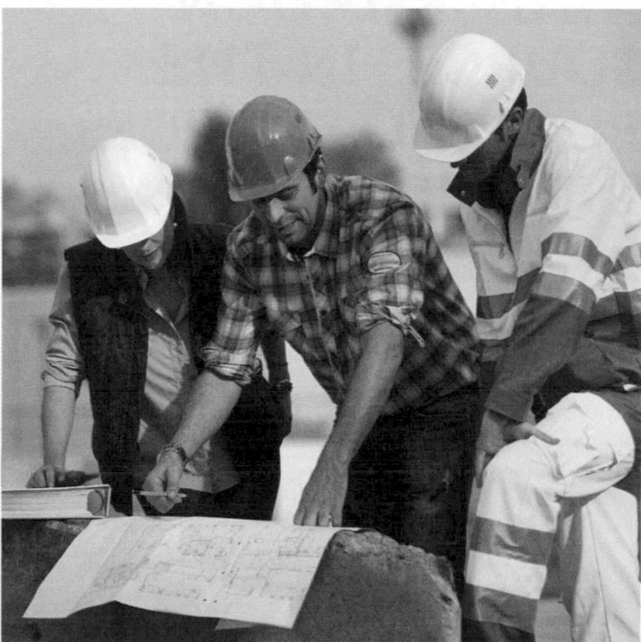

▲ Figure 4.5 Trades working together on-site

- **General building operatives and groundworkers:** usually mix concrete, lay drains, offload materials and generally assist the craft operatives
- **Specialist building operatives:** scaffolders, glaziers, suspended ceiling installers.

> **INDUSTRY TIP**
>
> Craft operatives, such as plumbers, electricians, joiners and bricklayers, have served a formal apprenticeship. This usually takes around three years to become fully qualified, with a formal City & Guilds (or equivalent) competency qualification being achieved. Specialist building operatives are often trained 'in-house' by the company that employs them, or they may have undergone formal training courses. These operatives quite often do not serve a formal apprenticeship.

The inspectors

There are other outside visitors to the construction site whose sole aim is health and safety. These are the inspectors. Their role is to check that we are complying with the rules and regulations, to ensure that the structure, the people who work in it and on it, and the services that the eventual occupiers will use, are safe and without risk.

There are four types of inspector:

1. health and safety inspector
2. building control inspector
3. water inspector
4. electrical services inspector.

The health and safety inspector

Also known as the 'factory inspector', the health and safety inspector usually works for the HSE, but can also be employed by the local authority. It is the inspector's duty to ensure that all health and safety law is fully implemented by the building contractor (this is covered in Chapter 1, Health and safety practices and systems, page 15).

The building control inspector

The building control inspector (now more generally known as the building control surveyor) works for the local authority and makes sure that each of the Building Regulations documents is observed in the planning and construction stages of new buildings. The Building Regulations are the statutory rules by which buildings are constructed, covering aspects such as drainage, energy efficiency, disabled access, etc.

Building control surveyors need to know the Building Regulations and how to interpret them accurately as they have the power to reject plans that fail to meet the Regulations. They may also have to use their professional judgement and skill to offer advice on acceptable solutions to meet statutory requirements should any problems arise. They will make site visits at different stages of construction to ensure that all construction work is being properly carried out.

The water inspector

Water inspectors are employed by the local water undertaker. The key objective of the water inspector's role is to reduce the risk of contamination of the public water supply from backflow of any fluid. They provide advice and guidance on regulation compliance in new and existing premises. The water inspector enforces the Water Regulations by inspecting a range of plumbing installations, as follows.

- Hands-on inspections:
  - in a percentage of new domestic premises
  - in all new non-domestic premises/connections
  - targeted inspections based on potential risk in existing premises.
- Reactive inspections:
  - requests to inspect due to water quality problems
  - requests from customers for advice and resolution of plumbing problems with old or new systems.

The electrical services inspector

Electrical inspections must be made on all new electrical installations, but more especially on commercial/industrial properties. They are undertaken by the local electrical supply company but, because these are now privately owned, the electrical supply companies usually employ private subcontractors to inspect the installations and issue test certificates on their behalf. The fees for these services are paid by the customer.

Site visitors

Construction sites occasionally get visits from people with little or no construction site experience. To many, construction sites are dangerous places with many different activities happening at once. To the experienced person, these activities seem perfectly normal, but to the uninitiated, construction sites can be confusing, noisy and daunting. Generally, there are three types of visitor to construction sites:

1 the frequent visitor with no construction site skills
2 the inexperienced visitor, including the general public
3 the experienced visitor, such as delivery drivers.

All visitors, regardless of the reasons for their visit, must follow the same rules as all other construction workers. They must:

- check in at the appropriate place, usually the general site office; often it is a requirement to sign in the visitor's book and wear a visitor's ID; visitors must also sign out again when leaving the site
- undergo a site health and safety induction
- wear the proper attire, such as hard hats, eye protection, high-visibility vest, hard-soled shoes (no high heels, sandals, sports shoes or open-toed shoes; no shorts or sleeveless tops); construction sites are often damp, dusty and dirty places, and the clothing should reflect this.

2 INFORMATION SOURCES IN THE BUILDING SERVICES INDUSTRY

Documentation on-site

No construction site can function without certain documents and a certain amount of day-to-day paperwork. Each of these documents has an important function:

- **Job specification:** a description of the installation that is being quoted for, complete with the types of materials and appliances that the installation must contain. Occasionally, it may specify the manufacturer or British Standard of the materials the installation is to use.
- **Working drawings (also known as building services drawings):** all plans, elevations and details needed by the contractor, along with the specifications, so that an estimate can be obtained and then the building can be constructed. These need to show all dimensions and be properly scaled.
- **Work programme:** another name for a work programme is a Gantt chart, and it has proved to be an excellent method of communication. Its purpose is to:
  - establish dates for work to start and finish
  - illustrate the labour and plant required for the duration of the contract
  - show the order of operations
  - provide information for monitoring work progress.

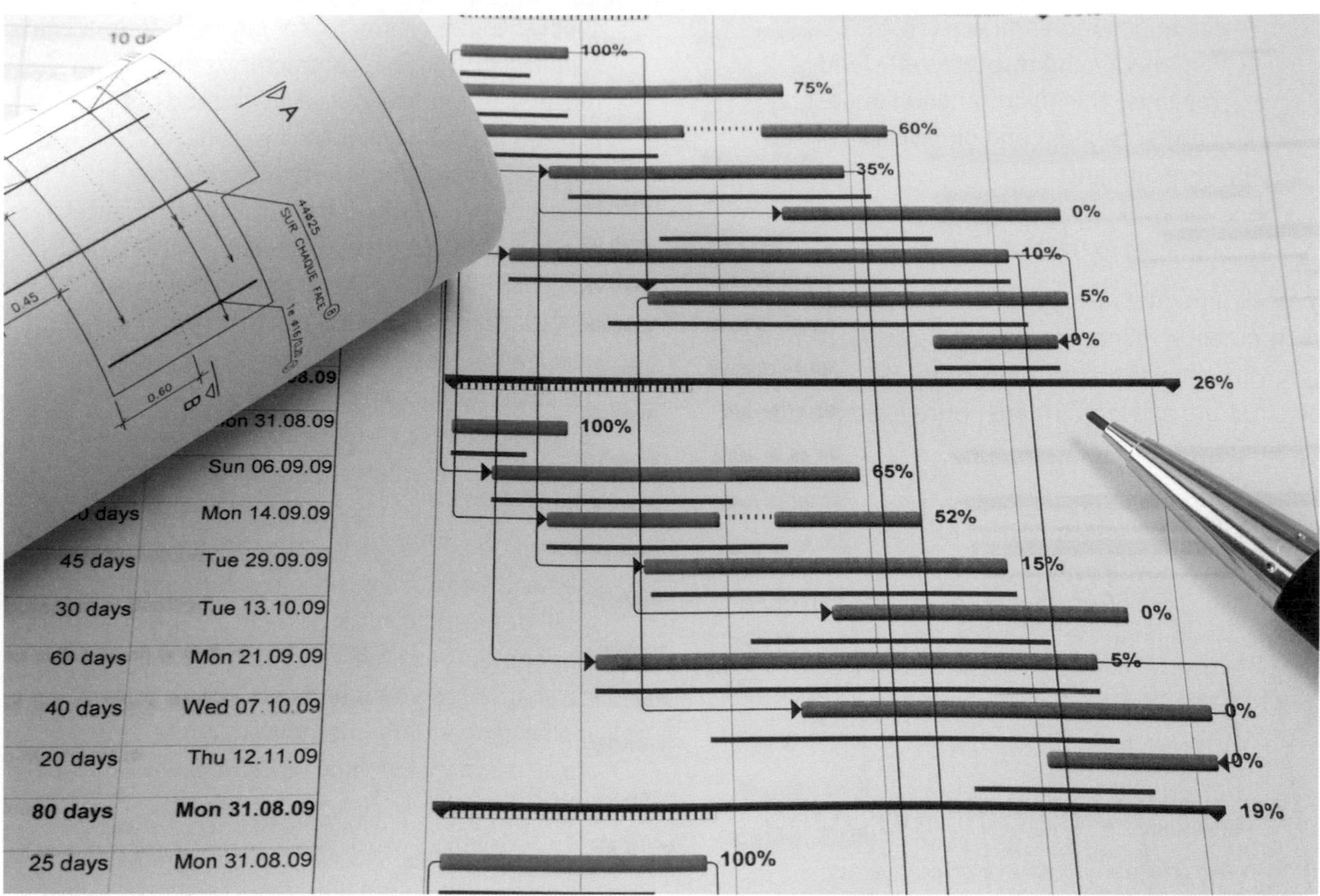

▲ Figure 4.6 Example of a Gantt chart

- **Delivery note:** also known as delivery advice note, this is a document that lists the type and amount of materials that are delivered to the site. It should be checked against the actual materials delivered and should be signed only if the materials on the note and the materials delivered are the same. A copy should be retained for administration purposes.
- **Time sheets:** these are completed by each employee on a weekly basis, on which they give details of hours worked and a description of the jobs they carried out. Time sheets are used by employers to calculate wages and provide information for planning future estimates. They are sometimes completed by the trade foreman.
- **Policy documents:** these include those listed below.
  - **Health and safety file:** a document held by the client in which health and safety information is recorded and kept for future use at the end of a construction project. It is a legal requirement of the CDM Regulations 2015. The type of information contained in the health and safety file is designed to help those in positions of responsibility to identify key health and safety risks that may be encountered on-site, and provide operating and maintenance manuals for the building and any equipment installed.
  - **Customer care charter:** also known as the customer service document. Good customer care makes for loyal customers, and loyal customers are a good source of positive advertising.
  - **Environmental policies:** an environmental policy can be described as a statement of intent to manage human activities to prevent, reduce or remove any harmful effects on the environment and the Earth's natural resources, while ensuring that any man-made changes to the environment do not have any harmful effects on humans. Environmental statements often make commitments to:

- lower pollution and waste
- use energy and resources efficiently
- minimise the environmental impact on natural habitats and biodiversity of new developments
- minimise the environmental impact of raw material extraction.

An environmental policy is implemented through an environmental management system (EMS). Writing an environmental policy is voluntary in the UK, and the structure and content are not regulated under UK legislation.

ACTIVITY

Think about how long it would take you to install a bathroom suite. Break the job down into days and what you do on those days (i.e. day 1 – first fix; day 2 – dress sanitary ware), then produce a simple work programme for the job.

VALUES AND BEHAVIOURS

Does the company you work for have an environmental policy? What does it contain? How does it affect the way that you work as a plumber? Find out what measures your company is taking to protect the environment.

Customer information

Communication between the company and the customer takes place at every stage of the contract, from the initial contact to customer care at the contract's completion. Written communication can take the following forms:

- **Quotations and estimates:** both of these are written prices as to how much the work will cost to complete. A quotation is a fixed price and cannot vary. An estimate, by comparison, is not a fixed price but can go up or down if the estimate was not accurate or the work was completed ahead of schedule. Most contractors opt for estimates because of this flexibility.
- **Invoices/statements:** documents that are issued at the end of any contract as a demand for final payment. Invoices and statements can be from the supplier to the contractor for payment for materials supplied, or from the contractor to the customer for services rendered. Usually, a period of time is allowed for the payment to be made.
- **Statutory cancellation rights:** a number of laws give the customer the legal right to cancel a contract after they have signed it. There is usually no penalty for cancellation provided that the cancellation is confirmed in writing within a specific time frame. Most cancellation periods start when the customer receives notification of their right to cancel up to seven days before work commences.
- **Handover information:** at the end of any contract, the customer must be given certain information. For large contracts, this includes the health and safety file already mentioned. For small domestic contracts, a file should be made that contains any manufacturer's information, installation, servicing and user instructions, the appliance warranty information, contact numbers of key personnel within the company, and a letter of thanks for their custom.

During the handover process, the customer should be shown where all control valves are, and how to use any appliances and controls that have been installed.

As with all data that contains private customer information, caution should be exercised to protect this information, as dictated in the General Data Protection Regulation 2018.

Company policies and procedures

Company working policies/procedures highlight what is important for the company and link this to its daily operations. Well-written policies and procedures allow employees to understand their roles and responsibilities, and management to guide operations without needing to constantly intervene because employees know what is expected of them.

Companies may have policies and procedures relating to the following aspects:

- **Behaviour:** companies and organisations demand a certain behaviour and professionalism towards their customers and management. Customers demand a certain respect, efficiency and professional attitude towards the work and themselves.

▲ Figure 4.7 Plumber presenting a professional image

- **Timekeeping:** customers do not expect and will not tolerate lateness, unless it is unavoidable. If lateness cannot be helped, then the customer should be informed at the earliest opportunity.
- **Dress code:** a company uniform or dress code presents a positive, professional image that the customer comes to recognise. Many companies and organisations have a set company uniform that is expected to be kept in reasonable order. Some companies have a laundry policy, where uniforms or company work wear are cleaned free of charge.
- **Contract of employment:** a contract of employment is a mutual agreement between the employer and the employee, which is the basis of the employment relationship. A contract is made when an offer of employment is accepted.

VALUES AND BEHAVIOURS

Remember, presenting a professional image is key to attracting and retaining customers, who will associate a uniform and well-organised work site with a skilled and competent plumber.

Limits to personal authority

As with most trades, plumbing follows a set pattern with regard to the roles and responsibilities of the qualified operatives. Each member of the team will have certain expectations placed on them by the management of the company or organisation. It follows, therefore, that the higher the qualification, the more responsibility will be given, as described in Table 4.2.

▼ Table 4.2 Limits of authority by professional level

Apprentice plumber	Has very little responsibility with regard to plumbing installations. Initially under constant supervision from the plumber they work with. As they gain experience they may work on simple installations and maintenance tasks. Responsible for maintaining the company image with regard to timekeeping, appearance and customer care. Their main task is learning their trade to the best of their ability.
Trained plumber	Domestic plumbers qualified up to NVQ Diploma Level 2 are able to install 'non-complex' hot and cold water systems, as well as domestic sanitation pipework and basic central heating pipework, only under regular supervision. They may also have some responsibility for improving business products and services.
Advanced plumber	Domestic plumbers qualified up to NVQ Diploma Level 3 have much more responsibility than those at Level 2. At Level 3 they will be gas qualified and may be included on the company Gas Safe registration. They are capable of running their own jobs, taking responsibility for domestic hot and cold water, and domestic heating installations, and working on their own initiative without supervision. They will be able to undertake unvented hot water installations and work to the Water, Gas and Building Regulations. They may also have responsibility for improving business products and services, and initiating some basic system design.
Plumbing supervisor	Plumbing supervisors will have many years' experience. They are capable of design and installation across a broad spectrum of systems, and have knowledge of the Regulations and British Standards. They will have good managerial and organisational skills, and will hold at least a Level 3 in Plumbing and sometimes a Level 5 qualification, e.g. an HNC in Building Services Engineering. They will also have responsibility for improving business products and services, and overall responsibility for the operatives and installations under their supervision.

Legislation and guidance information

There are many sources of information and forms of legislation that your employer (and you, the employee) must be aware of and follow. (All relevant health and safety legislation is covered in detail in Chapter 1, Health and safety practices and systems). The main legislation, regulations and guidance are summarised below.

The Equality Act 2010

The Equality Act came into force in October 2010. This Act provides a single legal framework with clear, simplified law in order to be more effective at tackling disadvantage and discrimination. It was implemented by the Equality and Human Rights Commission (EHRC). The EHRC was formed on 1 October 2007 and has responsibility for the promotion and enforcement of equality and non-discrimination laws in England, Scotland and Wales.

This Act brought together a range of previous laws covered in the Sex Discrimination Act 1975, Equal Pay Act 1970, Race Relations Act 1976 and Disability Discrimination Act 1995.

It protects people from discrimination in employment due to their race, sex, gender, age, sexual orientation, disability, marital status, maternity status or faith.

The Data Protection Act 2018

The Data Protection Act was updated in 2018 and sets out rules to ensure that everyone has the right to know what information is held about them, and that information is handled properly. The Act implements the General Data Protection Regulation into UK law. It is expected that the GDPR will continue to apply after the UK leaves the EU.

The Freedom of Information Act 2000

This Act gives you the right to ask any public body for all the information they have on any subject you choose. Unless there's a good reason, they have to provide it within a month. You can also ask for all the personal information they hold about you.

Regulations

Plumbing is one of the most regulated trades within building services engineering. Failure to comply with regulation often results in prosecution. Regulations in the plumbing industry include:

- Water Supply (Water Fittings) Regulations
- Gas Safety (Installation and Use) Regulations
- Building Regulations.

These are the main regulations that workers in the plumbing industry must comply with.

British Standards and approved codes of practice

These provide guidance on interpreting and following regulations. The British Standards are not enforceable, but they set out a series of recommendations so that the minimum standard to comply with the regulations can be achieved. By adhering to the recommendations within the British Standards, the regulations will be seen to be satisfied. Often the regulations and the British Standards will make reference to one another and it may even be the case that the regulations make reference to more than one British Standard.

However important the regulations and the British Standards are, they are not our primary source of information when installing equipment and appliances. Manufacturers' guidance overrides both of these.

Manufacturers' guidance

Manufacturers' installation, servicing/maintenance and user instructions are the most important documents you will have access to when installing, servicing and maintaining equipment and appliances. They tell us in basic installation language what we must do for correct and safe operation of their equipment. This guidance must be followed, otherwise:

- the terms of the warranty will be void
- the installation may be dangerous
- we may inadvertently be breaking the regulations.

In some instances, it may seem that the instructions contradict the regulations or the British Standards. This is because regulations are reviewed only periodically, whereas manufacturers are moving forward all the time with new, more efficient products, so their information may be more up to date. In these cases, follow a simple but effective rule: the manufacturers' guidance must be followed at all times.

3 COMMUNICATING WITH OTHERS

A company cannot function properly without proper methods of communication, whether a formal letter, an email, memo, fax or verbal instructions. Formal and informal communications take place in the workplace every day. Most people believe that formal communication is written communication but this is not the case. In a work context, communication in any form that is about your job should always be regarded as formal communication.

Methods of communication at work

There are several ways that companies communicate with customers, staff and suppliers, and other companies, such as:

- written communication (letters, email, faxes)
- verbal communication (face to face, telephone).

Verbal communication should always be backed up with written confirmation, to avoid confusion.

Written communication

Letters

Letters are an official method of communication and are usually easier to understand than verbal communication. Good written communication can help towards the success of any company by portraying a professional image and building goodwill.

Official company business should always be in written form, usually on the company's headed paper, and should have a clear layout. The content of the letter must be well written, using good English and correct grammar, and divided into logical paragraphs. Examples of business letters are sales letters, information letters, general enquiry or problem-solving letters, and so on.

Email

Emails have emerged as a hugely popular form of communication because of the speed that the information they contain is transferred to the recipient. As with letters, they should be well written and laid out, using correct grammar and spelling to convey professionalism, whether the recipient is a client, customer or colleague.

Faxes

Faxes are another useful form of communication for businesses. They are used mainly for conveying documents such as orders, invoices, statements and contracts, where the recipient may wish to see an authorising signature. Again, the basic rules apply with regard to layout, grammar and content. Remember to always use a cover page that is appropriate for your company. This is an external communication that reflects the business and company.

Verbal communication

The general rule of good, effective communication is that you should think beforehand about the kind of information you will need to give and what information

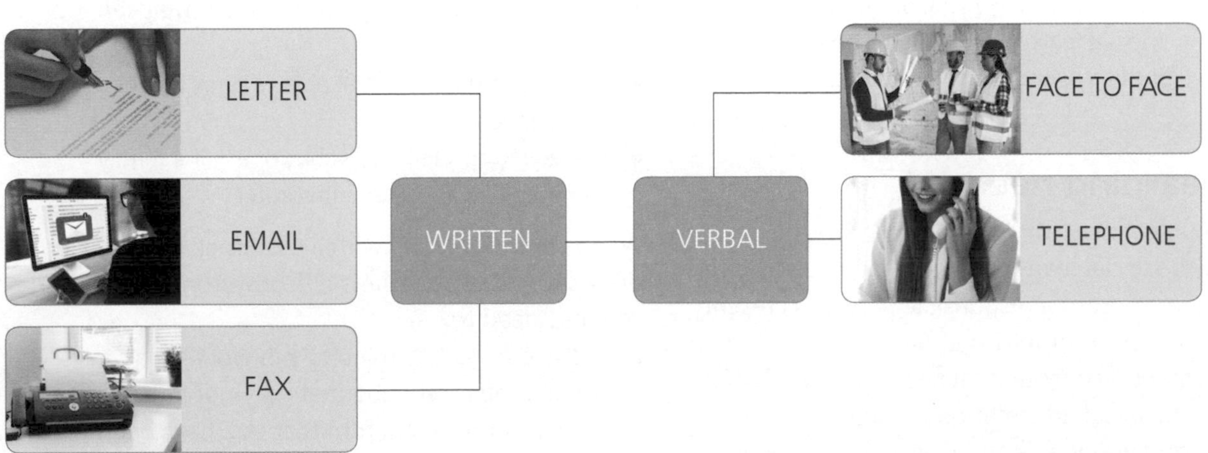

▲ Figure 4.8 Methods of communication

you will expect to receive. You should always make sure that your language, tone and body language are appropriate in terms of what you are saying and to whom you are saying it. Good communication is crucial if you are to carry out your job safely and efficiently, and you may need to adapt your communication skills to deal effectively with some individuals and groups. The principle behind effective communication is making sure that both parties completely understand each other. You may need to take into account the following factors.

Physical disabilities

When communicating with a customer, colleague or site visitor with a hearing impairment, you may need to:

- use written notes or drawings to reinforce verbal information
- use appropriate written information (such as a sales leaflet, manufacturer's literature or guides) to make sure that both you and the other person understand what is being referred to; if available and appropriate, use other means of technology (telephone amplifiers, etc.) to help communication.

When communicating with a customer, colleague or site visitor with a visual impairment, you may need to:

- give more verbal detail than you would usually use
- describe any diagrams or visual aids that you are using
- keep the person informed of his or her surroundings, e.g. who is present, who has left the room.

VALUES AND BEHAVIOURS

Remember: you should adjust your manner of communication to suit the individual needs of others, which may include a visual or hearing impairment, special learning needs, or those who do not have English as their first language.

Special learning needs

When communicating with a customer, colleague or site visitor with special learning needs, you may need to:

- if possible, make sure a responsible person is present to hear any important information
- keep information short and relevant, and avoid using too much technical information or jargon
- consider using visual aids and diagrams to back up information.

Language differences

Construction sites are often multicultural places, so you may be dealing with colleagues, clients and site visitors for whom English is not their first language, or who speak with a different **accent** or **dialect**.

KEY TERMS

Accent: the way in which people pronounce their words.

Dialect: a combination of the way people pronounce words, the vocabulary they use and the grammatical structures they use.

A person's accent and dialect are often a result of where in the country they live or were brought up, but other factors such as social class and gender may also play a part.

IMPROVE YOUR ENGLISH

Different names for tools, systems and so on from around the UK include:

- handi-bender vs scissor bender
- lump hammer vs club hammer
- troffins vs guttering
- tin snips vs shears.

When communicating with a customer, colleague or site visitor whose first language is not English, you may need to:

- speak clearly and avoid using slang words (words or phrases that are considered very informal and are often specific to certain geographical regions)
- use short sentences and simple words
- ask questions to confirm that you have been understood
- use diagrams and visual aids to back up verbal information
- use an interpreter, if possible, or ask if a family member can act as an interpreter.

When communicating with a customer, colleague or site visitor whose accent is different from your own, you may need to:

- use the correct terminology in work situations – avoid using local slang that may not be understood
- ask questions to confirm that you have been understood

- refer to product catalogues or manufacturer's literature to make sure that you are both talking about the same thing.

In all these cases, you should always show respect for the other person in the way in which you communicate. Keeping your body language open and engaged, with good eye contact, will help you to judge how the information is being received.

Conflicts in the workplace

When people work together in groups, there will be occasions when individuals disagree and conflicts occur. Whether these disagreements become full-blown feuds or instead fuel creative problem solving is, in large part, up to the person in charge. Conflicts can occur for many reasons, such as:

- unfair working conditions
- unfair pay structures
- clashes of personality
- language differences
- ethnic differences.

VALUES AND BEHAVIOURS

It is important to deal with workplace conflicts quickly and effectively as, if left unchecked, they can affect morale, motivation and productivity, and potentially cause stress and even serious accidents.

Conflicts may occur between:

- employer and employee – may need union involvement or some form of mediation
- two or more employees – will need employer intervention
- customer and employer – may need intervention by a professional body
- customer and employee – will need employer intervention.

IMPROVE YOUR ENGLISH

Reaching an agreement through discussion is known as 'negotiation'.

Dealing with workplace conflicts

There are several ways in which your employer may deal with disagreements. They should:

- identify the problem – make sure everyone involved knows exactly what the issue is, and why they are arguing; talking through the problem helps everyone to understand that there is a problem and what the issues are
- allow every person involved to clarify their perspectives and opinions about the problem – they should make sure that everyone has an opportunity to express their opinion; they may even establish a time limit for each person to state their case; all participants should feel safe and supported
- identify and clarify the ideal end result from each person's point of view
- work out what can reasonably be done to achieve each person's objectives
- find an area of compromise to see if there is some part of the issue on which everyone agrees; if not, they may try to identify long-term goals that mean something to all parties.

Informal counselling is one method that helps managers and supervisors to address and manage conflict in the workplace. This may be in the form of:

- meetings
- negotiation/mediation sessions
- other dispute-resolving methods.

VALUES AND BEHAVIOURS

It is important that employees know there is someone they can go to if a conflict develops. If an employee has a conflict with another member of staff, then they should first discuss the problem with their immediate supervisor.

In extreme cases where the matter cannot be resolved, then mediation or union involvement may be required (see Table 4.3).

▼ Table 4.3 Methods of resolving workplace conflict

Type of help	Mutually agreed solutions	Recommendations by an expert	Legally binding decisions	Key features
Mediation (sometimes referred to as 'collective **conciliation**' when used with a group of employees)	Yes	Not usually, but parties can ask for them		Helps to maintain ongoing working relationships. Develops problem-solving skills. Tackles conflict early.
Arbitration		Yes	Yes	Simpler, faster alternative to tribunal hearing (legal meeting with a judge presiding over it). Available only for cases involving unfair dismissal or flexible working.
Individual conciliation	Yes		Yes	Success rate of ACAS service: 70% of cases settled or withdrawn before they get to a tribunal hearing. Often conducted on the phone: parties may not talk to each other.

Source: Acas (2014) Advisory booklet: *Managing conflict at work*

KEY TERM

Conciliation: an alternative dispute resolution process whereby the parties to a dispute agree to use the services of a conciliator, who then meets with the parties separately in an attempt to resolve their differences. Collective conciliation is when a group of employees is involved, and individual conciliation is when there is only one employee involved in the dispute.

In the plumbing industry, workplace conflicts can usually be resolved by the Joint Industry Board (JIB), thus avoiding the need to approach the **Advisory, Conciliation and Arbitration Service (ACAS)** in all but the most severe disputes.

KEY TERM

Advisory, Conciliation and Arbitration Service (ACAS): an organisation that provides free and impartial information and advice to employers and employees on all aspects of workplace relations and employment law.

The effects of poor communication at work

The effects of poor communication can be extremely harmful to both businesses and personnel. If poor communication exists, then goals will not be achieved and this could develop into problems within the company. It can lead to de-motivation of the workforce and the business will not function as a unit. The effects are obviously negative:

- employees become mistrustful of management and, often, one another
- employees argue and reject their manager's opinions and input
- employees file more grievances (cause for complaint) related to performance issues
- employees don't keep their manager informed and avoid talking to management
- employees do their best to hide their professional deficiencies (lack of or gaps in skills) or performance problems
- employees refuse to take responsibility.

Poor communication in the workplace can disrupt the organisation and cause strained employee relations and lower productivity, which can often result in the following problems:

- Time may be lost as instructions may be misunderstood and jobs may have to be repeated.
- Frustration may develop, as people are not sure of what to do or how to carry out a task.
- Materials may be wasted.
- People may feel left out if communication is not open and effective.
- Messages may be misinterpreted or misunderstood, causing bad feelings.
- People's safety may be at risk.

All of these problems will eventually filter down to existing and potential customers, and when that happens, customer confidence will disappear, leading to a possible collapse of the company.

VALUES AND BEHAVIOURS

For more information and advice on ways of resolving disputes and avoiding conflict, visit the ACAS website: www.acas.org.uk

4 THE RESPONSIBILITIES OF RELEVANT PEOPLE IN THE BUILDING SERVICES INDUSTRY

Site responsibilities: communicating with the client

When working as an apprentice, you will meet many different types of clients. These may include a private customer in a domestic dwelling, a representative of a customer or managing agent, a contracting customer, or an internal customer who works within your company.

VALUES AND BEHAVIOURS

Excellent communication skills and good manners all contribute to a strong professional image and lead to a foundation of respect between the two parties.

Private customers are the people that most plumbers meet on a daily basis and first impressions can mean a great deal. A private customer can employ a company with specialist skills to work in their home, such as installers of solar panels or contractors who drill boreholes for private water supplies. A landlord of a dwelling is a private customer and can hire the services of plumbers and gas engineers.

The customer will need to trust the plumber and have faith in their competence. They will see them as representative of the company that they have hired to carry out the work. Direct communication, in the form of clearly thought out conversations, supported by plans and manufacturers' brochures, will help establish a good working relationship with the client.

Types of customer

Sometimes plumbing apprentices may be required to deal with a customer's representative such as a managing agent when engaged on a maintenance contract or if involved in minor electrical works. Access arrangements may need to be finalised and a timescale for the completion of a section of work will need to be agreed upon.

When dealing with a contracting customer, particularly on-site, a **CSCS card** will usually be required and they will expect an organised and efficient workforce to carry out a specific contract, often requiring a company to include a method statement for the job. A contract customer can hire a business with specialist equipment to carry out work on their behalf.

KEY TERM

CSCS card: this stands for Construction Skills Certification Scheme card. Its purpose is to confirm that people who work in the built environment have the necessary competence, and identifies their qualifications. For example, a trainee plumber would carry a small plastic ID craft or operative card that identifies them as a person enrolled on an NVQ programme but not yet qualified.

In larger companies, internal customers may be involved in contracts and the plumber could be included in a special project or even basic maintenance of sanitary appliances.

In all work situations, the image, performance and conduct of plumbers is paramount in creating a good impression and helping their company gain further work.

Communication throughout the progress of a job is very important in order to develop good and effective working relationships with a client.

VALUES AND BEHAVIOURS

Initial feedback to a customer on a job will be verbal and is essential for keeping them up to date with developments, and informing them of expected completion times and any likely changes to the schedule.

IMPROVE YOUR ENGLISH

The first point of written communication with a customer will be the quotation for the work and the last will be the invoice from the contractor.

Site responsibilities for plumbers

The level of personal authority regarding making decisions, solving technical problems and communications with customers will generally relate to a plumber's stage of progression within a company and their qualification status. For example, an apprentice, when working in a customer's home, would not usually communicate directly with the customer and should talk to their supervisor about any problems or issues they encounter. A situation could arise where an apprentice is asked by a customer to carry out a job that the company has not quoted for; this could lead to a number of problems, including not being able to complete the agreed specified contract for the work on time. They will usually be guided and supervised by a qualified plumber, who will help them to work efficiently and to the correct standard.

Once a plumber is qualified to craft Level 2, they become more useful to the company and will be expected to work with less supervision but not take on overall responsibility for a job. They will be able to impart information about a specific job to a customer, but if additional work is requested they will be expected to forward this information, along with details of any problems or complaints, to their supervisor.

When a plumber becomes fully qualified at plumbing craft Level 3 they can take full responsibility for a job, which will include dealing with direct enquiries from clients, including any complaints. At this level of competence, they would be expected to deal with requests for extra work from a client and be familiar with the pricing arrangements for the work.

Supervision

A plumber's supervisor will be the main point of contact for apprentices at work. They will usually be a fully qualified and experienced plumber who is entrusted to ensure that the work is carried out safely, efficiently and to the correct standard. One of their roles could be dealing with the hiring of subcontractors.

It is important that positive and motivating methods of supervision are employed in the day-to-day work on-site. For example, in the event of a building flood as a result of an apprentice's error, a positive approach could be to talk alone with the apprentice once the flooding problem is resolved and ask them where they think they made a mistake. An action plan, which could take the form of a checklist, for when they next carry out the same task could be drawn up and agreed upon. The impact of their error could be discussed but, in the end, there is a way forward to achieve an improved outcome the next time the apprentice takes on a similar job.

ACTIVITY

What would motivate you to improve your work? Make a note and discuss with your team to see what motivates them.

Punishing someone by deducting pay or making negative comments would be demoralising and extremely demotivating. Other ways of motivating staff can include inspiring workers to strive for a higher standard and helping them to realise that they are capable of such an improvement. Prompting can help them to remember their aims on how to improve, and some types of positive reinforcement can provoke people into a reaction, which in turn can improve their performance and perceptions about themselves. In essence, a supervisor is required to learn what makes each individual member of their team respond, so that they can improve and enjoy their work.

▲ Figure 4.9 Good and poor examples of supervision

VALUES AND BEHAVIOURS

Threats and bullying are negative, demotivating and create a poor working environment, with an often subdued workforce.

IMPROVE YOUR ENGLISH

Praise and encouragement are excellent methods of motivation, which can be more easily combined with positive discussions to help solve technical issues or staff problems with beneficial and productive outcomes.

Responsibilities when supervising staff

The role of a supervisor is quite involved and includes many responsibilities, such as:

- defining the overall team workload for a specific job
- allocating the daily work priorities and specifying the workforce for day-to-day tasks
- control and monitoring of work patterns and shift rotas
- explaining and communicating operational information to the team, and relaying feedback from them to management
- initiating and leading incident investigations and providing leadership in emergencies
- maintaining and updating procedures on-site
- finding solutions to problems
- identifying the competence levels for specific work tasks and assessing the training requirements of individual team members
- measuring the team performance and carrying out appraisals, as well as the implementation of first-level discipline
- identifying unacceptable or poor performance on jobs
- communicating with the team members concerned, with the aim of agreeing on ways of improvement.

INDUSTRY TIP

When a supervisor must intervene, they should support staff with a detailed plan of how to improve, and by creating records of improvement priorities and critical activities, which should then be identified in programmes and schedules for future reference.

IMPROVE YOUR ENGLISH

Sometimes competency discussions can be difficult to deliver, and excellent interpersonal skills are required.

The degree of supervision needed varies according to the apprentice's level:

- at Level 1, you will receive a high level of hands-on supervision
- at Level 2, you will have more autonomy to carry out tasks, with less strict supervision
- at Level 3, you are likely to work on your own without much supervision, depending on the type of work.

INDUSTRY TIP

The delegation of work tasks should be based on the competence of the person, not how fast they can do a job or how much money they charge.

A supervisor must be qualified at Level 3 and will usually have more experience than someone who has just qualified, as they may be required to explain installation details to someone less experienced.

KEY POINT

At apprentice level, it is common to have difficulty in comprehending complex manufacturers' instructions. You should receive coaching from your supervisor to help you.

A supervisor should communicate any problems as soon as they are identified. These could include:

- incorrect specification of pipework and materials
- any section of pipework that does not comply with the Water Regulations
- any safety risks as a result of how components were installed.

Adjusting work schedules

The supervisor must be able to know how to adjust work schedules when health and safety problems delay works. By referring to the work programme, they can produce a method statement and involve the team in its execution so that barriers to progress can be removed without compromising the ongoing safety of a given task. For example, if asbestos is encountered, the work in that area may have to stop immediately and an alternative phase of the job started until the problem is resolved. By referring to the work programme, a supervisor could manage this problem and use the diverse skills of their workforce to create a solution.

HEALTH AND SAFETY

Poor supervision can contribute to accidents. A recent report written for the HSE pointed out that the heavy workload of supervisors reduces their opportunities to recognise and respond to unsafe practices, and concluded that the lack of supervision in the workplace is a management failure.

Sometimes there are delays in the second fix and, in this instance, the building services supervisor should inform the construction manager. Good supervision is at the heart of the successful execution and smooth operation of the daily installation and servicing work of a plumbing team.

5 WORK PROGRAMMES IN THE PLUMBING AND HEATING INDUSTRY

What is a work programme?

The principle of a **work programme** is to plan work activity against the time frame of a job or contract so it is completed to the agreed schedule. The design of the work programme for larger projects, such as new-build work, can be very detailed. It may even include a separate plan for the individual trades built into a larger overall programme of work for all activities on-site. The main contractor for the project will generally oversee the work programme and, on a well-designed work programme, they will be able to identify whether everything is going to plan. Inevitably, problems will occur, but contingency plans can be put in place to keep the project on time and within the budget. Precise monitoring, timely intervention and good communication with other trades are key to managing a successful project.

It is therefore important for subcontractors to be organised when taking on such work within larger projects as they could lose out to competitors or face penalties if they fail to complete within the time allocated without the agreement of the main contractor.

KEY TERM

Work programme: a very detailed document used on projects to record and assess activity against expected time to complete the project. For example, it might highlight that poor quality of work and low safety standards could apply to someone completing work ahead of schedule. It could also demonstrate that very slow progress on a job would impact on labour costs. The competence of the plumber is very important and their performance must be assessed carefully.

▲ Figure 4.10 Supervisor showing an apprentice an area of work to be reviewed

Work programmes can be used for private installation work such as the removal of a bathroom and installing a wet room, where there could be quite a lot of disruption to people living in the property. A discussion with the customers at the outset will help to customise the work plan and organise different trades to carry out their specific tasks in an effective and efficient way. When working on-site, there are fewer restrictions than when working in a house. When a site is occupied it is essential that the customer's needs, requirements and lifestyle are carefully considered when planning. There will invariably have to be some compromises on both sides but the initial discussions are paramount to identify times where rapid progress can be made and where restrictions exist.

Using work programmes to arrange and co-ordinate maintenance activities

In the same way, service and maintenance contract work requires a work programme to ensure that appliances and components are kept functional, safe and in good working order. A plumbing company could be involved in regular maintenance of a range of appliances in a large building – for example, in a doctor's practice or an office block, to monitor the chlorination of water supplies and testing, and checking flow rates and temperatures of water outlets to ensure compliance with the Water Regulations and current British Standards.

INDUSTRY TIP

The Building Regulations now require thermostatic control of hot water temperature.

The heat source for the building could be a commercial gas boiler and these will need to be serviced in accordance with the manufacturer's instructions. The radiator circuit must also be checked for effective operation and leaks. Usually on a maintenance contract, the sanitary appliances will have to be inspected and the air admittance valves (AAVs) may have to be changed, as well as WC siphons. Performance testing of appliance traps' seals may be required if there have been complaints of foul smells within the building.

There is a large range of maintenance tasks to carry out, and accurate records of past events or risk assessments help plan for an effective work programme. Private service and maintenance work in houses will involve planning and record keeping to the same standard as larger projects but on a smaller scale. Yearly boiler servicing is recommended by manufacturers, but appliance safety checks are a legal requirement for landlords – the plumber's record keeping and planning will help them to work effectively. A heating system service will require the plumber to test the safety controls on a sealed system and inspect the float-operated valve in the feed and expansion cistern located in a loft, where safe access and good lighting needs to be provided.

VALUES AND BEHAVIOURS

Communication with the customer is essential to ensure minimum disruption, as some tasks, such as chlorination, may have to be carried out after hours.

ACTIVITY

If you are involved with installing a bathroom suite, measure the temperature of the hot water tap and see if it complies with the limit range set out by the Building Regulations.

Job specifications

Job specifications will identify precise details of a job and will normally complement services and site drawings for large projects. For example, if a bathroom suite is to be installed, then the type, model, quality and associated fittings will be clearly stated. Even the type of support for pipework can be mentioned. The timely delivery of the specified appliances and components is therefore important as delays will cost money and hold up other work associated with the job.

Testing procedures will be identified, such as soundness testing and sanitary installation, or pressure testing a hot and cold water pipework installation. The specifications can state who will notch the joints or make openings for first-fix pipework. Because the job specification is part of the overall contract, any changes must be carried out only after an agreement with the management has been made. Installers cannot take it upon themselves to make any changes without permission.

A job specification will also indicate what the documentation requirements are for power tools on-site and what plant is required to carry out specialist tasks. Where special vehicles are required to complete specified tasks, then this information will be stated in a job specification. Installation drawings and job specification can be compared with the work programme.

KEY POINT

Careful planning and monitoring against delivery times is important to avoid losing money on a project. It is also important to avoid theft from a site, which is best solved by arranging delivery on-site early in the morning when operatives are there.

Delivery of materials

Because a work programme can include a range of information, such as the progress of work and the strategy of start times, for multiple trades on a building project, the information must be clearly laid out on a simple bar chart. With domestic properties the non-arrival of goods and materials to a site can heavily impact a planned day's work. On larger sites the delivery of materials 'just in time' means that they are delivered to the exact location of the work at an agreed time that coincides with the plumbers commencing work. If material is not available this is likely to increase labour costs. In the same way, specialist plant and machinery should be delivered to coincide with specialist contractors' arrival to begin work on-site.

ACTIVITY

When a delivery of plumbing fixings and fittings arrives on-site, ask your supervisor if you can help with checking the goods delivered against the delivery note.

Other delays can occur, even if a delivery arrives on time, such as the wrong specification of goods being sent, or items missing or arriving damaged. This is why it is important to carefully check the delivery note to make sure that everything that has been ordered is there, before signing off any documentation.

Time allocation to work activities

The **delegation** of work tasks should be based on the competence of the person and not how fast they can do a job or how much money they charge. Good-quality work completed ahead of time is what everyone desires and this will lead to an increase in the profit margin for the company. Labour resources are best allocated after careful planning of when the work opportunity will arise in the work programme. When planning work with plumbing and other trades it is essential that the work is executed safely. If there are any delays because of health and safety issues, then the best solution would be to work around the problem safely until it is resolved.

KEY TERM 

Delegation: sharing or transfer of authority and responsibility, from an employer or supervisor to an employee.

First fix

The first fix comprises all the work required to take a building from foundation to plastering and painting the internal surfaces. For a plumber, this includes the installation of pipework in joists and in walls, and routed to the planned location of appliances such as radiators, boiler and sinks.

Second fix

The second fix includes all the installation work required once the plastering has finished, which means the appliances can then be connected to the first-fix pipework and commissioned.

ACTIVITY

When on-site, ask to see the bar chart to check where your particular work is located in the work programme. Identify the first- and second-fix stages of the plumbing team.

Once a **tender** for a project has been attained, the next step is to devise an efficient and timely way to execute the plan in order to complete the task. A simple bar progress chart will provide the essential information required to organise and monitor the progress of each individual trade on-site. The plan will include start and end dates, and even costs, for each section. The visual display will give a quick indication of the progress of a project and allow for changes to be made. Some trades, such as plumbing, will be involved in the first- and second-fix works of an installation. If the job specification states that carpenters must cut notches in joists for the

Construction team			**2014**											
			JAN				FEB				MAR			
			1	2	3	4	1	2	3	4	1	2	3	4
Ref	**Project**	**Roles**												
SC 1	Sports Centre	Carpenter		1										
		Plumber			1	1	1				2	2		
		Electrician						1	1				2	
		Plasterer												
		Painter												
Key: 1 first fix, 2 second fix														

▲ Figure 4.11 Excerpt from a simple bar progress chart

pipework, then the plumbing team can begin the first fix when that is completed. The job specification may also require that the carpenters insert battens in the studwork in the wall to support appliances and radiators. Once the dry lining has been fitted and any finish to the surface applied, then the plumber could come and complete the second fix, and testing and commissioning can follow. All of this information can be represented on a bar chart like the one shown in Figure 4.11.

KEY TERM

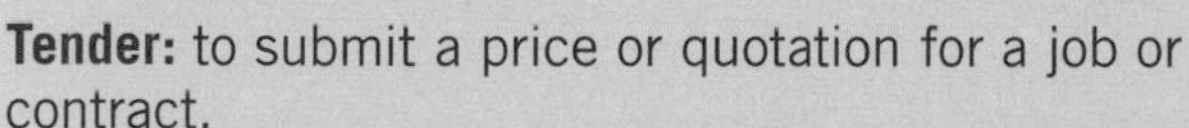

Tender: to submit a price or quotation for a job or contract.

Variations in work

Nearly all construction projects will encounter changes or variations during the design and construction process. Because of this, many construction contracts include provision for a variation clause. The term variation usually means a change, modification, alteration, revision or amendment to the original contract and how works are to be carried out.

In order to solve problems related to changes to a project, the project team must be equipped to analyse the variation, anticipate its immediate effect on other parts of the work programme and then effectively manage the new work.

Variation order

A variation order is a document that records any agreement made with a client to alter the existing work specified in a building contract.

Variation in the work that would involve any change to the agreed contract price for the work must be agreed and approved in writing by the owner before a variation order can be put into action. A variation can impact on timescales for completion and any order must include details of both cost and time changes.

If a variation of work is caused, for example, by poor installation or lack of ability to complete a task properly, then the contractor could be liable for any subsequent costs. A problem with the installation may have been identified by the clerk of works and, if it is found that the routing of exposed pipework is not acceptable, it would then alter the agreed design. A discussion with the contractors must take place and the reasons for the change, and its impact on time and cost, must be carried out with the customer.

INDUSTRY TIP

Producing the variation order as early as possible will save money, as one of the problems with reaching agreement is the time the whole process takes. Sometimes an independent company can be brought in to quickly resolve variation issues if there is poor communication and co-ordination on a project.

If the customer suggests changes after the contract has been signed, then they will have to bear the costs related to delays or additional material and design to complete the work. All changes are to be confirmed in writing as written communications have the advantage of providing a permanent record.

Causes of variation orders

The chart in Figure 4.12 shows a range of causes for variations and is helpful when assessing where liability rests.

Causes of variation orders

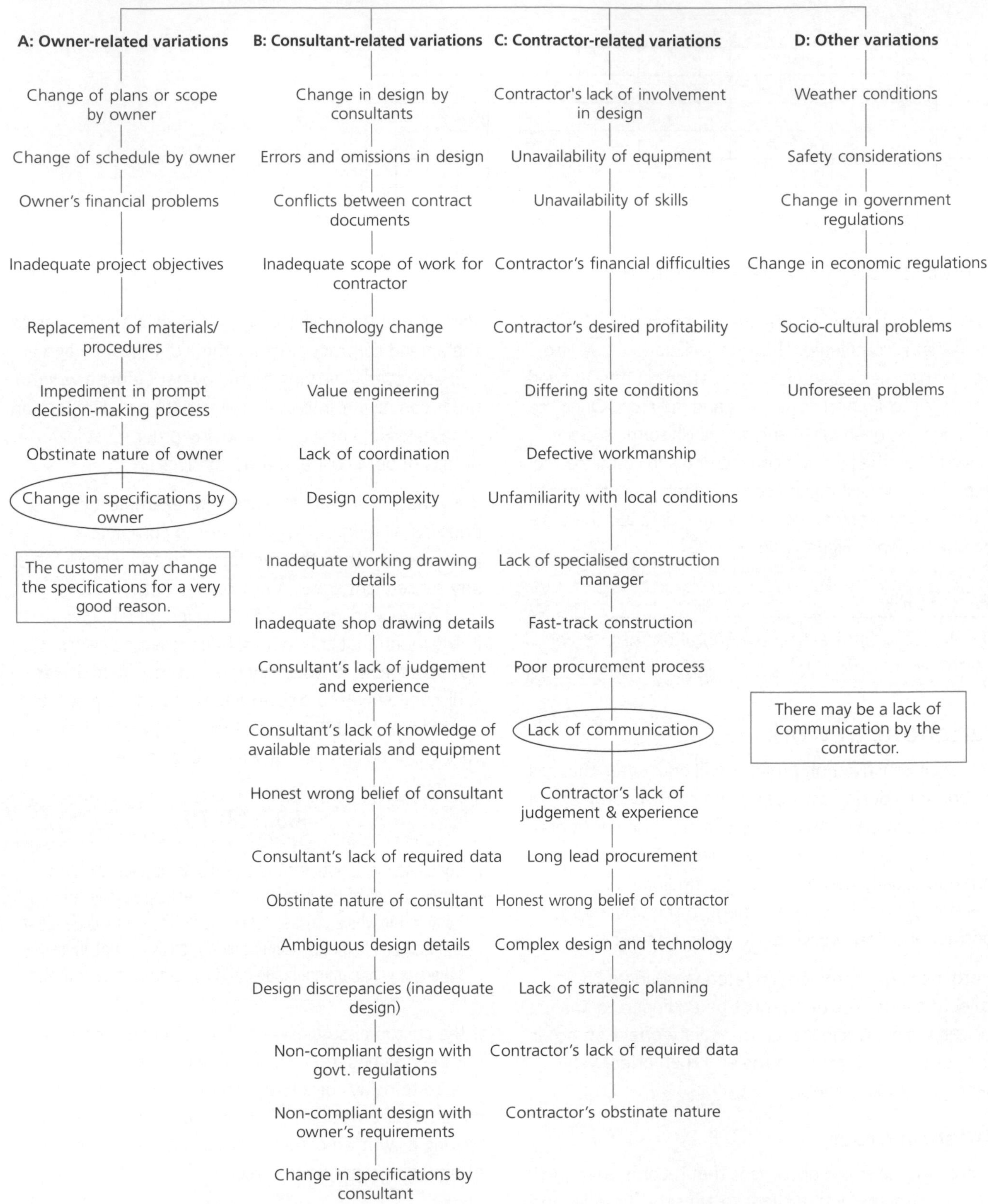
A: Owner-related variations
Change of plans or scope by owner
Change of schedule by owner
Owner's financial problems
Inadequate project objectives
Replacement of materials/ procedures
Impediment in prompt decision-making process
Obstinate nature of owner
Change in specifications by owner
The customer may change the specifications for a very good reason.
B: Consultant-related variations
Change in design by consultants
Errors and omissions in design
Conflicts between contract documents
Inadequate scope of work for contractor
Technology change
Value engineering
Lack of coordination
Design complexity
Inadequate working drawing details
Inadequate shop drawing details
Consultant's lack of judgement and experience
Consultant's lack of knowledge of available materials and equipment
Honest wrong belief of consultant
Consultant's lack of required data
Obstinate nature of consultant
Ambiguous design details
Design discrepancies (inadequate design)
Non-compliant design with govt. regulations
Non-compliant design with owner's requirements
Change in specifications by consultant
C: Contractor-related variations
Contractor's lack of involvement in design
Unavailability of equipment
Unavailability of skills
Contractor's financial difficulties
Contractor's desired profitability
Differing site conditions
Defective workmanship
Unfamiliarity with local conditions
Lack of specialised construction manager
Fast-track construction
Poor procurement process
Lack of communication
Contractor's lack of judgement & experience
Long lead procurement
Honest wrong belief of contractor
Complex design and technology
Lack of strategic planning
Contractor's lack of required data
Contractor's obstinate nature
D: Other variations
Weather conditions
Safety considerations
Change in government regulations
Change in economic regulations
Socio-cultural problems
Unforeseen problems
There may be a lack of communication by the contractor.

▲ Figure 4.12 Causes of variation orders

Monitoring of progress and identifying deficiencies in work performance

By referring to bar charts of the work programme, work activity, week numbers, expected completion dates, price of materials and week commencing dates can be identified. Installation drawings and job specifications can also be compared with the work programme. Labour resources can be best employed and organised after careful planning of when the work opportunity will arise in the work programme, and this will help with the cost effectiveness of the project.

A clear, well-designed work programme will help a building service supervisor to be able to know the exact time of a specific job, such as making a connection to an existing sewer from a new estate. They will be able to organise a **toolbox talk** before the work begins, to ensure that safe systems of work are observed because the plumbers will be in contact with waste matter from humans. The supervisor can emphasise that, when working on sanitary installations, rubber gloves should be worn, but point out that this work should not be undertaken if the plumber has any open wound.

The work programme would also show who was responsible for any work at a given time and what materials were being used. This helps managers assess the progress of a job against agreed timescales and anticipate any changes to costs. Monitoring the progress of the job will help when confirming delivery times for fixings and fittings because if material is not available this is likely to increase labour costs.

Deficiencies

Deficiencies in the context of the work performance relate to problems that could affect safety, quality and cost effectiveness.

It is important that a supervisor should communicate any installation problems as soon as they are identified while they are monitoring work. Examples of deficiencies can include incorrect specification of pipework and materials, any section of pipework that does not comply with the Water Regulations, or if there is a safety risk because of how components have been installed.

KEY TERM

Toolbox talk: a toolbox talk is an informal meeting to deal with matters of health and safety in the workplace and safe working practices. They are normally short meetings conducted on-site before the commencement of the day's work activities. Toolbox talks are an effective way of refreshing operatives' knowledge and communicating the company's health and safety culture.

6 RISK ASSESSMENTS AND METHOD STATEMENTS FOR THE PLUMBING AND HEATING INDUSTRY

Risk assessment

A risk assessment is a document drawn up after an evaluation of existing or potential hazards on a particular job that is about to start. Any such hazards are identified and precautions devised to reduce the risk. Its aim is to provide information that will help keep a worker safe and protect others from being injured or even suffering illness. Although an apprentice would not be responsible for completing the risk assessment, they must know how to follow it and be aware of its purpose. The law does not expect all risks to be eliminated but there is a requirement to protect people as far as is reasonably practicable.

As well as complying with the law, a risk assessment enables a worker to focus on those risks in the workplace with the potential to cause real harm. Straightforward measures can usually control risks, such as making sure that spillages are cleaned up quickly to help prevent the apprentice from slipping. Untidy work areas can also lead to accidents caused by people tripping over pipework and leads. Accidents and ill health can seriously affect lives and have a negative impact on business, especially as a result of court action. Therefore, there is a legal requirement to assess the risks in the workplace and have a plan in place to control the risks.

▲ Figure 4.13 Wet floor sign to warn people about a leak from an appliance

The HSE produces a document called 'Risk assessment: A brief guide to controlling risks in the workplace', which outlines five steps to risk assessment. These are:

1 Identify the hazards.
2 Decide who might be harmed and how.
3 Evaluate the risks and decide on precautions.
4 Record your findings and implement them.
5 Review your assessment and update if necessary.

Step 1: Identifying hazards in the workplace

Defining the risks in work situations involves being able to identify hazards, which is anything that may cause harm, such as chemicals, electricity, working from ladders or an obstruction. The risk is then defined by the possibility that a person could be harmed by these hazards, along with an indication of how harmful or serious they could be.

INDUSTRY TIP

Sometimes a conversation with other trades working in the same area may help to add important information that could improve the assessment.

Step 2: Deciding who might be harmed and how

A walk around the proposed job location and carrying out a visual inspection will help to make an assessment of what could reasonably be expected to cause anyone harm.

If dealing with, for example, gas appliances or plumbing equipment connected to an electrical supply, it is recommended that the manufacturer's instructions are consulted to ensure that these are properly installed and operating correctly. If chlorination of a cold water storage cistern (CWSC) is going to be carried out at the same time as other work, then COSHH data sheets for the chemicals should be referred to. Another example of a high-risk situation would be if a work environment is noisy. Noise can be a distraction and creates a problem when close communication is required for people working in pairs. Noise and exposure to chemicals can produce long-term hazards.

HEALTH AND SAFETY

The HSE publishes practical guidance on where hazards occur and how to control them, on its website at: www.hse.gov.uk

It is important to be clear about who could potentially be harmed by each hazard as this approach will help identify the best way to manage the risk. Identify precisely how people might be harmed for each different case, and state what type of injury could occur or how health could be impacted. For example, an engineer working in an oil- or gas-fired boiler room could be exposed to carbon monoxide poisoning if the open flue failed to remove the products of combustion effectively.

▲ Figure 4.14 COSHH hazard pictograms

▲ Figure 4.15 A typical safety helmet – check label to ensure it is in date; it is always a good idea to inspect the helmet for damage before use

Step 3: Evaluating the risks and deciding on precautions

Once hazards have been identified, then decisions must be made about what to do about them. The law states that you must do everything that is reasonably practicable to protect people from harm. A **risk assessment** of what type of work you will be doing will have to be carried out, as well as information on what controls you have in place and how the work will be organised. Work out if the problem can be removed completely and, if not, consider how the risks will be effectively controlled. There are several ways to reduce a risk, such as using barriers around a particular work area to prevent unauthorised access to a potentially hazardous area. The provision of an up-to-date first aid kit in the work area is important, as is appropriate personal protective equipment (PPE).

▲ Figure 4.16 Plumber's trousers – make sure kneepads are fitted before starting work

Step 4: Recording your findings and implementing them

It is important to write down the results of your risk assessment and share them with your team. Doing this will help to encourage all involved to put what is written into practice. The results need to be written simply and clearly. However, if there are fewer than five employees in a company, then nothing needs to be written down.

For example, a boiler room has many potential hazards, so carefully compiling a list of hazardous results is essential. You could begin implementing your findings by emphasising that work areas around the boiler are to be kept clear of any tripping hazards and that the area should be checked throughout the day. Perhaps in this same boiler room, it could be that natural gas will be released when installing new pipework. This risk can then be contained by writing step-by-step procedures, which should be stated in a method statement, especially if there is a risk to others. Finally, it is a possibility that some of the existing pipework insulation contains asbestos, so care and attention should be taken when working on pipework. Therefore, as you complete each action on the job, it should be recorded and then implemented into the plan.

Step 5: Reviewing your risk assessment and updating it if necessary

Quite often a plumber has to return to the same area of work on a servicing and maintenance contract, and circumstances could have changed. If the example of the boiler room is used again, then there could be new hazards such as the storage of combustible materials or even chemicals in the vicinity of the appliances. Someone may have damaged the main equipotential bonding at the gas meter, or a contractor could have inadvertently partially blocked a temperature relief discharge pipe.

It is therefore advisable that you look at the existing risk assessment for the job location, and make a note of any new changes and actions required. It is essential that the risk assessment stays up to date. In a maintenance work plan it is good practice to plan and review dates for risk assessments.

Risk calculation formulas

A **risk calculation formula** is also known as 'ranking the risk', and results in assigning a number to each risk. Start with an assessment of the likelihood of an accident, then proceed with listing the consequences. This is called the **quantitative approach**.

KEY TERMS

Risk calculation formula: this is a method of using a formula of multiplying likelihood by consequences to provide a number that quantifies the level of risk for a particular job.

Quantitative approach: ranking a risk with a number.

The formula for ranking risk with the quantitative approach is shown in Table 4.4.

▼ Table 4.4 Risk calculation formula

Likelihood of an accident occurring		Consequences of an accident occurring	
	Scale value		**Scale value**
No likelihood	0	No injury or loss	0
Very unlikely	1	Treated by first aid	1
Unlikely	2	Up to 3 days off work	2
Likely	3	More than 3 days off	3
Very likely	4	Specified major injury	4
Certainty	5	Fatality	5
Calculation of risk factors			
Likelihood × Consequence			
Calculated figure		**Action**	
Figures between 1 and 6		Minor, but monitor closely	
Figures between 8 and 15		Significant, immediate control action	
Figures between 16 and 25		Critical, all activities must stop until risk reduced	

Worked calculation

Imagine an apprentice is asked to solder pipework in the loft space of a house that was built in 1960. They are required to remove the galvanised CWSC. What precautions should be taken, and what are the risks to the plumber and others?

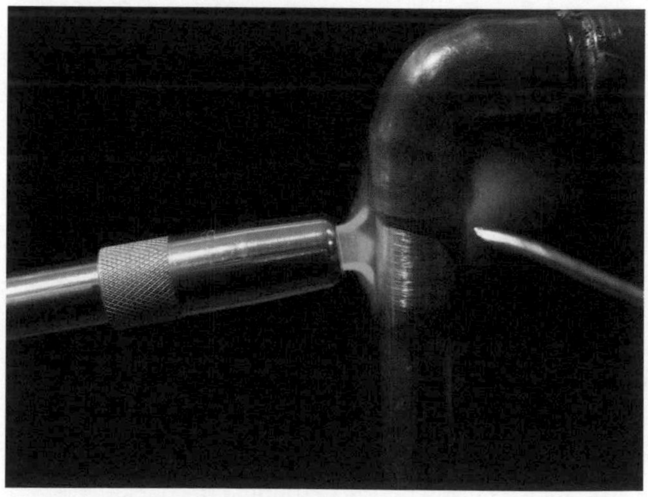

▲ Figure 4.17 Soldering pipework

There are several risks associated with this scenario. Access to the work area is to be carefully considered, along with the likelihood and consequences of an accident occurring, and the provision to be put in place to reduce the risk.

There is also a fire risk because of the use of naked flames from using a blowtorch in a confined area with combustible materials in the vicinity. The removal of the galvanised CWSC may require the use of a power tool to cut it into sections if the access to the loft space is too small to allow it to be removed in one piece. Finally, because the house was built in 1960 there is a possibility of the presence of asbestos in the building fabric, such as the fascia boards or even loft insulation.

There are several risks and by taking them one at a time, a picture of what is required to reduce them can be compiled. If access to the work area is looked at, then an assessment of the likelihood and the consequences can be made.

The likelihood of an accident from a fall or slip on a freestanding portable ladder could be **3** and the consequences could be **4**. By multiplying these figures, 3 × 4, we get a figure of **12**, therefore the outcome is significant (according to the formula in Table 4.4).

Task Accessing loft space with ladder		
Equipment Portable ladder		
Hazards Slipping and falling		
Likelihood 3	**Consequence** 4	**Risk factor** 12
Risk exposure Employees		
Control measures: ■ Ensure basic training is carried out for apprentices. ■ Inspect ladder for condition and correct grade. ■ Secure ladder at top and at base. ■ Ensure ladder extends sufficiently into loft space to enable safe descent. ■ Provide safe floor area in loft space to manoeuvre. ■ Provide lighting. ■ Use correct PPE for task.		

▲ Figure 4.18 Example of a basic risk assessment form for a typical task in a customer's home

A reduction of this risk is required. A possible solution could be the use of a secured ladder, with a protected area at the base or providing another trained person to stand at the base. Ascending a ladder can be easier than descending, so a clear, secure area at the top of the ladder is also required, in addition to adequate lighting. An exposed floor with only joists for foot support in the loft at the top of the ladder could be made safer with robust boards that cover the area of joists and make it safe to stand without the risk of a plumber falling through the ceiling.

The risk with these provisions in place could reduce the likelihood to **1** and the consequence to **0**. The risk is then reduced and safe access can be achieved.

ACTIVITY

Assess and rank the risk of soldering a 15 mm copper pipe above a classroom doorway in a primary school. What actions would you put in place to reduce the level of risks identified?

There is another way to carry out a risk assessment that does not use numbers. This is referred to as the **qualitative method**. The qualitative approach divides risks into priority categories – low, medium and high – based on a range of factors. An example of how this is applied is shown in the qualitative example of a risk assessment which involves gas work (Figure 4.19). This risk assessment form uses the qualitative approach of ranking risks.

KEY TERM

Qualitative method: divides risks into categories such as low, medium and high.

General Risk Assessment Form

Location: Boiler room no 2, Looming Towers, Scare Street, Wick, UK Company Sector: Plumbing

Environment/Activity/Equipment Live. Tightness testing and purging of gas installation in boiler room.

Who is at risk: plumbers public contractors others (state) ....

Hazard/s	Leading to a risk of	Existing measures to control risk	Risk rating			Action required Yes/No Ref No
			L	S	RR	
Gas leakage	Gas in air	System tested before use	1	1	1	no
		Apprentices monitored and supervised				
		System tested after any alteration				
Purging of gas	Gas in air / LEL lower Explosive Limit	Sectioning off areas	2	3	6	
		No naked flames or switching / unless intrinsically safe (IS)				yes

If young apprentices will be involved with this gas testing equipment, have the following been considered in the assessment: Is the installation work required as part of their on site training? **Y / N**; The need to be supervised by a competent person **Y / N**; Their lack of experience **Y / N**; If new to work, their lack of awareness of workplace risks **Y / N**; Their gender **Y / N**; Their manual dexterity **Y / N**; Their physical abilities **Y / N**.

Action plan				
Ref No	**Further action required**	**Action by whom**	**Action by when**	**Date completed**
1	Only persons deemed safe, responsible and conscientious will be allowed to touch the equipment even if supervised.	Supervisor / Foreman	Immediate effect	
2	Use of gas co seeker to search leaks plus leak detection fluid.	Senior gas engineer/ Plumber	Immediate effect	
3	Whole system to be commissioned after successful completion of documentation.	Senior gas engineer/ Plumber	Immediate effect	

Sign off procedure

Work Completed by:	Signature	Name	Date ....
Checked by Supervisor:	Signature	Name	Date ....
Approved By	Signature	Name	Date ....
Risk Number	Signature	Name	Date ....

Risk Rating Calculation

Likelihood		Severity	
3	Probable: Likely to occur each year; Has occurred recently	3	Death/hospitalisation
2	Possible: Likely to occur in 10 years; History of it happening	2	> 3 day injury
1	Remote: Not likely to occur in 10 years; Has not occurred	1	< 3 day injury

Risk Rating (RR) = Likelihood of Realisation of Hazard (L) × Severity of Hazard Being Realised (S)

Low = 1–2 Leave until last Medium = 3–4 Leave until later (set firm completion date) High = 6–9 **ACT NOW**

▲ Figure 4.19 Risk assessment form

Referring to the template above it can be seen that, when a tightness test takes place, the likelihood of a hazard is low and so is the severity of hazard; therefore, 1 × 1 = **low**.

However, as gas is released into air there is a high risk of flame ignition at the point of purging the pipework, and in more extreme situations the risk of explosion if the escape were left to continue and allowed to reach an explosive mixture. Therefore, for this part of the job, the likelihood is high and so is the severity of hazard. There are basic systems and processes used to reduce this risk and qualified Gas Safe engineers apply these on a daily basis.

Method statement (plans of work)

A **method statement** is often used for a high-risk situation. Its aim is to prevent accidents or dangerous situations from occurring. A suitable and sufficient method statement should be a practical and useful document that clearly describes a safe working method for a work team to follow easily. Senior management require this document to help them manage and control a job as it verifies that any significant risks have been considered and specific instructions to show how to deal with such risks have been provided. It should provide sufficient detail so that anyone inspecting the instructions, such as the HSE or even clients, can assess the effectiveness of the plan. The method statement is meant to help managers, supervisors and workers to carry out their work safely and efficiently.

KEY TERM

Method statement: the record of how management wants the job to be done. Its main purpose is to guide site work and it must always be available on-site as a live document with an aim to prevent accidents or dangerous situations from occurring.

A method statement must be clear and precise, otherwise it will be considered unsuitable or insufficient for the task. Typical information to be included in a method statement includes drawings, plans, specifications, schedules, risk assessments, site inspection reports, manufacturers' information, current regulations and official guidance associated with protecting work areas.

The purpose of the method statement is to compile into one document the control measures and findings for a range of risk assessments associated with a specific job. The information is then handed to employees so that they have clear and detailed guidance on how to safely carry out the task.

ACTIVITY

How would a method statement help you when carrying out work in a customer's home? Consider whether the extra planning would encourage a more enjoyable and productive working day.

For example, providing a method and system of working for employees investigating public health issues related to leaks from a domestic septic tank installation (like those shown in Figure 4.20) is essential as it is likely that there will be a risk assessment relating to the exposure of raw sewage, and another to cover the lifting and repair of low-level pipework, including possible trench and excavation work. It is important that a method statement is followed precisely to complete a job safely, as this will protect the workers and, in this particular case, the public.

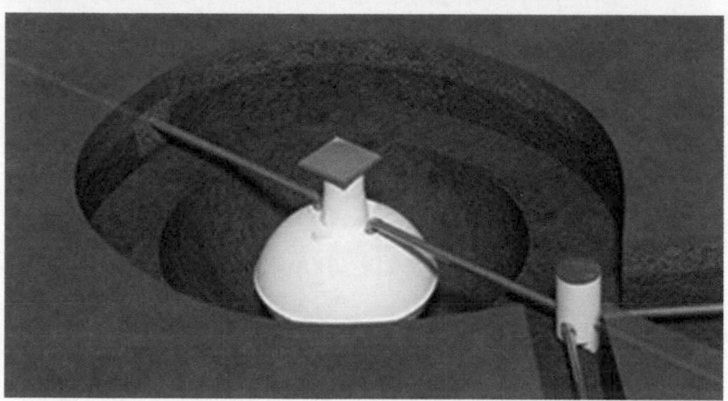

▲ Figure 4.20 Septic tank installations

It is also important that a method statement be provided when installing a septic tank because of the risk to and possible impact on others in the vicinity. There will be specialist contractors present to install the tank, and to prepare and dig the required excavations, as well as plumbers installing sanitary pipework. Local authorities will be involved and the management of a company needs to arrange a good level of co-ordination to ensure the job runs safely and on time. A method statement helps with the planning and execution of such tasks.

Presenting a method statement

A method statement could include headings such as details of the contract and the scope of work. It is difficult to be entirely prescriptive about the precise form a method statement will take, but anyone reading the document should quickly be able to establish the nature of the work. The sample in Table 4.5 shows how a method statement can be laid out, with examples of typical headings and details of how the plan will be managed.

▼ Table 4.5 Method statement examples

Typical heading	Example of what the details will include
Details of contract and scope of the work	**Work location** and attendance times/dates. **Contact details:** supervisor, client, survey details, principal contractor/co-ordinator (where relevant), senior manager responsible for the contract, local authority. **The scope of the work:** what does the team have to do and what are the constraints of the site.
Equipment, materials and controls	**Lifting, drilling and excavation equipment:** hand tools, sprays, gels, fencing, barriers and signage, etc. **Location** and access arrangements for water and power supplies. **Methods of storage and transport.** **Location and access to welfare facilities.**
Other relevant site-specific information	**Clear responsibilities and lines of communication** with relevant third parties (customers and other specialist contractors). **Adequate pre-planning** will ensure that others will not have a negative impact on the neighbours. **Emergency arrangements** and procedures. Any other **significant risks** (including how they will be controlled).
Method of work	**Site-specific sequence of work** and actual methods to keep work areas safe and tidy. The method will give details of **safe working practice**: working at height, working in excavations, confined spaces, live electrical installations.
Management arrangements	The **plan** is meant to be a management tool as well as a guide for site teams, so it should be clear how the supervisors/managers are expected to ensure that the working methods are followed. Regular **team meetings** to include any variations to the job specifications or working methods.

ACTIVITY

Complete the risk assessment form for the replacement of a WC suite in a customer's property. You must think of the hazards you might encounter while doing the work, such as causing a fire from the use of a blowtorch, and how you would prevent the risk.

Company name: **Date of risk assessment:**

Job description: *Installing a WC pan and cistern*

What are the hazards?	Who might be harmed and how?	What are you already doing?	Do you need to do anything else to control this risk?

▲ Figure 4.21 Risk assessment form

Source: adapted from the Health and Safety Executive's risk assessment template

SUMMARY

During this chapter, we have looked at the varied personnel of the construction industry, from the client through to the building contractor and the workers on-site. We have seen how the relationships between the trades are interwoven, with everyone working towards a common objective – a successful, quality building and a happy client.

Effective working relationships are crucial if the construction process is to be successful, but they are also often fragile, and it is important to know that help is at hand if these relationships, for whatever reason, break down.

The interaction between the many members of the construction team and the smoothness of the construction process is the most visible testament that effective working relationships at all levels of construction management, tradesperson and labourer are just that – effective and working!

Test your knowledge

1 In line with the Construction (Design and Management) Regulations 2015, a construction project is notifiable when:

- a The project lasts more than 500 days and involves 30 or more operatives
- b The project involves 500 or more person-days or lasts more than 30 days
- c The site includes three or more trades at any one time
- d The client is classified as a 'non-domestic client'

2 Which member of the site management team advises on how a project can be constructed within the client's financial budget?

- a The building surveyor
- b The estimator
- c The clerk of works
- d The quantity surveyor

3 Which of the following is the Building Control Inspector employed by?

- a The HSE
- b The local authority
- c The client
- d The main contractor

4 As a plumber working on a large domestic new-build project, which document would be consulted to confirm the type and quality of terminal fittings to be installed?

- a The specification
- b The work programme
- c Working drawings
- d Variation orders

5 Which of the following tasks is classed as 'first fix'?

- a Installing traps to wash hand basins
- b Installing flexible connectors to terminal fittings
- c Hanging radiators on finished walls
- d Notching joists to install heating pipework

6 Which of the following is the main role of the clerk of works?

- a To ensure that all work by contractors is completed in line with agreed plans and relevant standards on behalf of the client
- b To quantify the materials required for the installation and order as required
- c To oversee the craft operatives on-site and plan the daily activities
- d To manage the finances of the project, including materials and labour, ensuring the project is completed within budget

7 According to HSE guidelines, what is the third step to risk assessment?

- a Identify the hazard
- b Review the assessment and update as required
- c Decide who may be harmed and how
- d Evaluate the risks and decide on precautions

8 Plumbers that are able to install 'non-complex' hot and cold water systems, as well as domestic sanitation pipework and basic central heating pipework under regular supervision, are usually qualified to which level?

- a Level 1
- b Level 2
- c Level 3
- d Level 4

9 An apprentice plumber finds a crack in a fitted washbasin in a new property. Who should he/she report the problem to first?

- a The client
- b The clerk of works
- c The supervisor
- d The quantity surveyor

10 What is the <u>most likely</u> outcome of materials not being delivered to site on time?

- a Increased labour costs
- b Increased profit
- c Improved company reputation
- d Improved punctuality

11 Which person in the construction team would obtain quotations from the suppliers?
- a The clerk of works
- b The main contractor
- c The procurement officer
- d The client

12 Who is allowed to sign off a variation order?
- a The estimator
- b The quantity surveyor
- c The client
- d The clerk of works

13 Which trade would install conduit?
- a Electrician
- b Plumber
- c Carpenter
- d Bricklayer

14 Which document shows the order of construction and allows the monitoring of progress?
- a Delivery note
- b Work programme
- c Time sheet
- d Job specification

15 Which piece of legislation gives you the right to ask a public body for information they have on a subject?
- a Data Protection Act 2018
- b The Equality Act 2010
- c The Freedom of Information Act 2000
- d GDPR

16 Which of the following is NOT enforceable by law?
- a BS EN 806
- b The Data Protection Act 2018
- c The Water Regulations
- d Building regulations

17 If you were to produce a formal quotation for a customer, which method of communication should be used?
- a Text
- b Face to face
- c Written
- d Telephone

18 What is the correct process if a conflict occurs on-site?
- a Resolve it quickly
- b Take it to arbitration
- c Inform the unions
- d Seek mediation

19 Who would be an apprentice's main point of contact on-site?
- a The customer
- b The supervisor
- c The site manager
- d The clerk of works

20 Who does the subcontractor report to?
- a The architect
- b The building control officer
- c The main contractor
- d The trade supervisor

21 Which sentence best describes the first fix process?
- a The installation of pipework into joist and walls
- b The installation of appliances to pipework
- c The handing over of an installation to a customer
- d The design stage of an installation

22 What could conflict on-site between fellow workers result in?
- a A reduced level of accidents
- b An increase in output
- c A reduction in output
- d An increase in wages

23 What is the main advantage of communicating by email rather than posting a letter?
- a A more formal method of communication
- b More likely to be understood
- c There are less words used
- d It is a quicker form of communicating

24 An apprentice plumber has flooded part of a building. How should the supervisor deal with the situation?

25 Give three important points to remember when communicating with someone who has a visual impairment.

26 List the four on-site inspectors.

27 Who are ACAS and what is their role in workplace disputes?

28 What calculation is required to produce the risk factor of a task when following the quantitative risk assessment method?

29 Describe the role of the clerk of works.

30 The building control officer (BCO) is expected to visit the site. Describe what the role of the BCO is.

31 List the regulations that a plumber MUST adhere to while installing a system.

32 Describe what a 'works programme' is used for.

Answers can be found online at www.hoddereducation.co.uk/construction

COLD WATER SYSTEMS

CHAPTER 5

INTRODUCTION

The supply of fresh, wholesome cold water to people's homes is a basic human need. As a plumber, it is your job to get the water from the main external stop valve to the taps so that it is clean and fit for human consumption. Most people take for granted the supply of cold fresh water to their homes and few would probably appreciate the degree of work necessary to provide this service.

In this chapter, we will look at the subject of water from the cloud to the tap, the cleaning process that makes it fit for human consumption, the distribution of water and the systems that you will install. We will also explore the regulations that govern our industry and the processes you will need to understand to enable you to work safely and correctly on domestic cold water systems, from installation planning to testing, maintenance and fault finding.

By the end of this chapter, you will have knowledge and understanding of the following:

- the sources and properties of water
- the types of water supply to dwellings
- the treatment and distribution of water
- the sources of information relating to cold water systems
- the water service pipework to dwellings
- how to select cold water systems
- the system layout features of cold water systems fed from private water supplies
- the components used in boosted (pumped) cold water supply systems from private sources for single-occupancy dwellings
- backflow protection
- the Water Regulations
- how to install cold water systems and components
- how to carry out commissioning procedures
- how to replace or repair defective components
- how to decommission cold water systems.

1 SOURCES AND PROPERTIES OF WATER

The rainwater cycle

Water is a simple compound made up of two hydrogen atoms attached to a single atom of oxygen, with the chemical symbol H_2O. Water is tasteless and odourless and, in small quantities, it is colourless, while in large quantities it possesses a light blue hue.

There is no new water on Earth – all water is about 4.2 billion years old, whether it is sea water (saline), river or stream water, groundwater, fossilised water or water from the polar ice caps.

▲ Figure 5.1 Water molecule

Water moves constantly in what is scientifically called the hydrological cycle. We know it by its more common name: the rainwater cycle.

Simply explained, the rainwater cycle is a natural process where water is continually exchanged between the atmosphere, surface water, groundwater, soil water and plants. It can be divided into three main transfer processes:

1. evaporation from oceans and other water bodies into the air
2. transpiration from land plants and animals into the air
3. precipitation from water vapour condensing from the air and falling back to Earth or into the ocean.

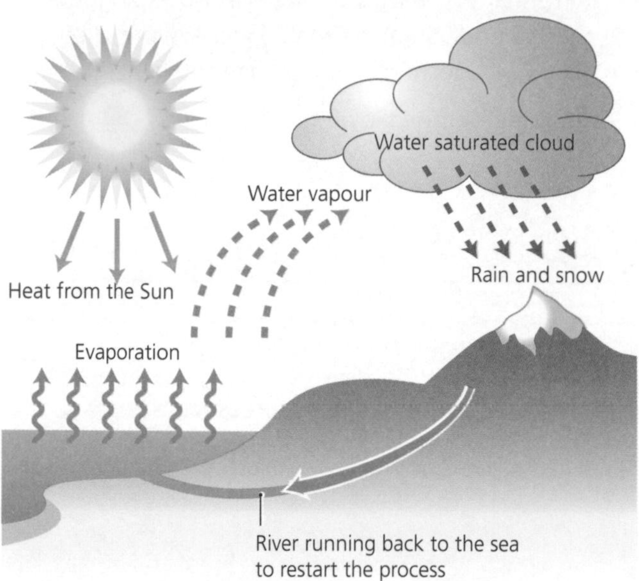

▲ Figure 5.2 The rainwater cycle

As the Sun warms the Earth, water on its surface evaporates. This vapour rises with the air and is carried by the prevailing winds. If the vapour passes over land, some of it condenses to form clouds and, as more water vapour is attracted or the ground rises (hills or mountains), the cloud becomes saturated to the point where it can no longer hold the moisture and the vapour is released in the form of rain, sleet, snow or hail.

On reaching the ground, there are many paths it may follow. Some of it may be re-evaporated back into the atmosphere; it may be absorbed by the ground, where it will travel towards the water table or aquifer; or it may remain on the surface, where it will eventually find its way into rivers, streams, lakes or the oceans. Here, the process begins again, an example of the Earth's natural recycling process.

Sources of water

If we look at all of the water on Earth, 97 per cent is saline (sea) water and only 3 per cent is fresh water. Of fresh water, nearly 69 per cent (or 2.07 per cent of the Earth's total water resources) is trapped in the polar ice caps and glaciers, and 30.7 per cent (0.9 per cent of the total water resources) is groundwater. It is groundwater that the population of the Earth relies on for its drinking water supply.

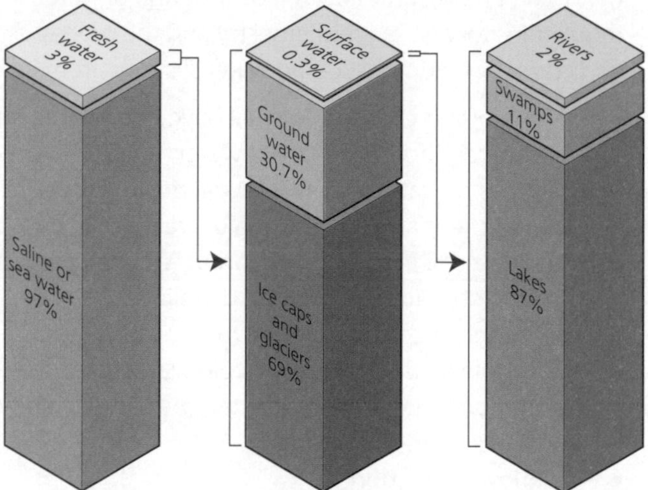

▲ Figure 5.3 Distribution of the Earth's water

The total freshwater supply for the world is in the region of 1350 trillion litres, the majority of which is stored on the ground, where it is available in reservoirs, streams, rivers, lakes, etc., with a further 13,650 trillion litres in the form of water vapour, which will eventually fall as rain. Conversely, about 1100 trillion litres of water evaporates into the atmosphere worldwide every day.

Sources of water in the UK

Of the rain that falls on the UK annually, only 5 per cent is collected and stored in reservoirs for the drinking water supply. The rest flows in rivers to the sea or is filtered down to the natural water table or aquifers that exist below the ground surface. The main sources of water in the UK are shown in Table 5.1.

▼ Table 5.1 Sources of water in the UK

Source of water	Description	Properties of water from this source
Deep wells	Machine-dug wells that draw their water from below the shallow impermeable strata (see Figure 5.4).	Usually good-quality water, as extracted from below the Earth's surface.
Shallow wells	Wells dug by hand or excavator that penetrate only the first water-bearing strata, or aquifer, in the Earth's surface (see Figure 5.4).	Must be considered dangerous because it can be contaminated with water from cesspits or broken drains, etc.
Upland surface water	Water that has collected in upland lakes and rivers without passing through the Earth's strata.	The main water source for the north-west of England. It is not contaminated with salts and minerals, and is naturally soft and acidic.
Spring	A naturally occurring flow of water from the Earth's surface.	The purity of the water is highly dependent on the distance it has travelled from the source.
River	A large natural flow of water, usually starting as a small stream on high ground, which enlarges with distance travelled. Usually terminates at the sea and can be tidal, such as the River Avon.	Usually poor quality due to industrial pollution. The cost of treatment is high.
Canals	Most canals are a product of the Industrial Revolution and for many years fell into dereliction. Many, though, have been cleaned and re-opened, and are now sites of natural beauty.	Very poor quality, generally used only for industrial purposes and irrigation.
Aquifers	Naturally occurring water-bearing strata, often deep beneath the Earth's surface. Mostly consist of permeable rock, such as sandstone, gravel silt or clay, which soaks up water like a sponge (see Figure 5.4).	Very high quality, but prone to contamination by nitrates from farming.
Artesian wells and springs	Water that rises from underground water-bearing rock layers under its own pressure, but only if the well head is below the level of the water table (see Figure 5.5).	Usually very good quality as the water is filtered naturally through layers of rock.
Boreholes	Man-made wells that are drilled directly to a below-groundwater source and the water extracted for use if connection to a water main is extremely difficult.	Very high-quality water that, in most cases, is cleaner than the water undertaker's water main. Filtering and chlorination are not necessary, although the quality should be monitored.

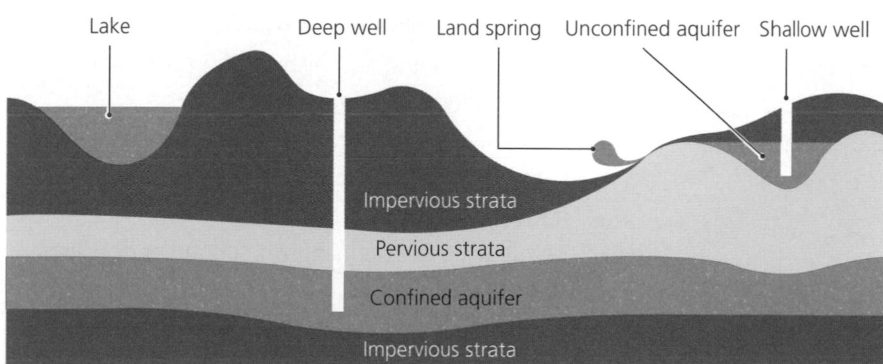

▲ Figure 5.4 Deep and shallow wells

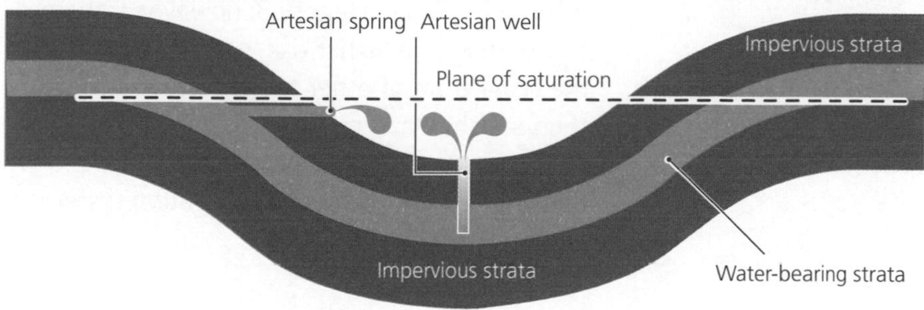

▲ Figure 5.5 Artesian wells and springs

INDUSTRY TIP

The Environment Agency is the overseeing authority for all watercourses in the UK. It samples about 7000 river and canal sites 12 times a year to test their chemistry and nutrients so it can see whether there are any pollutants and whether it needs to target areas for improvement.

INDUSTRY TIP

The information in this section on fluid categories is taken from the Water Supply (Water Fittings) Regulations 1999 document, accessible at: www.legislation.gov.uk/uksi/1999/1148/contents/made

The UK fluid categories

Fluid category means a category of fluid as described in Schedule 1 of the Water Supply (Water Fittings) Regulations and/or Scottish Water Byelaws. Any water that is not cold wholesome drinking water supplied by a water undertaker can be classed as a potential hazard. The Water Supply (Water Fittings) Regulations 1999 list five fluid categories, as described below.

Fluid category 1

Fluid category 1 is wholesome water supplied by a water undertaker, complying with the Water Quality Regulations made under Section 67 of the Water Industry Act 1991. It must be clean, cold and **potable**. All water undertakers have a duty to supply water that conforms to these regulations, which ensure wholesome water suitable for domestic use or food production purposes. Whenever practicable, water for drinking water purposes should be supplied direct from the water undertaker's mains without any intervening storage.

KEY TERM

Potable: drinkable, from the Latin word 'potabilis' meaning drinkable.

INDUSTRY TIP

The three words commonly used to describe fluid category 1 are wholesome water, potable water and drinking water.

Fluid category 2

Fluid category 2 is water that would normally be classified as fluid category 1 but whose aesthetic quality has been impaired because of:

- a change in temperature
- a change in appearance, taste or odour owing to the presence of substances or organisms.

These changes are aesthetic only and do not constitute a health risk. Typical situations where this may occur in domestic properties are:

- water heated in a hot water secondary system
- mixed fluid categories 1 and 2 water discharged from combination taps or showers
- water that has been softened by a domestic common salt regeneration process.

Fluid category 3

Fluid category 3 is water that constitutes a slight health hazard because of the concentration of low-toxicity substances. Fluids in this category are not suitable for drinking or any other domestic purpose or application. The substances might include:

- ethylene glycol (anti-freeze), copper sulphate or similar chemical additives such as heating inhibitors, cleansers and descalers
- sodium hypochlorite and other common disinfectants used in domestic properties.

Typical fluid category 3 situations occur in the following.

- In houses, apartments and other domestic dwellings:
  - water in the primary circuits of heating systems, whether chemicals have been administered or not
  - water in washbasins, baths and shower trays
  - clothes-washing and dishwashing machines
  - home dialysis machines
  - hand-held garden hoses with a flow-controlled spray or shut-off valve
  - hand-held fertilisers.
- In premises other than single-occupancy domestic dwellings:
  - domestic fittings and appliances such as washbasins, baths or showers installed in commercial, industrial or other premises may be regarded as fluid category 3; however, if there is a potential for a higher risk, such as a hospital, medical centre or other similar establishment, then a higher fluid category risk should be applied in accordance with the regulations
  - house-garden or commercial irrigation systems without insecticides.

Fluid category 4

Fluid category 4 is water that constitutes a significant health hazard because of the concentration of toxic substances, which can include:

- chemical, carcinogenic substances or pesticides (including insecticides and herbicides)
- environmental organisms of potential health significance.

Typical fluid category 4 situations are as follows.

- General:
  - primary circuits of heating systems in properties other than a single-occupancy dwelling (commercial systems)
  - fire sprinkler systems using anti-freeze chemicals
  - house gardens
  - mini irrigation systems without fertilisers or insecticides, including pop-up sprinkler systems and permeable hoses.
- Food processing:
  - food preparation
  - dairies
  - bottle washing plants.
- Catering:
  - commercial dishwashers
  - refrigerating equipment.
- Industrial and commercial installations:
  - dyeing equipment
  - industrial disinfection equipment
  - photographic and printing applications
  - car-washing and degreasing plant
  - brewery and distilling processes
  - water treatment plant or softeners that use methods other than salt
  - pressurised fire-fighting systems.

Fluid category 5

Fluid category 5 represents a serious health risk because of the concentration of pathogenic organisms, radioactive material or very toxic substances. These include water that contains:

- faecal material or any other human waste
- butchery or any other animal waste
- pathogens from any source.

Typical fluid category 5 situations are as follows.

- General:
  - industrial cisterns and tanks
  - hose union bib taps in a non-domestic installation
  - sinks, WC pans, urinals and bidets
  - permeable pipes in any non-domestic garden, whether laid at or below ground level
  - grey-water recycling systems.
- Medical:
  - laboratories
  - any medical or dental equipment with submerged inlets
  - bedpan washers and slop hoppers
  - mortuary and embalming equipment
  - hospital dialysis machines
  - commercial clothes-washing equipment in care homes and similar premises
  - baths, washbasins, kitchen sinks and other appliances that are in non-domestic installations.
- Food processing:
  - butchery and meat trade establishments
  - slaughterhouse equipment
  - vegetable washing.
- Catering:
  - dishwashing machines in healthcare premises and similar establishments
  - vegetable washing.
- Industrial/commercial:
  - industrial and chemical plants
  - laboratories
  - any mobile tanker- or gulley-cleaning vehicles.
- Sewage treatment works and sewer cleaning:
  - drain-cleaning plant
  - water storage for agricultural applications
  - water storage for fire-fighting systems.
- Commercial agricultural:
  - commercial irrigation outlets below or at ground level, and/or permeable pipes, with or without chemical additives
  - insecticide or fertiliser applications
  - commercial hydroponic systems.

The list of examples of applications outlined above for each fluid category is not exhaustive.

The distinction between fluid category 4 and fluid category 5 is often difficult to interpret. In general, we can assume that fluid category 4 is such that the risk to health, because of the level of toxicity or the concentration of substances, is such that harm will occur over a prolonged period of days to weeks to months, whereas the risk from fluid category 5, because of the high concentration of substances or the level of toxicity, is such that serious harm could occur after a very short exposure of minutes to hours to days, or even a single exposure.

KEY POINT

We must remember that fluid category 1 is clean, cold, wholesome water direct from the water undertaker's main and no other fluid category must come into contact with it or contamination may occur.

2 THE TYPES OF WATER SUPPLY TO DWELLINGS

Types of water supply in the UK

There are two types of water supply in the UK:

1. water supplied by a water authority, known as a water undertaker, under Section 67 of the Water Act
2. water supplied from a private source, such as a borehole, river or stream.

The Water Act 2003 (Water Industry Act 1991)

INDUSTRY TIP

Access the Water Act 2003 at: www.legislation.gov.uk/ukpga/2003/37/contents

The Water Act 2003 amalgamates and amends two previous pieces of legislation: the Water Industry Act 1991 and the Water Resources Act 1991. The Water Act 2003 introduced changes to the regulation of the water industry in England and Wales originally made under the Water Industry Act 1991. It is enforced by the Environment Agency and deals with such matters as:

- the appointment and regulation of water and sewerage companies and licensed water suppliers by the Water Services Regulation Authority (Ofwat)
- water supply and sewage disposal powers, and duties of the water companies and suppliers
- the obligations of the water companies and licensed water suppliers to supply water that is fit for human consumption, and the enforcement of those obligations by the Department for Environment, Food & Rural Affairs (Defra) and the Drinking Water Inspectorate
- charging powers of water companies and suppliers, and the control of those charges by Ofwat
- protection of customers and consumers by Ofwat and the Consumer Council for Water.

Under the provisions laid down by the Water Act 2003, the UK Government introduced two documents that regulate how plumbers install, commission and maintain water supplies within domestic buildings. These are:

1. the Water Supply (Water Fittings) Regulations 1999
2. the Private Water Supplies Regulations 2016.

These will be discussed later in the chapter.

Sources of recycled, unwholesome water supply in domestic dwellings

Over the past 20 years, demand for water has increased dramatically in the UK. Each of us now uses an average of 150 litres of water every day for washing, flushing the WC, drinking, cooking, gardening and other household tasks. With the climate changing and frequent periods of drought becoming a possibility, the need to save water is becoming more apparent.

There are many ways in which water usage can be reduced in a dwelling, from simple rainwater collection in water butts for garden use, to more complex systems for clothes washing and WC flushing. We will look at these here. It must be remembered that this type of water is not fit for human consumption and must be marked as such. Any installation in a dwelling must not cross-connect with the mains cold water supply.

There are three types of unwholesome water:

1. grey water
2. rainwater harvesting
3. black water.

Grey water

Waste water from baths, showers, washing machines, dishwashers and sinks is often referred to as grey water.

About a third of all water used in the average household is used for WC flushing. The water used for bathing from baths, showers and washbasins can be collected, cleaned and reused for this purpose.

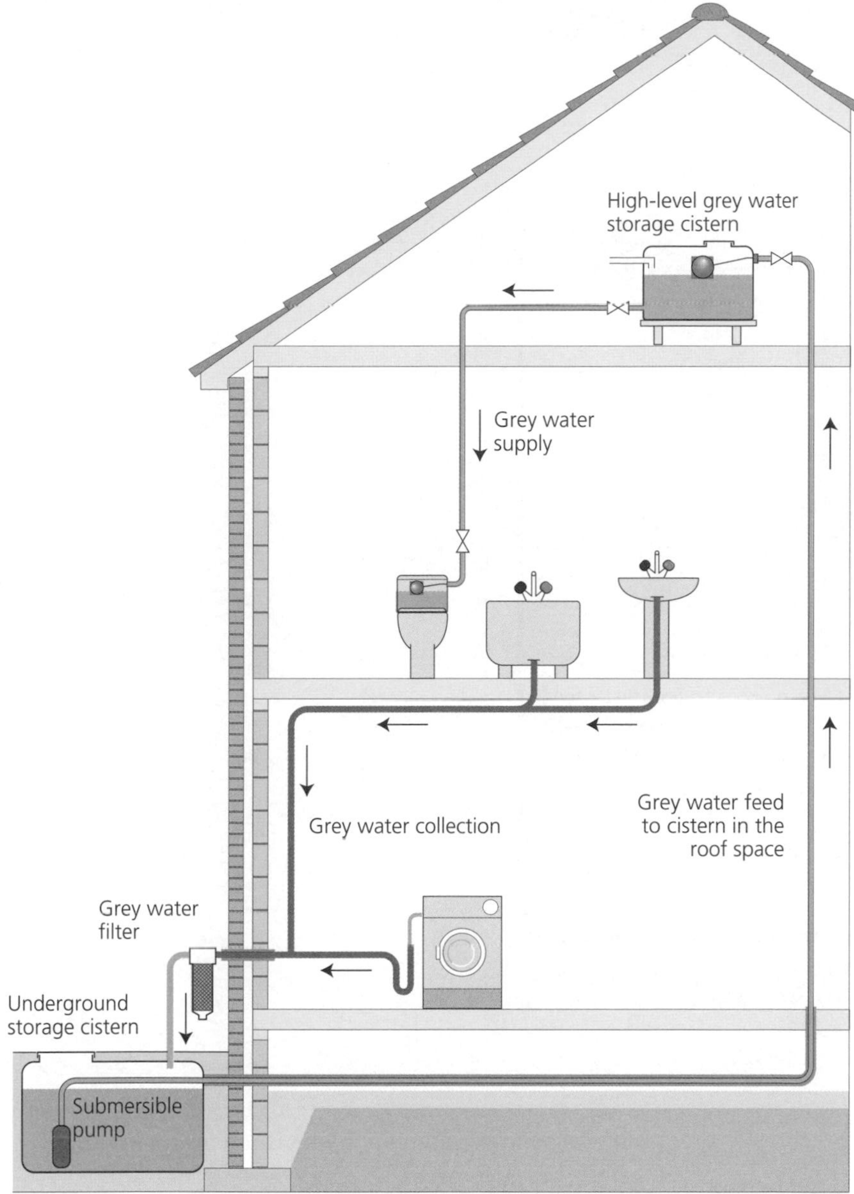

▲ Figure 5.6 Indirect grey water system feeding a WC

Grey water is usually clean enough for use in WCs with only minimal disinfection or micro-biological treatment. Problems can arise when the warm grey water deteriorates when stored, as the bacteria it contains rapidly multiply, making the water smell. This can be overcome by filtration and treatment with chemicals. There must also be a means of protecting the mains water against contamination by backflow from a grey water system, in order to comply with the Water Supply (Water Fittings) Regulations 1999.

Rainwater harvesting

Rainwater harvesting has the potential to save a large volume of mains water and reduce pressure on resources because water that would otherwise be lost can be used to flush toilets, water gardens and feed washing machines, instead of using water direct from the mains supply for such purposes.

Rainwater harvesters can be installed at domestic or commercial sites, and average households can expect to save up to 50 per cent of their water consumption by installing a rainwater harvesting system.

Harvesters are usually installed beneath the ground in an underground storage cistern or on the roof of a flat-roofed building. A typical four-bedroom house will capture enough water to keep a 5000-litre cistern in use throughout most of the year.

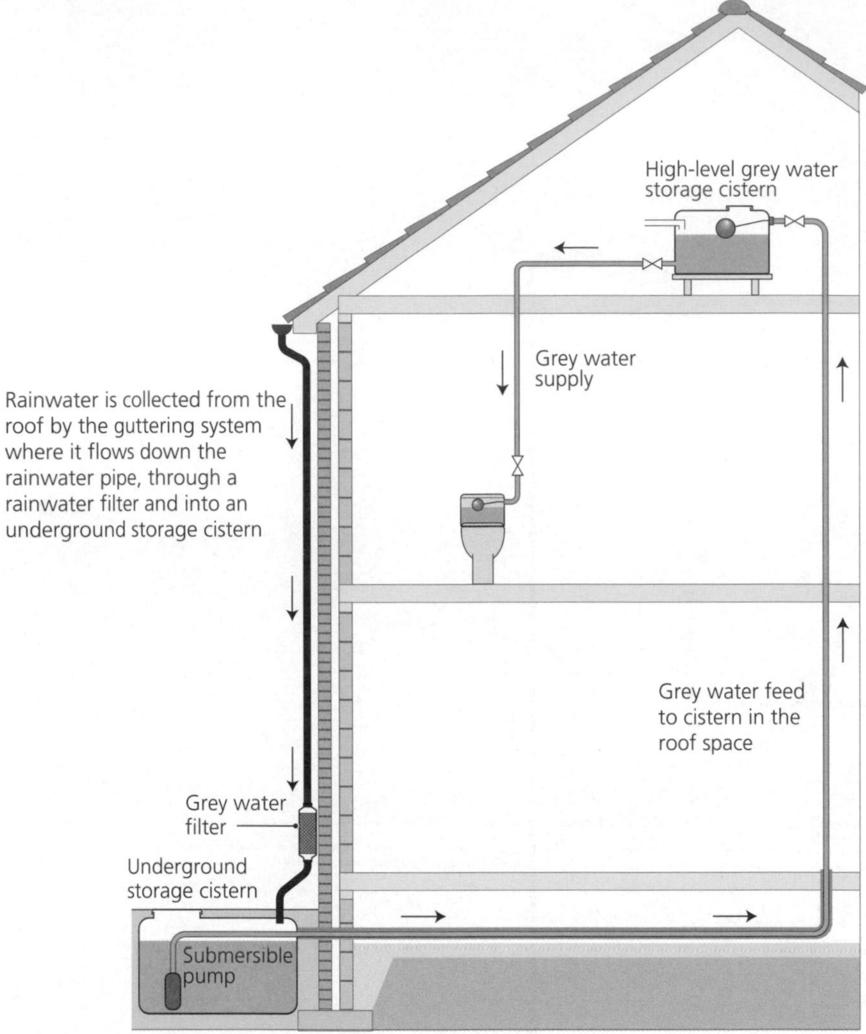

▲ Figure 5.7 Indirect rainwater harvesting system

Black water

Black water is water and effluent from WCs and kitchen sinks that can be treated only by a water undertaker at a sewage works.

Unwholesome water will be revisited in Chapter 9, Sanitation systems.

3 THE WATER TREATMENT PROCESS AND DISTRIBUTION OF WATER

In this section, we will look at the way the water we use every day for drinking, washing and cooking is filtered, cleaned and sterilised to ensure that it is fit for human consumption. The word we use to describe fresh, clean water is 'wholesome'.

Sedimentation, filtration, sterilisation and aeration of water

Before it is considered wholesome, the water undergoes several stages of treatment to ensure its cleanliness and quality. These stages are the responsibility of the water undertaker and are known as:

- sedimentation
- filtration
- sterilisation
- aeration.

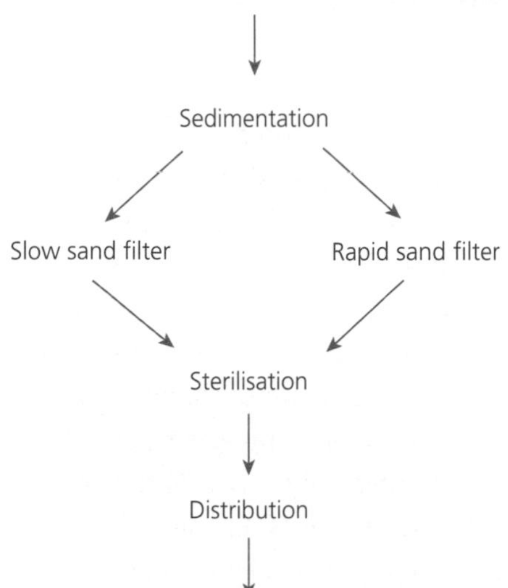

▲ Figure 5.8 The water treatment process

Sedimentation

Sedimentation tanks are designed to slow down the water velocity to allow the solids that the water contains to sink to the bottom and settle under gravity. Simple sedimentation may also be used to reduce **turbidity**.

> **KEY TERM**
>
> **Turbidity:** the cloudiness or haziness of water caused by particles that are usually invisible to the naked eye. Turbidity is a key test of water quality.

Sedimentation tanks are usually rectangular in shape, with a length to width ratio of 2:1, and are usually 1.5–2 m deep. The inlet and outlet must be on opposite sides of the tank, and the inlet designed to distribute the incoming flow as evenly across the tank as possible. The outlet should be designed to collect the cleared water across the entire width of the tank. The tank will also require covering to prevent external contamination.

Sedimentation tanks require cleaning when their performance begins to deteriorate; a 12-monthly period between cleaning operations is normally sufficient.

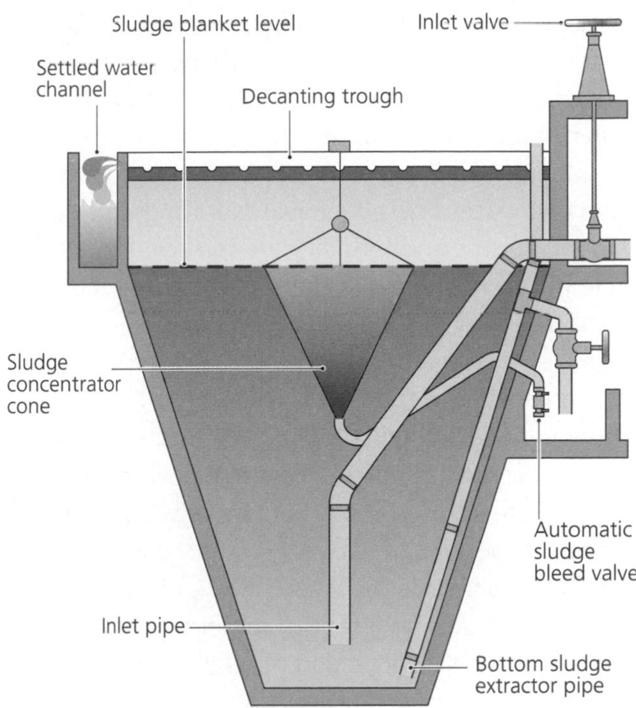

▲ Figure 5.9 Sedimentation tank

Filtration

Filtration is an important process that removes turbidity and algae from the raw, untreated water. There are many different types of filter, including screens, gravel filters, slow sand filters, rapid sand filters and pressure filters. We will concentrate on three of these:

1. slow sand filters
2. rapid sand filters
3. pressure filters.

The difference between these types is not just a matter of the speed of the filtration process, but the underlying principle of the method. Slow sand filtration is a biological process and rapid sand filtration is a physical treatment process.

Slow sand filters

These are often preceded by micro-straining or coarse filtration. These filters are used primarily to remove micro-organisms, algae and turbidity. It is a slow but very reliable method of water treatment, often suited to small supplies, provided that there is sufficient area to properly construct the filtration tanks.

Slow sand filters consist of tanks containing sand with a size range of 0.15 mm to 0.30 mm, and to a depth of around 0.5 m to 1.5 m. For single dwellings, circular modular units, usually used in tandem, are available. These have a diameter of around 1.25 m. As the raw water flows downwards through the sand, micro-organisms and turbidity are removed by a simple filtration process in the top few centimetres of sand. Eventually, a biological layer of sludge develops, which is extremely effective at removing micro-organisms in the water. This layer of sludge is known as the '**schmutzdecke**'. The treated water is then collected in underdrains and pipework at the bottom of the tank. The schmutzdecke will require removing at periods of between 2 and 10 weeks as the filtration process slows. The use of tandem filters means that one filter can remain in service while the other is cleaned and time allowed for the schmutzdecke to re-establish.

Slow sand filters should be sized to deliver between 0.1 m^3 and 0.3 m^3 of water for every 1 m^2 of filter per hour.

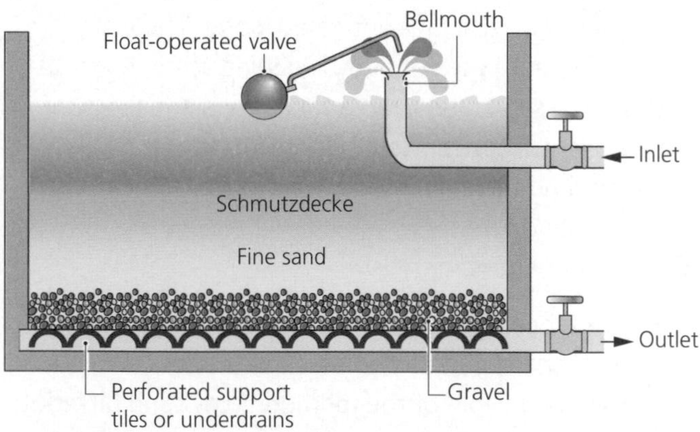

▲ Figure 5.10 Slow sand filter

KEY TERM

Schmutzdecke: 'schmutzdecke' comes from the German word meaning 'dirt cover'.

Rapid (gravity) sand filters

Rapid sand filters are predominantly used to remove the **floc** from coagulated water, but they can also be used to successfully remove algae, iron, manganese and water turbidity from raw water.

KEY TERM

Floc: a collection of loosely bound particles or materials. These are bound together by the coagulation process for easy removal from the water.

Rapid sand filters are usually constructed from rectangular tanks containing coarse silica sand with a size range of 0.5 mm to 1 mm laid to a depth of between 0.6 m and 1 m. As the water flows downwards through the filter, the solids remain in the upper part of the sand bed where they become concentrated. The treated water collects at the bottom of the filter and flows through nozzles in the floor. The accumulated solids are removed either manually every 24 hours or automatically when the head loss reaches a predetermined level. This is achieved by backwashing.

A variety of proprietary units are available containing filtering media of different types and sizes. In some filters, the water flows upwards, improving the efficiency.

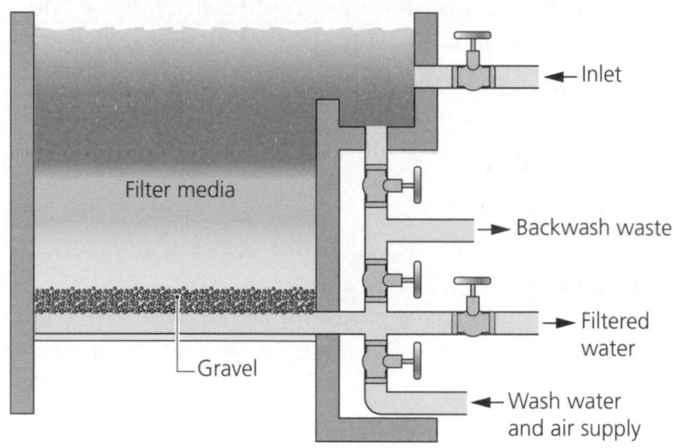

▲ Figure 5.11 Rapid gravity filter

Pressure filters

These are sometimes used where it is important to maintain a head of pressure to remove the need to pump the water into the supply. The filter bed is enclosed in a cylindrical pressure vessel. Some small pressure filters are capable of delivering as much as 15 m^3/h. The cylinder is typically made of specially coated steel, and smaller units can be manufactured from glass-reinforced plastic. They operate in a similar way to the rapid sand filter.

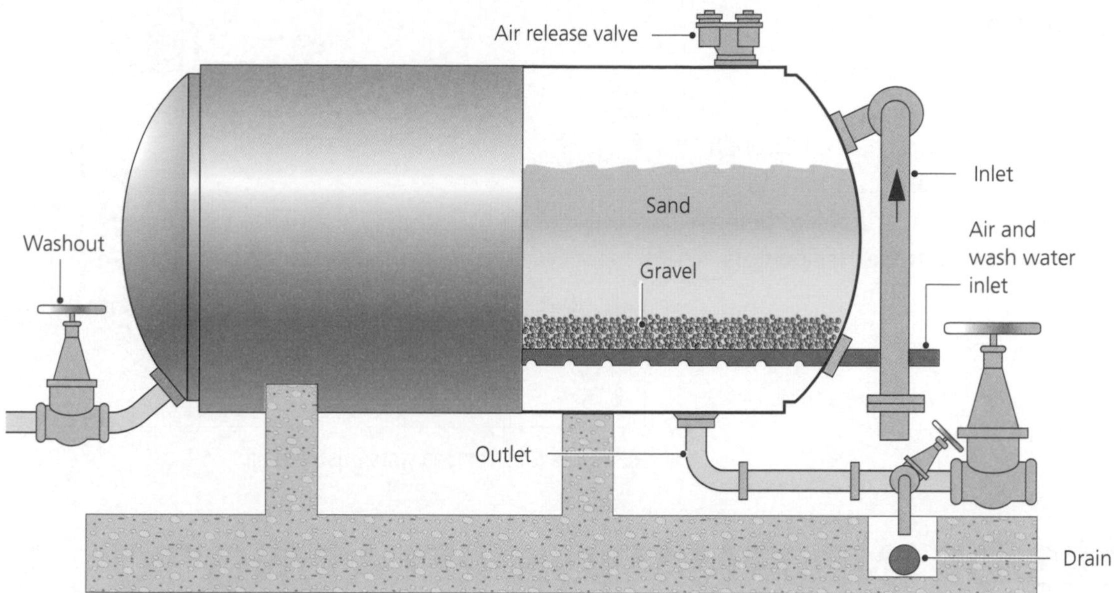

▲ Figure 5.12 Pressure filter

Sterilisation

Finally, water is treated with chlorine and ammonia before being allowed to enter the water supply. This will kill off any bacteria missed by the water filters. Fluoride is still added in some parts of the UK, but only in minute quantities. Ultraviolet (UV) water treatment uses a simple concept where water passes through a clear chamber where it is exposed to UV light. This UV light destroys the majority of bacteria and viruses that are present in the water.

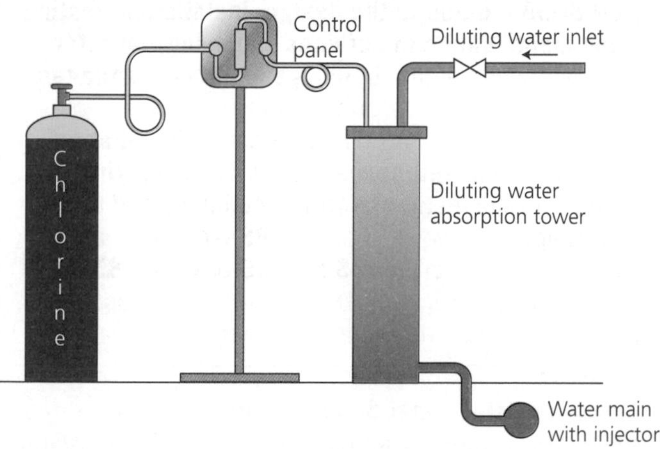

▲ Figure 5.13 Sterilisation by injection of chlorine

Aeration

Aeration is commonly used to treat groundwater by mixing water with air. This removes dissolved metals and gases like CO_2 through chemical reactions and physical mixing.

Distribution of cold water

There are two methods of water supply distribution used in the UK. These are:

1 gravity distribution
2 pumped distribution.

KEY POINT

There are more than 2000 reservoirs used for drinking water in the UK. These are the responsibility of the Environment Agency.

Gravity distribution

The water from the collection of upland surface water is impounded in reservoirs on high ground. Here, the water is filtered by slow sand filters, and chlorinated before being fed to homes and factories by gravity. No pumping is required.

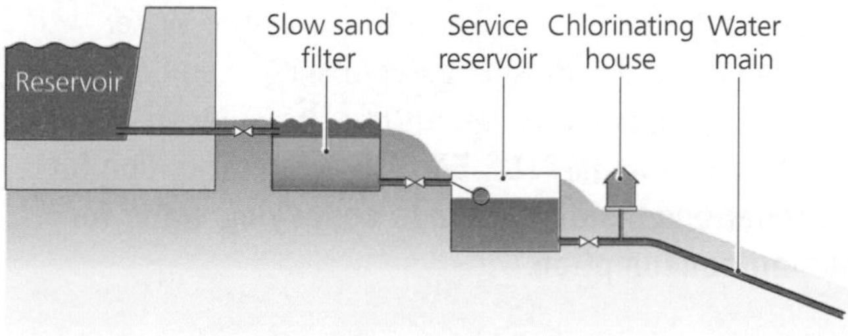

▲ Figure 5.14 Gravity water distribution

Pumped distribution

When water is taken from a river, it is pumped direct to a settlement tank where all of the heavier impurities sink to the bottom. It is then passed through a slow sand filter to remove any organic matter and chlorinated to wholesome water standard before being pumped to a water tower. From the tower, it flows via gravity to the water main.

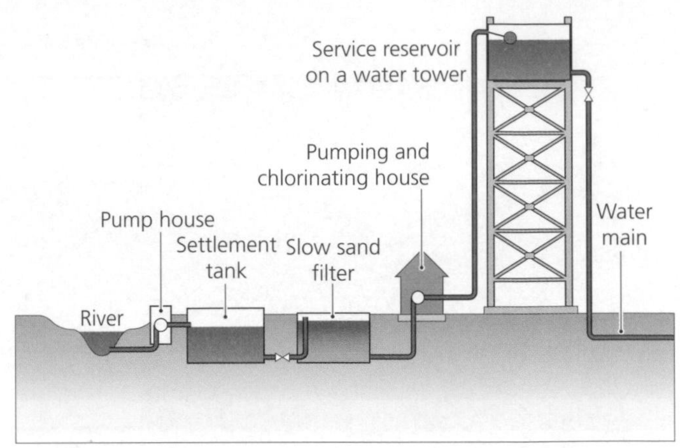

▲ Figure 5.15 Pumped water distribution

4 SOURCES OF INFORMATION RELATING TO COLD WATER SYSTEMS

The sources of information to be used when undertaking work on cold water systems are:

- statutory regulations
- industry standards
- manufacturers' instructions.

Statutory regulations

The Water Supply (Water Fittings) Regulations 1999

Before 1999, each water authority had its own set of water bye-laws that were based upon the 101 Model Water Bye-laws issued by the UK Government in 1986. The problem was that each water undertaker had local variations, which caused much confusion as there was no 'common' standard throughout the UK.

On 1 July 1999, the Office of the Deputy Prime Minister issued the first ever water regulations to be enforced in the UK. They are known as the Water Supply (Water Fittings) Regulations 1999 and offer a common practice throughout the UK. They are linked to a British Standard, **BS EN 806 – Specification for installations inside buildings conveying water for human consumption**.

> **KEY POINT**
>
> **BS EN 806 – Specification for installations inside buildings conveying water for human consumption** is a relatively new British Standard that came into force in 2011. It is divided into five parts:
>
> 1. General recommendations
> 2. Design
> 3. Pipe sizing
> 4. Installation
> 5. Operation and maintenance.
>
> Linked with this is a second British Standard, **BS EN 8558 – Guide to the design, installation, testing and maintenance of services supplying water for domestic use within buildings and their curtilages**. Together, these two publications replace the old British Standard, **BS 6700 – Design, installation, testing and maintenance of services supplying water for domestic use within buildings and their curtilages**. However, parts of **BS 6700** that are not covered in either **BS EN 806** or **BS EN 8558** are retained. For further information contact The Water Regulations Advisory Scheme.

Simply put, the Water Supply (Water Fittings) Regulations were made under Section 74 of the Water Industry Act 1991 and have been put in place to ensure that the plumbing systems we install and maintain prevent the following:

- contamination of water
- wastage of water
- misuse of water
- undue consumption of water
- **erroneous** metering of water.

KEY TERM

Erroneous: wrong; incorrect.

An important factor here is that these Regulations cover only those installations where the water is supplied from a water undertaker's water main, and are enforced by the water undertaker in your area. They are not, however, enforceable where the water is supplied from a private water source.

A free copy of the Water Supply (Water Fittings) Regulations 1999 can be downloaded from the relevant government pages at: www.legislation.gov.uk/uksi/1999/1148/contents/made

The Private Water Supplies Regulations 2016

A private water supply is defined as any water supply that is not provided by a water undertaker. It is not connected to any part of the water mains network and, as such, water rates are not charged, although the owner of any such supply may make a charge for any water used. Private supplies are commonly used in rural areas where connection to water mains is difficult. A private supply may serve one property or many properties on a private network. The water may be supplied from a borehole, spring, well, river, stream or pond.

Under the Water Act 2003, the local authority in the area where the private water supply is located is responsible for the inspection and testing of the water supply to ensure that it is maintained to a quality that is fit for human consumption. These inspections and tests are made in accordance with the Private Water Supplies Regulations 2016. Generally speaking, the more people that use the supply, the more detailed the tests and the more regular the inspections have to be. Supplies for commercial properties and activities, or food production and preparation, have to be tested more frequently and meet more stringent requirements than domestic supplies.

The Private Water Supplies Regulations 2016 stipulate that a risk assessment must be made of all private water supplies including the source, storage tanks, any treatment systems and the premises using the water supply.

INDUSTRY TIP

A free copy of the Private Water Supplies Regulations 2016 can be downloaded here: www.legislation.gov.uk/uksi/2016/618/contents/made

Industry standards

The British Standards (**BS EN 806 – 1 to 5** and **BS 8558**)

The main British Standard for design, installation, commissioning, testing, flushing and disinfection of systems is **BS EN 806:2010 Specifications for installations inside buildings conveying water for human consumption** (in conjunction with guidance document **BS 8558:2011 Guide to the design, installation, testing and maintenance of services supplying water for domestic use within buildings and their curtilages**).

In reality, the information has changed very little from the previous British Standard, **BS 6700: 2006+A1: 2009 Design, installation, testing and maintenance of services supplying water for domestic use within buildings and their curtilages**, and this document should be referenced where alternative information is not available.

The Building Regulations

The Building Regulations make reference to cold water services and systems. These are mentioned briefly in Approved Document G1 – Cold Water Supply and Approved Document G2 – Water Efficiency. Additional recommendations can be found in Annex 1 – Wholesome Water and Annex 2 – Competent Person Self-certification Schemes.

Manufacturers' instructions

Where appliances and equipment are installed on a system, the manufacturer's instructions are a key document when undertaking testing and commissioning procedures, and it is important that these are used correctly at both installation and commissioning operations. Only the manufacturers will know the correct procedures that should be used to safely put the equipment into operation so that it performs to its maximum specification. Remember:

- always read the instructions before operations begin
- always follow the procedures in the correct order
- always hand the instructions over to the customer upon completion
- failure to follow the instructions may invalidate the manufacturer's warranty.

5 THE WATER SERVICE PIPEWORK TO DWELLINGS

Distribution of water in cities, towns and villages

Water is supplied to our homes via a grid system network of pipes known as trunk mains, a phrase dating back to when the mains were constructed from hollowed-out tree trunks.

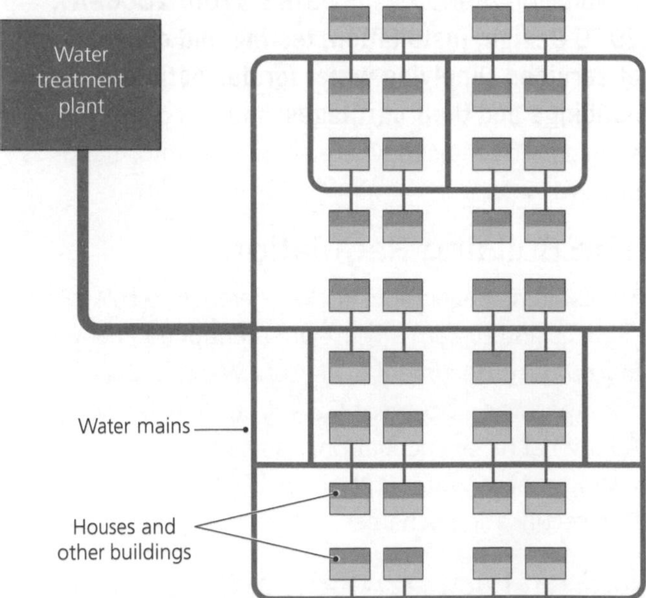

▲ Figure 5.16 Water supply grid system

Trunk mains will vary in diameter depending on the purpose of the main and the likely demand for the supply. Pipes that transfer water to the various points in the distribution system can vary in diameter from 75 mm to 2.3 m. The size of the water main depends upon the size of the community that it serves.

▼ Table 5.2 The size of water main required depends on the size of the community it is to serve

Town population	Size of main (metres diameter)
500,000	1.05 m to 1.20 m
200,000	0.75 m
5000–20,000	0.2 m to 0.3 m

When a new house has to be connected to the water supply, the supply pipes are usually 25 mm in diameter. At the boundary to the dwelling, a 'screw-down stop valve' is installed, so that the supply to the house can be isolated if necessary while any repairs are carried out.

Methods of connection to the water main

Underneath the road is the water main. The connection for the water supply to the dwelling is made by a brass ferrule, which is the responsibility of the water undertaker. The ferrule is a type of shut-off valve that allows the water supply connection to be isolated for maintenance and repair.

The connection to the water main can be made in a number of ways depending upon the material from which the water main is made. For instance, if the water main is made from cast iron, then a self-drilling and tapping machine is used. With this tool, the water main is drilled, threaded and a ferrule inserted while the main is still under pressure so that the supply to other properties is not disrupted. If the main is made from PVCu or cementitious lined asbestos, then a brass strap-type ferrule is used.

▲ Figure 5.17 A strap-type ferrule water main connection

▲ Figure 5.18 A standard gunmetal ferrule

From the water main to the building

The water supply from the water main into the building comprises two separate pipes:

1 the communication pipe, owned and maintained by the water undertaker
2 the supply pipe, owned and maintained by the owner of the building.

The communication pipe is installed by the water undertaker from the ferrule on the water main to the main external stop valve (also known as the boundary stop valve because it is usually located at the boundary of the property). It incorporates a gooseneck bend to allow for any settlement of the roadway or pavement. It is the sole responsibility of the water undertaker to install, repair and maintain the communication pipe and main external stop valve.

The supply pipe runs from the main external stop valve to the dwelling and is the responsibility of the house owner. It must be installed at a minimum depth of 750 mm and a maximum depth of 1350 mm. It must terminate within the building with a screw down-type stop valve with a drain-off valve installed immediately above the stop valve.

Together, the communication pipe and the supply pipe make up the service pipe to the building.

> **KEY POINT**
>
> Figure 5.19 identifies a lot of important factors: pipe names, pipe depth and three points of isolation.

The water supply to buildings can be arranged in numerous ways. In each case, separate dwellings supplied must have a controlling stop valve in a position that will allow the water supply to be turned off in an emergency without affecting any other property.

Most water supplies in modern dwellings and industrial premises are piped in medium-density polyethylene pipe (MDPE), which is coloured blue for easy identification to show mains cold (potable) water. This is generally known as 'blue poly'. The minimum pipe size for modern dwellings is 25 mm. Soft copper to **BS EN 1057 R220** can also be used (in older properties the cold mains may even be 15 mm copper).

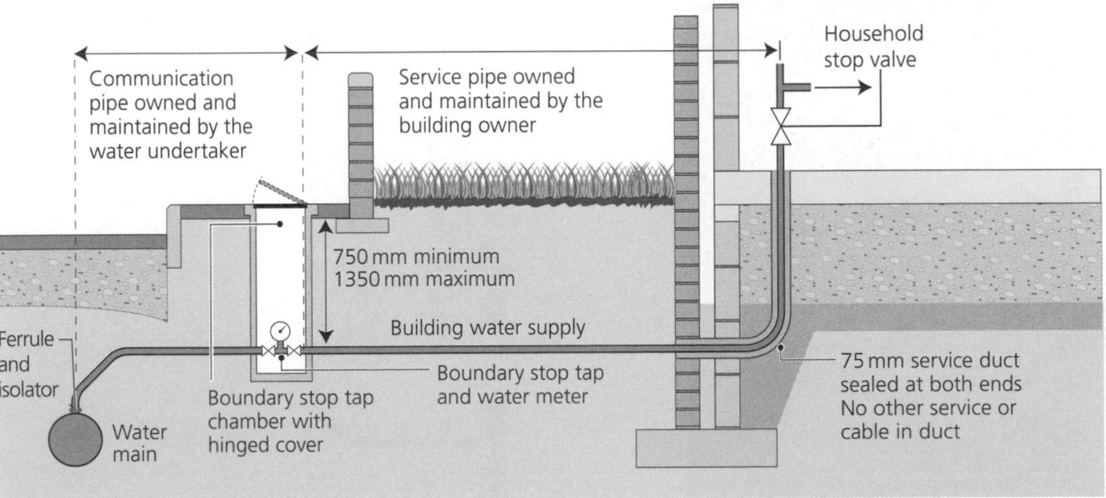

▲ Figure 5.19 The entry of the water supply into the building

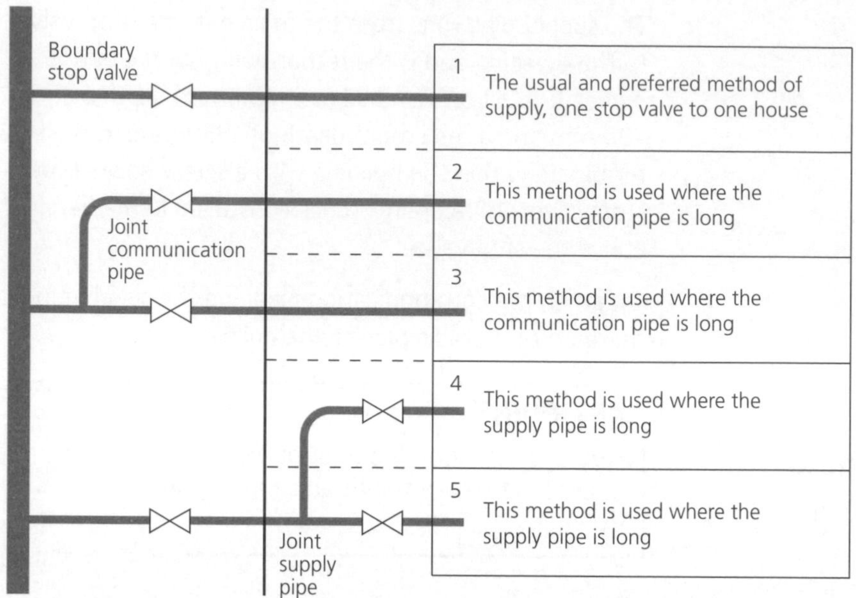
Boundary stop valve
Joint communication pipe
Joint supply pipe
1 The usual and preferred method of supply, one stop valve to one house
2 This method is used where the communication pipe is long
3 This method is used where the communication pipe is long
4 This method is used where the supply pipe is long
5 This method is used where the supply pipe is long

Water suppliers will normally insist on individual supplies to properties and DO NOT favour joint supplies (commonly called communal supplies).

▲ Figure 5.20 Alternative methods of supplying more than one dwelling

With new installations, a water meter is either fitted at the boundary to the property or in an external Groundbreaker-type meter box. This is so the customer does not have to be present when the meter is being read and to prevent illegal tampering with the water meter.

Water meters inside the dwelling are usually fitted to existing water supplies. They must be fitted between two stop taps with a drain-off valve fitted after the meter but before the upper stop tap.

▲ Figure 5.21 Groundbreaker-type meter box

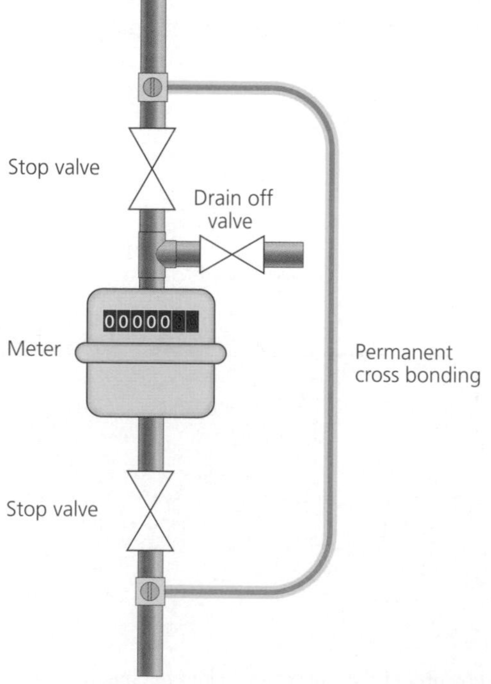
Stop valve
Drain off valve
00000
Meter
Permanent cross bonding
Stop valve

▲ Figure 5.22 Installation of an internal water meter

> **KEY POINT**
>
> From the external main stop valve is where the plumber's involvement with the Water Supply (Water Fittings) Regulations 1999 begins. From here on, everything that we do to the cold and hot water system in the building is regulated to ensure that the water supplied is fit for its intended purpose.

Entering the property

When the water supply enters the property, it should terminate with a screw-down stop tap/valve complying with the Water Regulations and **BS EN 806**. The Defra guidance to the Water Supply (Water Fittings) Regulations 1999 is very specific:

> As far as is reasonably practicable:
>
> 1 A stop valve should be located inside the building; and,
>
> 2 Be located above floor level; and,
>
> 3 As near as possible to the point where the supply enters the building; and,
>
> 4 Be so installed that its closure will prevent the supply of water to any point in the premises.

Source: Section 4, G10.5

The diagram in Figure 5.23 illustrates the point.

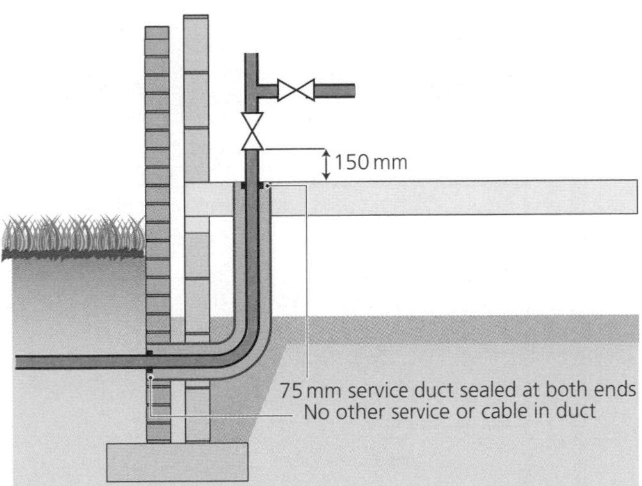

▲ Figure 5.23 The entry of the water supply to a property

The water authorities recommend that no more than 150 mm of blue MDPE pipe be exposed above the floor level of the building. This is to minimise the amount of MDPE pipework visible because MDPE decomposes under persistent exposure to the ultraviolet (UV) light present in daylight.

There are many different stop valve/tap styles available, but they must comply with the Water Regulations and **BS EN 806**. This, however, would not be acceptable when used as the lower stop tap on an internal water meter installation as it would be possible to draw water from the main before the water meter, leading to erroneous metering.

Any stop tap used above or below ground must be made from either gunmetal or corrosion-resistant brass, to prevent de-zincification of the stop tap. The Water Regulations state:

> Every water fitting shall be immune to or protected from corrosion by galvanic action or by any other process which is likely to result in contamination or waste of water.

Source: Schedule 2, Reg. 3

All fittings that are made of a copper alloy, such as brass or gunmetal, should carry either 'CR' or 'GM' markings on the fitting body to show that they are corrosion resistant.

A drain-off valve conforming to **BS EN 1254** should be installed immediately above any stop tap/valve to allow draining of the system.

6 SELECTING COLD WATER SYSTEMS

So far, we have looked at how water is collected, cleaned and distributed to houses and industry. We will now move on to look at the cold water systems we install in dwellings, their components, testing and maintenance.

Each dwelling should have a wholesome (often called 'potable', meaning 'drinkable') water supply, the most important place being at the kitchen sink. In most domestic premises, it is likely that people will drink water from most of the taps. This means that water to all taps should be connected to the mains supply or come from a protected storage cistern.

Drinking water should also be provided in convenient locations in offices and other buildings, especially where food is being eaten or prepared.

Domestic systems of cold water supply

There are two basic systems of cold water used in domestic dwellings:

1. the direct system of cold water
2. the indirect system of cold water.

The direct system of cold water supply

With this system, all cold water taps are fed direct from the mains supply. This means that all taps are provided with a supply of drinking water. Storage is required only for supplying cold water to the hot water cylinder via a 150-litre cistern. A feed cistern will not be necessary if the hot water is supplied via an instantaneous hot water heater or 'combi' boiler.

The direct system is the most commonly installed type of cold water system in domestic properties because its installation is cost effective and there is usually a relatively high-pressure supply available.

▼ Table 5.3 Advantages and disadvantages of the direct system of cold water supply

Advantages	Disadvantages
Cheaper to install Drinking water at all fittings Less pipework Less structural support required in roof space for the cold feed cistern More suitable for instantaneous showers, hose taps and mixer fittings. Used in conjunction with a high-pressure (unvented) hot water supply Smaller pipe sizes may be used in most cases Good pressure at all cold water outlets	At times of peak demand, the pressure may drop If the mains are under repair, the property has no water If there is a leak in the premises, there will be a great deal of damage due to high pressure Can be noisy Greater risk of contamination to mains Greater wear on taps and valves More problems with water hammer Greater risk of condensation build-up on the pipework, which can easily be mistaken for a leak

Pipe sizes for the direct system

Pipe size depends on the system design but, generally speaking:

- a 15 mm rising main will be large enough to supply most cold water demands for a three- to four-bedroom house with all cold water outlets being supplied in 15 mm, including the bath
- if a hot water storage vessel is to be installed, then a minimum of a 22 mm cold feed pipe is required, increasing in size to reflect increased demand to the hot water storage vessel supplied from a 100–150-litre cold water feed cistern in the roof space
- on larger installations, a 22 mm rising main may be required, but this will depend on the water needs of the household.

If a combination boiler or instantaneous water heater is installed, then a 15 mm mains cold water supply should, in most cases, be sufficient, depending upon the supply pressure and flow rate.

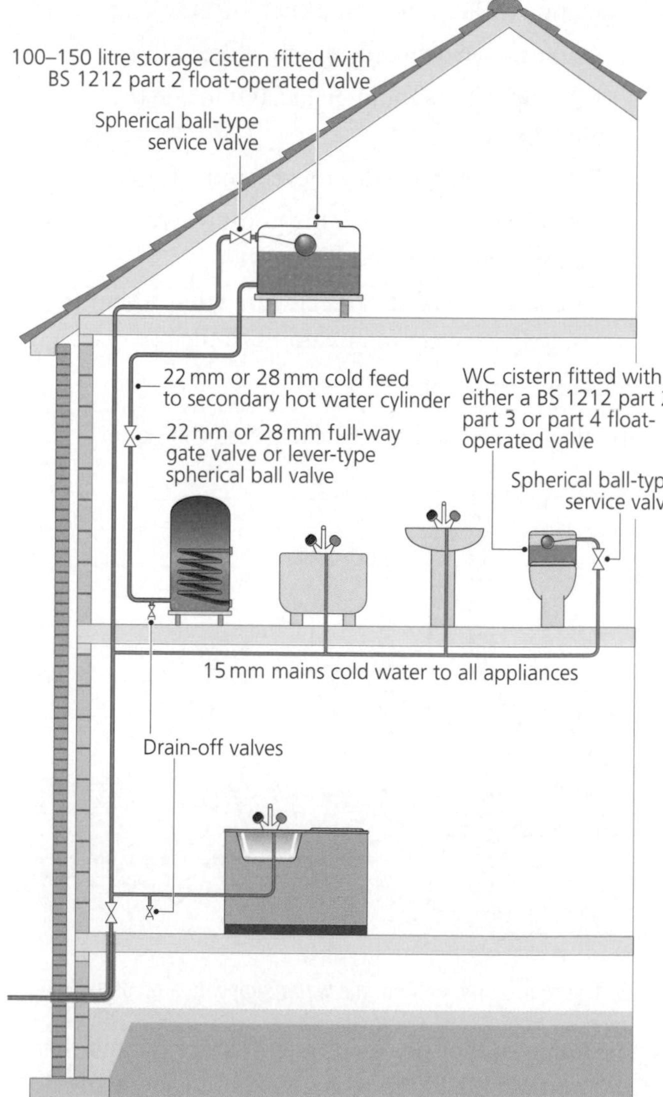

▲ Figure 5.24 The direct system of cold water supply

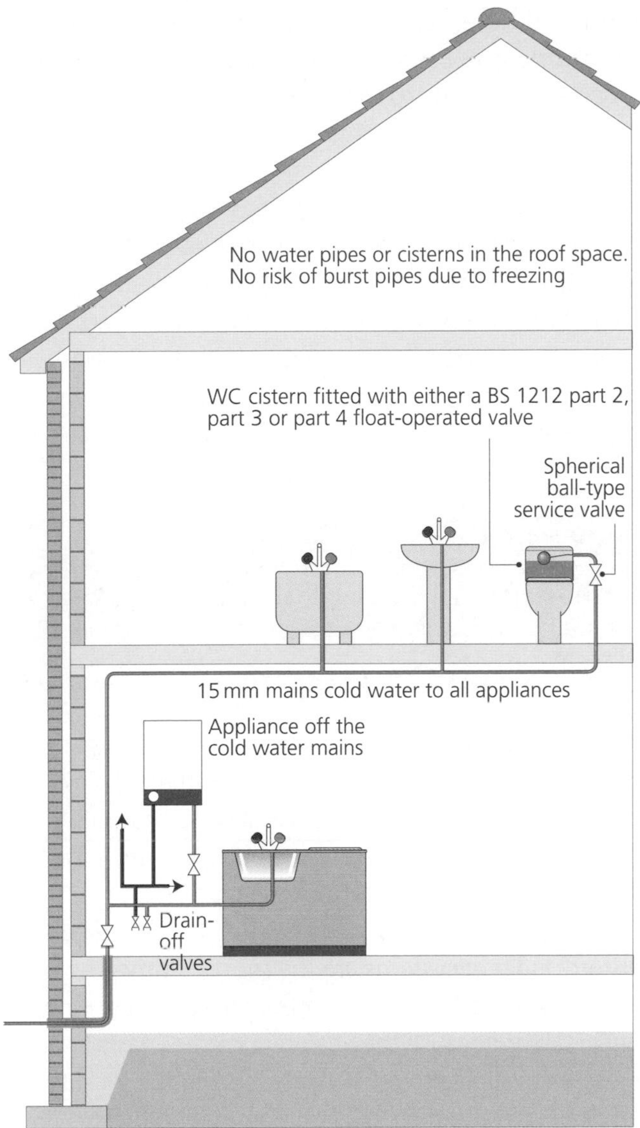

▲ Figure 5.25 Direct cold water system with combi boiler or instantaneous hot water heater

The indirect system of cold water supply

With this system of cold water supply, only the kitchen sink and the cold water storage cistern are fed directly from the mains cold water supply. The other appliances are fed indirectly via the cold water storage cistern in the roof space. A large amount of water will, therefore, need to be stored to supply both cold water and hot water to appliances and fittings from a minimum of 230 litres of water stored in the cistern. The system is designed to be used in low-pressure water areas where the mains supply pipework is not capable of supplying the full requirement of the system. This type of system also has a reserve of stored water for use in the event of mains failure.

The cistern should be installed as high as possible to increase the system pressure.

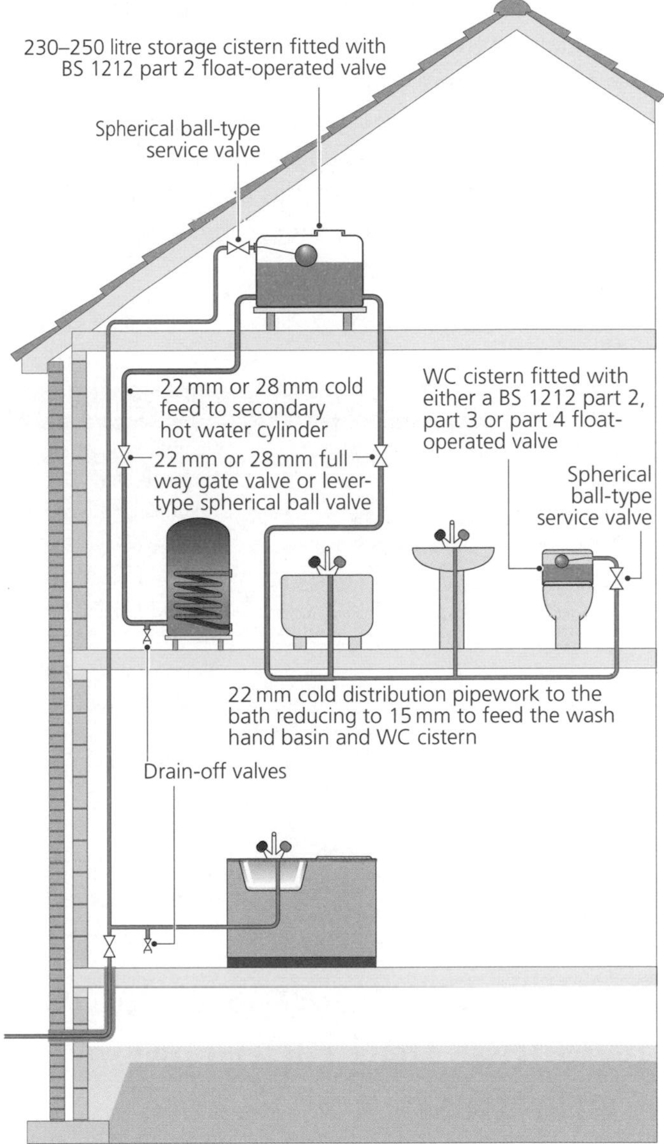

▲ Figure 5.26 The indirect system of cold water supply

Pipe sizes for the indirect system

Pipe size depends on the system design but, generally speaking:

- a 15 mm rising cold water main will be large enough to supply most cold water demands for a three- to four-bedroom house
- the kitchen sink should be supplied with water direct from the cold water main and 15 mm pipework is adequate for this; the cold water storage cistern can also be supplied via 15 mm pipework
- a cold water distribution pipe (22 mm minimum) distributes cold water from the cistern to the

washbasin, WC and bath; the bath should be supplied from 22 mm pipework because of the lack of pressure, but all other appliances can effectively be supplied from 15 mm pipework

- a 28 mm (22 mm minimum) cold feed pipe is needed to supply the hot water storage vessel; this system is ideal when mixing valves and taps require equal pressure and flow rate as both hot and cold supplies are fed from the same source, this being the cold water storage cistern.

▼ Table 5.4 Advantages and disadvantages of the indirect system of cold water supply

Advantages	Disadvantages
Reduced risk of water hammer and noise	Supply pipe must be protected against backflow from cistern
Constant low pressure supply reduces the risk and rate of leakage	Risk of frost damage in the roof space
Suitable for supply to mixer fittings for vented hot water supply	Structural support is needed for the cistern
Reserve supply of water available in case of mains failure	Space taken up
Less risk of backflow – fewer fittings supplied directly	Increased cost of installation
Showers may be supplied at equal head of pressure	Reduced pressure at terminal fittings
Reduces demand on main at peak periods	
Can be sized to give greater flow rate	

Cold water systems in larger dwellings and high-rise properties

For larger buildings (office blocks, factories, hotels, etc.), it is preferable for all water, except drinking water, to be supplied indirectly via a protected storage cistern, or cisterns.

Cold water systems in multi-storey buildings

In plumbing systems, the term **multi-storey** applies to buildings that are simply too tall to be supplied totally using just the pressure of the water main. Because of their design, these buildings have particular cold water system requirements that can be satisfied only by pumping or 'boosting' the cold water supply either in part or in total.

Most cold water supplies that are delivered from the mains cold water supply arrive at a building at a 3 to 7 bar pressure (30–70 metres head). A 30 m head is equivalent to around eight storeys in height. When taking into account a two-storey margin to allow for frictional losses, it becomes obvious that the height of the building will often outstrip the head of pressure available. In some parts of the UK, it is not unusual to find premises with pressures lower than 2 bars and flow rates of below 15 litres/minute. In these cases, the water undertaker should be consulted as to where supply pressures can be relied upon to ensure the correct operation of the cold water system.

KEY TERM

Multi-storey: tall building that requires boosting or pumping of the water supply pressure given its height.

If the public supply is inadequate or the building too high, then the water supply within the building must be boosted. There are several ways that this can be achieved and these can be divided into 'direct boosting' systems, direct from the cold water mains supply, and 'indirect boosting' systems from a break cistern. Indirect systems are the most common as direct boosting systems are often forbidden by water undertakers because they can reduce the mains pressure available to other consumers in the locality and can increase the risk of contamination by backflow. However, where insufficient water pressure exists and the demand is below 0.2 litres/second, then drinking water may be boosted directly from the supply pipe, provided that the water undertaker agrees. With indirect systems, a series of float switches in the break cistern starts and stops the pumps depending upon the water levels in the cistern.

Boosting pumps can create excessive aeration of the water, which, although causing no deterioration of water quality, can cause concern to the consumer because of the opaque, milky appearance of the water. There are several common examples of these systems:

- direct boosting systems
- direct boosting to a drinking water header and duplicate storage cisterns
- indirect boosting to a storage cistern
- indirect boosting with a pressure vessel.

Direct boosting systems

Where permission from the water undertaker has been granted, pumps can be directly fitted to the incoming supply pipe to enable the head of pressure to be increased.

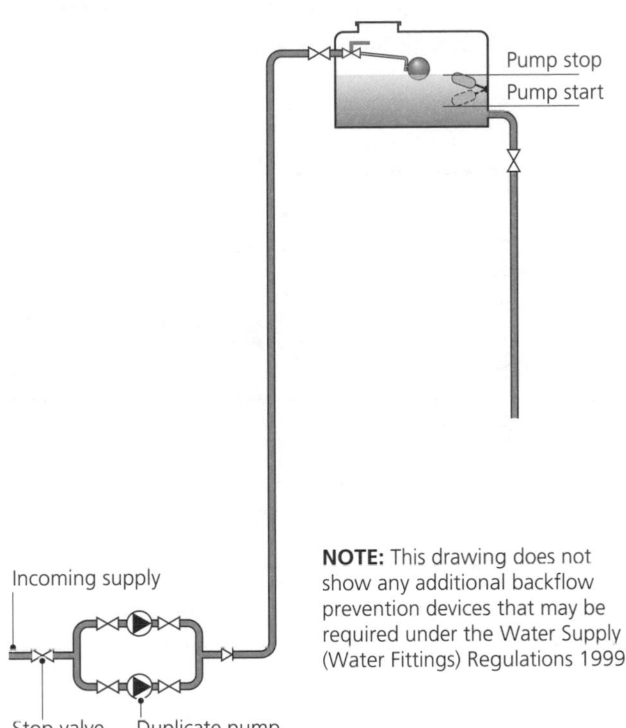

▲ Figure 5.27 Direct boosting system

A float switch or some other no less effective device situated inside the high-level cistern controls the pumps. The pumps either switch on or off depending upon the water level in the cistern. The pumps are activated when the water drops to a depth normally equal to about half the cistern capacity and switch off again when the water level reaches a depth approximately 50 mm below the shut-off level of the float-operated valve.

If the cistern is to be used for drinking water, then it must be of the protected type.

Direct boosting to a drinking water header and duplicate storage cisterns

This system is used mainly for large and multi-storey installations. With this system, the cisterns at high level are for supplying non-drinking water only; a drinking water header sited on the boosted supply pipe provides limited storage of 5 to 7 litres of drinking water to sinks in each dwelling when the pump is not running. Excessive pressure should be avoided as this can lead to an increase in the wastage of water at the sink taps, along with the nuisance of excessive splashing.

A pipeline switch on the header bypass starts the pumps when the water level falls to a predetermined level. The pumps can be time controlled or activated to shut down by a pressure switch. When filling the cisterns, the pumps should shut down when the water levels in the cisterns are approximately 50 mm below the shut-off level of the float-operated valve.

Secondary backflow devices may be required at the drinking water outlets on each floor.

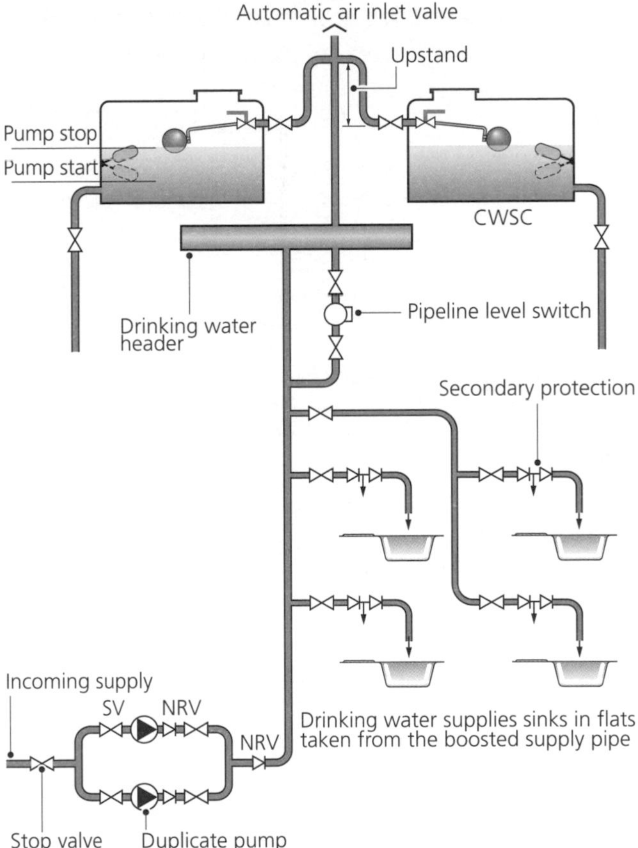

▲ Figure 5.28 Direct boosting to a drinking water header and duplicate cisterns

Indirect boosting to a storage cistern

This system incorporates a break cistern to store the water before it is pumped via a boosting pump (known as a booster set) to a storage cistern at high level. The pumps should be fitted to the outlet of the break cistern. The capacity of the break cistern needs careful consideration and will depend upon the total water storage requirements and the cistern's location within the building, but it should not be less than 15 minutes of the pump's maximum output. However, the cistern must not be oversized as this may result in water stagnation within the cistern.

The water level in the storage cistern (or cisterns) is usually controlled by means of water level switches that control the pumps. When the water drops to a predetermined level, the pumps start to fill the storage cisterns. The pumps are then switched off when the water level reaches a point about 50 mm from the shut-off level of the float-operated valve. A water level switch should also be positioned in the break cistern to automatically shut off the pumps if the water level drops to within 225 mm of the suction connection near the bottom of the break cistern. This is simply to ensure the pumps do not run dry.

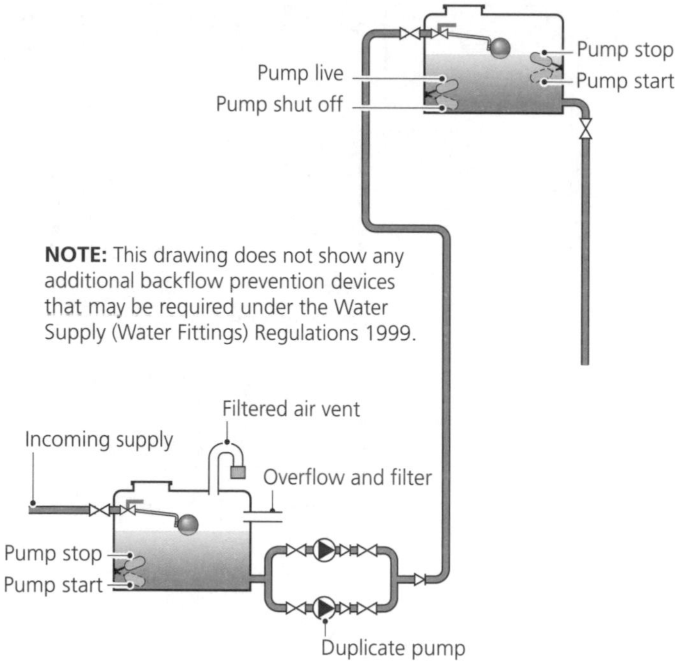

Pump stop
Pump live
Pump start
Pump shut off

NOTE: This drawing does not show any additional backflow prevention devices that may be required under the Water Supply (Water Fittings) Regulations 1999.

Filtered air vent
Incoming supply
Overflow and filter
Pump stop
Pump start
Duplicate pump

▲ Figure 5.29 Indirect boosting to a storage cistern

Indirect boosting with a pressure vessel

This rather complicated system is used mainly in buildings where a number of storage cisterns are fed at various floor levels, making it impractical to control pumps by water level switches. It utilises a pneumatic pressure vessel to maintain the pressure boost to the higher levels of the building.

The pneumatic pressure vessel comprises a small water reservoir with a cushion of compressed air. The water pumps and the compressed air operate intermittently. The pumps replenish the water level and the pressure vessel maintains the system pressure. Since the system may be supplying drinking water, the vessel capacity is purposely kept low to ensure a rapid and regular turnover of water. The compressed air must be filtered to ensure that dust and insects are eliminated.

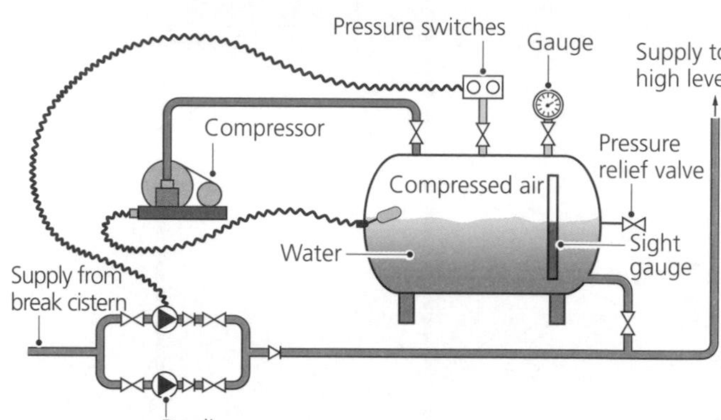

▲ Figure 5.30 Auto-pneumatic pressure vessel

Normally, the controls, including the pressure vessel, pumps, air compressor and control equipment, are purchased as a package, although self-assembly booster sets are available.

▲ Figure 5.31 A typical booster set with pressure vessel and control boards

As can be seen from Figure 5.32, some of the floors below the limit of the mains cold water supply pressure are supplied un-boosted direct from the cold water main, with the floors above the mains pressure limit being supplied via the break cistern and booster set. Drinking water supplies must be from a protected cistern.

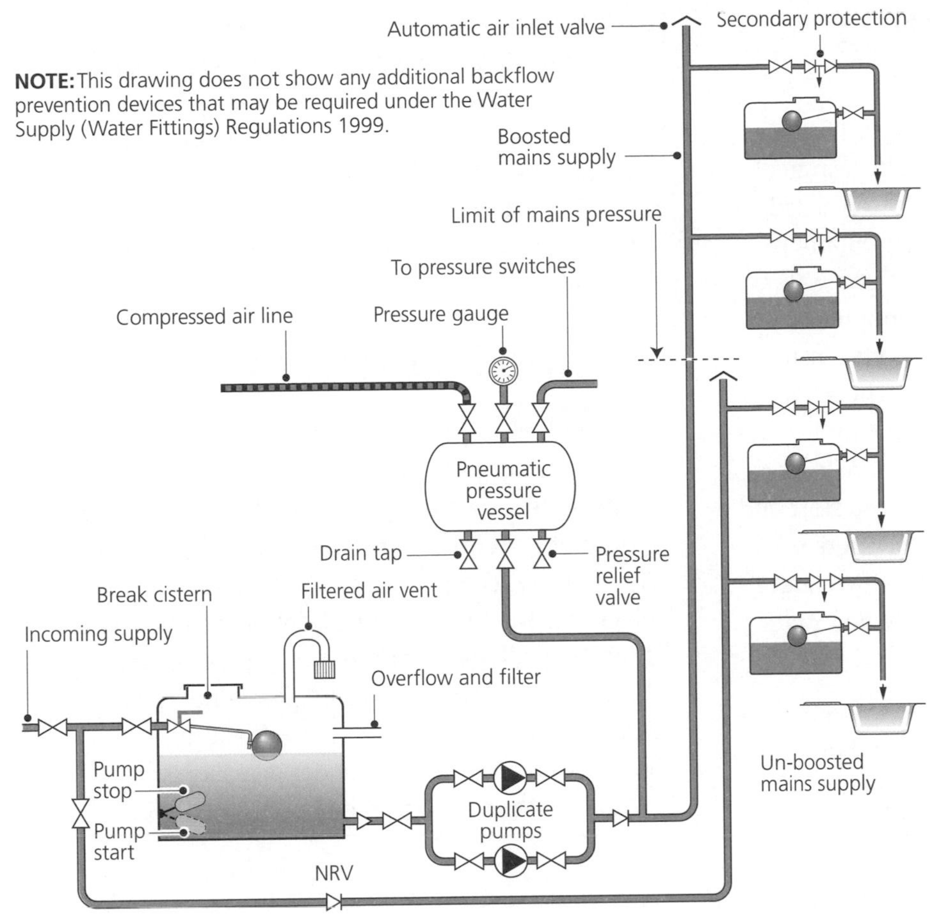

▲ Figure 5.32 Indirect boosting with pressure vessel

Summary of cold water systems

- In some cases, a combination of both methods of supply may be the best arrangement. In a dwelling or a house, for example, the ground-floor outlets and any outside tap could be supplied under mains pressure, while all other cold water outlet fittings could be fed from a storage cistern.
- The performance of any cold water system is dependent upon the pressure of the incoming supply and its flow rate.
- Direct systems require a good pressure and flow rate because all of the appliances use mains cold water supply and, in some cases, mains-fed instantaneous hot water supply too.
- Indirect cold water systems, where low-pressure supply is used, must be pipe-sized correctly to ensure that the system meets the design specification as the lack of pressure is compensated for by an increase in pipe size and, therefore, flow rate.

7 THE SYSTEM LAYOUT FEATURES OF COLD WATER SYSTEMS FED FROM PRIVATE WATER SUPPLIES

The UK has more than 500,000 people whose only source of potable drinking water is from a private supply. There are two methods of pumped supply from a well or a borehole:

1. pumped supply with pressure control
2. pumped supply with level control.

We will look at each of these methods separately.

Pumped supply with pressure control

This type of system provides directly drawn water at the point of use. Pressure is maintained within the system by the use of an accumulator (often called a pressure vessel) and a pump. The accumulator is a vessel that contains air under pressure and water. The water is contained within a neoprene rubber bag inside the accumulator, which expands when water is pumped into it under pressure. The air is then compressed and the pressure rises. As the water within the accumulator is used, the pressure will drop. At a predetermined pressure, the pump will start and the accumulator is refilled, raising the pressure to its operating level. These systems generally operate at 1.5 to 3 bar. This system is preferred when water treatment is being considered.

Control of the system is automatic. The system contains a submersible or surface-mounted pump to bring the water to the surface, filtration and sterilisation equipment (usually UV), a pressure transducer to sense pressure drop across the installation, a pressure gauge and an accumulator. The kitchen sink is usually installed with water under pressure direct from the accumulator. All other outlets are supplied from a low-pressure supply from a storage cistern situated in the roof space. A non-return or check valve must be fitted upstream of the accumulator.

Pumped supply with level control

This system uses a float switch to monitor the level of the water in a storage cistern. The storage cistern is normally situated in the roof space of a dwelling. The float switch operates a surface-mounted pump, which fills the tank until the level of the float switch is reached. All water for the dwelling passes through the storage cistern and this supplies all outlets with a low-pressure supply. Water fed direct from the borehole to a kitchen sink under pressure is not possible with this installation.

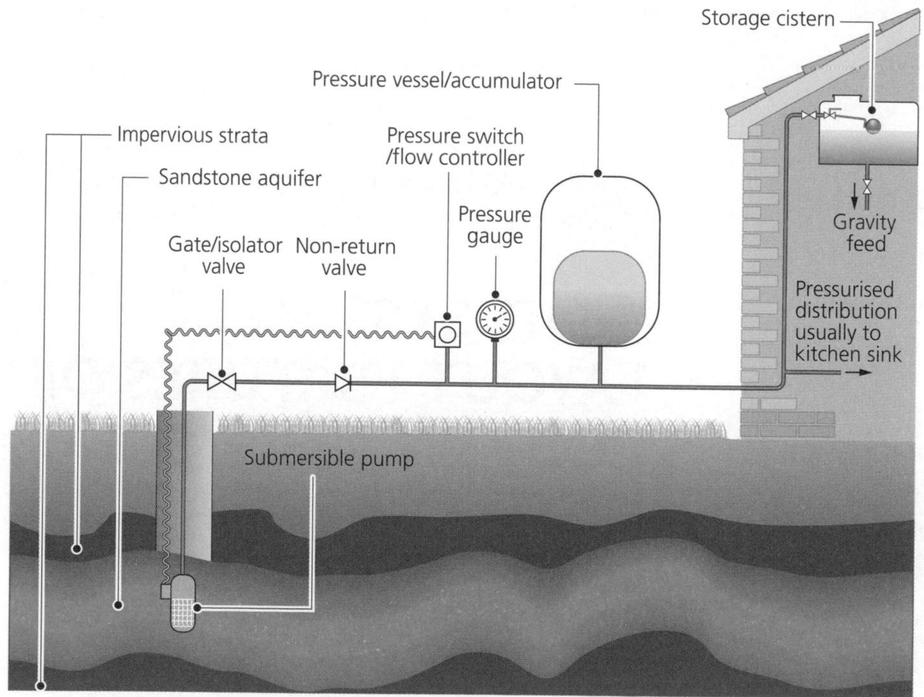

▲ Figure 5.33 A typical borehole installation with pressure control

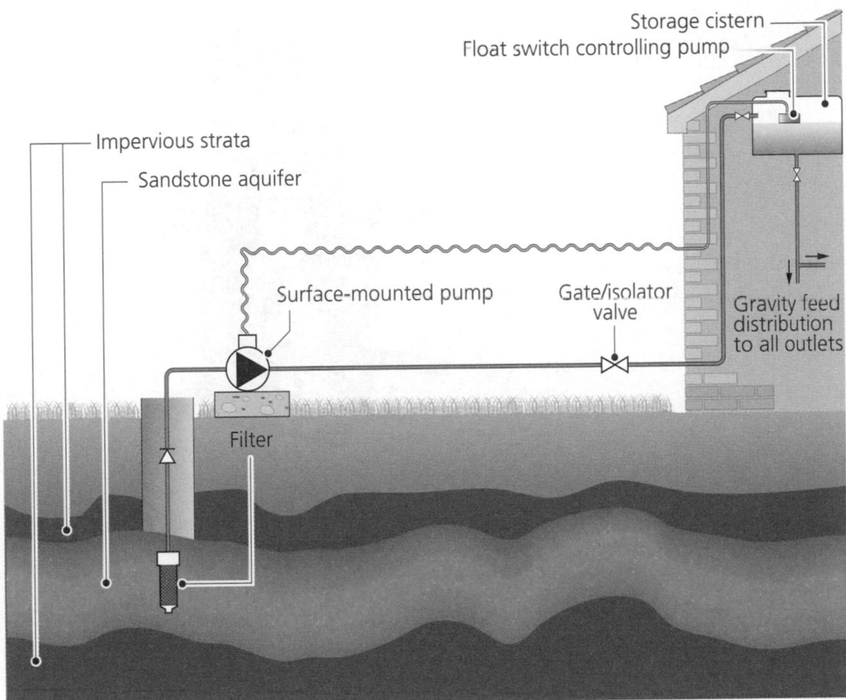

▲ Figure 5.34 A typical borehole installation with level control

Because all of the water for the dwelling is supplied at low pressure, this system can also be used with supplies that are fed via a catchment tank in a stream or spring via an external break/storage cistern. It is also possible to use water direct from a catchment tank without the use of a pump, provided that the source of water is higher than the dwelling. It must be remembered, however, that some form of filtration and sterilisation of the water is necessary. A non-return or check valve must be fitted upstream of the pump.

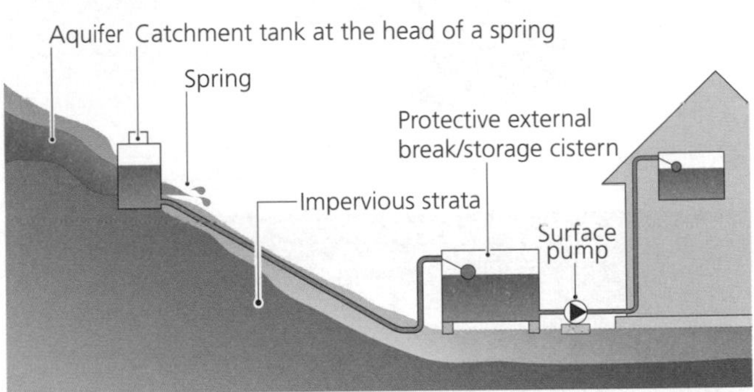

▲ Figure 5.35 A typical spring catchment tank installation with level control

⑧ THE COMPONENTS USED IN BOOSTED (PUMPED) COLD WATER SUPPLY SYSTEMS FROM PRIVATE SOURCES FOR SINGLE-OCCUPANCY DWELLINGS

In this, the final part of the unit dealing with private water supplies, we will investigate the components used with private water supplies to single domestic dwellings:

1. small booster pump sets, which incorporate all controls and components
2. boosted system with separate controls and components
3. use of accumulators in increasing system flow rate.

Vertical, horizontal and submersible pumps

There are two different types of pump that can be used with private water supplies and, more specifically, boreholes and springs:

1 surface pumps, such as:
   - horizontal single-stage types
   - vertical multi-stage types
2 submersible pumps.

Surface pumps for private water supplies are available either as single components or as packaged units containing all the necessary equipment pre-fitted. The latter are the easiest to install and require only the final plumbing and electrical connections.

Submersible pumps may be purchased as separate components or in packs with all the separately matched equipment supplied together ready to assemble.

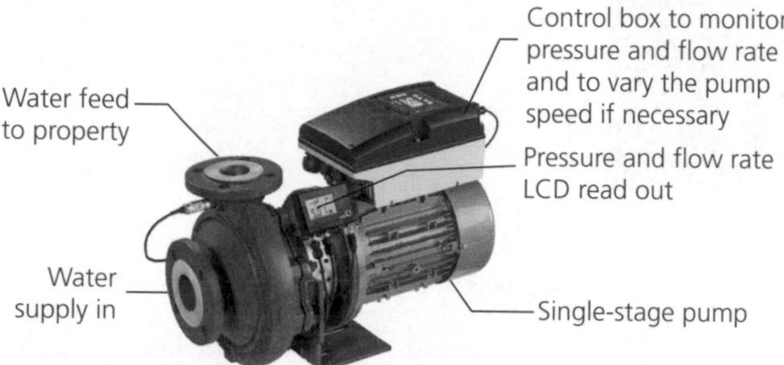

▲ Figure 5.36 Components of a horizontal pump

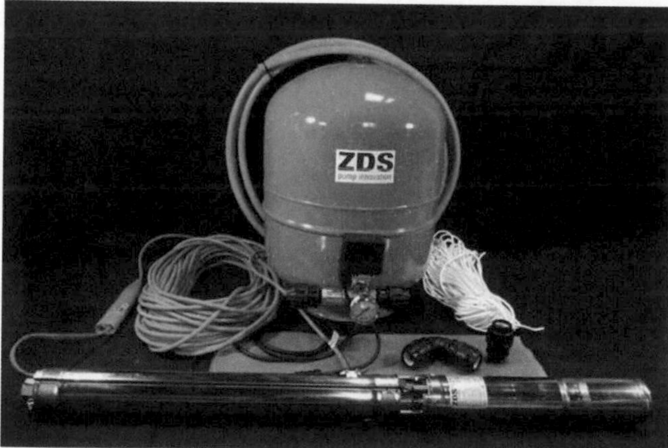

▲ Figure 5.37 A typical submersible pump kit

A typical pump package would normally consist of the following components:

- the pump
- a transducer to sense pressure and flow
- a control box to monitor pressure differentials and flow rate
- an accumulator to assist in providing sufficient system pressure for the installation
- a float switch to prevent the pumps running dry.

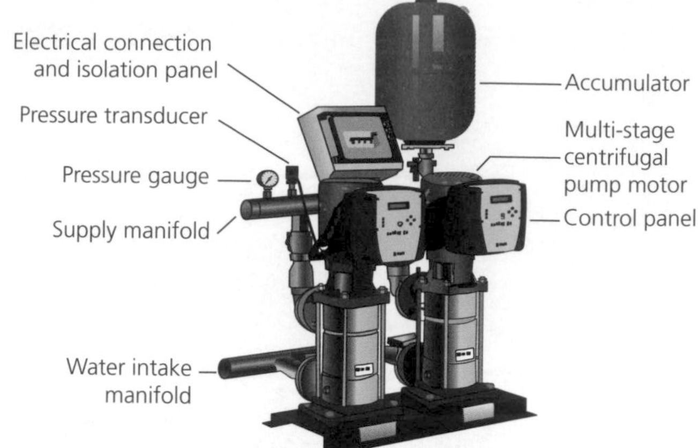

▲ Figure 5.38 Components of a vertical multi-stage pump set

The accumulator

The accumulator is a pressurised vessel that holds a small amount of water for distribution within the installation. It is designed to maintain mains operating pressure when the pump is not working, and to reduce pump usage. Small accumulators can also be used to suppress water hammer.

Small domestic installations use bladder-type accumulators. These consist of a synthetic rubber bladder or bag within a coated steel cylinder or vessel.

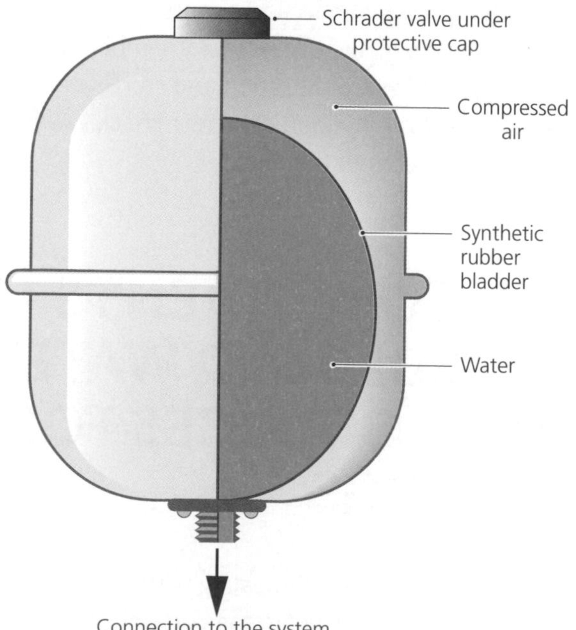

▲ Figure 5.39 A bladder-type accumulator

The operation of an accumulator can be broken down into three stages, as follows.

1 When the pump operates, it forces water into the accumulator bladder, compressing the air surrounding it to a pressure greater than the vessel's pre-charge pressure. This is the source of the stored energy.
2 When the bladder expands due to water being forced in by the pump, it deforms in shape and the pressure within the accumulator increases. Bladder deformation stops when the water and the now compressed air charge become balanced.
3 When a tap is opened, the pressure within the system drops and the compressed air forces the water out of the accumulator. When all of the water inside the accumulator is used and the pressure falls to a predetermined level, the pump energises to recharge the accumulator water storage and pressure, and the cycle begins again.

Probably the most important consideration when applying an accumulator is calculating the correct pre-charge pressure. The following points must be considered:

- the type of accumulator being used
- the work to be done
- the system operating limits.

IMPROVE YOUR MATHS

The pre-charge pressure is usually 80–90 per cent of the minimum system cut-in pressure (the pressure at which the pump energises), to allow a small amount of water to remain in the vessel at all times. This prevents the bladder from collapsing totally. To calculate the pre-charge pressure, follow this simple procedure:

If the minimum working pressure of a cold water system is 2 bar, then:

$$2 \times 0.9\ (90\%) = 1.8 \text{ bar}$$

Pre-charge pressure = 1.8 bar

The accumulator air charge must be lower than the mains pressure for water to enter the vessel and, on average, a pressure differential of around 1.5 bar lower than the supply pressure would be acceptable (but no more than 2 bar and no less than 0.8 bar). This means that, if the supply pressure is 3.5 bar, then the air charge within the accumulator must be around 2 bar; a supply pressure of 4.5 bar would require a 3 bar air charge, and so on. Air pressure can be checked and topped up as necessary at the Schrader valve (a tyre valve where you put the air in) situated at the top of the accumulator.

Float switches, transducers and temperature sensors

Float switches, transducers and temperature sensors play a vital part in modern boosted large-scale cold water systems. The problems encountered are not just those of how to install them but also where to install them. Installations of large cisterns are often undertaken in tight and restricted spaces. Difficulties arise in positioning these components while providing access for maintenance and inspection.

Here, we will look at these important components.

▲ Figure 5.40 Magnetic toggle float switch

Float switches

Float switches, often called level switches, provide detection of water levels within the cistern to activate various other pieces of remote equipment, such as start/stop functions on boosting pumps, open/close functions on solenoid valves, water level alarms and water level indicators.

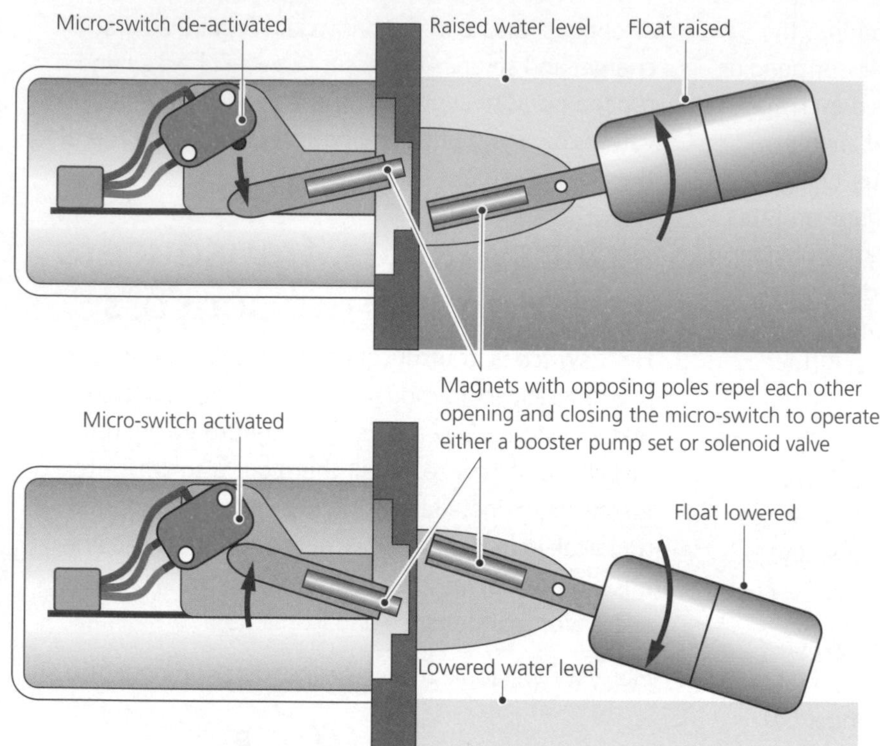

▲ Figure 5.41 A magnetic toggle float switch and how it works

INDUSTRY TIP

Float switches are commonly used in many types of systems to activate and de-activate pumps.

There are many different types of float switch available, and these can vary in sophistication from simple magnetic toggle switches to ultrasonic and electronic types. Popular types include:

- magnetic toggle – a simple float switch that uses the opposing forces of magnets to activate a micro-switch
- sealed float
- pressure-activated diaphragm
- electronic
- ultrasonic.

Transducers

A transducer is an electronic sensor that converts a signal from one form to another. In large-scale, multi-storey water systems, it senses system pressure variations and converts a pre-set low pressure into voltage to activate either the boosting pumps or the compressor feeding the pressure vessel to boost the pressure to normal operating pressure. Transducers may also be used to sense over-pressurisation.

▲ Figure 5.42 Water pressure transducer with pressure gauge

Temperature sensors

Temperature sensors are often used to monitor the temperature of large volumes of stored wholesome, potable water where the installation is of major importance, such as in a hospital, prison or any place where there is a **duty of care**.

KEY TERM

Duty of care: in British law, the moral and legal obligation imposed on an organisation or individual, which necessitates that a standard of reasonable care is adhered to. If the standard of care is not met, then the acts are considered to be negligent and damages may be claimed for in a court of law.

The Water Supply (Water Fittings) Regulations advise that stored wholesome water should not exceed 20°C in order to minimise the risk of micro-bacterial growth.

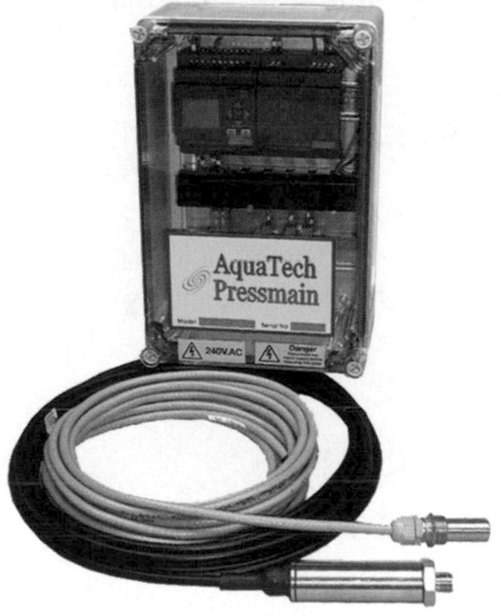

▲ Figure 5.43 Water temperature sensor

Cold water storage cisterns

Storage cisterns and the Water Regulations

Schedule 2, Paragraph 16, of the Water Regulations tells us that a storage cistern supplying low-pressure cold water supply to sanitary appliances, or feeding a hot water storage system, should be capable of supplying potable, wholesome water. Various protection measures must, therefore, be included in the design of the cistern to ensure the water supply does not become contaminated or unwholesome. To comply with the Water Regulations, cisterns must:

- be fitted with an effective inlet control device to maintain the correct water level, i.e. a float-operated valve
- be fitted with service valves on inlet and outlet pipework connections
- be fitted with screened warning/overflow pipes
- be covered with a rigid, close-fitting lid, which is not airtight but excludes light and insects
- be insulated against freezing or undue warming
- be installed so that the risk of contamination is minimised
- be arranged so that water can circulate, preventing stagnation
- be supported to avoid distortion or damage that could lead to leaks
- be readily accessible for maintenance and cleaning.

ACTIVITY

Look in a customer's loft or your loft area at home and identify each of these installation factors.

KEY POINT

Make sure you relate these factors to Figure 5.44.

Cisterns must be installed to these requirements if the problems of contamination are to be avoided. In the past, when cisterns were fitted with unscreened overflows and poorly fitting lids, insects and small mammals could easily gain access to the water the cisterns contained. Some insects, like mosquito larvae, need water to complete their life cycle and these must be avoided.

Schedule 2, Paragraph 16, therefore sets out to prevent this, both in the manufacture and the installation of the cistern.

Correct installation will, in most cases, eliminate the problems mentioned, especially when avoiding the problem of stagnation of water.

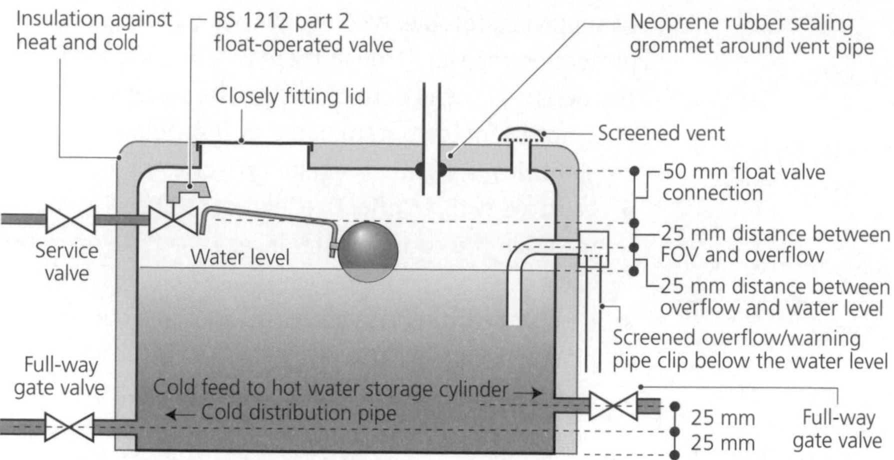

▲ Figure 5.44 Cistern complying with Schedule 2, Paragraph 16, of the Water Regulations

Types of domestic cistern, up to 1000 litres

- **Storage cistern:** this is designed to hold a supply of cold water to feed appliances fitted to the system. On indirect cold water systems, storage cisterns are used. It supplies cold water only.
- **Feed cistern:** this can be identical to the storage cistern. However, it holds only the water required to supply the hot water storage vessel. In other words, it supplies cold water to a hot water storage system.
- **Combined storage and feed cistern:** this is a combination of the previous two examples. It is used on an indirect system of cold water where only the drinking supply is taken direct from the main and the rest of the water is supplied from a cistern. It stores water for the domestic hot water system and the indirect system of cold water to the appliances, wash hand basin, bath, WC, washing machine, etc.
- **Feed and expansion cistern:** used to feed a vented central heating system; also allows expansion of water into the cistern when the system is hot.
- **WC and automatic urinal flushing cisterns:** used to clear the contents of a WC or urinal. The water they contain is not considered wholesome. They will be covered in more detail in Chapter 9, Sanitation systems.

Capacities of domestic cisterns

The British Standards no longer make reference to the minimum requirement as to the amount of water that is stored in a cold water cistern. Instead, **BS EN 806** and **BS 8558** recommend that each dwelling be taken separately, and the amount of water calculated to suit the needs of the dwelling and its occupants.

Kitchen sinks cannot be supplied from a protected cistern. Their supply must come direct from the mains cold water supply.

INDUSTRY TIP

It is recommended that the Water Efficiency Calculator be used to calculate the storage requirements of a dwelling. This can be found here: www.thewatercalculator.org.uk/calculator.asp

General installation requirements for domestic cisterns

Water is heavy. At 4°C it weighs 1 kg per litre, so 230 litres will weigh 230 kg – almost a quarter of a tonne! From this, we can appreciate that a cistern full of water will need adequate support, especially if placed in a roof space. Normal practice would be to try to place the cistern over a load-bearing supporting wall, as shown in Figure 5.45, but if this is not possible, then the platform (or stillage) must be big enough to support the weight of the cistern and the water it contains by spreading the load across the roof joists.

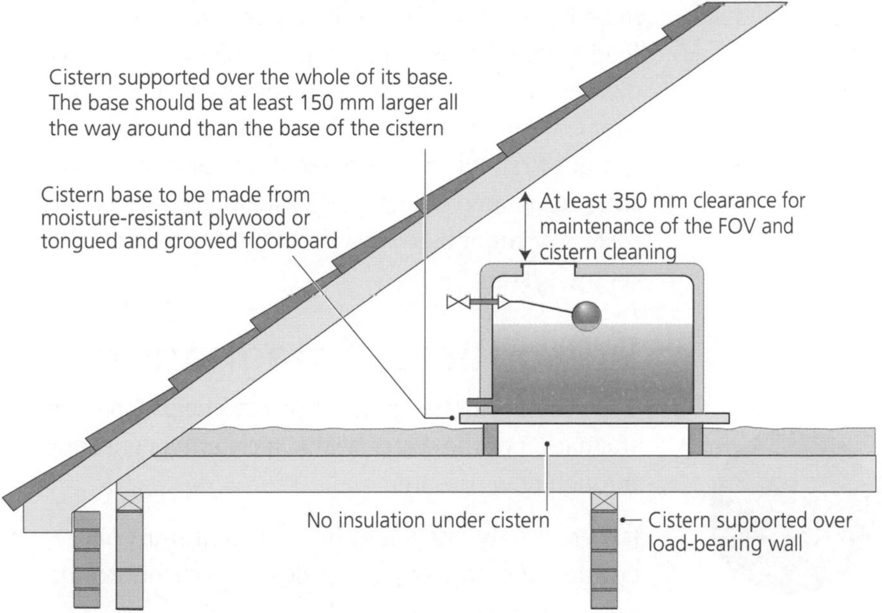
Cistern supported over the whole of its base. The base should be at least 150 mm larger all the way around than the base of the cistern
Cistern base to be made from moisture-resistant plywood or tongued and grooved floorboard
At least 350 mm clearance for maintenance of the FOV and cistern cleaning
No insulation under cistern
Cistern supported over load-bearing wall

▲ Figure 5.45 Cistern shown positioned over a load-bearing supporting wall

The platform that the cistern sits on should be covered with 21 mm tongue and groove boarding or moisture-resistant marine-grade plywood. The platform should be at least as big as the base area of the cistern and, if possible, 150 mm larger all the way around.

Access to the cistern, once it has been installed, is vital for cleaning, inspection and maintenance. The minimum access allowance will depend on the size of the cistern. For cisterns of less than 1000 litres, 350 mm must be allowed to permit access to the float-operated valve before removal and replacement.

Inlet requirements for domestic cisterns

The inlet requirements state that all cisterns will be fitted with an adjustable water inlet control device. These devices are usually float-operated valves that must conform to **BS 1212**. The British Standard defines four types of float-operated valve that are suitable for use in cold water cisterns:

- Part 1 – Portsmouth type (permitted only with a backflow prevention device)
- Part 2 – brass diaphragm type
- Part 3 – plastic diaphragm type
- Part 4 – diaphragm equilibrium type (used only on WC cisterns).

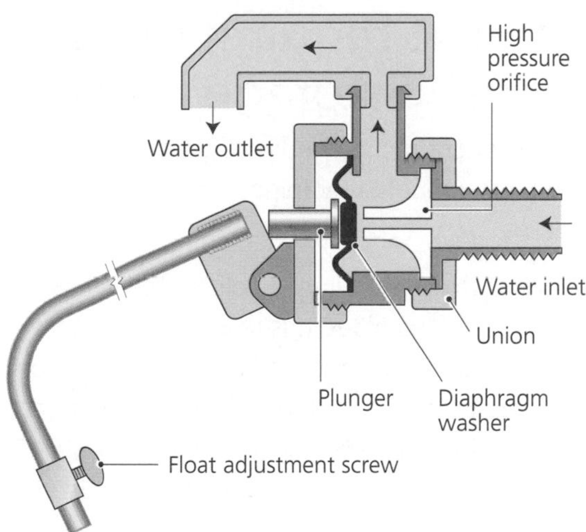
High pressure orifice
Water outlet
Water inlet
Union
Plunger
Diaphragm washer
Float adjustment screw

▲ Figure 5.46 A **BS 1212 Part 2 and 3** float-operated valve with a high-pressure orifice fitted

Figure 5.46 shows a cross-section of a float-operated valve, which can be made of brass (**BS 1212 Part 2**) or plastic (**BS 1212 Part 3**). The adjustment screw allows for the valve to be set at the correct water level. It is important to note that **BS 1212** only covers valves up to 54 mm in size. Any float-operated valve fitted that exceeds this size must be authorised by one of the following bodies:

- Water Regulations Advisory Service
- Water Fittings and Materials Directory
- your local water authority.

A low-pressure orifice, generally coloured red, would be used if the supply pressure was low or the float-operated valve was being fed from another cistern, say, on an indirect cold water system to a WC. The difference between an HP (white) and LP (red) orifice is simply that the LP orifice has a wider opening to allow greater flow of water.

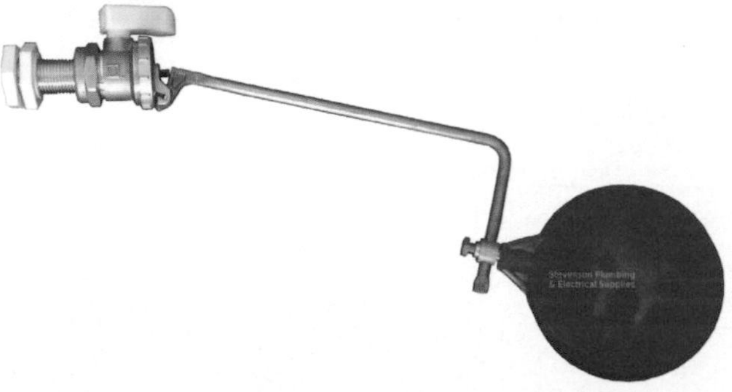

▲ Figure 5.47 A **BS 1212 Part 2** float-operated valve

Water outlet requirements for domestic storage cisterns

Outlets from a cistern include indirect cold water distribution pipes and cold feed pipes to hot water storage systems.

Figure 5.48 shows the positioning of cold water distribution and cold feed pipes. It is recommended that the cold water distribution pipe be taken from the bottom of the cistern. This is to prevent the build-up of sediment on the bottom of the tank; alternatively, the cold water distribution pipe can be located on the side of the cistern.

The distance between the cold water distribution outlet and the cold feed to the hot water system should be not less than 25 mm.

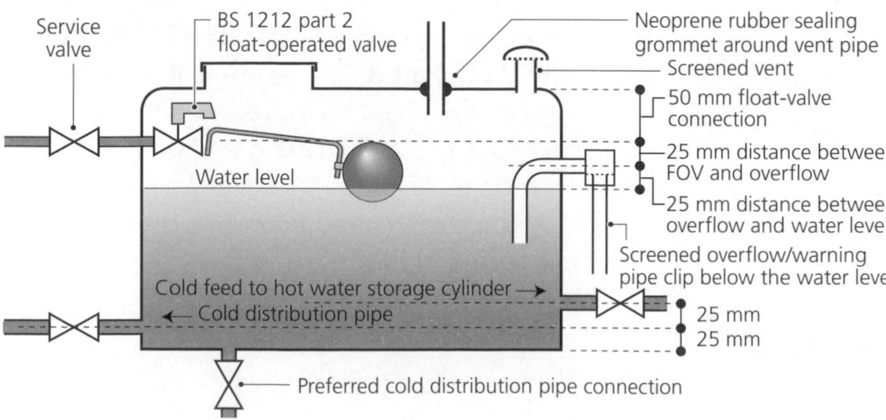

▲ Figure 5.48 Cistern connections

If we look at Figure 5.48, it can be seen that the cold feed for the hot water system is higher than the cold distribution pipe connection. This is so that, in the event of mains cold water failure, the hot water will run out first, which will prevent any potential scalding situation if any mixing valves, such as showers, bath mixers or monobloc washbasin mixers, are installed on the system.

Prevention of stagnation

Correctly positioned outlet pipes can help to prevent stagnation of the water held in a cistern by ensuring a through flow of water.

If there is only one outlet fitted, then it must be positioned on the opposite side of the cistern to the float-operated valve.

If there are two outlets fitted, they should be positioned on opposite sides of the cistern, with one higher than the other. The higher outlet should be on the opposite side of the cistern from the float-operated valve. This arrangement ensures circulation of the water within the cistern, which in turn helps to prevent stagnation.

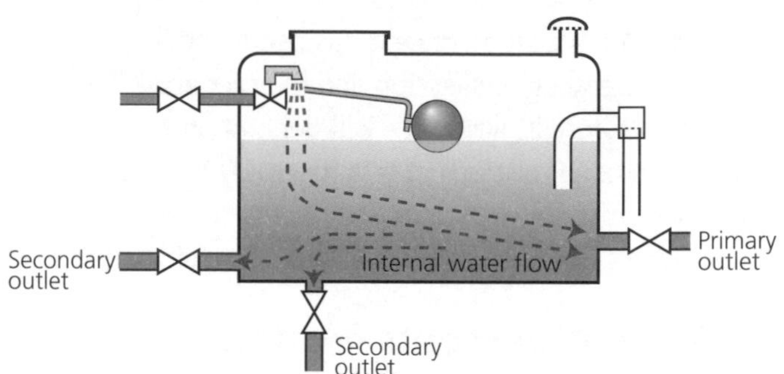

▲ Figure 5.49 Flow of water to prevent stagnation

Materials for domestic cisterns

Almost all new installations use cisterns made from plastics such as polyethylene, polypropylene and glass-reinforced plastic (GRP).

Most cisterns manufactured today are made from polypropylene because this allows:

- lightweight construction
- strength
- hygiene
- resistance to corrosion
- flexibility, as they can easily be handled through roof space openings.

Cisterns are available either square, rectangular or circular in shape, and are produced in black to prevent the growth of algae. However, because they are flexible, the base of the cistern must be fully supported throughout its entire length and width.

Holes for pipe connections should be cut out using a hole saw, and not by using a heated section of copper pipe and using it to make a hole in the cistern. Doing the latter alters the molecular structure of the plastic and will result in the cistern cracking. The joint between the cistern wall and fitting should be made using plastic or rubber washers.

Galvanised steel cisterns were used for many years, but these were notorious for corrosion. They are still manufactured and can still be used, provided the inside of the cistern is protected by the use of a special paint that is registered by the Water Fittings and Materials Directory as safe to use with potable water.

HEALTH AND SAFETY

On no account must any linseed oil-based jointing compounds be used as this also breaks down the plastic and provides a culture where micro-biological growth such as *Legionella pneumophila* (Legionnaires' disease) can occur.

Warning and overflow pipes

Simply put, the difference between a warning pipe and an overflow pipe lies in the fact a warning pipe has a smaller diameter than an overflow pipe. It is intended to act as a warning that the float-operated valve has malfunctioned and the cistern is about to overflow.

An overflow pipe has a larger diameter than a warning pipe and should be able to carry the excess water that would be present if the inlet valve (float-operated valve) fails completely and lets in the maximum amount of water possible. This should ensure that the inlet valve will never become submerged in water and the cistern will not flood the area in which it is situated.

Warning and overflow pipes should run to a point outside of the building that is clearly visible and below the level of the storage cistern. Warning pipes should be situated below the overflow pipe so it is obvious which pipe is which.

Overflow and warning pipe requirements for cold water cisterns vary with the storage volume of the cistern installed.

Cisterns up to 1000-litre capacity

Cisterns below 1000-litre capacity require a single combined warning and overflow pipe. The bottom of the combined warning and overflow pipe should be a minimum of 25 mm above the water level of the cistern.

Layout features for large-scale storage cisterns used in multi-storey cold water systems

The installation of large-scale cisterns differs somewhat from the cisterns you have already been introduced to. Large cisterns must be installed in accordance with the Water Supply (Water Fittings) Regulations 1999 (and the Scottish Water Byelaws 2004). Regulation 5 states that the water undertaker must be notified before the installation of large cisterns begins, and it is important to remember that the correct backflow protection must be present in relation to the fluid category of the contents of the cistern.

In this section, we will look at the general requirements of large-scale cisterns.

Materials for large-scale cisterns

Large cisterns can be made from several materials, and can be either one piece or sectional. Sectional cisterns are constructed, usually on-site, from 1 m^2 sections, which are bolted together and can be made to suit literally any capacity and tailored to fit any space. Sectional cisterns can be internally or externally flanged and are bolted together with stainless steel bolts. The main materials are described below.

For one-piece cisterns:

- glass-reinforced plastic (GRP) **BS EN 13280:2001**
- plastic **BS 4213:2004** and **BS EN 12573–1:2000**
- polypropylene (PP)
- polyethylene (PE)
- polyvinyl chloride (PVC).

For sectional cisterns:

- GRP **BS EN 13280:2001**
- steel, with protection against corrosion and subsequent water contamination in the form of:
  - protection with a paint that is listed in the Water Materials and Fittings Directory
  - glass coated
  - galvanised
  - rubber lined
  - aluminium–rubber lined.

Overflow and warning pipe requirements of large-scale cisterns

Overflows for large cisterns are quite different from those fitted to cisterns for domestic purposes. The objective is the same – to warn that the float-operated valve is malfunctioning and to remove water that may otherwise damage the premises. However, with larger cisterns, the potential for water wastage and water damage is far greater. Therefore, the layout is different.

The overflow/warning pipe on large-scale cisterns must:

- contain a vermin screen to prevent the ingress of insects and vermin
- be capable of draining the maximum inlet flow without compromising the inlet air gap
- contain an air break before connection to a drain
- not be of such a length that it will restrict the flow of water, causing the air gap to be compromised
- discharge in a visible, conspicuous position.

The warning pipe invert needs to be located a minimum of 25 mm above the maximum water level of the cistern, and the air gap not less than 20 mm or twice the internal diameter of the inlet pipe, whichever is the greater.

The general features of larger cisterns are as follows.

- Cisterns with an **actual capacity** of 1000 litres to 5000 litres:
  - the discharge level of the inlet device must be positioned at least twice the diameter of the inlet pipe above the top of the overflow pipe

KEY TERMS

Actual capacity: (of a cistern) the maximum volume it could hold when filled to its overflowing level.

Nominal capacity: (of a cistern) the total volume it could hold when filled to the top of the cistern.

  - the overflow pipe invert must be located at least 25 mm above the invert of the warning pipe (or warning level if an alternative warning device is fitted)
  - the warning pipe invert must be located at least 25 mm above the water level in the cistern and must be at least 25 mm diameter.

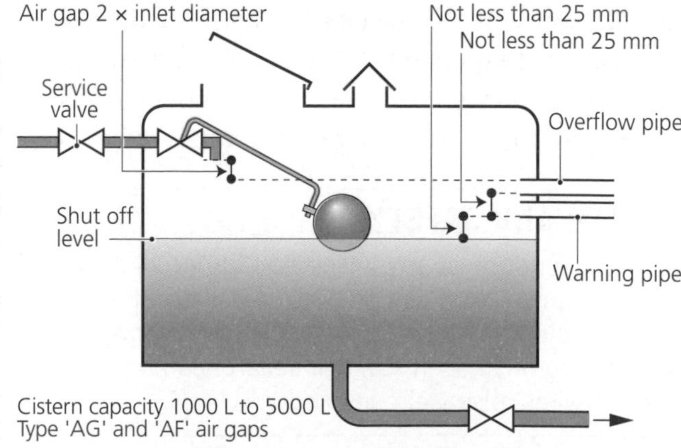

▲ Figure 5.50 Cistern with a capacity of 1000 litres to 5000 litres

- Cisterns with an actual capacity greater than 5000 litres:
  - the discharge level of the inlet device must be positioned at least twice the diameter of the inlet pipe above the top of the overflow pipe

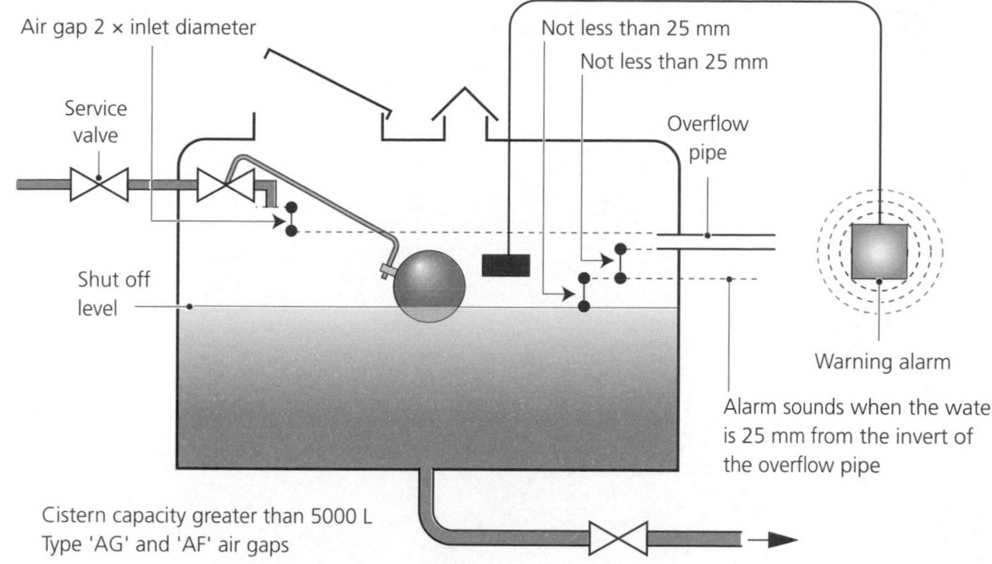

▲ Figure 5.51 Cistern with a capacity greater than 5000 litres

- the overflow pipe invert must be located at least 25 mm above the invert of the warning pipe (or warning level if an alternative warning device is fitted)
- the warning pipe invert must be located at least 25 mm above the water level in the cistern and must be at least 25 mm diameter
- alternatively, the warning pipe may be discarded provided a water level indicator with an audible or visual alarm is installed that operates when the water level reaches 25 mm below the invert of the overflow pipe.

In both cases, the size of the overflow pipe will depend upon the type of air gap incorporated into the cistern (we will look at air gaps and backflow protection a little later in this unit) and this will depend upon the fluid category of the cistern contents. It must be remembered that:

- if a type AG air gap (fluid category 3) is fitted, the overflow diameter shall be a minimum of twice the inlet diameter
- if a type AF air gap (fluid category 4) is fitted, the minimum cross-sectional area of the overflow pipe must be, throughout its entire length, four times the cross-sectional area of the inlet pipe
- for all cisterns greater than 1000 litres, the invert of the overflow must not be less than 50 mm above the working level of the cistern.

Multiple cistern installations: interconnection of two or more cisterns

Where large quantities of water are required but space is limited, cisterns can be interlinked, provided the cisterns are of the same size and capacity. Problems can occur if the cisterns are not linked correctly, especially where the cisterns are to supply drinking water. Stagnation of the water in some parts of the cistern may cause the quality of the water to deteriorate. It should be remembered that the number of cisterns to be linked should be kept to a minimum.

Stagnation can be avoided by following some basic rules. Connection must be arranged to encourage the flow of water through each cistern. This can be achieved by:

- keeping the cistern volumes to a minimum to ensure rapid turnover of water and thus prevent stagnation
- connecting the cisterns in parallel wherever possible
- connecting the inlets and the outlets at opposite ends of the cistern
- using delayed-action float-operated valves to limit stratification.

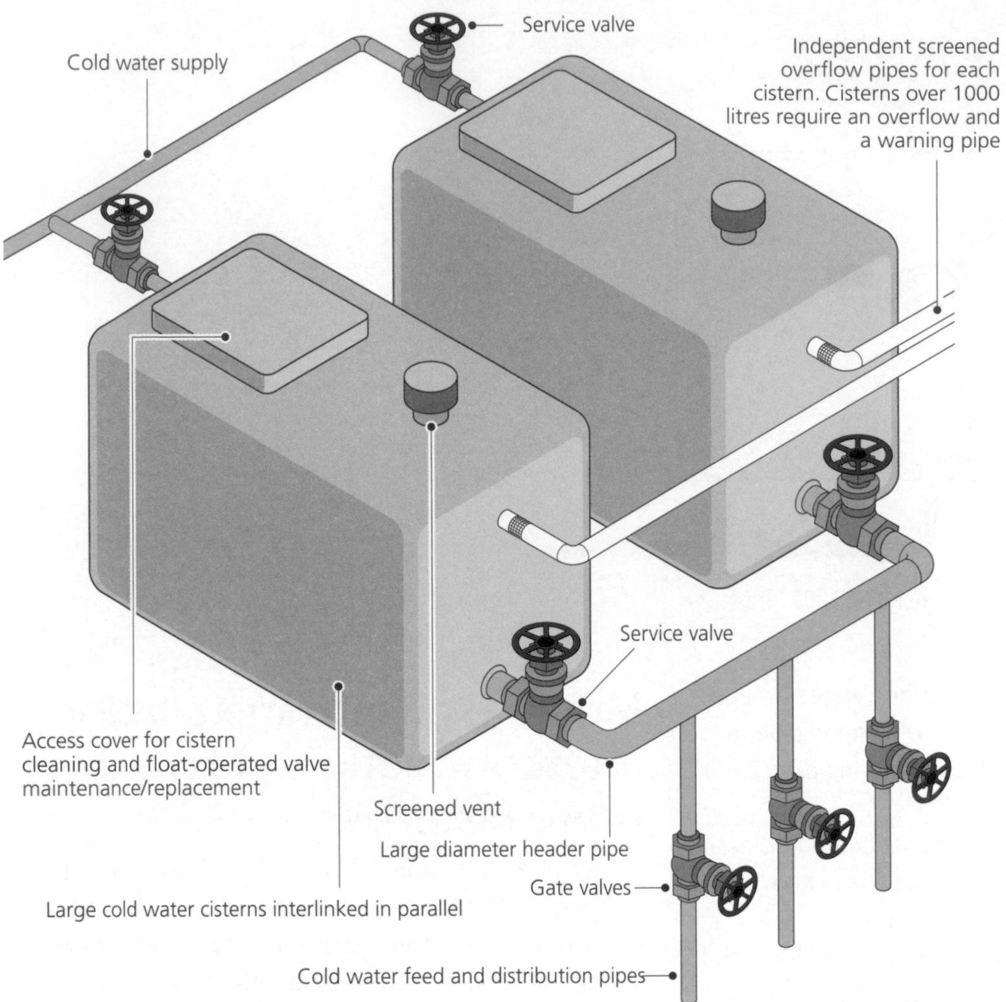

▲ Figure 5.52 Cisterns in parallel

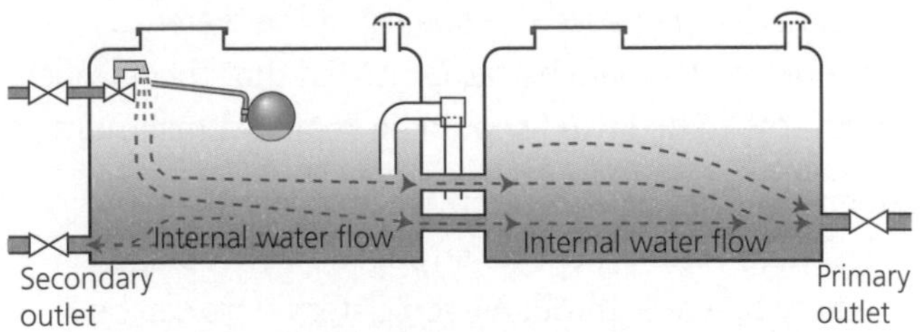

▲ Figure 5.53 Cisterns in series

Where it is not possible to connect cisterns in parallel, cisterns may be connected in series.

In practice, cisterns in series should be interconnected to allow free movement of water from one cistern to the other. They should be connected at the bottom and the middle so that water passes evenly through them. The primary outlet connection should be made on the opposite cistern to the float-operated valve to encourage water movement, with the secondary connection made on the cistern with the float-operated valve installed. The overflow/warning pipe should be fitted onto the same cistern as the float-operated valve. Both cisterns must be of the same size and capacity.

When connecting two or more cisterns, care should be taken to ensure that the water movement is regular and even across all cisterns. In this situation, it is a good idea to install float-operated valves on all cisterns with appropriate service valves, as detailed in the Defra guidance to the Water Supply (Water Fittings) Regulations 1999:

> Service valves should be fitted as close as is reasonably practical to float-operated valves.

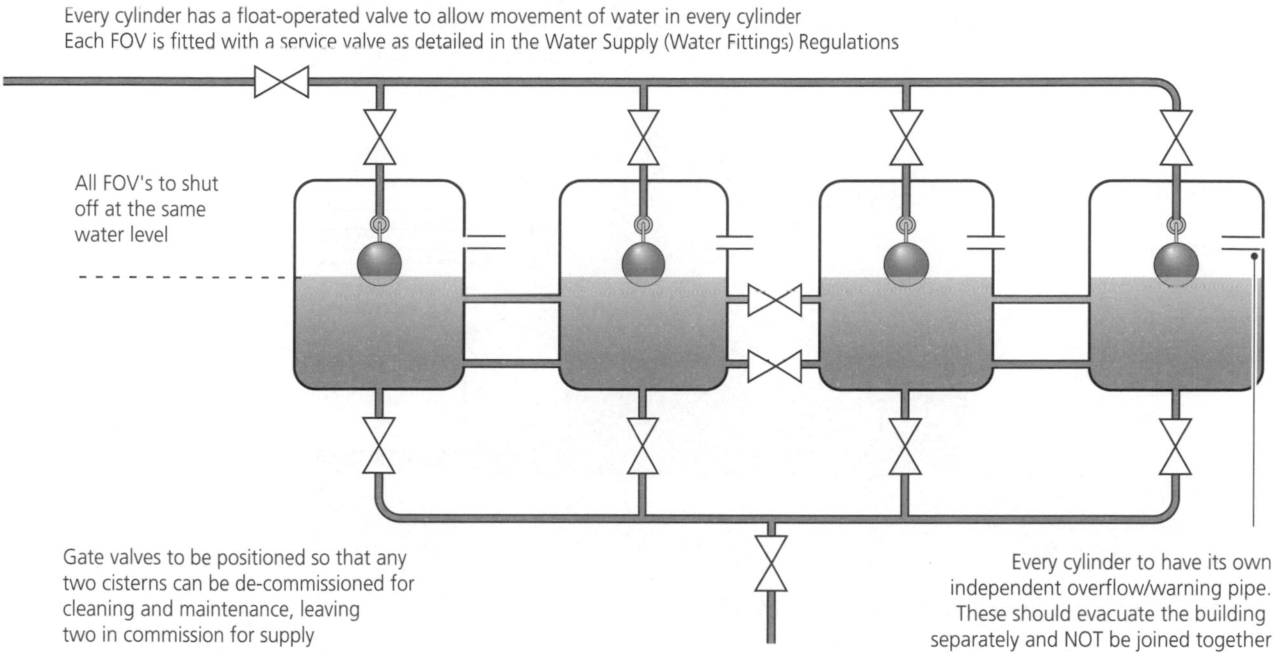

Every cylinder has a float-operated valve to allow movement of water in every cylinder
Each FOV is fitted with a service valve as detailed in the Water Supply (Water Fittings) Regulations

All FOV's to shut off at the same water level

Gate valves to be positioned so that any two cisterns can be de-commissioned for cleaning and maintenance, leaving two in commission for supply

Every cylinder to have its own independent overflow/warning pipe. These should evacuate the building separately and NOT be joined together

▲ Figure 5.54 Installing three or more cisterns

INDUSTRY TIP

A free copy of the Water Supply (Water Fittings) Regulations 1999 can be downloaded from the relevant government pages at: www.legislation.gov.uk/uksi/1999/1148/contents/made

Wherever a float-operated valve is fitted, then an overflow/warning pipe must accompany it. These should terminate in a conspicuous, visible position outside the building. On no account should they be coupled together.

There should be service/gate valves positioned to allow for isolation and maintenance of the cisterns without interrupting the supply. In Figure 5.54, you will see that any two of the four cisterns can be decommissioned, leaving two in operation. This ensures continuation of supply.

Break cisterns

Break cisterns (often called break tanks) are used in large cold water installations in order to supply the system with water via a set of boosting pumps when the mains supply is insufficient. They provide a 'break' in the supply between the mains supply and the installation. This has several advantages over pumping direct from the mains supply:

- Using break cisterns ensures that there is no surge on the mains supply when the boosting pumps either start or stop.
- Break cisterns ensure that contamination of the mains cold water supply from multi-storey installations does not occur.
- Break cisterns ensure that there is sufficient supply for the installation requirements at peak demand.
- Break cisterns safeguard the water supply to other users by not drawing large amounts of water from the mains supply through the boosting pumps.

Break cisterns are often used in very tall buildings as intermediate cisterns on nominated service floors, thus dividing the system into a number of manageable pressure zones. The break cisterns provide water to both user outlets and other break cisterns higher up, where the water is then boosted to other pressure zones further up the building.

As with all cistern installations, break cisterns must be fitted with an appropriate air gap that ensures zero backflow into any part of the system.

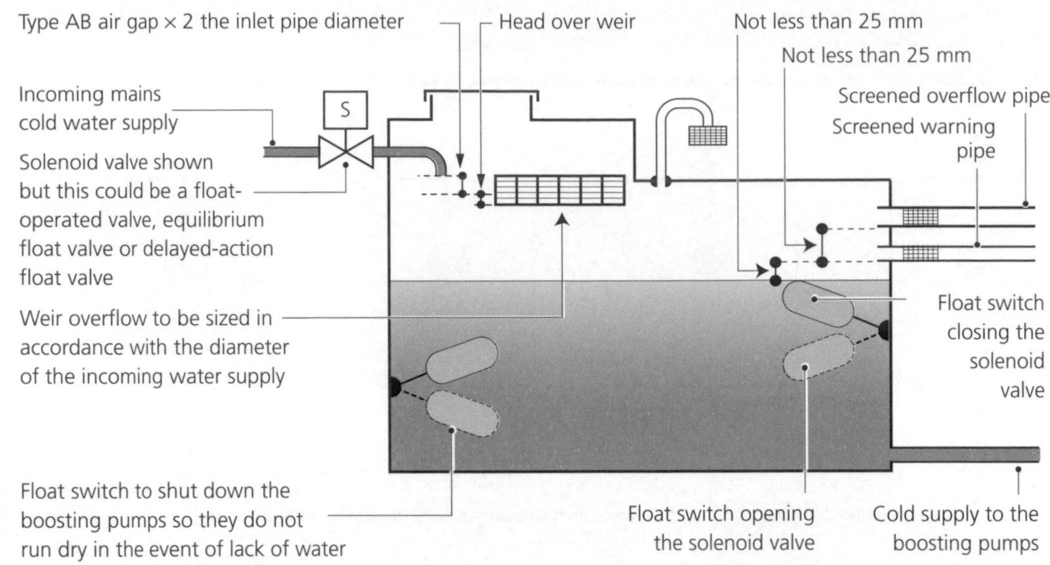

▲ Figure 5.55 The layout of a break cistern with a raised chamber

Frost protection of pipes, fittings and cisterns

You can never fully protect against freezing temperatures. No matter how much insulation we wrap around pipes and fittings, if the weather gets cold enough the pipes will freeze. Therefore, we merely delay the freezing process as long as we possibly can by insulating.

When we insulate pipes, we are not attempting to 'keep the cold out'. The idea of insulation is to keep in the heat that is already there. In other words, we are attempting to retain the 'heat energy' already present in the water for as long as possible. This means insulation is important even under normal conditions as it maximises energy efficiency. The greater the thickness of insulation, the longer the heat energy is retained. This is illustrated in Figure 5.56, where you will see that, for a greater volume of water, i.e. larger pipe sizes, less insulation will be needed. Trace heating is sometimes used. This is covered in Plumbing Book 2, also published by Hodder Education.

The Defra guidance to the Water Supply (Water Fittings) Regulations 1999 (G4.2) states:

> All cold water fittings located within a building but outside the thermal envelope, or those outside the building must be protected against damage by freezing.

▲ Figure 5.56 Insulation around pipes

KEY POINT

The thermal envelope is defined as that part of a building that is enclosed within walls, floor and roof, which is thermally insulated in accordance with the requirements of the Building Regulations.

The Defra guidance to the Water Supply (Water Fittings) Regulations 1999 (G4.3) states:

> If the frost protection provided is insufficient for exceptional freezing conditions or the premises are left unoccupied without adequate heating, damage and leakage can often be avoided by shutting off the water supply and draining down the system before the onset of freezing.

The Defra guidance to the Water Supply (Water Fittings) Regulations 1999 (G4.4) states:

> Where low temperatures persist, insulation will only delay the onset of freezing. Its efficiency is dependent upon its thickness and thermal

conductivity in relation to the pipe size, the time of exposure, the location and, possibly, the wind chill factor.

INDUSTRY TIP

There are two conditions used to determine the thickness of pipe insulation required:

- Normal condition refers to the insulating of pipework in unheated rooms of a building which has a working heating system. All unheated rooms that contain pipework should be protected from freezing, e.g. cloakrooms, store rooms, utility rooms etc.
- Extreme conditions means pipework that will require thicker insulation to give better protection against freezing, e.g. under suspended floor, in garages, pipework externally run above ground and buildings that have no form of heating.

In general, all pipes and fittings that are installed in vulnerable or exposed locations inside and outside a building, such as unheated cellars, roof spaces, under ventilated suspended floors, garages and outbuildings, must be insulated. Where pipework is installed in a roof space, the pipes should still be insulated, even if they are placed below the roof insulation. This is to avoid unnecessary warming by heat from the rooms below. The thickness of the insulation will, as we have already seen, be dependent upon the size of the pipe. Where pipes are located outside the dwelling, the insulation should be to external standards and waterproof.

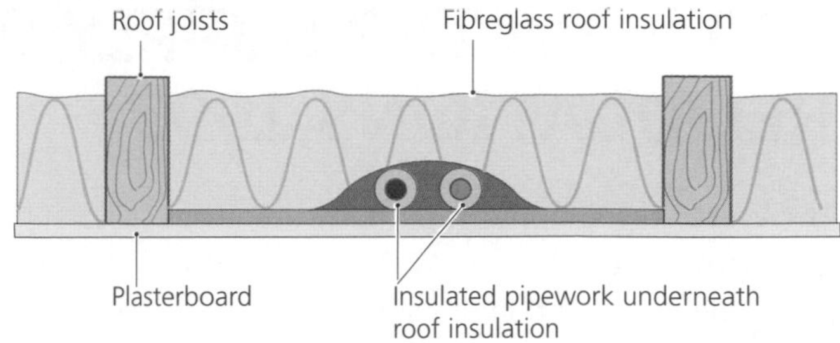

▲ Figure 5.57 Pipes in roof spaces

INDUSTRY TIP

The same process of frost protection also applies to hot water systems, which are covered in Chapter 6.

Insulation materials and their effectiveness

Pipework insulation should be of the closed-cell type complying with **BS 5422** and installed in accordance with **BS 5970**.

The recommended materials for pipe insulation are:

- rigid phenolic foam (less than 0.020 W/m²K)
- polyisocyanurate foam (0.020–0.025 W/m²K)
- PVC foam (0.025–0.030 W/m²K)
- expanded polystyrene, extruded polystyrene, cross-linked polyethylene foam and expanded nitrile rubber (0.030–0.035 W/m²K)
- expanded synthetic rubber, cellular glass and standard polyethylene foam (0.035–0.040 W/m²K).

The wall thickness of the insulation is shown in Table 5.5.

▼ Table 5.5 Thermal properties of pipe insulation

External diameter of pipe	Thermal conductivity of insulation material at 0°C in W/m²K				
	0.020	0.025	0.030	0.035	0.040
mm	mm	mm	mm	mm	mm
15	20 (20)	30 (30)	25* (45)	25* (70)	32* (91)
22	15 (9)	15 (12)	19 (15)	19 (19)	25 (24)
28	15 (6)	15 (8)	13 (10)	19 (12)	22 (14)
35	15 (4)	15 (6)	9 (7)	9 (8)	13 (10)
42 and over	15 (3)	15 (5)	9 (5)	9 (5)	9 (8)

Source: Defra guidance to the Water Supply (Water Fittings) Regulations, Section 3, Schedule 2.

Note: 15 mm pipes with thermal conductivities of 0.030, 0.035 and 0.040 W/(m.K), shown with a *, are limited to 50% ice formation after 9, 8 and 7 hours respectively. The figure in brackets indicates minimum thickness for 12 hours frost protection.

9 THE WATER SUPPLY (WATER FITTINGS) REGULATIONS 1999

The Water Regulations set out to prevent five factors:

- waste of water
- undue consumption of water
- misuse of water
- contamination of water
- erroneous measurement of water.

The water supplier, known as the water undertaker, has the responsibility to supply water fit for human consumption (wholesome water) to properties in the UK. This is covered by the Water Act 2003. Once the water is inside the property, plumbers have the responsibility to install systems and appliances that maintain the quality of that water. This is covered by the Water Supply (Water Fittings) Regulations. This a legal requirement which could lead to serious consequences if not followed.

It is a vital step for plumbers to become an 'approved contractor'. This means the plumber has studied the Water Regulations, taken a test and received a certificate to prove their knowledge about the Water Regulations.

In brief, the Water Regulations make sure that:

- All fittings and materials are of an approved quality and standard and will be installed correctly.
- Correct notice is given for certain installations (from which approved contractors are exempt).
- Certificates of compliance are issued when required.
- Systems must be tested for water tightness.
- Stored water and pipework must be protected.
- Cold water is protected from freezing or being heated to above 25°C.
- Systems are flushed out correctly when required.
- Pipework is labelled when required.
- Crossflow, backflow and back pressure (back siphonage) must be prevented.
- The five categories of water are listed.
- The mechanical and non-mechanical backflow prevention methods are outlined.
- Factors for unvented hot water are outlined.
- The flushing requirements for WCs and urinals are detailed.
- Correct valves and isolators must be installed and pipework must be accessible.

Becoming an approved contractor is a separate course.

The items outlined above are covered in this cold water chapter and are all based on the Water Regulations.

INDUSTRY TIP

Understanding the requirements of the Water Regulations is vital to becoming a professional plumber.

10 BACKFLOW PROTECTION

Backflow and back siphonage risks in the home

There are many instances in the home where backflow and back siphonage could present contamination risks. These will need to be considered during any planning, design and installation of hot and cold water supplies and central heating systems. Let us look first at some of the appliances and systems we use, and consider the risks. This will give you some idea of how the fluid categories occur.

▼ Table 5.6 Appliances and fluid category risk

Appliance or system	Content of the water	Risk
Kitchen sink	May contain animal remains from food preparation	Fluid cat. 5
WC	Contains human waste	
Bidet (over rim type)	May contain human waste	
Grey water and rainwater harvesting systems	May contain bacteria and disinfectants	
Washing machines and dishwashers	Contains soap and other detergents, and chemicals from dish washing and clothes cleaning	Fluid cat. 3
Bath	May contain soap and other detergents from personal hygiene	
Wash hand basin		
Shower valves and instantaneous showers	At risk from soap and other detergents from personal hygiene	
Hose union bib taps (outside tap)	At risk from gardening and other activities such as watering, weed killing, car washing, irrigation, etc.	
Combination boilers	The water in the heating system is often contaminated with dissolved metals, flux and some form of chemical inhibitor	Fluid cat. 3 or 4 (depending on boiler size)
Hot water system	Contains hot water	Fluid cat. 2

Note: This table is designed to give a brief overview of how and where fluid categories occur in the home and should not be viewed as exhaustive.

As you can see from Table 5.6, there are many potential contamination risks in every dwelling, and the bigger the building the more risks there are likely to be.

Whole-site, zone and point-of-use protection

There are many commercial and industrial processes where the whole or part of a plumbing system can present a high risk of backflow to other parts of the installation, or even the water undertaker's mains supply, despite the fact that the installation is installed to the required standards. In these circumstances, whole-site or zone protection must be installed on those parts that are deemed to be high risk.

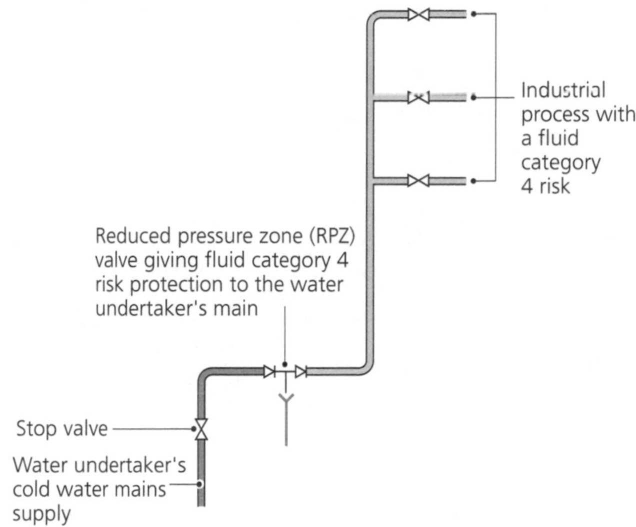

▲ Figure 5.58 Whole-site protection

Whole-site protection

The term 'whole-site protection' simply means that the water undertaker's main is protected at all times from backflow or back siphonage from any fluid category that is not fluid category 1 by a suitable backflow device. Protection should be at the point of entry of the cold water supply.

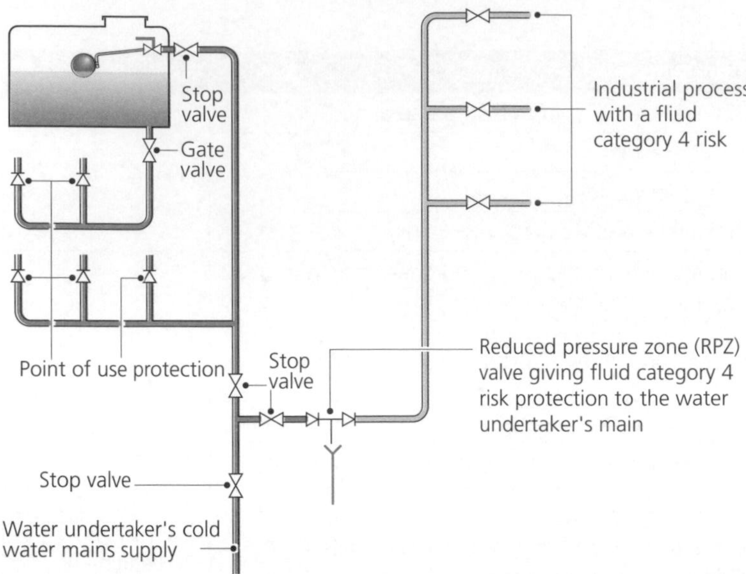

▲ Figure 5.59 Zoned protection

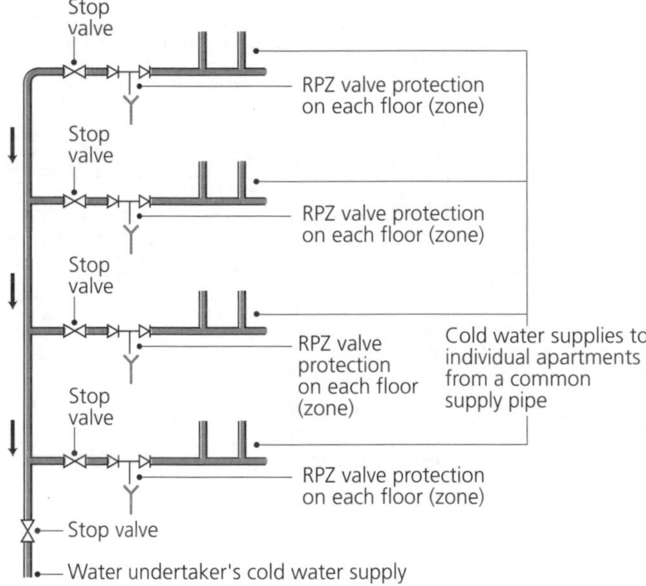

▲ Figure 5.60 Zoned protection for domestic premises

If whole-site protection is required, it is important that the water undertaker is informed at the application/notification for water supply stage. They will assess the application for a water supply and advise on what fluid category of backflow protection device must be installed to comply with the Water Supply (Water Fittings) Regulations. The backflow protection device must be installed before the system is commissioned.

Zoned protection

Zoned backflow protection simply means that, where different fluid categories exist within the same building, premises or complex, these have their own backflow protection devices to protect any part of the system that is fluid category 1. Zoned protection is also required where any water supply pipe is supplying more than one separately occupied premises.

Point-of-use protection

This is the simplest form of backflow protection. Point-of-use backflow protection devices are used to protect an individual fitting or outlet against backflow and are usually located close to the fitting they protect, such as a single check valve on a mixer tap to protect against fluid category 2, or a double check valve on a domestic hose union bib tap as protection against fluid category 3.

Eliminating the risk of contamination of wholesome water

The Water Regulations and, more specifically, the Water Regulations Guide can help us to choose the right course of action based upon the risk. Manufacturers, too, help in this regard by designing and manufacturing their appliances, taps and valves to conform to the Water Regulations. For example, most kitchen and bidet taps are designed and made with fluid category 5 risk in mind, and most bath and washbasin taps are designed and made with fluid category 3 in mind.

In most cases, where baths, washbasins, bidets and kitchen sinks are concerned, a simple air gap will protect the mains cold water supply. The size of the air gap, however, is dependent on the size of the tap, appliance type and its likely contents.

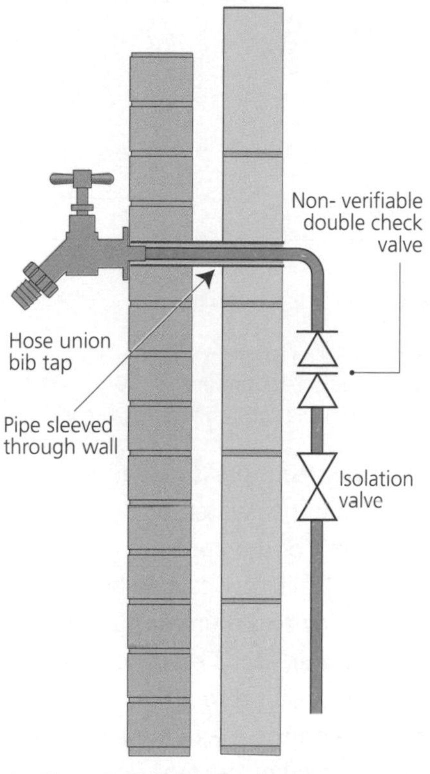

▲ Figure 5.61 Point-of-use protection

Air gaps used as a method of backflow prevention

An air gap is simply a physical unrestricted open space between the wholesome water and the possible contamination; the greater the air gap, the greater the level of protection that is offered. It does not require the use of a mechanical backflow prevention device. Here, we will consider the most important air gaps and how we can apply them. We will look at those listed in Table 5.7.

INDUSTRY TIP

Verifiable means the valve has a test nipple situated on the upstream side of the valve so it can be tested while in position to verify it is working correctly. A non-verifiable valve does not have this test point.

▼ Table 5.7 Schedule of non-mechanical backflow prevention arrangements and their respective fluid category protection

<table>
<tr><th colspan="2">Type</th><th colspan="2">Description of backflow prevention arrangements and devices</th><th colspan="2">Suitable for protection against fluid category</th></tr>
<tr><th></th><th></th><th colspan="2"></th><th>Back pressure</th><th>Back siphonage</th></tr>
<tr><td>a</td><td>AA</td><td colspan="2">Air gap with unrestricted discharge above spill-over level</td><td>5</td><td>5</td></tr>
<tr><td>b</td><td>AB</td><td colspan="2">Air gap with weir overflow</td><td>5</td><td>5</td></tr>
<tr><td>c</td><td>AD</td><td colspan="2">Air gap with injector</td><td>5</td><td>5</td></tr>
<tr><td>d</td><td>AG</td><td colspan="2">Air gap with minimum size circular overflow determined by measure or vacuum test</td><td>3</td><td>3</td></tr>
<tr><td>e</td><td>AUK1</td><td colspan="2">Air gap with interposed cistern (e.g. a WC suite)</td><td>3</td><td>5</td></tr>
<tr><td rowspan="5">f</td><td rowspan="5">AUK2</td><td colspan="2">Air gaps for taps and combination fittings (tap gaps) discharging over domestic sanitary appliances, such as a washbasin, bidet, bath or shower tray, shall not be less than the following:</td><td rowspan="5">X</td><td rowspan="5">3</td></tr>
<tr><td>Size of tap or combination fitting</td><td>Vertical distance of bottom of tap outlet above spill-over level of receiving appliance</td></tr>
<tr><td>Not exceeding G ½</td><td>20 mm</td></tr>
<tr><td>Exceeding G ½ but not exceeding G ¾</td><td>25 mm</td></tr>
<tr><td>Exceeding G ¾</td><td>70 mm</td></tr>
</table>

➜

Type		Description of backflow prevention arrangements and devices	Suitable for protection against fluid category	
			Back pressure	Back siphonage
g	AUK3	Air gaps for taps or combination fittings (tap gaps) discharging over any higher-risk domestic sanitary appliances where a fluid category 4 or 5 is present, such as: • any domestic or non-domestic sink or other appliance, or • any appliances in premises where a higher level of protection is required, such as some appliances in hospitals or other healthcare premises shall be not less than 20 mm or twice the diameter of the inlet pipe to the fitting, whichever is the greater.	X	5
h	DC	Pipe interrupter with permanent atmospheric vent	X	5

Notes:

1. 'X' indicates that the backflow prevention arrangement or device is not applicable or not acceptable for protection against back pressure for any fluid category within water installations in the UK.
2. Arrangements incorporating type DC devices shall have no control valves on the outlet of the device; they shall be fitted not less than 300 mm above the spill-over level of a WC pan, or 150 mm above the sparge pipe outlet of a urinal, and discharge vertically downwards.
3. Overflows and warning pipes shall discharge through, or terminate with, an air gap, the dimension of which should satisfy a type AA air gap.

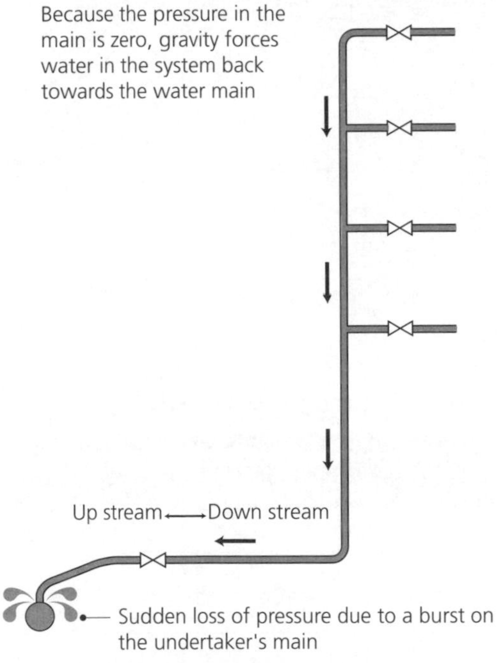

Because the pressure in the main is zero, gravity forces water in the system back towards the water main
Up stream ⟵⟶ Down stream
Sudden loss of pressure due to a burst on the undertaker's main

▲ Figure 5.62 Back pressure

Each of the air gaps described in Table 5.7 will have two fluid categories attached to it: one for back pressure and one for back siphonage. The difference between the two is simple to explain.

- **Back pressure:** this is caused when a downstream pressure is greater than the upstream or supply pressure in the water undertaker's main or the consumer's potable water supply. Back pressure can be caused by:
  - a sudden loss of upstream pressure, i.e. a burst pipe on a water undertaker's mains supply
  - an increase in downstream pressure caused by pumps or expansion of hot water
  - a combination of both of the above.
- **Back siphonage:** this is backflow caused by a negative pressure creating a vacuum or partial vacuum in the water undertaker's mains cold water supply. It is similar to drinking through a straw. If a sudden loss of pressure on the mains supply were to occur while a submerged outlet was flowing, then water would flow back upwards through the submerged outlet and down into the water undertaker's main.

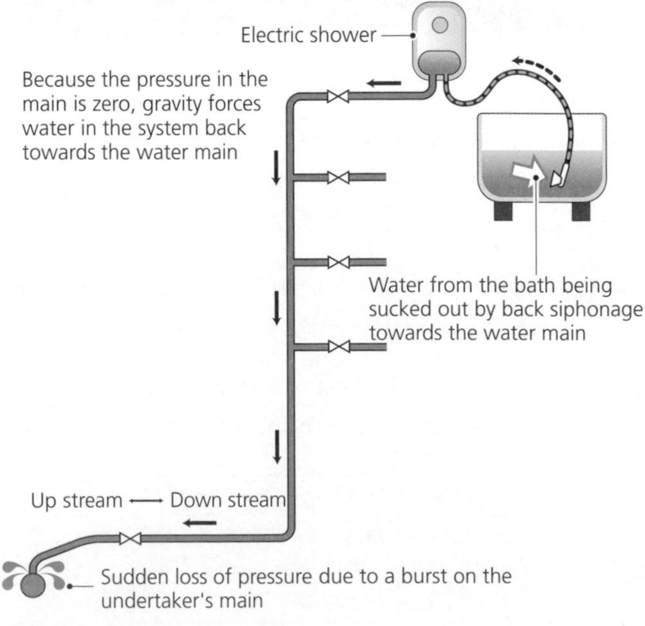

Electric shower
Because the pressure in the main is zero, gravity forces water in the system back towards the water main
Water from the bath being sucked out by back siphonage towards the water main
Up stream ⟵⟶ Down stream
Sudden loss of pressure due to a burst on the undertaker's main

▲ Figure 5.63 Back siphonage

Type AA air gap with unrestricted discharge above spill-over level

This gives protection against fluid category 5 and is a non-mechanical backflow prevention arrangement of water fittings, where water is discharged through an air gap into a cistern, which has, at all times, an unrestricted spill-over to the atmosphere. The air gap is measured vertically downwards from the lowest point of the inlet discharge orifice to the spill-over level. It should be remembered that:

- the type AA air gap is suitable for all fluid categories
- the size of the air gap is subject to the size of the inlet (see Table 5.8)
- the flow from the inlet into the cistern must not be more than 15° from the vertical.

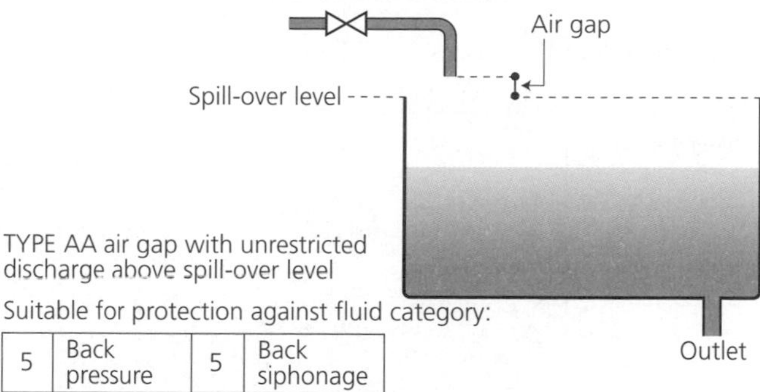

▲ Figure 5.64 Type AA air gap

▼ Table 5.8 Air gaps at taps, valves, fittings and cisterns

Situation	Nominal size of inlet, tap, valve or fitting	Vertical distance between tap or valve outlet and the spill-over level of the receiving appliance or cistern
Domestic situation with fluid categories 2 and 3 (AUK2)	Up to and including G ½	20 mm
	Over G ½ and up to G ¾	25 mm
	Over G ¾	70 mm
Non-domestic situation with fluid categories 4 and 5 (AUK3)	Any size of inlet pipe	Minimum diameter of 20 mm or twice the diameter of the inlet pipe, whichever is the greater of the two

▲ Figure 5.65 Animal trough

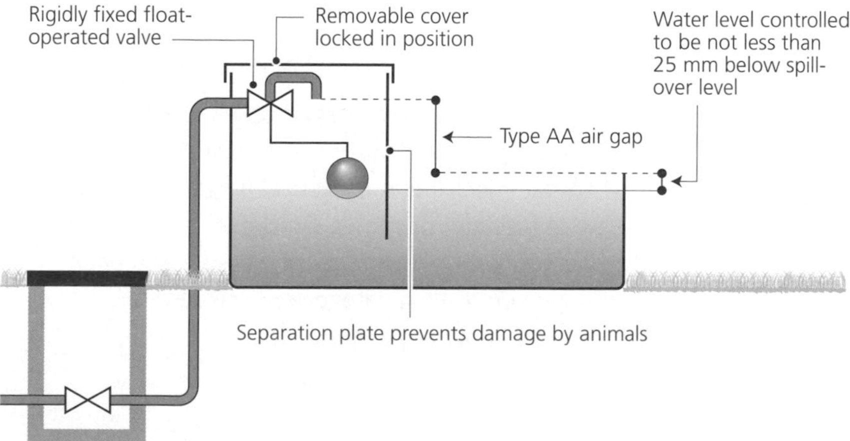

▲ Figure 5.66 Animal trough schematic

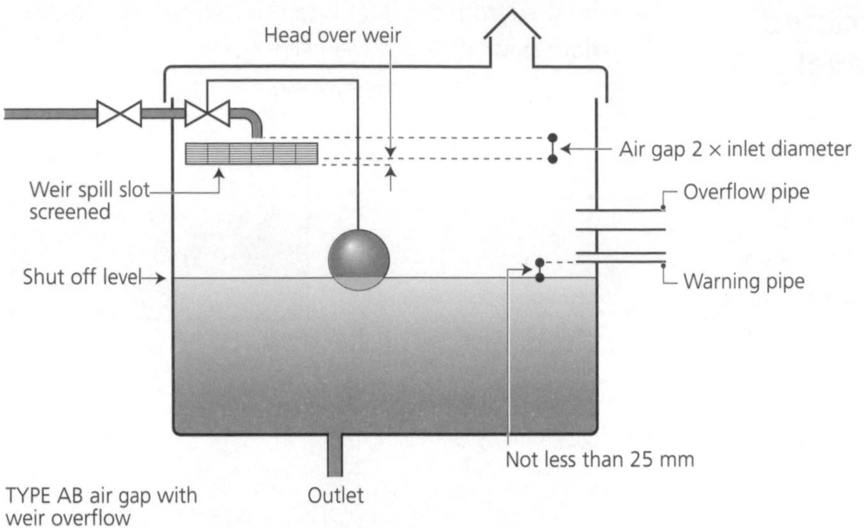

Suitable for protection against fluid category:

5	Back pressure	5	Back siphonage

▲ Figure 5.67 Type AB air gap with weir overflow

A good example of the use of a type AA air gap is in the form of animal drinking troughs, where the discharge of water into the trough is in a raised housing on the edge of the trough. The housing is covered to prevent the animals from having access to the water supply.

Type AB air gap with weir overflow

This gives protection against fluid category 5. It is a non-mechanical backflow prevention arrangement of water fittings complying with type AA, except that the air gap is the vertical distance from the lowest point of the discharge orifice, which discharges into the receptacle, to the critical level of the rectangular weir overflow.

The type AB air gap is suitable for high-risk fluid category 5 situations and is particularly suited to installations where the contents of the cistern need to be protected from contaminants such as insects, vermin and dust. A good example of this is feed and expansion cisterns in industrial/commercial installations, or where high-quality water is required, such as in dental surgeries.

INDUSTRY TIP

The size of the weir needs to be calculated based on the inlet size. This is usually completed using a weir overflow calculator.

▲ Figure 5.68 Type AB air gap with weir overflow on a cistern

Type AD air gap with injector

This is defined as a non-mechanical backflow prevention arrangement of water fittings with a horizontal injector and a physical air gap of 20 mm, or twice the inlet diameter, whichever is the greater. It gives protection against back pressure and back siphonage up to fluid category 5. This device is commonly known as a 'jump jet'.

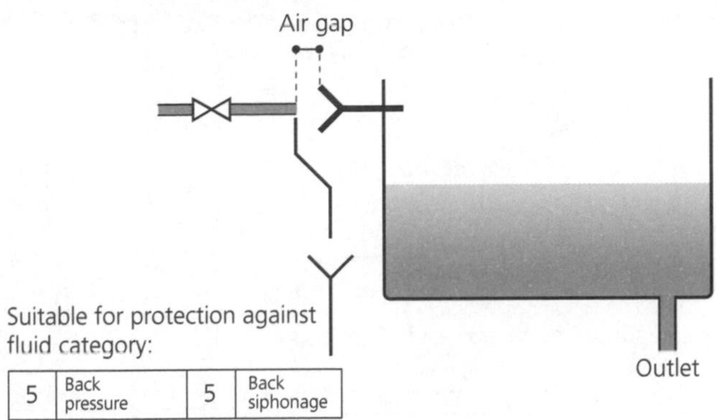

▲ Figure 5.69 Type AD air gap with injector

The principal uses of this type of air gap arrangement are in commercial clothes washing and dishwashing machines. It also has the potential to be used in catering equipment such as steaming ovens.

Type AG air gap with minimum size circular overflow determined by measure or vacuum test

This is a non-mechanical backflow prevention arrangement of water fittings with an air gap, together with an overflow, the size of which is determined by measure or a vacuum test. This arrangement gives protection against fluid category 3.

The type AG air gap fulfils the requirements of **BS EN 14623:2005 Devices without moving parts for the prevention of contamination of water by backflow. Specification for type B air gaps.** In a cistern that is open to the atmosphere, the vertical distance between the lowest point of discharge and the critical water level should comply with one of the following requirements:

- It should be sufficient to prevent back siphonage.
- It should not be less than the distances specified in Table 5.7, depending on cistern type.

The following points about type AG air gaps should be noted.

- The air gap is related to the size of the inlet supply and is the minimum vertical distance between the critical water level and the lowest part of the discharge outlet of the float-operated valve, as specified in Table 5.7.
- The critical water level is the level that is reached when the float-operated valve has failed completely and the water is running freely at maximum full-bore flow rate and pressure.
- AG air gaps must comply with the requirements of **BS EN 13076:2003**.

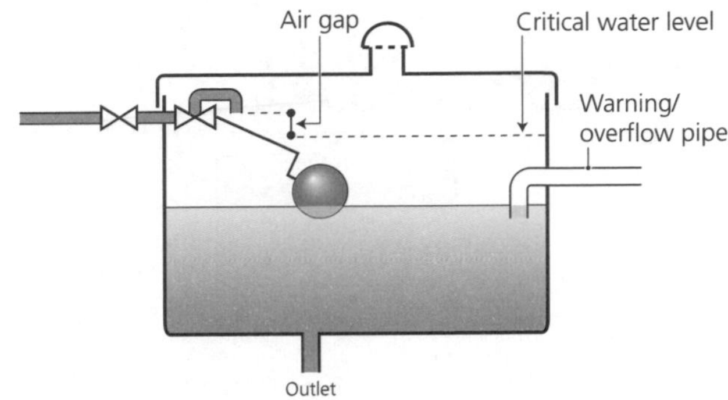

▲ Figure 5.70 Type AG air gap

Where storage cisterns are installed, it is likely that the critical water level would differ from installation to installation because of the varying flow rates and pressures of the incoming supply, and the differing lengths and gradients of the overflow pipe. With this type of installation, the type AG air gap is not practical because the critical water level cannot be calculated accurately. It is the critical water level that would determine the position on the cistern of the float-operated valve and the distance between the float-operated valve and the overflow.

Type AUK1 air gap with interposed cistern

This is a non-mechanical backflow prevention arrangement consisting of a cistern incorporating a type AG overflow and an air gap. The spill-over level of the receiving vessel is located not less than 300 mm below the overflow pipe and not less than 15 mm below the lowest level of the interposed cistern. It is suitable for protection against fluid categories 5 for back siphonage and 3 for back pressure.

This arrangement is most commonly found on WC installations, with the WC pan being the receiving vessel containing fluid category 5 water. A conventional domestic WC suite consists of a 6 l/4 l dual flushing cistern, a part 2, 3 or 4 float-operated valve with an AG air gap and overflow arrangement. This creates an AUK1 interposed cistern or, in other words, a cistern that can be supplied from a mains supply or another protected cistern without the need for additional backflow protection.

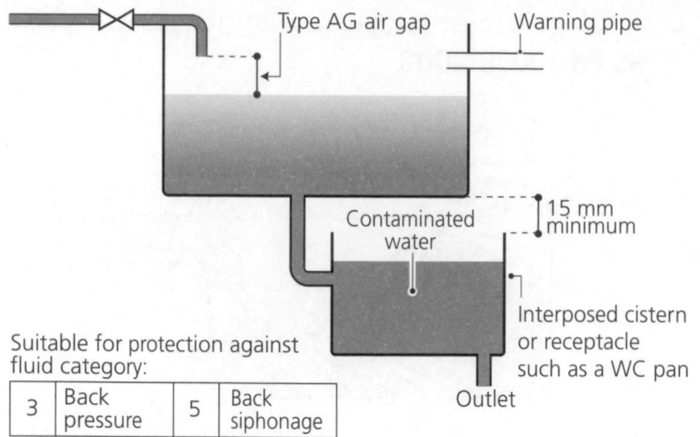

Suitable for protection against fluid category:

3	Back pressure	5	Back siphonage

▲ Figure 5.71 AUK1 air gap

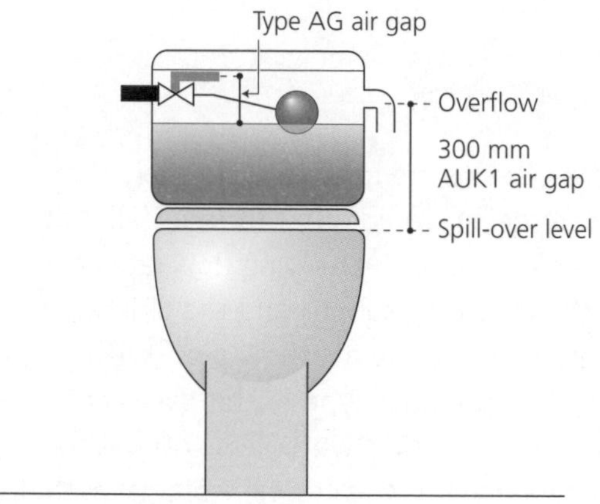

▲ Figure 5.72 AUK1 air gap on a WC cistern

Type AUK2 air gaps for taps and combination fittings (tap gaps) discharging over domestic sanitary appliances

This refers to the height of the air gap between the lowest part of the outlet of a tap, combination fitting, showerhead or other fitting discharging over a domestic sanitary appliance or other receptacle, and the spill-over level of that appliance, where a fluid category 2 or 3 risk is present downstream. An AUK2 air gap is suitable only for back siphonage up to fluid category 3 and must comply with the distances stated in Table 5.7.

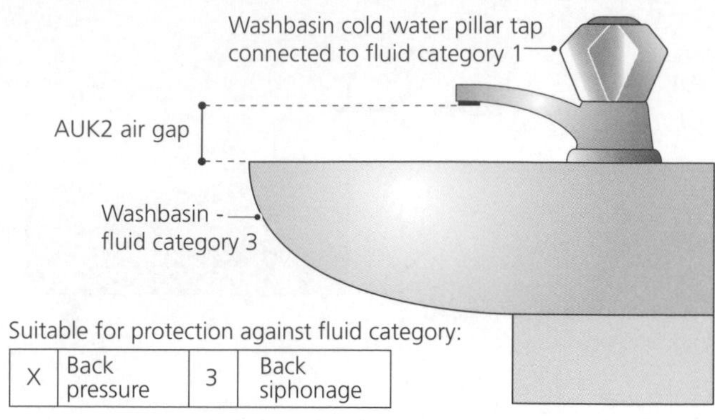

Suitable for protection against fluid category:

X	Back pressure	3	Back siphonage

▲ Figure 5.73 AUK2 air gap (tap gaps)

AUK3 air gaps for taps or combination fittings (tap gaps) discharging over any higher-risk domestic sanitary appliances where a fluid category 4 or 5 is present

Type AUK3 – higher-risk tap gap means the height of an air gap between the lowest part of the outlet of a tap, combination fitting, showerhead or other fitting discharging over any appliance or other receptacle, and the spill-over level of that appliance, where a fluid category 4 or 5 risk is present downstream.

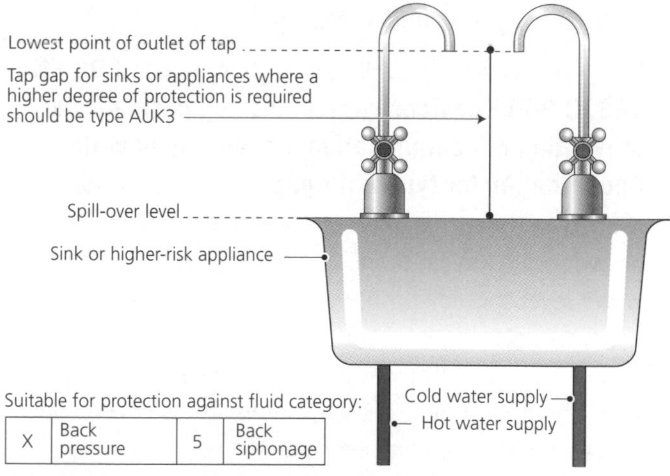

Suitable for protection against fluid category:

X	Back pressure	5	Back siphonage

▲ Figure 5.74 AUK3 air gap (higher-risk tap gaps)

In a domestic dwelling, AUK3 air gaps are most common at the kitchen sink in the form of high-necked pillar taps, sink mixer taps or sink monobloc taps. Sink mixers and monoblocs have a swivel spout. If a cleaners' sink, Belfast sink or London sink is being installed, it is important that any bib taps installed are positioned so as to maintain an AUK3 air gap.

Taps and combination fittings discharging on non-domestic appliances and any appliances in premises where a higher level of protection is required, such as appliances in hospitals or other healthcare premises, require a type AUK3 tap gap.

DC pipe interrupter with a permanent atmospheric vent

This refers to a non-mechanical backflow prevention device with a permanent unrestricted air inlet, the device being installed so that the flow of water is in a vertical downward direction. They are used where there is a threat of back siphonage from a fluid category 5.

The idea behind the DC pipe interrupter is to create an air inlet should a back-siphonage situation occur. When water begins to backflow upwards due to back siphonage, the DC pipe interrupter allows air into the system to break the siphonic action, thus preventing contamination.

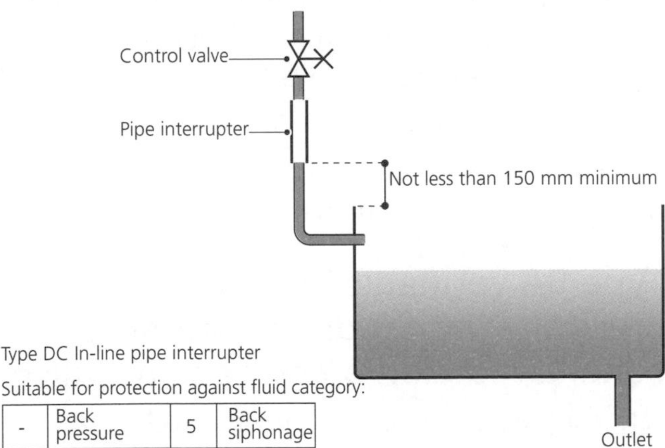

-	Back pressure	5	Back siphonage

▲ Figure 5.75 Installation of a DC pipe interrupter

Type DC pipe interrupter, this device must be fitted with the lowest point of the air aperture not less than 150 mm above the free discharge or spill-over level of an appliance and have no valve, flow restrictor or tap on its outlet.

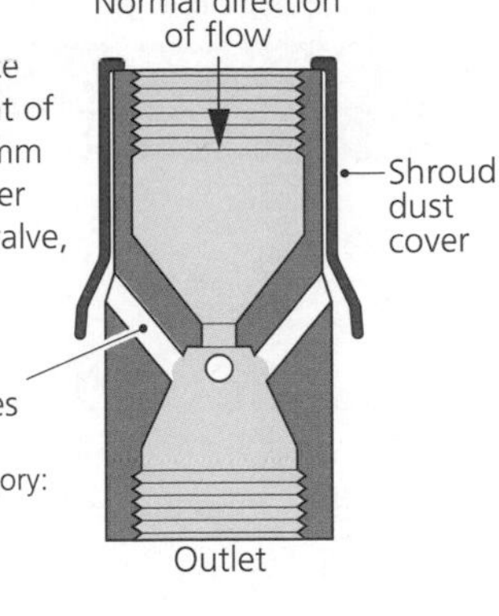

Suitable for protection against fluid category:

-	Back pressure	5	Back siphonage

▲ Figure 5.76 DC pipe interrupter

▲ Figure 5.77 Typical DC pipe interrupter

▼ Table 5.9 Schedule of mechanical backflow prevention arrangements and fittings and their respective fluid category protection

Type		Description of backflow prevention arrangements and devices	Suitable for protection against fluid category	
			Back pressure	Back siphonage
a	BA	Verifiable backflow preventer with reduced pressure zone	4	4
b	CA	Non-verifiable disconnector with difference between pressure zones not greater than 10%	3	3
c	DB	Pipe interrupter with atmospheric vent and moving element	X	3
d	EA/EB	Verifiable and non-verifiable single check valves	2	2
e	EC/ED	Verifiable and non-verifiable double check valves	3	3
f	HA	Hose union backflow preventer; permitted only for use on existing hose union bib tap in house installations	2	3
g	HUK1	Hose union bib tap incorporating a double check valve arrangement; permitted only as a replacement for existing bib taps in house installations	3	3
h	HC	Diverter with automatic return (normally integral with some domestic appliance applications only)	X	3

Notes:

'X' indicates that the backflow prevention device is not acceptable for protection against back pressure for any fluid category.

Arrangements incorporating a type DB device shall have no control valves on the outlet of the device. The device shall not be fitted less than 300 mm above the spill-over level of an appliance and must discharge vertically downwards.

Relief ports from BA and CA devices should terminate with an air gap, the dimension of which should satisfy a type AA air gap.

The DC pipe interrupter is a non-mechanical fitting. It does not contain any moving parts. They are manufactured from corrosion-resistant brass. Typical uses include WCs and urinal installations. The following points should be noted.

- The valve should be fitted in the vertical position, discharging downwards.
- It must be installed at least 300 mm above the overflowing level, or 150 mm if fitted above a urinal.
- No tap or valve should be installed downstream of the interrupter.
- Pipe size reductions downstream of the interrupter are not allowed.
- The length of the pipe downstream after the interrupter should be as short as possible.
- The interrupter should be accessible for replacement and repair.
- DC pipe interrupters must comply with **BS EN 14453:2005**.

Mechanical backflow prevention devices

An air gap is the most effective method of preventing contamination of the water supply; most installers will try to achieve this within their installations and designs but there are many cases where air gaps are not practical as a method of protection. In these instances, installers may opt to install a mechanical backflow prevention device. These provide a physical barrier to backflow. However, it must be remembered that mechanical backflow prevention devices have limitations and can be subject to failure.

In this section, we will look at some of the more common mechanical backflow prevention devices and where we can install them (see Table 5.9).

Type BA verifiable backflow preventer with reduced pressure zone (a reduced pressure zone valve)

Better known as an RPZ valve, this is a mechanical, verifiable, backflow prevention device, offering protection to water supplies up to and including fluid category 4. Verifiable simply means that the valve can be checked via test points to see if it is working correctly (verified).

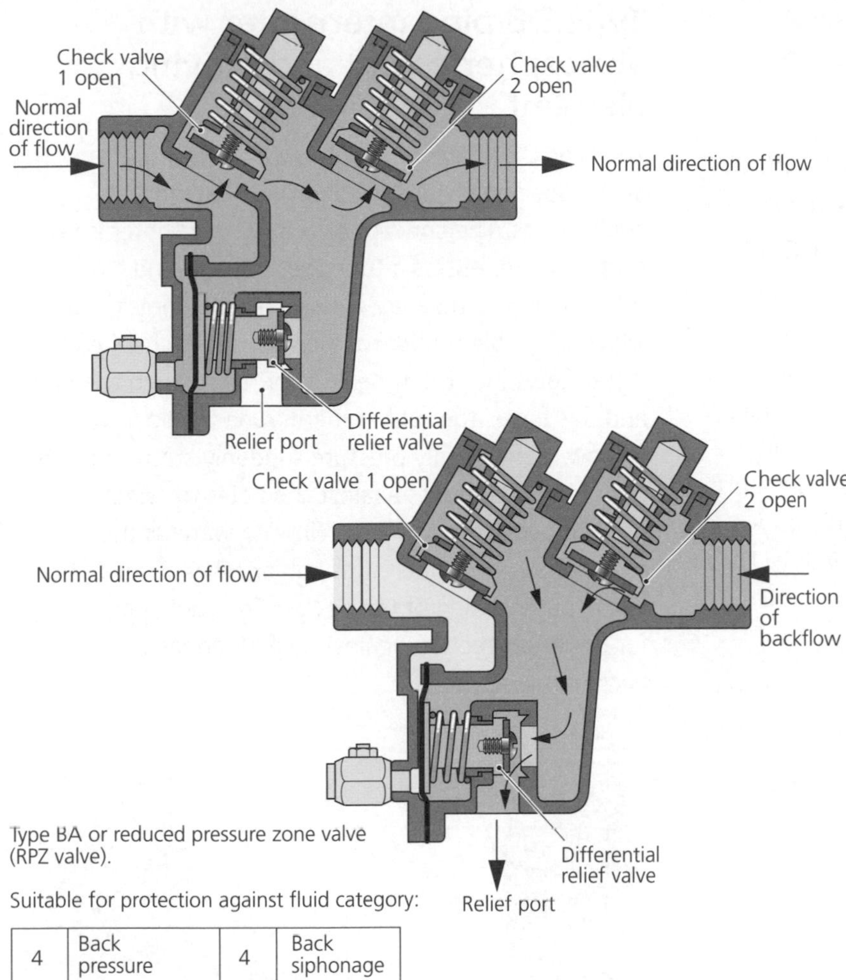

▲ Figure 5.78 An RPZ valve cut-away

Most RPZ valves consist of three separate elements:

1 two check valves
2 a differential relief valve
3 three test points.

The first check valve is spring loaded to generate a specific pressure drop across this part of the valve. This creates a reduced pressure zone downstream in the middle chamber of the valve and on the downstream side of the differential relief valve. The incoming mains supply maintains supply pressure on the upstream side of the differential valve and, as long as the mains pressure is higher, the differential relief valve will remain closed.

If, under static conditions, the mains pressure reduces to where it is just 0.14 bar above the pressure in the reduced pressure zone, the differential relief valve will open and release the contents of the middle chamber to drain. Should backflow occur past the first check valve element, the pressure on both sides of the differential valve will equalise and the differential relief valve will open to discharge the water.

If complete mains failure occurs, the contents of the middle chamber are discharged to drain, provided that both check valve elements are functioning correctly. However, should the upstream check valve become faulty, the pressure in the middle chamber will equalise to that of mains pressure and the differential relief valve will open and continuously discharge water at a steady rate. If the downstream check valve fails under zero mains pressure conditions, the differential relief valve will open and water will discharge from the downstream side of the system until the pressure here also becomes zero.

Testing, commissioning, maintenance and annual inspection can be carried out only by a trained and approved installer. Anyone who tests RPZ valves must be certificated. Specialist training is available from various test centres across the UK. Further recommended reading is the Water Regulations Advisory Scheme Information and Guidance Note No. 9-03-02.

Type CA non-verifiable disconnector with difference between pressure zones not greater than 10 per cent

These are very similar to BA devices (RPZ valves) in that they provide a positive disconnection chamber between the downstream water and the upstream water. The disconnection area between the two main check valves is open to the atmosphere under fault conditions, thereby maintaining an air gap should a loss of upstream pressure occur. Like the RPZ valve, any water discharged would run to drain via a tundish. They are suitable for fluid category 3.

A typical use of a type CA disconnector is as a permanent connection between a sealed central heating system and the water undertaker's cold water supply.

▲ Figure 5.79 A type CA backflow preventer

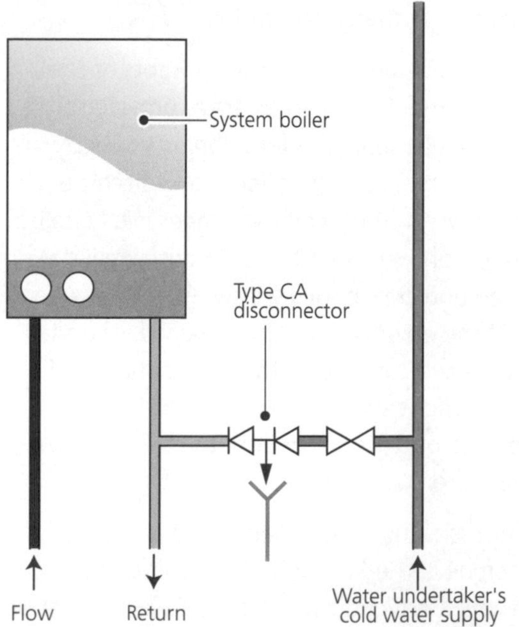

▲ Figure 5.80 Use of a type CA backflow preventer

Type DB pipe interrupter with atmospheric vent and moving element

The type DB pipe interrupter is a backflow prevention device specifically designed for fluid category 4 applications. The concept of the DB interrupter is very simple. Water enters a tube that has one end blanked off. Around the tube are a series of small holes over which a flexible rubber membrane is stretched. As the water flows into the tube, it is forced through the holes and this flexes the rubber membrane to allow water to flow. If the supply pressure suddenly stops, then the membrane contracts against the holes to effectively prevent backflow. Any back-flowing water is then released to atmosphere through another series of holes in the outer casing of the device. They are approved for use as protection against back siphonage but not back pressure.

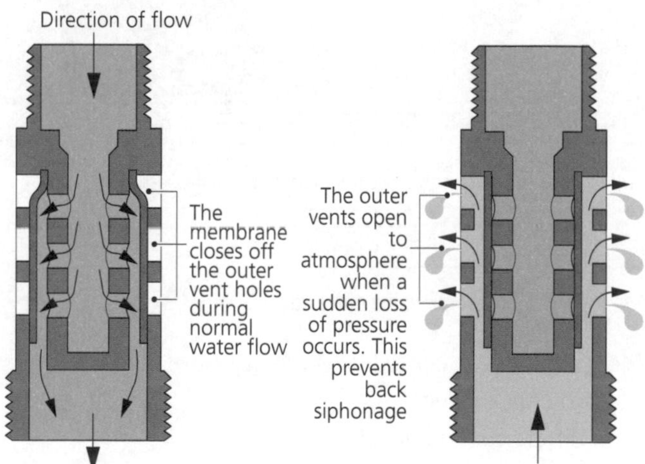

▲ Figure 5.81 Type DB pipe interrupter schematic

DB pipe interrupters are generally used externally as attachments to hose union bib taps and must not be used on appliances that have a control valve restriction, such as a washing machine. They are resistant to frost damage. They must be fitted vertically and have no valves fitted downstream of the device.

▲ Figure 5.82 Type DB pipe interrupter

Some DB interrupters are manufactured with bayonet-type attachments for domestic garden perforated hose irrigation systems.

Type EA and type EB verifiable and non-verifiable single check valves

These two valves are the simplest of all mechanical backflow prevention devices and can be used to protect against fluid category 2 for both back pressure and back siphonage. Generally regarded as point-of-use protection, they consist of a spring-loaded one-way valve that will allow water to flow from upstream to downstream only. If back siphonage or back pressure occurs, the valve will shut to prevent a reverse water flow. When no water is flowing, the valve remains in the closed position. Both types are almost identical in appearance. The difference between them is that the type EA device has a test nipple situated on the upstream side of the valve so that it can be tested while in position to verify that it is working correctly. The type EB non-verifiable single check valve does not have a test point but can be used in the same way as the type EA single check valve.

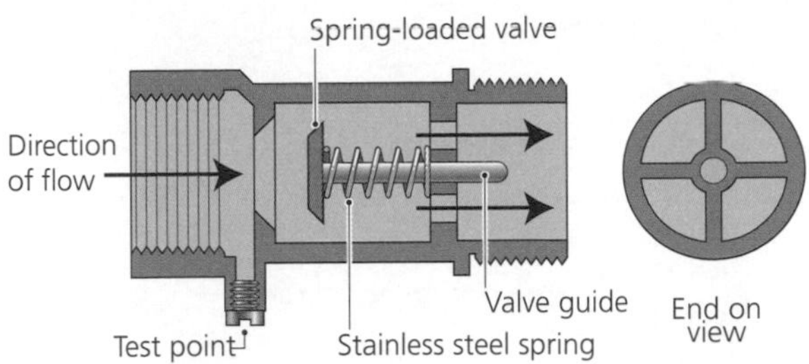

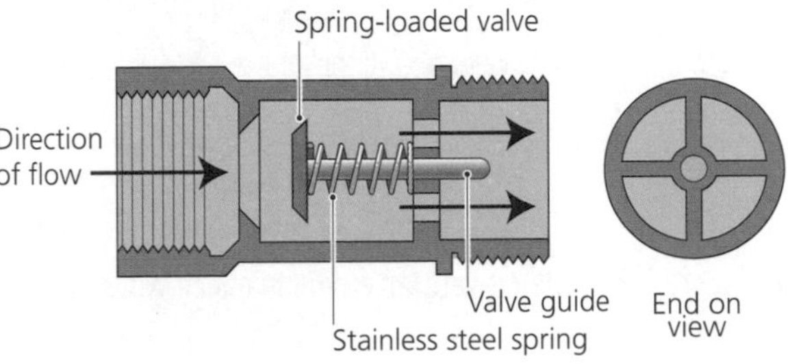

Suitable for protection against fluid category:

Type EA/EB single check valve			
2	Back pressure	2	Back siphonage

▲ Figure 5.83 Type EA verifiable single check valve and type EB non-verifiable single check valve

Both valves are manufactured from DZR resistant brass and have either type A compression fittings or female BSP threads for connection to the pipework. The valves should conform to **BS EN 13959:2004** for use in hot or cold water systems up to 90°C.

In domestic premises the risk from fluid category 2 generally occurs where the hot and cold supplies are taken to a single terminal fitting such as mixer taps or shower valves. This is known as a cross-connection. However, care must be taken when installing single check valves to hot water supplies as the expansion of the water can cause excessive pressure on the check valve causing it to fail. Other uses include the cold water connections to drinks machines.

Type EC and type ED verifiable and non-verifiable double check valves

These are mechanical backflow prevention devices consisting of two single check valves in series, which will permit water to flow from upstream to downstream, but not in the reverse direction. They are used primarily to protect against fluid category 3 for both back pressure and back siphonage.

Suitable for protection against fluid category:

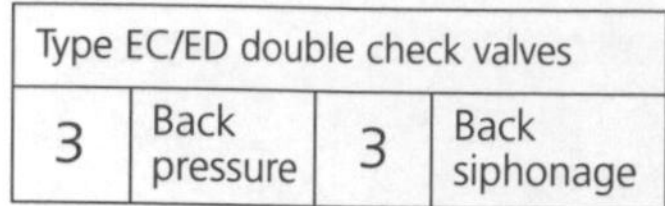

Type EC/ED double check valves			
3	Back pressure	3	Back siphonage

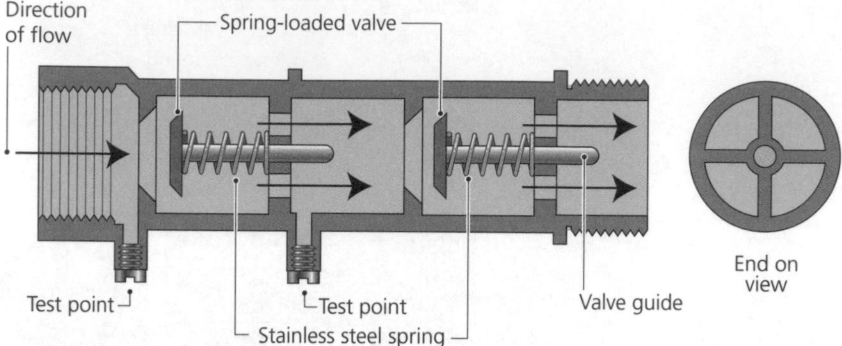

▲ Figure 5.84 Type EC verifiable double check valve

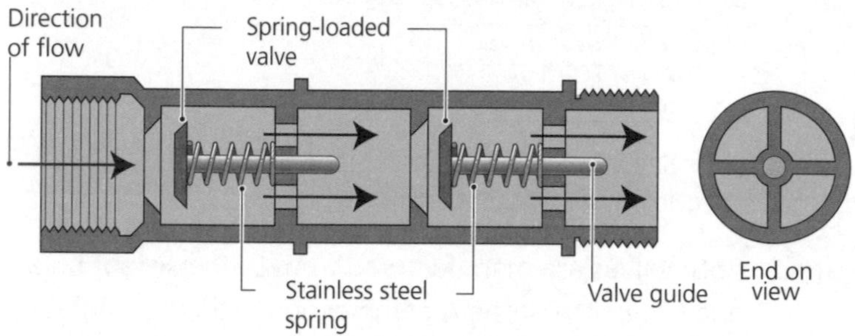

▲ Figure 5.85 Type ED non-verifiable double check valve

The type EC verifiable double check valve has two test nipples, one on the upstream side of the first check valve and another in the chamber between the first and second check valves. These are used to verify that the valve is working correctly. The type ED non-verifiable double check valve does not have a test point but can be used in the same way as the type EA single check valve.

Typical uses in domestic installations include garden hose union bib taps and sealed heating systems fitted in conjunction with a temporary filling loop. When used with sealed heating systems, the double check valve <u>must</u> be fitted to the cold water supply connection to the filling loop and <u>not</u> to the sealed heating connection.

Type HA hose union backflow preventer (permitted for use only on existing hose union bib taps in house installations)

As the name suggests, this mechanical backflow prevention device screws onto the outlet thread of a hose union bib tap. It is specifically for use with existing hose union bib taps that do not have any form of backflow protection. It is used to protect against back pressure at fluid category 2 and back siphonage at fluid category 3.

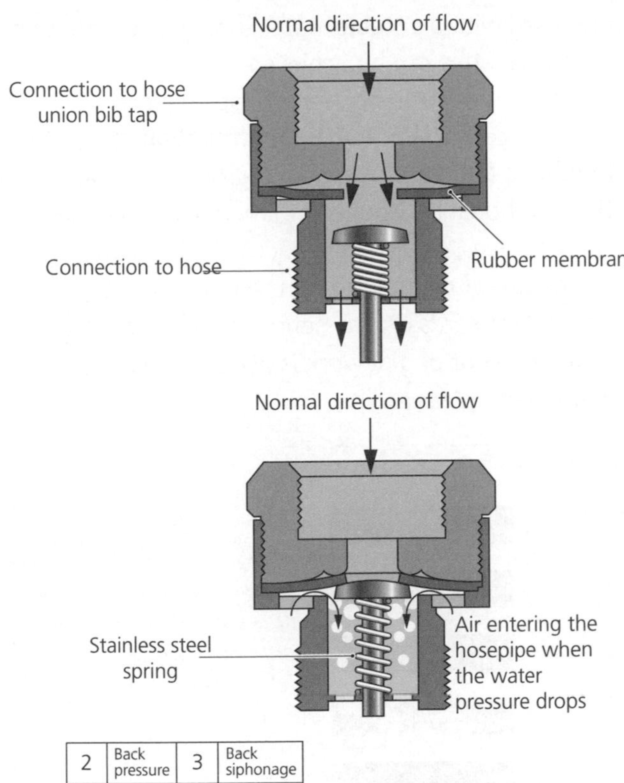

2	Back pressure	3	Back siphonage

▲ Figure 5.86 Type HA hose union backflow preventer

Type HUK1 hose union bib tap incorporating a double check valve arrangement

This hose union bib tap incorporates two single check valves, one situated at the inlet to the tap and one at the outlet. A screw-type test point is also included on the tap body. They are fitted in the same way as a normal HU bib tap. However, they are not suitable for new installations and can only be used as replacements where a hose union bib tap already exists. This is simply because the Water Supply (Water Fittings) Regulations state that any mechanical backflow prevention device should be fitted within the envelope of the building to prevent damage by freezing. They are suitable as protection against fluid category 3 for both back pressure and back siphonage.

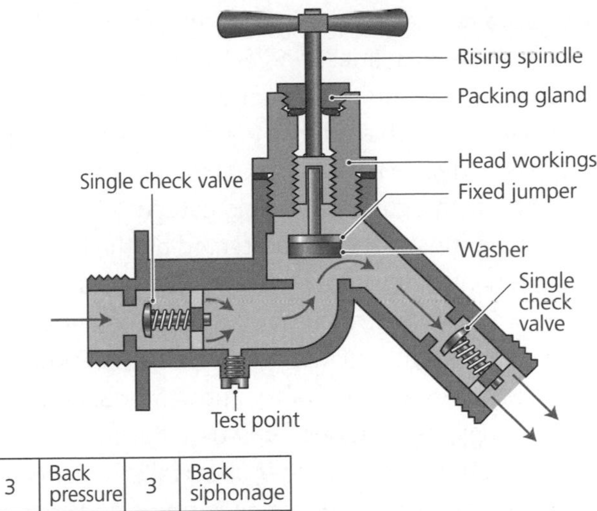

3	Back pressure	3	Back siphonage

▲ Figure 5.87 Type HUK1 hose union bib tap with double check valve arrangement

Type HC diverter with automatic return

This refers to a mechanical backflow prevention device used in bath/shower combination tap assemblies, which automatically returns the bath outlet open to atmosphere if a vacuum occurs at the inlet to the device.

The type HC diverter with automatic return is usually incorporated into the design of an appliance or fitting. It is not a 'stand alone' fitting that can be added to the installation. A good example of a type HC diverter would be a bath/shower mixing valve with a diverter valve to operate the shower. While pressure is maintained, the diverter valve remains open and the water is fed to the shower hose. Should loss of pressure occur, the diverter valve closes and any excess water in the shower hose returns to the bath through the open tap, thus preventing the water from back-flowing down the cold supply pipe. They are suitable for fluid category 3 to prevent back siphonage only.

Methods of preventing cross-connection in systems that contain non-wholesome water sources

A cross-connection is a direct, physical connection between wholesome, potable water and water that is considered non-potable, such as recycled water or harvested rainwater. In extreme circumstances, this can result in serious illness and even death. Cross-connections occur during correct plumbing design and installation, such as the hot and cold connections to a shower valve or a mixer tap (cross-connection between fluid category 1 and fluid category 2) and these, for the most part, are protected by the correct use of mechanical backflow prevention devices. However, some modern plumbing systems require much more thought and planning, rather than simply the installation of a check valve. The Water Supply (Water Fittings) Regulations 1999 demand that cross-connections from a water undertaker's mains to recycled and rainwater harvesting systems and even connections to private water supplies are eliminated completely in order to safeguard the wholesome water supply. There are several ways in which we can do this:

- correct design of systems, taking into account the requirements of the regulations in place
- careful planning and routing of pipework and fittings
- careful use of mechanical backflow prevention devices and air gaps
- using the correct methods of marking and colour coding pipework and systems.

Of these, identification of pipework is most important, especially when additions to the system are required or during routine and emergency maintenance operations.

Colour coding pipework

All pipes, cisterns and control valves that are used for conveying water that is not considered to be wholesome must be readily identifiable from pipes or fittings used with a potable water supply. There are two ways in which this can be achieved:

1 by the use of labels or colour banding pipes in accordance with **BS 1710: Identification of pipelines and services**; above-ground pipes and fittings should be labelled at junctions, and either side of valves, service appliances and bulkheads
2 by the use of pigmented materials and pipes – British Standards recommend that a contrasting type or colour of pipework is used to make identification easier.

Pipeline colour codes to BS 1710		
Wholesome drinking water		Green - Blue - Green
Hot water supply		Green - White - Crimson - White - Green
Reclaimed water		Green - Black - Green
Effluent		Black
Chemical dosing		Violet
Fire fighting		Green - Red - Green

▲ Figure 5.88 Pipeline colour codes to **BS 1710**

> **KEY POINT**
>
> Blue medium-density polyethylene (MDPE) water supply pipe must not be used under any circumstances to convey anything other than wholesome drinking water, nor must it be used to form ducts for conveying pipes with any other fluids or cables.

11 INSTALLING COLD WATER SYSTEMS AND COMPONENTS

The working principles of taps, valves and other terminal fittings

Taps and valves can be divided into five separate categories:

1 isolation valves
2 float-operated valves
3 terminal fittings
4 drain-off valves
5 shower mixer valves.

In this part of the chapter, we will look at each one in turn, including the types available, their uses and the working principles.

Isolation valves and drain valves

As already stated, the main use of an isolation valve is to turn off (isolate) either complete systems, or parts of a system or appliances. They can be divided into four distinct types:

1. those that isolate high-pressure systems, such as stop taps
2. those that isolate low-pressure systems, such as full-way gate valves
3. those that isolate appliances and terminal fittings on either high- or low-pressure systems
4. those that are used for draining down systems.

Stop taps/valves (high-pressure isolation) to BS 5433 or BS 1010

Stop taps are designed for isolation of high-pressure cold water systems and, because of their restrictive internal design, should not be used on low-pressure supplies. They are manufactured to either **BS 5433** or **BS 1010** for domestic use.

They consist of a brass valve body, a head gear with a rising spindle, a packing gland and a re-washerable loose jumper plate. Stop taps have an arrow on the valve body that shows the direction flow of the water.

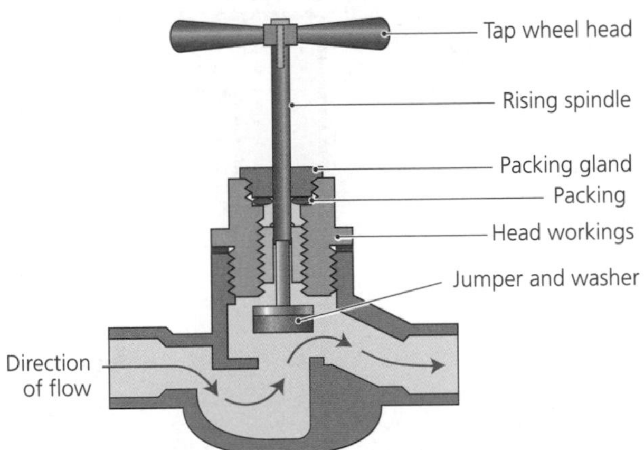

▲ Figure 5.89 Internal workings of a **BS 1010** tap

Some stop taps have a drain valve built in to the stop tap body, but care should be taken with this type when installing internal water meters as the drain valve position may allow water to be drawn from the main without being metered.

Stop taps are available with either capillary or compression connections to suit copper tubes to **BS EN 1057**, compression connections for MDPE and push-fit connections for polybutylene pipe.

Full-way gate valves (low-pressure isolation) to BS 5154

Gate valves are used on low-pressure installations such as the cold feed to vented hot water storage cylinders and the cold distribution pipework for indirect cold water systems. They do not have a washer, instead using a brass, wedge-shaped gate that rises inside the valve.

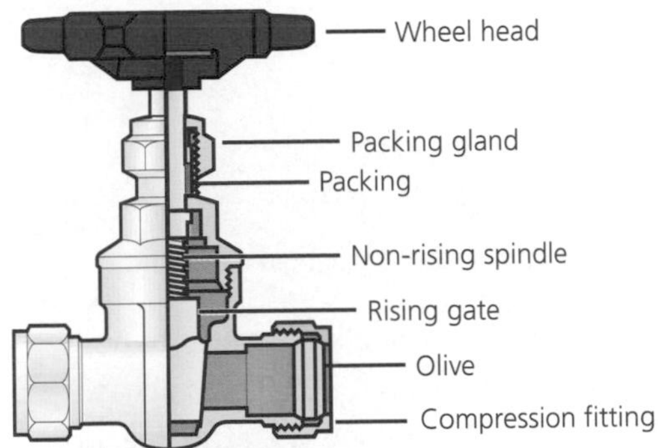

▲ Figure 5.90 Internal workings of a full-way gate valve

▲ Figure 5.91 Full-way gate valve

They are known as 'full-way' gate valves because the design allows water to flow at full bore without much restriction to the flow rate. However, they should not be used on high-pressure supplies as they tend to allow water to pass by the gate when the valve is under pressure. They consist of a brass valve body and a head gear with a non-rising spindle.

Gate valves are also available with a lockshield head to prevent the valve being tampered with.

Spherical plug valves (isolation)

Spherical plug valves are used for isolation of appliances and terminal fittings such as taps and float-operated valves. A variety of styles are available, including with or without a handle (these use a screwdriver slot to isolate the water), or for use with an appliance such as a washing machine.

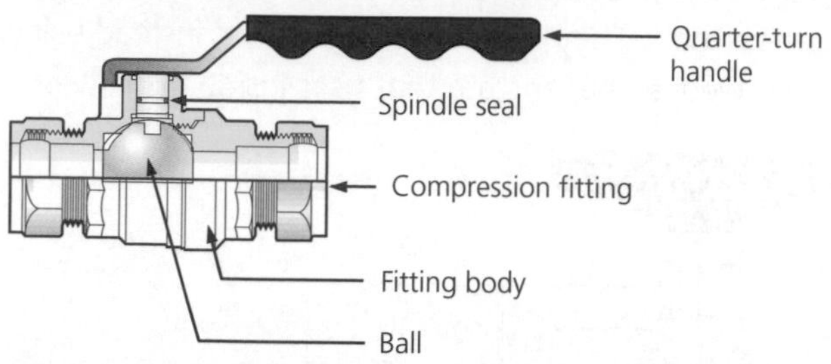

▲ Figure 5.92 Internal workings of a spherical plug valve

The internal design of the valve allows water to be isolated by turning a ball through 90°. The ball has a hole through it, which, when in line with the direction of water flow, allows water to pass through it. It is isolated when the hole is at 90° to the flow of water.

▲ Figure 5.93 A spherical plug valve

Drain-off valves

Drain-off valves are small valves that are strategically placed at low points in the installation to allow draining down of the system. Several types are available:

- with a male thread to allow connection to low carbon steel pipes and fittings
- with a spigot end to facilitate connection to either copper capillary fittings or compression fittings
- with or without a packing gland.

Drain-off valves should be positioned in accordance with the Water Supply (Water Fittings) Regulations, which state that:

> No drain valve should be placed below ground or in any position which allows the valve to become submerged in water.

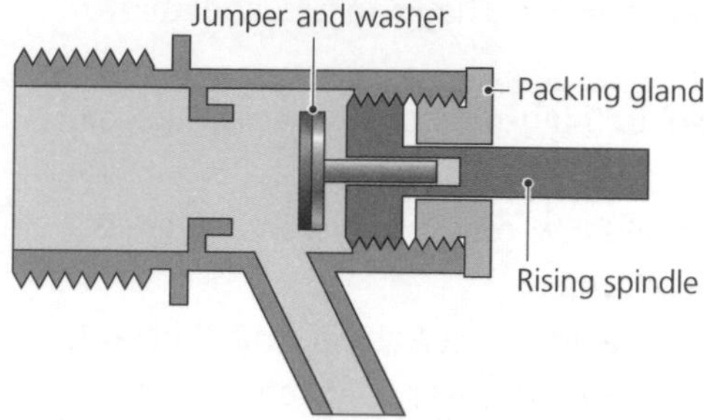

▲ Figure 5.94 Internal workings of a drain-off valve

The types of isolation valve and their use

Figure 5.95 shows the types of isolation valve that can be used on cold water installations and the position to which they are best suited. As a general rule of thumb, it should be remembered that:

- stop taps/valves are high-pressure only valves and should not be used on low-pressure supplies
- gate valves are for low-pressure installations
- service/Ballofix/ISO valves are primarily for terminal fixture isolation.

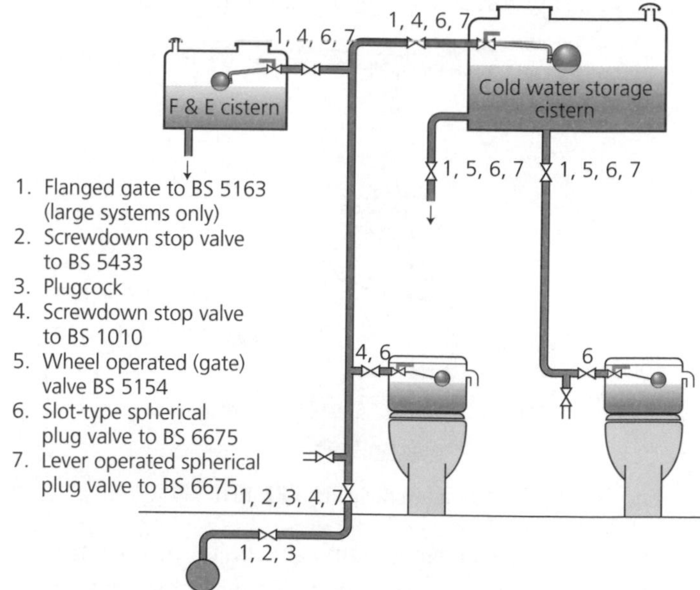

▲ Figure 5.95 The type and position of valves

Float-operated valves to BS 1212

Float-operated valves are used to control the flow of water into cold water storage and feed cisterns, feed and expansion cisterns, and WC cisterns. They are designed to close when the water reaches a pre-set level. They are made to **BS 1212** and it is important that plumbers recognise the different types. There are four basic float-operated valve types:

1. **BS 1212 Part 1**: Portsmouth pattern and Croydon pattern
2. **BS 1212 Part 2**: Diaphragm type
3. **BS 1212 Part 3**: Diaphragm type (plastic)
4. **BS 1212 Part 4**: Torbeck equilibrium type (WC cisterns only).

Float-operated valves can either be high pressure or low pressure depending on the type of orifice fitted. The orifice is the part of the valve that the water passes through. A high-pressure orifice is white in colour and has a small hole for the water to flow through, whereas the low-pressure orifice is coloured red with a larger hole. The orifice is universal for Parts 1, 2 and 3 float-operated valves.

BS 1212 Part 1 Portsmouth pattern float-operated valve

The Portsmouth-type float-operated valve discharges water from the bottom of the valve, which makes it susceptible to back siphonage should the valve become submerged in water. It should not be fitted on new installations without some form of backflow protection device, although existing Portsmouth-type valves can be repaired and maintained.

Portsmouth float-operated valves have moving parts that will come into contact with water, and this makes them vulnerable to failure and noise.

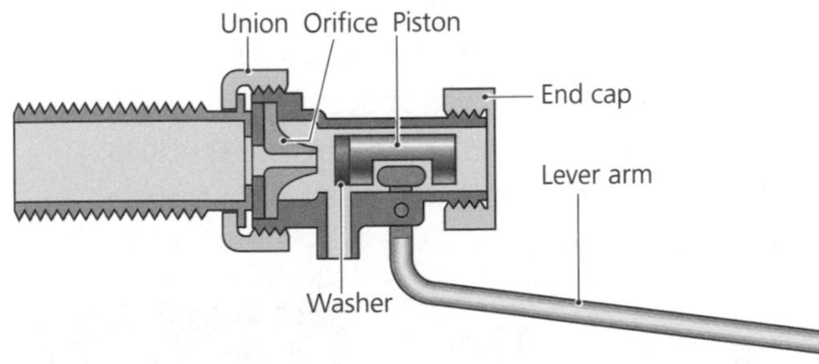

▲ Figure 5.96 A **BS 1212 Part 1** Portsmouth-type float-operated valve

BS 1212 Part 1 Croydon pattern float-operated valve

The Croydon-type float-operated valve is less common than the Portsmouth type. Like the Portsmouth, it discharges water from the bottom of the valve, but it is easily recognisable by its vertical piston and by the fact it delivers water into the cistern in two streams. This type of float-operated valve is very noisy and no longer manufactured, but it may still be used in some older WC cisterns.

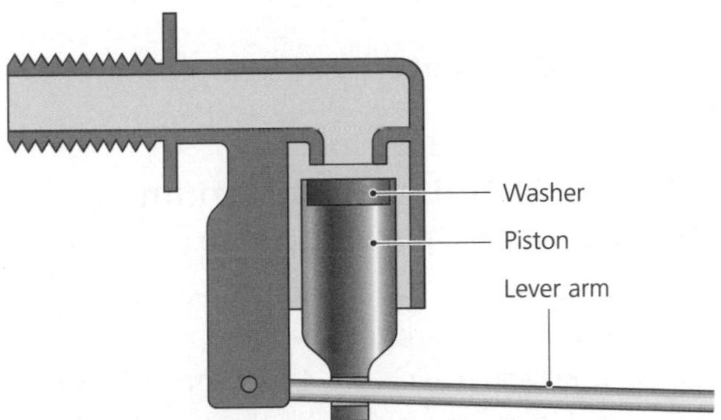

▲ Figure 5.97 A **BS 1212 Part 1** Croydon-type float-operated valve

BS 1212 Part 2 and 3 (plastic) diaphragm float-operated valves

These float-operated valves use a diaphragm rather than a washer to control the flow of water and, unlike Part 1 float-operated valves, they discharge water over the top of the valve. This makes them less susceptible to being submerged in water when the overflow runs and so less likely to cause a contamination issue. They also have fewer moving parts, which makes the valve quieter in operation and less likely to cause **water hammer** and reverberation of the pipework.

KEY TERM

Water hammer: caused by a rapid opening and closing of the float-operated valve. As the water nears the water level in the cistern, the ball valve can begin to bounce quickly up and down and from side to side. This causes the noise to travel down the pipework, resulting in reverberation or a whining noise. It can also be caused by a faulty washer or diaphragm.

The main difference between a Part 2 float-operated valve and a Part 3 float-operated valve is that the Part 2 is made of brass and the Part 3 is made of plastic. They are almost identical in all other respects. It should be noted that plastic float-operated valves are not recommended for cisterns other than WC cisterns because of the risk of freezing and subsequent splitting of the plastic.

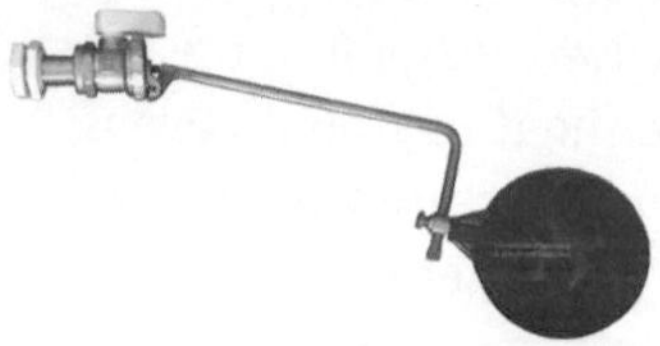

▲ Figure 5.98 A **BS 1212 Part 2** diaphragm-type float-operated valve

BS 1212 Part 4 Torbeck equilibrium diaphragm float-operated valve

The Torbeck equilibrium float-operated valve is a diaphragm valve that works on the principle of equal pressure in front and behind the diaphragm when the valve is open. No moving parts come into contact with the diaphragm. It closes the valve when a build-up of pressure occurs in front of the diaphragm due to the float arm closing the pressure relief orifice on the front of the valve. Although quieter in operation than other float-operated valves, the positive 'snap' closing action can lead to problems of banging and reverberation in some systems.

The valve can be used on either high or low pressures by the insertion of either a low-pressure or high-pressure flow restrictor in the valve stem. Some valves also have a filter, to filter out any minute solid impurities in the water, which could cause malfunction.

The Torbeck valve must be used only on WC cisterns.

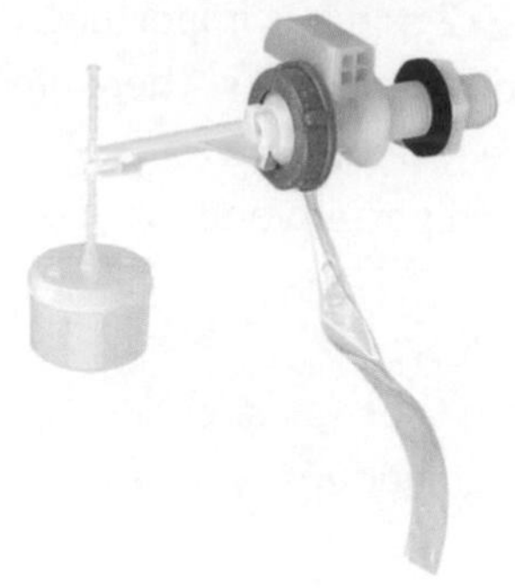

▲ Figure 5.99 A **BS 1212 Part 4** Torbeck-type float-operated valve

Terminal fittings

Terminal fittings are those that are fitted to sanitary appliances, such as baths and washbasins. There are several different types, which are:

- Pillar taps for baths, washbasins and bidets – these are available for baths (¾-inch tails), washbasins and bidets (½-inch).
- High-necked pillar taps for kitchen sinks – similar internal design to pillar taps but designed with a high stem to provide an AUK3 air gap at kitchen sinks.
- Bi-flow mixer taps including monobloc mixers – these are two taps in a single body. A bi-flow mixer has a single spout that is divided down the middle so that the water does not mix until it has exited the tap. It is not a true mixer tap.
- True mixer taps – allow the hot and cold water supplies to be mixed inside the body of the tap. Caution should be exercised as these taps can provide a cross-connection between low-pressure hot (fluid cat. 2) and high-pressure cold (fluid cat. 1).
- Bib taps and hose union bib taps – bib taps are mostly fitted to the wall above cleaners' sinks and Belfast sinks. Hose union bib taps are specifically designed for garden use so that a hose may be connected.

▲ Figure 5.100 Pillar taps

▲ Figure 5.101 Bi-flow mixer taps

▲ Figure 5.102 True mixer taps

▲ Figure 5.103 Bib taps

They fall into three categories:

1 taps with a rising spindle
2 taps with a non-rising spindle to **BS EN 200:2008**
3 ceramic disc taps.

Taps with a rising spindle

These are referred to as **BS 1010** type taps, but **BS 1010** has now been withdrawn. They have a rising spindle attached to a jumper plate and a washer. When the tap is turned on, the spindle rises, allowing the pressure of the water to push the jumper plate and washer upwards to start the flow of water. Originally, there were two different types:

1 those taps with loose jumper plates for high-pressure supplies such as mains cold water
2 those taps with fixed jumper plates for low-pressure supplies such as indirect cold water installations and vented hot water supplies.

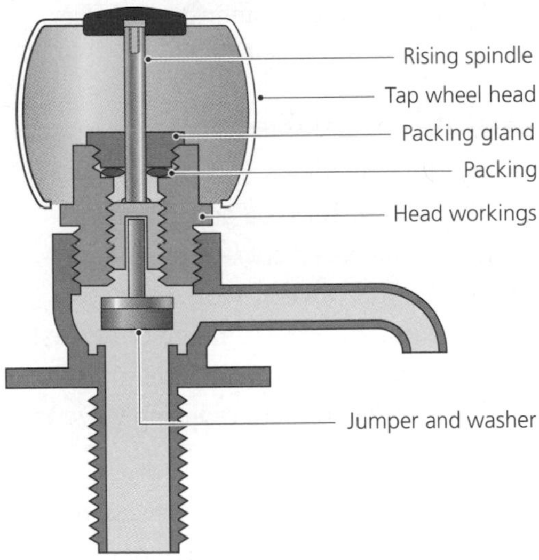

▲ Figure 5.104 Rising spindle pillar tap

Both types have a packing gland designed to stop water leaking through the spindle. The design of these taps is generic across most manufacturers. This means that the head workings of one tap will almost certainly fit the tap body of another manufacturer, including stop tap heads.

These taps are available as stop taps, pillar taps for washbasins and bidets, high-necked pillar taps for kitchen sinks, mixer taps for baths (¾-inch thread) and kitchen sinks (½-inch thread), and bib taps.

Taps with a non-rising spindle to BS EN 200:2008

These taps do not have a rising spindle. Instead, the spindle has a thread at the end that lifts a hexagonal barrel, with a rubber washer attached, inside the valve head workings. The spindle is fixed in the head workings by a circlip.

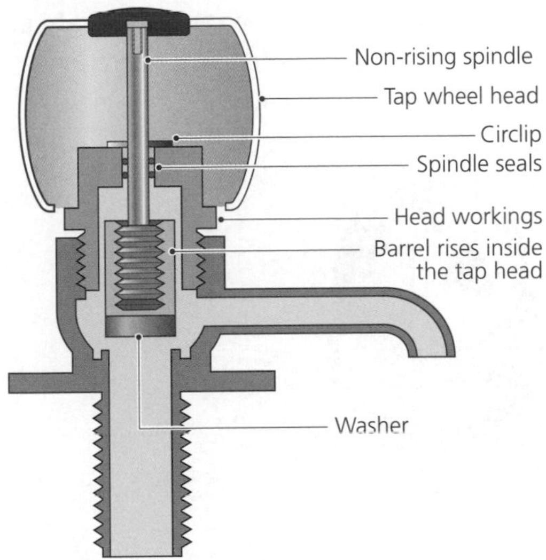

▲ Figure 5.105 Non-rising spindle tap

▲ Figure 5.106 Non-rising spindle pillar tap head workings

There are many different styles and types of **BS EN 200:2008** tap and each manufacturer has its own style of conforming to the British Standard. The result of this is that very few of the head workings are interchangeable between manufacturers.

These taps are available as pillar taps for washbasins and bidets, high-necked pillar taps for kitchen sinks, mixer taps for baths (¾-inch thread) and kitchen sinks (½-inch threads), monobloc mixer taps and bib taps.

Ceramic disc taps

Unlike washer-type taps, ceramic disc taps use two thin, close-fitting, slotted ceramic discs in place of rubber washers. One of the discs is fixed, while the other is turned by the handle of the tap a quarter of a turn through 90°.

Ceramic disc tap heads are 'handed'. In other words, there are specific hot tap head workings, which turn to the left, and specific cold tap head workings, which turn to the right, and they are usually colour coded for easy identification.

Ceramic disc taps are not universal. If replacement head workings are required during maintenance operations, the correct type for the make of tap will be needed.

▲ Figure 5.107 Ceramic discs

Shower mixer valves

Shower mixer valves mix water from both the cold water and hot water installations, and discharge the mixed water from a showerhead. They can be either:

- manual mixing valves
- venturi boost mixing valves
- pressure compensating mixing valves
- thermostatic mixing valves – for example:
  - wax capsule type
  - bi-metal coil type.

Shower valves are available in three styles:

1 exposed, surface-mounted valves – mounted on the surface, generally with concealed pipework
2 concealed valves – all the valve and pipework is concealed with only the controls on show
3 bar valves – a recent addition, an exposed-type shower valve designed to be thin and modern looking.

Externally, all mixing valves appear very similar in style and most have common distances of 150 mm between the hot and cold connections. The difference is in the internal workings of the shower. In this part of the chapter, we will look at those differences.

The requirement of all shower valves is that they blend hot and cold water to the required temperature.

Manual mixing valves

Manual mixing valves do not have thermostatic control. They rely wholly on the hot and cold supplies being balanced in terms of pressure and flow rate. Once the temperature of the blended water has been adjusted, it remains fixed and does not adjust to fluctuations in flow rate, pressure or temperature. For this reason, the temperature of the hot water needs to be stable.

Although manual mixing valves can be used on high-pressure supplies, they are best suited to low-pressure installations to avoid pressure fluctuations. They should not be fitted to systems that contain instantaneous water heaters or combination boilers.

Venturi boost mixing valves

The venturi mixing shower valve is specifically designed for installations that do not have balanced hot and cold supplies, such as mains-fed cold water and low-pressure hot water. For the valve to work correctly, the mains cold water must have a pressure of at least 1 bar and a maximum pressure of 3 bar. Pressures in excess of this will require a pressure reducing valve to be fitted.

The venturi mixing valve uses the extra pressure of the cold water supply to increase the pressure of the hot water supply by using the venturi principle. The operating principle is as follows.

As the cold water passes through the venturi tube within the valve, its velocity increases and its pressure is slightly reduced. At this point, the hot water is drawn in to the cold supply and mixed. As the mixed water leaves the venturi, the pressure reverts to almost as high as the initial cold supply, giving a fairly powerful shower.

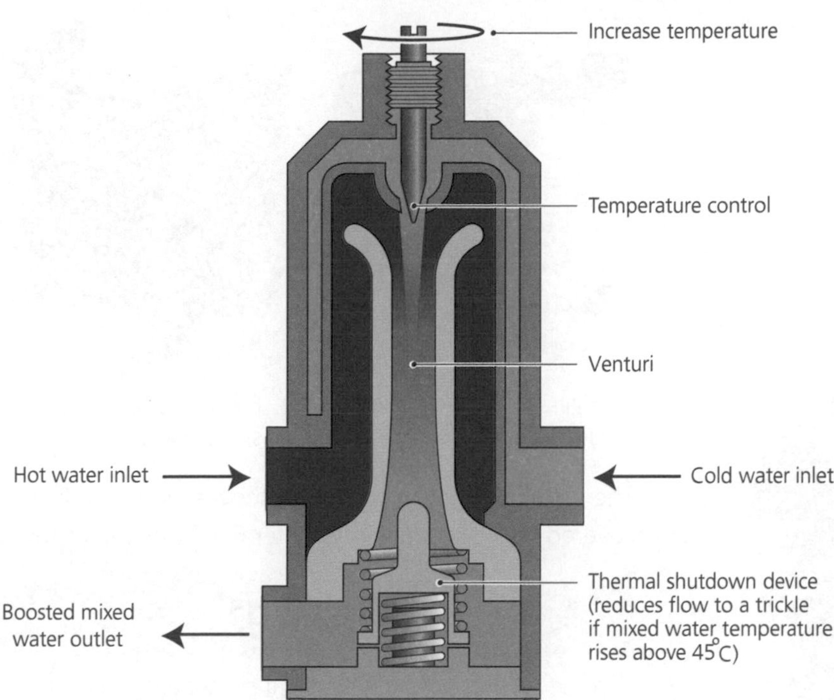

▲ Figure 5.108 The working principles of a venturi boost shower mixing valve

Pressure-compensating mixing valves

This type of mixing valve gives greater temperature stability compared to manual mixing valves. Some valves can be used on both high- and low-pressure systems, while others are specifically designed for high-pressure system use.

Pressure-compensating mixing valves are manufactured with either of the following two types of control.

1. **Sequential control:** starting the shower at a low temperature and progressively turning the control towards hot gradually increases the temperature and maintains a steady flow rate. When the temperature is set, a balancing diaphragm reacts to subtle changes in water pressure and maintains the correct hot/cold mix.
2. **Dual control:** these have a separate flow control and temperature control mechanism. The temperature control mechanism consists of a metallic shuttle that moves backwards and forwards inside a plastic mixing tube. The hot and cold water is regulated as the water flows through the tube. If there is a drop in pressure on either supply, the shuttle is moved inside the mixing tube, increasing the flow on the reduced pressure side and decreasing the flow on the opposite supply. This maintains an even showering temperature when pressure fluctuations occur. They do not, however, react to changes in temperature.

Thermostatic mixing valves

Thermostatic mixing valves give the best overall temperature control of all the shower valves currently available. In most cases, the maximum temperature is pre-set by the manufacturer with a manual override for the end user. Incoming hot water temperature averages about 55°C and the cold supply at 15°C, giving a showering temperature of between 38°C and 42°C. There are two different types:

1. **Wax capsule type:** a copper capsule containing a mixture of fine metal particles and a heat-sensitive wax is positioned in the mixing chamber of the valve. The wax expands with heat. As the wax expands, it forces a metal piston to activate a shuttle, which effectively controls the flow of hot and cold water into the valve by restricting the flow rate of the hot and cold water. If the temperature of the hot water is very hot then the hot flow is restricted, allowing more cold water into the valve to compensate for the high temperature. When a cooler shower is required, then the reverse happens and the shuttle moves backwards as the wax contracts, aided by a spring pushing against it.

2 **Bi-metal coil type:** these work on the bi-metallic coil principle where two metals with differing expansion rates are bonded together. When heat is applied, the two metals expand but one faster than the other, causing the metal coil to distort. In the case of a shower valve, the bi-metallic coil is fastened at one end to a shuttle that controls the in-flow of hot and cold water to the mixing chamber of the valve.

▲ Figure 5.109 Wax capsule

▲ Figure 5.110 Bi-metallic coil

Shower pumps

Low-pressure shower valves can have boosted hot and cold supplies by the use of shower boosting pumps. There are two types available:

1 **Single impeller outlet pumps:** this type of pump is designed to pump hot or cold water to individual outlets such as hot or cold water taps throughout the property. They were commonly installed after the shower mixer valve to boost the mixed water to the showerhead, however it can sometimes be difficult to install them as per the manufacturer's guidelines in this way; as such twin impeller pumps have become more common to boost the water to the mixer valve.

2 **Twin impeller inlet pumps:** these are fitted before the mixing valve, and boost the individual hot and cold supplies to the valve where the water is mixed, or if the correct pump is selected, can also supply hot and cold water to the whole house. They have a single electric motor, which drives two impellers (hot and cold). Care should be taken when installing the pipework to ensure that it meets the manufacturer's specific requirements with regard to pump position, pipe size and minimum head of water required.

▲ Figure 5.111 Single impeller shower pump

▲ Figure 5.112 Twin impeller shower pump

The installation of showers and shower pumps will be covered in more detail in Chapter 6, Hot water systems.

INDUSTRY TIP

Remember: it is not permitted to fit shower pumps on mains cold water installations. The Water Supply (Water Fittings) Regulations prohibit the use of pumps on mains cold water except when special permission has been given by the water undertaker.

Scale reduction and water treatment in domestic properties

As well as the treatment given to the water by the water undertaker, many domestic properties, especially in hard water areas, employ alternative methods to condition the water so that scaling problems do not occur.

Scaling occurs in hot and cold water systems and central heating installations when the water contains salts and minerals, such as calcium carbonate, that re-forms in the water as a hard limescale that sticks to the inside of pipes and appliances. This process is known as precipitation. The resulting limescale reduces the appliance's efficiency and can, in some cases, make the appliance unusable.

There are several methods we can use to prevent precipitation from occurring; these include the use of:

- water conditioners
- water softeners
- water filters.

In this part of the chapter, we will look at the use of these appliances and how they work.

Water conditioners

The term 'water softener' is used to describe a variety of products that are designed to prevent the build-up of limescale. Water conditioners (also known as limescale inhibitors) work by altering the chemistry of the precipitation process by suppressing limescale formation and thereby reducing the rate of scaling. There are many different types of water conditioner using a wide variety of conditioning methods, including those described below.

- **Magnetic:** prevents scale build-up by influencing the type of calcium crystals precipitated, which ensures that only needle-like aragonite crystals are formed. These find it harder to stick to smooth surfaces than the normal calcium crystals. These are for individual appliance protection only, such as combi boilers. They are installed on the cold main to the appliance.
- **Electrolytic:** these work by adding a minute amount of zinc to the water, which suppresses the formation of calcium crystals. Any crystals that are formed are washed away by the flow of water. Can be used for whole-house protection.

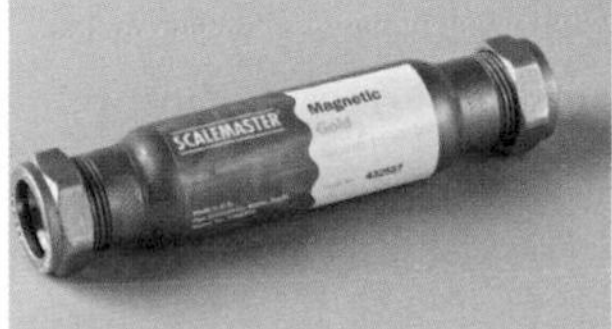

▲ Figure 5.113 Magnetic scale inhibitor

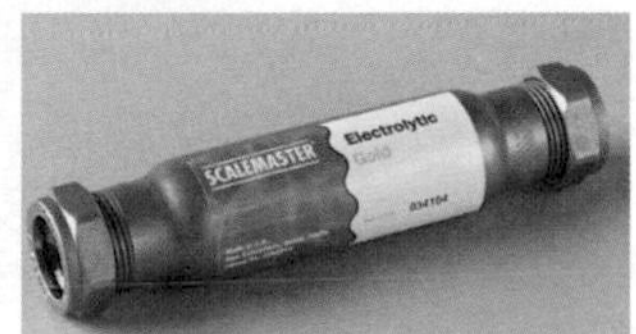

▲ Figure 5.114 Electrolytic scale inhibitor

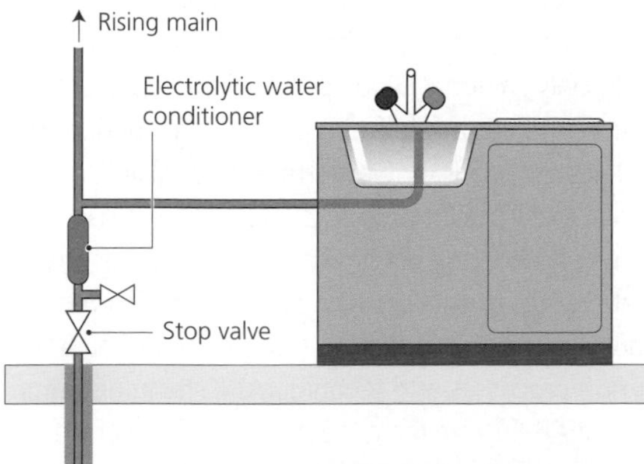

▲ Figure 5.115 Fitting an electrolytic water conditioner

- **Electronic (electromagnetic):** these cause dissolved hardness salts and minerals to cluster together rather than form on surfaces.

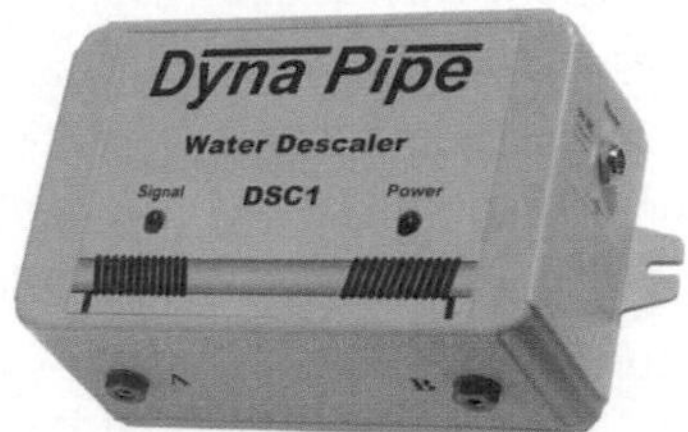

▲ Figure 5.116 Electromagnetic water conditioner

- **Electrochemical:** typically German or Austrian made, these conditioners contain a cartridge filled with ceramic beads that cause the magnesium and calcium crystals to precipitate. The conditioner units are usually quite large, requiring an electrical supply.

The benefits of installing a water conditioner include:

- reduction in the scale formation in pipes
- reduction of limescale on taps
- easier cleaning of showerheads and places where limescale may form.

Water softeners

A water softener is an appliance that is fitted directly to the water supply to a domestic dwelling or a commercial building, specifically designed to remove the water hardness. They are usually installed as close to water main entry into the building as possible. Most modern softeners are very compact and can easily be fitted under a kitchen sink.

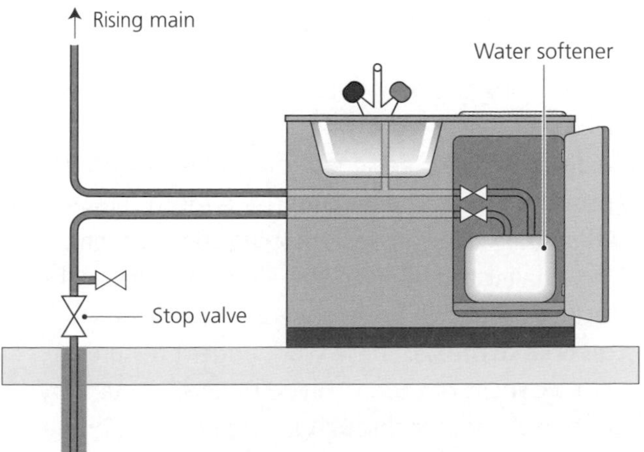

▲ Figure 5.117 Fitting a water softener

Water softeners use a process called ion exchange. The softener contains a column that is filled with special resin beads. These remove the dissolved calcium and magnesium salts by replacing them with sodium as the water passes through them. Once a day, the unit automatically washes the beads with brine (salt water) to remove calcium and magnesium ions, taking the solution to drain. Every month, the unit has to be refilled with salt in the form of granules, tablets or blocks.

Use of a water softener generally reduces the hardness of the water from 350 mg/l (milligrams per litre) to less than 10 mg/l.

When installing a water softener, there must be at least one unsoftened cold water outlet in the dwelling.

Drinking water filters

Drinking water filters alter the water composition to improve its taste, odour and appearance for drinking and cooking purposes. There are two common types:

1 jug filters – filled from a tap and stored in a fridge
2 plumbed-in filters – usually sited underneath the kitchen sink with a separate drinking water tap installed at the kitchen sink.

These can usually be supplied in six different forms:

1 **Activated carbon filter:** used to reduce taste and odour such as chlorine. The carbon filter has a large surface area that attracts and absorbs organic substances from the water. The carbon is usually in powder, granular or block form.
2 **Ion exchange:** used to reduce limescale formation and other metal-ion contaminants such as lead. It takes the form of tiny granules, which work by replacing the mineral or contaminant ions with hydrogen ions.
3 **Sediment filter:** designed to remove fine particles from the water. These comprise a mesh through which the water passes, trapping the sediment. The smaller the holes in the filter, the smaller the particles that can be removed.
4 **Reverse osmosis:** these work under pressure to remove most of the dissolved mineral content by passing the water through a very fine membrane.
5 **Distillation:** removes the mineral content of the water by boiling it and condensing the steam back to water vapour.
6 **Disinfection:** used to reduce the bacteria content and other micro-organisms by either UV light or a very fine sediment filter (usually ceramic or membrane).

▲ Figure 5.118 Fitting a water filter

Installation of cold water pipework

Many of the requirements for pipework installation are covered in Chapter 2, Common processes and techniques. Here, we will look at those techniques specific to cold water installations.

Choosing the right materials

Cold water supply in domestic dwellings, as we have already discussed, is strictly regulated by the Water Supply (Water Fittings) Regulations. This means that the choice of materials for cold water installations is limited to the following.

- Copper tubes and fittings (refer back to Chapter 2, Common processes and techniques) – copper has a proven record for cold water installations. It is light, rigid, has many jointing techniques available and requires only minimal clipping. It is highly resistant to corrosion and has a minimum life, in ideal conditions, of 150 years. It does, however, take great skill to fabricate and install it properly. There is a fire risk when using soldering equipment and it requires many specialist tools to successfully complete an installation.
- Polybutylene pipe and fittings (refer back to Chapter 2, Common processes and techniques) – manufacturers state that PB-1 has a life expectancy of 50 years. It is light and extremely flexible, and requires regular clipping when fixed on the surface. It is easier to install and can be cabled through joists easily and quickly. Push-fit joints make installation quicker and so

installation time can be reduced by 40 per cent with no fire risk. Testing techniques are more complicated and time consuming than for copper tube.

Preparation, planning and positioning of pipes

The installation of cold water systems needs to comply with the Water Supply (Water Fittings) Regulations and we must always consider the recommendations of **BS EN 806** and **BS 8558**. The manufacturer's instructions have to be followed with regard to the appliances installed and materials used.

The installation procedures will vary depending on the property. For instance, the methods used on new buildings will differ from those in an occupied dwelling where the customer's possessions will need to be taken into account.

Irrespective of the property type, pipework runs need to be planned carefully. It is advisable to avoid positions where frost and heat could cause a problem, such as outside walls, in cellars and unheated roof spaces. Wherever possible, pipework should be positioned out of sight and boxed in where appropriate. It should be remembered, however, that pipework should not be buried in walls or floors unless provision can be made to make it accessible.

Pipes in suspended timber floors

Pipes have been installed in timber floors for many years. Notching or drilling of joists should not be carried out in joists or rafters 100 mm deep or less. Notches should not be too tight for the pipes or creaking and 'ticking' noises may become a problem as the pipes expand and contract. Pipes in notches should be covered with joist clips to prevent excessive movement, and floorboards should be screwed (not nailed) when they are repositioned.

There are many different styles of suspended floor, including engineered timber joists, lightweight fabricated steel joists, lightweight cellular steel joists, and concrete block and beam systems. Figure 5.119 shows the installation requirements for these systems.

INDUSTRY TIP

Notching and drilling of joists should be done carefully, taking care to follow the recommendations mentioned in Chapter 2, Common processes and techniques, page 105.

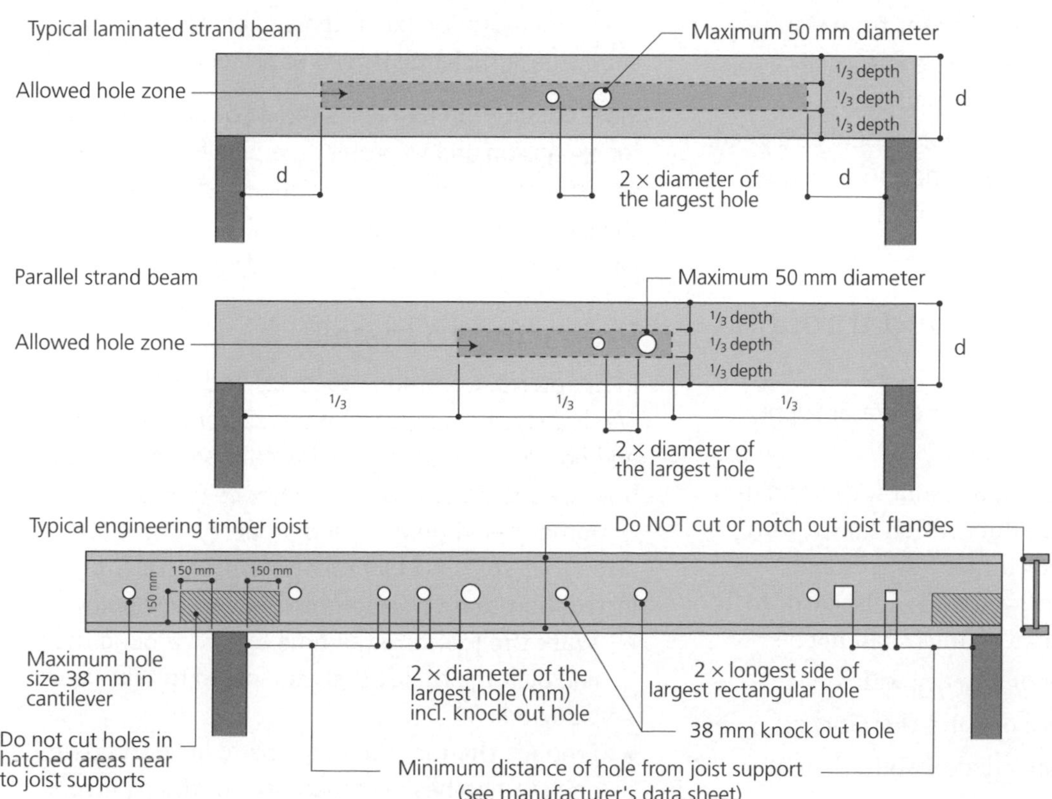

▲ Figure 5.119 Pipe installation requirements of typical joist systems

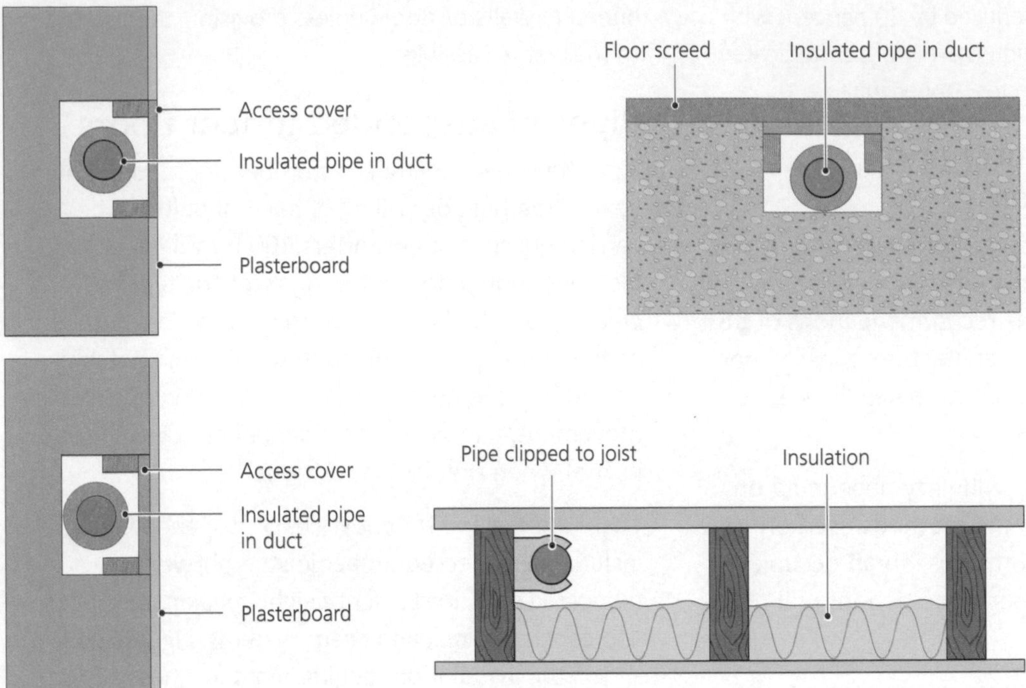

▲ Figure 5.120 Pipe in walls and floors

Cold water pipes should not be installed in the same notch as hot water and central heating pipes. There must be a minimum horizontal distance of 300 mm between cold water pipes and any hot water/central heating pipes to prevent radiated heat from warming the cold drinking water. Where there is a significant risk of cold water pipes being warmed by other pipework, the cold water installation should be lagged. To eliminate the risk of contamination from undue warming, the cold water pipework must never be allowed to exceed a temperature of 25°C.

Pipes in walls and sleeved through walls

According to Defra's guidance on the Water Supply (Water Fittings) Regulations 1999:

Unless they are located in an internal wall which is not a solid wall, a chase or duct which may readily be removed or exposed, or under a suspended floor which may be readily removed and replaced, or to which there is access, water fittings shall not be:

1 Located in the cavity of a cavity wall, or;

2 Embedded in any wall or solid floor, or;

3 Installed below a suspended or solid floor at ground level.

Where the laying of pipes in walls and floors is unavoidable, they should be placed in purpose-made ducts that have an accessible, removable cover, as shown in Figure 5.120. Pipes laid in chases must have adequate room for expansion and contraction, and should be sufficiently lagged or protected.

Pipes passing through walls should be sleeved to allow for expansion and to protect the pipe from building settlement and the corrosive effects of the masonry on the pipe. The sleeve should be sealed at both ends. The pipe should be thermally insulated where necessary.

Preparing to install

In Chapter 2, Common processes and techniques, we discussed taking care of the customer's property and possessions during the installation process, and how we should use various methods to protect the customer's environment and property. There are also other steps we can take before we start the installation to help save time, as described below.

- **Walk the job:** take the time to walk around the job and plan the routes that you intend to take your pipework.
- **Prepare the job:** use this time to lift floorboards and cut notches in preparation for the pipework installation. The floorboards can be replaced

temporarily so that the customer is not inconvenienced by holes in the floors. Remember to clear any mess as you go along. Don't leave it all to the last minute and never leave cleaning to the customer.

- **Mark out:** if you have decided on the routes that you intend to take, then mark out any surface-mounted pipework and drill any holes that you need to drill. Chases in walls and floors can also be marked at this point. Don't forget to use the correct PPE, such as protective goggles, when carrying out drilling and chasing procedures. It may be necessary to perform a risk assessment first.
- **Keep the customer informed:** let the customer know where you are going to be working and how long you plan to be in this area.
- **Keep entrances and exits clear:** don't leave trip hazards, such as cables and tools, lying around the work area.

Installation, testing and commissioning of cold water systems

The fabrication of pipework, installation techniques, commissioning and testing are dealt with elsewhere in this book. These, however, are important subjects that require reinforcement of your learning. The important aspects of installation are as follows.

- Keep all exposed pipework as neat as possible. Use the recommended clipping distances and protect the building fabric when making soldered joints.
- Prefabrication techniques for copper tube can save time and money on installations. Try to use machine bends wherever possible as these help with the flow rates in the finished installation.
- If the installation is an existing system, leave the final connections to the system until last. This will help to keep the decommissioning and turnover time of the system as short as possible.
- Pipework installed in floors and walls should be placed in properly prepared and accessible chases and ducts. Protect the customer's property at all times with dust sheets when cutting in chases.
- Hot and cold pipework should not be installed together unless the cold water can be protected from undue warming from the other surrounding services. If possible, when pipework is to be fixed on wall surfaces, the hot water pipework should be installed above the cold water pipework and, when installed in a floor cavity, a gap of at least 300 mm should be maintained horizontally.
- Cisterns should be marked and drilled for pipe connections in accordance with **BS EN 8558 4.3.10** and all holes drilled with a hole saw. Installation requirements should be in accordance with the Water Supply (Water Fittings) Regulations.
- Holes and notches in joists must be carried out in line with the building regulations.
- A water dead leg refers to any pipework that is no longer in use and there is a risk of the water turning stagnant which could contaminate the system. These are normally redundant branches and should be removed in order to prevent this from happening.

Connections to bathroom equipment and other common components

When connecting bathroom equipment, the manufacturer's installation instructions should be referred to. The design of the installation will dictate the size of the pipe required to deliver the flow rate, but the connection size to the tap will be dictated by the tap itself (see Table 5.10).

▼ Table 5.10

Baths	Bath taps usually require 22 mm pipework unless the system water is to be delivered at high pressure, then 15 mm pipework usually suffices.
Washbasins	Washbasins are usually connected with 15 mm pipe, but many new monobloc washbasin taps have 10 mm connections. However, 10 mm pipe should be restricted to the last 1 m of pipework, otherwise the flow rates required may be affected.
WCs	WCs must be fitted with an isolation valve prior to connection to the float-operated valve for maintenance and replacement purposes. They are usually connected in 15 mm pipe.
Bidets	The kind of connection to bidets is dictated by bidet type. Over rim-type bidets can be connected in the same manner as washbasins. However, ascending spray-type bidets must be connected only to low-pressure supplies fed from a storage cistern. The recommendations of **BS EN 806** and the Water Supply (Water Fittings) Regulations 1999 must be adhered to, as ascending spray bidets are a backflow risk.

Cold water cisterns	The size of pipe connecting to the float-operated valve in a cold water cistern will depend on the size of the float-operated valve. Most cisterns are connected in 15 mm pipe for domestic cisterns. However, on rare occasions, a 22 mm connection is required, especially on large domestic installations, where a ¾-inch float-operated valve has been installed. Cold feed pipes to hot water storage cylinders and cold distribution pipework on indirect cold water installations from the cold water cistern should be sized in accordance with the demands of the system. The more appliances installed, the greater the flow rate and the larger the pipe. Cisterns must have isolation valves on both the inlet and outlet pipework.
Boiler jigs	Boiler jigs should be installed in accordance with the manufacturer's instructions. Copper pipe must be installed from the jig for at least 1 m.
Boosting sets and pumps	Boosting sets, again, must be installed in accordance with the manufacturer's instructions. Most booster sets are now supplied with variable-speed pumps so the flow rate and pressure can be set by the installer to match the system design. This must be adjusted carefully to the required pressure and flow rate, and checked at commissioning stage. Booster sets are usually installed with a cold water accumulator.
Cold water accumulator	The accumulator is basically a pressurised water storage vessel, designed to limit the use of the pump and maintain system pressure. These must be installed after the pump but before the first appliance.

Working on existing systems

Existing systems can be notoriously difficult to work on and, the older the system, the more difficult it can be. Over the years, a variety of materials have been used for the installation of cold (and hot) water systems, and each of them brings its own unique set of problems (see Table 5.11).

▼ Table 5.11

Lead pipes	There are still hundreds of installations that contain lead pipe and there are situations where making a joint on lead pipe is unavoidable. Joints using leaded grade D solder were banned in 1986. This means that proprietary joints, such as leadlocks and Philmac fittings, can be used only to convert the old lead pipe to workable copper tubes or polybutylene pipes. Even so, we must still exercise caution as brass fittings such as leadlocks can cause galvanic corrosion to occur, which could lead to water contamination downstream of the fitting. Wherever possible, lead pipe should be removed and replaced.
20 thread copper tube	Occasionally, you may come across an installation that contains thick-walled screwed copper pipe jointed using screwed brass fittings. This is known as 20 thread copper tube because the threads on the pipe measure 20 threads to the inch. Sizes 3/8, ½ and ¾-inch pipes were generally used in domestic installations and are not compatible with modern **BS EN 1057** copper tubes or polybutylene pipes. Capillary converter fittings are available but these are becoming increasingly rare.
Copper tube	This type of copper tube was introduced in the 1950s and has a much thicker pipe wall compared with modern copper tube. Jointing techniques were very similar to those of today, with both compression and capillary fittings being used. However, the tube sizes are imperial and so converters are required for some sizes. ½-inch tube will fit modern 15 mm, although it is a tight fit; ¾-inch is much smaller than modern 22 mm tube and so must be converted; and 1-inch tube is extremely tight when used with 28 mm fittings, so a converter fitting is, again, recommended. Both capillary and compression converter fittings are available.
Red band thin wall copper tube	This kind of copper tube is identifiable by a red line running down the length of the tube and is mostly of German origin. It was used in the early 1970s when copper tube was scarce due to a copper shortage. It is very susceptible to pin hole corrosion. Only capillary joints should be made on this type of tube. The sizes of tube are imperial.
Stainless steel	Again, stainless steel tube was used extensively in the early 1970s due to a copper shortage. Unfortunately, the tube was manufactured from low-grade stainless steel, which has led to many problems of corrosion. Compression joints can be made onto this type of tube but care should be taken as it requires harder tightening because stainless steel is a much harder metal than copper. Again, tube sizes are imperial.
High-density polyethylene (HDPE)	HDPE was used for underground service pipes from the external stop valve (boundary stop valve) to the dwelling. It is black in colour and comes in four grades (A, B, C and D). Compression fittings are still available for this type of pipe but it should be noted that the grades have different wall thicknesses and so it is important that the correct type of pipe insert is used when making joints. Conversion to blue MDPE is a fairly simple task when the correct fitting is used.

➜

CuPVC (chlorinated unplasticised polyvinyl chloride)	Better known as 'PolyYork', this is a plastic pipe that is suitable for cold water supplies only. It was again used extensively in some parts of the UK during the early 1970s for cold water systems inside a domestic dwelling. Fittings used a solvent cement system that, once a joint was made, had to be left for 24 hours before testing could take place. It is very susceptible to fracture and fitting blow-off. Care should be taken when this pipe is encountered as it is extremely easy to fracture a fitting just by turning the water supply off!
Acorn (polybutylene)	An early version of polybutylene pipe that first appeared in the mid-1980s. It is compatible with all new polybutylene pipes and fittings, and copper tubes and compression fittings; however, a special pipe insert is required.

Testing cold water systems

- Before testing takes place, walk around the job and check that all joints have been made correctly, that there are a sufficient number of pipe supports and clips, and that you are happy that the installation conforms to the regulations.
- Close any open ends of pipes with cap ends.
- Pressure testing of the completed installation will depend upon the materials used:
  - Copper tubes – testing as detailed in **BS EN 806** and in Chapter 2, Common processes and techniques.
  - Plastic (polybutylene PB-1) – this will depend on which test is being performed. The requirements for both test A and test B are detailed in **BS EN 806** and in Chapter 2, Common processes and techniques.
- Testing should be performed using a hydraulic test pump like the one shown in Figure 5.121.

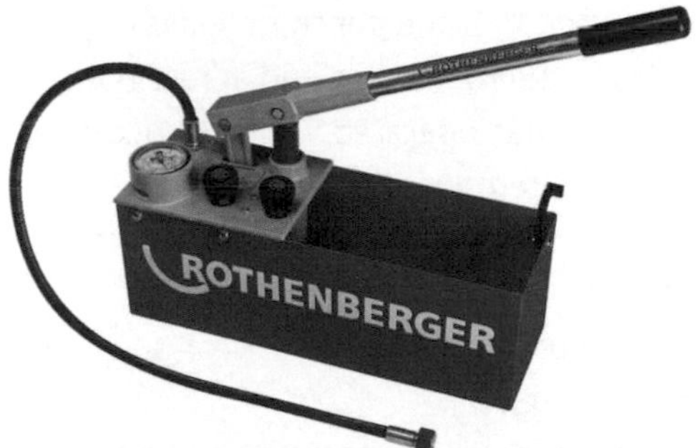

▲ Figure 5.121 Hydraulic pressure test pump

Noise

System noise can take many forms, from a squealing tap washer to violent pipework reverberation and water hammer. Most noise within direct systems of cold water is a direct result of the high pressure and flow rate that can occur within this system. It should be remembered that whenever a system has a mixture of high pressure and high flow rate, there will always be a certain amount of noise within that system. Sometimes, however, the noise can be excessive and this may be attributable to:

- **Faulty tap washers:** these tend to make a humming or squealing noise when the tap is opened. It is usually because the tap washer is either worn or split, and re-washering the tap cures the problem in most cases.
- **Faulty FOV washers:** this can cause a very loud hum throughout the pipework. Unfortunately, the noise is amplified if the cistern is in the roof space. Re-washering the FOV generally cures the fault. One way of testing to see if it is the FOV washer is to turn on a cold tap when the noise begins. If the noise stops or goes quieter, it is probably the FOV washer.
- **Loose or incorrectly supported pipework:** this can be the cause of very violent banging within the system. Every time the pipework reverberates, it is equal to twice the incoming mains pressure. If the supply is at 3 bar, then each bang is the equivalent of 6 bar. This can eventually lead to fittings failure and leakage. The best course of action is to try to find where the pipework is loose, and re-fix it. If this is not possible, the installation of a water hammer arrester fitted near to the main stop valve inside the property may cure the problem.

Inadequate water supply

Airlocks on low-pressure systems can be a constant nuisance, especially during the commissioning stage. Airlocks stop the flow of water due to air trapped in the pipework, and there is insufficient water pressure from the cistern to push the air out. They usually occur because the cold distribution pipe rises as it leaves the cistern rather than falling towards the appliances, and this causes a high spot where air collects. It is often a result of poor installation or design.

Curing an airlock is not easy. Usually, the best course of action is to leave the system to settle. Most airlocks eventually move, allowing water to flow. This can be problematic if the system is new and at the commissioning stage, because the system cannot be tested properly until the airlock clears. To be sure that airlocks do not occur, ensure that distribution pipes from cisterns have a slight but constant fall towards the appliances.

Leakage

Leakage is a common problem in cold water systems. It can take three main forms:

- **Leakage from the cold water service pipe below ground before it enters the property:** this is quite difficult to detect. The main signs of leakage are loss of water pressure and flow rate and a constant distant sound of running water. To find out whether the leak is before or after the external (boundary) stop valve, the external stop valve must be turned off; if the water supply has stopped but the sound of running water remains, the communication pipe is leaking and this must be repaired by the water undertaker. If the sound of running water stops when the external stop valve is turned off, the leak is on the service pipe to the property and this is the responsibility of the property owner.
- **Leakage from the internal cold water system pipes and fittings:** this can cause a lot of damage to the property. It is fairly easy to detect the source by isolating the mains cold water stop valve. If the water stops, it is on the mains cold water supply. If the water continues to run, it is on the distribution pipework. By isolating the mains internal stop valve and opening the hot and cold water taps in the property, the system will drain quickly, allowing repairs to be carried out.
- **Leakage from taps and FOVs:** dripping taps are an annoyance but they can also waste quite a lot of water if they are dripping for a long time. If the property is on a water meter, they can make a significant impact on the water bill. Dripping FOVs are detected when the overflow to the cold water storage cistern or the WC cistern begins to run. This can first show itself by the overflow running only at night when the pressure of the water main rises. Gradually, it will start to run all the time and will need to be repaired.

Commissioning cold water systems

Refer to Chapter 2, Common processes and techniques, for further information on the commissioning of cold water systems.

- The commissioning and flushing procedure should be undertaken with fresh wholesome water direct from the water undertaker's main.
- Check that all pipework is secure and check that all tap connectors and tank connectors are fully tightened, and all drain-off valves turned off.
- Check the inside of any cisterns installed to ensure that they are free of debris and that all connections are tight.
- Ensure that all isolation valves and terminal fittings are off.
- Open the kitchen cold tap and slowly open the mains cold water stop valve. Allow the water to flow into the kitchen sink to clear any debris that may have collected in the pipework.
- Close the cold tap on the kitchen sink and allow the system to fill to full standing pressure.
- Turn on the cold taps one at a time until the water runs clear, and check for leaks.
- Turn on the isolation valves to the float-operated valve in the WC cistern and allow the cistern to fill to the water line. Adjust the water level as necessary. Flush the WC and check for leaks.
- Fill any cisterns in the roof space and adjust the water level at the float-operated valve as necessary.
- Open any taps and terminal fittings fed from the cistern and clear any air in the system. Allow the water to run, to clear any debris.
- Allow the system to stand, then check for any leaks throughout the system.
- Isolate at the mains cold water stop valve and completely drain the system to flush the system through. This should clear any flux residue and swarf from the system.
- Refill the system and test for standing and running pressure at all mains outlets using a pressure gauge.
- Check that all flow rates meet the specification and any manufacturers' instructions, using a flow meter or a weir gauge.
- Re-check the system for leaks.

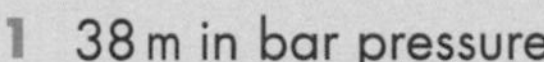

ACTIVITY

Remember: water pressure is measured in bar pressure – 1 bar is the equivalent of 10 m head of water or 100 kPa. Now, using the above figures, calculate:

1. 38 m in bar pressure
2. 4.5 bar in kilopascals (kPa)
3. 150 kPa in bar pressure

ACTIVITY

Water flow rate is measured in litres per second or litres per minute. To convert from litres per second (l/s) to litres per minute (l/m), simply multiply the l/s by 60. For example:

$$0.3 \text{ l/s} = 0.3 \times 60 = 18 \text{ l/m}$$

To convert from litres per minute (l/m) to litres per second (l/s), simply divide the l/m by 60. For example:

$$25 \text{ l/m} = 25 \div 60 = 0.41 \text{ l/s}$$

Now attempt the following calculations:

1. 30 l/m into l/s
2. 0.25 l/s into l/m
3. 12 l/m into l/s
4. 0.12 l/s into l/m

12 REPLACING OR REPAIRING DEFECTIVE COMPONENTS: PLANNED AND UNPLANNED MAINTENANCE

Maintenance tasks on cold water services, appliances and valves are essential to ensure the continuing correct operation of the system. The term used when isolating a water supply during maintenance operations is 'temporary decommissioning'.

Before undertaking the repair or replacement of components, we must first ascertain what the problem is. The customer will be able to tell you what is happening with the component. They may not know the technical language but they will be able to explain the problem well enough for you to understand.

IMPROVE YOUR ENGLISH

Communication is key when it comes to both resolving issues and also ensuring that your customer feels informed and confident in your ability to complete the job. You need to remember that most of your customers will not have the technical vocabulary that you are accustomed to. Remember, this may be their home, so reassure them by explaining the problem and the processes necessary to fix (or not) the problem. Avoid overly technical terms, clearly stating the stages of the job so they know what to expect and when.

VALUES AND BEHAVIOURS

Always try to maintain and return the customer's property as it was. If a component requires replacement, we must ensure that we get as near to a like-for-like replacement as possible, that we have the correct tools available, and that the customer's property is either removed or protected with dust sheets and other coverings before we begin.

The manufacturer's instructions

When repairing or replacing components, the manufacturer's instructions give step-by-step methods. These should be followed wherever possible. In some instances where the component is old or the customer has lost the original instructions, a copy may be available on the manufacturer's website.

Maintenance

There are basically two types of maintenance:

1 planned preventative maintenance
2 unplanned/emergency maintenance.

Planned preventative maintenance

Planned preventative maintenance is usually performed on larger systems and commercial/industrial installations. It is performed to a pre-arranged maintenance schedule, which may mean out-of-hours working if the supply of water cannot be disrupted during normal working hours. It is designed to stop problems from occurring by catching faults in their early stages. Planned preventative maintenance could include:

- periodic system inspection – checking for leaks
- re-washering of float-operated valves
- re-washering and re-seating of terminal fittings and taps
- inspection and cleaning of cisterns
- readjustment of water levels in cisterns
- re-washering of drain valves
- cleaning of filters and strainers
- maintenance of water softeners
- checking the correct operation of stop valves
- checking flow rates at all outlets.

HEALTH AND SAFETY

When a maintenance task involves isolating the cold water supply, a notice will need to be placed at the point of isolation, stating 'System off – do not turn on', to prevent accidental turning on of the system. In most systems, it will be possible to isolate specific parts of the installation without the need to have the whole supply turned off. Where no such isolation exists, it may be of benefit to use a pipe-freezing kit so that total system isolation is not undertaken.

A record of all repairs and maintenance tasks completed will need to be recorded on the maintenance schedule at the time of completion, including their location, the date when they were carried out and the types of test performed. This will ensure that a record of past problems is kept for future reference.

Where appliance servicing is carried out, the manufacturer's installation and servicing instructions should be consulted. Any replacement parts may be obtained from the manufacturer.

VALUES AND BEHAVIOURS

Do not forget to keep the householder/responsible person informed of the areas that are going to be isolated during maintenance tasks and operations, and always ask the customer if they need to 'draw off' a temporary supply of water (kettle, saucepans, bucket, etc.) to cover a short period of system isolation.

Unplanned and emergency maintenance

Unplanned and emergency maintenance occur when a fault suddenly develops, such as a burst pipe, or a small problem suddenly becomes a larger issue, such as a dripping tap or sudden loss of water. Unplanned and emergency maintenance can include:

- burst pipes and leaks
- running overflows
- dripping taps
- loss of low-pressure, cistern-fed cold water supply due to faulty float-operated valves
- poor past installation practices, such as incorrectly positioned overflow pipes
- complete component breakdown necessitating the replacement of the component.

Many of the maintenance practices we use involve the decommissioning of systems so that parts and pipes can be replaced.

Maintenance tasks

In this part of the chapter, we will look at some of the basic maintenance tasks we have to perform, including:

- re-washering and re-seating taps
- maintaining a ceramic disc tap
- maintaining **BS 1212 Part 1, 2 and 3** float-operated valves.

Re-washering and re-seating a rising spindle tap

During the maintenance operation, taps should be re-seated as well as re-washered. This involves using a special tool, called a tap re-seating tool, which grinds the seat of the tap to remove any pits that have occurred due to water passing between the seat and the tap, ensuring that the washer sits evenly on the tap seat.

▲ Figure 5.122 Tap re-seating tool

The procedure for re-washering a tap is as follows.

First, ensure that the water supply is isolated, open the tap to relieve the pressure and put the plug into the sink. This will ensure that any dropped small screws and nuts do not disappear down the sink waste and into the waste pipe trap.

STEP 1 Locate the screw that holds the tap head onto the spindle and carefully remove with a small screwdriver.

STEP 2 Carefully remove the tap head. Invariably, taps are cross-top heads, which can prove difficult to remove. Care must taken here to prevent damage to the appliance that the tap is fixed to.

STEP 3 With the head removed, we can now break the joint between the tap head workings and the tap body using an adjustable spanner. This may involve using a pair of water pump pliers to counteract the force of the adjustable spanner on the head workings. Ensure that a cloth is used to protect the tap body from the effects of the jaws of the water pump pliers on the tap body.

STEP 4 Remove the jumper plate and washer from the spindle. A little force may be needed from the flat blade of a screwdriver if the jumper plate is fixed.

STEP 5 Some rubber tap washers are held onto the jumper plate by a small brass nut. Carefully remove the nut and replace the existing rubber washer with a new rubber washer of the correct size, then replace the washer nut. Do not over-tighten the washer nut as it may break.

STEP 6 Remove the packing gland nut and remove the spindle by fully winding in a clockwise direction and pushing the spindle through the packing gland.

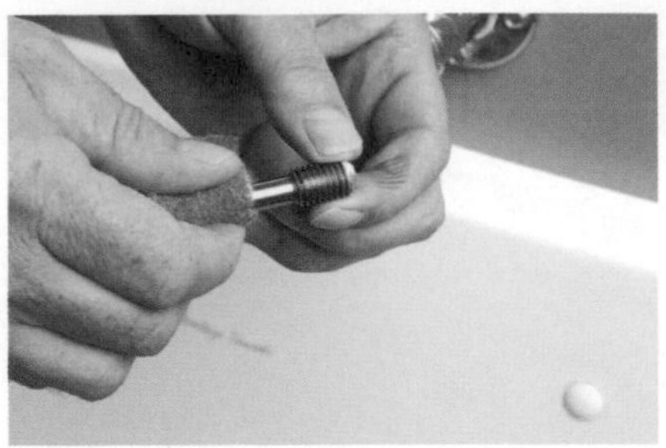

STEP 7 Check the spindle for any signs of wear and remove any scale that may have gathered on the spindle shaft. A non-metallic fittings cleaning pad is ideal for this. Re-grease the spindle using silicone grease.

STEP 8 Push the spindle back through the packing gland and fully wind until the tap spindle is in the fully open position.

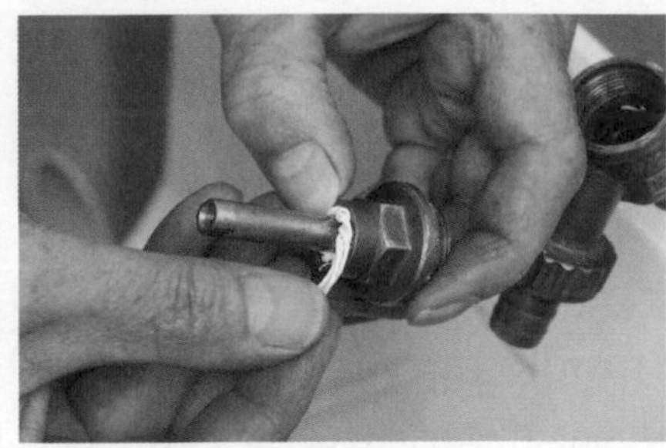

STEP 9 Check the packing in the packing gland and replace with a PTFE grommet where necessary.

STEP 10 Re-insert the jumper plate into the spindle.

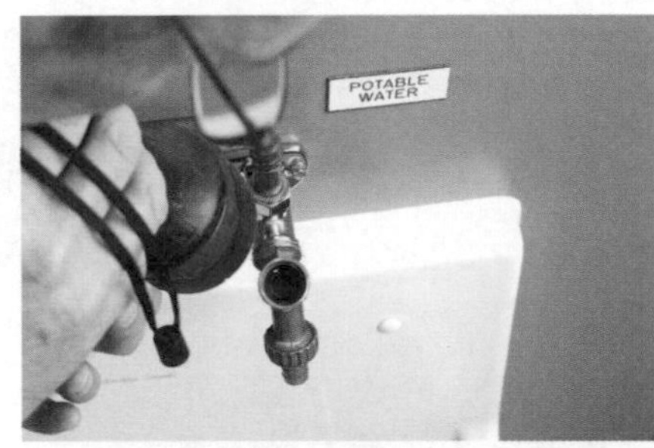

STEP 11 Check the seat of the tap by shining a torch into the tap body. If the tap requires re-seating, use the tap re-seating tool with correct size grinding head and re-seat as necessary.

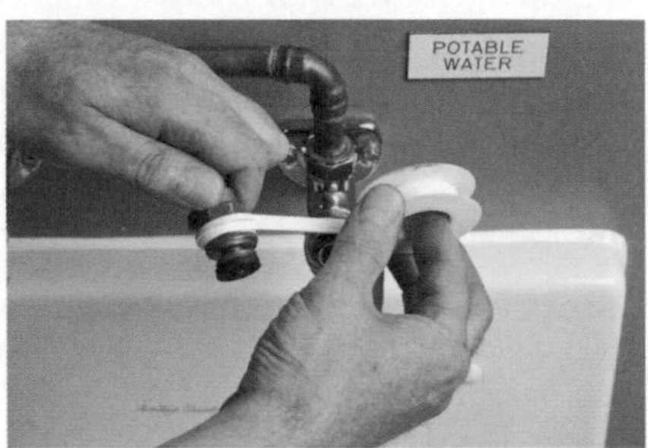

STEP 12 Check the fibre sealing washer on the head workings. These tend to break when the tap head is removed. If the fibre sealing washer needs replacing, this can be done using PTFE tape.

STEP 13 Replace the head workings into the tap body (ensuring the head workings are fully open) and re-tighten into the tap.

STEP 14 Tighten the packing gland nut, taking care not to over-tighten as the tap will be difficult to open.

STEP 15 Replace the tap head but do not secure with the screw at this point. Turn on the water with the tap open. This will ensure that any debris from re-seating will be washed out of the tap. Turn off the tap and check for any drips. Replace the tap head securing screw.

▲ Figure 5.123 Re-washering a tap

Re-washering and re-seating a non-rising spindle tap

Some of the problems that can occur with these taps are as follows.

- The barrel, which rises inside the tap head workings, can become dislodged causing the tap to seize in the closed position and prevent the tap being opened. Often, this is a result of its having been over-tightened, compressing the washer.
- The circlip, which holds the non-rising spindle in position, can very often break.

There is no packing gland with non-rising spindle taps, so maintenance is a little easier. The maintenance procedure for non-rising spindle taps is as follows.

First, ensure that the water supply is isolated, open the tap to relieve the pressure and put the plug into the sink. This will ensure that any dropped small screws and nuts do not disappear down the sink waste and into the waste pipe trap.

STEP 1 Carefully remove the cap on the tap head to gain access to the screw.

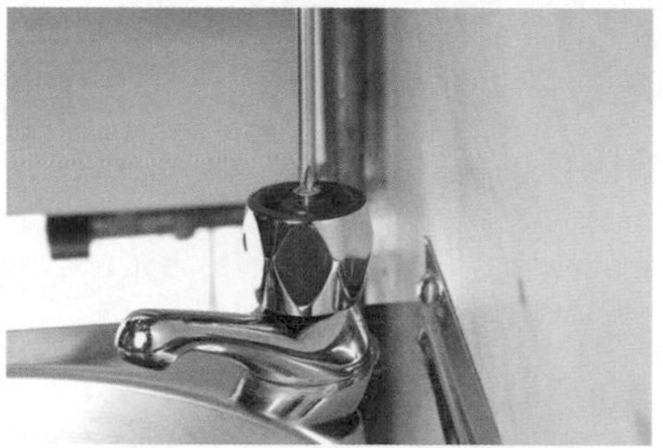

STEP 2 Locate the screw that holds the tap head on to the spindle and carefully remove with a small screwdriver. Some tap heads simply pull off the spindle. Carefully remove the tap head.

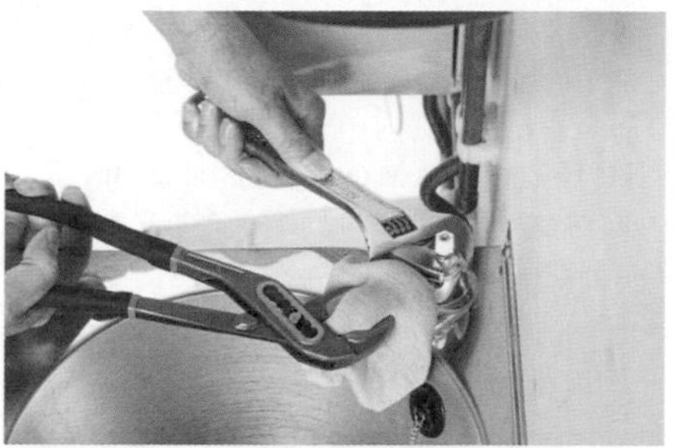

STEP 3 With the head removed, you can now break the joint between the tap head workings and the tap body using an adjustable spanner. This may involve using a pair of water pump pliers to counteract the force of the adjustable spanner on the head workings. Ensure that a cloth is used to protect the tap body from the effects of the jaws of the water pump pliers on the tap body.

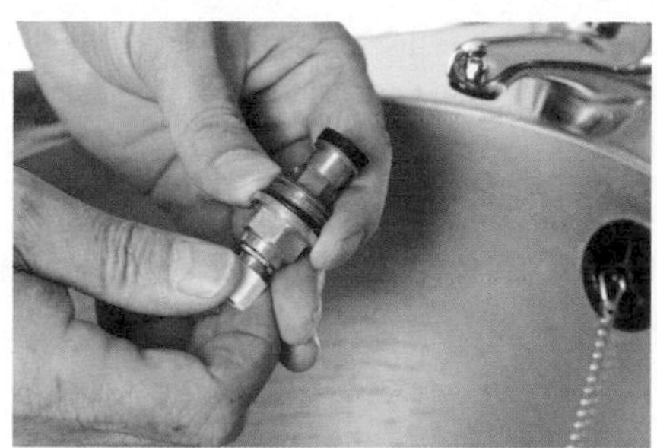

STEP 4 Fully unwind the spindle until the hexagonal barrel can be removed from the head workings.

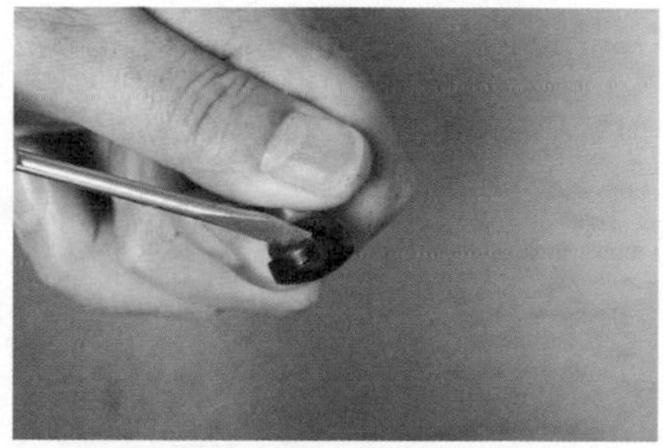

STEP 5 Carefully remove the rubber washer and replace with the correct size washer. A tap washer kit may be of benefit here, as there are many different sizes and styles of washer for a non-rising spindle tap.

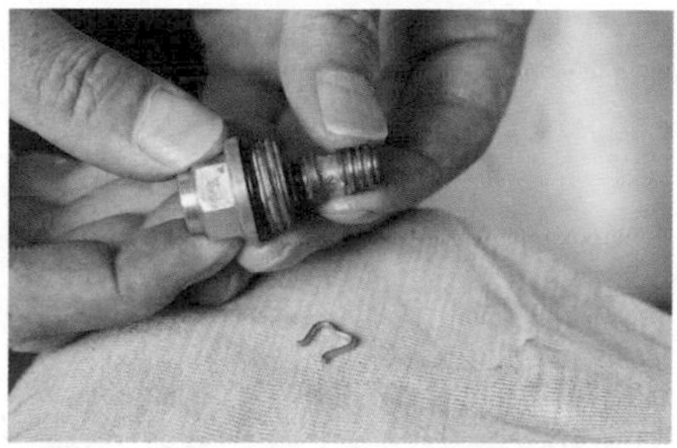

STEP 6 Carefully remove the circlip with circlip pliers and push the spindle downwards and out of the head workings.

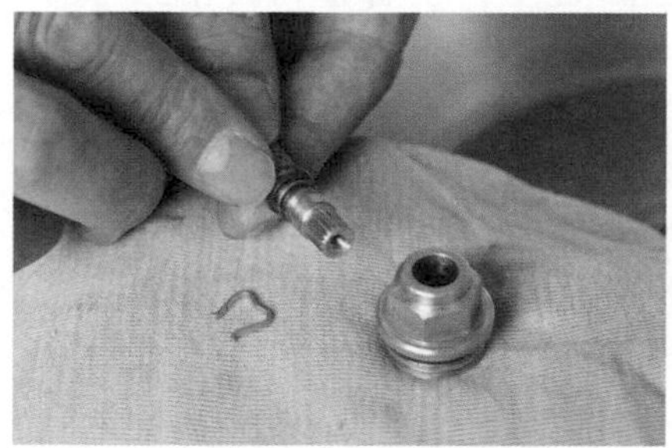

STEP 7 Check and replace the spindle 'O' ring seals as necessary.

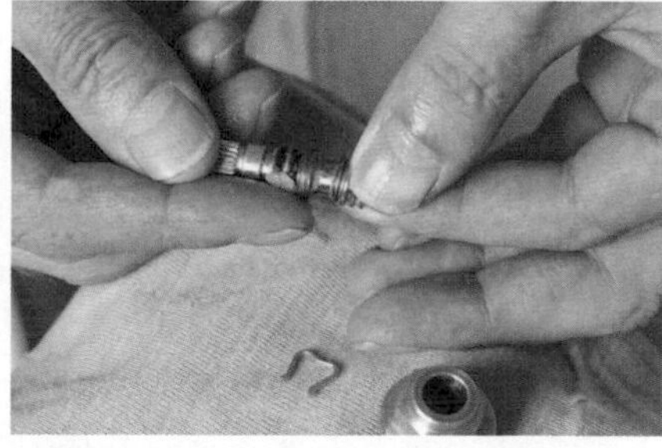

STEP 8 Re-grease the spindle with silicone grease.

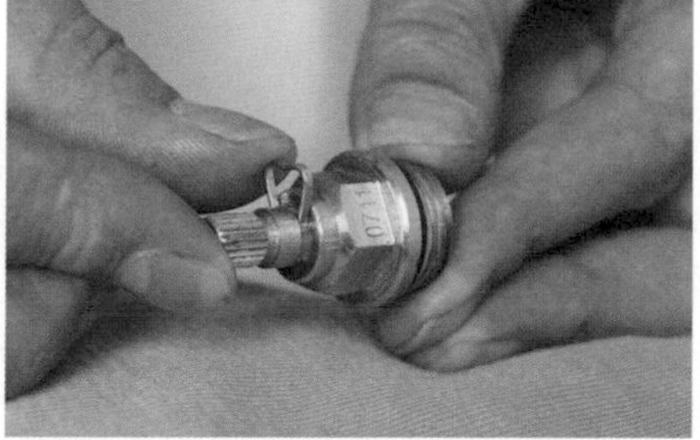

STEP 9 Re-insert the spindle into the head workings and replace the circlip.

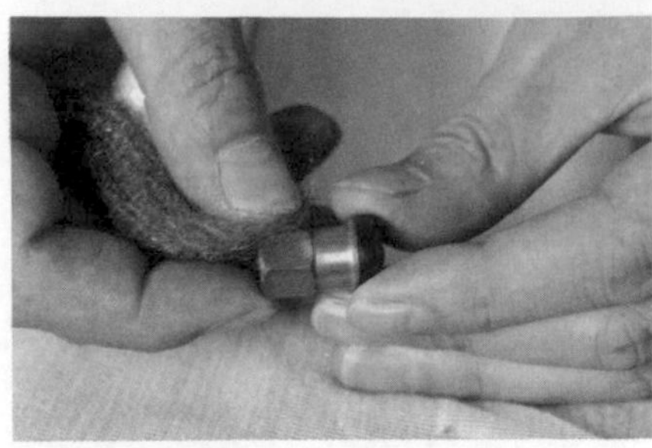

STEP 10 Check the hexagonal barrel for any signs of scale and clean with a cleaning pad as necessary.

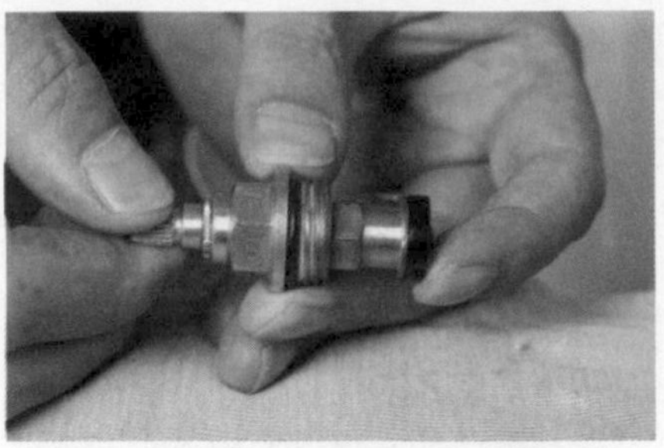

STEP 11 Re-grease the barrel using silicone grease and very carefully rewind back into the head workings. Ensure that the tap head is in the fully open position.

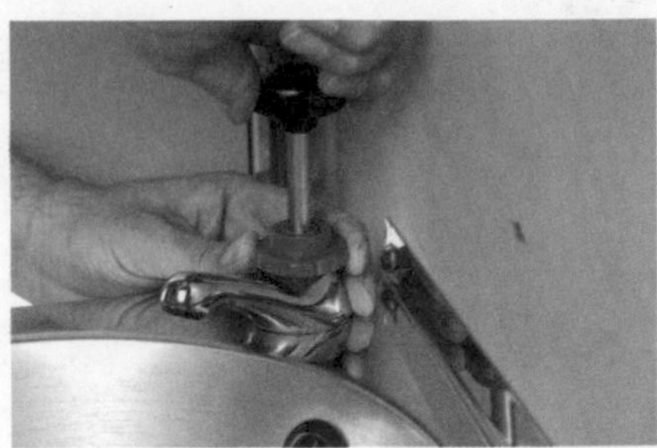

STEP 12 Check the tap seating and re-seat using the tap re-seating tool with the correct size grinding head as required.

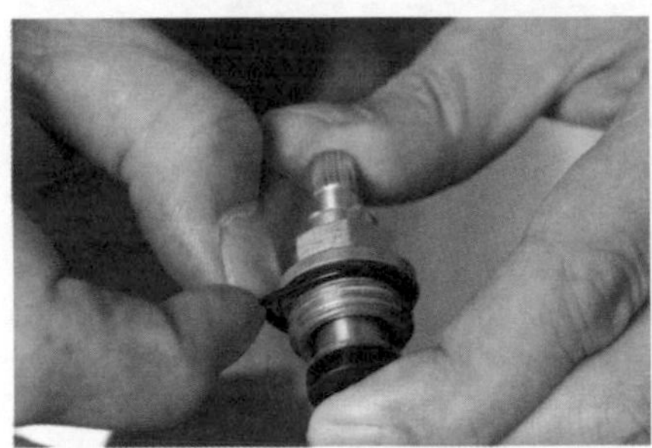

STEP 13 Check the rubber 'O' ring on the tap head workings. This washer seals the head workings to the tap body. Replace as required.

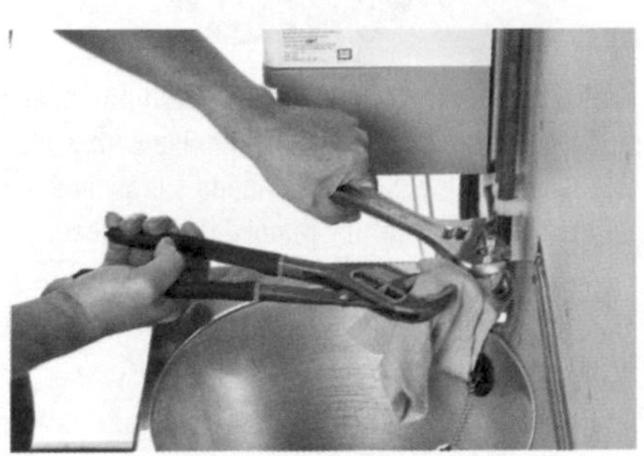

STEP 14 Replace the head workings into the tap body (ensuring the head workings are fully open) and re-tighten into the tap.

STEP 15 Replace the tap head but do not secure with the screw at this point. Turn on the water with the tap open. This will ensure that any debris from re-seating will be washed out of the tap. Turn off the tap and check for any drips. Replace the tap head securing screw.

▲ Figure 5.124 Re-washering a non-rising spindle tap

Maintaining a ceramic disc tap

Ceramic disc taps do not have a washer to replace. Instead, they use two very thin plates or discs of a ceramic material to allow water to flow through the tap. Most ceramic disc taps are not repairable. The tap head workings will need to be replaced with a like-for-like unit, which can be obtained from the manufacturer or from the local merchant or stockist. There are a wide variety of ceramic disc sets available and the correct one for the tap must be obtained. When ordering the part, the type of head workings, i.e. hot or cold, will need to be stated as they open and close in different directions.

Replacing taps

Modern taps can be replaced easily, as the threaded tap-connecting tail is of a generic length, irrespective of the manufacturer.

- Ensure that the water supply is isolated. Open the tap to relieve the pressure and drain the pipework.
- Using a crows-foot spanner or a tap spanner, twist the tap locking nut holding the tap to the appliance counter-clockwise half a turn. This is to release the tap from the appliance. By not loosening the tap connector first, the tap connector will prevent the tap from spinning in the appliance.
- Attach the tap spanner to the tap connector and turn counter-clockwise. Ensure that you hold the tap during this process as it may turn, damaging the appliance.
- Remove the old tap and clean the hole.
- Take off the locking nut to the new tap and make sure that the rubber grab washer is in place on the tap.
- Insert the tap into the appliance and wind up the new locking nut clockwise by hand, then finally tighten with the tap spanner.

- Replace the fibre sealing washer on the tap connector and wind the tap connector onto the tap. Tighten the tap connector.
- Turn on the water, check for leaks and test the tap.

When replacing old taps, the threaded tap-connecting tail is longer than more modern taps. This means that a tap extender fitting is used to lengthen the thread to the correct length for it to fit the existing pipework. Tap extender fittings are available from most good plumbers' merchants.

Repairing a BS 1212 Part 1 float-operated valve (Portsmouth type)

Portsmouth-type float-operated valves are allowed to be fitted on new installations only if some form of backflow prevention device is installed before the float-operated valve; usually this would be a double check valve. However, if a Portsmouth valve is part of an existing installation, then repair is permissible. To repair a Portsmouth valve, follow the steps listed below.

1. Turn off the water supply at the isolation valve to the float-operated valve.
2. Remove the float-operated valve from the cistern by unscrewing the union nut.
3. Remove the end cap on the valve body.
4. Remove the cotter (split) pin holding the float arm to the valve body and remove the float arm.
5. Remove the piston from the valve body.
6. The piston is generally made from one of two materials. It can either be brass or nylon.
7. For brass pistons, the float-operated valve washer is held in the end of the piston by a retaining cap, which will need to be unscrewed to allow the washer to be removed. To remove the retaining cap:
   - place a flat-blade screwdriver in the slot for the float arm and unscrew the retaining cap using a pair of pliers
   - remove the washer and replace with a like-for-like washer
   - replace the retaining cap and tighten
   - check the piston for any signs of scale and remove these with a cleaning pad.
8. For nylon pistons, simply push the washer out of the gap in the side of the washer housing and replace the washer.
9. Remove the orifice from the float-operated valve body and check to ensure that there are no cracks or splits visible. Replace as necessary.
10. Reassemble the valve, making sure that the washer is towards the spindle.
11. Replace the cotter pin and open to ensure that it does not fall out.
12. Re-install the valve into the cistern, making sure the fibre sealing washer is in place.
13. Re-tighten the union and turn on the water.
14. Check the operation of the valve, adjusting the water level as necessary.

Repairing a BS 1212 Part 2 and 3 float-operated valve (diaphragm type)

Diaphragm-type float-operated valves discharge water over the top of the valve. They have a large diaphragm-type washer that is easily accessible for repair and replacement. To replace the diaphragm washer, follow the steps listed below.

1. Turn off the water supply at the isolation valve to the float-operated valve.
2. Remove the float-operated valve from the cistern by unscrewing the union nut.
3. Unscrew the large washer-retaining union and float arm arrangement at the front of the valve, and withdraw the washer.
4. Replace the washer, ensuring that it is fitted the correct way. These washers must be inserted correctly for the float-operated valve to operate as normal.
5. Replace the large washer-retaining union and float arm arrangement, ensuring that is engaged into the retaining notch at the top of the front plate and hand tighten the union.
6. Check that the orifice is in good order, with no cracks or splits. Replace as necessary.
7. Re-install the valve into the cistern, making sure the fibre sealing washer is in place.
8. Re-tighten the union and turn on the water.
9. Check the operation of the valve, adjusting the water level as necessary with the float arm adjustment screw.

Replacing float-operated valves

Float-operated valves can be replaced easily, as the threaded connecting tail is of a generic length, irrespective of manufacturer.

- Ensure that the water supply is isolated. Open the float-operated valve to relieve the pressure and drain the pipework.
- Using an adjustable spanner, twist the float-operated valve locking nut holding the float-operated valve to the cistern counter-clockwise half a turn. This is to release the tap from the appliance. By not loosening the tap connector first, the tap connector will prevent the float-operated valve from spinning in the cistern.
- Attach the spanner to the tap connector and turn counter-clockwise, ensuring that the float-operated valve is held firm by a pair of water pump pliers.
- Remove the old float-operated valve.
- Take off the locking nut to the new float-operated valve and make sure that the rubber grab washer is in place on the locking nut.
- Insert the float-operated valve into the cistern and wind up the new locking nut clockwise by hand, then finally tighten with the tap spanner.
- Replace the fibre sealing washer on the tap connector and wind the tap connector onto the float-operated valve. Tighten the tap connector.
- Turn on the water, check for leaks and test the float-operated valve.

Replacing pumps in boosting sets

When replacing any electrical component, first ensure that the electricity supply is totally isolated by initiating the safe isolation procedure for electricity supplies, as follows.

- Remove the fuse from the consumer unit.
- Check that the electrical circuit is dead using a GS38 tester and proving unit or some other effective electrical testing device.
- Make a simple drawing of the live/neutral/earth connections on the pump and disconnect the cable.
- Turn off the isolating valves on the inlet and outlet to the pump.
- Carefully loosen the unions on the pump by turning them anti-clockwise using water pump pliers. It may be a good idea to have some old towels handy to catch any water.
- Once both unions have been disconnected, remove the pump. The pump unions should have the old washers removed and the union faces cleaned. The new pump should include flat rubber washers.
- Position the new pump, with the sealing washers in place, between the valves and hand tighten the unions. Take care to ensure that the pump is facing in the right direction for the system.
- Fully tighten the unions with the water pump pliers.
- Turn on the pump valves and check for leaks.
- Carefully reconnect the electrics to the pump: live to the L point, neutral to the N point and earth to the E point. Make sure that all electrical connections are tight.
- Reinstate the fuse in the consumer unit. Switch on and test for correct operation.

13 DECOMMISSIONING OF SYSTEMS

Occasionally, systems will require isolation for repairs, renewal of appliances and extensions to systems, or when systems or appliances are being permanently removed. This is known as decommissioning.
Decommissioning takes two forms:

1. **Temporary decommissioning:** this is where systems are isolated for a period of time so that work can be performed. Eventually, the system will be recommissioned and put back into normal operation.
2. **Permanent decommissioning:** when a system or an appliance is taken out of use, it has to be permanently decommissioned. This will require that the system is isolated and drained, the appliance(s) removed, and the pipework cut back, removed and capped to the nearest live line to prevent stagnation of water in a live cold water supply.

Information to be provided to other users before decommissioning

Whether permanently decommissioning a system or temporarily decommissioning a section of pipework to allow repairs or the replacement of appliances or components, the end user or customer should be informed of:

- which part of the system is likely to be out of service
- approximately how long for
- where they can get water from in the meantime.

Consider ways that can reduce periods when facilities are not available

To lessen the inconvenience of the water system being off, the installation can be separated into zones by installing isolation valves at key points. This would ensure that not all of a system is out of commission and that parts remain in service.

The work, rather than being done when the property is occupied, could be completed at those times when the building is closed, outside of normal operating hours.

Always remember to isolate the fuel system/electricity supply to the hot water system, to prevent accidental heating of a decommissioned system.

Preventing the end user from using an appliance or system

To prevent the end user from using the cold water system once it has been turned off for decommissioning, a notice should be placed at the point of isolation informing other water users that part of the system is out of commission and that it should not be turned back on. It is also a good idea to put your telephone number on the notice so that people can contact you for information. The pipework should be cut at the nearest tee piece or live line to prevent stagnation of water in the decommissioned section of pipework. If the decommissioning is temporary, then the appliance can be disconnected and the pipework capped off.

SUMMARY

During this chapter, we have investigated water supply from the cloud to the tap and we have seen the correct practice of system installation, materials and components. It is an almost impossible task to attempt to describe every aspect of cold water supply and the best way to gain experience in this field is by working on the systems themselves and seeing the different systems, both new and existing, in operation. It will soon become apparent that there is a multitude of different systems, materials and fittings based upon those we have looked at. By seeing these different systems in operation, you will soon become proficient at identifying the correct methods of working. And as long as we can do that, we will enhance and develop our knowledge.

Test your knowledge

1 How many UK fluid categories are there?

a 4
b 5
c 6
d 7

2 Water within a food preparation area such as a kitchen sink falls within which UK fluid category?

a 4
b 5
c 6
d 7

3 Waste water collected from baths, showers and washing machines that is then reused for WC flushing is known as:

a Rainwater harvesting
b Wholesome water
c Black water
d Grey water

4 With regard to water treatment, what does the passage below describe?

'... designed to slow down the water velocity to allow the solids that the water contains to sink to the bottom and settle under gravity. It may also be used to reduce turbidity.'

a Sedimentation
b Filtration
c Sterilisation
d Chlorination

5 Which document is split into five parts and gives the specification for installations inside the buildings conveying water for human consumption?

a BS EN 806
b BS 6700
c The Water Supply (Water Fittings) Regulations
d The Private Water Supplies Regulations

6 What is the minimum supply pipe size to a modern dwelling?

a 20 mm
b 25 mm
c 32 mm
d 40 mm

7 What is the minimum recommended storage capacity of a CWSC within an indirect cold water system?

a 150 litres
b 230 litres
c 250 litres
d 310 litres

8 What type of **BS 1212** float-operated valve is shown below?

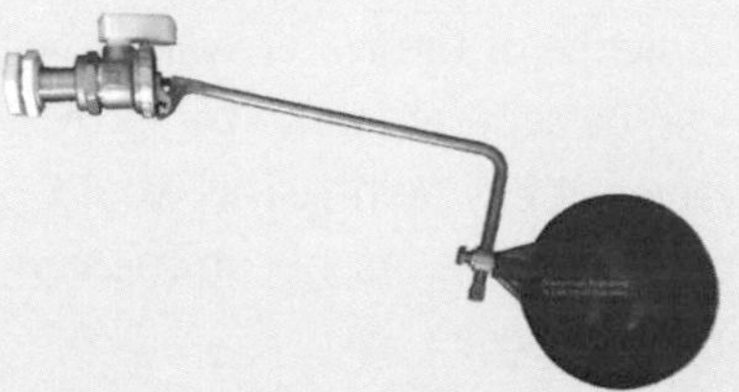

a Part 1
b Part 2
c Part 3
d Part 4

9 Where is the most suitable location for the component in the image below to be installed?

a On the outlet of a CWSC
b On the inlet of a WC cistern within an indirect system
c On the incoming water main as it enters the property
d On the cold supply to a washing machine

10 What type of air gap is incorporated over a kitchen sink?

a AG
b AUK1
c AUK2
d AUK3

11 What temperature does the Water Regulations state that cold water must never reach?

a 10°C
b 25°C
c 15°C
d 5°C

12 What are the two main considerations for the incoming cold water supply?

a Flow and colour

b Temperature and pressure

c Taste and temperature

d Pressure and flow

13 Following the commissioning of a cold water system, you find the flow rate from a float-operated valve to be poor. Which of the following could be the cause?

a The FOV was wrongly connected to a direct system

b The system wasn't flushed and debris is blocking the orifice

c The FOV can only be used on an indirect system

d Incorrect chemicals were used to flush the system out

14 Where should a drain-off valve be installed?

a Immediately above the internal stop valve and at every low point

b At every appliance to aid system draining

c Below the cold water storage cistern to avoid wasting water

d Immediately below any appliance supplied by the mains

15 Between what depths should the service pipe be laid?

a 650–1300 mm

b 750–1350 mm

c 800–1600 mm

d 350 –700 mm

16 If a tap washer in a tap is damaged, what would be the likely effect?

a The tap would drip from the outlet

b The tap would drip from the body

c The tap would drip from the spindle

d The tap would drip from the handle

17 What material is a modern service pipe made from?

a R250 half hard copper

b Lead

c MDPE

d ABS

18 According to the Water Regulations and **BS EN 806**, how should a cold water storage cistern be installed in a loft area?

a Next to an outside wall

b Supported over its entire base

c Have 750 mm clearance on all sides

d Insulated under the whole base area

19 Which British Standard outlines the required capacity of a cold water storage cistern?

a BS EN 806

b BS 14336

c BS 5422

d BS EN 1057

20 Which of the following is regarded as a 'surface water' source?

a Underground reservoir

b Cold water storage cistern

c Rising main

d Lake

21 Why is a **BS 1212 Part 1** float-operated valve not allowed to be installed in a WC cistern?

a The material it is made of could cause contamination

b The float is too big for a modern cistern

c The AG air gap cannot be maintained

d It can only be used on low pressure systems

22 What colour is a HIGH pressure orifice used in a Part 2 float-operated valve?

a Green

b Red

c Blue

d White

23 What is fluid category 3 described as in the Water Regulations?

a Slight health risk

b Wholesome water

c Wholesome water except for colour, odour or temperature

d Significant health risk

24 What is the reason plumbers are required to use 'approved' fittings?

a They are cheaper

b They are legal and avoid contamination

c They can be used on both copper and plastic pipework

d They have universal joints

25 Who is responsible to ensure that water is fit for human consumption in the mains?
  a The customer
  b The plumber
  c The local health authority
  d The water undertaker

26 Explain fluid category 2 and give an example from within a dwelling.

27 Give two advantages of a rainwater harvesting system.

28 Consider the table at the bottom of the page. Tick the boxes that indicate advantages when comparing either direct or indirect system selection.

29 What can be provided to prevent pumps running dry within a boosted old water system?

30 Explain the difference between a true mixer tap and a bi-flow mixer tap.

31 Describe where a grey water re-use system collects water from.

32 List the advantages of a direct cold water system in a property.

33 A cold water storage cistern you are installing has a cold feed and a cold distribution connection. Explain why care needs to be taken when positioning these connections.

34 Outline why a Part 1 float-operated valve is not allowed to be installed today.

Answers can be found online at www.hoddereducation.co.uk/construction.

Practical activity

At your place of work or training centre, why not ask your supervisor or tutor if you could prepare to install a CWS cistern? Prepare by marking with a pen the correct outlet and inlet positions in accordance with WRAS guidance, ensuring that all dimensions are adhered to.

If it is convenient (ask permission), cut the holes using the correct tools and ensure that all necessary components are fitted in accordance with Schedule 2, Paragraph 16 of the Water Regulations. Once the cistern is completed, ask your supervisor or tutor to check. This may also be a good opportunity to practise replacing the washer within the float-operated valve while it is easily accessible.

Particular advantage	Indirect system	Direct system
Cheaper to install		
Drinking water to all fittings		
Less fluctuation of pressure during peak demand periods		
Less risk of leaks due to lower pressures		
Smaller pipe sizes may be used		
Good pressure at all outlets		
Less risk of backflow		

CHAPTER 6

HOT WATER SYSTEMS

INTRODUCTION

A supply of hot water is essential. We use it every day for personal hygiene, cooking and clothes washing. It is a vital resource for combating germs and bacteria, but it can also cause harm if the temperature of the water is not controlled.

In this chapter, will we investigate the many methods of supplying hot water in the home. We will look at the systems of hot water supply, the installation methods we should employ, the appliances we use to generate hot water and the ways in which we can control its temperature to safe, usable limits. We will also explore some of the common hot water-related faults that occur and look at ways of maintaining systems so that they give optimum performance.

By the end of this chapter, you will have knowledge and understanding of the following:

- sources of information relating to work on hot water systems
- hot water systems and components
- system safety and efficiency
- how to prepare for the installation of systems and components
- how to install and test systems and components
- how to decommission systems and components
- how to replace defective components.

ACTIVITY

As a starting point to this unit, consider the following questions:

- What are the recommended design temperatures for hot water systems?
- What safety devices are required in hot water systems?
- How is *Legionella* prevented in hot water systems?
- Where can faults occur in hot water systems?

KEY POINT

There are some areas that come under the subject of hot water – such as sanitary appliances, taps, valves, pumps and backflow protection – that are identical to those areas discussed within other chapters in this book. Where such duplication exists, you will be encouraged to read and research within those chapters. In most cases, further reading is encouraged within Chapter 2, Common processes and techniques, Chapter 3, Scientific principles, and Chapter 5, Cold water systems.

1 SOURCES OF INFORMATION RELATING TO WORK ON HOT WATER SYSTEMS

Here, you will learn to identify and use the information sources that should be referred to when designing hot water systems, including:

- statutory regulations
- industry standards
- manufacturer technical instructions.

Statutory regulations

The installation of hot water systems is governed strictly by various regulations:

- the Building Regulations Approved Document G3 2010

- the Building Regulations Approved Document L1A/B 2013 (with 2016 amendments)
- the Water Supply (Water Fittings) Regulations 1999
- the Gas Safety (Installation and Use) Regulations
- the IET (18th Edition) (**BS 7671:2008**) Wiring Regulations.

The Building Regulations Approved Document G3 2010

In the past, Building Regulations Approved Document G3 related only to unvented hot water supply systems. In 2010 it was updated to encompass all hot water delivery systems in domestic dwellings. It is divided into four parts:

1. Part 1 of G3 is a new requirement. It states that heated wholesome water must be supplied to any washbasin or bidet that is situated in or adjacent to a room containing a sanitary convenience, to any washbasins, bidets, fixed baths or showers installed in a bathroom, and any sink in an area where food is prepared.
2. Part 2 is an expanded requirement. It states that any hot water system, including associated storage (including any cold water storage cistern) or expansion vessel, must resist the effects of any temperature or pressure that may occur during normal use as a consequence of any reasonably anticipated fault or malfunction. This amendment was enforced after the failure of an immersion heater thermostat that caused the collapse of a storage cistern containing water almost at boiling point.
3. Part 3, again, is an amended requirement. It states that any part of a hot water system that incorporates a hot water storage vessel must include precautions to ensure that the temperature of the stored water does not exceed 100°C and that any discharge from such safety devices is safely conveyed to a point where it is visible without constituting a danger to persons in or about the building.
4. Part 4 states that any hot water supply to a fixed bath must include provision to limit the temperature of the discharged water from any bath tap to not in excess of 48°C. This requirement applies to any new-build or property conversions. It is a new requirement that is intended to prevent scalding.

It is interesting to note that Regulation G3 applies to all domestic dwellings, including greenhouses, small detached buildings, extensions and conservatories, but only if they are served with hot water supplied from a dwelling.

It should be noted that the local building control officer should be informed before commencing any installation of a hot water system.

INDUSTRY TIP

Remember that:

- Building Regulations Part G is about TEMPERATURE
- Building Regulations Part L is about EFFICIENCY.

The Building Regulations Approved Document L1A/B 2013

This document promotes the conservation of fuel and power. The basic outline to this document is that the building and services contained within a dwelling must be designed and installed to actively reduce the amount of CO_2 produced. The building fabric must contain insulation to limit heat loss and heating appliances, associated controls and equipment and lighting systems must all reduce the energy wasted. Pipes and storage vessels must also be insulated to reduce the waste of energy.

This document should be read in conjunction with the Domestic Building Compliance Guide.

INDUSTRY TIP

Copies of the Building Regulations Approved Documents G3 2010 and L 2013 can be downloaded free from these links:

www.gov.uk/government/publications/sanitation-hot-water-safety-and-water-efficiency-approved-document-g

www.gov.uk/government/publications/conservation-of-fuel-and-power-approved-document-l

The Water Supply (Water Fittings) Regulations 1999

In many respects, the Water Regulations mirror the Building Regulations, and these two documents should be consulted before undertaking any design or installation of hot water systems.

Hot water supply is covered in Section 8 of Schedule 2 of the Water Supply (Water Fittings) Regulations. The Document G Guidance for Hot Water Supply is reproduced in Table 6.1, complete with the guidance notes attached to the Regulations.

INDUSTRY TIP

A copy of the Water Supply (Water Fittings) Regulations can be downloaded free from: www.legislation.gov.uk/uksi/1999/1148/contents/made

▼ Table 6.1 Document G Guidance for Hot Water Supply

SECTION 8 Schedule 2: Paragraphs 17, 18, 19, 20, 21, 22, 23 and 24: Hot water services	Guidance
17 (1) Every unvented water heater, not being an instantaneous water heater with a capacity not greater than 15 litres, and every secondary coil contained in a primary system shall: a) Be fitted with a temperature control device and either a temperature relief valve or a combined pressure and temperature relief valve; or b) Be capable of accommodating expansion within the secondary hot water system. (2) An expansion valve shall be fitted with provision to ensure that water is discharged in a correct manner in the event of a malfunction of the expansion vessel or system.	**Unvented hot water systems** **G17.1** a A temperature control device; and either a temperature relief valve or combined temperature and pressure relief valve; and b An expansion valve; and c Unless the expanded water is returned to the supply pipe in accordance with Regulation 15(2)(a), either; i An expansion vessel; or ii Contain an integral expansion system, such that the expansion water is contained within the secondary system to prevent waste of water. **G17.2** An expansion valve should be fitted to all unvented hot water storage systems, with a capacity in excess of 15 litres, to ensure that expansion water is discharged in a correct manner in the event of a malfunction of the expansion vessel or system. **G17.3** Where expansion water is accommodated separately the expansion vessel should preferably be of an approved 'flow through type' and should comply with the requirements of **BS 6144** and **BS 6920**.
18 Appropriate vent pipes, temperature control devices and combined temperature pressure and relief valves shall be provided to prevent the temperature of the water within a secondary hot water system from exceeding 100°C.	**Temperature of hot water within a storage system** **G18.1** Irrespective of the type of fuel used for heating, the temperature of the water at any point within a hot water storage system should not exceed 100°C and appropriate vent pipes, temperature control devices and other safety devices should be provided to prevent this occurring. **Hot water distribution temperatures** **G18.2** Hot water should be stored at a temperature of not less than 60°C and distributed at a temperature of not less than 55°C. This water distribution temperature may not be achievable where hot water is provided by instantaneous or combination boilers. **G18.3** The maintenance of acceptable water temperatures may be achieved by efficient routing of pipes, reducing the lengths of pipes serving individual appliances and the application of good insulation practices to minimise freezing of cold water pipes and to promote energy conservation for hot water pipes. For references, see Comments and Recommendations of Clause 4.3.32.2 in **BS 8558**. **Temperature of hot water supplies at terminal fittings and on surfaces of hot water pipes** **G18.4** Where practicable the hot water distribution system should be designed and installed to provide the required flow of water at terminal fittings to sanitary and other appliances at a water temperature of not less than 50°C and within 30 seconds after fully opening the tap. This criteria may not be achievable where hot water is provided by instantaneous or combination boilers. **G18.5** Terminal fittings or communal showers in schools or public buildings, and in other facilities used by the public, should be supplied with water through thermostatic mixing valves so that the temperature of the water discharged at the outlets does not exceed 43°C. **G18.6** The temperature of water discharged from terminal fittings and the surface temperature of any fittings in healthcare premises should not exceed the temperatures recommended in HS(G)104 – Safe hot water and surface temperatures.

→

SECTION 8 Schedule 2: Paragraphs 17, 18, 19, 20, 21, 22, 23 and 24: Hot water services	Guidance
	Energy conservation **G18.7** All water fittings forming part of a primary or secondary hot water circulation system and all pipes carrying hot water to a tap that are longer than the maximum length given in the table below should be thermally insulated in accordance with **BS 5422**. ▼ Table 6.1a Maximum recommended lengths of uninsulated hot water pipes Outside diameter (mm) — Max. length (m) 12 — 20 Over 12 and up to 22 — 12 Over 22 and up to 28 — 8 Over 28 — 3
19 Discharges from temperature relief valves, combined temperature pressure and relief valves and expansion valves shall be made in a safe and conspicuous manner.	**Discharge pipes from safety devices** **G19.1** Discharge pipes from expansion valves, temperature relief valves and combined temperature and pressure relief valves should be installed in accordance with the guidance given in this document and should also comply with the requirements of Building Regulation G3. **G19.2** Where discharge pipes pass through environments outside the thermal envelope of the building they should be thermally insulated against the effects of frost.
	G19.3 The discharge pipe from a temperature relief valve or combined temperature and pressure relief valve should: a Be through a readily visible air gap discharging over a tundish located in the same room or internal space and vertically as near as possible and in any case within 600 mm of the point of outlet of the valve; and, b Be of non-ferrous material, such as copper or stainless steel, capable of withstanding any temperatures arising from a malfunction of the system; and, c Have a vertical drop of 300 mm below the tundish outlet, and thereafter be laid to a self-draining gradient; and, d Be at least one size larger than the nominal outlet size of the valve, unless its total equivalent hydraulic resistance exceeds that of a straight pipe 9 metres long. Where the total length of the pipe exceeds 9 metres equivalent resistance, the pipe shall be increased in size by one nominal diameter for each additional, or part of, equivalent 9 metres resistance length. The flow resistance of bends in the pipe should be taken into consideration when determining the equivalent length of pipe; and, e Terminate in a safe place where there is no risk to persons in the vicinity of the point of discharge. See Building Regulation G3. **Note:** Pipe sizing is now covered by **BS EN 806-3**. **Discharge pipes from expansion valves** **G19.4** The discharge pipe from an expansion valve may discharge into the tundish used for the discharge from a temperature relief valve or from a combined temperature and pressure relief valve as described in G19.1; or: a Discharge through a readily visible air gap over a tundish located in the same room or internal space and vertically as near as possible and in any case within 600 mm of the point of outlet of the valve; and, b Be of non-ferrous material, such as copper or stainless steel; and, c Discharge from the tundish through a vertical drop outlet and thereafter be laid to a self draining gradient; and, d Not be less than the nominal outlet size of the expansion valve and discharge external to the building at a safe and visible location.

→

SECTION 8 Schedule 2: Paragraphs 17, 18, 19, 20, 21, 22, 23 and 24: Hot water services	Guidance
20 (1) No vent pipe from a primary circuit shall terminate over a storage cistern containing wholesome water for domestic supply or for supplying water to a secondary system. (2) No vent pipe from a secondary circuit shall terminate over any combined feed and expansion cistern connected to a primary circuit.	**Vent pipes** **G20.1** Vent pipes from primary water systems should be of adequate size but not less than 19 mm internal diameter. They may terminate over their respective cold water feed and expansion cisterns, or elsewhere providing there is a physical air gap, at least equivalent to the size of the vent pipe, above the top of the warning pipe, or overflow if there is one, at the point of termination. **G20.2** Vent pipes from hot water secondary storage systems should be of adequate size but not less than 19 mm internal diameter and be insulated against freezing. **G20.3** Where vent pipes, from either a primary or secondary system, terminate over their respective cold water feed cisterns, they should rise to a height above the top water level in the cistern sufficient to prevent any discharge occurring under normal operating conditions. **Hot water systems supplied with water from storage cisterns** **G20.4** In any cistern-fed vented or unvented hot water storage system the storage vessel should: a be capable of accommodating any expansion water; or b be connected to a separate expansion cistern or vessel; or c be so arranged that expansion water can pass back through a feed pipe to the cold water storage cistern from which the apparatus or cylinder is supplied with water.
	G20.5 Where the cold water storage cistern supplying water to the hot water storage vessel is also used to supply wholesome water to sanitary or other appliances, any expansion water entering the cistern through the feed pipe should preferably not raise the temperature of the wholesome water in the cistern to more than 20°C. **Vented systems requiring dedicated storage cisterns or mechanical safety devices** **G20.6** Every vented and directly heated hot water storage vessel, single feed indirectly heated hot water storage vessel, or any directly or indirectly heated storage vessel where an electrical immersion heater is installed, should be supplied with water from a dedicated storage cistern unless: a Where the energy source is gas, oil or electricity, a non-self-setting thermal energy cut-out device is provided in addition to the normal temperature-operated automatic-reset cut-out; or, b Where the energy source is solid fuel, a temperature relief valve complying with **BS EN 1490:2000**, or a combined temperature and pressure relief valve complying with **BS EN 1490:2000**, is provided complete with a readily visible air-break to drain device and discharge pipe as described in G19.3. **G20.7** Every double feed indirectly heated hot water storage system which is heated by a sealed (unvented) primary circuit, or the primary circuit heating medium is steam or high temperature hot water, or where an electric immersion heater is installed, should: a Be supplied with water for the secondary circuit from a dedicated cold water storage cistern; or, b Be provided with a non-self-setting thermal energy cut-out device to control the primary circuit, and any electric immersion heaters, in addition to any temperature-operated automatic-reset cut-out. **G20.8** No water in the primary circuit of a double feed indirect hot water storage vessel should connect hydraulically to any part of a hot water secondary storage system. **G20.9** Vent pipes from primary circuits should not terminate over cold water storage cisterns containing wholesome water for supply to sanitary appliances or secondary hot water systems. **G20.10** Vent pipes from secondary hot water systems should not terminate over feed and expansion cisterns supplying water to primary circuits. **G20.11** No water in the primary circuit of a single feed indirect hot water storage vessel, under normal operating conditions, should mix with water in the secondary circuit. Single feed indirect hot water storage vessels should be installed with a permanent vent to the atmosphere.

SECTION 8 Schedule 2: Paragraphs 17, 18, 19, 20, 21, 22, 23 and 24: Hot water services	Guidance
21 Every expansion cistern or expansion vessel, and every cold water combined feed and expansion cistern connected to a primary circuit, shall be such as to accommodate any expansion water from that circuit during normal operation.	**Primary feed and expansion cisterns** **G21.1** Every expansion cistern, and every cold water combined feed and expansion cistern connected to a primary or heating circuit should be capable of accommodating any expansion water from the circuit and installed so that the water level is not less than 25 mm below the overflowing level of the warning pipe when the primary or heating circuit is in use.
22 (1) Every expansion valve, temperature relief valve or combined temperature and pressure relief valve connected to any fitting or appliance shall close automatically after a discharge of water.	**Expansion and safety devices** **G22.1** Expansion valves, temperature relief valves or combined temperature and pressure relief valves connected to any fitting or appliance should close automatically after an operational discharge of water and be watertight when closed.
(2) Every expansion valve shall: a Be fitted on the supply pipe close to the hot water vessel and without any intervening valves; and b Only discharge water when subjected to a water pressure of not less than 0.5 bar (50 kPa) above the pressure to which the hot water vessel is, or is likely to be, subjected in normal operation.	**G22.2** Expansion valves should comply with **BS EN 1491:2000**. They should be fitted on the supply pipe close to the hot water vessel and without any intervening valves, and only discharge water when subjected to a water pressure of not less than 0.5 bar (50 kPa) above the pressure to which the hot water vessel is, or is likely to be, subjected to in normal operation.
23 (1) A temperature relief valve or combined temperature and pressure relief valve shall be provided on every unvented hot water storage vessel with a capacity greater than 15 litres. (2) The valve shall: a Be located directly on the vessel in an appropriate location, and have a sufficient discharge capacity, to ensure that the temperature of the stored water does not exceed 100°C; and b Only discharge water at below its operating temperature when subjected to a pressure of not less than 0.5 bar (50 kPa) in excess of the greater of the following: i The maximum working pressure in the vessel in which it is fitted, or ii The operating pressure of the expansion valve. (3) In this paragraph 'unvented hot water storage vessel' means a hot water storage vessel that does not have a vent pipe to the atmosphere.	**Temperature and combined temperature relief valves** **G23.1** Except for unvented hot water storage vessels of a capacity of 15 litres or less, a temperature relief valve complying with **BS EN 1490:2000**, or a combined temperature and pressure relief valve complying with **BS EN 1490:2000**, should be provided on every unvented hot water storage vessel. The valve should: a Be located directly on the storage vessel, such that the temperature of the stored water does not exceed 100°C; and, b Only discharge water at below its operating temperature when subjected to a pressure not less than 0.5 bar (50 kPa) greater than the maximum working pressure in the vessel to which it is fitted, or 0.5 bar (50 kPa) greater than the operating pressure of the expansion valve, whichever is the greater. **Non-mechanical safety devices** **G23.2** If a non-mechanical safety device such as a fusible plug is fitted to any hot water storage vessel, that vessel requires a temperature relief valve or combined temperature and pressure relief valve designed to operate at a temperature not less than 5°C below that at which the non-mechanical device operates or is designed to operate.

→

SECTION 8 Schedule 2: Paragraphs 17, 18, 19, 20, 21, 22, 23 and 24: Hot water services	Guidance
24 No supply pipe or secondary circuit shall be permanently connected to a closed circuit for filling a heating system unless it incorporates a backflow prevention device in accordance with a specification approved by the regulator for the purposes of this Schedule.	**Filling of closed circuits** **G24.1** No primary or other closed circuit should be directly and permanently connected to a supply pipe unless it incorporates an approved backflow prevention arrangement. **G24.2** A connection may be made to a supply pipe for filling or replenishing a closed circuit by providing a servicing valve and an appropriate backflow prevention device, the type of which will depend on the degree of risk arising from the category of fluid contained within the closed circuit, providing that the connection between the backflow prevention device and the closed circuit is made by: a A temporary connecting pipe which must be completely disconnected from the outlet of the backflow prevention device and the connection to the primary circuit after completion of the filling or replenishing procedure; or b A device which in addition to the backflow prevention device incorporates an air gap or break in the pipeline which cannot be physically closed while the primary circuit is functioning; or c An approved backflow prevention arrangement.

Source: Water Supply (Water Fittings) Regulations 1999, Section 8

The Gas Safety (Installation and Use) Regulations 1998

Many hot water supply appliances use gas as their main fuel source for both direct and indirect domestic hot water heating. This, obviously, means that the Gas Regulations play an important part in any hot water installation.

The Gas Safety (Installation and Use) Regulations deal with the safe installation, maintenance and use of these appliances, and any gas pipework and fittings connected to them in both domestic and industrial/commercial premises. The main requirement of the Regulations is that only a competent person (deemed by the HSE to be any person that is a member of an approved body) must carry out work on any gas fitting. In this case, installers of gas appliances, pipework and fittings must by registered with Gas Safe.

INDUSTRY TIP

A copy of the Gas Safety (Installation and Use) Regulations 1998 can be downloaded free from this link: www.hse.gov.uk/pubns/priced/l56.pdf

The IET (18th Edition) (BS 7671) Wiring Regulations

As with the Gas Regulations, heating hot water often uses electricity either as a direct or indirect fuel source. All domestic and industrial electrical installations must conform to the IET Wiring Regulations. In England and Wales, the Building Regulations Approved Document P 2010 requires that domestic installations be designed and installed according to **BS 7671, Chapter 13**. This document was written to standardise electrical installations in line with international document IEC60364-1 and equivalent standards from other countries. Guidance is given in installation manuals such as the IET on-site guide and IET Guidance notes 1 to 7.

Installations in industrial and commercial premises must also satisfy various other legislative documents, such as the Electricity at Work Regulations 1989. Again, the recognised standards and practices contained in **BS 7671** will help meet these requirements.

INDUSTRY TIP

A copy of the Building Regulations 2010 Approved Document P can be downloaded free from this link: www.planningportal.gov.uk/uploads/br/BR_PDF_AD_P_2010.pdf

Industry standards

There are a number of industry standards that we can reference to ensure that we conform to the regulations when installing hot water systems. Some of these share a commonality with those discussed in Chapter 5, Cold water systems.

- **British Standard BS EN 806 Parts 1 to 5:** again, this standard contains extensive information regarding the design and installation of hot water supply systems.
- **British Standard BS 8558:2011:** this provides complementary guidance to **BS EN 806**. It is a guide to the design, installation, testing, operation and maintenance of services supplying water for domestic use.
- **The Domestic Building Services Compliance Guide:** this guide provides guidance to the Building Regulations Approved Documents L1 and L2 when installing fixed building services within new and existing dwellings to help them comply with the Building Regulations. The guide specifically targets space heating, domestic hot water services, mechanical ventilation, comfort cooling and interior lighting. New technologies such as heat pumps, solar thermal panels and micro-combined heat and power systems are also discussed. The guide also refers to other publications that refer to techniques to assist in the design and installation of systems that are over and above the standard that is required by the Building Regulations.

INDUSTRY TIP

British Standards, **BS EN 806 Parts 1 to 5** and **BS 8558** should be read in conjunction with each other. Although **BS 6700** has been superseded by **BS EN 806** and **BS 8558**, there are still parts of the document that remain relevant and it should still be consulted when either designing or installing hot water systems.

Manufacturer technical installation and maintenance instructions

Unvented hot water storage systems must be fitted, commissioned and maintained strictly in accordance with the manufacturer's instructions. These contain vital information for the correct and safe installation, operation and maintenance of the system and its components, such as:

- the minimum required pressure and flow rate of the incoming supply, for satisfactory operation of the system
- the minimum size of the incoming cold water supply
- the minimum size of any hot water distribution pipework
- the required heat input and heat recovery time
- any electrical installation requirements
- the operation of any controls
- the calculation required to ascertain the correct size of the discharge pipework
- fault-finding techniques.

INDUSTRY TIP

If the manufacturer's instructions are not available or have been misplaced, most manufacturers now offer the facility to download the instructions from their website.

Factors affecting hot water systems

You will learn about the factors that affect the selection of hot water systems for dwellings, with consideration of:

- customer needs/occupancy and purpose
- building layout and features
- energy efficiency
- environmental impact
- appliance location
- cost
- storage type/location
- legislation.

The type of system we choose will depend on the following points.

The customer's needs/occupancy and purpose

This concerns the number of occupants and the amount of hot water required. Larger households will require more hot water, which can be supplied in a number of ways, i.e. an instantaneous water heater giving unlimited hot water amounts or a large hot water storage cylinder, although other factors must also be considered before a decision is made.

Building layout and features: the size of the property and the distance from the outlets

The Water Supply (Water Fittings) Regulations stipulate the maximum distance that a hot water supply pipe may run without constituting wastage of water. This is because of the amount of cold water that is drawn off before hot water arrives at the taps. This 'dead' cold water must be limited. Large properties may exceed the maximum distances for hot water dead legs, which excludes some hot water systems. In these cases, only systems that can incorporate secondary circulation should be considered.

Running costs and energy efficiency

New, more efficient methods of heating water are constantly being developed. Perhaps the most important recent development is that of solar hot water heating, which can, theoretically, offer a 60 per cent saving on domestic hot water heating costs, despite its initial costly installation. The development of fuel-efficient condensing oil and gas boilers and storage cylinders with fast heat recovery times have also helped in terms of energy efficiency.

Environmental impact: the type(s) of fuel to be used

With most storage hot water systems, multiple fuels may be used in one system, i.e. utilising gas, oil or solid fuel as the main fuel source, with an electrical alternative (immersion heater) as back-up or for summer use. Multipoint heaters do not have this capability and so fuel type usage is very limited. The environmental impact of fuels like gas and oil are now important considerations, especially in buildings where a low/zero carbon footprint is preferred.

Appliance location: the number of hot water outlets

Again, an important point because this may automatically exclude such appliances as combi boilers and instantaneous heaters because, although classed as multipoint heaters, only one outlet at a time may be opened satisfactorily, whereas other types of hot water system may allow multiple open taps with a good flow rate. This becomes important where there are long distances between the appliance and the hot water source.

Installation and maintenance costs

This is also a very important point because of the size of the system, and initial cost of the appliance and materials. Add to this the installation costs and any maintenance costs over the lifetime of the system.

Storage type/location

The type of hot water storage system used (vented or unvented) will play a vital part in its location within the dwelling. While a vented system may be fitted in an airing cupboard, an unvented system generally requires much more space than that available in an airing cupboard, because of the need for safety and functional controls. The distance from the outlets also may create a problem because the longer the run of hot water pipework, the greater the need for secondary circulation. Location of the storage vessel is, therefore, of great importance.

Legislation

As with all plumbing systems, the legislation that covers the installation of hot water systems must be considered. Some systems, such as unvented hot water installations, are governed much more rigidly because of the safety features that must be installed. Other systems, such as vented hot water installations, do not require such tight regulation but may not meet the specification of the appliances fitted.

Whichever system is fitted, the regulations that cover its installation must be adhered to at all times.

Choosing the right hot water system

When the above points are considered, the choice of hot water system should be quite a straightforward affair. Certain dwellings almost dictate the system that should be fitted. For example, it would be foolhardy to install a combi boiler in a dwelling with three bathrooms, a kitchen, utility room and downstairs washroom. The hot water demand would be more than the boiler could cope with. By far the main considerations that must be taken into account are the type and number of appliances, and their pattern and frequency of use. Knowing this will indicate the correct choice of system to install and the customer can then be advised accordingly.

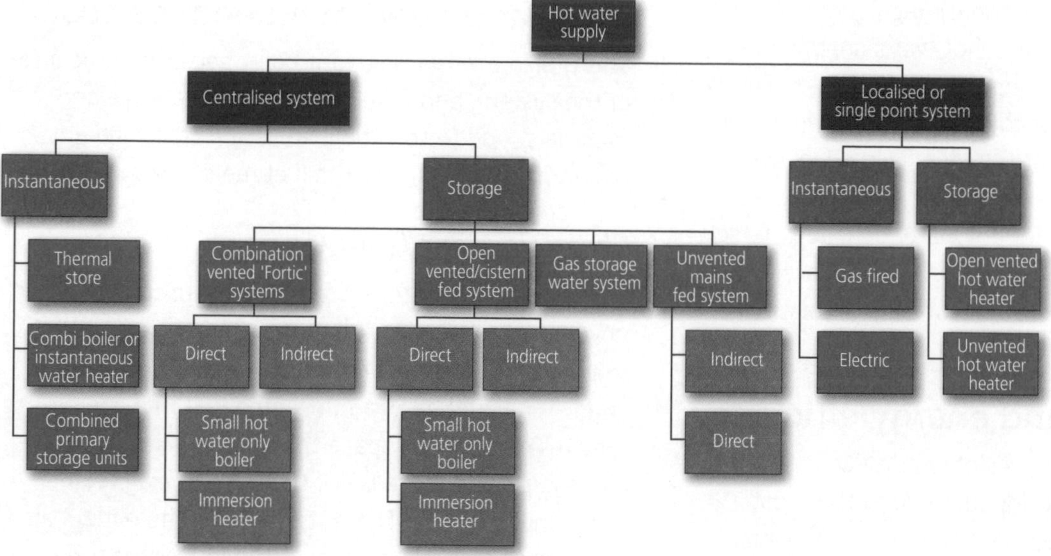

▲ Figure 6.1 Types of hot water system

2 HOT WATER SYSTEMS AND COMPONENTS

Identify types and layout features of hot water systems

In this section, you will compare the types of hot water supply systems used in dwellings and describe their applicable pipework layout features.

Hot water systems include:

- centralised systems – unvented hot water systems and vented hot water systems
- localised systems – unvented point-of-use heaters and instantaneous heaters
- indirect storage systems
- direct storage systems, such as:
  - electrically heated
  - gas or oil fired
- small point of use (under sink)
- bulk storage heaters (combination tank)
- solar thermal hot water systems
- combination boilers.

Pipework layout features include:

- unvented hot water
- secondary circulation
- solar thermal
- thermal stores
- combination boilers.

Hot water system types

Hot water systems can be divided into two categories, as follows.

1 **Centralised systems**, where hot water is delivered from a central point to all hot water outlets in the dwelling. The water may be heated by a boiler or immersion heater. Centralised systems are those where the source of hot water supply is sited centrally in the property for distribution to all of the hot water outlets. They are usually installed in medium to large domestic dwellings such as a three-bedroomed house.
2 **Localised systems**, often called single-point or point-of-use systems. With these systems, the hot water is delivered by a small water heater at the point where it is needed.

Centralised hot water storage systems

These are divided into the following types.

- **Open vented systems:** those hot water storage systems that are fed from a cistern in the roof space and contain a vent pipe that is open to the atmosphere.
- **Unvented systems:** those hot water storage systems that are fed directly from the cold water main, and utilise an expansion vessel or an internal air bubble to allow for expansion.
- **Gas-fired instantaneous multipoint hot water heaters:** those heaters that heat the water instantaneously.
- **Gas- or oil-fired combination boilers:** operate in a similar fashion to instantaneous hot water heaters, but also have a central heating capability.
- **Thermal stores:** sometimes referred to as water-jacketed tube heaters.
- **Gas- or oil-fired combined primary storage units:** these are very similar in operation to the thermal store (see above).

Open vented hot water storage systems

In an open vented storage hot water system, water is heated, generally by a boiler or an immersion heater, and stored in a hot water storage vessel sited in a central location in the property, usually in the airing cupboard. Open vented systems contain a vent pipe, which remains open to the atmosphere ensuring that the hot water cannot exceed 100°C. The vent pipe acts as a safety relief should the system become overheated. It must be sited over the cold feed cistern in the roof space.

> **ACTIVITY**
>
> To refresh yourself as to the types and make-up of hard and soft water, check out Chapter 5, Cold water systems, and Chapter 3, Scientific principles.

The cylinder is fed with water from the cold feed cistern. The capacity of the cistern will depend upon the capacity of the hot water storage vessel, and **BS EN 806** and **BS EN 8558** recommend that the capacity of the cistern feeding cold water to a hot water storage vessel must be at least equal to that of the hot water storage vessel.

Below are some important points to note about open vented hot water systems.

- The open vent pipe must not be smaller than 22 mm pipe and must terminate over the cold feed cistern.
- The open vent pipe must not be taken directly from the top of the hot water storage vessel.
- The hot water draw-off pipe should rise slowly from the top of the cylinder to the open vent pipe and incorporate at least 450 mm of pipe between the storage cylinder and its connection point to the open vent. This is to prevent **parasitic circulation** (also known as one pipe circulation) from occurring.
- The cold feed pipe should be sized in accordance with **BS EN 806**. The cold feed is the main path for expansion of water to take place within the cylinder when the water is heated. The heated water from the cylinder expands up the cold feed pipe, raising the water level in the cold feed cistern.
- The cistern should be placed as high as possible to ensure good supply pressure. The higher the cistern, then the greater the pressure at the taps. Poor pressure can be increased by raising the height of the cistern.

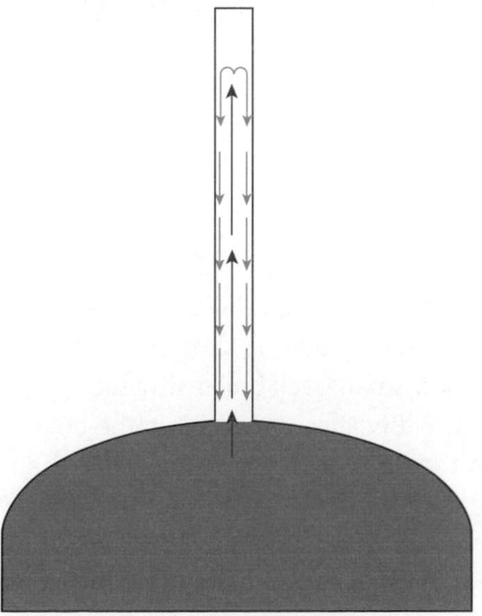

▲ Figure 6.2 Parasitic circulation

- All pipes should be laid with a slight fall (except the hot water draw-off), to prevent air locks within the system.
- The cold feed pipe from the storage cistern must feed only the hot water storage cylinder.
- A drain-off valve should be fitted at the lowest point of the cold feed pipe.

KEY TERM

Parasitic circulation: circulation that occurs within the same pipe; often called one pipe circulation. It generally occurs in open vent pipes that rise vertically from the open vented hot water storage cylinder. The hotter middle water rises up the vent pipe, and the cooler water, towards the wall of the pipe, falls back to the cylinder. It can be a major source of heat loss from hot water storage cylinders.

There are two types of open vented hot water storage system. These are:

1. the open vented direct hot water storage system
2. the open vented indirect hot water storage system:
   - double feed type
   - single feed type.

The open vented direct hot water storage system

Direct systems use a direct-type cylinder that is heated by either a small hot water only boiler or an immersion heater. The direct cylinder contains no form of heat exchanger and so is not suitable for use with central heating systems. The connections for the cold feed and draw-off are usually male thread connections, while the primary flow and return connections have female threads.

The boiler can either be a small gas-fired hot water heater (often called a gas circulator), designed to heat the water directly, or a small back boiler situated behind a solid fuel fire. Because the water in the boiler comes direct from the hot water storage cylinder, the boiler must be made of a material that does not rust. This is to prevent rusty water being drawn off at the taps. Suitable boiler materials are:

- copper
- stainless steel
- bronze.

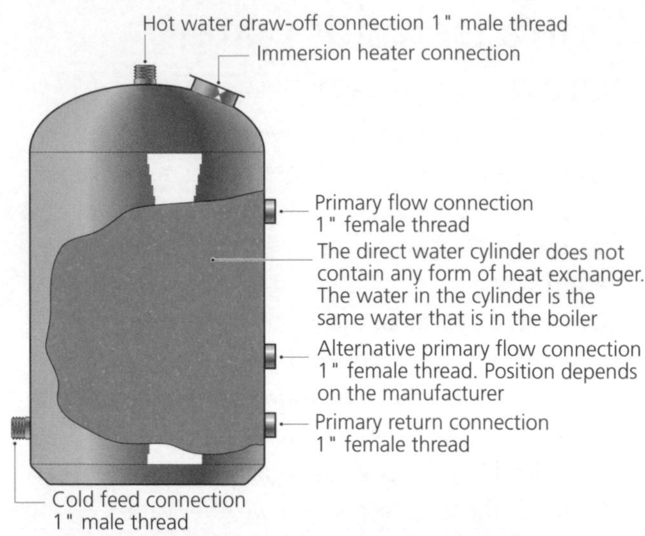

▲ Figure 6.3 Old gravity direct cylinder

The hot water circulates from the boiler or circulator by the principle of convection. This is known as gravity circulation (see Chapter 3, Scientific principles). The hot water rises in the primary flow pipe, directly heating the stored water in the cylinder before the cooler water returns to the boiler. The water in the cylinder does not heat uniformly. The water at the top of the cylinder is usually 10°C hotter than at the bottom (generally 60°C at the top, 50°C at the bottom). This is known as stratification and is desirable in stored hot water systems.

The primary flow and return pipes to and from the boiler/circulator should be a minimum of 28 mm in older gravity systems regardless of pipe length, unless stated differently in the manufacturer's instructions.

ACTIVITY

Look back at Chapter 5, Cold water systems and remind yourself how a cold water storage cistern should be installed.

ACTIVITY

Look back at Chapter 3 and remind yourself as to how the 'head height' of a system will affect the pressure of water at each outlet.

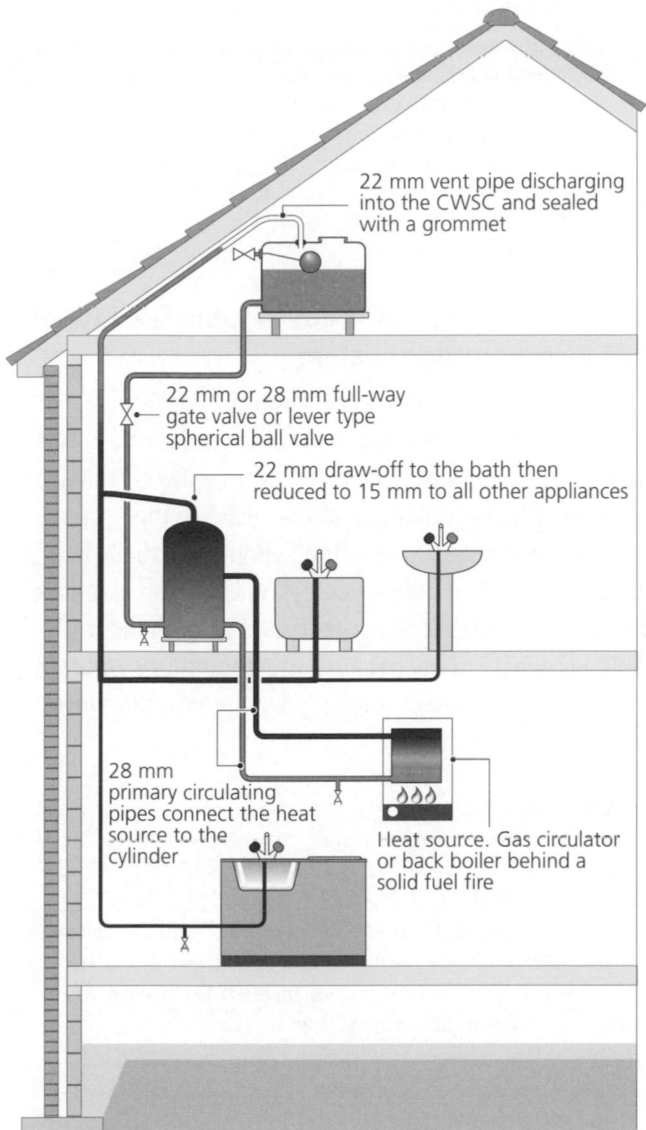

▲ Figure 6.4a Old direct system of hot water, relying on gravity circulation

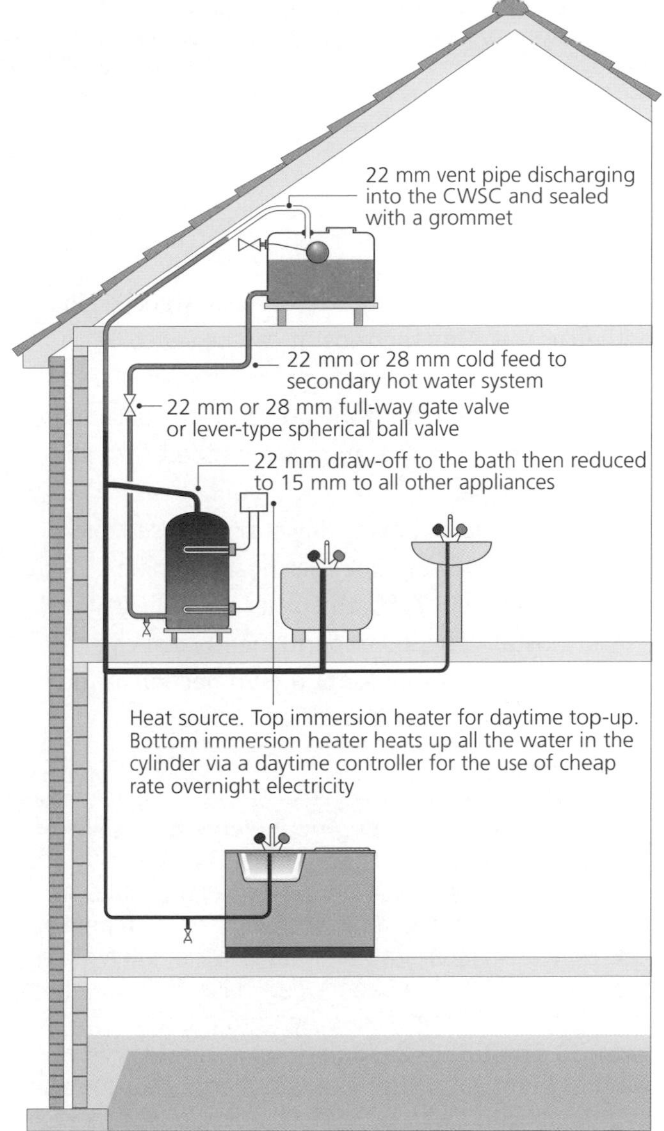

▲ Figure 6.4b Current direct hot water system, using immersion heaters

Direct cylinders, when connected to solid fuel back boilers, are susceptible to boiling because there is no effective method of temperature control.

▼ Table 6.2 Advantages and disadvantages of direct systems

Advantages	Disadvantages
Quick heat up time of the water Cheap to install	Risk of rusty water being drawn off at the taps if the wrong type of boiler is used High risk of scale build-up in hard water areas if the water temperature exceeds 65°C High risk of scalding because of the lack of thermostatic control

Alternative direct systems using immersion heaters

As an alternative to direct systems with a circulator/back boiler, some direct systems use two 3 kW **immersion heaters** placed in the side of the cylinder to heat the water. One immersion heater is placed at the bottom of the cylinder to heat all the contents and a second immersion heater is placed halfway down the cylinder for daytime top-up. The immersion heaters are wired to a time controller for use with cheap-rate overnight electricity. The temperature of the immersion heaters should be limited to 55°C to prevent build-up of scale.

▼ Table 6.3 Criteria of choice for direct systems

Property size	Storage capacity	Fuel type	Installation cost	Fuel efficiency
Suitable for most houses.	Varies with occupancy. Generally, 210 litres for four people.	Mostly used with Economy 7 electricity but can also be used with some solid fuel boilers and gas circulators. Gravity hot water circulation only.	The least expensive of all storage systems for houses when boilers and circulators are not fitted.	Economy 7 electricity is 100% efficient but the tariffs can be very expensive.

- Some immersion heaters have a resettable double thermostat. One thermostat can be set to 50–70°C, the other is a resettable high limit thermostat designed to switch off the power to the unit when the maximum temperature is exceeded. It can be manually reset.
- Some immersion heaters have a non-resettable double thermostat. One thermostat can be set between 50–70°C; the other is a high limit thermostat designed to permanently switch off the power to the unit until the immersion heater is replaced and the fault rectified.

KEY TERM

Immersion heater: an electrical element that sits in a body of water, just like in a kettle. When switched on, the electrical current causes the electrical element to heat up, which in turn heats up the water. Most immersion heater elements are rated at 3 kW but cylinders can have 1, 2, 3 or 4 elements. All immersion heaters must comply with **BS EN 60335-2-73** and have a resettable double thermostat (RDT) as standard. This enables problems with overheating to be recognised quickly.

The indirect system

An indirect system uses an indirect-type hot water storage cylinder, which contains some form of heat exchanger to heat the secondary water. There are two distinct types:

1 the double feed indirect hot water storage cylinder
2 the single feed, self venting indirect hot water storage cylinder.

The heat exchanger contains primary water and is classed as part of the central heating system to the dwelling.

The open vented indirect (double feed type) hot water storage system

This is probably the most common of all hot water delivery systems installed in domestic properties. It uses a double feed indirect hot water storage cylinder, which contains a heat exchanger, at the heart of the system. The heat exchanger within the cylinder is usually a copper coil but, in older-type cylinders, it can also take the form of a smaller cylinder called an annular. It is called indirect simply because the **secondary water** in the cylinder is heated indirectly by the **primary water** via the heat exchanger.

KEY TERM

Primary and secondary water: the primary water is the water that is in the boiler, central heating system and the heat exchanger of an indirect-type hot water storage cylinder/vessel. It is called the primary water because it is heated by the primary source of heat and hot water in the dwelling, namely the boiler. The pipes that connect the boiler to the heat exchanger are called the primary flow and the primary return. The secondary water is the stored water in the cylinder itself that is delivered to the hot water outlets and taps. The primary water heats the secondary water indirectly via the heat exchanger.

The double feed indirect cylinder

The double feed indirect cylinder contains a heat exchanger in the form of a coil and so is suitable for use with central heating systems. The connections for the cold feed, draw-off and the primary flow and return are usually 1-inch male thread connections.

In a double feed indirect system, two cisterns are used: a large cistern for the domestic hot water and a smaller one for the heating. It is now general practice to install indirect cylinders in preference to direct types, even if the indirect flow and return are capped off.

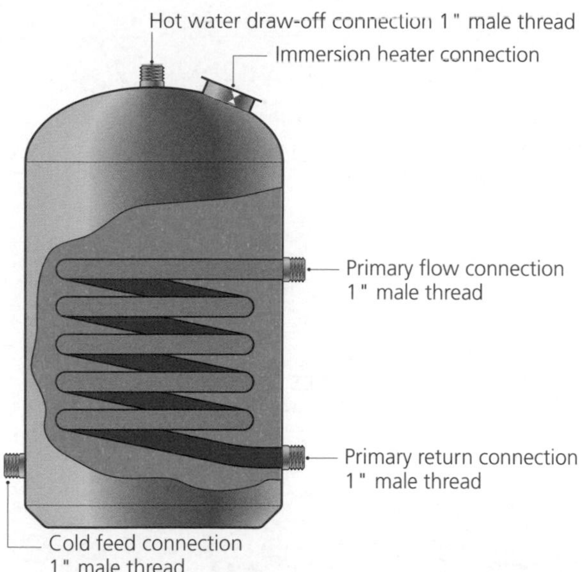

▲ Figure 6.5 An indirect cylinder

The double feed indirect hot water storage cylinder allows the use of boilers and central heating systems that contain a variety of metals, such as steel and aluminium because the water in the cylinder is totally separate from the water in the heat exchanger. This means that there is no risk of dirty or rusty water being drawn off at the taps. The system is designed in such a way that the water in the boiler and primary pipework is hardly ever changed, the only loss of water being in the feed and expansion cistern through evaporation.

The secondary water is that which is drawn from the hot water storage cylinder to supply the hot taps. It is heated by conduction as the water in the cylinder is in contact with the **heat exchanger**.

KEY TERM

Heat exchanger: a device or vessel that allows heat to be transferred from one water system to another without the two water systems being allowed to come into contact with each other. The transfer of heat between the two systems takes place via conduction (see Chapter 3, Scientific principles).

A feed and expansion cistern feeds the primary part of the system, and this must be large enough to accommodate the expansion of the water in the system when it is heated. The vent pipe from the primary system must terminate over the feed and expansion cistern. An alternative method would be to use a sealed heating system, which is fed with water from the cold water main via a filling loop. Expansion of water is accommodated in an expansion vessel.

Hot water storage cylinders must conform to **BS 1566**, which specifies the minimum heating surface area of the heat exchanger.

- **Existing double feed indirect systems:** existing double feed indirect systems use gravity circulation via 28 mm gravity primary flow and return pipes to heat the water in the cylinder. This type of system can no longer be installed as they are extremely wasteful in terms of energy usage. Document L1B of the Building Regulations recommends that these systems should be replaced with fully pumped systems wherever possible (see Chapter 7, Central heating systems) or they must be updated to include a cylinder thermostat and a motorised zone valve arrangement, as stated in the Domestic Heating Compliance Guide. This is to limit the amount of energy wastage.

INDUSTRY TIP

Look at a copy of the Domestic Heating Compliance Guide and list the necessary controls for a central heating system.

ACTIVITY

Using Figure 6.6, identify:

- the two feeds
- where the expansion is taken up
- where the gravity circulation happens
- primary water
- secondary water
- open vents.

- **New double feed indirect systems:** on new installations, double feed indirect cylinders must incorporate pumped circulation to the heat exchanger. Document L1A of the Building Regulations dictates that all new installations must have pumped primary circulation with controls that prohibit energy wastage. This is achieved by installing thermostatic control over the hot water storage cylinder via a cylinder thermostat and a motorised zone valve arrangement, as stated in the Domestic Heating Compliance Guide. Because the primary flow and return are pumped, the pipe size, in most cases, can be reduced to 22 mm. This subject will be covered in greater detail in Chapter 7, Central heating systems.

ACTIVITY

Relate Building Regulations Part L and G to Figure 6.7 and compare the potential efficiency of this system to Figure 6.6.

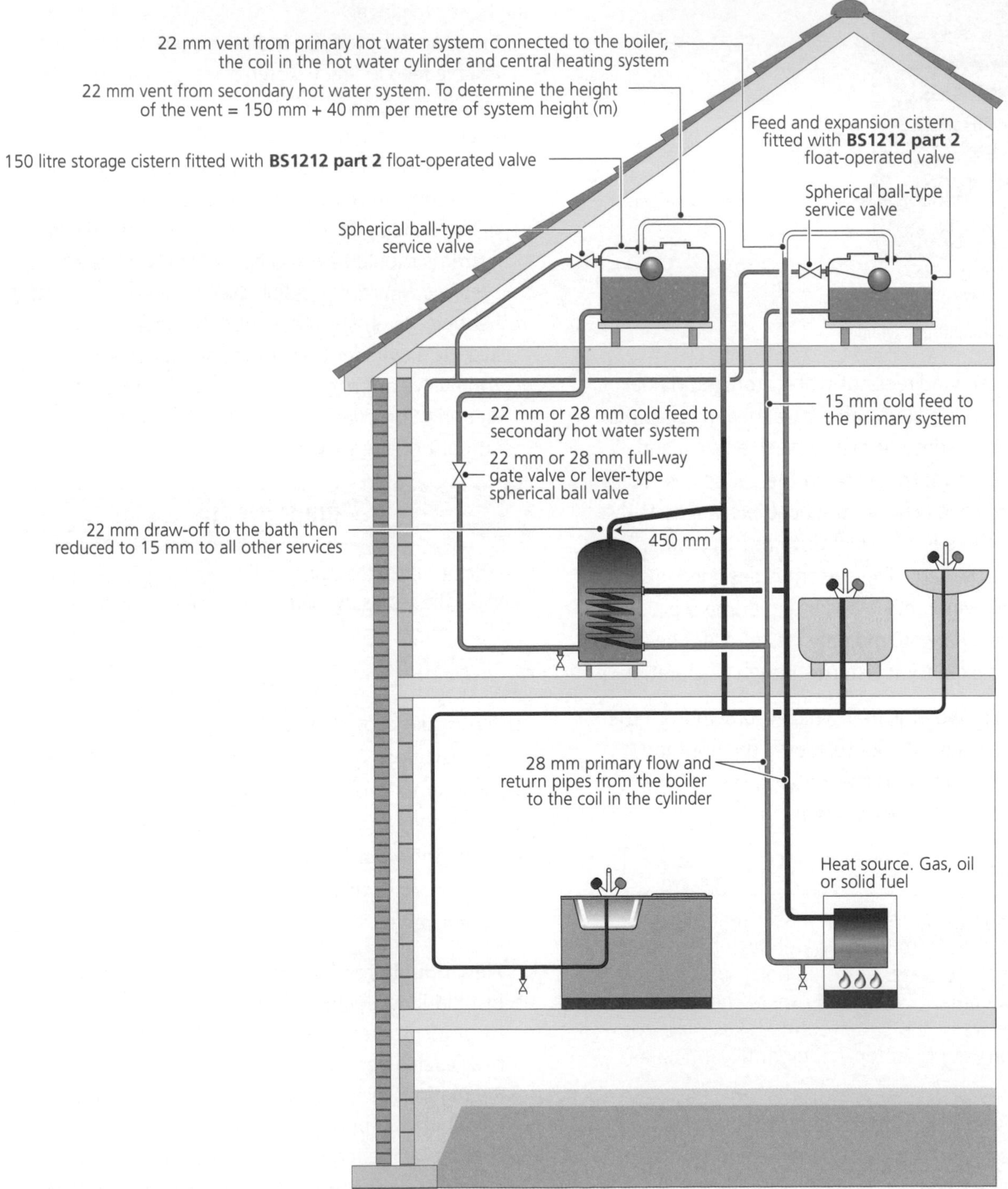

▲ Figure 6.6 An old indirect gravity open vented (double feed) hot water storage system

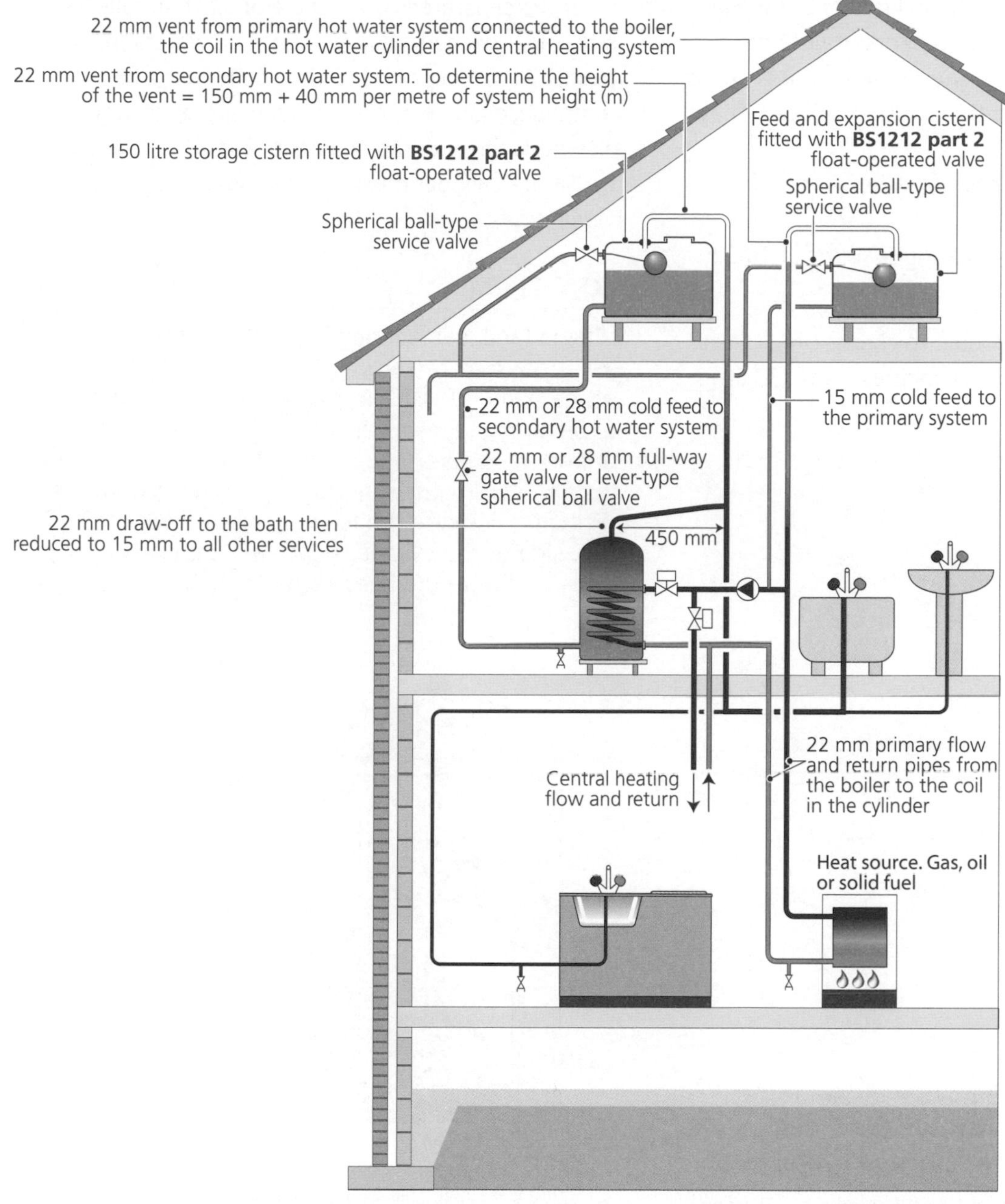

▲ Figure 6.7 A modern indirect open vented (double feed) hot water storage system with pumped primary circulation

▼ Table 6.4 Criteria of choice for double feed indirect systems

Property size	Storage capacity	Fuel type	Installation cost	Fuel efficiency
Suitable for all domestic properties.	Varies with occupancy. Generally, 210 litres for four people.	Can be used with gas, oil, solid fuel and electricity. Suitable for fully pumped heating systems. Conforms to Doc. L of the Building Regulations.	More expensive than direct systems due to the extra pipework for the feed and expansion cistern and associated pipework.	Gas and oil appliances must be energy-efficient condensing types. Can also be used with Economy 7 electricity.

Indirect cylinders for renewable energy hot water supply

Open vented cylinders have been developed for installation onto renewable energy hot water supply systems, such as solar, geothermal and ground-source heat pumps. The cylinder contains two heat exchanger coils. The first coil is used with a conventional fuel source such as gas or oil and this accounts for 70 per cent of the cylinder's hot water volume. The second coil has 30 per cent volume dedicated to the renewable energy heat source and is usually situated in the top third of the cylinder. They are suitable for:

- modern fully pumped heating systems (see Chapter 7)
- both vented and sealed heating systems up to 3.5 bar pressure.

They are supplied with a double thickness of CFC-free polyurethane insulation and capacities from 130 litres to 300 litres.

▲ Figure 6.8 Double coil cylinder

Open vented indirect (single feed, self-venting type) hot water storage systems

This is an old system that is rarely found now. It uses a **single feed, self-venting indirect cylinder**, often referred to by its trade name: the 'Primatic' cylinder. It contains a special heat exchanger, which uses air entrapment to separate the primary water from the secondary water.

It is fitted in the same way as a direct system, with only one cold feed cistern in the roof space but, unlike the direct system, it allows a boiler and central heating to be installed. It does not require a separate feed and expansion cistern. The heat exchanger works in such a way that the primary and secondary water are separated by a bubble of air that collects in the heat exchanger, preventing the waters from mixing. According to the Domestic Building Services Compliance Guide, these cylinders are no longer allowed for new or replacement cylinders. A 'double feed'-type cylinder must be used on all replacement installations.

KEY TERM

Single feed, self-venting indirect cylinder: often referred to as the 'Primatic' cylinder, which is a trade name of IMI Ltd. Another version of this type of cylinder was also available and may be found in some existing installations. It was known as the 'Aeromatic'. It is slightly different from the Primatic because it has an air release valve on the side of the cylinder near the heat exchanger to bleed air from the heat exchanger.

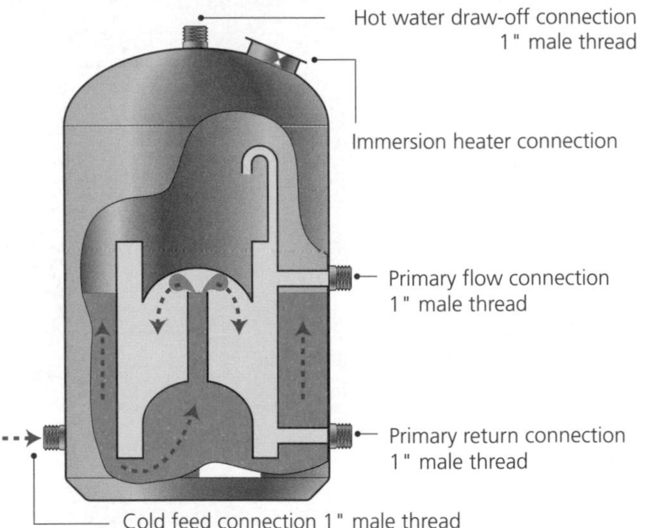

▲ Figure 6.9 Single feed, self-venting indirect cylinder

A typical open vented indirect (single feed, self-venting type) hot water storage system utilising gravity circulation is shown in Figure 6.10.

INDUSTRY TIP

On no account must central heating inhibitors be used in the primary water if a single feed cylinder is installed as this would cause contamination of the water if the air bubbles were to rupture.

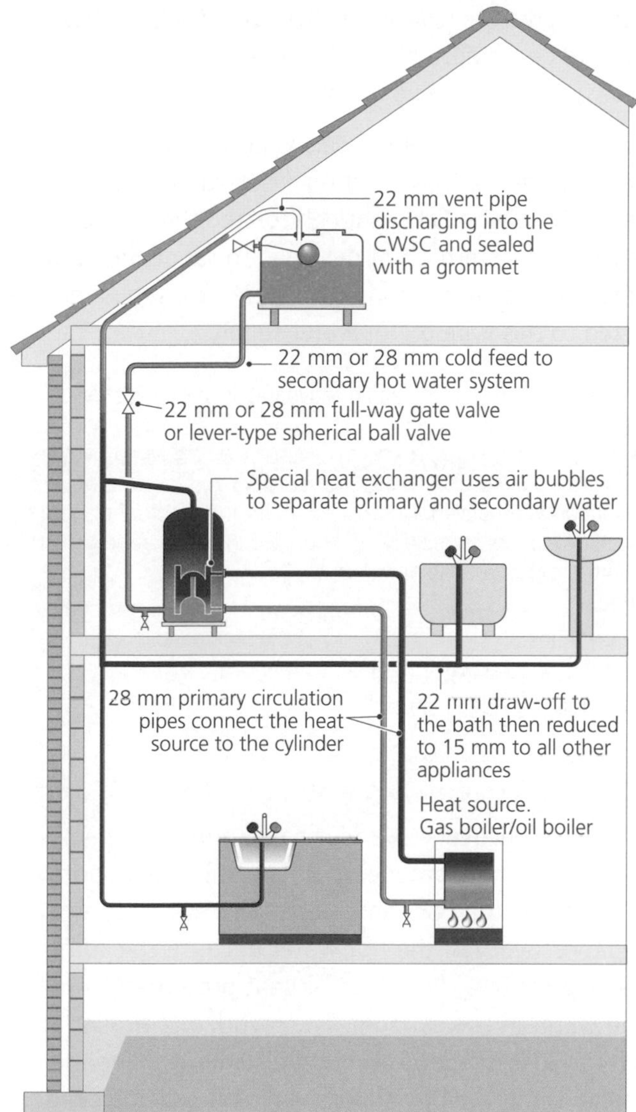

▲ Figure 6.10 Single feed, self-venting indirect system

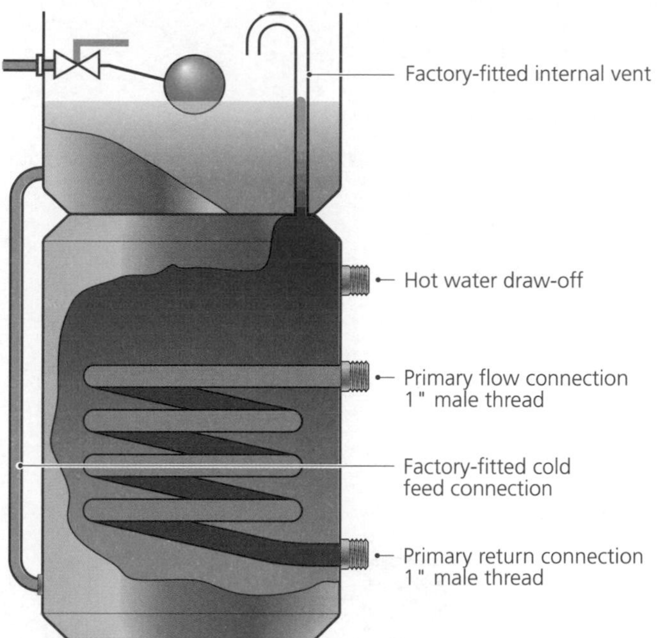

▲ Figure 6.11 The layout of a combination open vented hot water storage cylinder

▲ Figure 6.12 Combination open vented hot water storage cylinder

Other types of open vented cylinder

Combination centralised open vented hot water storage systems

A combination cylinder has its own cold feed cistern attached on the top of the cylinder, and factory-fitted cold feed and vent pipes. The connections for the draw-off and the primary flow and return are usually 1-inch male thread connections. Isolation of the hot water for maintenance should be via a full-way gate valve installed on the hot water draw-off pipe.

They are known as 'Fortic' cylinders and are available in both direct (immersion heater only and circulator/boiler types) and double feed indirect types. They come in a variety of sizes, 115 litres of hot water storage being the most common with 20–115 litres of cold storage above.

Integrating a cold water feed cistern and a hot water cylinder in a compact all-copper unit is an effective way of providing adequate supplies of hot water when storage space is limited. It also enables a dry roof space, eliminating the risk of freezing. It is an ideal system for rented accommodation due to its low maintenance requirement.

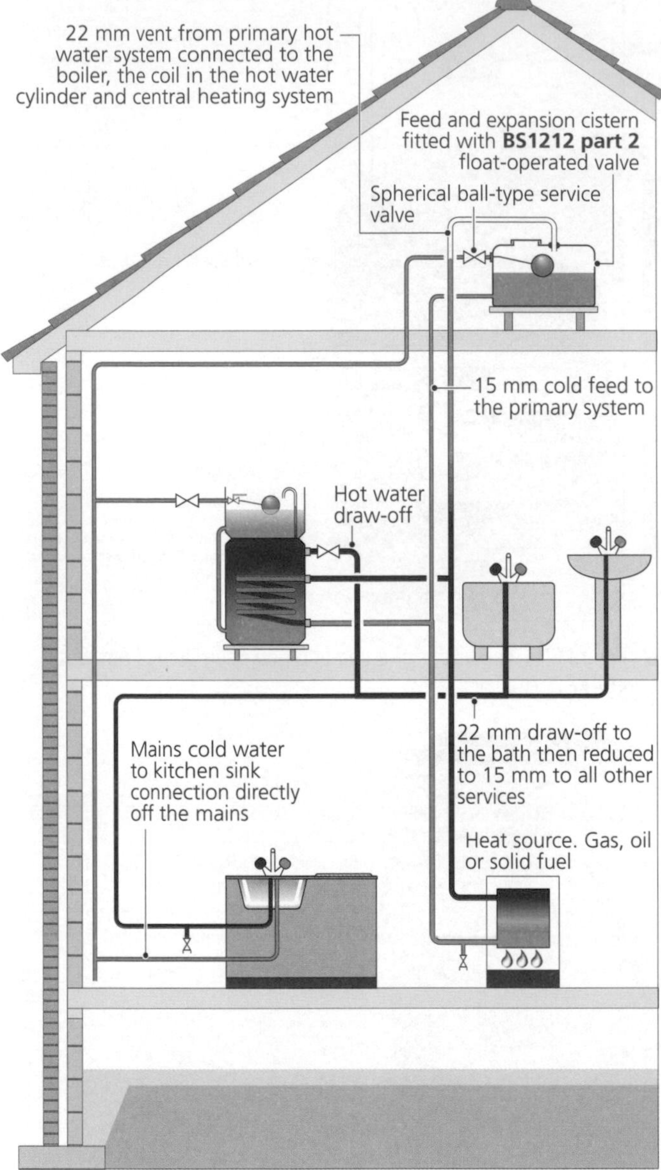

▲ Figure 6.13 A combination cylinder open vented hot water storage system

The main problem with this type of cylinder is the lack of water pressure at the taps. The cold water storage is very close to the hot water cylinder and so the static head of pressure is very low. Because of this, Fortics need to be installed as high as possible (but not in the roof space) to improve the pressure at the outlets.

Power shower pumps may also cause a problem because the cold water storage cistern cannot replenish itself quickly enough to feed the shower pump. The cistern is not designed to supply hot and cold water systems, and so shower pumps must not be fitted to this type of hot water supply.

▼ Table 6.5 Advantages and disadvantages of combination cylinder systems

Advantages	Disadvantages
A cheap alternative for hot water systems, especially suited to flats and small houses Easy to install Fully compliant with Doc. L of the Building Regulations	Suffers from lack of pressure unless installed at height Not suitable for pumped shower installations because of the lack of cold water storage

'Quick recovery' hot water storage cylinders

The quick recovery cylinder has a multi-coil heat exchanger that is made up of several smaller bore coils rather than one large one. This encourages a rapid recovery of the hot water because the coil has a greater surface area of heat presented to the water in the cylinder. They work in a similar way to instantaneous hot water cylinders and are replenishing the hot water even as it is being used. This reduces the amount of storage required and can save up to 40 per cent on fuel bills when compared to the standard cylinder type. When cheap-rate electricity is used for heating the water, it is most economical to heat the entire contents of the cylinder overnight. This reduces the need to use the immersion heater during the day when electricity is more expensive.

▼ Table 6.6 Criteria of choice for combination cylinder systems

Property size	Storage capacity	Fuel type	Installation cost	Fuel efficiency
Small properties and flats.	Usually 114 litres for flats but larger capacities are available.	Best used with Economy 7 electricity but can be double feed type for use with fully pumped heating systems.	Ideally suited to flats and small houses because of the low installation and materials costs.	Economy 7 electricity is 100% efficient but the tariffs can be very expensive.

Key points are:

- rapid heated water recovery; generally, recovery times are 15 minutes for 45 litres and 45 minutes for 210 litres (assuming a boiler output of 9 kW); 4 minutes for 45 litres and 19 minutes for 210 litres (assuming a boiler output of 30 kW)
- multi-coil heat exchanger
- smaller storage cylinder means more space in the airing cupboard
- reduces boiler cycling
- saves on fuel bills
- can be used with conventional fully pumped systems.

Quick recovery cylinders work at their most efficient when installed alongside condensing boilers running at maximum temperature. This will ensure that recovery times are at their absolute minimum and the lower return temperature of the condensing boiler will maximise the time the boiler spends in condensing mode.

INDUSTRY TIP

The high heat recovery cylinder relies on the coil having an increased surface area. This increased surface area is able to transfer the heat from the primary water at a quicker rate, so in a property the cylinder can stay the same size if space is limited but produce hot water quicker.

▲ Figure 6.14 Shower outlet fed from a domestic hot water cylinder

▼ Table 6.7 Advantages and disadvantages of Superduty cylinder systems

Advantages	Disadvantages
Quick turnaround of hot water Only small storage capacity needed Very energy efficient Fully compliant with Doc. L of the Building Regulations	Initial cost of the cylinder

▼ Table 6.8 Criteria of choice for quick recovery cylinder systems

Property size	Storage capacity	Fuel type	Installation cost	Fuel efficiency
All domestic properties.	Usually 80 litres.	Can be used with Economy 7 electricity and fully pumped heating systems.	Initial cost of the cylinder is expensive, but is installed as a double feed indirect cylinder with comparable costs.	Extremely energy efficient when used with condensing boilers due to fast heating of the water. Can cut fuel costs by up to 40%.

Storage cylinder insulation

Cylinders are insulated with polyurethane foam, which is sprayed on to a predetermined thickness. The thickness of the insulation is covered by Building Regulations Document L: Conservation of heat and power, which was updated in October 2010. The insulation thicknesses have been modified to deliver low standing heat loss and keep CO_2 emissions to a minimum, in line with the Regulations. The new thicknesses are:

- Part L1A (new build and replacement cylinders) have 50 mm insulation
- Part L1B (replacement only cylinders) have 35 mm insulation.

Cylinder insulation jackets are also available for uninsulated cylinders. They are made from fibreglass insulation with a PVC jacket. They are tied with a lace at the top and kept in place by either aluminium bands or plastic straps.

Grades of storage cylinder

Open vented hot water storage cylinders are manufactured to **BS 1566–1:2002 – Copper Indirect Cylinders for Domestic Purposes. Open Vented Copper Cylinders. Requirements and Test Methods.**

BS 1566 specifies three grades of cylinder, with each grade indicating the pressure the cylinder will withstand. The grades of cylinder are:

1. Grade 1: 25 metres head
2. Grade 2: 15 metres head
3. Grade 3: 10 metres head.

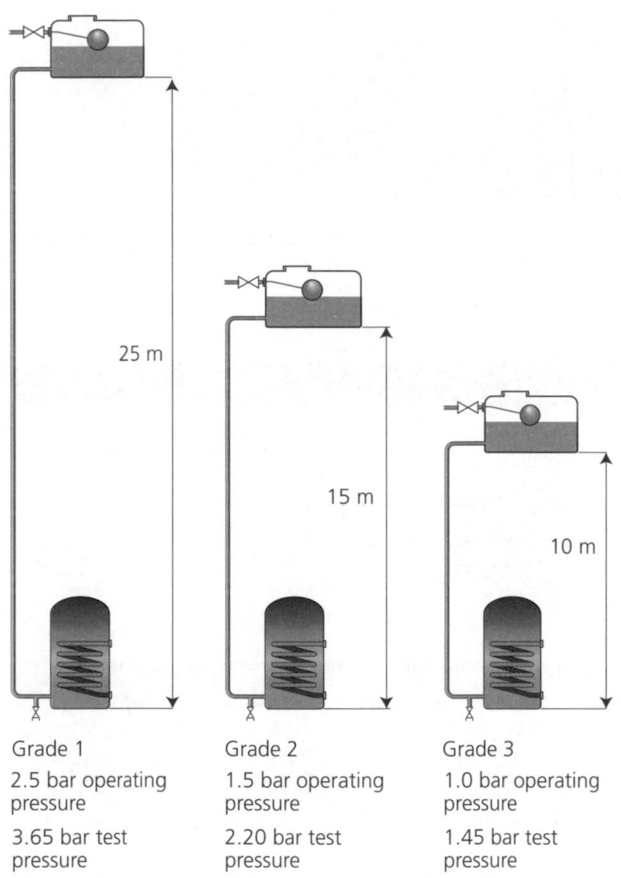

▲ Figure 6.15 The grades of cylinder and their maximum working pressures

Storage cylinder sizes and capacities

Open vented hot water storage cylinders are available in a wide range of sizes and capacities. The more common sizes are listed in Table 6.9.

▼ Table 6.9 Common sizes of open vented hot water storage cylinders

Size	Capacity
900 mm × 350 mm	74 litres
900 mm × 400 mm	98 litres
1050 mm × 400 mm	116 litres
900 mm × 450 mm	120 litres
1050 mm × 450 mm	144 litres
1200 mm × 450 mm	166 litres
1500 mm × 450 mm	210 litres

ACTIVITY

Working out the capacity of a cylinder calls for a relatively simple calculation that involves the use of Pi (π). Take π as being 3.142.

The formula for calculating the capacity of a cylinder is:

$$\pi r^2 \times h \times 1000$$

where:

π = 3.142

r = radius

h = height

Example:

A cylinder has a diameter of 500 mm and a height of 1000 mm. What is its capacity in litres?

Answer:

First, we will need to convert mm to m. Therefore, 500 mm becomes 0.5 m and 1000 mm becomes 1 m. The diameter is 0.5 m so the radius will be half of that. Therefore, the calculation will read:

$$3.142 \times (0.250 \times 0.250) \times 1 \times 1000 = 196.375 \text{ litres}$$

Now attempt the following calculations:

1. A cylinder measures 300 mm × 1050 mm. What is its capacity?
2. A cylinder measures 400 mm × 850 mm. What is its capacity?
3. A cylinder measures 500 mm × 1500 mm. What is its capacity?

Anodic corrosion protection of hot water storage cylinders

Hot water storage cylinders can suffer from electrolytic corrosion where there are two or more dissimilar metals present, especially in areas where the water is soft as this is aggressive to certain metals.

Placing two dissimilar metals in aggressive water produces a very small electric current, which flows from the weaker (anodic) metal to the noble (cathodic) metal where the anodic metal is gradually eaten away. This occurs commonly when some types of brass fittings are used.

Hot water cylinders can be protected from electrolytic corrosion by the use of a magnesium rod, which is either fastened to the bottom of the storage cylinder during manufacture or by simply dropping the magnesium rod in the draw-off connection during installation. This magnesium rod is known as the sacrificial anode. It works by distracting the corrosion away from the weaker anodic metal in the installation to be eaten away itself. If necessary, it can be replaced once the anode has been completely destroyed.

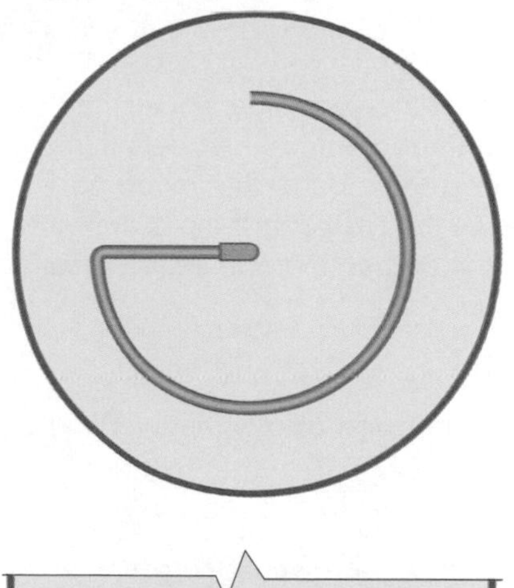

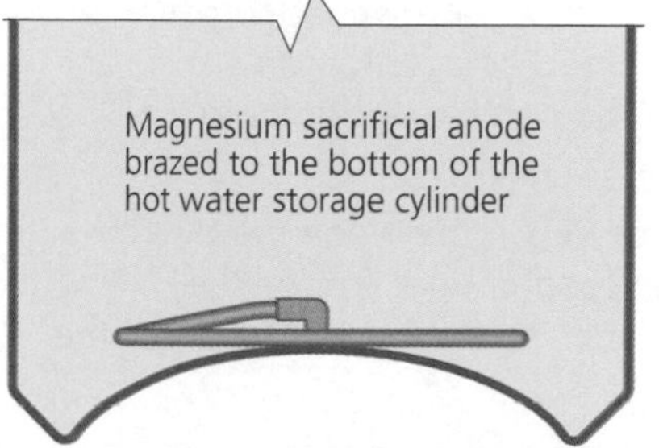

▲ Figure 6.16 Sacrificial anodes

Pipe sizes for open vented hot water storage systems

Pipe sizes are critical if the correct flow rate is to be achieved at the outlets. For open vented hot water systems fed from a cistern in the roof space, the size of the pipework would generally depend on the size of the system. A minimum 22 mm cold feed pipe to the cylinder should be installed, with a full-way gate valve to provide isolation of the hot water system. Occasionally, the cold feed may be 28 mm if there is more than one bathroom in the property. The cold feed should be fitted with a drain-off valve at the lowest point to allow complete drain down of the hot water storage cylinder. The connection of the cold feed to the cistern must be at least 25 mm above any cold distribution pipework to ensure that, in the event of mains cold water failure, the hot water runs out first.

The hot water draw-off should have a gradual rise towards the vent and must be a minimum of 450 mm in length to prevent parasitic circulation occurring. The vent pipe must rise vertically, terminate inside the cold water storage cistern and be sealed by means of a rubber grommet; it should have no valve installed anywhere along its length. The vent pipe and draw-off must be installed in a minimum of 22 mm size pipe.

The hot distribution pipework must be a minimum of 22 mm pipework to any large-volume appliances such as baths, but can be reduced in size to 15 mm to supply kitchen sinks, washbasins and shower valves. It is good practice to install isolating valves at the appliances, although it is not a requirement of the Water Supply Regulations. The pipework should have a gradual incline towards drain-off valves to permit total draining of the system for maintenance and repair.

Where double feed indirect cylinders are installed, the primary system must contain a separate feed and expansion cistern or expansion vessel and disconnectable filling loop, which separates the primary water from the secondary water. The cold feed to the primary system from the F and E (feed and expansion) cistern can be installed in 15 mm pipework and must not contain any form of isolation valve. The vent from the primary system must be installed in 22 mm pipework. It should rise vertically and terminate over the F and E cistern. The height of the vent pipe above the F and E cistern should not be less than 150 mm plus 40 mm for every metre in height from the overflow level to the lowest point of the cold feed pipe.

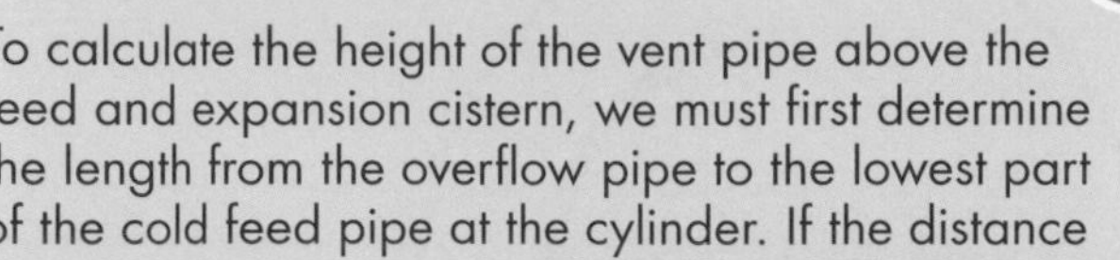

ACTIVITY

To calculate the height of the vent pipe above the feed and expansion cistern, we must first determine the length from the overflow pipe to the lowest part of the cold feed pipe at the cylinder. If the distance between them is, say, 4 m then the calculation is as follows:

$$4 \times 40 + 150 = 310$$

So, the vent pipe must be taken above the overflow level 310 mm.

Now try it for yourself:

1 There is a distance of 6 m between the overflow level and the cold feed connection on the cylinder. What is the recommended height of the vent pipe?
2 There is a distance of 3 m between the overflow level and the cold feed connection on the cylinder. What is the recommended height of the vent pipe?
3 The height of the vent pipe above the F and E cistern is 350 mm. What is the distance between the overflow level and the cold feed connection to the cylinder?

The F and E cistern must be capable of accommodating an expansion of 4 per cent of the total amount of water contained in the primary system and any heating system installed.

Unvented hot water storage systems

An unvented hot water storage system is simply a sealed system of pipework and components that is supplied with water above atmospheric pressure. The system does not require the use of a feed cistern. Instead, it is fed with water direct from a water undertaker's mains supply, or with water supplied by a booster pump and a cold water accumulator if the mains pressure is low.

An unvented hot water system differs from open vented types because there is no vent pipe. Expansion of water due to the water being heated is accommodated in either an external expansion vessel or an expansion bubble within the storage cylinder.

The system also requires other mechanical safety devices for the safe control of the expansion of water and to ensure that the water within the storage cylinder does not exceed 100°C. There are two categories of centralised unvented hot water storage systems:

1 directly fired/heated storage systems
2 indirectly fired/heated storage systems.

KEY POINT

Unvented systems require safety discharge pipework that must be correctly sized and positioned in accordance with Building Regulations Document G3. This is discussed later in the chapter.

The various types of unvented hot water system

There are three basic types of unvented hot water system. They are defined by how the water is heated. These are:

1 indirect storage systems
2 direct storage systems:
   - electrically heated
   - gas or oil fired
3 small point of use (under sink).

Indirect storage systems

Indirect unvented hot water storage systems utilise an indirect unvented hot water storage cylinder at the heart of the system. As with open vented systems, the cylinder contains a coiled heat exchanger to transfer the heat indirectly from the primary system to the secondary system. This can be done in one of two ways:

1 by the use of a gas-fired condensing boiler
2 by the use of an oil-fired condensing boiler.

Older, non-condensing boilers may be used if the boiler is an existing appliance, provided that the boiler contains both a control thermostat and a high energy cut-out (high-limit) thermostat to limit the water temperature at the coil should the control thermostat fail. On no account must solid fuel appliances and boilers be used to provide heat to the coil. The primary hot water system may either be an open vented or sealed system.

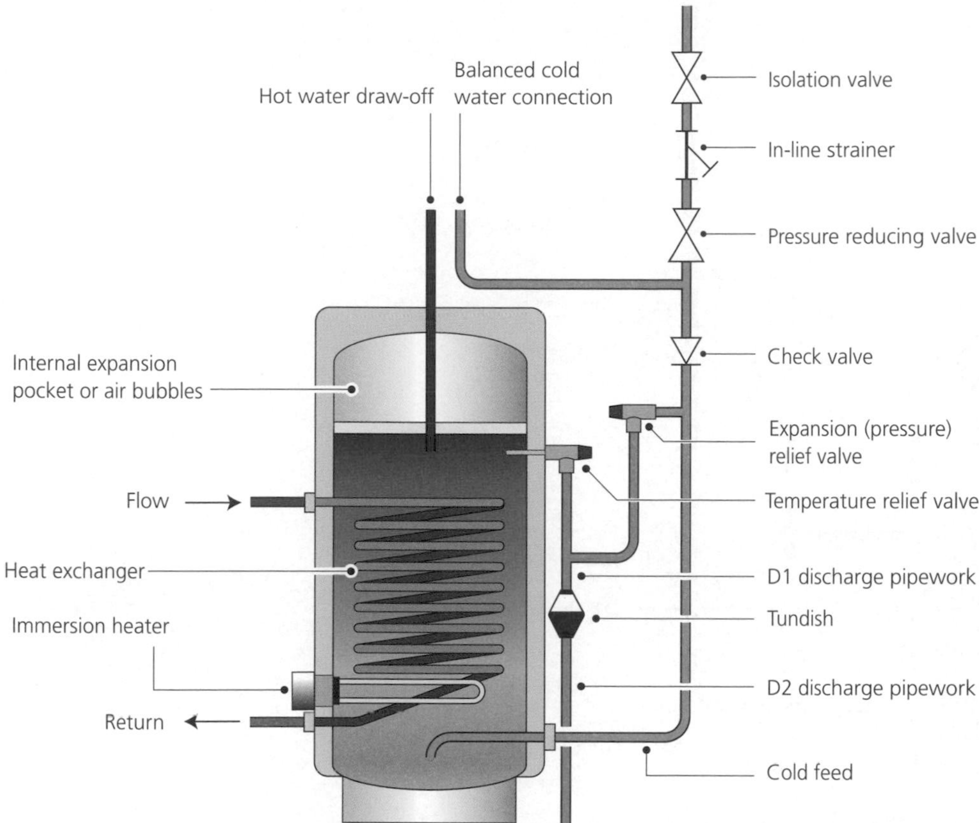

▲ Figure 6.17 Indirect-type unvented hot water storage cylinder with internal expansion

An immersion heater provides back-up hot water heating for use during the summer or for when the boiler malfunctions.

Direct storage systems

The direct system uses a direct-type unvented hot water cylinder that does not contain any form of heat exchanger. There are two very different types, as described below.

- **Electrically heated:** this type of cylinder does not contain a heat exchanger. Instead, the water is heated directly by two immersion heaters controlled by a time switch. One immersion heater is located close to the bottom of the cylinder to heat all of the contents of the cylinder at night and another located in the top third to top up the hot water during the day if required via a one-hour boost button on the time switch. Both immersion heaters are independently controlled and cannot be used simultaneously. The immersion heaters are manufactured to **BS EN 60335–2–73** and must contain a user thermostat usually set to 60°C and a non-resetting thermal cut-out (high limit stat).
- **Gas or oil fired:** the design of these water heaters originated in North America. They consist of a hot water storage vessel with a flue pipe that passes through the centre. Expansion of the water is catered for by the use of an external expansion vessel. Below the storage vessel is a burner to heat the water; this can be fuelled by either gas or oil, depending on the type. The burner is controlled by a thermostat and a gas/oil valve. An energy cut-out prevents the water exceeding the maximum of 90°C. The safety and functional controls and components layout is almost identical to other unvented hot water storage systems.

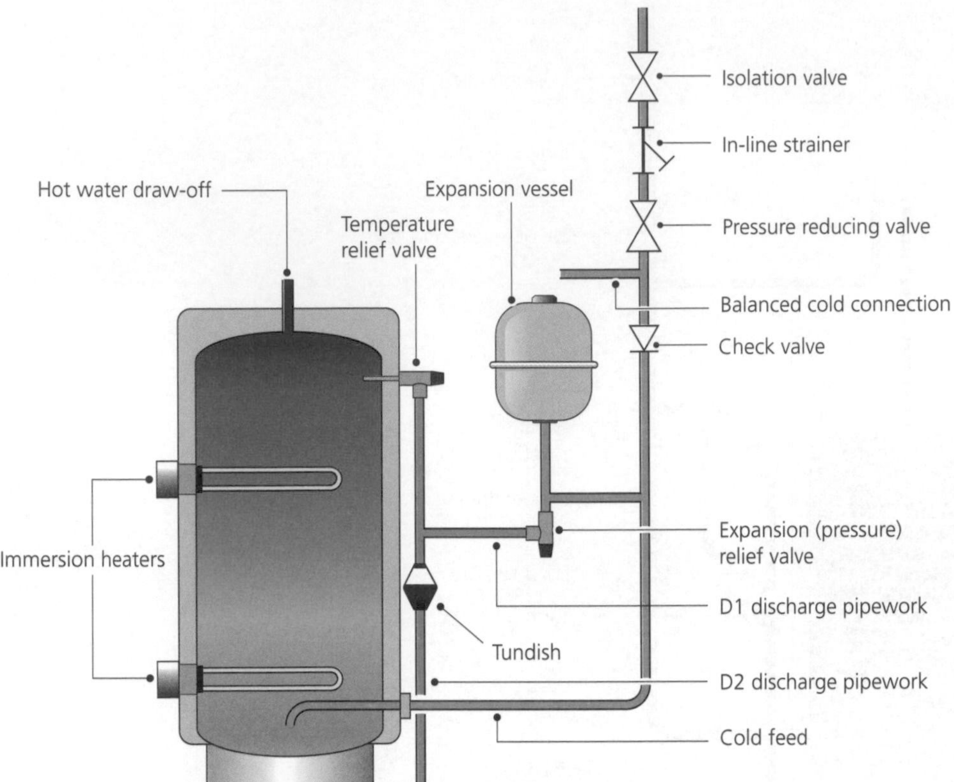

▲ Figure 6.18 Direct-type unvented hot water storage cylinder with external expansion vessel

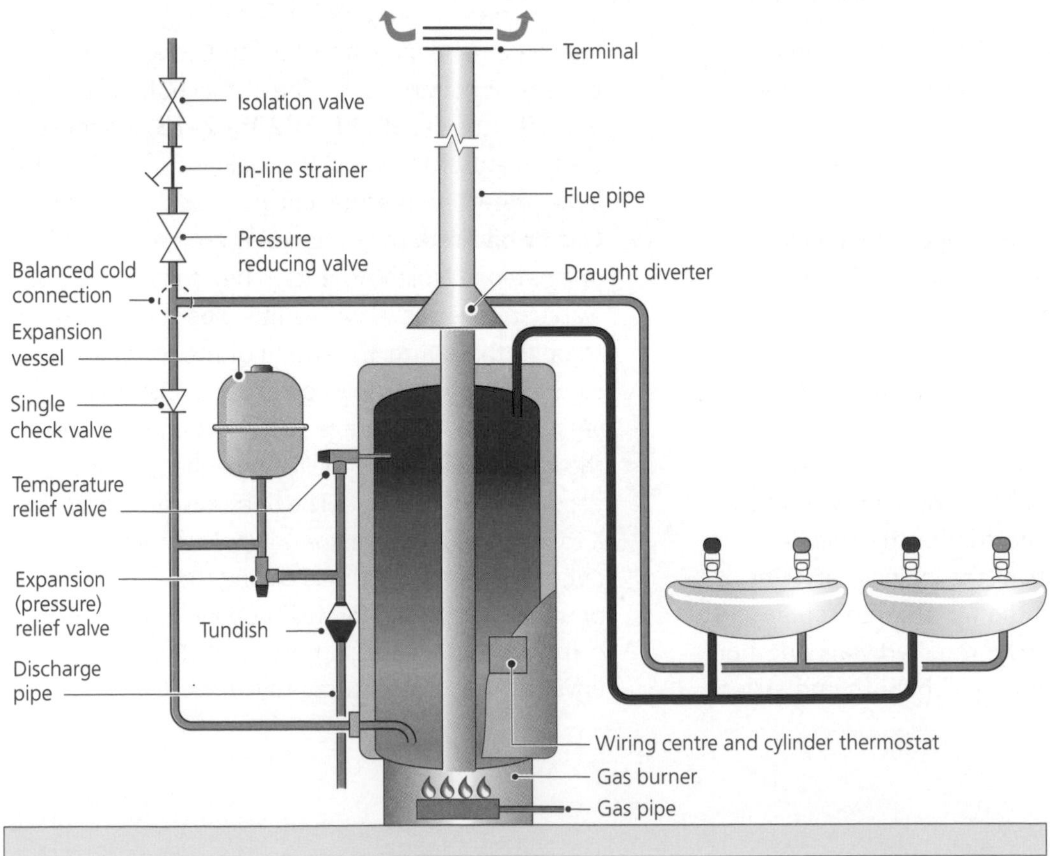

▲ Figure 6.19 Gas-fired direct-type unvented hot water storage cylinder with external expansion vessel

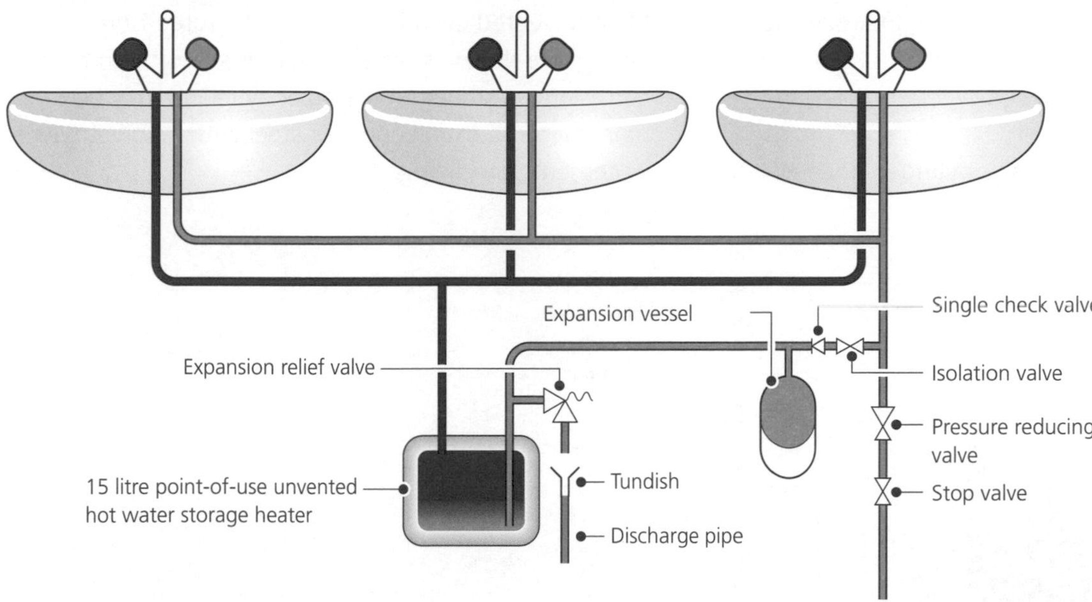

▲ Figure 6.20 Direct-type under-sink type unvented hot water storage cylinder with external expansion vessel

Direct unvented under-sink storage heaters

Unvented under-sink hot water storage heaters are connected direct to the mains cold water supply and deliver hot water at near mains cold water pressure. Because they have less than 15 litres of storage, they are not subject to the stringent regulations that surround the installation of larger unvented hot water storage units.

The expansion of water may be taken up within the pipework, provided the pipework is of sufficient size to cope with the water expansion. If not, then an external expansion vessel will be required.

▲ Figure 6.21 Unvented-type under-sink storage water heater with expansion vessel

Unvented hot water storage systems and pipework arrangements

Many installers claim that an unvented hot water storage system is the best type of system for any domestic situation, but this is far from the case. There are many factors that must be considered before this arrangement is installed into a property:

- Available pressure and flow rate – this is probably the most important factor, simply because poor pressure and flow rate will affect the operating performance of the installation. Pressure and flow rate readings should be taken at peak times to ensure adequate water supply before recommending this type of system.
- The route of the discharge pipework, termination and discharge pipework size.
- The type of terminal fittings to be used. This is especially important when retro-fitting unvented installations onto existing hot water systems as the existing taps etc. may not be suitable.
- Cost – unvented systems tend to be very expensive.

The types of unvented hot water storage cylinder

There are two types of unvented hot water storage cylinder; both are manufactured to **BS EN 12897:2016 +A1:2020 – Specification for Indirectly Heated Unvented (Closed) Storage Water Heaters** and available as direct fired/heated or indirectly heated vessels:

- unvented hot water storage cylinders using an external expansion vessel
- unvented hot water storage cylinders incorporating an internal expansion air gap.

Most unvented cylinders are manufactured from high-grade duplex stainless steel for strength and corrosion resistance. Some older cylinders may be manufactured from copper or steel with a polyethylene or cementitious lining.

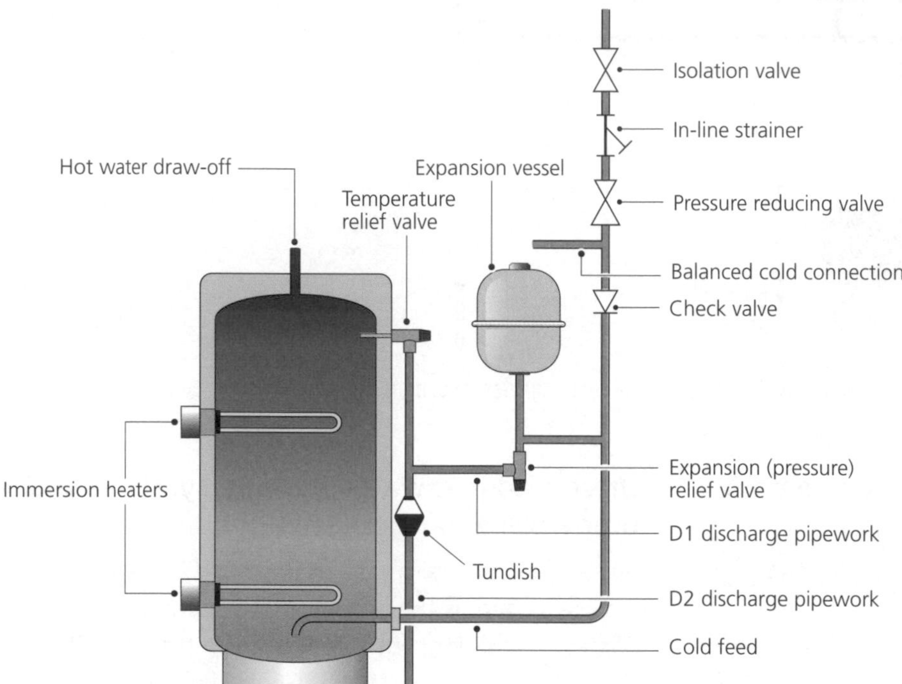

▲ Figure 6.22 A typical unvented cylinder with external expansion vessel

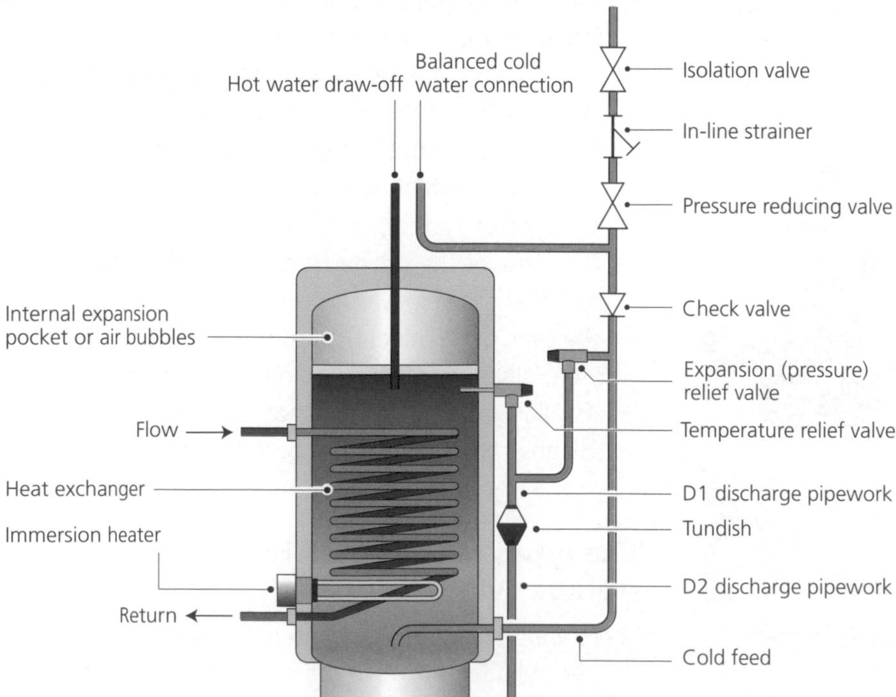

▲ Figure 6.23 A typical unvented cylinder with internal expansion

Unvented hot water storage cylinders can be purchased as 'units' or 'packages':

- units are delivered with all the components already factory fitted and require less installation time
- packages are delivered with all components separately packaged (except those required for safety, such as temperature relief valves); these have to be fitted by the installer in line with the manufacturer's instructions.

The installation of unvented hot water storage cylinders

The installation of unvented hot water storage systems (UHWSS) is subject to the strict requirements of Building Regulations Approved Documents G3 and L, and the Water Supply (Water Fittings) Regulations. Typical pipework layouts are shown in Figures 6.24 and 6.25.

The unit or package must be installed in accordance with the manufacturer's instructions supplied with the vessel. There may be special instructions from the manufacturer regarding the installation requirements of that particular vessel.

The floor on which the vessel is to be sited must be substantial enough to accommodate the weight of the vessel and its water contents.

The pipework must be fitted in accordance with **BS EN 806** and **BS 8558**. Unvented hot water storage systems require at least a 22 mm cold water feed supplied by a water undertaker because of the high flow rate and pressure that the vessels operate at. Water can be supplied through a boosting pump and cold water accumulator if necessary (this will be discussed later in the chapter). A 22 mm hot water draw-off is required in all installations, but this may be reduced for particular appliances such as washbasins, sinks and bidets. Isolation valves should be fitted at all appliances in line with good practice.

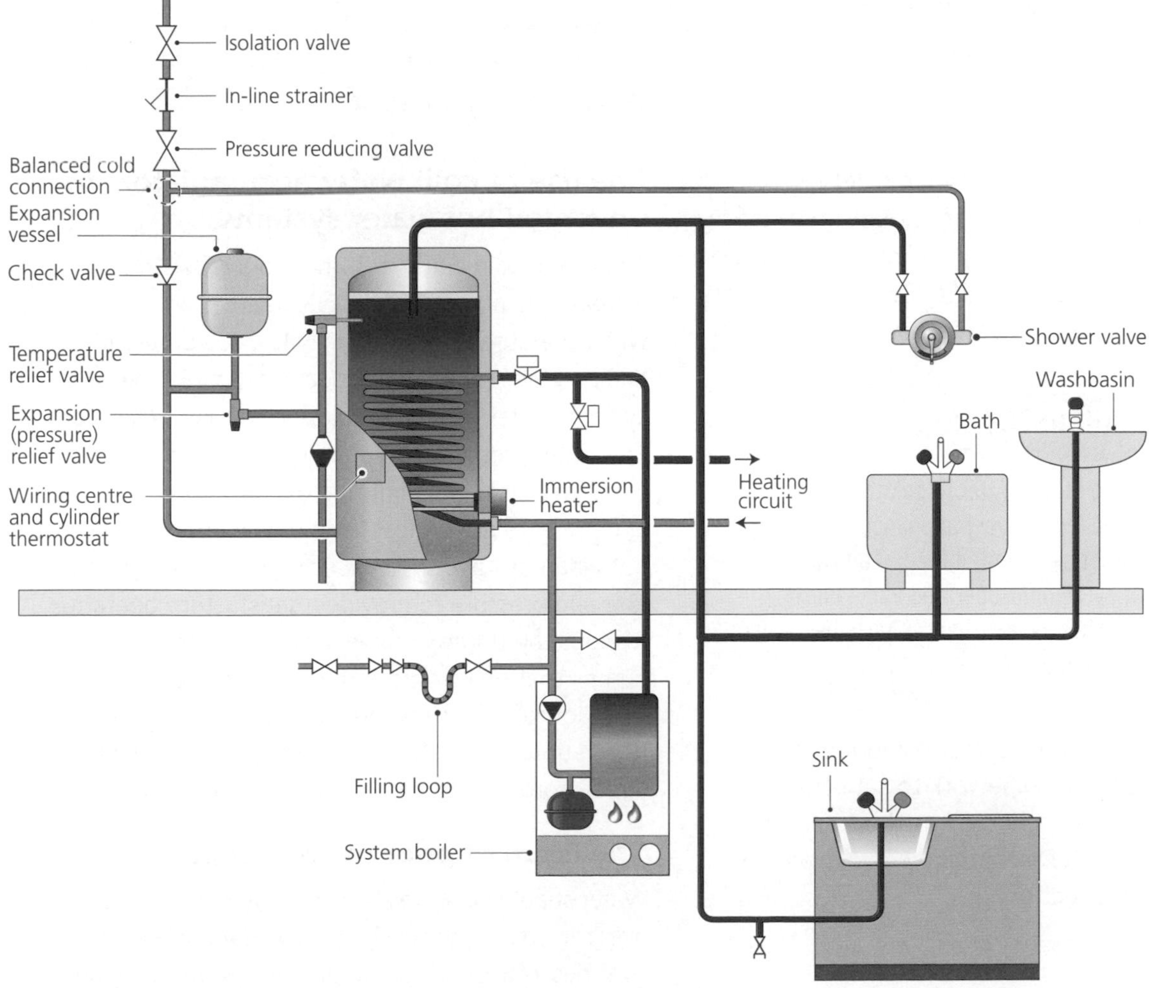

▲ Figure 6.24 Installation of an indirectly heated UHWSS with a system boiler

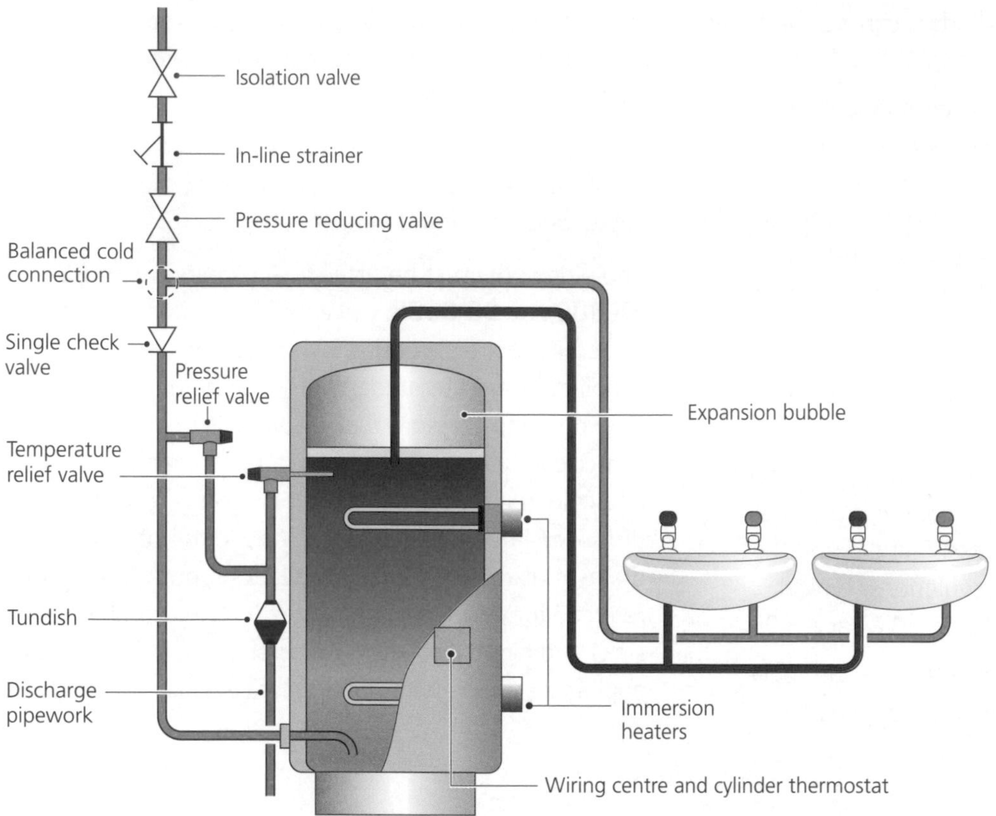

▲ Figure 6.25 Installation of a directly fired UHWSS with immersion heaters as the primary heat source

The order in which the functional and safety components are installed is of paramount importance if the system is to operate safely and efficiently, which can be seen in Figure 6.24 for indirectly heated vessels and Figure 6.25 for directly heated vessels.

INDUSTRY TIP

It doesn't actually state in the Water Regulations that isolation valves are needed at every appliance but it is considered good practice to install them. The Water Regulations say that every float-operated valve must have a service valve fitted as near to it as possible. Other appliances aren't mentioned.

Unvented hot water storage systems require the installation of a discharge pipework to safely convey any water that may be discharged as the result of a defect or malfunction. Discharge pipework will be discussed later in this section.

The use of cold water accumulators in unvented hot water systems

The use of cold water accumulators is becoming increasingly popular, especially in areas where the water pressure is exceptionally low. Accumulators and boosting pumps, as we saw in Unit 5, Cold water systems, offer a positive solution to the problem of low water pressure and poor low flow rate by storing water at night for use during the day. Both flow rate and pressure are critical factors when fitting unvented hot water storage systems as these rely on a good flow rate and pressure to provide a satisfactory operation. It should be borne in mind, however, that boosting pumps that deliver more than 12 litres per minute are not allowed under the Water Supply (Water Fittings) Regulations when the cold water supply is being taken direct from a water undertaker's mains supply.

The issue of poor mains supply

Water supply pressures have consistently diminished over the past 30 years. As more and more homes, factories, offices and shops are built, the loading on the UK water system has increased, with little or no upgrading of the water mains supply network. The pipework that serves our towns and cities is now supplying more properties than ever before, and this has resulted in a gradual degradation of both pressure and flow rate. In some areas of the UK, the supply pressure can be as little as 1 bar, which is unsatisfactory for an unvented hot water storage system.

Pressure of water takes two forms:

1 static pressure – this is the water pressure when no flow is occurring; this is always greater than the dynamic pressure
2 dynamic pressure (also known as 'running pressure') – this is the water pressure when outlets are open and water is flowing.

During periods of peak use, both static and dynamic pressures will decrease. If, during this time, a property has a static pressure of, say, 2 bar, then the dynamic pressure could drop to below 1 bar. At off-peak times, say, during the night, this could rise significantly, to 3 bar static and 2 bar dynamic, simply because less water is being used in the surrounding area. An accumulator would take advantage of the night-time rise in pressure to replenish its storage capacity while the mains pressure is at its highest. With the accumulator fully replenished, a good pressure and flow rate would be available throughout the day, provided that the accumulator has been sized correctly.

Figure 6.26 shows a typical unvented hot water storage system with an accumulator installed to increase both the pressure and the flow rate. An important factor here is the use of two pressure reducing valves (PRVs). The first PRV regulates the pressure entering the property so that any pressure fluctuations can be controlled to a predetermined pressure at night when the accumulator is replenishing. The second PRV reduces the pressure to that of the UHWSS manufacturer's recommendations.

Accumulators require a minimum incoming supply pressure to replenish successfully, usually around 2 bar. If the incoming supply cannot deliver this, even at off-peak periods, then a booster pump should also be installed, as shown in Figure 6.27.

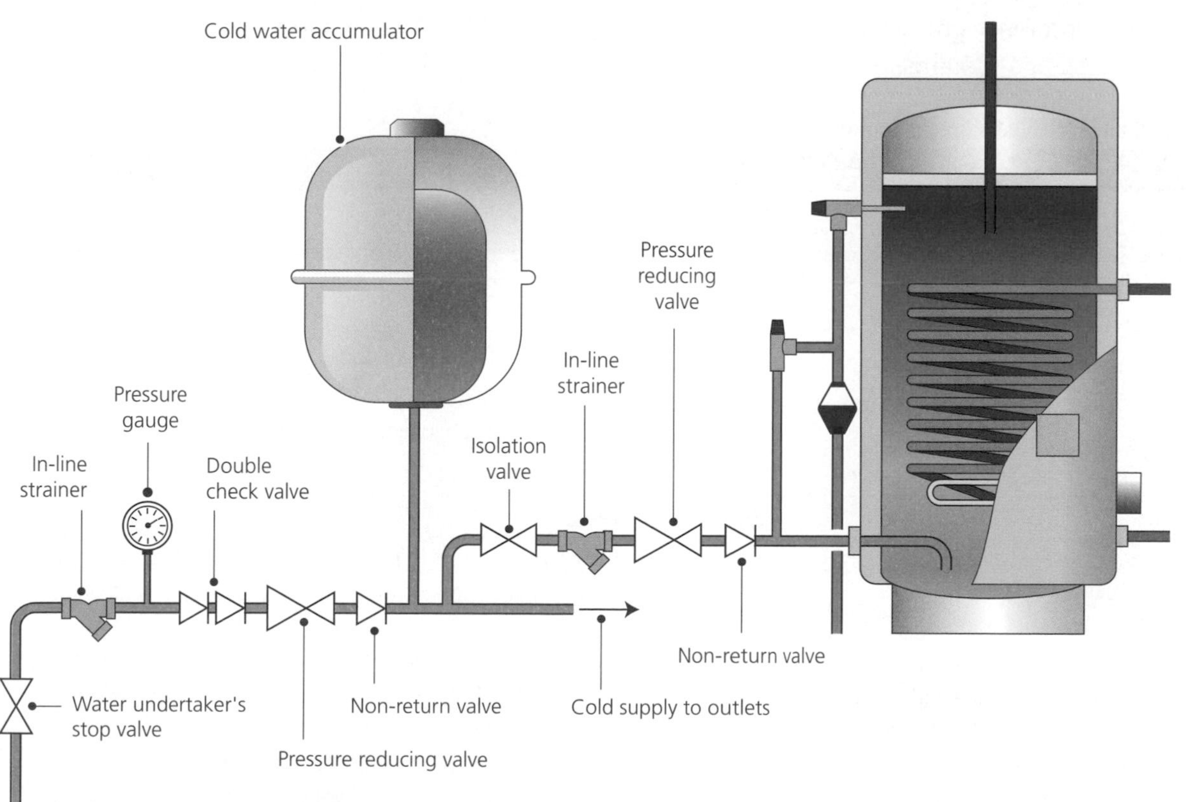

▲ Figure 6.26 An accumulator installed on an unvented system

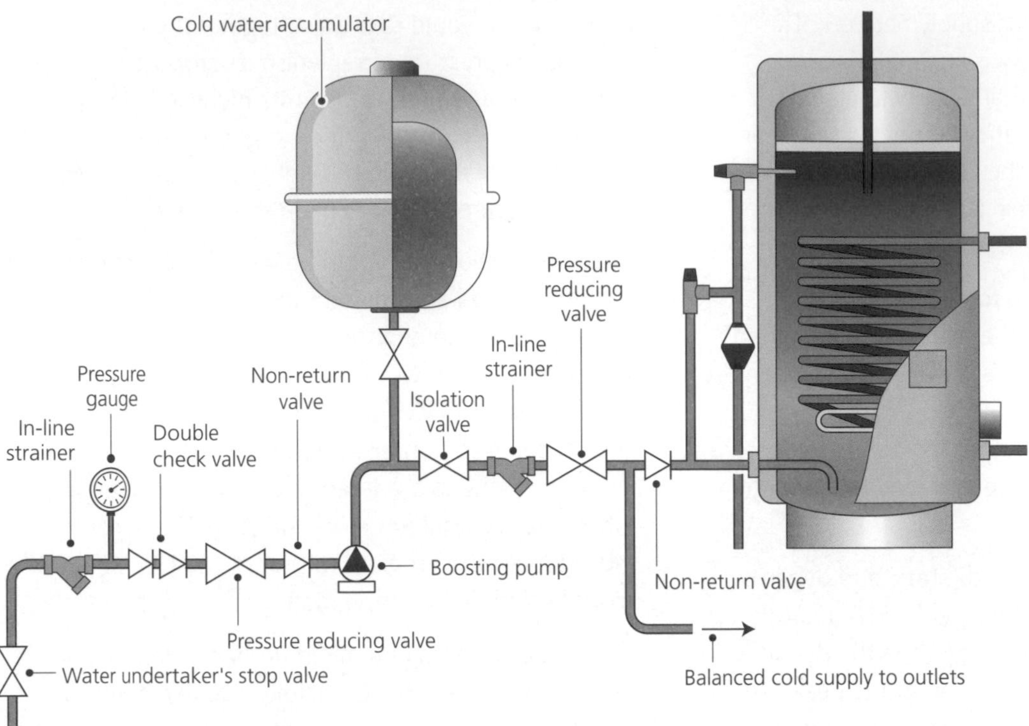

▲ Figure 6.27 An accumulator with a boosting pump installed on an unvented system

Comparisons between open vented and unvented hot water storage systems

There are important differences between these two types of system. Table 6.10 compares open vented and unvented hot water storage systems.

▼ Table 6.10 Vented and unvented storage hot water systems: a comparison

Advantages	Disadvantages
Open vented systems	
Storage is available to meet demand at peak times Low noise levels Always open to the atmosphere Water temperature can never exceed 100°C Reserve of water available if the mains supply is interrupted Low maintenance Low installation costs	Space needed for both the hot water storage vessel and the cold water storage Risk of freezing Increased risk of contamination Low pressure and, often, poor flow rate Outlet fittings can be limited because of the low pressure
Unvented systems	
Higher pressure and flow rates at all outlets, giving a larger choice of outlet fittings Balanced pressures at both hot and cold taps Low risk of contamination The hot water storage vessel can be sited almost anywhere in the property, making it a suitable choice for houses and flats alike The risk from frost damage is reduced Less space required because cold water storage is not needed Installation is quicker as less pipework is required Smaller-diameter pipework may be used in some circumstances	No back-up of water should the water supply be isolated If the cold water supply suffers from low pressure or flow rate, the system will not operate satisfactorily There is the need for discharge pipes that will be able to accept very hot water and there will be restrictions on their length A high level of maintenance is required Higher risk of noise in the system pipework Initial cost of the unvented hot water storage vessel is high

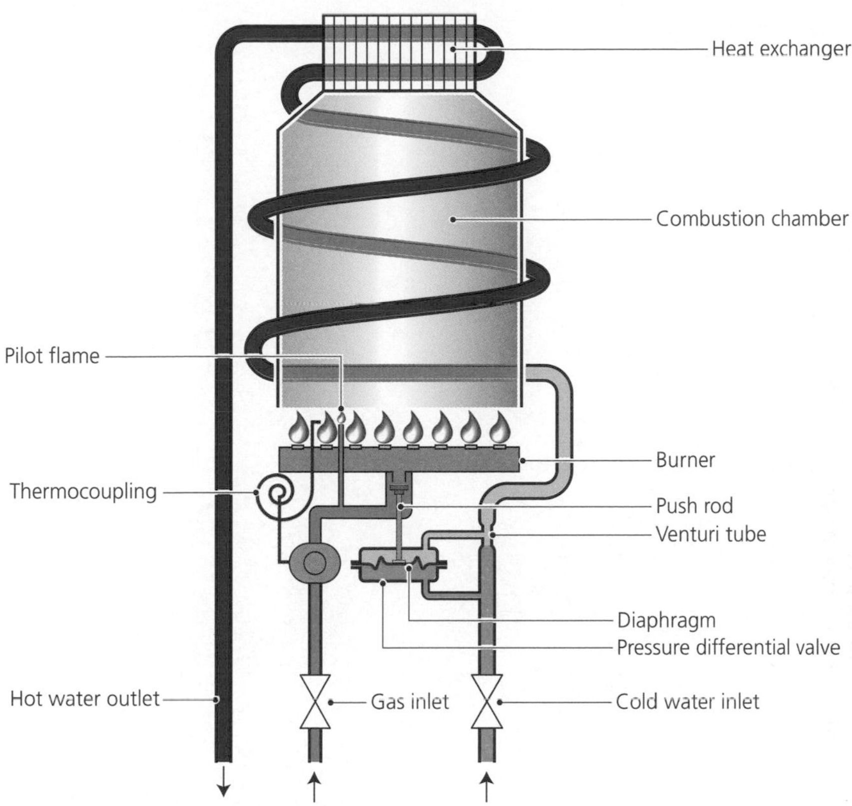

▲ Figure 6.28 Gas instantaneous hot water heater

Gas-fired instantaneous multipoint hot water heaters

With this type of hot water heater, cold water is taken from the water undertaker's main and heated in a heat exchanger as demand requires before being distributed to the outlets. As long as a tap is running, hot water will be delivered to it. There is no limit to the amount of hot water that can be delivered. There is no storage capacity.

Expansion of water due to being heated is accommodated by back pressure within the cold water main. However, if this is not adequate or the cold water system contains pressure reducing valves or check valves, then an expansion vessel must be fitted.

The heater works on **Bernoulli's principle** by using a venturi tube to create a pressure differential across the gas valve when the cold water is flowing into the heater.

KEY TERM

Bernoulli's principle: when a pipe reduces in size, the pressure of the water will drop but the velocity of the water increases. When the pipe increases back to its original size, then the velocity will decrease and the pressure will increase almost to its original pressure.

Gas- or oil-fired combination boilers

Combi boilers are dual-function appliances. They provide instantaneous hot water and central heating within the same appliance. In normal working mode, combination boilers are central heating appliances, supplying a proportion of their available heat capacity to heat the central heating water. When a hot tap is opened, a diverter valve diverts the boiler water around a second heat exchanger, which heats cold water from the water undertaker's cold water mains to supply instantaneous hot water at the hot taps. In this mode, the entire heat output is used to heat the water. Temperature control is electronic and this automatically adjusts the burner to suit the output required. Typical flow rates are around 9 litres per minute (35°C temperature rise). Some combination boilers incorporate a small amount of storage and this can double the flow rate to around 18 litres per minute.

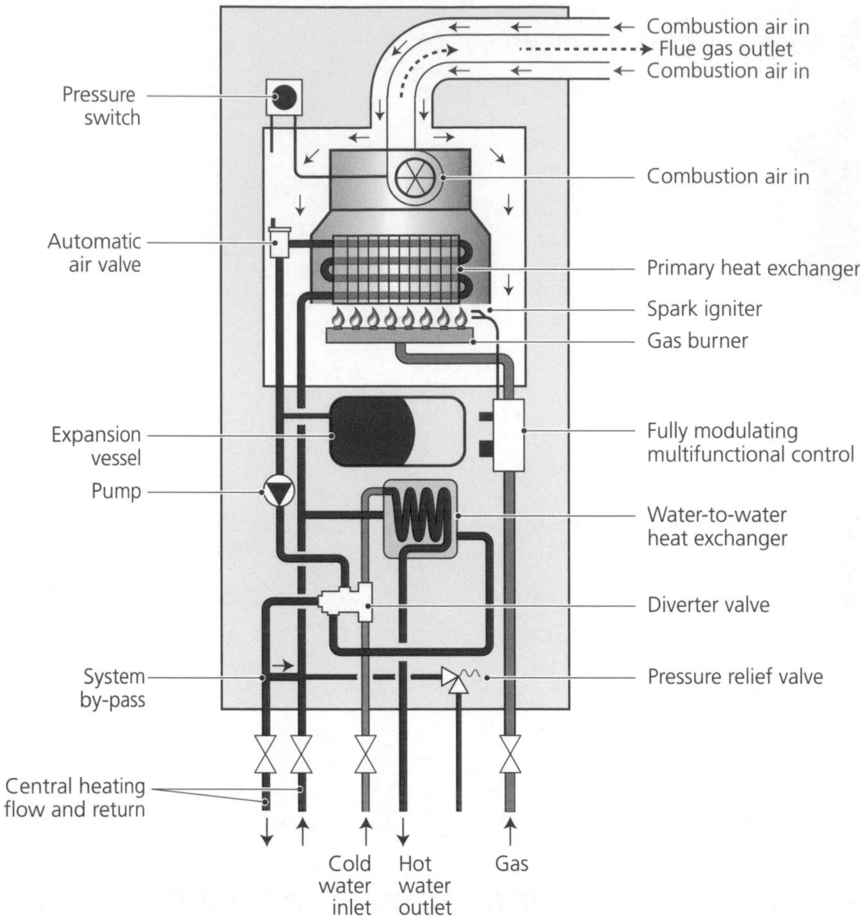

▲ Figure 6.29 Combination boiler

Thermal stores

Sometimes called water-jacketed tube heaters, thermal stores work by passing mains cold water through two heat exchangers that are encased in a large storage vessel of primary hot water fed from a boiler. They are very similar to an indirect system but work in reverse.

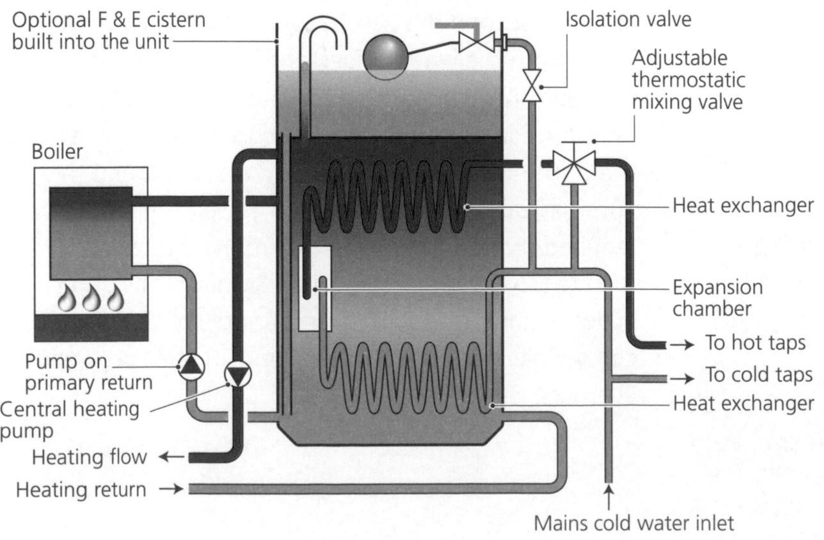

▲ Figure 6.30 Thermal store

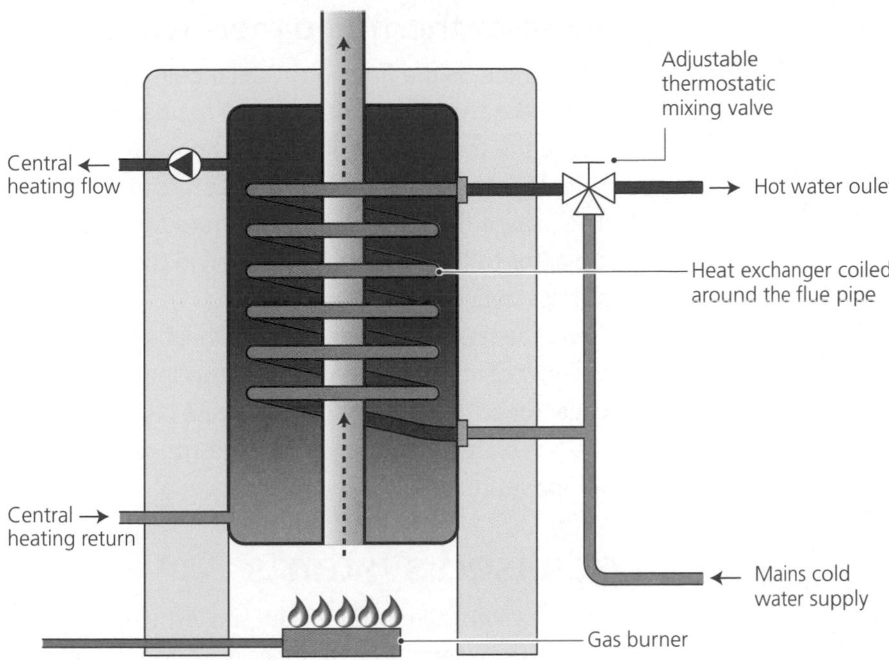

▲ Figure 6.31 Combined primary storage unit

Inside the unit are two heat exchangers, which the mains cold water passes through, and a small expansion chamber. The expansion chamber allows for the small amount of expansion of the secondary water. The primary water can reach temperatures of up to 82°C, which can, potentially, be transferred into the secondary water. Because of this, an adjustable thermostatic mixing valve blends the secondary hot water with mains cold water so that the water does not exceed 60°C.

Gas- or oil-fired combined primary storage units

These are very similar in design to thermal stores and work in exactly the same way, in that cold water from the mains supply is passed through a heat exchanger. The difference here is that the unit has its own heat source, in the form of a gas burner, to heat the primary water, eliminating the need for a separate boiler.

Solar thermal systems

A solar water heating system uses roof-mounted solar collectors aligned to face south to capture the heat generated by the Sun. The solar collector can be either a series of vacuum tubes or a flat panel, both of which are filled with a heating fluid (usually a mixture of water and anti-freeze). On average during the summer months, 1 m^2 of solar panel will deliver around 1 kW of energy, therefore 1 m^2 is needed for every occupant of the dwelling, with a minimum recommended area of 2.5 m^2. This will supply about 80 per cent of the hot water demand during the summer and around 20 per cent over the winter season, an average of 60 per cent over the whole year. A conventional gas or oil boiler, or an electric immersion heater, will be required for the remaining 40 per cent heating requirement or in case the solar system should fail.

The components of a solar thermal hot water system

Solar hot water systems require certain components, some of them specialised, to enable the system to work effectively. These are as follows.

- **Collector:** these can either be:
    - **Flat-plate collectors:** these are the simplest form of collector. They are tubes that run through shallow metal boxes with a front of thick black glass to trap the heat in a greenhouse effect. As the heating fluid is pumped through the tubes, it collects the Sun's heat, which is then pumped through the heat exchanger where the heat is transferred to the water inside the storage cylinder.

  - **Evacuated tubes:** these are a little more complicated but, in essence, are tubes that have a vacuum inside. These collect the heat from the Sun, passing it to a manifold through which the heating fluid runs. The heated fluid is then pumped to the coil in a similar way to the flat plate collector.
- **Hot water storage cylinder to store the hot water.** The cylinder should contain two coils: one to transfer the heat from the solar collector and the other to transfer the heat from a conventional boiler/water heater.
- **Heat exchanger**, usually in the form of a coil that transfers the heat from the solar collector to the water stored in the hot water storage cylinder.
- **Circulating pump** to circulate the hot fluid from the solar collector to the heat exchanger and back.
- **Control system:** the control system is used to prevent freezing fluid being circulated through the coil during the winter or at night when the Sun goes down. A typical control system will incorporate a pump, flow meter, pressure gauge, a thermometer and a thermostat.

How solar thermal panels work

1. The Sun heats the fluid in the solar collector.
2. When the thermostat senses that the panel is 6°C above the temperature inside the hot water storage cylinder, the circulation pump will start to run.
3. The heated fluid is then pumped from the solar collector to the heat exchanger coil in the hot water storage cylinder.
4. Here, the heated fluid gives off its heat into the cylinder of stored water before returning to the collector to be reheated. This process continues until the hot water storage cylinder is at the required temperature.

Localised systems

Localised systems are often called single-point or point-of-use systems. They are designed to serve one outlet at the position where it is needed and are usually installed where the appliance is some distance away from the fuelled hot water supply.

Again, these can be divided into two categories:

1. instantaneous-type heaters
2. storage-type heaters.

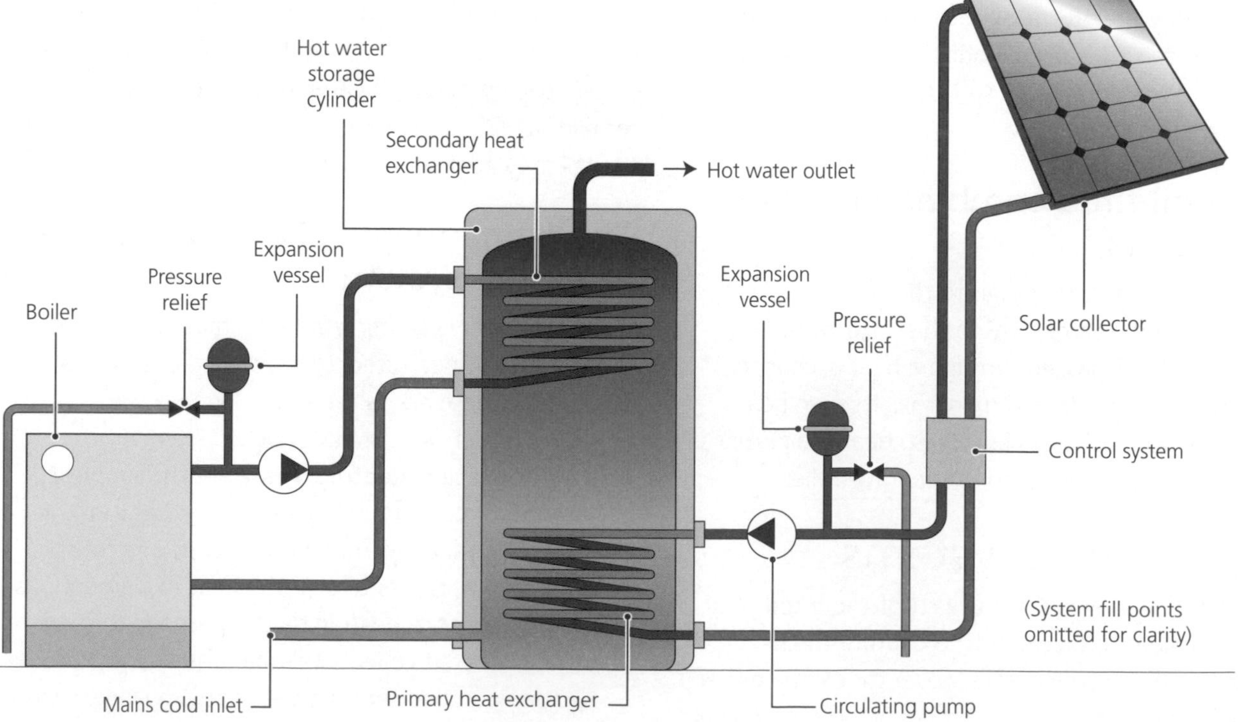

▲ Figure 6.32 Working principles of solar thermal hot water

Instantaneous-type fuelled water heaters

These can either be fuelled by gas or electricity, and are generally described as inlet controlled. This simply means that the water supply is controlled at the inlet to the heater. The water is heated as it flows through the heater and will continue to be heated as long as the water is flowing. When the control valve is closed, the water flow stops and the heat source shuts down.

This type of heater is generally used to supply small quantities of hot water such as washbasins and showers. Typical minimum water pressure is 1 bar. There are many different types of electric shower with varying outputs from 8.5 kW to 11 kW. The higher the kW output, the better the overall flow rate at a showering temperature. All electric showers feature a low-pressure heater element cut-off so that the temperature of the water does not cause harm if the supply pressure/flow rate is low.

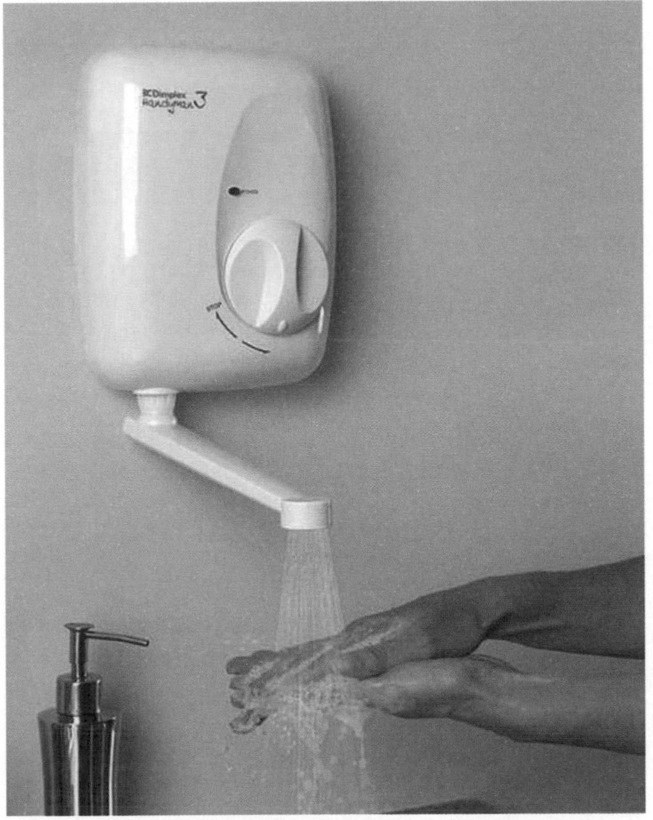

▲ Figure 6.33 Localised instantaneous hot water heater

HEALTH AND SAFETY

You must not attempt to install any electrical appliance such as showers, immersion heaters or hand wash heaters unless you are competent to do so and have the correct qualifications. Remember: electricity can kill!

INDUSTRY TIP

Most localised gas instantaneous water heaters do not contain a flue. The gases from the appliance simply disperse in the room where they are installed. They are known as 'flueless appliances'.

Storage-type localised water heaters

This type of heater is often referred to as the displacement type heater, as the hot water is displaced from the heater by cold water entering the unit. Typical storage capacities are between 7 litres and 10 litres (for the over-sink type). They can be divided into the following categories.

- **Over-sink heaters:** as the name suggests, these are fitted over an appliance such as a sink. The water is delivered from a spout on the heater. A common complaint with this type of heater is that they constantly drip water from the spout. This is normal as the heater must be open to the atmosphere at all times to accommodate the expansion when the water is heated. The dripping water is the expansion taking place and will stop once the heater has reached its operating temperature.
- **Under-sink heaters:** the under-sink heater works in exactly the same way as the over-sink heater. The main difference is that these heaters usually require a special tap or mixer tap that permits the outlet to be open to the atmosphere at all times to allow for expansion. The inlet of water to the heater is still controlled from the tap. Typical capacities are up to 15 litres.

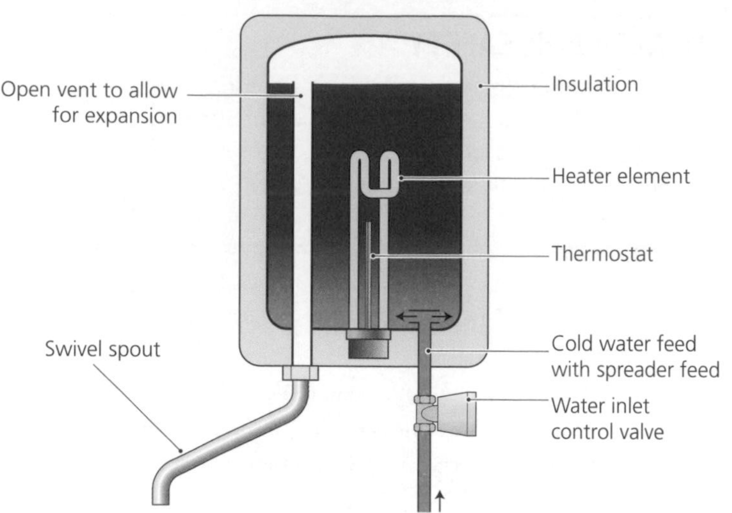

▲ Figure 6.34 A typical over-sink storage water heater

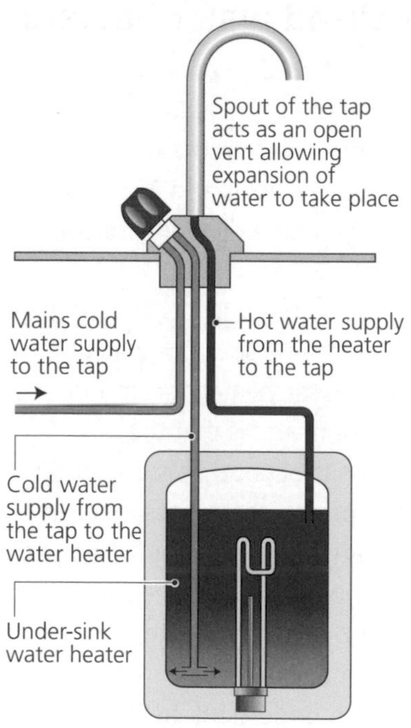

▲ Figure 6.35 A typical under-sink storage water heater

Secondary circulation

Secondary circulation is necessary to prevent the wastage of water due to the excessive lengths of hot water draw-off from the storage vessel to the outlet. Here, we will look at the various methods of providing secondary circulation in hot water systems, including larger domestic systems.

Secondary circulation in domestic dwellings

Secondary circulation is required where the length of any draw-off pipework is excessive. British Standard **BS EN 806** (and **BS 6700**) and the Water Supply (Water Fittings) Regulations give the maximum length a hot water draw-off pipe may travel without a secondary circulation system being installed. These lengths are reproduced in Table 6.11.

▼ Table 6.11 Maximum length of pipe before it becomes a dead leg and requires secondary circulation

Outside diameter (mm)	Maximum length (m)
12	20
Over 12 and up to 22	12
Over 22 and up to 28	8
Over 28	3

Secondary circulation is a method of returning the hot water draw-off back to the storage cylinder in a continuous loop, to eliminate cold water 'dead legs' by reducing the distance the hot water must travel before it arrives at the taps.

INDUSTRY TIP

What is a 'dead leg'?

When a hot tap is opened, a certain amount of cold water is usually drawn off and allowed to run to drain before hot water arrives at the tap. This wasted, cold water is known as a dead leg. Under the Water Regulations, dead legs must be restricted to the lengths given in Table 6.11. If this is not possible, then secondary circulation is required. Dead legs are a potential source of *Legionella* and noise.

In all installations, secondary circulation must use forced circulation via a bronze- or stainless steel-bodied circulating pump to circulate the water to and from the storage cylinder. The position of the pump will depend on the type of hot water system installed.

INDUSTRY TIP

A secondary circulation pump is very similar in design to a central heating circulator. The difference is that the secondary circulating pump has a body cast from bronze to eliminate discoloured water and subsequent contamination. Bronze, as we discovered in Chapter 3, is a non-ferrous metal that does not rust. Central heating circulators have a cast iron or steel body that would rust if used on a secondary circulation system. Under no circumstances should a central heating circulator be used.

Secondary circulation installations on unvented hot water storage systems

In most cases, a secondary circulation connection is not fitted on an unvented hot water storage vessel and, unlike open vented hot water storage vessels, it is not possible to install a connection on the vessel itself. Where secondary circulation is required, this must be taken to the cold water feed connection using a swept tee just before the cold feed enters the unit. To safeguard against reverse circulation, a non-return valve or single check valve must be fitted after the circulating pump and just before the swept tee branch. The pump should be fitted on the secondary return, close to the hot water storage vessel.

Secondary circulation installations on open vented hot water storage systems

With secondary circulation on open vented systems, the return pipe runs from the furthest hot tap back to the cylinder, where it enters at about a quarter of the way down. A circulating pump is placed on the return, close to the hot water cylinder, pumping into the vessel. As with all secondary circulation systems, the pump must be made from bronze or stainless steel to ensure that corrosion does not pose a problem. Isolation valves must be installed either side of the pump so that the pump may be replaced or repaired. The system is shown in Figure 6.37.

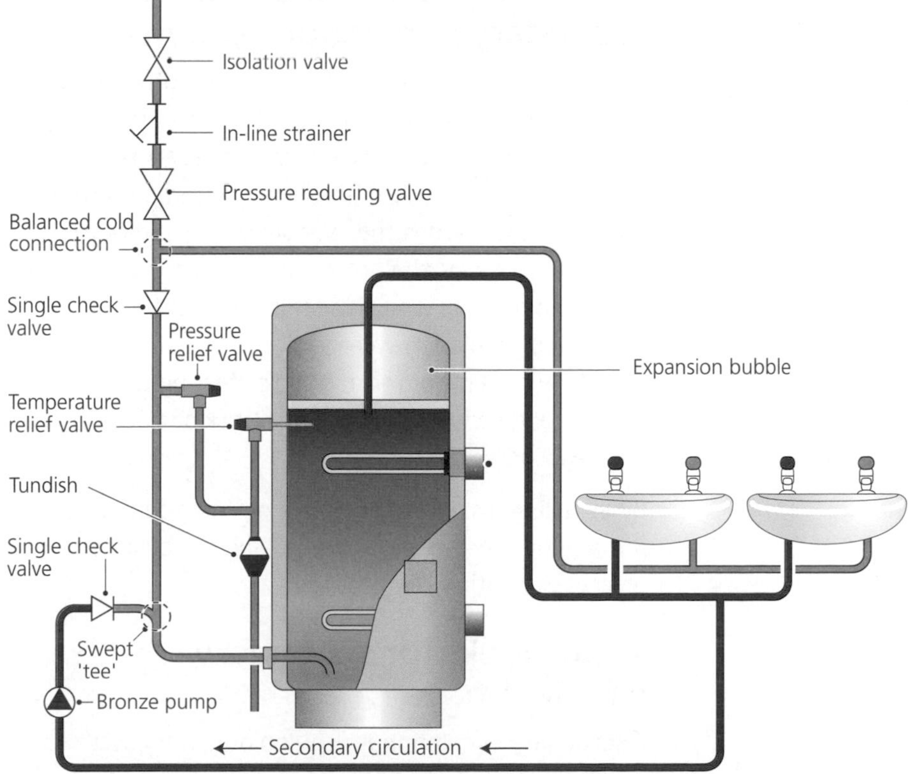

▲ Figure 6.36 Secondary circulation on an unvented hot water storage installation

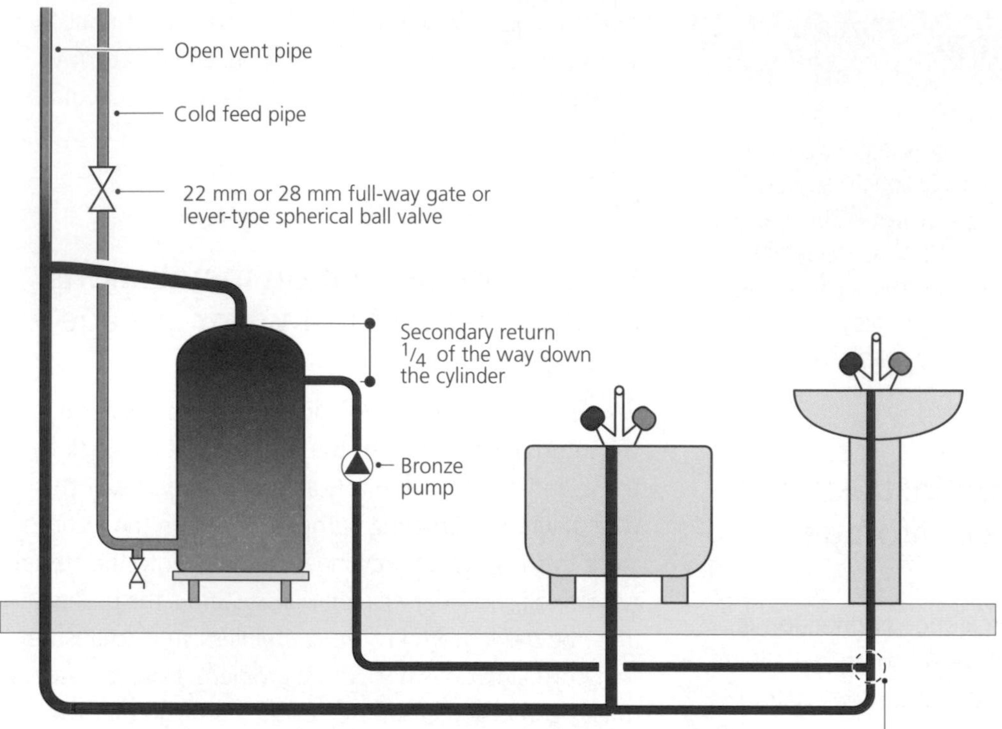

▲ Figure 6.37 Secondary circulation on an open vented hot water storage installation

Some open vented cylinders can be purchased with a secondary return connection already installed on the cylinder. Alternatively, an Essex flange (Figure 6.38) can be used on cylinders where no connection exists.

▲ Figure 6.38 An Essex flange

Preventing reversed circulation in secondary circulation systems

The secondary flow (the hot water draw-off), as we have already seen, should have a temperature of at least 60°C. The secondary return of the secondary circulation circuit should have a return temperature of 50°C when it reaches the cylinder at the end of the circuit. In this way, the hottest part of the cylinder will always be the top, where the hot water is drawn off. If reversed circulation were to occur, the water in the cylinder would never reach the disinfecting temperature of 60°C and so would always be at risk of a *Legionella* outbreak, however remote.

By installing a single check valve on the return, and positioning it between the pump and the cylinder, reverse circulation is prevented.

Time clocks for secondary circulation

If secondary circulation is used on hot water systems, it should be controlled by a time clock so that the circulating pump is not running 24 hours a day. The time clock should be set to operate only during periods of demand, and should be wired in conjunction with pipe thermostats (also known as aquastats) to switch off the pump when the system is up to the correct temperature and circulation is not required and to activate the pump when the water temperature drops.

Insulating secondary circulation pipework

If secondary circulation systems are installed, they should be insulated for the entire length of the system. This is to prevent excessive heat loss through the extended pipework due to the water being circulated by a circulating pump. The insulation should be thick enough to maintain the heat loss below the values shown in Table 6.12.

▼ Table 6.12 Insulation thickness for secondary circulation pipework

Tube/pipe size	Maximum heat loss per metre
15 mm pipe	7.89 w/m
22 mm pipe	9.12 w/m
28 mm pipe	10.07 w/m

Secondary circulation on large open vented hot water storage systems

Figure 6.39 shows a large domestic hot water system with secondary circulation. As can be seen, there are some significant differences from other secondary circulation systems, as discussed below.

- The hot water vessel includes a shunt pump. This is to circulate the water within the cylinder to ensure that the varying temperature (stratification) of the water inside is kept to a minimum, and to ensure an even heat distribution throughout, thereby preventing the growth of *Legionella* bacteria. Stratification is desirable during the day so that the draw-off water is maintained at its hottest for the longest period of time. Because of this, the shunt pump should operate only during periods of low demand, i.e. at night.
- The secondary circulation pump (component **5** on the drawing) is installed on the secondary flow and not the secondary return as with other, smaller systems.
- A non-return valve (component **6** on the drawing) is installed on the secondary flow to ensure that reverse circulation does not occur.
- A cylinder thermostat (component **3** on the drawing) is provided to maintain the temperature within the cylinder at a maximum of 60°C.
- A pipe stat (component **2** on the drawing) installed on the secondary flow maintains the temperature at a minimum of at least 50°C.
- A motorised valve (component **4** on the drawing) is installed on the secondary return close to the hot water storage vessel, to prevent water being drawn from the secondary return when the pump is not operating.
- Lockshield gate valves (components **7** and **8** on the drawing) are provided to balance the system to ensure even circulation throughout the secondary water system.
- The secondary circulation system, shunt pumps and thermostats are controlled through a control box (component **1** on the drawing).

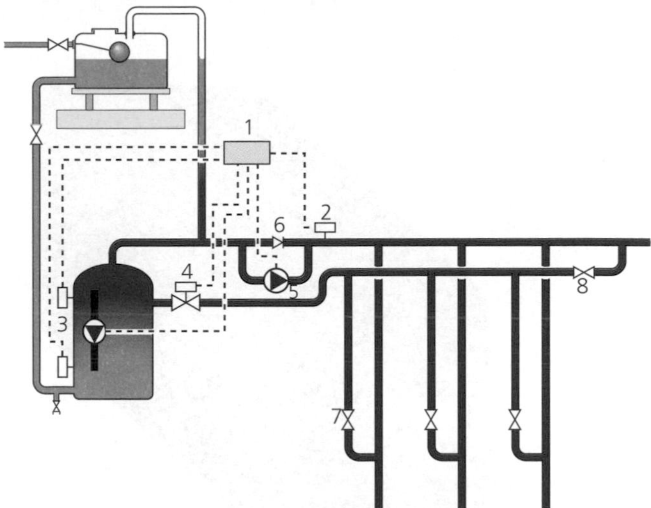

▲ Figure 6.39 Secondary circulation on a large domestic open vented hot water storage installation

Here are some points to remember regarding large centralised hot water systems:

- The pipework should be carefully designed to prevent dead legs as this is a major concern with regard to *Legionella pneumophila*.
- The hot water storage vessel should be capable of being heated to 70°C, again to kill any *Legionella* that may be present.

- There should be easy access for draining, cleaning, inspection and maintenance.
- If a shunt pump is installed, the storage vessel should be insulated on its underside to prevent excessive heat loss.

The use of trace heating instead of secondary circulation

Electric trace heating uses an electric cable that forms a heating element. It is positioned directly in contact with the pipe along the whole length of the pipe. The pipe is then covered in thermal insulation. The heat generated by the element keeps the pipe at a specific temperature.

The operation of the trace heating element should be timed to a period when the hot water system is in most use, i.e. early in the morning and in the evening. If the pipe is well insulated and installed with a timer, the amount of energy usage will be minimal.

By using trace heating, the additional cost of the extra pipework for the secondary return and its associated pump and running costs is removed.

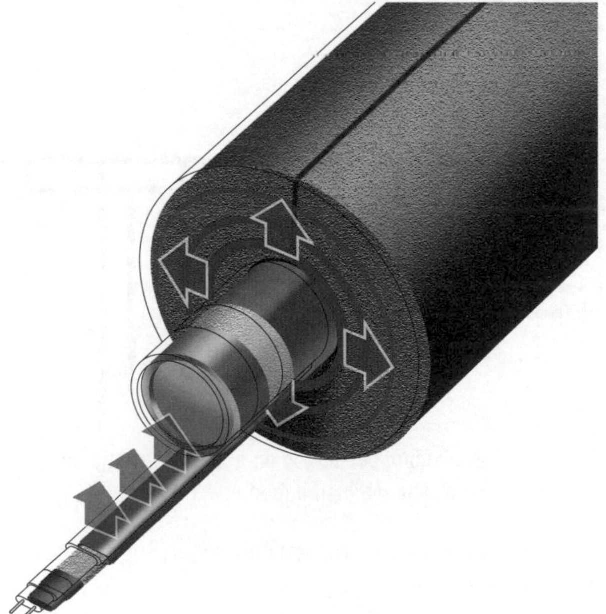

▲ Figure 6.40 Trace heating

INDUSTRY TIP

Trace heating can also be used as frost protection on cold water systems.

System components and controls

You will be able to describe the layout and operational requirements of hot water system components including the location and safety features for unvented/vented hot water systems, with consideration of standard components:

- line strainers
- pressure reducing valves
- single check valves
- expansion devices (vessel or integral to cylinder)
- expansion relief valves
- tundish arrangements
- application of composite valves
- safety features – including expansion and temperature relief pipework, vent pipes
- thermostatic mixing valves (TMV2 and TMV3).

You will be able to explain the working principles of specialist components used in systems including:

- infrared-operated taps
- concussive taps
- combination bath tap and showerhead
- flow-limiting valves
- spray taps
- shower pumps – single and twin impeller
- pressure reducing valves
- shock arrestors/mini expansion vessels.

Note: the eight specialist components listed above were covered in Chapter 5, Cold water systems.

The controls for vented/unvented hot water storage systems fall into two categories:

1. safety
2. functional.

In this part of the chapter, we will look at the various controls and components for unvented hot water storage systems, their function and the position that they occupy within the system.

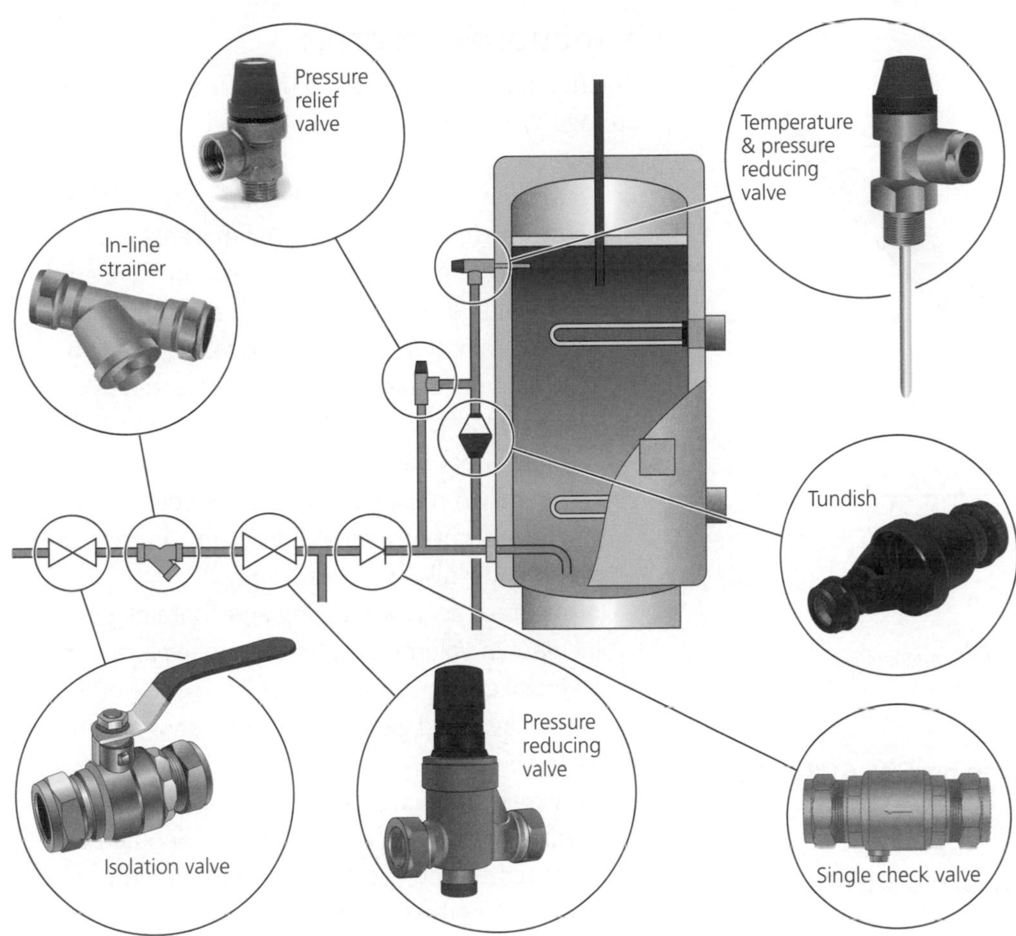

▲ Figure 6.41 The controls on a modern UHWSS

Safety controls

With the water inside the storage vessel at a pressure above atmospheric pressure, the control of the water temperature becomes vitally important. This is because, as the pressure of the water rises, so the boiling point of the water rises. In simple terms, if total temperature control failure were to occur, the water inside the vessel would eventually exceed 100°C, with disastrous consequences. The graph in Figure 6.42 demonstrates the pressure/temperature relationship.

On the graph it can be seen that at the relatively low pressure of 1 bar the boiling point of the water has risen to 120.2°C! If a sudden loss of pressure at the hot water storage vessel were to occur due to vessel fracture, at 120.2°C the entire contents of the cylinder would instantly flash to steam with explosive results, causing structural damage to the property. Calculating how much steam would be produced illustrates the point further.

1 cm^3 of water creates 1600 cm^3 of steam; if the storage vessel contains 200 litres of water and each litre of water contains 1000 cm^3, then the amount of steam produced would be 200 × 1000 × 1600 = 320,000,000 cm^3 of steam!

The Building Regulations Approved Document G3 states that unvented hot water storage systems must have a three-tier level of safety built in to the system. This takes the form of three components that are fitted to the storage vessel. The aim of these components is to ensure that the water within the system never exceeds 100°C.

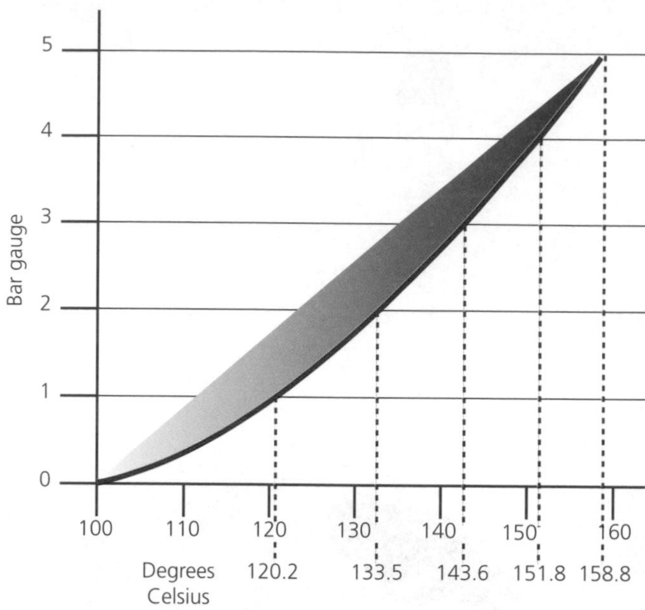

▲ Figure 6.42 Boiling point/pressure relationship

These components are:

1 **Control thermostat (set to 60°C to 65°C):** this can take two forms depending on the type of storage vessel:
   - with direct heated vessels, this is the immersion heater user thermostat
   - with indirectly heated vessels, it is the cylinder thermostat wired to the central heating wiring centre. Indirectly fired systems are also controlled, in part, by the boiler thermostat (82°C maximum setting) and the boiler high limit stat, designed to operate at typically 90°C.

2 **Overheat thermostat (thermal cut-out 90°C maximum but more usually factory set at between 85°C and 89°C):** again, this can take two forms:
   - with direct heated systems, it is incorporated into the immersion heater thermostat
   - with indirectly heated systems, it is a separate component factory wired into the vessel and designed to operate the motorised valve at the primary hot water coil.

3 **Temperature/pressure relief valve (95°C):** a standard component used on most vessels that is designed to discharge water when the temperature exceeds 95°C. Most types have a secondary pressure relief function.

Functional controls

The functional controls of an unvented hot water storage system are designed to protect the water supply.

- To avoid contamination, the storage cylinder or vessel must be of an approved material, such as copper or duplex stainless steel, or have an appropriate lining that will not cause corrosion or contamination of the water contained within it. Where necessary it must be protected by a sacrificial anode.
- A single check valve (often referred to as a non-return valve) must be fitted to the cold water inlet to prevent hot or warm water from entering the water undertaker's mains supply.
- A means of accommodating and containing the increase in volume of water due to expansion must be installed. This can either be by the use of an externally fitted expansion vessel or via an integral air bubble.
- An expansion valve (also known as a pressure relief valve) must be installed, and should be designed to operate should a malfunction occur with either the pressure reducing valve or the means of accommodating the expanded water. The expansion valve must be manufactured to **BS EN 1491:2000 – Building Valves. Expansion Valves. Tests and Requirements**.

The Water Supply (Water Fittings) Regulations also state that:

> Water supply systems shall be capable of being drained down and fitted with an adequate number of servicing valves and drain taps so as to minimise the discharge of water when water fittings are maintained or replaced.

To comply with this requirement, a servicing valve should be fitted on the cold supply close to the storage vessel, but before any other control. The valve may be a full-bore spherical plug, lever action-type isolation valve or a screw-down stop valve. Any drain valves fitted should be manufactured to **BS 2879** and be 'type A' drain valves with a locking nut and an 'O' ring seal on the spindle.

The functional controls of an unvented hot water storage system are listed below. We will look at each one in turn and identify its position within the system.

In-line strainer

The in-line strainer is basically a filter designed to prevent any solid matter within the water from entering and fouling the pressure reducing valve and any other mechanical components sited downstream. In modern storage systems, this is incorporated into the composite valve, which will be discussed later in this section.

Pressure reducing valve

Pressure reducing valves (PRVs) were looked at in detail in Chapter 5, Cold water systems; however, they are of sufficient importance to warrant an explanation here too.

The PRV of an unvented hot water storage system reduces the pressure of the incoming water supply to the operating pressure of the system. In all cases this will be set by the manufacturer and sealed at the factory. The outlet pressure will remain constant even during periods of fluctuating pressures. Should the pressure of the water supply drop below that of the operating pressure of the PRV, it will remain fully open to allow the available pressure to be used.

Replacement internal cartridges are available and easily fitted without changing the valve body should a malfunction occur.

Modern PRVs for unvented hot water storage systems are supplied with a balanced cold connection already fitted.

Single check valve

The single check valve (also known as a non-return valve) is fitted to prevent hot water from back-flowing from the hot water storage vessel, causing possible fluid category 2 contamination of the cold water supply. The single check valve also ensures that the expansion of water when it is heated is taken up within the system's expansion components or expansion bubble. Single check valves are classified as either type EA or EB backflow prevention devices.

KEY POINT

Backflow prevention devices were discussed in detail in Chapter 5, Cold water systems.

In most cases, the check valve will be part of the composite valve, to be discussed later in this section.

Expansion device (vessel or integral to cylinder)

Water expands when heated. Between 4°C and 100°C it will expand by approximately 4 per cent. Therefore, 100 litres of water at 4°C becomes 104 litres at 100°C. It is this expansion of water that must be accommodated in an unvented hot water storage system. This can be achieved in one of two ways:

1. by the use of an externally fitted expansion vessel, or
2. by the use of a purpose-designed internal expansion space or 'expansion bubble'.

Expansion vessels

An expansion vessel is a cylindrical-shaped vessel that is used to accommodate the thermal expansion of water to protect the system from excessive pressures. It is installed as close to the storage vessel as possible and preferably higher. There are two basic types.

The bladder (bag) type expansion vessel

Also known as the bag-type expansion vessel, this is usually made from steel and contains a neoprene rubber bladder to accept the expanded water. At no time does the water come into contact with the steel vessel as it is contained at all times within the bladder.

The inside of the steel vessel is filled with either air or nitrogen to a predetermined pressure. The initial pressure charge from the manufacturer is usually made with nitrogen to negate the corrosive effects on the steel vessel's interior. A Schrader valve is fitted to allow the pressures to be checked and to allow an air 'top-up' if this becomes necessary. Figure 6.43 shows the workings of a bladder-type expansion vessel.

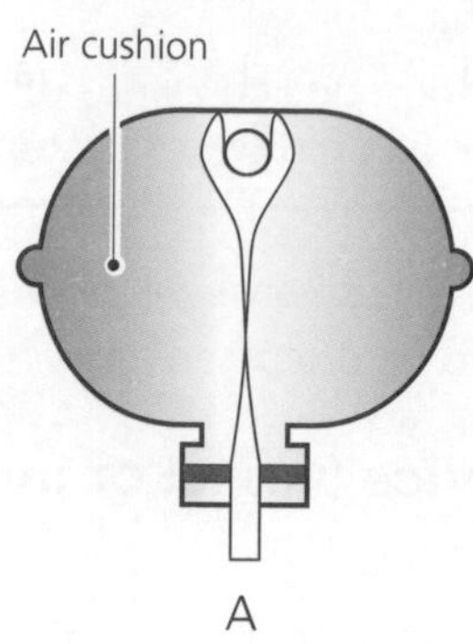

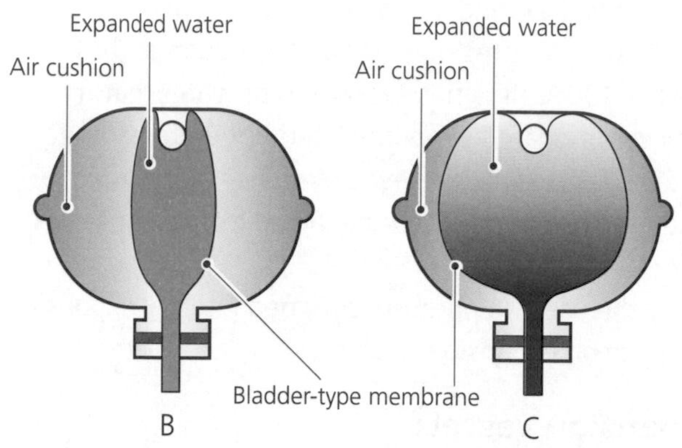

▲ Figure 6.43 Workings of a bladder (bag) type expansion vessel

- Diagram A shows the bladder in its collapsed state; this is because the only pressure is the air/nitrogen charge compressing the empty bladder. There is no water in the bladder.
- Diagram B shows that water under pressure has entered the bladder during the initial cold fill of the storage cylinder, causing the bladder to expand and pressurising the air in relation to the water pressure. The bladder has expanded because the water pressure is greater than the pressure of the air.
- Diagram C shows the bladder fully expanded due to the hot water expansion when the system is heated.

▲ Figure 6.44 Bladder (bag) type expansion vessel

With some bladder expansion vessels, the bladder is replaceable in the event of bladder failure. A flange at the base of the vessel holds the bladder in place. By releasing the air and removing the bolts, the bladder can be withdrawn and replaced.

The diaphragm-type expansion vessel

Diaphragm expansion vessels are used where the water has been deoxygenated by the use of inhibitors or because the water has been repeatedly heated, such as in a sealed central heating system. They must not be used with UHWSS because the water is always oxygenated and comes into direct contact with the steel of the vessel.

They are made in two parts with a neoprene rubber diaphragm separating the water from the air charge. Again, like the bladder-type expansion vessel, a Schrader valve is fitted to allow top-up and testing of the air pressure. Figure 6.45 shows the workings of a diaphragm-type expansion vessel.

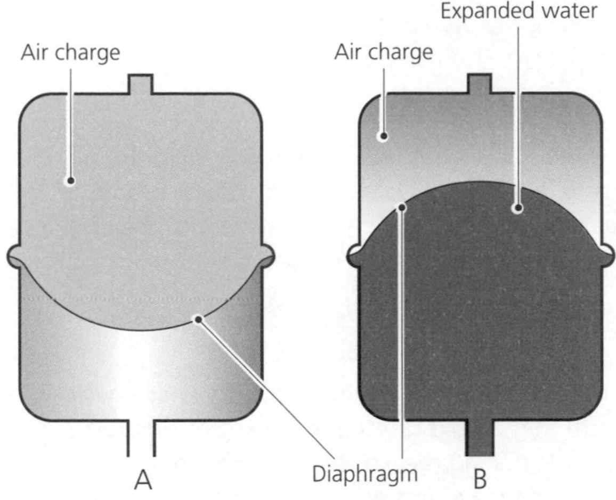

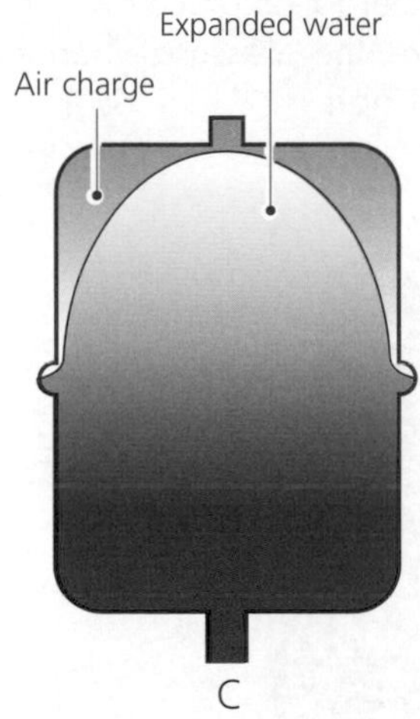

▲ Figure 6.45 The workings of a diaphragm-type expansion vessel

Internal expansion

With internal expansion, an air pocket is formed as the hot water storage vessel is filled. A floating baffle plate provides a barrier between the air and the water so that there is minimum contact between the air and the water in the cylinder. When the water is heated, the expansion pushes the baffle plate upwards in a similar manner to an expansion vessel.

Over a period of time, the air within the air bubble will dissipate as it is leeched into the water. When this happens, expansion cannot take place and the pressure relief valve will start to discharge water. However, this will only occur as the water heats up. Once the cylinder is at its full temperature, the pressure relief valve will close and will only begin to discharge water again when expansion is taking place. Because of this, manufacturers of bubble top units and packages recommend that the cylinder is drained down completely and refilled to recharge the air bubble. This should be done on an annual basis or as and when required.

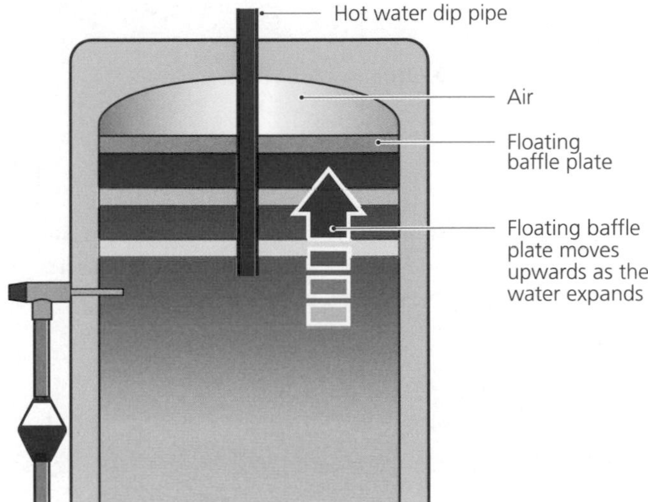

▲ Figure 6.46 Integral air bubble

The scientific principles of expansion vessels

The principle of an expansion vessel is that a gas is compressible but liquids are not. That principle is based upon Boyle's law. In this case, the gas is air or nitrogen and the liquid is water.

IMPROVE YOUR MATHS

Boyle's law states:

> The volume of a gas is inversely proportional to its absolute pressure provided that the temperature remains constant.

In other words, if the volume is halved, the pressure is doubled.

Mathematically, Boyle's law is expressed as $P_1V_1 = P_2V_2$

Where:

P_1 = Initial pressure = 1 bar

V_1 = Initial volume = 20 litres

P_2 = Final pressure = to be found

V_2 = Final volume = 20 litres – 10 litres of expanded water

→

So, to find the pressure in the vessel, the formula must be transposed:

$$P_2 = \frac{P_1 \times V_1}{V_2}$$

Therefore:

$$P_2 = \frac{1\ \text{bar} \times 20\ \text{litres}}{10\ \text{litres}}$$

$$= 2\ \text{bar final cold pressure}$$

If, on the initial cold fill of the system, the vessel required, say, 5 litres of water to be taken in, the air pressure to apply to the vessel can be calculated. We can assume a water pressure of 1 bar.

P_1 = 1 bar

V_1 = 20 litres

V_2 = 20 litres – 5 litres = 15 litres

P_2 = Pressure to be calculated

$$P_2 = \frac{P_1 \times V_1}{V_2} = \frac{1\ \text{bar} \times 20\ \text{litres}}{15\ \text{litres}} = 1.33\ \text{bar}$$

The capacity left in the vessel after the initial fill is 15 litres with a cold fill pressure of 1.33 bar and, if 10 litres of water are to expand inside the vessel, the final pressure of the system will be:

$$P_2 = \frac{P_1 \times V_1}{V_2} = \frac{1.33 \times 15}{15 - 10} = \frac{20}{5} = 4\ \text{bar}$$

The initial pressure of the empty 20-litre vessel was 1 bar. On initial cold fill, 5 litres of water entered the vessel, reducing the capacity to 15 litres. As a result, the air was compressed even more when the expansion of water takes place and, instead of 1.33 bar final pressure, the pressure when the water is heated will be 4 bar.

ACTIVITY

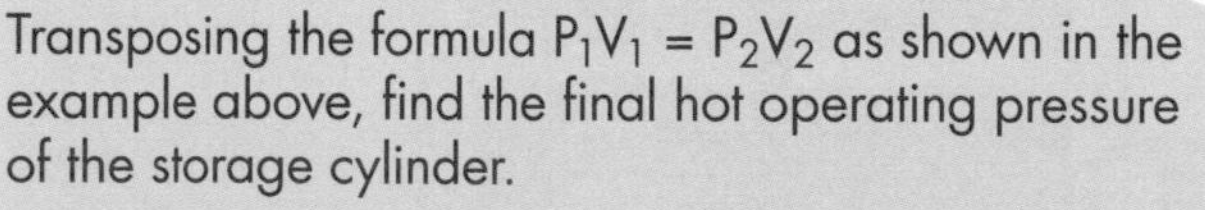

Transposing the formula $P_1V_1 = P_2V_2$ as shown in the example above, find the final hot operating pressure of the storage cylinder.

Where:

P_1 = Initial pressure = 1.5 bar

V_1 = Initial volume = 18 litres

P_2 = Final pressure = to be found

V_2 = Final volume = 18 litres – 9 litres of expanded water

Pressure relief valve

Often referred to as the expansion relief valve, the pressure relief valve is designed to automatically discharge water in the event of excessive mains pressure or malfunction of the expansion device (expansion vessel or air bubble). It is important that no valve is positioned between the pressure relief valve and the storage cylinder.

The pressure at which the pressure relief valve operates is determined by the operating pressure of the storage vessel and the working pressure of the pressure relief valve. The valve is pre-set by the manufacturer and must not be altered.

The pressure relief valve will not prevent the storage vessel from exploding should a temperature fault occur and, as such, is not regarded as a safety control.

▲ Figure 6.47 Pressure relief valve

Tundish arrangements

The tundish is part of the discharge pipework and is supplied with every unvented hot water storage system. It is the link between the D1 and D2 pipework arrangements. It has three main functions:

1 to provide a visual indication that either the pressure relief or temperature relief valves are discharging water due to a malfunction
2 to provide a physical, type A air gap between the discharge pipework and the pressure relief/temperature relief valves

3 to give a means of releasing water through the opening in the tundish in the event of a blockage in the discharge pipework.

The tundish must always be fitted in the upright position in a visible place close to the storage vessel. The tundish will be looked at in more detail when discharge pipework arrangements are discussed later in this section.

Composite valves

These days, it is very rare to see individual controls fitted on an unvented hot water storage system unless it is an early type manufactured in the 1990s. Most manufacturers now prefer to supply composite valves, which incorporate many components into one 'multi-valve'. A typical composite valve will contain:

- a strainer
- a pressure reducing or pressure limiting valve, followed immediately by
- a balanced cold take off, and finally
- a pressure relief valve.

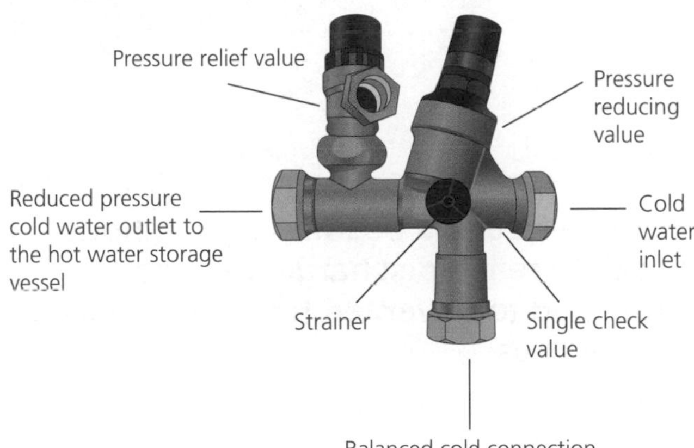

▲ Figure 6.48 A typical composite valve

Some composite valves may also contain an isolation valve. With all controls contained in a single valve, making the connection to an unvented hot water storage vessel is a simple matter of just connecting the cold supply, without the need to ensure that the controls have been fitted in the correct order.

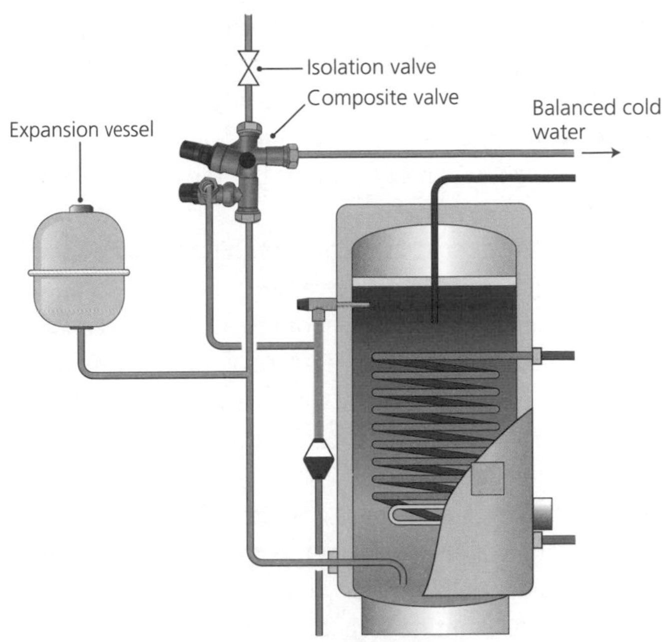

▲ Figure 6.49 Position of a composite valve

Unvented hot water systems: the discharge pipework

The layout features for temperature and expansion relief (discharge) pipework

With unvented hot water systems, there is always the possibility, however undesirable, that the pressure relief and temperature relief valves may discharge water. The discharge pipework is designed specifically to remove the discharged water away from the building safely. It is, therefore, very important that it is installed correctly with the correct size of pipe and that the pipework is made from the correct material, especially since the water discharged may be at near boiling point.

There are three sections to the discharge pipework:

1 D1 pipework arrangement
2 the tundish
3 D2 pipework arrangement.

As we have already established the role of the tundish earlier in the chapter, we will concentrate specifically here on the D1 and D2 sections of the discharge pipework.

To ensure that there is no damage to the property, the discharge pipework should be positioned in a safe but visible position, and should conform to the following.

- The discharge must be via an air break (tundish) positioned within 600 mm of the temperature relief valve.
- The tundish must be located within the same space as the hot water storage vessel.
- It should be made of metal or other material capable of withstanding the temperature of the discharged water. The pipe should be clearly and permanently marked to identify the type of product and its performance standards.*
- The discharge pipe must not exceed the hydraulic resistance of a 9 m straight length of pipe without increasing the pipe size.
- It must fall continuously throughout its entire length with a minimum fall of 1 in 200.
- The D2 pipework from the tundish must be at least one pipe size larger than the D1 pipework.
- The discharge pipe should not connect to a soil discharge pipe unless the pipe material can withstand the high temperatures of discharge water, in which case it should:
  - contain a mechanical seal (such as a Hepworth HepvO valve), not incorporating a water trap, to prevent foul air from venting through the tundish in the event of trap evaporation
  - be a separate branch pipe with no sanitary appliances connected to it
  - where branch pipes are to be installed in plastic pipe, be either polybutylene (PB) to class S of **BS 7291–2:2010** or cross-linked polyethylene (PE-X) to Class S of **BS 7291–3:2010**
  - be marked along the entire length with a warning that no sanitary appliances can be connected to the pipe.
- The D1 pipework must not be smaller than the outlet of the temperature relief valve.
- The D1 discharge from both the pressure relief and temperature relief valves may be joined by a tee piece, provided that all of the points above have been complied with.
- There must be at least 300 mm of vertical pipe from the tundish to any bend in the D2 pipework.

Figure 6.50 illustrates some of the requirements mentioned above.

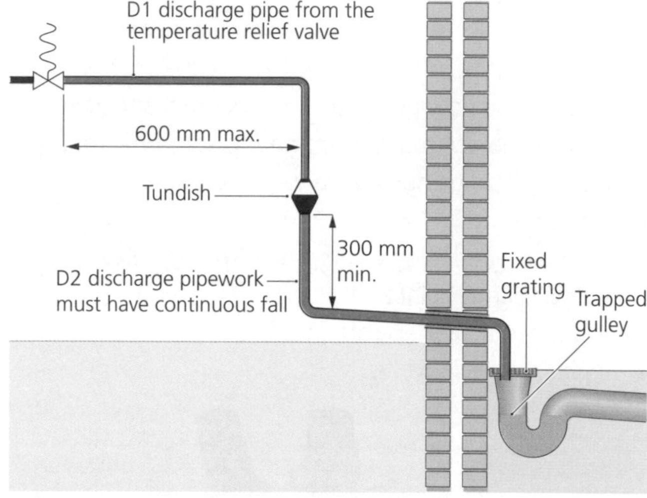

▲ Figure 6.50 The layout of the discharge pipework

The pipe size and positioning methods for safety relief (discharge) pipework connected to unvented hot water cylinder safety valves

As we have already seen, the discharge pipework must not exceed the hydraulic resistance of a 9 m straight length of pipe without increasing the pipe size. Where the discharge pipework exceeds 9 m, the size of the discharge pipe will require calculating, including the resistance of any bends and elbows. Table 6.13 can be used.

* Paragraph 3.9 of Approved Document G3 Guidance specifies metal pipe for the discharge pipework. However, G3 itself states only that hot water discharged from a safety device should be safely conveyed to where it is visible but will not cause a danger to persons in or about the building. Since many types of plastic pipe are now able to withstand the heat of the discharge water, the responsibility for the choice of material rests with the installer, the commissioning engineer and the local Building Control Officer to ensure that G3 is complied with. It is also important that, if plastic pipes are used, the type of plastic is clearly indicated for future reference when inspections and servicing are carried out.

▼ Table 6.13 Discharge pipework D1/D2 pipe sizing chart

Valve outlet size	Maximum size of discharge to tundish (D1)	Maximum size of discharge pipe from tundish (D2)	Maximum resistance allowed, expressed as a length of straight pipe without bends or elbow	Resistance created by each bend or elbow
G ½	15 mm	22 mm 28 mm 35 mm	Up to 9 m Up to 18 m Up to 27 m	0.8 m 1.0 m 1.4 m
G ¾	22 mm	28 mm 35 mm 42 mm	Up to 9 m Up to 18 m Up to 27 m	1.0 m 1.4 m 1.7 m
G 1	28 mm	35 mm 42 mm 54 mm	Up to 9 m Up to 18 m Up to 27 m	1.4 m 1.7 m 2.3 m

IMPROVE YOUR MATHS

Let's look at how Table 6.13 works.

The temperature and pressure relief valves both have ½-inch BSP outlets. Therefore, the D1 pipework, as can be seen from the table, can be installed in 15 mm tube. The discharge pipe run is 6 m long to the final termination and there are six elbows installed in the run of pipe.

Using the first row in the table, the first option has to be 22 mm because the D2 pipework must be at least one pipe size larger than the D1 pipework. The maximum length of 22 mm pipe is 9 m but there are six elbows in the run and each of these has a resistance of 0.8 m.

$$6 \times 0.8 = 4.8 \text{ m}$$

If we add the original length of 6 m, we get:

$$4.8 + 6 = 10.8 \text{ m}$$

The maximum length of 22 mm discharge pipe, as we have already seen, is 9 m so, at 10.8 m, 22 mm pipe is not large enough for the discharge pipe run. Another pipe size will have to be chosen.

Looking at 28 mm, we see that the maximum run of pipe is 18 m but the 28 mm elbows now have a resistance of 1 m and there are six of them. Therefore:

$$6 \times 1 = 6 \text{ m}$$

Add this to the original length of 6 m:

$$6 + 6 = 12 \text{ m}$$

In this case, the discharge pipework is well within the 18 m limit and so 28 mm discharge pipework can be installed.

ACTIVITY

A customer wishes to have an unvented hot water storage system installed. You have been asked by the site supervisor to size the discharge pipework. The temperature and pressure relief valves both have ½-inch BSP outlets. Therefore, the D1 pipework can be installed in 15 mm tube. The discharge pipe run is 9 m long to the final termination and there are 5 elbows installed in the run of pipe.

What size of discharge pipework should be installed?

Correct termination of the discharge pipework

A risk assessment is likely to be needed where any termination point for the discharge pipework is to be considered. This will determine whether any special requirements are needed in relation to the termination point and its access. Points to be considered here are:

- areas where the public may be close by or to which they have access
- areas where children are likely to play or to which they have access

- areas where the discharge may cause a nuisance or a danger
- termination at height
- the provision for warning notices in vulnerable areas.

The Building Regulations Approved Document G3 states that the discharge pipe (D2) from the tundish must terminate in a safe place, with no risk to any person in the discharge vicinity. Acceptable discharge arrangements are:

(a) To trapped gully with pipe below gully grate but above the water seal.

(b) Downward discharges at low level up to a maximum 100 mm above external surfaces, such as car parks, hard standings and grassed areas, are acceptable provided a wire cage or similar guard is provided to prevent contact, whilst maintaining visibility.

(c) Discharges at high level, onto a flat metal roof or other material capable of withstanding the temperature of the water may be used provided that any plastic guttering system is at least 3 m away from the point of discharge to prevent damage to the guttering.

(d) Discharges at high level, into a metal hopper and metal downpipe may be used provided that the end of the discharge pipe is clearly visible. The number of discharge pipes terminating in a single metal hopper should be limited to 6 to ensure that the faulty system is traceable.

(e) Discharge pipes that turn back on themselves and terminate against a wall or other vertical surface should have a gap of at least 1 pipe diameter between the discharge pipe and the wall surface.

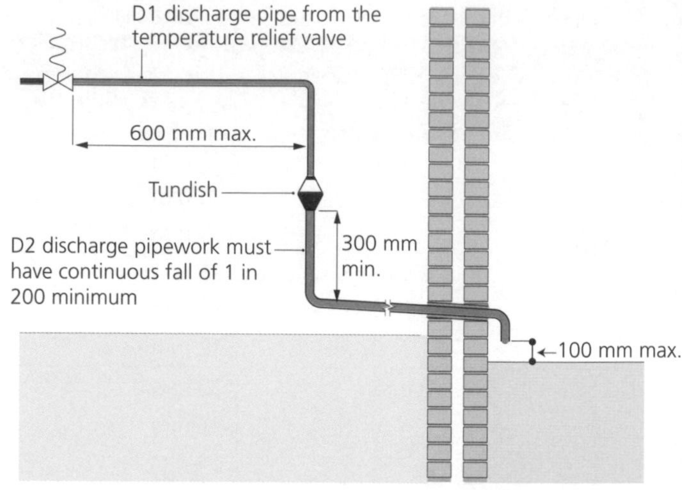

▲ Figure 6.51 The low-level termination of discharge pipework 1

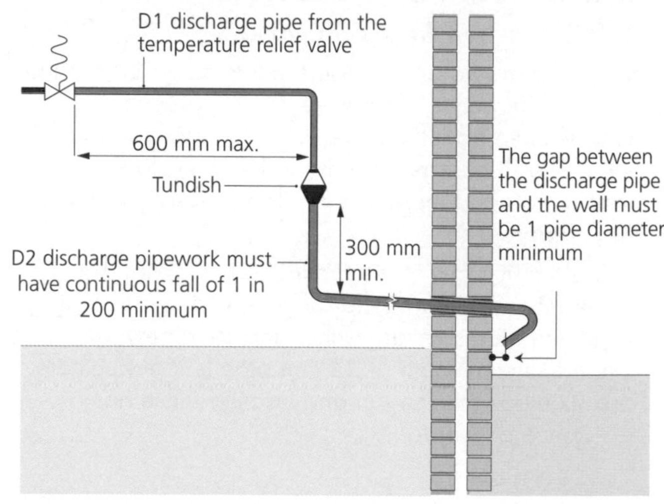

▲ Figure 6.52 The low-level termination of discharge pipework 2

Note: The discharge may consist of high temperature water and steam. Asphalt, roofing felt and other non-metallic rainwater goods may be damaged by very high temperature hot water discharges.

Termination of the discharge pipework where the storage vessel is sited below ground level

When storage vessels are sited below ground, such as in a cellar, the removal of the discharge becomes a problem because it cannot be discharged safely away from the building. However, with the approval of the local authority and the vessel manufacturer it may be possible to pump the discharge to a suitable external point. A constant temperature of 95°C should be allowed for when designing a suitable pumping arrangement. The pump should include a suitable switching arrangement installed in conjunction with a discharge collection vessel made from a material resistant to high temperature water. The vessel should be carefully sized in line with the predicted discharge rate, and should include an audible alarm to indicate discharge from either of the pressure or temperature relief valves is taking place.

3 SYSTEM SAFETY AND EFFICIENCY

INDUSTRY TIP

Frost protection is covered in detail in Chapter 5, Cold water systems

Hot water, by its nature, can be dangerous if:

- the temperature of the water is too high
- the delivery system does not contain a vent pipe to keep the system at atmospheric pressure
- there are no means to accommodate the expansion of the water due to the water being heated
- there are no means to relieve excessive pressure and/or temperature.

Because of this, various safety features must be built in to hot water systems to prevent the water from:

- exceeding 60°C at the point of use
- exceeding 100°C at the point of storage
- over-pressurising the water beyond safe limits.

In this part of the chapter, we will look at the safety features that prevent excessive pressure and water temperature.

Open vent pipe

Open vented systems contain a vent pipe, which remains open to the atmosphere, ensuring that the hot water cannot exceed 100°C. The vent pipe acts as a safety relief outlet should the system become overheated.

Water at atmospheric pressure boils at 100°C. However, once the water is pressurised, the boiling point temperature rises. The higher the pressure, the higher the boiling point. So, a system without a vent pipe to maintain 100°C maximum is classed as an unvented system and this means that the water will exceed this – often, as we will see later, with catastrophic consequences.

The vent pipe will also assist with the expansion of water by allowing the expanded water to rise within the pipe. The vent pipe must be sited over the cold feed cistern in the roof space.

Temperature relief valve

In an unvented hot water storage system, where there is a risk of the water temperature exceeding 100°C, a temperature relief valve must be fitted as part of the three-tier level of safety. To evacuate the hot water away from the building in the event of the water reaching 95°C, the temperature relief valve must be connected to the discharge pipework.

The discharge pipework

This pipework is connected to both the temperature and pressure relief valves via a tundish. It is designed to evacuate any discharged water quickly and safely away from the building to a drain.

Thermostatic mixing valve (TMV2 and TMV3)

The object of any hot water storage system is to store water at the relatively high temperature of 60°C to ensure that it is free from any bacteria, to distribute the water at 55°C and yet to deliver the water at the hot water outlets at the relatively low temperature of 35°C to 46°C, to ensure the safety of the end user. The most efficient way to do this is by the use of thermostatic mixing valves (TMVs).

TMVs (sometimes known as a thermostatic blending valves) are designed to mix hot and cold water to a predetermined temperature, to ensure that the water is delivered to the outlet at a temperature that will not cause injury but is hot enough to facilitate good personal hygiene. There are three methods of installing TMVs, as described below.

Single valve installations

This is probably the most common of all TMV installations. The maximum pipe length to a single appliance is 2 m from the TMV to the outlet. Back-to-back installations are acceptable from a single valve provided that the use of one appliance does not affect the other, and that both appliances have a similar flow rate requirement, e.g. two washbasins. Typical installations are those listed below.

- **Baths:** it is now a requirement of Building Regulations Approved Document G3 that all bath installations in new and refurbished properties incorporate the use of a TMV. This would normally be set to a temperature of between 41°C and 44°C, depending on personal comfort levels. Temperatures above this can be used only in exceptional circumstances.
- **Showers:** these installations usually require a temperature of not more than 43°C. In residential care homes and other medical facilities, a temperature of not more than 41°C should be used according to NHS guidelines.
- **Washbasins:** careful consideration must be applied to washbasin installations because this is probably the only appliance used in domestic dwellings where the user puts their hands directly in the running water without waiting for the water to get hot. When the water reaches maximum temperature, scalding can occur. Therefore, typical temperatures between 38°C and 41°C can be used, depending upon the application. Again, NHS guidelines recommend a temperature of no more than 41°C.
- **Bidets:** a maximum of 38°C should be used with bidet installations.
- **Kitchen sinks:** this is probably the area where the user is most at risk. The need to ensure that bacteria and germs are killed, and that grease is thoroughly removed, dictates that a water temperature of between 46°C and 48°C is used. However, as the kitchen is an area with no published recommendations on hot water temperature, a safe temperature similar to that of washbasins should be considered, to lessen the risk of scalding unless notices warning of very hot water are used.

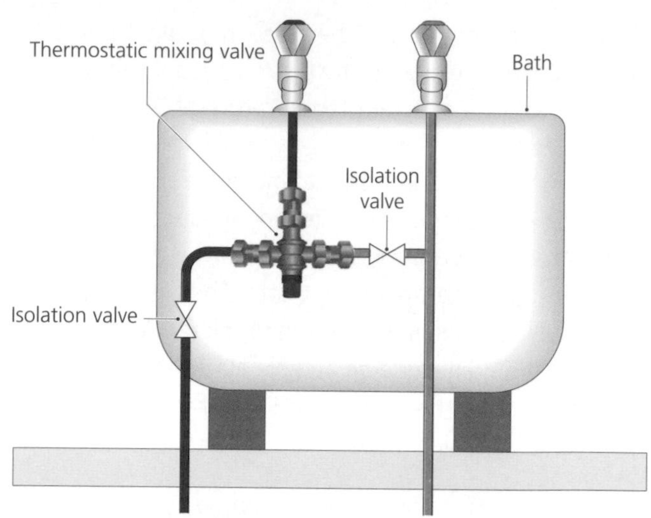

▲ Figure 6.53 A single thermostatic mixing valve installation

Group mixing

Installations where a number of appliances of a similar type are fed from a single TMV are allowed in certain installations. However, installations of this type are not recommended where the occupants are deemed to be at high risk, such as in nursing homes. If a group installation is to be considered, then the points listed below should be followed.

- The operation of any one appliance should not affect others on the run.
- When one TMV is used with a number of similar outlets, the length of the pipework from the valve to the outlets should be kept as short as possible so that the mixed water reaches the furthest tap within 30 seconds.
- With group shower installations, it is not unusual to see pipe runs in excess of 10 m. Pipework runs of this length carry an unacceptable *Legionella* risk. These situations can be dealt with by:
  - careful monitoring of the water at the showerheads and appropriate treatment should *Legionella* be detected
  - regular very hot water disinfection when the system is not in use.

Typical group installations are those listed below.

- **Group showers:** with the correct-sized TMV, a number of shower outlets may be served at a temperature of between 38°C and 40°C. For safety reasons, the temperature must not exceed 43°C.
- **Washbasins:** rows of washbasins may be served from a single TMV. Temperatures of between 38°C

and 40°C are typical, but should not exceed 43°C for safety reasons.

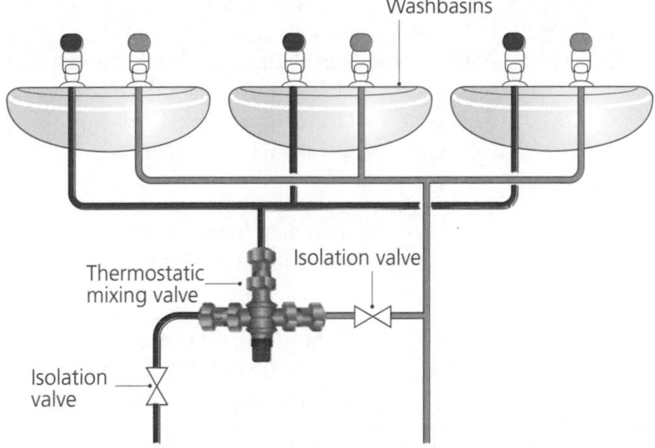

▲ Figure 6.54 A group thermostatic mixing valve installation

Centralised mixing

Centralised mixing is very similar to group mixing but occurs when there are groups of different hot water appliances to be served from a single TMV. The recommendations listed below should be followed.

- If the mixed water is recirculated within the *Legionella* growth temperature range, then anti-*Legionella* precautions similar to those recommended for group mixing will need to be implemented.
- If the mixed water is recirculated at about *Legionella* growth temperature regimes, then the recommendations for single TMV installations are appropriate.
- The operation of any one outlet should not affect other outlets.

The types of thermostatic mixing valve

Thermostatic mixing valves are certificated under a third-party certification scheme set up and administrated by BuildCert. Under the BuildCert scheme, thermostatic mixing valves are certificated and approved for use depending on their application. They are divided into two groups: TMV2 and TMV3.

TMV2

Approved Document G – Sanitation, hot water safety and water efficiency of the Building Regulations in England and Wales requires that the hot water outlet to a bath should not exceed 48°C. It also states that valves conforming to **BS EN 1111** or **BS EN 1287** are suitable for this purpose. Similar requirements exist in Scotland.

TMV2 approval is for the domestic thermostatic installations and uses **BS EN 1111** and **BS EN 1287** as a basis for the thermostatic valves' performance testing.

TMV3

These valves are manufactured and tested for healthcare and commercial thermostatic installations, and use the NHS specification D08 as a basis for the thermostatic valves' performance testing.

Table 6.14 presents a guide to the selection of TMVs for a given application.

▼ Table 6.14 Guide to TMVs for different applications

Environment	Appliance	Is a TMV required by legislation or authoritative guidance?	Is a TMV recommended by legislation or authoritative guidance?	Is a TMV suggested best practice?	Valve type	Reference documents
Private dwelling	Bath Basin Shower Bidet	Yes Yes		Yes Yes	TMV2 TMV2 TMV2 TMV2	Part G – building regulations
Housing Association dwelling	Bath Basin Shower Bidet	Yes Yes		Yes Yes	TMV2 TMV2 TMV2 TMV2	Part G – building regulations

→

Environment	Appliance	Is a TMV required by legislation or authoritative guidance?	Is a TMV recommended by legislation or authoritative guidance?	Is a TMV suggested best practice?	Valve type	Reference documents
Housing Association dwelling for the elderly	Bath Basin Shower Bidet	Yes Yes Yes Yes			TMV2 TMV2 TMV2 TMV2	Housing Corp Standard (1.2.1.58 and 1.2.1.59) & Part G – building regulations
Hotel	Bath Basin Shower			Yes Yes Yes	TMV2 TMV2 TMV2	Guidance to the Water Regulations (G18.5)
NHS nursing home	Bath Basin Shower		Yes Yes Yes		TMV3 TMV3 TMV3	HTM 04 01 & HTM 64. Care Standards Act 2000, Care Homes Regulation 2001, D08
Private nursing home	Bath Basin Shower		Yes Yes Yes		TMV3 TMV3 TMV3	Guidance to the Water Regulations (G18.6), Care Standards Act 2000, Care Homes Regulation 2001, HSE Care Homes Guidance
Young persons' care home	Bath Basin Shower	Yes Yes Yes			TMV3 TMV3 TMV3	DoH National Minimum Standards Children's Homes Regulations, Care Standards Act 2000, Care Homes Regulations 2001, HSE Care Homes Guidance
Schools, including nursery	Basin Shower Bath	Yes Yes, but 43°C maximum	Yes		TMV2 TMV2 TMV2	Building Bulletin 87, 2nd edition, The School Premises Regulations/National Minimum Care Standards Section 25.8
Schools for the severely disabled including nursery	Basin Shower Bath	Yes Yes, but 43°C maximum	Yes		TMV3 TMV3 TMV3	Building Bulletin 87, 2nd edition, The School Premises Regulations, if residential, Care Standards Act
NHS hospital	Bath Basin Shower	Yes Yes Yes			TMV3 TMV3 TMV3	HTM 04 01 & HTM 64, D08
Private hospital	Bath Basin Shower		Yes Yes Yes		TMV3 TMV3 TMV3	Guidance to the Water Regulations (G18.6)

4 PREPARE FOR THE INSTALLATION OF SYSTEMS AND COMPONENTS

Preparing for installation deals with the preparatory work required before an installation can commence:

- consulting drawings and specifications
- positioning components in line with the regulations, manufacturers' instructions and the customer's wishes
- marking out pipework runs
- making a fittings list.

This subject was dealt with in detail in Chapter 5, Cold water systems.

5 INSTALL AND TEST SYSTEMS AND COMPONENTS

Here, we will look at the general requirements for hot water systems within a dwelling, including:

- hot water pipework installation
- installing storage cylinders and cisterns
- temperature control
- the use of thermostatic blending valves
- insulation of pipework
- expansion of hot water pipework.

The installation of hot water pipework to **BS EN 806**

The installation of hot water pipework is covered in **BS EN 806**. Materials used are usually copper tubes to **BS EN 1057** and polybutylene pipes and fittings as these are the only materials that do not cause contamination of the water and can withstand the temperatures associated with hot water distribution pipework. The pipework should be capable of withstanding at least 1.5 times the normal operating pressure of the system and sustained temperatures of 95°C, with occasional temperature increases up to 100°C to allow for any malfunctions of any hot water heating appliances that may occur. All systems must be capable of accommodating thermal expansion and movement within the pipework. Care should be taken when pressure testing open vented cylinders to ensure that the maximum pressure that the cylinder can withstand is not exceeded. If necessary, the cylinder should be disconnected and the pipework capped before testing commences.

The installation methods for hot water systems are very similar to those for cold water installations. Care should be taken when installing hot and cold water pipework side by side so that any cold water installation is not adversely affected by the hot water pipework.

Installation techniques, such as installing pipework below timber floors, solid floors, within walls, marking out, cabling plastic pipework and so on, are discussed at length in Chapter 5, Cold water systems.

ACTIVITY

To refresh yourself as to the tools, materials and installation requirements of pipework within dwellings, check out Chapter 2, Common processes and techniques, and Chapter 5, Cold water systems.

Installing storage cylinders and cisterns

Where the storage of large amounts of water is required, such as in hot water storage cylinders and cold water storage cisterns, care must be taken to ensure that the substrate, where the component is to be installed, can withstand the weight of the stored water. Water is heavy! Every litre of water has a mass of 1 kg and, wherever possible, stored water should be positioned over load-bearing structures and walls. Where this is not possible, a suitable platform must be built that distributes the weight evenly to the entire structure or directs the weight to a load-bearing part of the building. This was discussed briefly in Chapter 5, Cold water systems.

Temperature control of hot water systems

According to **BS EN 806**, hot water systems must not be allowed to exceed 100°C at any time. A maximum normal operating temperature of 60°C is required to kill off *Legionella* bacteria. There are several methods by which we can maintain and control the temperature of hot water systems and prevent it from exceeding the maximum temperature specified. A thermostat should be installed and set to the temperature required. A second thermostat, called a high-limit thermostat, operates should the maximum temperature be exceeded. This is known as a second-tier level of temperature control.

- **Immersion heaters that have a re-settable double thermostat:** one thermostat can be set between 50°C and 70°C, the other is a re-settable high-limit thermostat designed to switch off the power to the unit when the maximum temperature is exceeded. It can be reset manually.
- **Immersion heaters with a non-resettable double thermostat:** one thermostat can be set between 50°C and 70°C, the other is a high-limit thermostat designed to permanently switch off the power to

the unit until the immersion heater is replaced and the fault rectified.

- **Open vented double feed indirect cylinders with gravity or pumped primary circulation:** must be fitted with a minimum of a cylinder thermostat and a motorised zone valve, which closes when the water in the cylinder reaches a pre-set level.
- **Open vented cylinders with no high-limit thermostat:** can be fitted with a temperature relief valve that opens automatically at a specified temperature to discharge water via a tundish and discharge pipework safely to outside the property.

The use of thermostatic mixing valves

As we have already seen, the maximum temperature of hot water in a dwelling should not exceed 60°C but this is far too hot for bathing and showering. Water with a temperature as low as 51.66°C can cause serious burns to a child if it is exposed to the skin for two minutes or more.

In April 2013, new legislation under Building Regulation Document G required that all new-build properties and renovations have temperature control to baths not exceeding 48°C and all hot water storage cylinders where the stored water may exceed 80°C (usually solid fuel-heated cylinders). All properties to which the public have access, such as schools, hospitals, nursing homes and so on, under the Care Standards Act 2000, require that the temperature of water delivered to all hot outlets, except where food preparation is carried out, be limited to 43°C. This is done by the use of thermostatic mixing valves for appliances and in-line blending valves for storage cylinders.

A thermostatic mixing valve mixes hot and cold water together and supplies it to an appliance at exactly the correct temperature. They use a temperature-sensitive element, usually a wax cartridge, that expands and contracts to maintain a specific temperature based on the temperatures of the hot and cold water entering the valve. The length of pipe from the mixing valve to the taps should be kept as short as possible.

The insulation of hot water pipework

When installing new hot water installations in domestic properties, pipes should be wrapped with thermal insulation that complies with the Domestic Heating Compliance Guide. There are four main considerations:

1. Primary circulation pipes for heating and hot water circuits should be insulated wherever they pass outside the heated living space, such as below ventilated suspended timber floors and unheated roof spaces. This is for protection against freezing.
2. Primary circulation pipes for domestic hot water circuits should be insulated throughout their entire length, except where they pass through floorboards, joists and other structural obstructions.
3. All pipes connected to hot water vessels, including the vent pipe, should be insulated for at least 1 m from their points of connection to the cylinder, or at least up to the point where they become concealed.
4. If secondary circulation, such as a pumped circuit feeding bath and basin taps in a large property, is installed, all pipes fed with hot water should be insulated to prevent excessive heat loss through the secondary circulation circuit.

Expansion of hot water pipework

When the pipework of the hot water system is filled with hot water, the heated pipework will expand. As the pipework cools down, it will contract. This expansion and contraction must be accommodated for during the installation process or noise within the installation will result. Pipes that pass through walls and floors where not enough room has been left for expansion will 'tick' and 'creak' as the expansion and contraction takes place.

The rate of expansion will depend upon the material the pipe is made from. It is known as the coefficient of linear expansion. Generally, pipework made from plastic materials tends to expand more than that made from copper. The coefficients of linear expansion for polybutylene and copper are as follows:

- the coefficient of linear expansion of plastic pipe is 0.00018 per metre per °C
- the coefficient of linear expansion of copper pipe is 0.000016 per metre per °C.

This means that, for every degree rise in temperature, polybutylene pipe will expand 0.00018 m in every metre and copper will expand 0.000016 m in every metre.

ACTIVITY

To calculate the amount of expansion that takes place on a given length of pipe:

Length of pipe (m) × coefficient of linear expansion × temperature rise

What is the expansion on a 15 mm copper pipe 6 m in length, when the pipe is heated from 10°C to 60°C?

$6 \times 50 \times 0.000016 = 0.0048$ m or 4.8 mm

Now attempt these examples:

1 What is the expansion on a 15 mm polybutylene pipe 6 m in length, when the pipe is heated from 10°C to 60°C?
2 What is the expansion on a 15 mm copper pipe 20 m in length, when the pipe is heated from 15°C to 50°C?
3 What is the expansion on a 15 mm copper pipe 30 m in length, when the pipe is heated from 12°C to 58°C?

Installation of shower mixing valves and shower boosting pumps

In Chapter 5, Cold water systems, we looked at shower mixing valves and the various types of shower boosting pumps. In the next section of this chapter we will take this a step further and see how we install these appliances within hot water systems.

As we have already seen, there are a number of different shower valves available, ranging from bath/shower mixer taps and simple shower mixing valves to thermostatic and pressure balancing shower valves. The method of installation is, in most cases, the same for each type of valve, with the requirement that equal pressure and flow rate exist on both the hot water and cold water installations. There are five methods of installation:

1 simple installations from a storage cistern in roof space supplying water to both hot and cold water systems, thus ensuring equal pressures across both systems
2 installations that include an inlet, twin impeller shower-boosting pump (often called a 'power shower')
3 installations that include a single impeller outlet pump
4 installations that use mains cold and mains-fed hot water systems
5 installations that use supplies where there is an imbalance in supply pressures, such as those systems that use a combination boiler/instantaneous hot water heater for the hot water supply.

We will look at each of these installations in turn.

Installation of shower mixing valves using cistern-fed supplies

Shower mixing valves fed from a storage cistern require equal pressures on both the hot and cold supplies to maintain the correct mixing ratio of hot and cold water. The safest type of valve to use is the thermostatic type, which maintains a constant temperature irrespective of the temperature of the incoming hot and cold supplies to the valve. Ordinary mixing valves also work well with cistern-fed supplies. It has to be remembered, though, that because ordinary mixing valves are not thermostatically controlled, the water will eventually become cooler the longer the shower is used. This is because of **stratification** within the cylinder.

KEY TERM

Stratification: in a hot water storage cylinder, water forms in layers of temperature from the top of the cylinder, where the water is at its hottest, to the base where it is at its coolest. Stratification is necessary if the cylinder is to perform to its maximum efficiency and manufacturers will purposely design storage vessels and cylinders with stratification in mind. Designers will generally design:

- a vessel that is cylindrical in shape
- a vessel that is designed to be installed upright rather than horizontal
- a vessel with the cold feed entering the cylinder horizontally.

To create enough pressure to give a reasonable shower, there has to be a minimum of 1 m from the bottom of the cistern to the showerhead at its highest position.

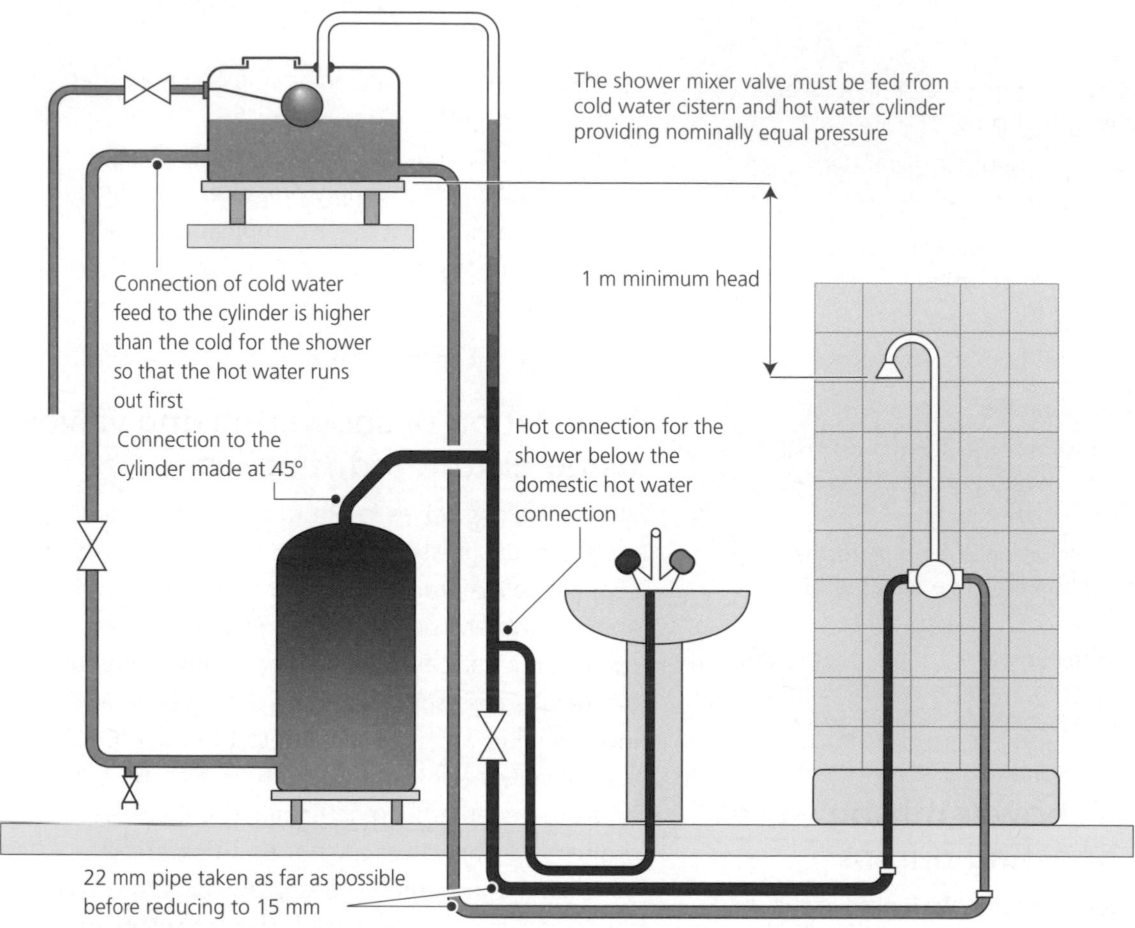
The shower mixer valve must be fed from cold water cistern and hot water cylinder providing nominally equal pressure
1 m minimum head
Connection of cold water feed to the cylinder is higher than the cold for the shower so that the hot water runs out first
Connection to the cylinder made at 45°
Hot connection for the shower below the domestic hot water connection
22 mm pipe taken as far as possible before reducing to 15 mm

▲ Figure 6.55 Gravity-fed shower installation

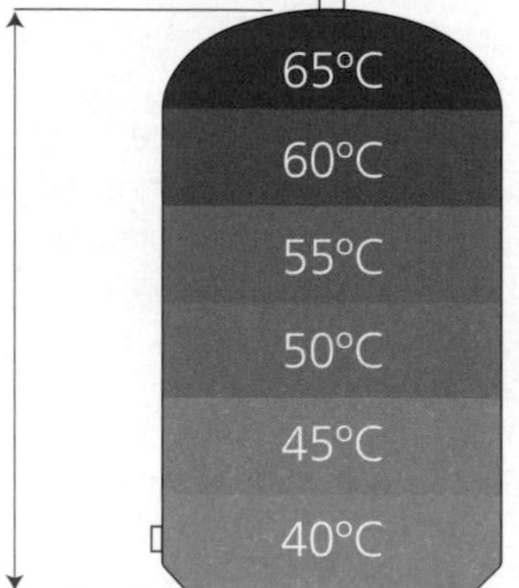
Hottest water at a max. temperature of 65°C is at the top of the cylinder
65°C
60°C
55°C
50°C
45°C
40°C
Hottest water at a max. temperature of 40°C is at the bottom of the cylinder

▲ Figure 6.56 Stratification

Installation of shower mixing valves using cistern-fed supplies and a booster pump: the 'power shower'

There are two systems that use a shower booster pump.

Systems that use a twin impeller pump on the inlet to the mixer valve

The pump increases the pressure of the hot and cold water supplies to the mixer valve independently. The water is then mixed to the correct temperature in the valve before flowing to the showerhead.

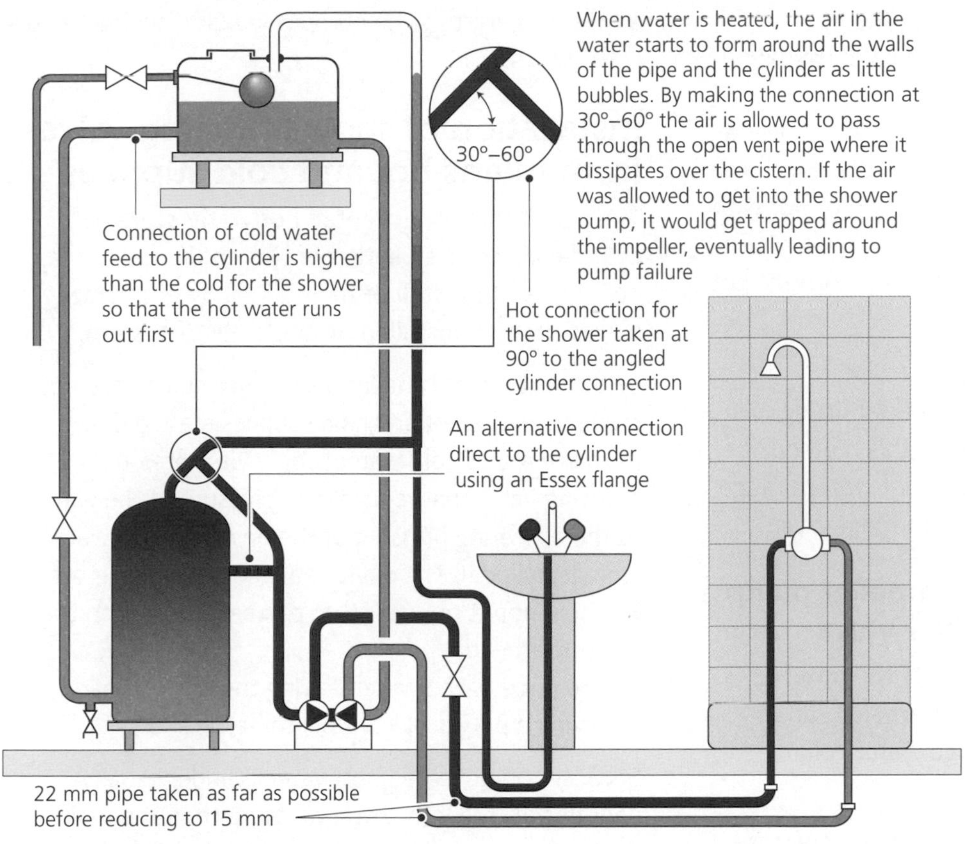
When water is heated, the air in the water starts to form around the walls of the pipe and the cylinder as little bubbles. By making the connection at 30°–60° the air is allowed to pass through the open vent pipe where it dissipates over the cistern. If the air was allowed to get into the shower pump, it would get trapped around the impeller, eventually leading to pump failure
30°–60°
Connection of cold water feed to the cylinder is higher than the cold for the shower so that the hot water runs out first
Hot connection for the shower taken at 90° to the angled cylinder connection
An alternative connection direct to the cylinder using an Essex flange
22 mm pipe taken as far as possible before reducing to 15 mm

▲ Figure 6.57 Pump-assisted shower installation with twin impeller, inlet shower booster pump

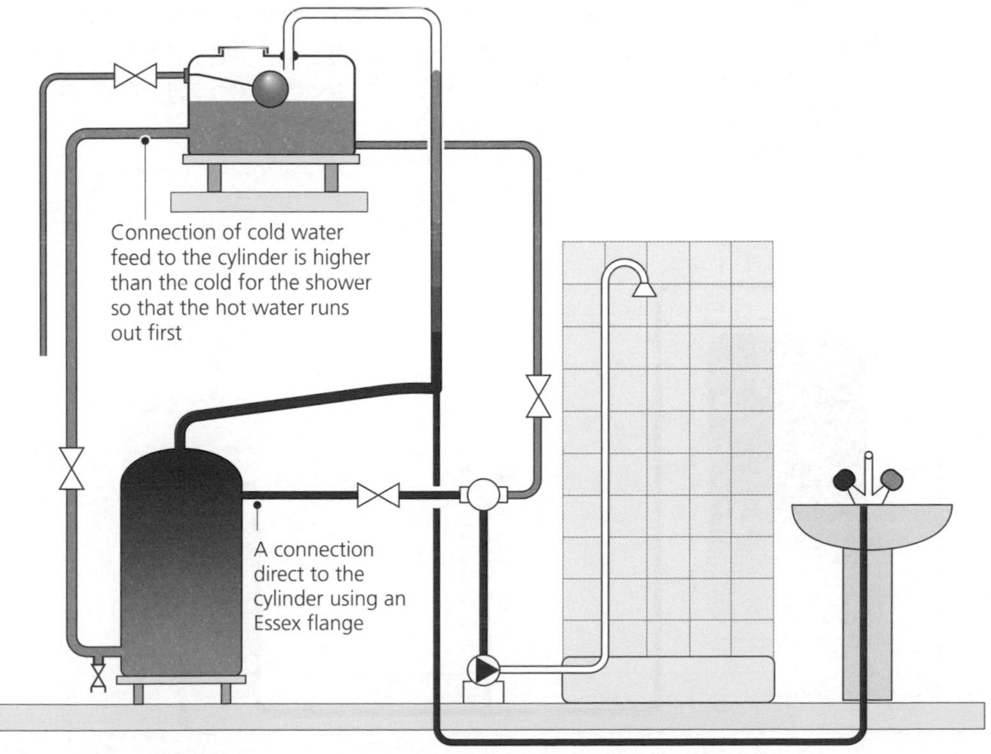
Connection of cold water feed to the cylinder is higher than the cold for the shower so that the hot water runs out first
A connection direct to the cylinder using an Essex flange

▲ Figure 6.58 Pump-assisted shower installation with single impeller, outlet shower booster pump

Care must be exercised when making the hot connection to the cylinder. There are two ways in which this can be done. The first method involves installing the hot water draw-off from the cylinder at an angle of between 30° and 60°, with the hot shower pump connection being made at an angle of 90° with a tee piece (see Figure 6.57). This allows any air in the system to filter up to the vent and away from the hot shower pump inlet.

The second method involves making a direct connection to the cylinder using a special fitting called an Essex flange (see page 324). With this method, the hot water is taken directly from the hot water storage vessel, avoiding any air problems that may occur.

Systems that use a single impeller pump off the outlet from the mixer valve

These boost the water after it has left the mixer valve. They are usually used with concealed shower valves and fixed 'deluge'-type, large water volume showerheads.

In both of these installations, the pump increases the pressure of the water, which means that the minimum 1 m head is not necessary. However, a minimum head of 150 mm is required to lift the flow switches as these switch the pump on. With some installations it is possible to install the pump with a negative head, where the cistern is lower than the pump, provided that a means of starting the pump is in place, such as a pull-cord switch.

Installation of shower mixing valves from mains hot and cold supplies

The installation of unvented hot water storage cylinders is covered earlier in this chapter. You may be required to install or maintain shower mixing valves that are installed on this type of system.

With this type of installation, shower pumps are not required as the hot and cold supplies are fed direct from the mains cold water supply via a pressure reducing valve that reduces the pressure of the water to the operating pressure of the unvented hot water storage cylinder. The obvious advantages of this are:

- the amount of water that can be delivered to the showerhead
- the force of the water leaving the showerhead, giving a powerful 'continental'-type shower.

Because the unvented hot water cylinder usually operates at a slightly lower pressure than the mains cold water supply, the cold water to the shower must be at the same pressure as the hot water supply. This means that the cold supply needs to be connected after the pressure reducing valve but before the single check valve on the unit (see Figure 6.59) to ensure equal hot and cold pressures.

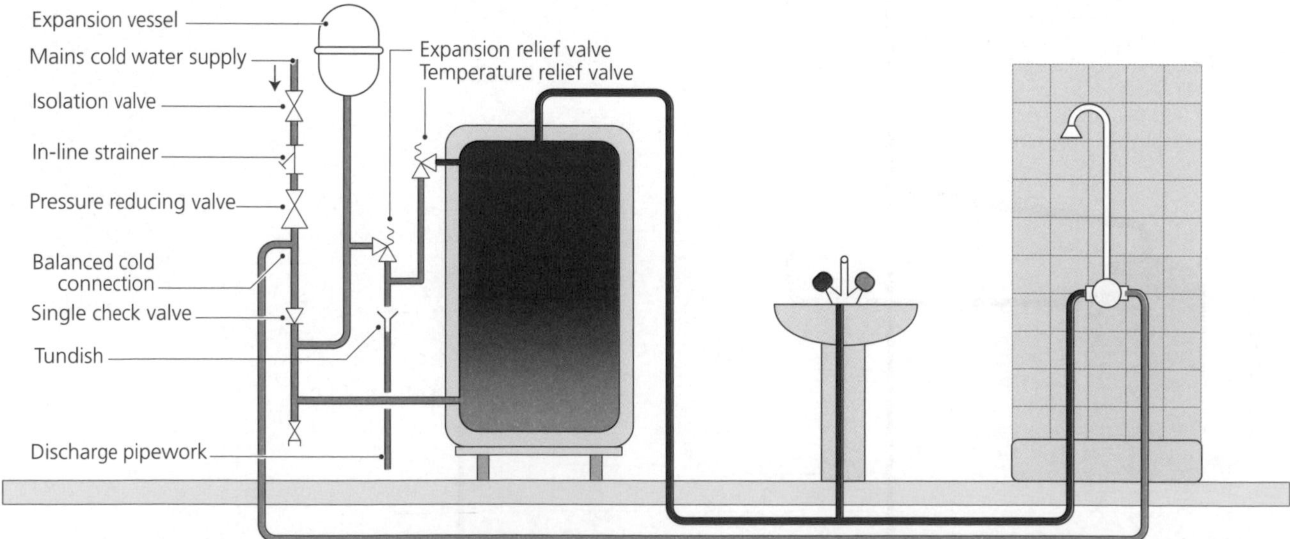

▲ Figure 6.59 Installation of shower mixing valves from an unvented hot water storage cylinder

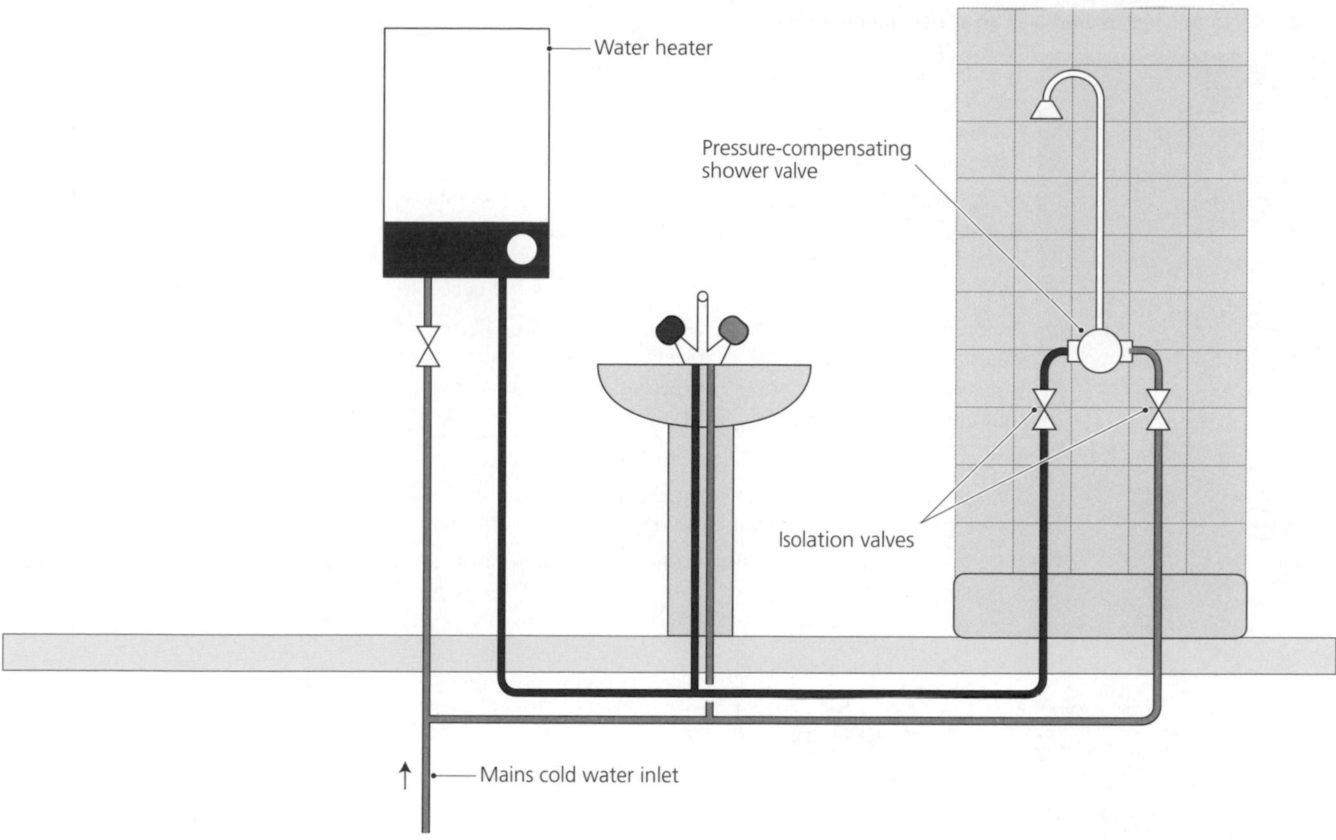

▲ Figure 6.60 Installation of a pressure-compensating shower mixing valve

Installations that use supplies where there is an imbalance in supply pressures (instantaneous hot water heaters/combi boilers)

Showers installed on instantaneous water heaters and combination boilers require a shower valve that is pressure compensating. This is because as the cold water passes through the hot water heater/combi boiler, it loses pressure and flow rate, and so an imbalance of pressure/flow rate between the mains cold water and the hot water from the heater occurs. The pressure-compensating shower mixer valve adjusts both pressure and flow rate within the shower valve body to give a reasonably powerful shower.

Protection against backflow and back siphonage

This subject is dealt with in Chapter 5, Cold water systems, where different types of basic backflow prevention devices and air gaps are discussed. That theme is continued in this chapter as hot water is itself categorised as fluid category 2 simply because heat has been added to the cold wholesome water. Other considerations here are that many of the bathroom appliances that are connected to the hot and cold supply are also at risk from fluid categories 3 and 5.

Appliances that may be at risk from backflow are listed in Table 6.15.

▼ Table 6.15 Appliances that may be at risk from backflow

Washbasins **Fluid cat. 2 and 3 risk**	Taps for use with washbasins should discharge at least 20 mm above the spill-over level of the appliance (AUK2 air gap). Mixer taps should be protected by the use of single check valves on the hot and cold supplies. Twin-flow mixer taps do not require any backflow protection as the water mixes on exit from the tap.
Kitchen sinks **Fluid cat. 5 risk**	No backflow protection is required as the height of the outlet is well above the spill-over level of the appliance. This is classified as an AUK3 air gap. If a mixer tap, where both hot and cold water mix in the tap body, is installed then single check valves must be fitted on both hot and cold supplies. Twin-flow mixer taps do not require any backflow protection as the water mixes on exit from the tap.
Baths **Fluid cat. 2, 3 and 5 risk**	As for washbasins, except that the air gap should be 25 mm. Bath/shower mixer taps, where the water is fed from the mains cold water supply and there is a risk of the showerhead being below the water level in the bath, should be protected by double check valves or a shower hose retaining ring that maintains an AUK2 air gap above the spill-over level of the bath.
Bidets **Fluid cat. 2, 3 and 5 risk**	There are two types of bidet that are at risk from backflow. These are: 1 the ascending spray type – special consideration must be made when fitting this type of bidet (see Figure 6.61); these cannot be used with mains-fed hot and cold water systems; fluid cat 5 risk. 2 the over rim with shower hose connection – with this installation there is fluid cat. 5 risk as well as a fluid cat. 2 risk. 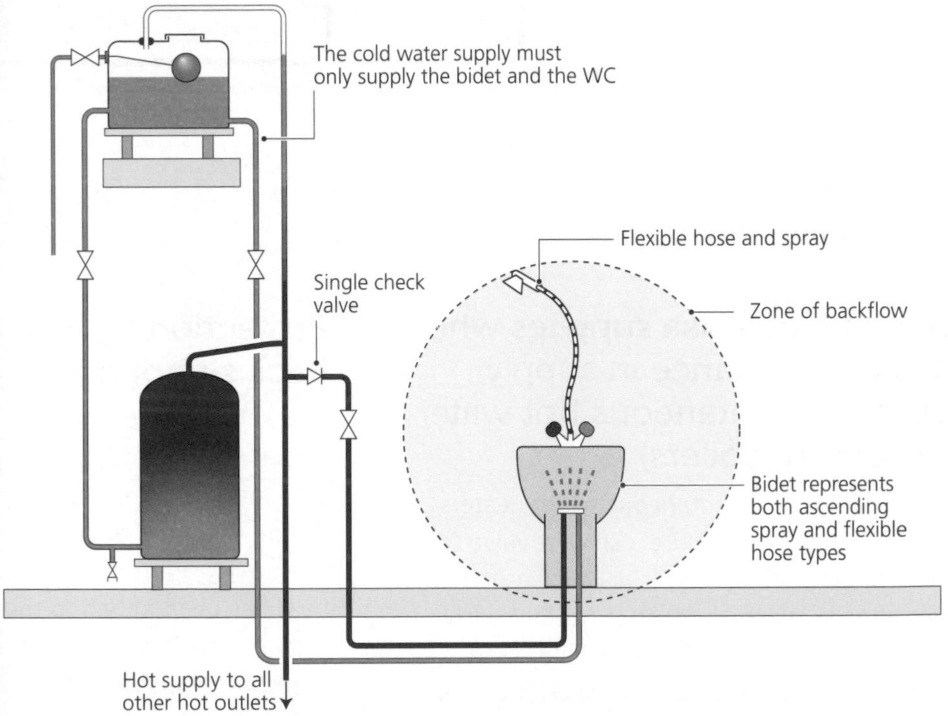 ▲ Figure 6.61 Installation of an ascending spray bidet and bidet with flexible hose
Shower valves **Fluid cat. 2 and 3 risk**	When both hot and cold supplies are fed from a cistern, no backflow protection is required. However, when both are fed from mains-fed supplies, then single check valves are required with a hose retaining ring to prevent the hose entering the water. If no retaining ring is fitted, then both hot and cold supplies should have a double check valve installed.
Electric shower units **Fluid cat. 2 and 3 risk**	A double check valve is required where a hose retaining ring is not fitted.

The installation of other common components (taps, valves, pumps, cisterns and appliances)

Many of the components and appliances installed on hot water systems are generic and share common installation techniques with cold water systems. These were dealt with in Chapter 5, Cold water systems.

Testing and commissioning of hot water systems

Testing and commissioning of hot water systems is probably the most important part of any installation, as it is here that the system design is finally put into operation. For an installation to be successful, it has to comply with both the manufacturer's installation instructions and the regulations in force. It also has to satisfy the design criteria and flow rates that have been calculated and the customer's specific requirements.

Testing and commissioning performs a vital role and its importance cannot be overstated. Correct commissioning procedures and system set-up often make the difference between a system working to the specification and failing to meet the required demands.

In this part of the chapter, we will look at the correct methods of testing and system commissioning.

Information sources required to complete commissioning work on hot water systems

Inadequate commissioning, system set-up, system flushing and maintenance operations can affect the performance of any hot water system, irrespective of the materials that have been used in the system installation. Building debris and swarf (pipe filings) can easily block pipes, and these can also promote bacteriological growth. In addition, excess flux used during installation can cause corrosion and may lead to the amount of copper that the water contains exceeding the permitted amount for drinking water. This could have serious health implications and, in severe cases, may cause corrosion of the pipework, fittings and any storage vessel installed.

It is obvious, then, that correct commissioning procedures must be adopted if the problems stated are to be avoided. There are four documents that must be consulted:

1. the Water Supply (Water Fittings) Regulations 1999
2. British Standard **BS 6700** and **BS EN 806** (in conjunction with **BS 8558**)
3. the Building Regulations Approved Document G3
4. the manufacturer's instructions of any equipment and appliances.

The documents required for correct testing and commissioning were investigated in Chapter 5, Cold water systems.

The checks to be carried out during a visual inspection of an unvented hot water storage system to confirm that it is ready to be filled with water

Before soundness testing a hot water system, visual inspections of the installation should take place. These should include:

- walking around the installation; check that you are happy that the installation is correct and meets installations standards
- check that all open ends are capped off and all valves isolated
- check that all capillary joints are soldered and that all compression joints are fully tightened
- check that sufficient pipe clips, supports and brackets are installed, and that all pipework is secure
- check that the equipment, i.e. unvented hot water storage cylinder, shower boosting pumps, expansion vessels and subsequent safety and functional controls, are installed correctly and that all joints and unions on and around the equipment are tight
- check that the pre-charge pressure in the expansion vessel is correct and in accordance with the manufacturer's data
- check that any cisterns installed on open vented hot water storage systems are supported correctly and that float-operated valves are provisionally set to the correct water level
- check that all appliances' isolation valves and taps are off; these can be turned on and tested when the system is filled with water

- check that the D1 and D2 discharge pipework complies with the Building Regulations and that it terminates in a safe but visible position.

The initial system fill

The initial system fill is always conducted at the normal operating pressure of the system. The system must be filled with fluid category 1 water direct from the water undertaker's mains cold water supply. It is usual to conduct the fill in stages so that the filling process can be managed comfortably. There are several reasons for this:

- Filling the system in a series of stages allows the operatives time to check for leaks stage by stage. Only when the stage being filled is leak free should the next stage be filled.
- **Open vented systems:** air locks from cistern-fed open vented systems are less likely to occur, as each stage is filled slowly and methodically. Any problems can be assessed and rectified as the filling progresses without the need to isolate the whole system and initiate a full drain down. Allowing cisterns to fill to capacity and then opening any gate valves is the best way to avoid air locks. This ensures that the full pressure of the water is available and the pipes are running at full bore. Trickle filling can encourage air locks to form, causing problems later during the fill stage.
- **Unvented systems:** before an unvented hot water storage system is filled, the pressure at the expansion vessel (if fitted) should be checked with a Bourdon pressure gauge to check the pre-charge pressure. Unvented hot water storage systems should be filled with all hot taps open. This is to ensure that pockets of air at high pressure are not trapped within the storage vessel as this can cause the system to splutter water, even after the system has filled. Water should be drawn from every hot water outlet to evacuate any air pockets from the system. The taps can be closed when the water runs freely without spluttering. The temperature and pressure relief valves should be opened briefly to ensure their correct operation and to test the discharge pipework arrangement.
- When the system has been filled with water it should be allowed to stabilise to full operating pressure. Any float-operated valves should be allowed to shut off. The system will then be deemed to be at normal operating pressure.

Once the filling process is complete, another thorough visual inspection should take place to check for any possible leakage. The system is then ready for pressure testing.

Soundness testing hot water systems

The procedure for soundness testing hot water systems is described in **BS EN 806** and the Water Supply (Water Fittings) Regulations. There are two types of test:

1. testing metallic pipework installations
2. testing plastic pipework systems.

Both of these test procedures are covered in detail in Chapter 5, Cold water systems.

Flushing procedures for hot water systems and components

Again, this subject was covered in detail in Chapter 5, Cold water systems, but differs slightly in this case because of the appliances and equipment installed on hot water systems.

Like cold water installations, the flushing of hot water systems is a requirement of the British Standards. All systems, irrespective of their size, must be thoroughly flushed with clean water direct from the water undertaker's main supply before being taken into service. This should be completed as soon as possible after the installation has been completed to remove potential contaminates, such as flux residues, PTFE, excess jointing compounds and swarf. Simply filling a system and draining down again does not constitute a thorough flushing. In most cases, this will only move any debris from one point in the system to another. In practice, the system should be filled and the water run at every outlet until the water runs completely clear and free of any discolouration. It is extremely important that any hot water storage vessels and cold water storage cisterns should be drained down completely.

It is generally accepted that systems should not be left charged with water once the flushing process has been completed, especially if the system is not going to be used immediately, as there is a very real risk that the water within the system could become stagnant. In practice, it is almost impossible to effect a complete drain down of a system, particularly large systems, where long horizontal pipe runs may hold water. This, in itself, is very detrimental as corrosion can often set in and this can also cause problems with water contamination. It is recommended therefore that, to minimise the risk of corrosion and water quality problems, systems should be left completely full and flushed through at regular intervals of no less than twice weekly, by opening all terminal fittings until the system has been taken permanently into operation. If this is the case, then provision for frost protection must be made.

Taking flow rate and pressure readings

Once the hot water system has been filled and flushed, the heat source should be put into operation and the system run to its operating temperature. Thermostats and high-limit thermostats should be checked to ensure that they are operating at their correct temperatures. When the system has reached full operating temperature and the thermostats have switched off, the flow rates, pressures and water temperatures can then be checked against the specification and the manufacturer's instructions. This can be completed in several ways:

- Flow rates can be checked using a weir gauge. This is sometimes known as a weir cup or a weir jug. The method of use is simple. The gauge has a slot running vertically down the side of the vessel, which is marked with various flow rates. When the gauge is held under running water, the water escapes out of the slot. The height that the water achieves before escaping from the slot determines the flow rate. Although the gauge is accurate, excessive flow rates will cause a false reading because the water will evacuate out of the top of the gauge rather than the side slot.
- System pressures (static) can be checked using a Bourdon pressure gauge at each outlet or terminal fitting. Bourdon pressure gauges can also be permanently installed either side of a boosting pump to indicate both inlet and outlet pressures.
- Both pressure (static and running) and flow rate can be checked at outlets and terminal fittings using a combined pressure and flow rate meter.
- The temperature should be checked using a thermometer at the hot water draw-off to ensure that it is at least 60°C but does not exceed 65°C. Each successive hot water outlet, moving away from the storage vessel, should be temperature checked to ensure that any thermostatic mixing valves are operating at the correct temperature and that the hot water reaches the outlet within the 30-second limit. If a secondary return system is installed, then the circulating pump should be running when the tests are conducted and the temperature of the return checked just before it re-enters the cylinder, to ensure that the temperature is no less than 10°C lower than the draw-off, 50°C minimum.

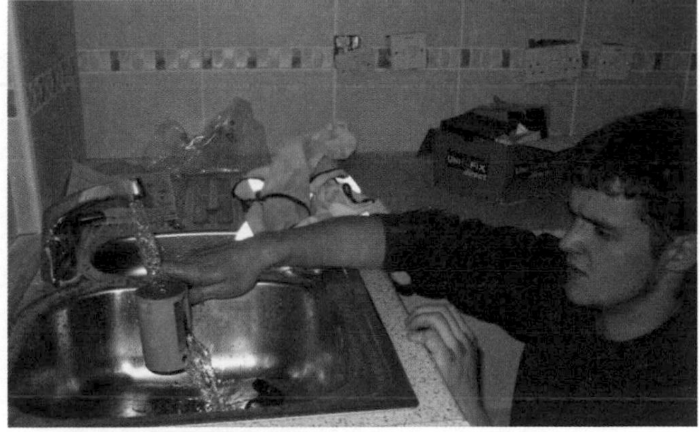

▲ Figure 6.62 Checking hot water flow rates

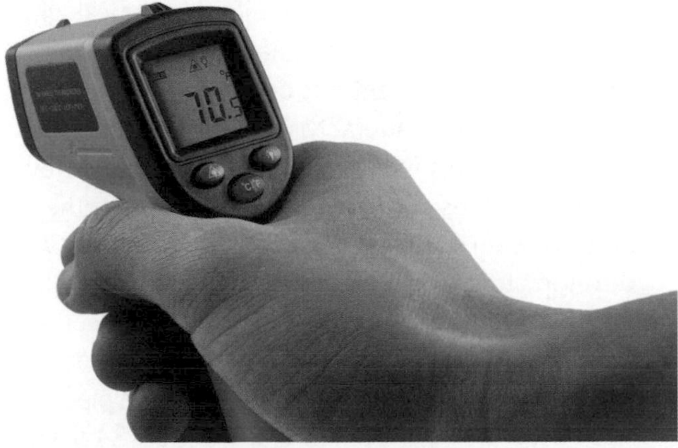

▲ Figure 6.63 Using an infrared thermometer

Balancing a secondary circulation system

Large secondary circulation systems should contain bronze lockshield valves on every return leg of the hot water secondary circuit. These should be fitted as close to the appliances as possible and are used to balance the system so that the flow rates to each leg are such that:

- heat loss through the circuit is kept to a minimum
- the temperature of each leg is constant
- the temperature of the return at the cylinder is not less than 50°C.

Correct balancing is achieved by opening the valves on the longest circuits and then successively closing the lockshield valves a little at a time, working towards the cylinder until the flow rates through each circuit are equal. The flow rate should be balanced so that all of the circuits achieve the same temperature at the same time. This is especially important with those systems that operate through a time clock.

Dealing with defects found during commissioning

Commissioning is the part of the installation where the system is filled and run for the first time. It is now that we see if it works as designed. Occasionally, problems will be discovered when the system is fully up and running, such as those described below.

Systems that do not meet correct installation requirements

This can take several forms, as follows.

- **Systems that do not meet the design specification:** problems such as incorrect flow rates and pressures are quite difficult to deal with. If the system has been calculated correctly and the correct equipment has been specified and installed to the manufacturer's instructions, then problems of this nature should not occur. However, if the pipe sizes are too small in any part of the system, then flow rate and pressure problems will develop almost immediately downstream of where the mistake has been made. In this instance, the drawings should be checked and confirmation with the design engineer sought that the pipe sizes used are correct before any action is taken. It may also be the case that too many fittings or incorrect valves have been used, causing pipework restrictions.
- Another cause of flow rate and pressure deficiency is the **incorrect set-up of equipment such as boosting pumps and accumulators**. In this instance, the manufacturer's data should be consulted and set-up procedures followed according to the installation instructions. It is here that mistakes are often made. If problems persist, then the manufacturer's technical support should be contacted for advice. In a very few cases, the equipment specified is at fault and will not meet the design specification. If this is the case, then the equipment must be replaced.
- **Poor installation techniques:** installation is the point where the design is transferred from the drawing to the building. Poor installation techniques account for problems such as noise. Incorrectly clipped pipework can often be a source of frustration within systems running at high pressures because of the noise it can generate. Incorrect clipping distances and, often, lack of clips and supports can put a strain on the fittings and cause the pipework to reverberate throughout the installation, even causing fitting failure and leakage. To prevent these occurrences, the installation should be checked as it progresses and any deficiencies brought to the attention of the installing engineer. Upon completion, the system should be visually checked before flushing and commissioning begins.
- **Leakage:** water causes a huge amount of damage to a building and can even compromise the building structure. Leakage from pipework, if left undetected, causes damp, mould growth and an unhealthy atmosphere. It is, therefore, important that leakage is detected and cured at a very early stage in the system's life. It is almost impossible to ensure that every joint on every system installed is leak free. Manufacturing defects on fittings and equipment, as well as damage, sometimes cause leaks. Leakage due to badly jointed fittings and poor installation practice are much more common, especially on large systems where literally thousands of joints have to be made until the system is complete. These can often be avoided by taking care when jointing tubes and fittings, using recognised jointing materials and compounds, and following manufacturers' recommended jointing techniques.

▲ Figure 6.64 A plumber's nightmare! A badly designed plumbing system makes fault finding almost impossible

The risk from *Legionella pneumophila* in hot water systems

According to the HSE, instances of Legionnaires' disease derived from hot water supply have diminished over recent years due to better installation techniques and greater awareness of sterilisation methods. However, large hot water systems can often be complex in their design and, therefore, still present a significant risk of exposure. The environments where *Legionella* bacteria proliferate are listed below.

- At the base of the cylinder or storage vessel where the cold feed enters and cold water mixes with the hot water within the vessel. The base of the storage vessel may well eventually contain sediments, which support the bacterial growth of *Legionella*.
- The water held in a secondary circulation system between the outlet and the branch to the secondary circulation system, as this may not be subject to the high temperature sterilisation process.

In general, hot water systems should be designed to aid safe operation by preventing or controlling conditions that allow the growth of *Legionella*. They should, however, permit easy access for cleaning and disinfection. The following points should be considered.

- Materials such as natural rubber, hemp, linseed oil-based jointing compounds and fibre washers should not be used in domestic water systems. Materials and fittings acceptable for use in water systems are listed in the directory published by the Water Research Centre.
- Low-corrosion materials (copper, plastic, stainless steel, etc.) should be used where possible.

Defective components and equipment

Defective components cause frustration and cost valuable installation time. If a component or piece of equipment is found to be defective, do not attempt a repair as this may invalidate any manufacturer's warranty. The manufacturer should first be contacted as they may wish to send a representative to inspect the component prior to replacement. The supplier should also be contacted to inform them of the faulty component. In some instances where it is proven that the component is defective and was not a result of poor installation, the manufacturer may reimburse the installation company for the time taken to replace the component.

The procedure for notifying works carried out to the relevant authority

At all stages of the installation, from design to commissioning, notification of the installation will need to be given so that the relevant authorities can check that it complies with the regulations and to ensure that the installation does not constitute a danger to health. It must be remembered that only operatives that are registered to do so can install unvented hot water storage systems. The operative's registration number must be given on any paperwork submitted to the local authority.

Under Building Regulations Approved Document G, hot water installations are notifiable to the local authority Building Control Office. Building Regulations approval can be sought from the local authority by submitting a 'building notice'. Plans are not required with this process so it's quicker and less detailed than the full plans application. It is designed to enable small building works to get under way quickly. Once a building notice has been submitted and the local authority has been informed that work is about to start, the work will be inspected as it progresses. The authority will notify if the work does not comply with the Building Regulations.

INDUSTRY TIP

Notice should be given to Building Control not later than five days after work completion and, until this is received, no completion certificates can be issued.

Building Regulations Compliance certificates

From 1 April 2005, the Building Regulations have demanded that all installations must be issued with a Building Regulations Compliance certificate. This is to ensure that all Building Regulations relevant to the installation have been followed and complied with.

Commissioning records for hot water systems

Commissioning records, such as benchmark certificates for hot water systems, should be kept for reference during maintenance and repair, and to ensure that the system meets the design specification. Typical information that should be included on the record is as follows:

- the date, time and the name(s) and ID numbers of the commissioning engineer(s)
- the location of the installation
- the amount of hot water storage and cold water storage (if any)
- the types and manufacturer of equipment and components installed
- the type of pressure test carried out and its duration
- the incoming static water pressure
- the flow rates and pressures at the outlets
- the expansion vessel pressure
- whether temperature and pressure relief valves have been fitted
- the results of tests on the discharge pipework.

The benchmark certificate should be signed by the operative and the customer, and kept in a file in a secure location.

Hand over to the customer or end user

When the system has been tested and commissioned, it can then be handed over to the customer. The customer will require all documentation regarding the installation and this should be presented to them in a file, which should contain:

- all manufacturers' installation, operation and servicing manuals for the unvented hot water storage vessel and associated controls
- the commissioning records and certificates
- the Building Regulations Compliance certificate
- an 'as fitted' drawing showing the position of all isolation valves, backflow prevention devices, etc.

The customer must be shown around the system and shown the operating principles of any controls. Emergency isolation points on the system should be pointed out and a demonstration given of the correct isolation procedure in the event of an emergency. Explain to the customer how the systems work and ask if they have any questions. Finally, highlight the need for regular servicing of the appliances and leave emergency contact numbers.

6 DECOMMISSION SYSTEMS AND COMPONENTS

Decommissioning hot water systems for maintenance and the replacement of components can be a delicate task. It is important to ensure that the heat source is totally isolated before work on the system begins. A notice should be placed next to the heat source informing people that the system is decommissioned and must not be turned on. Fuses to electric heaters, thermostats and motorised valves should be removed and retained. If appliances are removed, any open pipes should be capped off. The customer should be informed when the system is turned off.

The main components of hot water systems that require periodic maintenance are as follows.

- **The hot water storage vessel:** should be periodically checked for any signs of corrosion. Diminishing flow rates could indicate scale build-up in either the cold feed connection or the hot water draw-off connection. These can be removed and descaled as necessary. When replacing hot water cylinders, the cylinder should be pre-assembled as much as possible before installation begins, to reduce the time the hot water supply is off.

- **The hot water appliance:** this should be serviced annually in line with the manufacturer's instructions.
- **The cistern (for open vented systems):** cisterns should be checked periodically for sediment build-up on the bottom of the cistern. If a cistern is to be replaced, then the replacement cistern should be pre-assembled before decommissioning the system. This will reduce the length of decommissioning time.
- **Taps and terminal fittings such as float-operated valves:** taps should be re-washered and float-operated valves checked for correct shut-off, and water levels checked and adjusted as necessary.
- **Isolation valves such as full-way gate valves and service valves:** these should be checked to ensure that they shut off the flow of water fully.
- **Thermostats:** systems, such as immersion heaters and boilers, should be run to operating temperature to ensure the correct operation of any thermostats. They should be checked using digital thermometers.
- **Shower mixing valves and pumps:** these should be inspected to ensure that they are functioning in accordance with the manufacturer's specifications. Flow rates can be confirmed by using a weir cup. Filters can be removed and cleaned. The operation of the flow switch on shower boosting pumps should be checked, as these turn the shower pump on. Showerheads should be cleaned of any scale build-up as this can significantly reduce the flow of water.

7 REPLACE DEFECTIVE COMPONENTS

The replacement of float-operated valves, taps, pumps and valves is covered in Chapter 5, Cold water systems.

Faults with open vented hot water systems

There are many faults that can occur with open vented hot water systems. Some of these may be due to poor system design but most occur with use. Some of the more common faults are:

- **Loss of hot water:** this may be due to evaporation of the water in the feed and expansion (F&E) cistern installed on double-feed indirect cylinders, with gravity circulation to the heat exchanger. This is usually due to a sticking float-operated valve (FOV) that fails to top the water up as evaporation occurs. Because the FOV is stuck in the off position, the water evaporates down to the primary flow pipe and this stops circulation to the heat exchanger and prevents the cylinder getting hot. To rectify the fault, the FOV in the F&E cistern should be removed and repaired/replaced.
- **Immersion heater element failure:** this is usually due to corrosion of the immersion heater element sheath, allowing water to penetrate the heater element. This causes a short circuit, which usually blows the fuse. The immersion heater will need to be replaced.
- **Cylinder thermostat failure:** a very rare fault. The thermostat should first be tested to confirm that it has failed before replacing it.
- **Motorised valve failure:** this is a common occurrence with fully pumped systems. The valve should be tested to confirm whether it is the valve itself that has failed or just the motor in the actuator head.
- **Boiler failure:** this is a more serious fault that may mean specialist diagnosis and repair by an experienced plumber.
- **Airlocks:** these can usually be traced to long horizontal runs in the cold feed to the cylinder as it leaves the cistern. The closer the horizontal run is to the cistern, the less head of pressure there is on the cold feed. This can create an airlock before the cold feed drops vertically to the cylinder. Low-pressure systems always work better when the pipework exits the roof space quickly. Long horizontal runs create problems with flow rate when the head of pressure is low.
- **The cold feed has a backfall towards the cistern:** air collects in the high point in the pipework. The pipework leaving the cistern should fall away from the cistern to ensure a good flow rate.
- **Noise in the system:** this can be due to oscillation of the float-operated valve. This may

be because of a faulty float-operated valve or a missing cistern wall-strengthening plate, which prevents the cistern wall from vibrating. Vibration may also come from the immersion heater when the electricity is turned on. The heater element vibrates quickly, making a humming sound. The only action here is to replace the heater.

- **Overheating of the water:** this causes the water to boil and is a problem found in some older direct systems with a coal-fired back boiler.
- **Expansion of the pipework:** this causes ticking and creaking noises when not enough room has been allowed for expansion of the pipework. On new properties, this type of noise is not allowed and must be traced and rectified.
- **Excessively hot water:** this is usually caused by immersion heater thermostat failure. This will need testing and replacing with a thermostat that has a high limit stat cut-out.
- **Uncontrolled heat from a solid fuel appliance:** this may occur in direct systems.
- **Cylinder collapse:** due to the creation of a vacuum in the cylinder caused by the hot water dropping as soon as it leaves the cylinder before it enters the vent pipe; having no vent pipe installed; a blocked vent pipe; or an isolation valve installed on the vent pipe which is turned off.

SUMMARY

The choice of hot water system is a confusing task. There are so many systems to choose from, and each one has its advantages and disadvantages. In this chapter, we have investigated a sample of the most popular systems from simple point-of-use heaters to Building Regulations-compliant storage and non-storage systems for whole-house hot water distribution for a variety of property types and sizes. These systems should be considered carefully to give the best possible combination of initial cost, efficiency, hot water control, maintenance costs and eventual replacement.

Hot water is a necessity. How we deliver it is a matter of choice.

Test your knowledge

1 What is the minimum recommended distribution temperature of hot water supplied by a storage cylinder?

a 48°C

b 55°C

c 60°C

d 65°C

2 Which Building Regulation relates to the hot water delivery systems in dwellings?

a Part F

b Part G

c Part P

d Part L

3 Which one of the following is classed as a localised hot water system?

a Under-sink single point heater

b Thermal store

c Combination boiler

d Open vented

4 Within an open vented hot water system, which of the following are the correct connection methods for the open vent pipe off the draw-off pipe?

a Open vent a minimum of 15 mm diameter connected within 450 mm of the draw-off connection from the cylinder

b Open vent a minimum of 15 mm diameter connected within at least 450 mm of the draw-off point from the cylinder

c Open vent a minimum of 22 mm diameter connected within 450 mm of the draw-off connection from the cylinder

d Open vent a minimum of 22 mm diameter connected within at least 450 mm of the draw-off point from the cylinder

5 The safety devices within an unvented hot water cylinder are designed to prevent the water from exceeding which of the following temperatures?

a 60°C

b 65°C

c 95°C

d 100°C

6 Where should a secondary return be connected within a hot water storage cylinder?

a A quarter of the way down

b Halfway down

c Two-thirds of the way from the bottom

d A quarter of the way up

7 Identify the components in the unvented hot water system below.

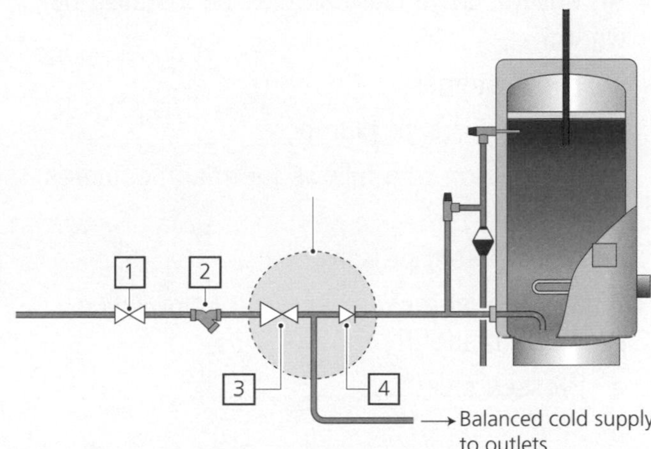

a 1: Strainer; 2: Isolation valve; 3: Expansion relief valve; 4: Check valve

b 1: Isolation valve; 2: Strainer; 3: Check valve; 4: Expansion relief valve

c 1: Pressure reducing valve; 2: Isolation valve; 3: Check valve; 4: Strainer

d 1: Isolation valve; 2: Strainer; 3: Pressure reducing valve; 4: Check valve

8 What temperature should the temperature relief valve fitted to an unvented hot water storage vessel be set to discharge at?

a 60°C

b 72°C

c 90°C

d 95°C

9 Which of the following types of thermostatic mixing valve (TMV) should be installed within NHS and healthcare properties?

a TMV1

b TMV2

c TMV3

d TMV4

10 When considering discharge pipework from an unvented storage cylinder, what is the minimum size for D2 pipework if the D1 pipework is connected to a G1/2 valve outlet?

a 15 mm
b 22 mm
c 28 mm
d 35 mm

11 An unvented cylinder can only be installed by whom?

a The customer
b IPHE registered plumber
c Any person who follows the manufacturer's instructions
d A competent person

12 What material must a secondary circulation pump be made from?

a Pressed steel
b Zinc
c Cast iron
d Bronze

13 The required controls for a hot water system must comply with:

a The Domestic Hot Water Guide
b The Domestic Heating Compliance Guide
c Approved Document H
d BS EN 806

14 What does the 450 mm leg from the hot water outlet to the base of the open vent prevent?

a One pipe circulation
b Back flow
c Back siphonage
d One pipe pull

15 What is the most likely cause of warm water coming out of the cold water tap when it is first opened up?

a The cold feed has been incorrectly installed
b The hot and cold pipe are touching
c The wrong pipe is insulated
d The hot distribution is wrongly sized

16 What can system noise can be caused by?

a Flow of water
b Water hammer
c Thermal expansion
d All of these

17 A new hot water cylinder must comply with what Building Regulation?

a H
b P
c L
d G

18 What component is installed inside showers to prevent backflow?

a Drain-off valve
b Double check valve
c Stop valve
d Single check valve

19 If a hot water cylinder does not have a heat exchanger coil and the water is being heated by an immersion heater, what type of cylinder is installed in the system?

a Indirect
b Radiated
c Instantaneous
d Direct

20 Where is the expansion in a hot water cylinder taken up?

a Cold feed from the cold water storage cistern
b Primary flow from the boiler
c Hot distribution to the outlets
d Primary return to the boiler

21 A hot water cylinder contains 100 litres of cold water. When heated the water will expand. Approximately how much will the water expand by when heated?

a 10 litres
b 2 litres
c 4 litres
d 12 litres

22 Where on a domestic hot water cylinder should the cylinder thermostat be located?

a Top quarter of the cylinder
b Middle of the cylinder
c 1/3rd way up the cylinder
d Bottom of the cylinder

23 What does the term 'heat recovery' refer to?

a The time taken for the boiler to get to temperature
b The time taken for a cylinder of water to get to temperature
c The time taken for a cylinder thermostat to activate the boiler
d The time taken for the primary waters to start heating the heat exchanger coil

24 On an unvented system, what safety item will be activated if the expansion vessel fails?

a Pressure relief valve
b Temperature/pressure relief valve
c Over heat button
d High limit cut out valve

25 Identify these three systems:

A

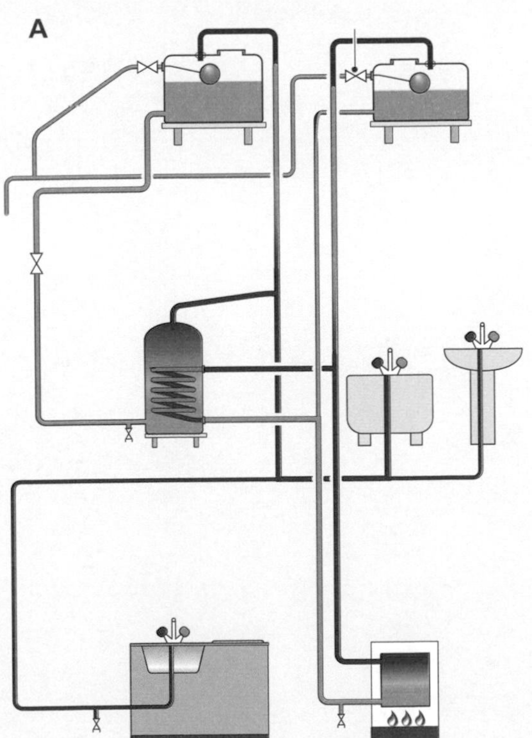

B

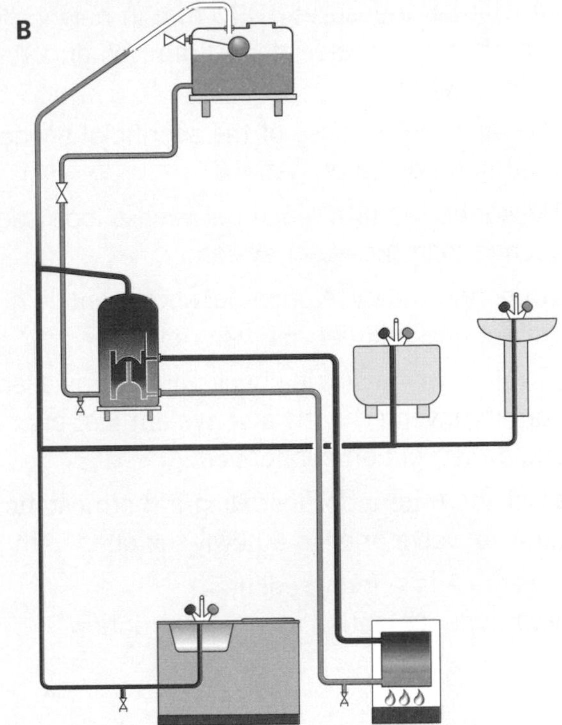

C

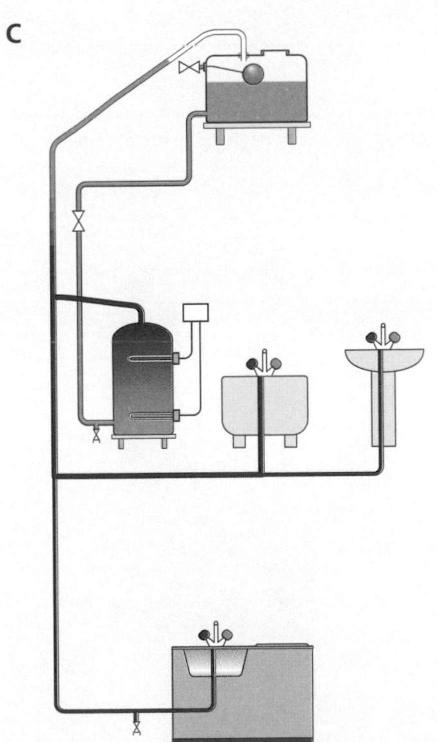

26 Explain the purpose of a secondary return circuit.

27 Name two types of shower pump.

28 A cylinder measures 1050 mm in height and has a diameter of 350 mm. What is its capacity?

29 What is the purpose of the sacrificial anode within a hot water system?

30 Describe the difference between a localised and centralised hot water system.

31 Describe the difference between a modern direct and indirect hot water cylinder.

32 Explain why a single check valve is installed on an unvented hot water system just after the cold distribution connection.

33 List the criteria for installing a thermostatic mixing valve prior to a newly installed bath.

Answers can be found online at www.hoddereducation.co.uk/construction

Practical activity

With permission, in your training centre or on-site, locate a thermostatic mixing valve. Locate any relevant maintenance documentation (often this information can be downloaded from manufacturers' websites) and safely decommission the valve ready for routine maintenance.

Follow the maintenance instructions to adequately clean the filters, check for operation and adjust to a suitable temperature. Ensure that the supervising person checks your work before leaving in operation.

CENTRAL HEATING SYSTEMS

CHAPTER 7

INTRODUCTION

Some 97 per cent of homes in the UK have a central heating system, and most of these are in the traditional form of a boiler and radiators. In the past ten years, central heating has developed into a sophisticated home heating system that incorporates energy-saving appliances and controls designed to heat the dwelling quickly and efficiently using as little fuel as possible and saving thousands of tonnes of CO_2 from being released into the atmosphere.

In this chapter, we will look at the subject of central heating from a domestic perspective. We will investigate existing and modern systems, their pipework layouts, methods of control, the various types of appliances and the fuels they use.

By the end of this chapter, you will have knowledge and understanding of the following:

- central heating systems and their layouts
- how to install central heating systems and components
- the decommissioning requirements of central heating systems and their components.

1 UNDERSTAND CENTRAL HEATING SYSTEMS AND THEIR LAYOUTS

The main purpose of central heating is to provide thermal comfort conditions within a building or dwelling. Central heating is preferable to open fires as it heats the whole property. Thermal comfort is achieved when a desirable heat balance between the body and surroundings is met. How we achieve this balance is down to the design of the central heating system and the way it is installed.

Sources of information

The recommendations for good central heating installations are set out in the British Standards and various other documents, some of which are legislative and take the form of regulations. In this first part of the chapter, we will look at the criteria used for efficient central heating design.

Regulations

These are as follows:

- The **Building Regulations:**
    - **Approved Document L1A:** conservation of fuel and power in new dwellings, 2013 edition with 2016 amendments
    - **Approved Document L1B:** conservation of fuel and power in existing dwellings, 2010 edition (incorporating 2010, 2011, 2013 and 2016 amendments)
    - **Approved Document L2A:** conservation of fuel and power in new buildings other than dwellings, 2013 edition with 2016 amendments
    - **Approved Document L2B:** conservation of fuel and power in existing buildings other than dwellings, 2010 edition (incorporating 2010, 2011, 2013 and 2016 amendments)

- **Approved Document J:** combustion appliances and fuel storage systems (incorporating 2010 and 2013 amendments)
- **Approved Document F:** ventilation (2010 edition incorporating 2010 and 2013 amendments)
- The **Water Supply (Water Fittings) Regulations 1999**
- The **Gas Safety (Installation and Use) Regulations 1998**
- The **IET 18th Edition Wiring Regulations BS 7671.**

INDUSTRY TIP

Remember, you can access all of these Building, Water, and Gas Regulations documents via the government's database at: www.gov.uk/search

The British Standards

These are as follows:

- **BS EN 12828:2012+A1:2014.** Heating systems in buildings. Design for water-based heating systems
- **BS EN 12831–1:2017.** Energy performance of buildings. Method for calculation of the design heat load. Space heating load, Module M3-3
- **BS EN 12831–3:2017.** Energy performance of buildings. Method for calculation of the design heat load. Domestic hot water systems heat load and characterisation of needs, Module M8-2, M8-3
- **BS EN 1264–1:2021.** Water based surface embedded heating and cooling systems. Definitions and symbols
- **BS EN 1264–2.** Water based surface embedded heating and cooling systems. Floor heating: Prove methods for the determination of the thermal output using calculation and test methods
- **BS EN 1264–4:2021.** Water based surface embedded heating and cooling systems. Installation
- **BS EN 1264–3:2021.** Water based surface embedded heating and cooling systems. Dimensioning
- **BS EN 1264–5:2021.** Water based surface embedded heating and cooling systems. Heating and cooling surfaces embedded in floors, ceilings and walls. Determination of the thermal output
- **BS EN 14336:2004.** Heating systems in buildings. Installation and commissioning of water based heating systems.
- **BS EN 442–1:2014.** Radiators and convectors. Technical specifications and requirements
- **BS EN 442–2:2014.** Radiators and convectors. Test methods and rating.

The recommendations

These are as follows:

- **Domestic Building Services Compliance Guide 2013** (amended in 2018) – this document offers practical assistance when designing and installing to Building Regulations requirements for space heating and hot water systems, mechanical ventilation, comfort cooling, fixed internal and external lighting, and renewable energy systems
- **Central Heating System Specifications (CHeSS) 2008** – this publication offers advice for compliance with good practice and best practice for the installation of central heating systems
- **HVDH Domestic Heating Design Guide (2021)** – this was produced to assist heating engineers to specify and design wet central heating systems.

INDUSTRY TIP

The documents are available from the following websites:

- Domestic Building Services Compliance Guide 2013 – https://assets.publishing.service.gov.uk/government/uploads/system/uploads/attachment_data?file/697525/DBSCG_secure.pdf
- Central Heating System Specifications (CHeSS) 2008 – http://bpec.org.uk/downloads/CE51%20CHeSS%20WEB%20FINAL%20JULY%2008.pdf
- HVDH Domestic Heating Design Guide (2021) – www.cibse.org/knowledge/knowledge-items/detail?id=a0q20000008I7odAAC (note: this document must be purchased)

Manufacturers' technical instructions

Central heating systems and components must be installed, commissioned and maintained strictly in accordance with manufacturers' instructions. If these are not available or have been misplaced, most manufacturers now have the facility to download the instructions from their websites.

Operating principles and system layouts of central heating systems and components

Central heating is a vast and complex subject. There are now more options with regard to system types, sources of heat, pipe materials and heat emitters than ever before. Environmentally friendly technology and the re-emergence of underfloor heating have meant that the customer can now afford to be selective about the system they have installed into their property. The advent of heat pumps and solar systems, with the accompanying savings on fuel and running costs, has dramatically lowered the carbon footprint of domestic properties. No longer does the customer have to rely on appliances that burn carbon-rich fuels such as gas and oil. Zero-carbon and carbon-neutral fuels have revolutionised domestic heating, while advances in technology have lowered the cost of the energy-saving appliances that formerly were available to only a select few.

By far the most popular heating system in the UK is the 'wet' system, whether supplying radiators, convectors or underfloor heating, but while wet systems have enjoyed the monopoly thus far, other systems, such as electric storage heaters and warm air, continue to be available. In some areas of the UK, district heating, supplied from a central source and serving many properties, is also commonplace.

Here, we will look at the following central heating options that are available for today's homeowner, their layouts and operating principles:

- wet central heating
- warm air systems
- electric storage heaters
- district heating installations.

Types of system

Wet central heating

Domestic wet central heating systems fall into two different categories, based upon the way the system is filled with water and the pressure at which it operates:

1. **Low pressure, open vented central heating systems**, fed from a feed and expansion cistern in the roof space. These can be both modern fully pumped systems and existing gravity hot water/pumped heating installations.
2. **Sealed, pressurised central heating systems**, fed direct from the mains cold water supply and incorporating an expansion vessel to take up the expansion of water due to the water being heated. These are generally more modern fully pumped and **combination ('combi') boiler** systems.

The water in low pressure open vented central heating systems is kept below 100°C. For existing systems the flow water from the boiler is usually about 80°C and the return water temperature is usually 12°C to 15°C lower.

Circulation of the water can be either by:

- **gravity circulation** to the heat exchanger in the hot water cylinder and pumped heating to the heat emitters, or
- by means of a **fully pumped system** where both the hot water heat exchanger and heat emitters are heated using a circulating pump.

KEY TERMS

Combination ('combi') boiler: a boiler that provides central heating and instantaneous hot water.

Gravity circulation: circulation that occurs because heat rises through the water. No pump is required.

Fully pumped system: a heating system that uses pumped circulation to both heating and hot water circuits.

Fully pumped systems have the advantage that system resistance created by the pipework, fittings and heat emitters can be overcome much more easily and this enables the system to heat up faster, giving the occupants a much more controllable system.

Sealed heating systems operate at a higher pressure, with modern systems incorporating condensing boilers operating at a slightly lower temperature of 65°C for the flow temperature with a return temperature 20°C lower at 45°C.

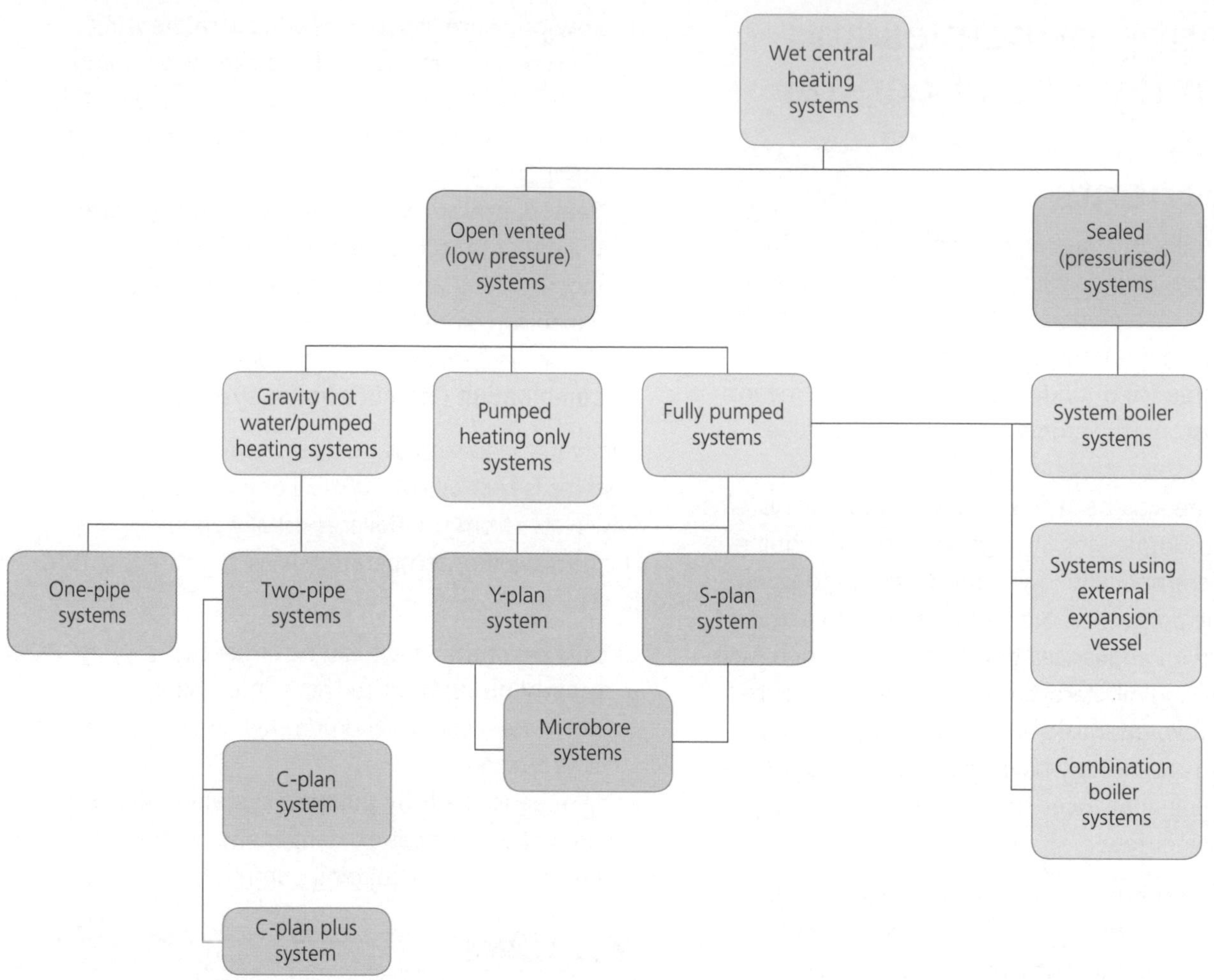

▲ Figure 7.1 The development of pumped central heating

In both cases the difference between the flow and return temperatures is the amount of heat lost to the heated areas.

Figure 7.1 illustrates the development of central heating, from the open vented one-pipe system through to the more modern sealed combination boiler systems and fully pumped systems using system boilers.

Low pressure, open vented central heating systems

Pumped central heating only systems

The simplest of all heating systems – pumped central heating only systems – do not contain any provision for heating the domestic hot water. They serve only the heat emitters, usually radiators/convectors, for domestic installations. The cold feed and the vent pipe can either be taken direct from the boiler or direct from the heating pipework. They are generally two-pipe systems, with the central heating circulating pump installed on the flow pipe.

Semi-gravity heating systems

Semi-gravity heating systems utilise gravity circulation to heat the domestic secondary water and pumped central heating circulation. The heat exchanger within the hot water storage cylinder is connected to the boiler by the primary flow and return pipes, usually 28 mm in diameter with a 22 mm vent pipe branched from the primary flow and a 15 mm cold feed pipe connected to the primary return. They may still be found as existing systems in older properties. There are three old basic semi-gravity systems that are not installed today, and each is an advance on the previous system. These are:

1 the one-pipe system
2 the two-pipe system (C-plan system)
3 the C-plan plus system.

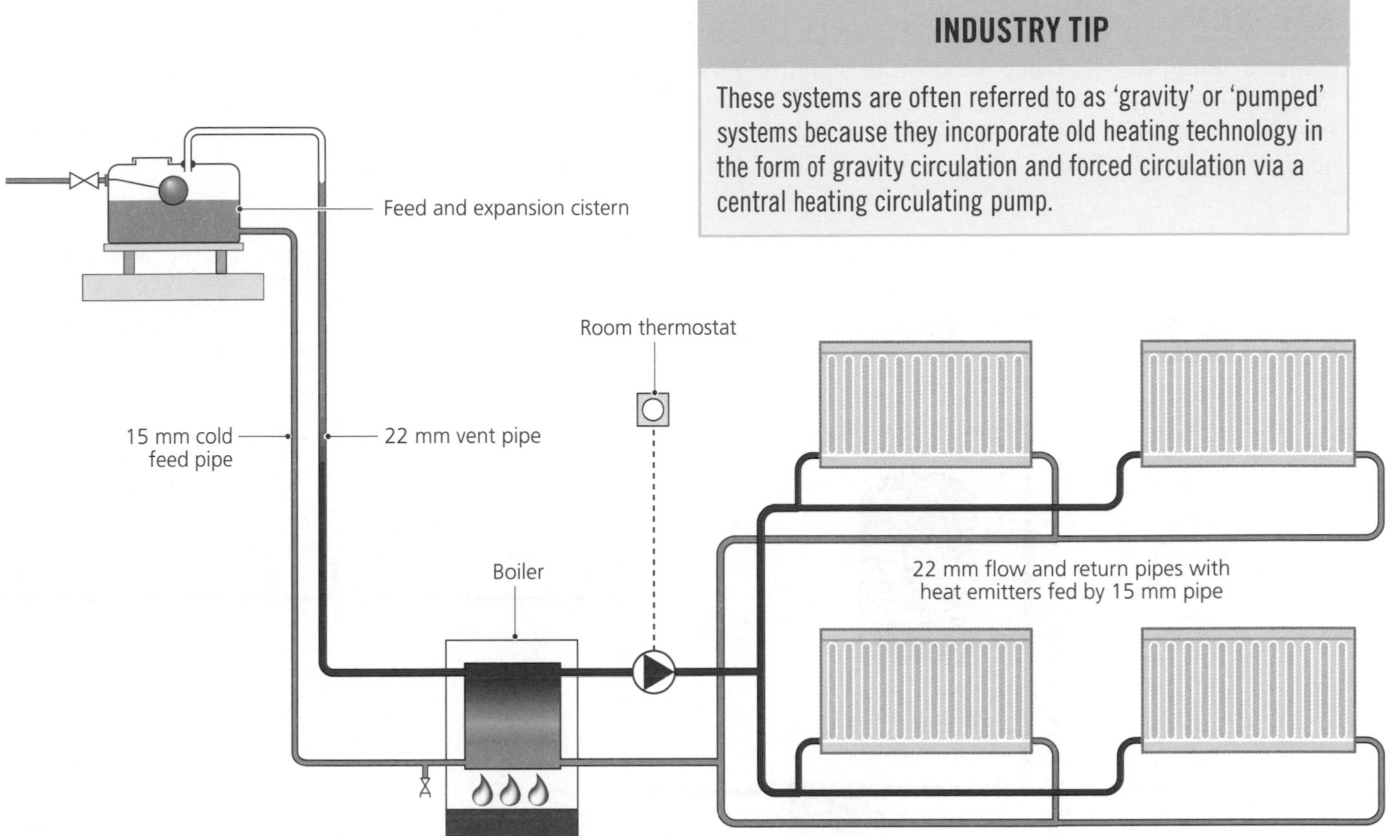

▲ Figure 7.2 Pumped central heating only

INDUSTRY TIP

These systems are often referred to as 'gravity' or 'pumped' systems because they incorporate old heating technology in the form of gravity circulation and forced circulation via a central heating circulating pump.

The one-pipe system

The one-pipe system is an old system that is rarely found today as it is an early design of central heating. It is a simple ring circuit of pipework to and from the boiler and, for this reason, there are no separate flow and return pipes. The main 'ring' is pumped and the water circulates through the radiators by gravity circulation. The size of the radiators is calculated from the temperature drop at each successive radiator, with the last radiator always being around 15°C cooler than the first. Balancing the flow of water to each radiator is a simple process by the use of radiator valves, but this increases system resistance and slows the heating process.

▼ Table 7.1 Advantages and disadvantages of the one-pipe system

Advantages	Disadvantages
Cheap to install because there is less pipework involved in the installation when compared to other heating systems	The water in the system cools as it travels from one heat emitter to the next, which has the effect of increasing the heat emitter sizes the further from the boiler they are
	The system tends to circulate within the main pipework ring; circulation within the heat emitters can be induced only by a difference in the density of the water entering and leaving the system (gravity circulation)
	Uncontrolled heating of the primary circuit leading to overheating of the domestic secondary hot water
	Constant **boiler cycling** even when the hot water and heating are up to temperature leads to wastage of fuel energy
	The system is not Building Regulations Document L compliant and must be updated
	The boilers fitted to this type of system are only about 78% efficient or less
	Condensing-type boilers cannot be fitted to this type of installation because of the gravity circulation needed by the hot water storage cylinder

KEY TERM

Boiler cycling: the constant firing up and shutting down as the system water cools slightly. When a heating system has reached temperature, the boiler shuts down on the boiler thermostat. A few minutes later the boiler will fire up again to top up the temperature as the system loses heat and, after a few seconds, shuts down again. This wastes a lot of fuel energy.

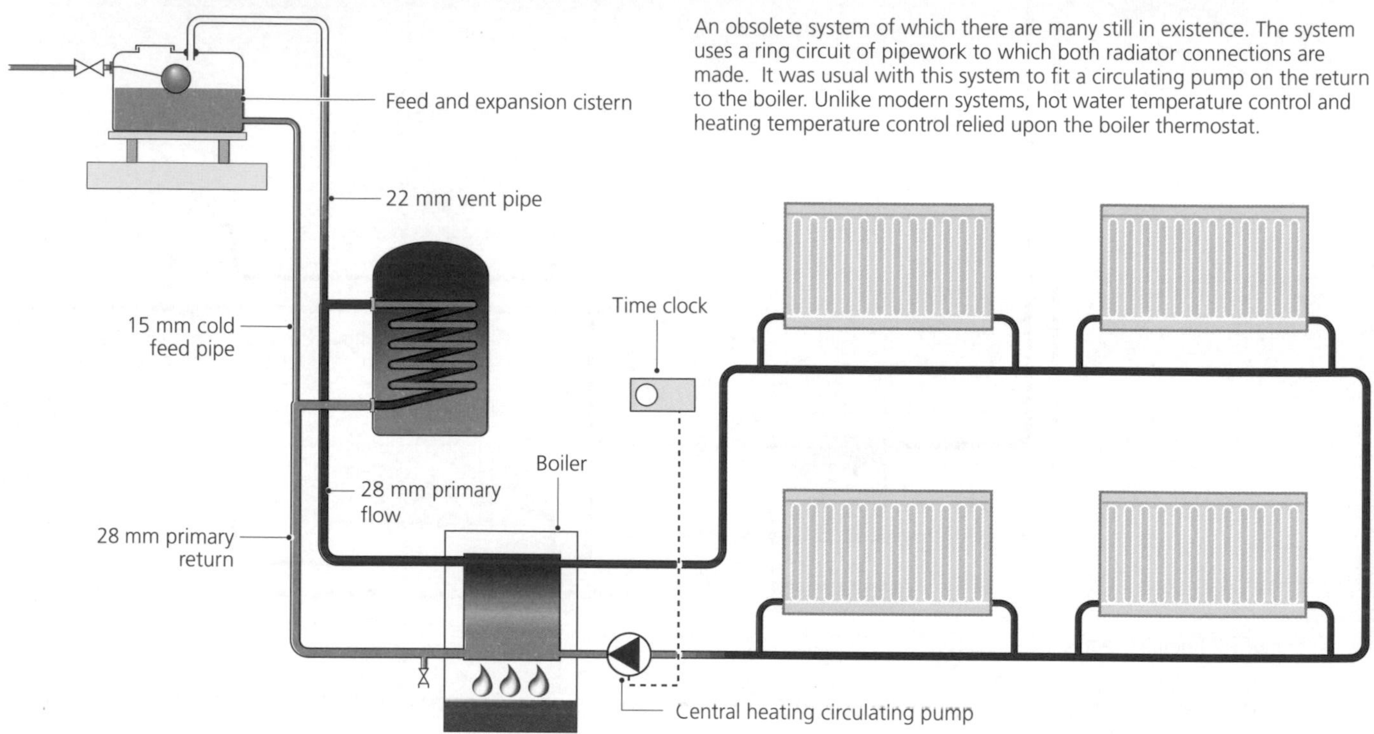

An obsolete system of which there are many still in existence. The system uses a ring circuit of pipework to which both radiator connections are made. It was usual with this system to fit a circulating pump on the return to the boiler. Unlike modern systems, hot water temperature control and heating temperature control relied upon the boiler thermostat.
Feed and expansion cistern
22 mm vent pipe
Time clock
15 mm cold feed pipe
Boiler
28 mm primary flow
28 mm primary return
Central heating circulating pump

▲ Figure 7.3 One-pipe system

One-pipe systems will not work effectively when installed with combination boilers, so these systems are not recommended for 'combi' boiler conversions.

The C-plan (two-pipe) system

Like the one-pipe system, the C-plan system has gravity circulation to the hot water circuit and pumped circulation to the central heating circuit. The system differs considerably from the one-pipe system by having two pipes: a flow pipe from the boiler to the heat emitters and a return pipe from the emitters back to the boiler. The heat emitters are connected to separate branches of the main flow and return pipes so, in effect, each heat emitter has its own flow and return pipework to the boiler. This means that all of the heat emitters achieve the same temperature and this negates the need to increase heat emitter size due to temperature loss. The temperature difference across each flow and return is usually 12°C to 15°C with a flow temperature of around 80°C.

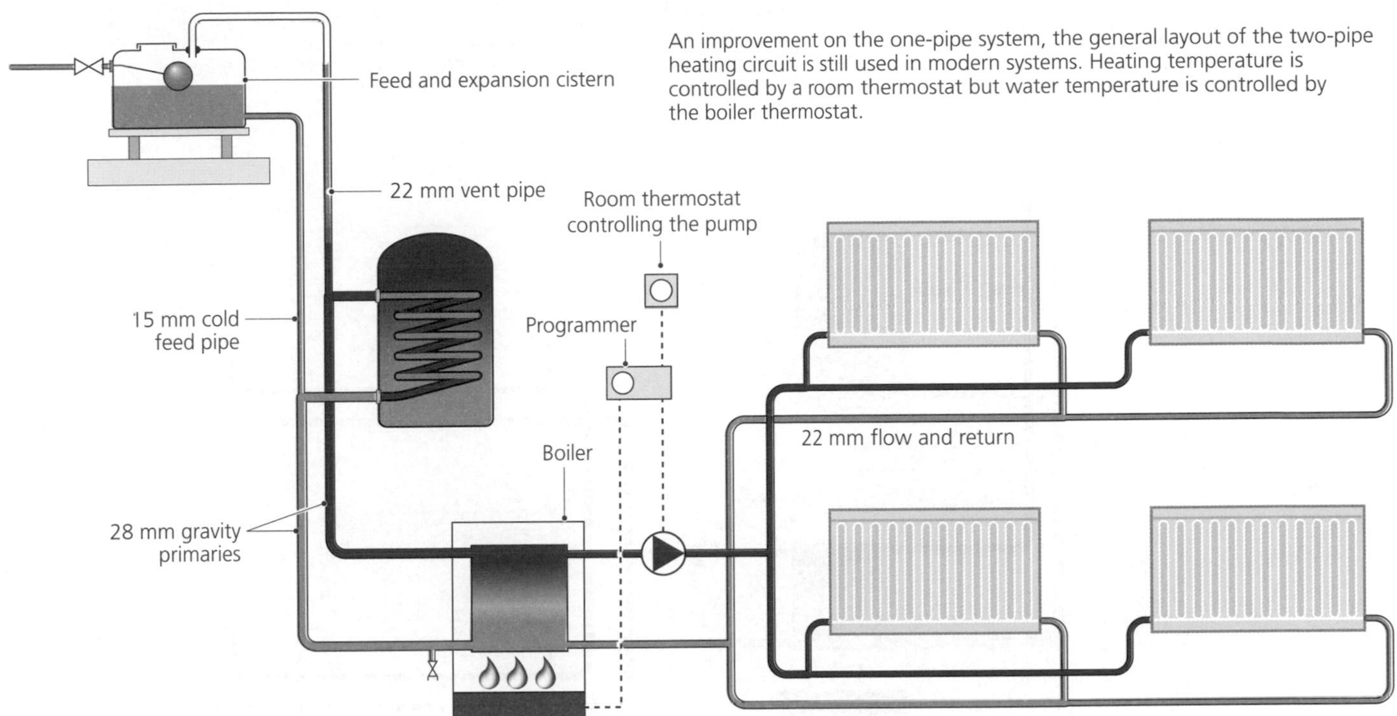

▲ Figure 7.4 Two-pipe semi-gravity system

An improvement on the one-pipe system, the general layout of the two-pipe heating circuit is still used in modern systems. Heating temperature is controlled by a room thermostat but water temperature is controlled by the boiler thermostat.

KEY POINT

The two-pipe **semi-gravity system** is no longer installed as it does not comply with Building Regulations Document L. Systems of this type must be updated to include full thermostatic control over both hot water temperature and room temperatures by the inclusion of separate controls. The updated system is known as the C-plan plus.

KEY TERM

Semi-gravity system: a central heating system that has pumped heating circulation but gravity hot water circulation.

One of the biggest problems with older central heating systems was the lack of temperature control on both the hot water and heating circuits, which meant that the hot water and the radiators became as hot as the water in the boiler. The C-plan went some way towards addressing this problem with the inclusion of a room thermostat that simply switched off the pump when the desired room temperature was reached. The secondary water, however, was still uncontrolled and was often too hot.

▼ Table 7.2 Advantages and disadvantages of the C-plan (two-pipe) system

Advantages	Disadvantages
All of the heat emitters reach the same temperature	Uncontrolled heating of the primary circuit leading to overheating of the domestic secondary hot water
	Constant boiler cycling even when the hot water and heating are up to temperature leads to wastage of fuel energy
The two-pipe system is much quicker at heating up than the one-pipe system; this saves on fuel usage	The system is not Building Regulations Document L compliant and must be updated to C-plan plus system as a minimum standard
	The boilers fitted to this type of system are only about 78% efficient or less
	Condensing-type boilers cannot be fitted to this type of installation because of the gravity circulation needed by the hot water storage cylinder

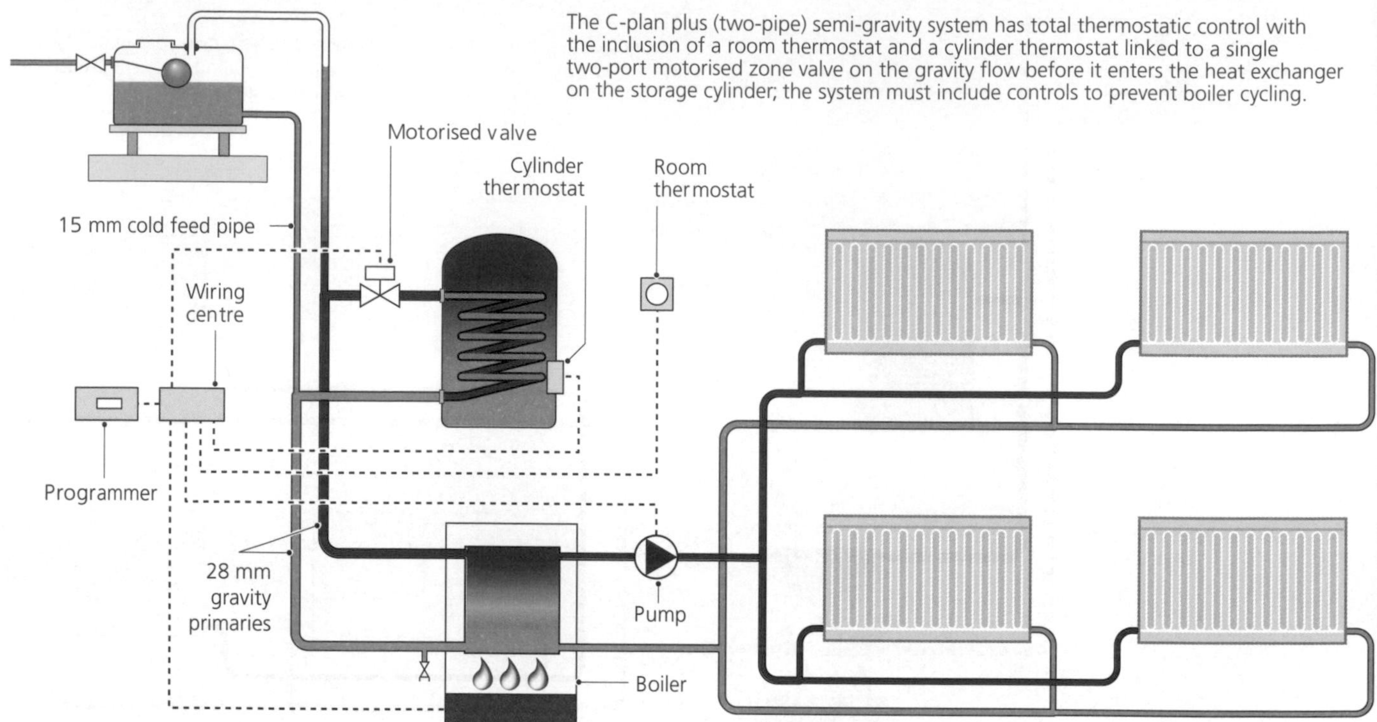
The C-plan plus (two-pipe) semi-gravity system has total thermostatic control with the inclusion of a room thermostat and a cylinder thermostat linked to a single two-port motorised zone valve on the gravity flow before it enters the heat exchanger on the storage cylinder; the system must include controls to prevent boiler cycling.
Motorised valve
Cylinder thermostat
Room thermostat
15 mm cold feed pipe
Wiring centre
Programmer
28 mm gravity primaries
Pump
Boiler

▲ Figure 7.5 C-plan plus system

▼ Table 7.3 Advantages and disadvantages of the C-plan plus system

Advantages	Disadvantages
All of the heat emitters reach the same temperature	The system is not as controllable as more modern fully pumped systems
The two-pipe system is much quicker at heating up than the one-pipe system; this saves on fuel usage	Condensing-type boilers cannot be fitted to this type of installation because of the gravity circulation needed by the hot water storage cylinder
The system is Building Regulations Document L compliant	The boilers fitted to this type of system are only about 78% efficient or less
Full control over both heating and hot water circuits is possible	

The C-plan plus (two-pipe) system

This is a two-pipe system that is basically an updated version of the C-plan system and incorporates full thermostatic control of both heating and hot water circuits. Room temperatures are controlled by a room thermostat and thermostatic radiator valves, while the hot water temperature is controlled by a cylinder thermostat linked to a single two-port motorised zone valve installed on the gravity flow before it enters the heat exchanger at the hot water storage cylinder.

The C-plan plus system is accepted as Building Regulations Document L1b compliant for the updating of existing systems. Some advantages and disadvantages of the C-plan plus system are listed in Table 7.3.

Fully pumped systems

Modern heating systems utilise pumped primary circuits as well as pumped heating circuits. By installing 2 × two-port zone valves or a three-port mid-position valve, the user can have hot water only, heating only or a combination of both. There are three basic types:

1 the Honeywell Y-plan, which uses one three-port motorised mid-position valve
2 the Honeywell W-plan, which uses one three-port motorised diverter valve
3 the Honeywell S-plan, which uses two two-port motorised zone valves.

Fully pumped systems offer a better choice of both system design and boiler choice and, because the need for gravity circulation has been eliminated, fully pumped systems give much greater scope for

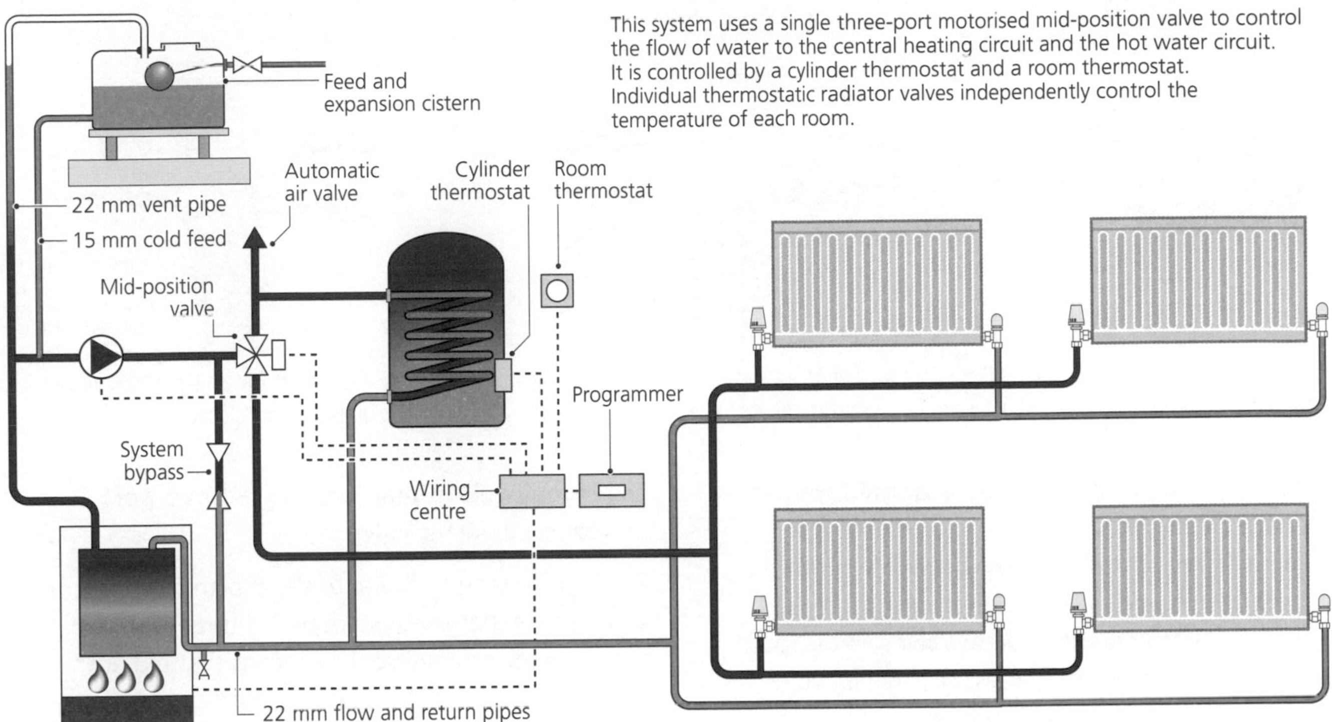
This system uses a single three-port motorised mid-position valve to control the flow of water to the central heating circuit and the hot water circuit. It is controlled by a cylinder thermostat and a room thermostat. Individual thermostatic radiator valves independently control the temperature of each room.
Feed and expansion cistern
Automatic air valve
Cylinder thermostat
Room thermostat
22 mm vent pipe
15 mm cold feed
Mid-position valve
Programmer
System bypass
Wiring centre
22 mm flow and return pipes

▲ Figure 7.6 Y-plan system

installation options, especially when positioning the boiler, as the need for the boiler to be lower than the storage cylinder is no longer a consideration.

Full thermostatic control is available to both hot water and heating circuits by means of a cylinder thermostat, a room thermostat and thermostatic radiator valves.

Fully pumped systems heat up much more quickly than semi-gravity systems, offering savings on fuel and operating costs, and both Y-plan and S-plan systems can be used with natural gas, liquid petroleum gas (LPG) and oil appliances.

The Honeywell Y-plan utilising a three-port motorised mid-position valve

The three-port motorised mid-position valve controls the flow of water to the primary (cylinder) circuit and the heating circuit. The valve reacts to the demands of the cylinder thermostat or the room thermostat.

An outline of the operating sequence of the Y-plan system

1. At a set time, the programmer activates the system calling for both hot water and heating.
2. With the motorised valve in the mid-position, water from the boiler circulates around both primary and heating circuits. The boiler fires up and the circulating pump begins to circulate the water.
3. a or b:
   a. When the cylinder reaches temperature, the valve is energised by the cylinder thermostat, which closes the hot water port, preventing water flowing to the hot water cylinder heat exchanger, or
   b. When the room reaches its set temperature, the valve is energised by the room thermostat, which closes the heating port, preventing water flowing to the heating circuit.
4. With both the room thermostat and the boiler thermostat satisfied, the pump and the boiler shut down and the valve returns to the mid-position. In this condition, the system will only operate should either the room thermostat or cylinder thermostat call for heat. This is known as **boiler interlock**.

KEY TERM

Boiler interlock: this is not a physical item, but the way in which a modern system is wired up to prevent the boiler from firing unless there is a demand for heat.

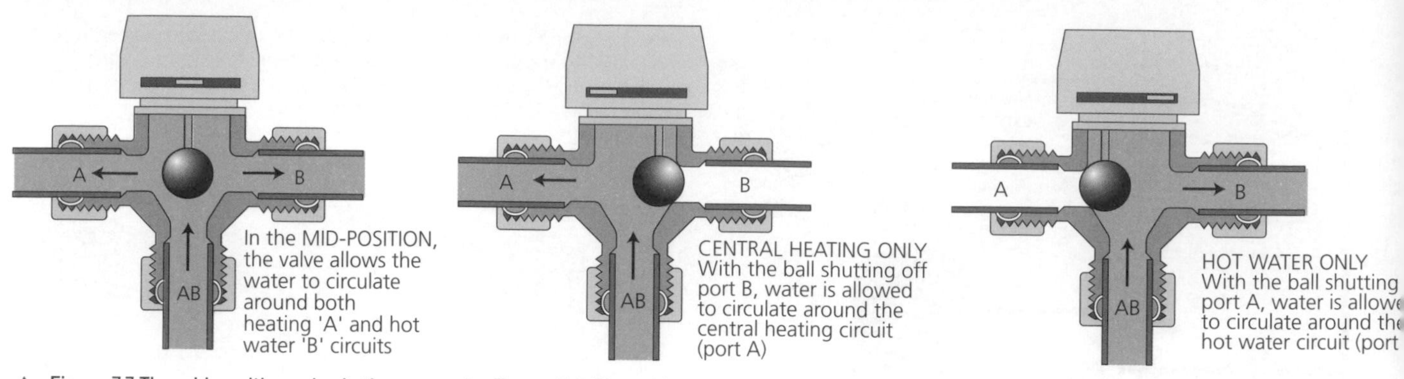

▲ Figure 7.7 The mid-position valve in the mid-position serving heating and hot water

▲ Figure 7.8 The mid-position valve with the hot water port closed

▲ Figure 7.9 The mid-position valve with the heating port closed

The system contains a system bypass fitted with an automatic bypass valve that simply connects the flow pipe to the return pipe. The bypass is required when all circuits are closed either by the motorised valve or the thermostatic radiator valves as the rooms reach their desired temperatures. The bypass valve opens automatically as the circuits close, to protect the boiler from overheating by allowing water to circulate through the boiler, keeping the boiler below its maximum high temperature. This prevents the boiler from 'locking out' on the overheat energy cut-out.

INDUSTRY TIP

On a three-port mid-position valve, it is important to pipe up the correct port:

- AB: Flow from the boiler
- A: Flow to the central heating circuit
- B: Flow to the hot water cylinder

(Remember 'B' for bath where you need hot water!).

KEY POINT

Locking out on the overheat high-limit thermostat

Modern boilers contain two thermostats. The first controls the temperature of the water inside the boiler and can be set by the user up to a maximum of 82°C. The second is for protection of the boiler and is known as the 'high- limit' thermostat. Its job is to protect the boiler from overheating by shutting it down or 'locking out' when a temperature of around 85°C is reached. High-limit thermostats are manually resettable by pushing a small button on the boiler itself.

The Honeywell W-plan utilising a three-port motorised diverter valve

The W-plan is very similar to the Y-plan. The main difference is that the system uses a three-port motorised diverter valve. This means that either the hot water circuit or the heating circuit can be opened but not both circuits at the same time. It is known as a hot water priority system. If both circuits are calling for heat, the heating circuit will not open until the hot water circuit is satisfied.

The W-plan system is not recommended where a high hot water demand is required as it would lead to the space heating temperature dropping below comfort levels. The pipework layout is identical to that of the Y-plan.

The Honeywell S-plan utilising two two-port motorised zone valves

The S-plan has two two-port motorised zone valves to control the primary and heating circuits separately by the cylinder and room thermostats respectively. This system is recommended for dwellings with a floor area greater than 150 m² because it allows the installation of additional two-port zone valves to zone the upstairs heating circuit from the downstairs circuit. A separate room thermostat and possibly a second time clock/programmer would also be required for upstairs **zoning**.

KEY TERM

Zoning: a process where living spaces and sleeping spaces are individually controlled via independent time clocks, room thermostats and motorised zone valves.

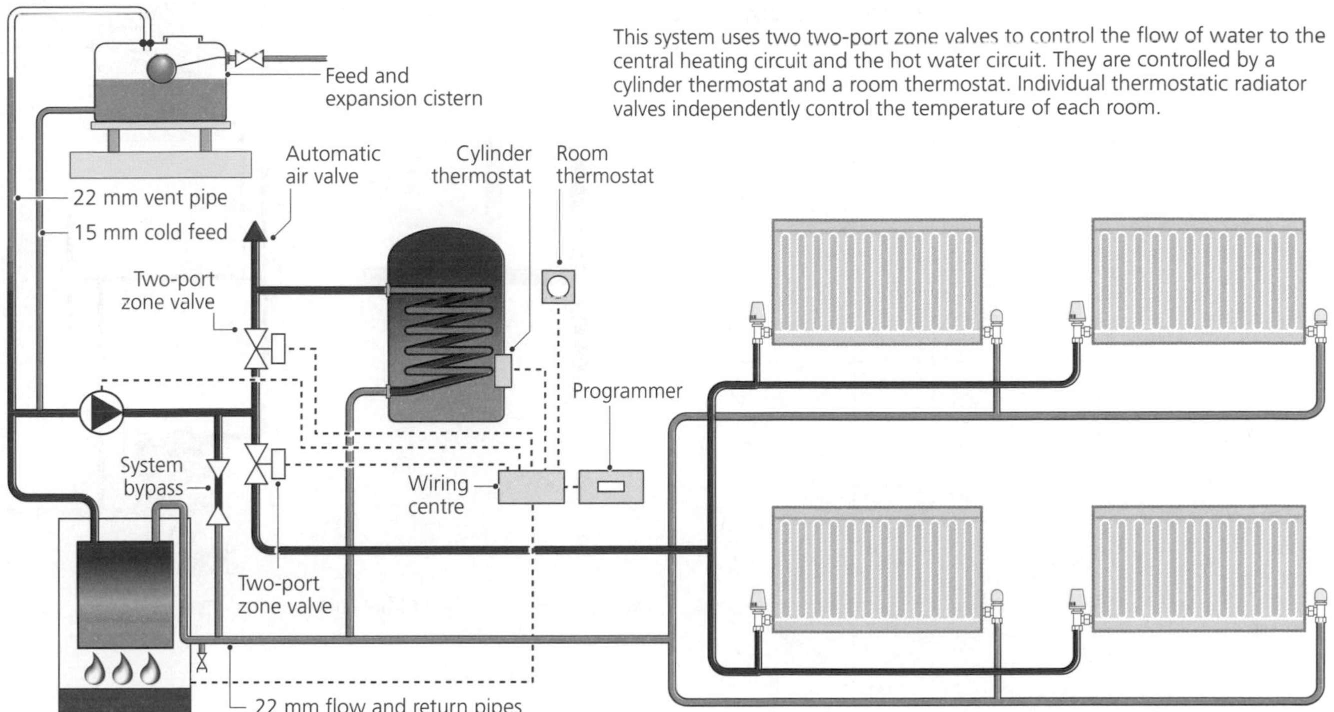
This system uses two two-port zone valves to control the flow of water to the central heating circuit and the hot water circuit. They are controlled by a cylinder thermostat and a room thermostat. Individual thermostatic radiator valves independently control the temperature of each room.
Feed and expansion cistern
Automatic air valve
Cylinder thermostat
Room thermostat
22 mm vent pipe
15 mm cold feed
Two-port zone valve
Programmer
System bypass
Wiring centre
Two-port zone valve
22 mm flow and return pipes

▲ Figure 7.10 S-plan system

As with the Y-plan, a system bypass is required for overheat protection of the boiler.

An outline of the operating sequence of the S-plan system

1 At a set time, the programmer activates the system calling for both hot water and heating.
2 Both of the two-port motorised zone valves open, and water from the boiler circulates around both primary and heating circuits. The boiler fires up and the circulating pump begins to circulate the water.
3 a or b:
   a When the cylinder reaches temperature, the two-port zone valve is energised by the cylinder thermostat, which closes the hot water zone valve preventing water flowing to the hot water cylinder heat exchanger, or
   b When the room reaches its set temperature, the two-port zone valve is energised by the room thermostat, which closes the valve preventing water flowing to the heating circuit.
4 With both the room thermostat and the boiler thermostat satisfied, the pump and the boiler shut down. In this condition, the system will operate only should either the room thermostat or cylinder thermostat call for heat. This is known as 'boiler interlock'.

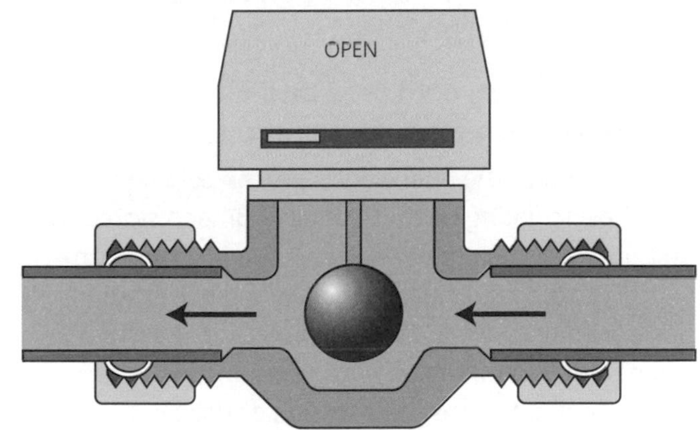
OPEN

▲ Figure 7.11 Zone valve open

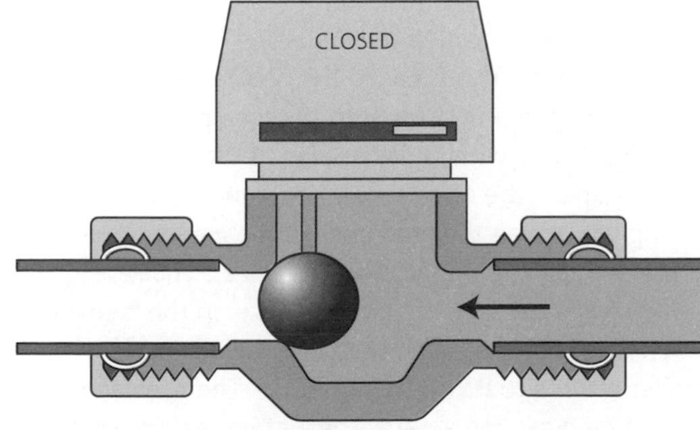
CLOSED

▲ Figure 7.12 Zone valve closed

▼ Table 7.4 The comparisons between the Y-plan, W-plan and S-plan systems

	Full thermostatic control	Building Regulations Document L compliant	Recommended for larger properties	Can be used with sealed (pressurised) systems	Can be used with system boilers	Can be zoned	Anti-cycling boiler interlock
Y-plan system	✓	✓	×	✓	✓	×	✓
W-plan system	✓	✓	×	✓	✓	×	✓
S-plan system	✓	✓	✓	✓	✓	✓	✓
S-plan plus system	✓	✓	✓	✓	✓	✓	✓

*The S plan plus system has the extra two-port valve to zone the downstairs and upstairs heating circuits separately (see Figure 7.18).

The open vent, cold feed and circulating pump position for fully pumped systems

The position of the open vent pipe, the cold feed pipe and the circulating pump to a fully pumped system is an important part of the system design. If the open vent pipe, the feed pipe and circulating pump are positioned onto the system incorrectly, the system will not work properly and may even induce system corrosion due to constant aeration of the system water.

The open vent and the cold feed should be positioned on the flow from the boiler on the suction side of the circulating pump with a maximum of 150 mm distance between them. This is called the neutral point, as the circulating pump acts on both the feed pipe and the open vent pipe with equal suction. If they are any further apart, the neutral point becomes weak and the pump will act on the feed pipe with a greater force than the open vent pipe. This creates an imbalance, which leads to a lowering of the water in the feed and expansion cistern. When the pump switches off, the water returns to its original position. The constant see-sawing motion aerates the water creating corrosion within the system.

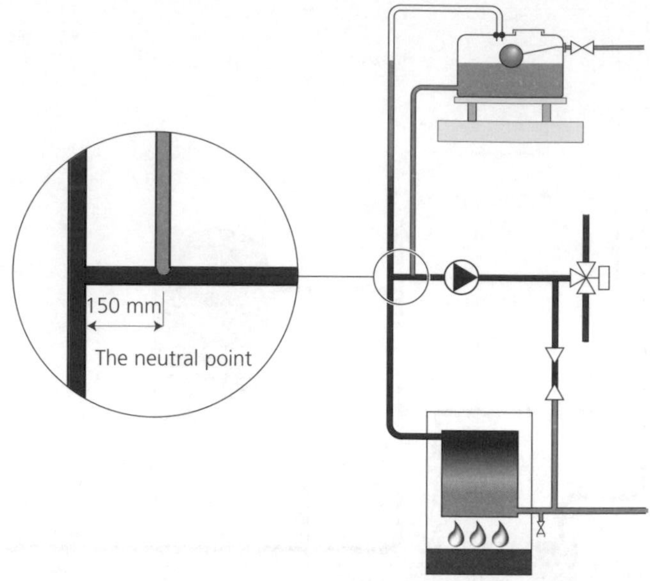

▲ Figure 7.13 Position of the cold feed and open vent pipes

The circulating pump

The circulating pump must also be positioned with care to avoid design faults that could lead to problems with corrosion by aeration of the water due to water movement in the feed and expansion cistern. This occurs when water is either pushed up the cold feed pipe and the open vent pipe or is circulated between the cold feed pipe and the open vent pipe.

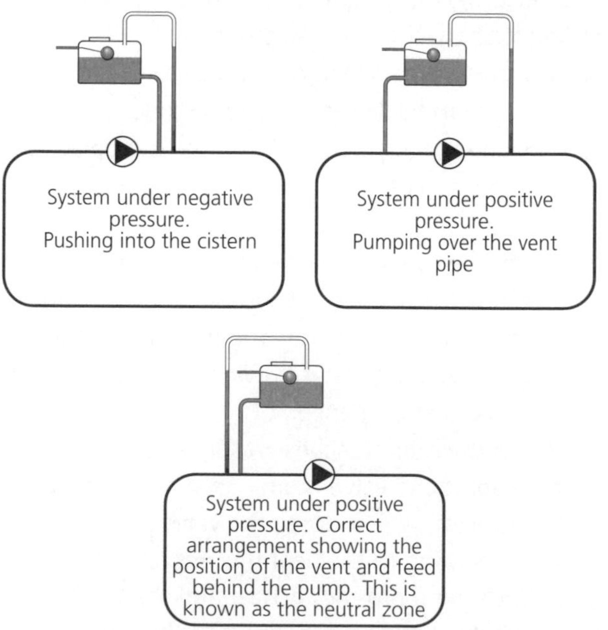

System under negative pressure. Pushing into the cistern
System under positive pressure. Pumping over the vent pipe
System under positive pressure. Correct arrangement showing the position of the vent and feed behind the pump. This is known as the neutral zone

▲ Figure 7.14 The position of the circulating pump

> **KEY POINT**
>
> **What is aeration of the water?**
>
> Central heating systems do not like air. Air is one of the biggest causes of corrosion in heating systems because the air in the water contributes to rust occurring throughout the system and the formation of red oxide sludge. Water alone will not cause corrosion, even with ferrous metals present, such as radiators and convectors. It is the air present in the water that causes metals to rust and constant water movement at the feed and expansion cistern will aerate the water in the system enough for corrosion to take place.

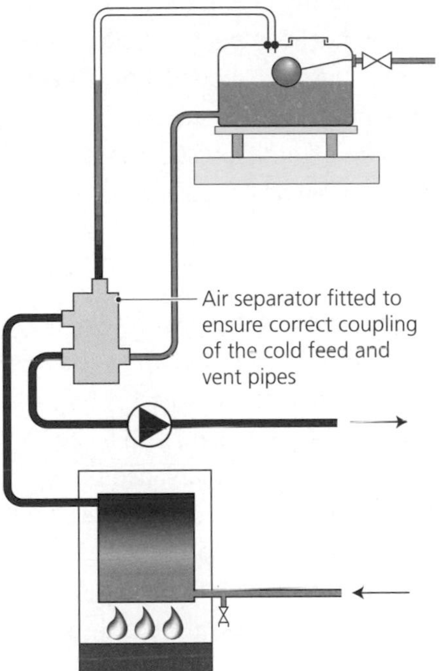
Air separator fitted to ensure correct coupling of the cold feed and vent pipes

▲ Figure 7.15 The use of an air separator

The circulating pump, or to give it its correct name the 'hydronic central heating circulator', is a simple electric motor with a fluted water wheel-like impeller that circulates the water around the system by centrifugal force. The faster the impeller rotates, the greater the circulation that occurs in the system. For quiet operation of the system, the flow rate should not exceed 1 litre per second and 1.5 litres per second for microbore systems (see page 371 of this chapter). Most domestic circulating pumps have three speeds, which correspond to varying circulatory pressures or 'heads'. Domestic circulating pumps have either a 6 m head to circulate up to a height of 6 m, or a 10 m head to circulate up to a height of 10 m.

The use of air separators

The use of an air separator helps in the positioning of the feed and vent by ensuring that the neutral point is built in to the system. The positioning of the pipework on an air separator creates a turbulent water flow in the separator body and this helps to remove air from the system, which makes the system quieter in operation and significantly reduces the risk of corrosion.

▲ Figure 7.16 An air separator; the top connection is the vent to release the air

> **ACTIVITY**
>
> Work out which pipes are connected to each of the air separator designs:
>
> - flow for the boiler
> - flow to the system
> - open vent
> - cold feed.

The feed and expansion cistern

Open vented systems contain a feed and expansion cistern, which fulfils three important functions:

1. it is the means by which water enters the system for filling and top-up
2. it allows space for the system water to expand into when it is heated
3. it provides a static head (or water pressure) to the system.

Generally, the size of the F and E cistern will depend on the size of the system, but for most domestic systems an 18-litre cistern is recommended. The bigger the system, the more water it will contain and so the water expansion will be greater. The water level in the cistern should, therefore, be set at a low level.

The cistern must be located at the highest part of the system and must not be affected by the operation of the circulating pump. For fully pumped systems, the cistern must be at least 1 metre above the highest part of the pumped primary flow to the heat exchanger in the hot water storage cylinder. For gravity systems, the minimum height of the cistern can be calculated by taking the maximum operating head of the pump and dividing it by 3.

The cold feed for the system for most domestic properties is 15 mm. The cold feed pipe should not contain any service or isolation valves. This is to ensure that there is a supply of cold water in the event of overheating and leakage, preventing the system from boiling.

Should the valve be inadvertently closed, a dangerous situation could develop, especially if the vent is also blocked as the pressure will build up in the system, raising the boiling point of the water to dangerous levels. Both the cistern and any float-operated valve it may contain must be capable of withstanding hot water at a temperature close to 100°C.

Primary open safety vent

The purpose of the open vent pipe is one of safety. The open vent is installed to:

- provide a safety outlet should the system overheat due to a component failure
- ensure that the system always remains at atmospheric pressure limiting the boiling point to 100°C.

In a fully pumped system, the height of the open vent should be a minimum of 450 mm from the water level in the cistern to the top of the open vent pipe. This is to allow for any pressure surges created by the circulating pump. The minimum size of pipe for the open vent is 22 mm and this, like the cold feed pipe, should not be fitted with any valves.

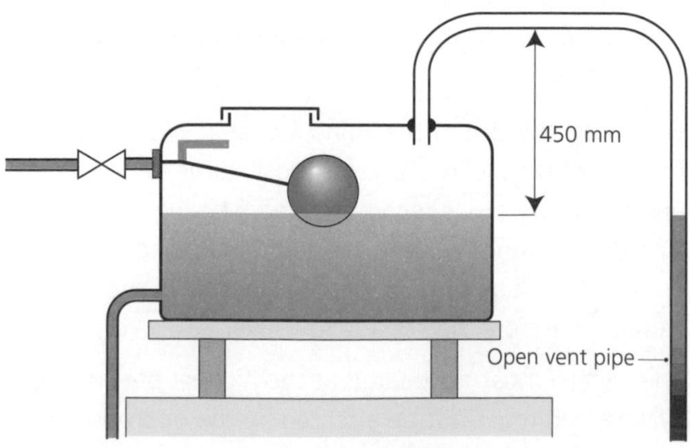

▲ Figure 7.17 Height of the open vent pipe

Sealed (pressurised) heating systems

Sealed heating systems are those that do not contain a feed and expansion cistern but are filled with water direct from the mains cold water supply via a temporary filling loop. Large systems would be filled via an automatic pressurisation unit. The expansion of water is taken up by the use of an expansion vessel and the open vent is replaced by a pressure relief valve, which is designed to relieve the excess pressure by releasing the system water and discharging safely to a drain point outside of the dwelling. This is vital as the water may be in excess of 80°C. A pressure gauge is also included so that the pressure can be set when the system is filled, and periodically checked for rises and falls in the pressure as these could indicate a potential component malfunction. The system is usually pressurised to around 1 bar. There are several types of fully pumped alternatives:

- sealed systems with an external pressure vessel
- system boilers that contain all necessary safety controls
- combination boilers.

All fully pumped systems, such as those with two, or three or more, two-port zone valves (known as the S-plan and the S-plan plus), or a three-port mid-position valve (known as the Y-plan) or a three-port diverter valve (known as the W-plan), can be installed as sealed systems or can be purpose-designed 'heating only' systems using a combination boiler with instantaneous hot water supply. All the pipework layouts described below can be used with the three boiler systems above.

> **KEY TERM**
>
> **Sealed heating systems:** heating systems that are sealed from the atmosphere and operate under pressure. They do not contain a feed and expansion cistern. Instead, they have an expansion vessel to take up water expansion and a filling loop to fill the system from the cold water main.

Fully pumped systems with three or more two-port zone valves (known as the S-plan plus)

The S-plan plus has three or more two-port motorised zone valves to control the primary and heating circuits separately by the cylinder and room thermostats respectively. This system is recommended for dwellings with a floor area greater than 150 m² because it allows the installation of additional two-port zone valves to zone the upstairs heating circuit from the downstairs circuit. A separate room thermostat, and possibly a second time clock/programmer, would also be required for upstairs zoning. A system bypass is required for overheat protection of the boiler.

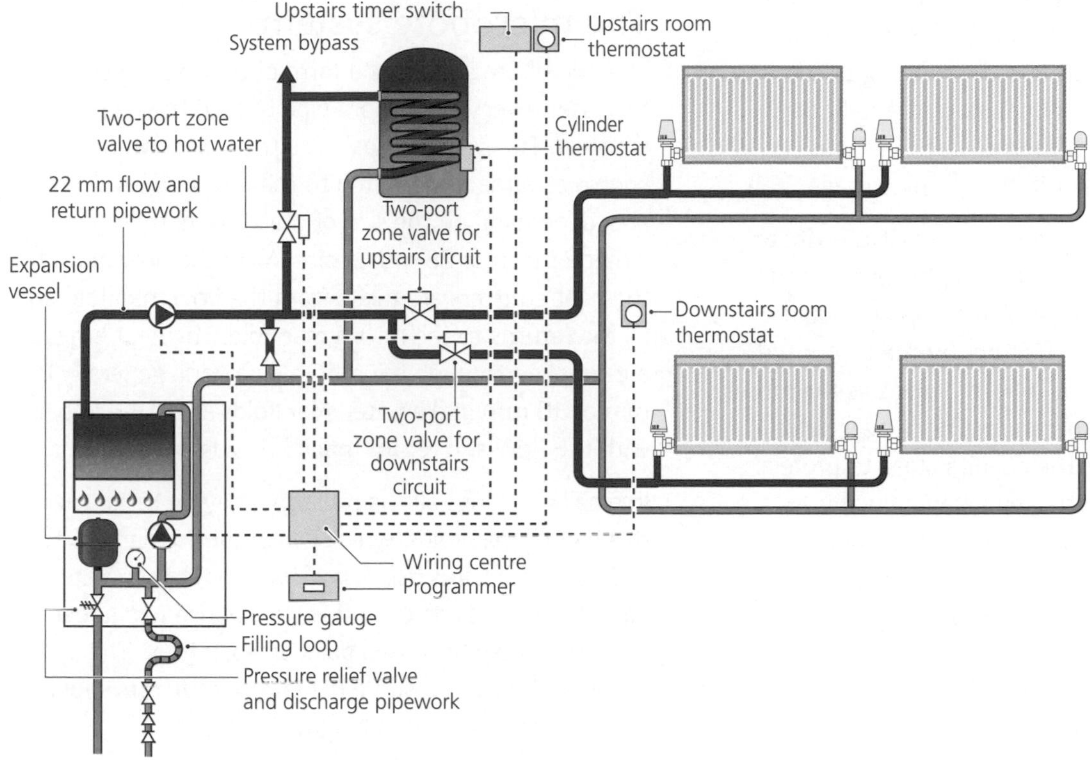

▲ Figure 7.18 The sealed S-plan plus system

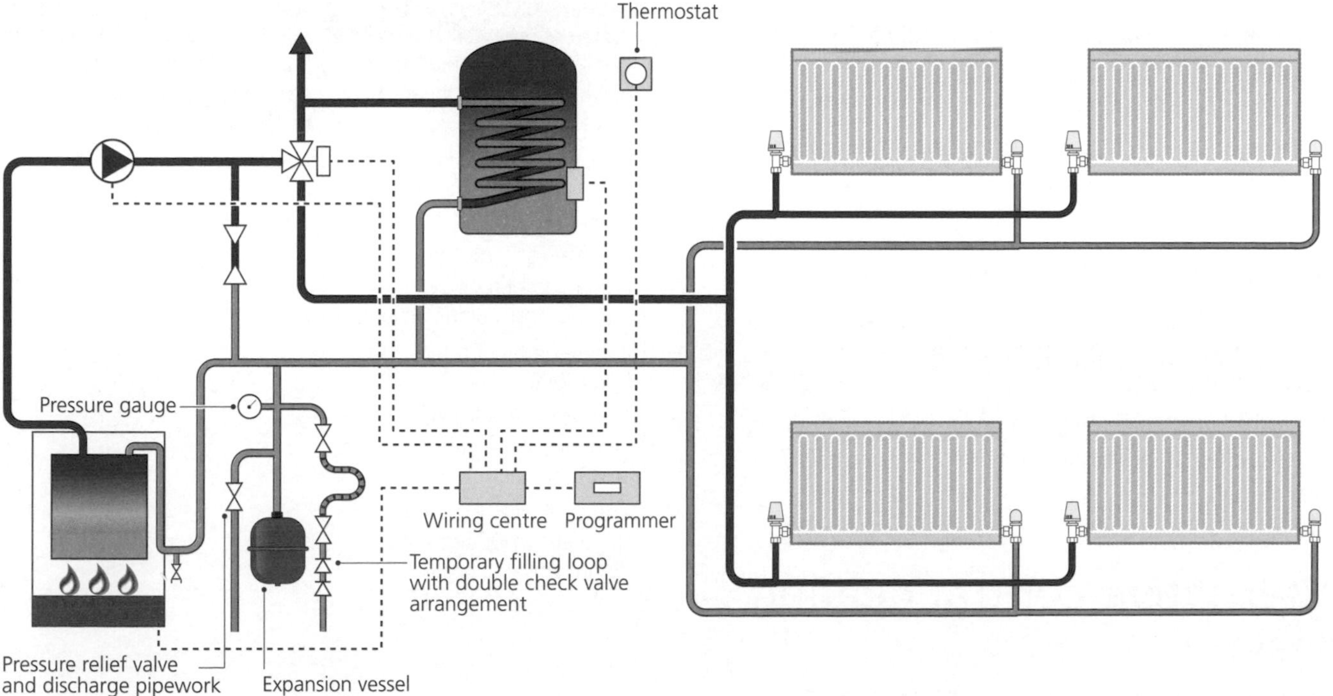

▲ Figure 7.19 The sealed Y-plan system

Fully pumped systems with three-port mid-position valve (known as the Y-plan) or a three-port diverter valve (known as the W-plan)

The three-port mid-position valve (Y-plan) or diverter valve (W-plan) controls the flow of water to the primary (cylinder) circuit and the heating circuit. The valve reacts to the demands of the cylinder thermostat or the room thermostat. This was discussed in detail earlier in the chapter (page 363).

The system contains a system bypass fitted with an automatic bypass valve, which simply connects the flow pipe to the return pipe. The bypass is required when all circuits are closed either by the motorised valve or the thermostatic radiator valves as the rooms reach their desired temperature. The bypass valve opens automatically as the circuits close to protect the boiler from overheating by allowing water to circulate through the boiler, keeping the boiler below its maximum high temperature. This prevents the boiler from 'locking out' on the overheat energy cut-out.

Sealed system components

As we have already seen, sealed systems do not contain a feed and expansion cistern, nor open vent pipe. Instead, these systems incorporate the following components:

- an external expansion vessel fitted to the system return
- a pressure relief valve
- the system is filled via a temporary filling loop or a CA disconnection device
- a pressure gauge.

These will be covered later in this chapter.

INDUSTRY TIP

Instead of an external filling loop, some boiler manufacturers have an integral CA disconnection device (see pages 247 and 391).

Alternative central heating designs

Apart from the central heating systems we have already looked at, there are two other pipework arrangements that can be installed in domestic premises. These are:

1. the microbore system
2. the reversed return system.

The microbore system

The microbore system is a form of two-pipe system that uses a very small-bore pipe to feed the heat emitters. The system uses a multi-connection fitting, known as a manifold, fitted to the flow and return pipes and, depending on the size of the system, these are either 22 mm or 28 mm in size. All of the flow pipes to the heat emitters are taken from the flow manifold and all the returns to the return manifold. The heat emitters are supplied through microbore pipework, generally 8 mm or 10 mm in diameter. Manifolds are fitted in pairs with the flow and return manifolds side by side.

In small dwellings all the radiators may be taken from one pair of manifolds, which can accommodate up to eight radiators. It is usual, however, to fit a separate pair of manifolds on each floor in a house and larger properties may have two pairs on each floor. The pipework loops that serve the largest radiators should be no more than 9 m in length.

▼ Table 7.5 Advantages and disadvantages of the microbore system

Advantages	Disadvantages
Contains only a small amount of water and so is heated quickly	Microbore piping is easily damaged and not very resistant to knocks
Microbore tubing comes in fully annealed coils, is easily bent by hand and is easily hidden	Microbore tubes can easily get blocked with sludge if the system is installed poorly
It can sometimes be a cheaper form of installation	
Long lengths of tubing mean fewer joints	
Can be used with sealed and open vented systems, Y-plan or S-plan	
The system is Building Regulations Document L compliant	

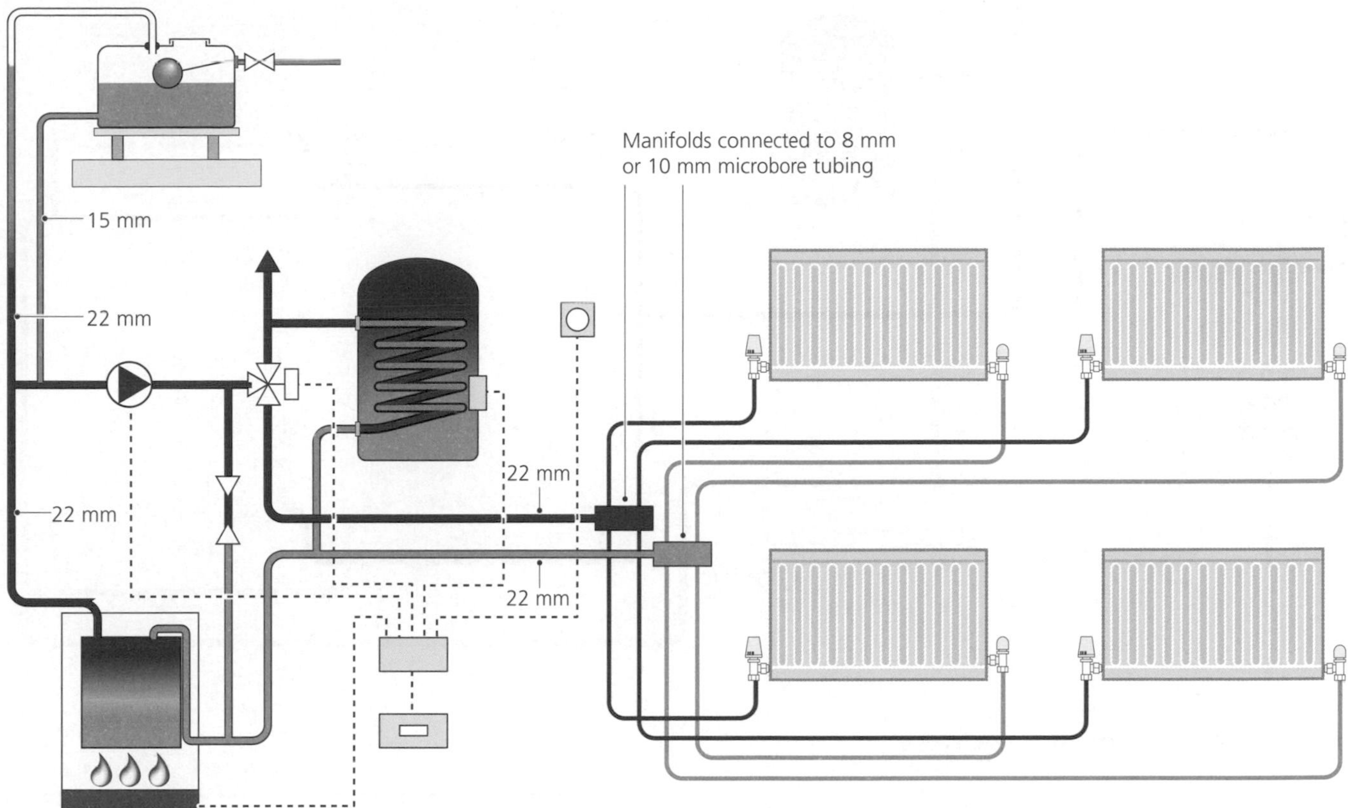

▲ Figure 7.20 Microbore system

▲ Figure 7.21 Microbore manifolds

The reversed return system

The reversed return system is designed for larger systems and is a variation on the two-pipe system. In the reversed return system, the flow and returns are connected, as before, to separate flow and return pipes, but the return travels away from the boiler in the same direction as the flow before looping around to be connected to the return at the boiler. By doing this, the amount of pipe used on both flow and the return is almost equal, which has the effect of ensuring that all of the heat emitters reach full temperature at about the same time. Reversing the return makes balancing the system much quicker and easier and, in some cases, balancing is eliminated completely.

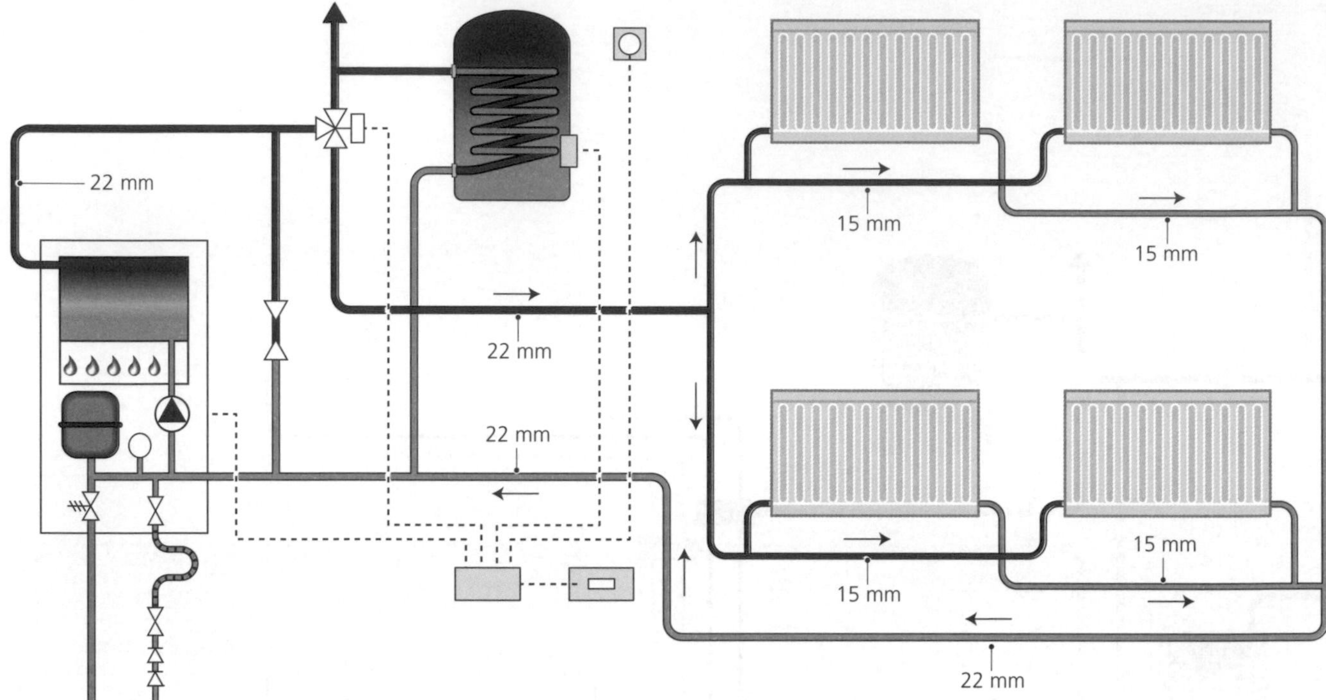

▲ Figure 7.22 The reversed return system

▼ Table 7.6 Advantages and disadvantages of the reversed return system

Advantages	Disadvantages
Eliminates the need for complex balancing procedures	It is difficult to install
Can be used with sealed and open vented systems, Y-plan or S-plan	It is a more expensive system due to the extra time taken on installation and the extra materials required
The system is Building Regulations Document L compliant	The system installation requires careful planning and design

KEY POINT

The legal requirements for the installation of solid fuel and oil heat-producing appliances, such as boilers, cookers and room heaters, are covered in Building Regulations Document J: Heat Producing Appliances. The legal requirements for the installation of gas appliances are given in the Gas Safety (Installation and Use) Regulations. In all cases, the manufacturer's instructions must always be followed when installing heat-producing appliances of any kind.

The governing bodies for the different fuels used with heating appliances are:

- gas – Gas Safe, www.gassaferegister.co.uk
- oil – OFTEC, www.oftec.org
- solid fuel – HETAS, www.hetas.co.uk

Heat-producing appliances

So far in this chapter, we have looked at the different central heating systems and their layouts. In this part of the chapter, we will investigate the different appliances we can use to generate the heat required to warm the systems and the different fuels they use.

Boilers used for central heating systems are generally heated by one of three different fuels. These are:

1 solid fuel
2 gas
3 oil.

Solid fuel appliances

The most common types of solid fuel appliances are:

- open fires with a high-output back boiler
- room heaters
- cookers (Aga type)
- independent boilers.

▼ Table 7.7 Comparison of appliance types to fuel types

	Open flued	Room sealed (natural draught)	Room sealed (fan assisted)	Freestanding / independent boilers	Wall mounted	Condensing	Non-condensing (traditional)	System boilers	Cookers	Open fire with high output back boiler	Room heaters
Solid fuel	✓	×	×	✓	×	×	✓	×	✓	✓	✓
Gas	✓	✓	✓	✓	✓	✓	✓	✓	✓	×	×
Oil	✓	×	✓	✓	✓	✓	✓	✓	✓	×	×

Open fires with a high-output back boiler

High-output back boilers are installed behind a real open coal fire. These appliances give their heat output in two forms:

1. radiation from the open fire for direct room heating
2. hot water from the boiler, which is available for domestic hot water supply and central heating.

▲ Figure 7.23 Solid fuel high-output back boiler

These appliances work on an open flue or chimney, and contain a manual flue damper to regulate the amount of updraught through the chimney. By regulating the updraught, a certain amount of control can be administered over the heat of the fire. Typically, with the damper open, a fire of this type will give around 6.8 kW to 10 kW of hot water heating output and, with the damper closed, outputs vary from 5.3 kW to 8.4 kW. Radiated heat outputs from the coal fire directly into the room peak at around 2.6 kW.

Room heaters

A solid fuel room heater is an enclosed appliance usually with a glass door so that the fire can be viewed. They are installed directly into a chimney or open flue capable of accepting solid fuel, and can either be stand alone or fitted into chimney breasts with a high-output back boiler capable of serving up to ten heat emitters. Room heaters provide radiant heat for direct warmth and a constant circulation of convected heat.

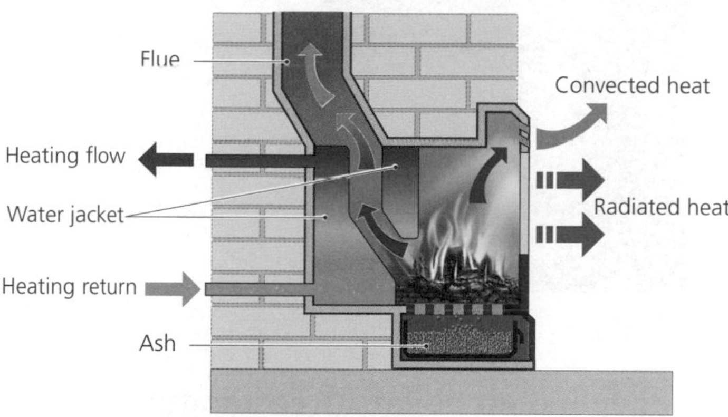

▲ Figure 7.24 A room heater cut-away

▲ Figure 7.25 A solid fuel room heater

Solid fuel cookers (Aga type)

Open-flued solid fuel cookers have been around for many years. The concept of the solid fuel cooker is very simple: a controllable fire, burning continuously, inside a well-insulated cast iron shell, which retains the heat. When cooking is required, the heat is transferred to the ovens. The hot plates, because they are always hot, are covered with insulated cast iron covers, which lift up when hot-plate cooking is required. Many models provide hot water and central heating as well as radiated heat in the room where they are fitted.

Solid fuel cookers burn a wide variety of solid fuels, including wood, and all have easy to empty ash pans so that the fire never goes out.

▲ Figure 7.26 An Aga-type solid fuel cooker

Independent boilers (freestanding)

Domestic open-flued independent solid fuel boilers are designed to provide both domestic hot water and central heating in a whole range of domestic premises, from the very large to the very small.

There are two main types of independent boiler for domestic use. These are as follows.

1. **Gravity feed boilers:** often called hopper-fed boilers, these appliances incorporate a large hopper, positioned above the firebox, which can hold two or three days' supply of small-sized anthracite. The fuel is fed automatically to the fire bed as required and an in-built, thermostatically controlled fan aids combustion. This provides a rapid response to an increase in demand. They are available in a wide range of sizes and outputs.
2. **Batch feed boilers:** these are 'hand fired' appliances requiring manual stoking. They require much more refuelling than hopper-fed boilers. They can, however, be less expensive to run in some cases and will often operate without the need for an electrical supply, thereby providing hot water and central heating during power failure.

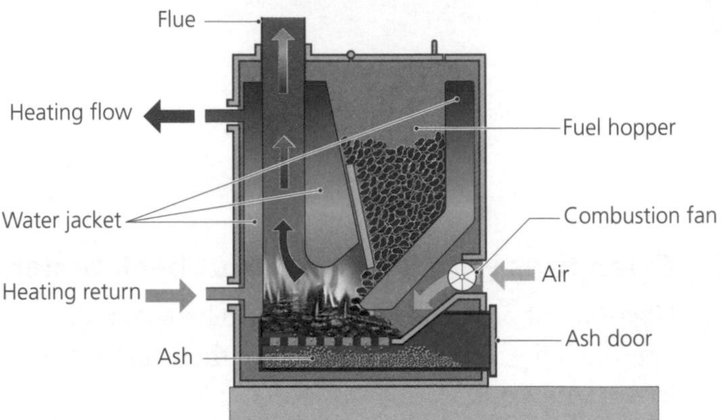

▲ Figure 7.27 Gravity-fed boiler

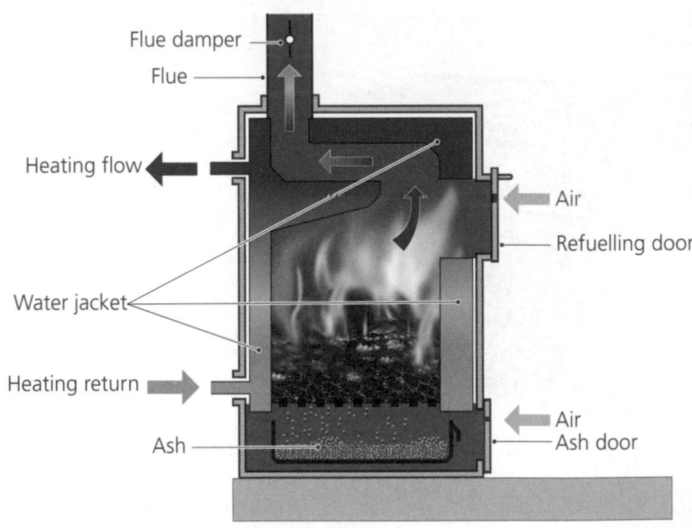

▲ Figure 7.28 Batch fed boiler

> **HEALTH AND SAFETY**
>
> The main danger with gravity feed boilers is the risk of fire in the hopper. The fuel fed to the fire bed needs to be regulated with care.

Gas central heating boilers

Gas central heating boilers are the most popular of all central heating appliances. Over the years there have been many different types, from large multi-sectional

▼ Table 7.8 The different kinds of boiler and their flue arrangements

	Energy efficient	Cast iron heat exchanger	Low water content	Open vented system	Sealed (pressurised) system	Open flue	Room sealed (natural draught)	Room sealed (fan assisted)	Wall mounted	Freestanding
Traditional boilers	×	✓	✓	✓	✓	✓	✓	✓	✓	✓
Condensing boilers	✓	×	✓	×	✓	×	×	✓	✓	✓
System boilers	✓	×	✓	×	✓	×	×	✓	✓	×
Combination boilers	✓	×	✓	×	✓	×	✓	✓	✓	✓

cast iron domestic boilers to small, low water content condensing types. Both natural gas (those that burn a methane-based gas) and LPG (those that burn propane) types are available. Gas burning boilers may be converted into hydrogen boilers in the future as environmental considerations are made.

Central heating boilers can be categorised as:

- traditional boilers (non-condensing)
- traditional (condensing)
- cast iron heat exchanger
- low water content heat exchanger
- combination boilers
- condensing boilers (system and combination boilers).

KEY POINT

The different types of commercially available gas are dealt with in Chapter 10, Domestic fuel systems.

Traditional boilers (non-condensing)

Traditional non-condensing boilers have been around for many years and in many different forms. Usually, they contain cast iron heat exchangers, although some models are low water content with copper or aluminium heat exchangers.

Traditional boilers (condensing)

A condensing traditional boiler does not give instantaneous hot water. It requires to be installed in conjunction with a hot water storage system and heating system. It contains an expansion vessel, filling loop and pressure relief valve, and does not require a feed and expansion cistern.

Boilers with cast iron heat exchangers

For many years, boilers were made with cast iron heat exchangers. They were often very large and heavy, even for small domestic systems. Some heat exchangers were made from cast iron, which was cast in a single block, while older types were made up of cast iron sections that were bolted together. The more sections a boiler had, the bigger the heat output.

Fuel efficiency was, typically, 55 to 78 per cent, with much wasted heat escaping through the flue. Most traditional boilers were fitted onto open vented systems but sealed (pressurised) systems could also be installed with the inclusion of an external expansion vessel and associated controls.

Cast iron boilers can be found either freestanding (floor mounted) or wall mounted, using a variety of flue types:

- open flued
- room sealed (natural draught)
- fan-assisted room sealed (forced draught).

▼ Table 7.9 Advantages and disadvantages of cast iron heat exchangers

Advantages	Disadvantages
Long lasting, typically 20 to 30 years Very robust	Heavy Not energy efficient Do not comply with Building Regulations Document L Noisy Very basic boiler controls

Boilers with low water content heat exchangers

Low water content heat exchangers were usually made from copper tube with aluminium fins or lightweight cast iron. They were an attempt to reduce the water content of the heating system, thus speeding up heating times and improving efficiency. Typical efficiencies for this type of boiler were around 82 per cent.

The boilers were always wall mounted, very light in weight and, as a consequence, often quite small in size, designed for fully pumped S- and Y-plan heating systems only. They were the first generation of central heating boilers to use a high temperature limiting thermostat (or energy cut-out) to guard against overheating, and often used a basic printed circuit board to initiate a pump overrun, which kept the pump running for a short period after the boiler had shut down. It was required to dissipate any latent heat build-up in the water in the heat exchanger as this could 'trip' the energy cut-out resulting in boiler lock-out.

Low water content boilers can be found with a variety of flue types:

- open flued
- room sealed (natural draught)
- fan-assisted room sealed (forced draught).

▼ Table 7.10 Advantages and disadvantages of low water content boilers

Advantages	Disadvantages
Light in weight Often a cheaper appliance Relatively fast water-heating times	Not energy efficient Do not comply with Building Regulations Document L Could be very noisy Relatively short working life High maintenance compared with other boilers

Combination boilers (non-condensing)

Combination boilers that supply instantaneous hot water as well as central heating have been around for many years. Early models, although wall mounted, were very large. Most had a sealed (pressurised) heating system but some were of the low pressure, open vented type. Hot water flow rates were often poor by comparison to modern condensing types.

Early combination boilers can be found with a variety of flue types:

- open flued
- room sealed (natural draught)
- fan-assisted room sealed (forced draught).

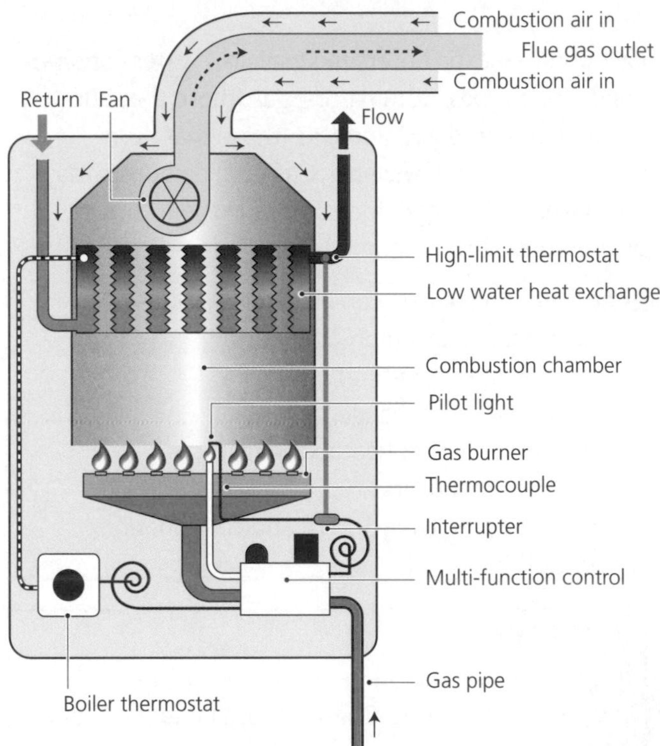

▲ Figure 7.29 A fan-assisted low water content boiler

▼ Table 7.11 Advantages and disadvantages of non-condensing combination boilers

Advantages	Disadvantages
Instantaneous hot water supply Sealed system means no F and E cistern required in the roof space	Not energy efficient Very large in size Do not comply with Building Regulations Document L Can be very noisy High maintenance compared with other boilers Poor hot water flow rates Difficult to maintain

Condensing boilers

A more recent addition to the gas central heating family is the condensing boiler. These work in a very different way from the traditional boiler.

Natural gas, when it is combusted, contains CO_2, nitrogen and water vapour. As the flue gases cool, the water vapour condenses to form water droplets. It is this process that condensing boilers use.

With a condensing boiler, the flue gases first pass over the primary heat exchanger, which extracts about 80 per cent of the heat. The flue gases, which still contain 20 per cent of latent heat, are then passed over a secondary heat exchanger where a further 12–14 per cent of the heat is extracted. When this happens, the gases cool to their '**dew point**', condensing the water vapour inside the boiler as water droplets, which are then collected in the condensate trap before being allowed to fall to drain via the condensate pipe. The process gives condensing boilers their distinctive 'plume' of water vapour during operation, which is often mistaken for steam.

KEY TERM

Dew point: the temperature at which the moisture within a gas is released to form water droplets. When a gas reaches its dew point, the temperature has been cooled to the point where the gas can no longer hold the water and it is released in the form of 'dew', or water droplets.

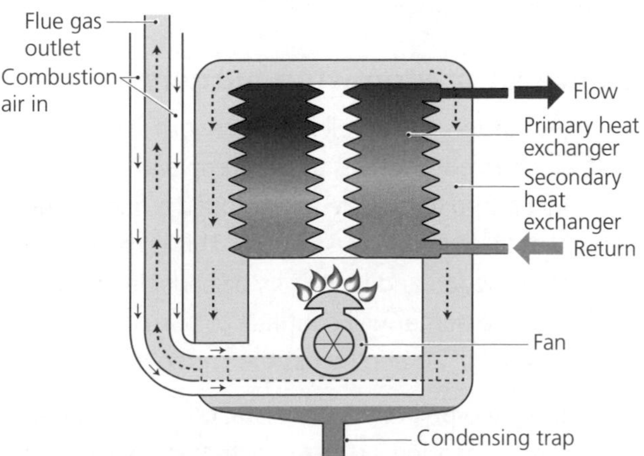

▲ Figure 7.30 How a condensing boiler works

Modern condensing boilers are around 94 per cent efficient, releasing only 6 per cent of wasted heat in the cooler flue gases to the atmosphere. Some advantages and disadvantages of condensing boilers are listed in Table 7.12.

▼ Table 7.12 Advantages and disadvantages of condensing boilers

Advantages	Disadvantages
Building Regulations Document L compliant Very high efficiency Sealed (pressurised) system gives better heating flow rates System corrosion can be reduced Very quiet in operation Can be used with all modern fully pumped heating systems (system boilers) No F and E cistern required in the roof space Very good flow rate on hot water supply (condensing combination boilers)	High maintenance compared with other boilers Siting of the condense pipework can often prove difficult Does not work if the condense line freezes during cold weather Use more gas when not in condensing mode

The types of boiler that can be used with wet central heating systems fall into three distinct categories:

1 system boilers
2 traditional boilers
3 combination boilers.

System boilers

A system boiler is an appliance where all necessary safety and operational controls are included and fitted directly to the boiler. There is no need for a separate expansion vessel, pressure relief valve or filling loop, and this makes the installation much simpler.

Traditional boilers

A traditional boiler does not contain any form of expansion vessel or operational controls, such as the pump or filling loop. It is, however, a condensing boiler.

Combination boilers

In recent years, combination boilers have become one of the most popular forms of central heating in the UK. A combination boiler provides central heating and

The system boiler has all the components for a sealed system contained within the boiler unit. It is filled directly from the mains cold water via a filling loop which is often fitted by the boiler manufacturer.

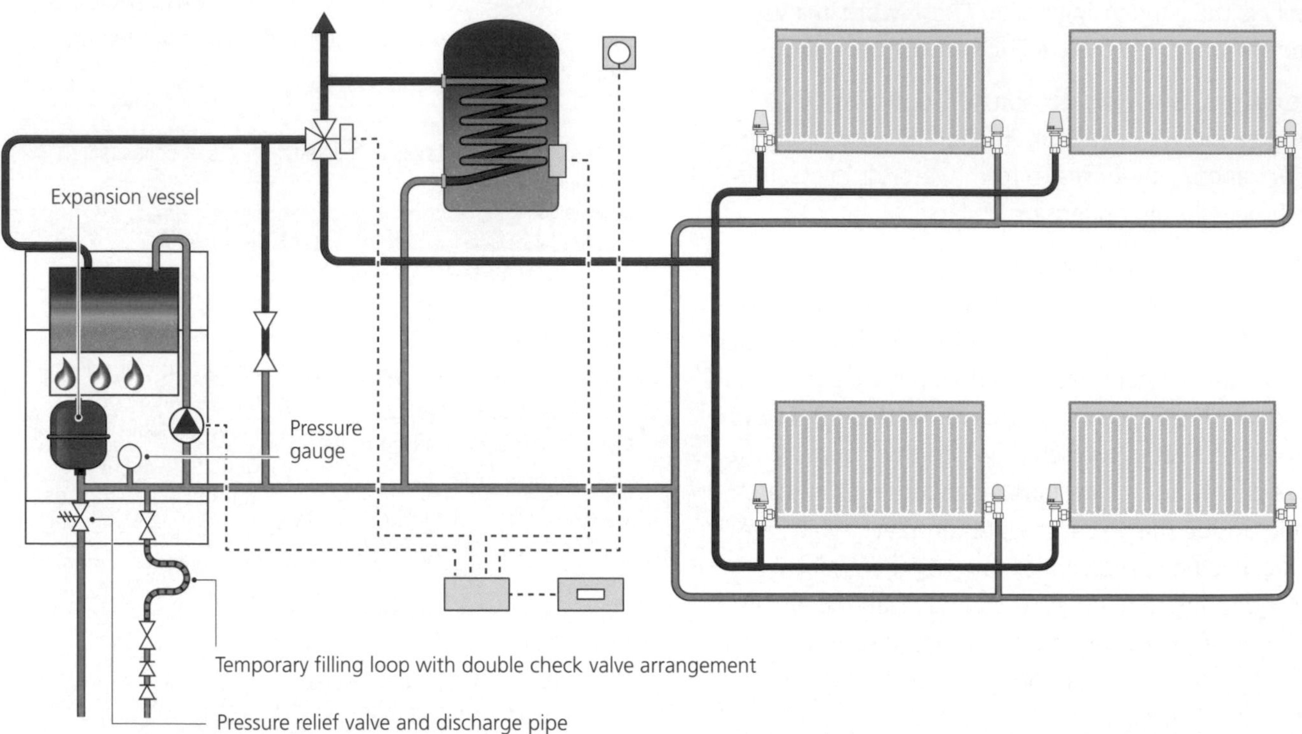

▲ Figure 7.31 A sealed system with a system boiler

instantaneous hot water supply from a single appliance. Modern combination boilers are very efficient and contain all the safety controls (i.e. expansion vessel, pressure relief valve) of a sealed system. Most 'combis' also have an integral filling loop.

Hydrogen boilers

In the UK, most domestic properties are connected to the gas mains and run their boiler on gas. Hydrogen is a low-carbon alternative that in the future could reach each home through the same gas network. In this way it could enable each householder to reduce their carbon footprint. It is currently not possible to purchase a hydrogen boiler, although manufacturers are already manufacturing protypes. They will be similar to natural gas boilers except for the fuel they burn. Engineers would need to be re-trained to install, service and maintain hydrogen boilers. Currently it is too expensive to produce large enough quantities of hydrogen to meet demand but technology will soon catch up.

Hydrogen is a very efficient fuel. For example, 1 kg of hydrogen has the same energy as 2.8 kg of petrol.

Oil-fired central heating appliances

Oil-fired boilers use two different firing methods:

1 pressure jets or atomising burners
2 vaporising burners.

Pressure jet or atomising burners

Pressure jet burners use an oil burner that mixes air and fuel. An electric motor drives a fuel pump and an air fan. The fuel pump forces the fuel through a fine nozzle, breaking the oil down into an oil mist. This is then mixed with air from the fan and ignited by a spark electrode. Once it is lit, the burner will continue to burn as long as there is a supply of air and fuel in the correct ratio.

Oil pressure jet-type boilers are installed on all modern oil-fired central heating systems, including condensing system boilers, condensing 'combi boilers' and wall-mounted types.

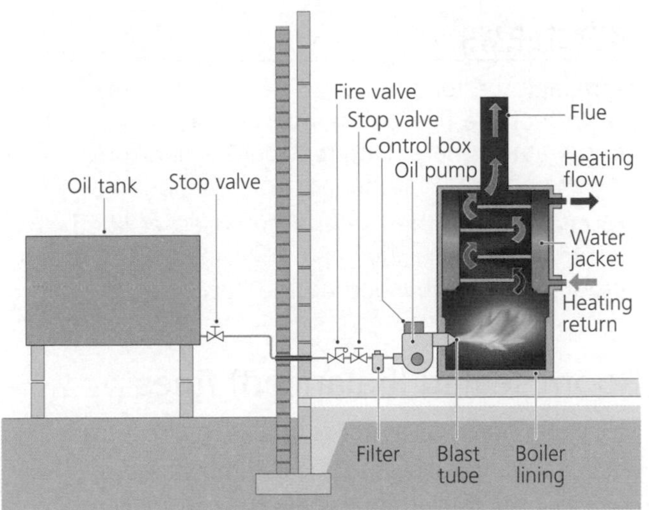

▲ Figure 7.32 A pressure jet oil burner installation

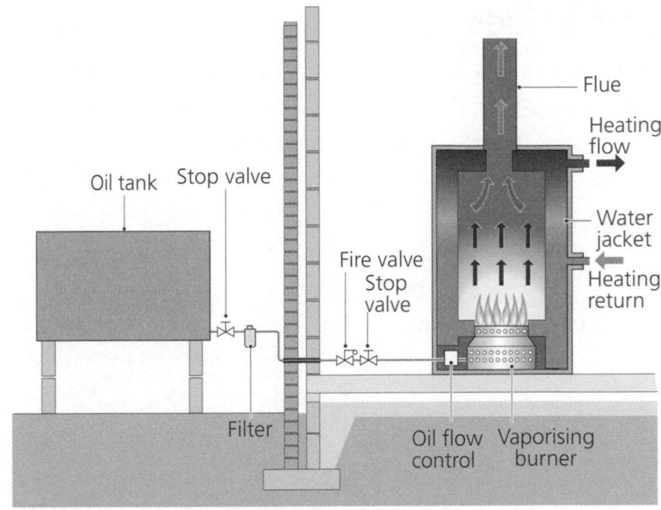

▲ Figure 7.33 A vaporising oil burner installation

▼ Table 7.13 Advantages and disadvantages of pressure jet oil burner-type boilers

Advantages	Disadvantages
Building Regulations Document L compliant Very high efficiency Sealed (pressurised) system gives better heating flow rates Can be used with all modern fully pumped heating systems (system boilers) No F and E cistern required in the roof space (system and combination types) Very good flow rate on hot water supply (condensing combination boilers) Can be used in areas where there is no gas supply	High maintenance compared with gas boilers Noisy in operation Needs an oil tank for fuel storage

Vaporising burners

Vaporising burners work on gravity oil feed. There is no pump. The oil flows to the burner, where a small oil heater warms the oil until vapour is given off and it is the vapour that is then ignited by a small electrode. As the oil burns, vapour is produced continually, which keeps the burner alight.

They are generally used only in oil-fired cookers.

▼ Table 7.14 Advantages and disadvantages of vaporising oil burner-type boilers

Advantages	Disadvantages
Very quiet in operation	Very limited use (cookers only)

Typical flue systems for central heating appliances

All central heating appliances need a flue to remove the products of combustion safely to the outside. The basic concept is to produce an updraught, whether by natural means or by the use of a fan, to eject the fumes away from the building. There are two flue concepts:

1 open flues
2 room sealed (balanced) flues.

▼ Table 7.15 Boiler/flue arrangements

	Open flue (natural draught)	Open flue (forced draught)	Room sealed (natural draught)	Room sealed (fan assisted)
Solid fuel boilers	✓	✓	×	×
Gas boilers	✓	✓	✓	✓
Pressure jet oil burners	×	✓	×	✓
Vaporising oil burners	✓	×	×	×

Open flues

The open flue is the simplest of all flues. Because heat rises, it relies on the heat of the flue gases to create an updraught. There are two different types:

1 natural draught
2 forced draught.

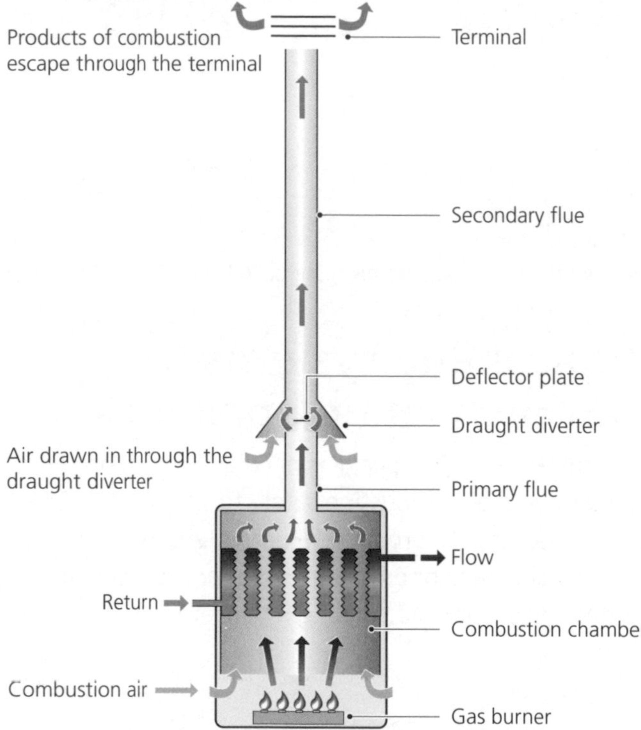

▲ Figure 7.34 The operation of an open flue

With a boiler having this type of flue, air for combustion is taken from the room in which the boiler is located. The products of combustion are removed vertically by natural draught into the atmosphere, through a suitable **terminal**. The room must have a route for combustion air direct from outside. This is usually supplied through an air brick on an outside wall. All natural draught open flue appliances work in this way. The material from which the flue is made, however, will differ depending on the type of fuel used.

Occasionally, an open flue may be **forced draught**. This is where a purpose-designed fan is positioned either before the combustion chamber or close to the primary flue. The fan helps to create a positive updraught by blowing the products of combustion up the flue. Forced draught open flues are not suitable for all open flue types and their use will depend upon the boiler manufacturer and the boiler/flue design.

KEY TERMS

Terminal: the terminal of a flue system is the last section of the flue before the flue gases evacuate to the atmosphere. Different boilers and fuels require different terminals.

Forced draught: the use of a purpose-designed fan to create a positive updraught by forcing the products of combustion up the flue.

Room sealed (balanced) flues

This boiler draws its air for combustion direct from outside through the same flue assembly used to discharge the flue products. This boiler is inherently safer than an 'open flue' type, since there is no direct route for flue products to spill back into the room. There are two basic types:

1 natural draught
2 fan assisted (forced draught).

Natural draught

Natural draught room sealed appliances have been around for many years and there are still many thousands in existence. The basic principle is very simple – both the combustion air (fresh air in) and the products of combustion (flue gases out) are situated in the same position outside the building. The products of combustion are evacuated from the boiler through a duct that runs through the combustion air duct, one inside the other.

The boiler terminals are either square or rectangular and quite large in size. Terminal position is critical to avoid fumes going back into the building through windows and doors.

Fan assisted (forced draught)

Fan-assisted room sealed appliances work in the same way as their natural draught cousins, with the products of combustion outlet positioned in the same place (generally) as the combustion air intake but there are two distinct differences:

1 the process is aided by a fan, which ensures the positive and safe evacuation of all combustion products and any unburnt gas that may escape
2 the flue terminal is circular, much smaller and can be positioned in many more places than its predecessors.

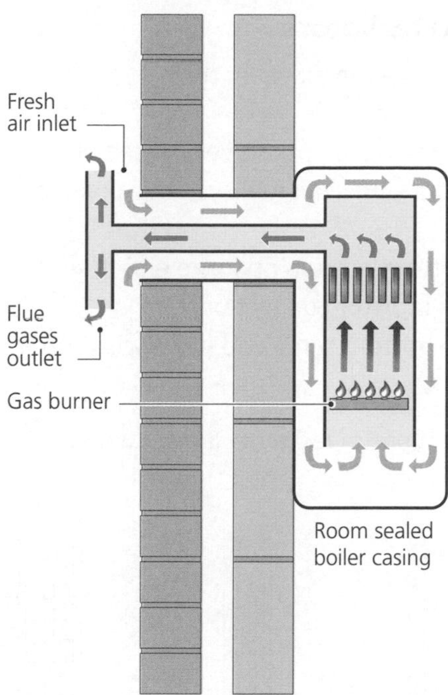

▲ Figure 7.35 The operation of a natural draught room sealed boiler

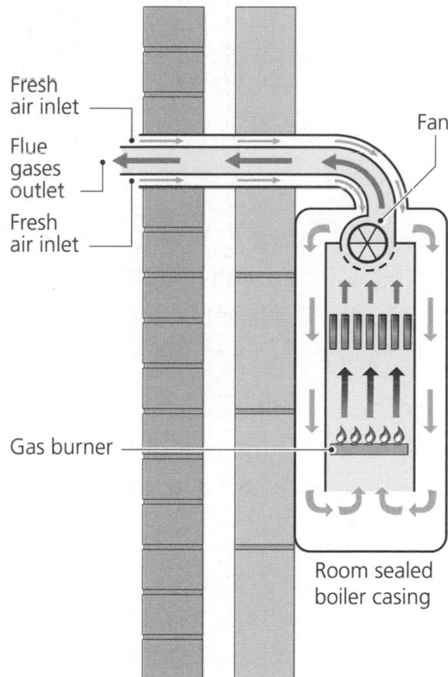

▲ Figure 7.36 The operation of a fan-assisted room sealed boiler

There are two very different versions of the fan-assisted room sealed boiler. These are:

1 the fan positioned on the combustion products outlet from the heat exchanger; this creates a desired negative pressure within the casing
2 the fan positioned on the fresh air inlet blowing a mixture of gas and air to the burner; this creates a positive pressure within the boiler casing. This positive pressure caused some problems with boiler seals, so the fan position was changed to the flue gas outlet sucking air through the boiler and creating a negative pressure within the boiler. Nearly all condensing boilers use this principle.

District heating

This is a form of heat that is produced centrally and then supplied to either individual houses or larger buildings. Often based on a combined heat and power (CHP) system, it is a form of heating that has been successfully used in many cities as it is most effective where there is a high density of properties or buildings.

There is one large heat source which is run very efficiently. This is linked to properties by a network of highly insulated pipes. Steam or high temperature water passes down the pipework to a set of plate heat exchangers or heat interface units (HIU), which in turn transfer the required heat to each property's system.

The overall effect is efficiency, an overall reduction in CO_2 emissions and a saving on running costs.

ACTIVITY

Read about the district heating project based in this UK city: www.theade.co.uk/news/market-news/how-a-hidden-london-power-plant-is-tackling-the-uks-energy-needs

Common heat emitters

So far in this chapter, we have looked at heating systems and the appliances that drive them. Here, we will look at the methods of getting the heat into the room or dwelling. For this, we need to look at the many different types of heat emitter that are available. These include:

- panel radiators
- column radiators
- low surface temperature radiators
- fan convectors
  - wall mounted
  - kick space
- towel warmers
- towel warmers with integral panel radiators
- skirting heating.

Panel radiators

Modern panel convectors/radiators are designed to emit heat by convection and radiation (refer to Chapter 3, Scientific principles, to read more about this); 70 per cent of the heat is convected. They have fins (often called a convector) welded to the back of the radiator, which serve to warm the cold air that passes through them, creating warm air currents, which flow into the room. This dramatically improves the efficiency of the radiator. Steel radiators that do not have fins rely on radiant heat alone and this leads to cold spots in the room. Positioning of the radiator is, therefore, critical. Radiators should be sited on a clear wall with no obstructions, such as window sills, above it. If this is not possible, enough space should be left between the top of the radiator and the obstruction to allow the warm air to circulate.

It is recommended by radiator manufacturers that radiators should be fitted at least 150 mm from finished floor level to the bottom of the radiator (depending on the height of the skirting board), to allow air circulation.

KEY POINT

Radiator standard BS EN 442

Radiators must now meet the above BS EN number, which supersedes the old British Standard which was **BS 3528**. After years of testing in Europe the certification of 'rads' was brought up to date to ensure that all radiators met the minimum thickness, pressure tolerance, treatment and paint quality which helps with limiting corrosion.

Under these tests, the technical committee responsible for the changes use something called Delta T (ΔT) to set the new standard. Delta T defines the difference between the water (delta) temperature in the radiator and the ambient air temperature (T) in a room. As water passes from the flow and returns through the radiator it will give away the energy it has obtained from the boiler. Delta T is set to certain data standards to ensure that the least amount of energy is lost from the radiator, helping to reduce the use of natural resources without reducing the radiator's performance.

Connections to radiators

Radiator connections are classified by their abbreviations. For example:

- TBOE means top bottom opposite end (used on heat sink radiators with solid fuel systems and one-pipe systems)
- BBOE means bottom bottom opposite end (the usual method of radiator connection)
- TBSE means top bottom same end (used with some one-pipe systems).

The most common types of radiators are shown In Figure 7.37.

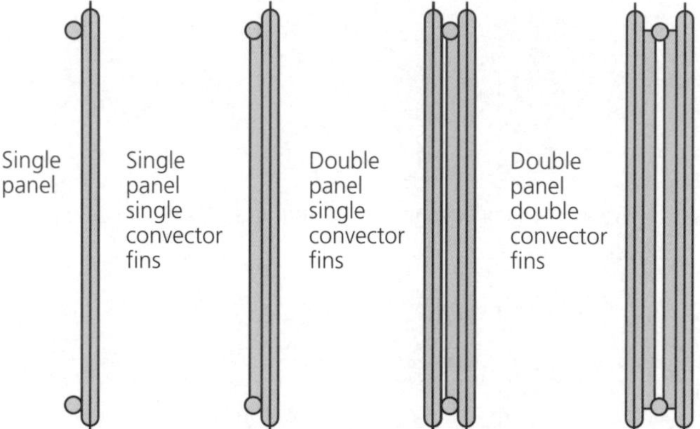

▲ Figure 7.37 Types of panel radiator

Manufacturers provide a wide range of heights, from 300 mm high through to 900 mm, and lengths from 400 mm increasing by 100/200 mm increments through to 3 m.

It is important that radiators are fitted according to the manufacturer's instructions if the best output performance is to be achieved. Outputs vary from manufacturer to manufacturer.

There are three different styles of radiator, as follows.

1. **Seamed top:** this is a very common style of radiator that was for many years the market leader. Top grilles and side panels are available for this radiator style.
2. **Compact:** these have factory-fitted top grilles and side panels, making them a more attractive radiator style. These are currently the most popular radiator style available.
3. **Rolled top:** the least popular of all radiator styles. They are somewhat old-fashioned looking, with exposed welded seams either side.

Hanging a radiator

1. Before hanging the radiator, you must decide how close you want it to be to the wall. Radiator brackets have two options – near and far – therefore, select the one that is best for the installation and the customer. Maximising the space between the radiator and the wall increases convection.
2. Mark the centre of the radiator and the position of the radiator brackets.
3. Place a radiator bracket into position on the radiator and measure from the bottom of the bracket to the bottom of the radiator. This is usually (depending on the manufacturer) 50 mm. This is measurement A.
4. Mark the centre of the position of the radiator on the wall where the radiator is to be hung.
5. Place the radiator against the wall on the centre line and mark the position of the brackets on the wall. Using a spirit level, draw two vertical lines where the brackets are to be fixed.
6. Radiators are best hung at 150 mm from the floor (depending on the skirting board height) to allow air circulation through the fins, so add measurement A to 150 mm and mark across the two bracket marks on the wall, using a spirit level.
7. Radiator brackets can usually be hung either with the radiator close to the wall, or with a larger gap. Decide which way the brackets are to be fixed, then place the bracket against the marked position on the wall, making sure that the bottom of the bracket is sitting on the bottom bracket mark. Mark the fixing position.
8. For masonry walls: using a suitable masonry drill bit, drill the four bracket holes (a 7 mm masonry drill bit and brown wall plugs are usually suitable). Screw the brackets to the wall using 50 mm × no. 10 screws.
9. For timber-studded walls: use plasterboard fixings that are capable of carrying the weight of the radiator plus the water inside.
10. Hang the radiator onto the brackets. Check that it is level using a spirit level and that it is 150 mm from the finished floor level.

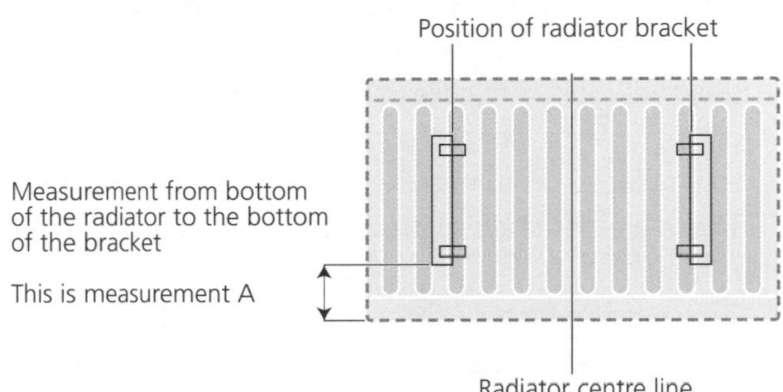

▲ Figure 7.38 Marking radiator bracket positions on the radiator

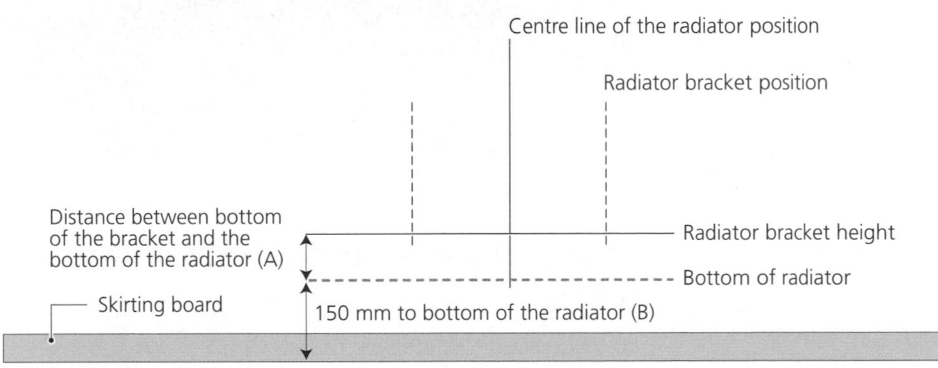

▲ Figure 7.39 Marking heights on the wall

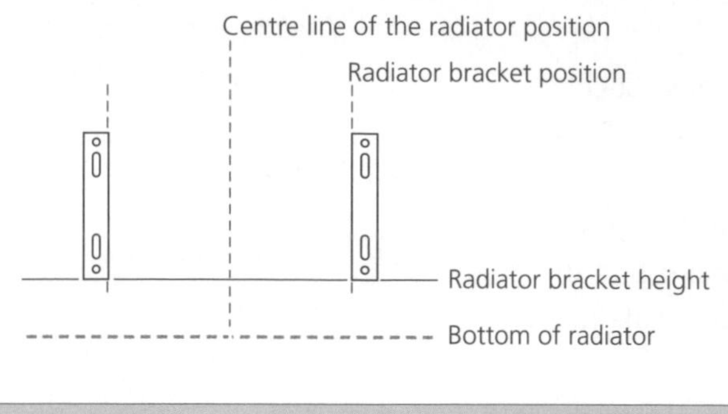

▲ Figure 7.40 Marking bracket positions on the wall

Dressing a radiator

Dressing a radiator involves getting the radiator ready for hanging by putting in the valves, the air release valve and the plug. The process is as follows.

1. Carefully remove the radiator from its packing. Inside the packing you will find the hanging brackets, the air release valve and the plug – and, often, small 'u'-shaped pieces of plastic, which are to be placed on the brackets where the radiator fits. These are designed to prevent the radiator from rattling.
2. Take out the factory-fitted plugs. Be careful here, especially if you are working in a furnished property, as the radiator often contains a small amount of water from when it was tested at the factory.
3. Split the valves at the valve unions and wrap PTFE tape around the valve tail. Between 10 and 15 wraps will ensure the joint between tail and radiator does not leak. This may seem an awful lot of PTFE but the ½-inch female sockets on radiators are notoriously slightly oversized and this leads to leaks.
4. Make the tail into the radiator using a radiator spanner.
5. Insert and tighten the air release valve and plug using an adjustable spanner.

Domestic panel radiators have ½-inch BSP female threads at either side, top and bottom, and these will accept a variety of radiator valves. One end of the radiator has an air release valve, with the other end being blanked by the use of a plug. These are usually supplied by the radiator manufacturer.

Column radiators

Column radiators (often known as hospital or church radiators) have been available for many years. As the name suggests, they are made up of columns; the more columns the radiator has, the better the heat output. They are increasingly being used with modern heating systems, especially on period refurbishments.

Column radiators can be made from three different metals, these being traditional cast iron, steel and aluminium, with many modern column radiator designs now being produced by a variety of manufacturers.

▲ Figure 7.41 Modern column radiator

Low surface temperature radiators (LSTs)

Low surface temperature radiators (LSTs) were specifically designed to conform to the NHS Estates guidance note 'Safe hot water and surface temperatures', which states that:

> Heating devices should not exceed 43°C when the system is running at maximum design output.

This has been adopted not only by the NHS but also local authorities and commercial buildings installations where the general public may have access, including residential care homes and schools. LSTs are also becoming popular in domestic installations, especially in children's bedrooms and nurseries, and where the elderly, infirm or disabled are likely to come into contact with radiators.

Fan convectors

Fan convectors work on the same basic principle as traditional finned radiators. A finned copper heat exchanger is housed in a casing, which also contains a low-volume electrically operated fan. As the heat exchanger becomes hot, a thermostat operates the fan and the warm air is blown into the room. Because the warm air is forced into the room, more heat can be extracted from the hot, circulating water. Once the desired temperature has been reached, the fan is again switched off by the thermostat.

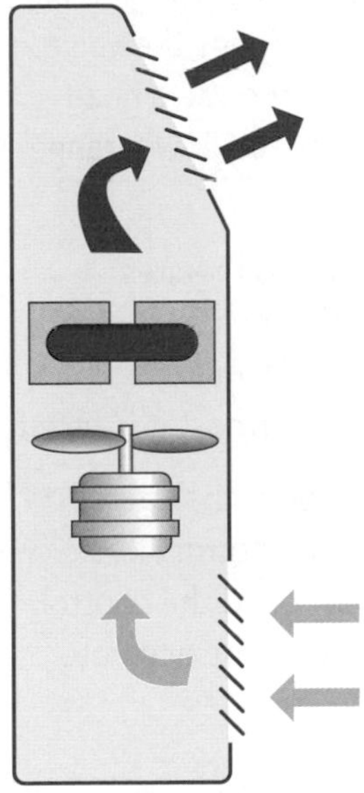

▲ Figure 7.42 The operation of a fan convector

Fan convectors tend to be larger than traditional radiators and they also require a mains electric connection, usually via a switched fuse spur. There are two separate types of fan convector, as follows.

1. **Wall mounted:** these tend to be quite large in size. The manufacturer's data should be consulted to allow the correct heat output to be selected.
2. **Kick-space heaters:** specifically designed for kitchen use where space to mount a radiator is limited. They are installed under a kitchen unit and blow warm air via a grille mounted on the kick plinth.

▲ Figure 7.43 Installation of a kick-space fan convector

Tubular towel warmers

These are available in a range of different designs and colours, and are often referred to as designer towel rails. They can be supplied for use with wet central heating systems with an electrical element option, for use during the summer when the heating system is not required. They are usually mounted vertically on the wall and can be installed in bathrooms and kitchens.

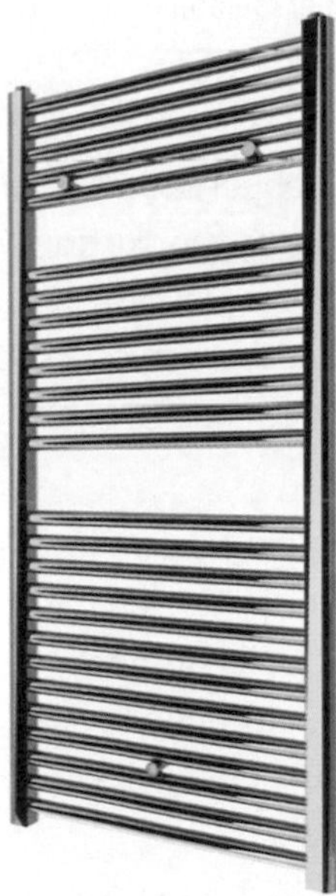

▲ Figure 7.44 A tubular towel rail

Towel warmers with integral panel radiators

Less popular than tubular towel rails, these heat emitters combine a towel rail and radiator into one unit. They allow a towel to be warmed without affecting the convection current from the radiator. They are generally installed only in bathrooms.

▲ Figure 7.45 A towel rail with integral panel radiator

Skirting heating

Skirting heating consists of a finned copper tubular heat exchanger in a metal casing that replaces the skirting boards in a room. It is usually used where unobtrusive heat emitters are required. Skirting heating can be used as perimeter heating below glazing or for background heat in some areas.

The heat output, at 450 watts per metre, is quite low, which means that, to be effective, the skirting heating would need to be at floor level on all walls of the room to off-set the room heat losses, although the heat coverage is very similar to that experienced with specialist underfloor heating.

One disadvantage is that efficiency is reduced by dust collecting in the fins.

Mechanical central heating components

Mechanical central heating components are those that do not use electricity but still play a vital role in helping to ensure the correct and efficient operation of the system.

In this section, we will look at the most common of the mechanical components and controls used on domestic central heating systems.

Radiator valves: thermostatic and manual valves

There is a wide selection of radiator valves available from many different manufacturers. There are three basic types of radiator valve (thermostatic, wheel head and lockshield), as described below.

Thermostatic radiator valves (TRVs)

These control the temperature of the room by controlling the flow of water through the radiator. They react to air temperature. TRVs have a heat-sensitive head that contains a cartridge, which is filled with either a liquid, gas or wax, and this expands and contracts with heat. As the room heats up, the wax/gas/liquid cartridge expands and pushes down on a pin on the valve body. The pin closes and opens the valve as the room heats up or cools down. The valve head has a number of temperature settings to allow a range of room temperatures to be selected.

Document L1 of the Building Regulations requires that TRVs are installed on new installations to control individual room temperatures, and on all radiators except the radiator where the room thermostat is fitted.

Most TRVs are bi-directional. This means that they can be fitted on either the flow or the return. TRVs are now available in a digital format to aid the control of water through the system and give more accurate temperature control.

Wheel head valves

These allow manual control of the radiator by being turned on or off. The valve is turned on by rotating the wheel head anti-clockwise and turned off by rotating clockwise.

Lockshield valves

These are designed to be operated only by a plumber and not by the householder. They are adjusted during system balancing to regulate the flow of water through the radiator. The lockshield head covers the valve mechanism. They can be turned off for radiator removal.

> **KEY POINT**
>
> **What is balancing a central heating system and what does it involve?**
>
> Balancing a system is ensuring an even heat distribution to all of the radiators in the system. Water will always take the line of least resistance and this means that the radiators nearer to the circulating pump will always get more than their fair share of heat. This is detrimental to the furthest radiators on the system and has the effect of slowing down the heating effect.
>
> Balancing is simply evening out the heat distribution by restricting the flow of water to the quickest radiators by simply adjusting the flow rate through the lockshield valves. This has the effect of slowing the heat circulation through the quick radiators and increasing the flow rate to the slower, furthest-away radiators, thus balancing the heat distribution throughout the system.

Automatic air valves

Automatic air valves are fitted where air is expected to collect in the system, usually at high points. They allow the collected air to escape from the system but seal themselves when water arrives at the valve.

When water reaches the valve, the float arm rises, closing the valve. As more air reaches the valve, the float momentarily drops, allowing the air out of the system. These valves are often used with a check valve that prevents air from being drawn into the system backwards through the valve.

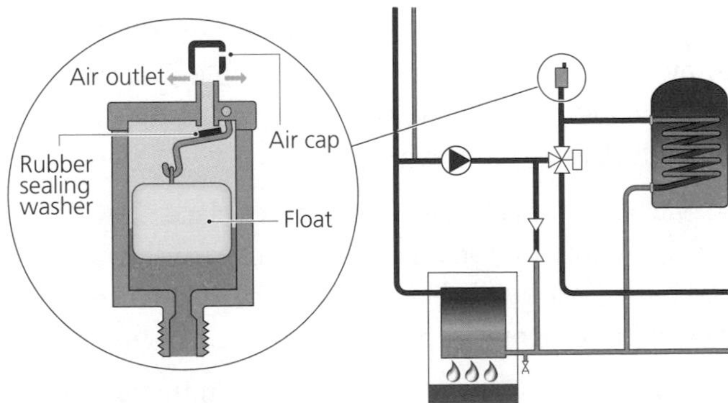

▲ Figure 7.46 Automatic air valve

Automatic bypass valves

The automatic bypass valve controls the flow of water across the flow and return circuit of fully pumped heating systems by opening automatically as other paths for the water close, such as circuits with motorised valves and radiator circuits with thermostatic radiator valves. This occurs as the hot water circuit and heating circuit/thermostatic radiator valves begin to reach their full temperature. As the circuits close, the bypass will gradually open, maintaining circulation through the boiler and reducing noise in the system due to water velocity. Most boiler manufacturers require a bypass to be fitted to maintain a minimum flow rate through the boiler, to prevent overheating.

Automatic bypass valves are much better than fixed bypass valves, as these, being permanently open, take the flow of hot water away from the critical parts of the system, which increases the heating time for both hot water and heating circuits. This reduces the efficiency of the system and increases fuel usage.

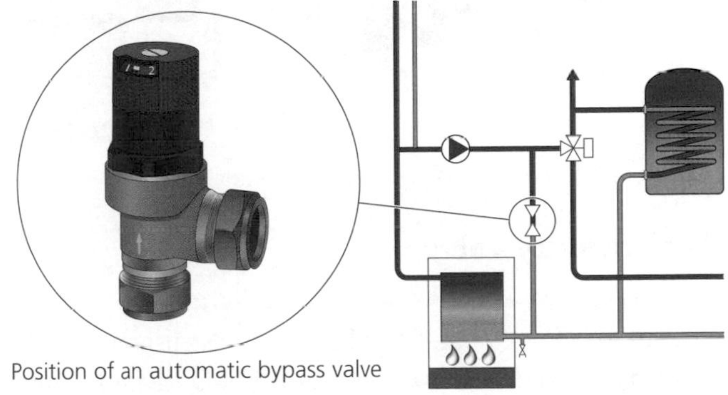

▲ Figure 7.47 Automatic bypass valve

Thermo-mechanical cylinder control valves

Thermo-mechanical cylinder control valves are non-electrical valves used to control the temperature of a hot water cylinder. They are mainly used with gravity primary circulation as part of an upgrade to give some control over secondary hot water temperature. The Domestic Heating Compliance Guide states that:

> For replacement systems where only the hot water cylinder is being replaced and where hot water is on a gravity circulation system, a thermo-mechanical cylinder thermostat should be installed as a minimum provision.

This means that if the hot water cylinder only is being replaced and no control over the hot water temperature exists, then a thermo-mechanical thermostat is the minimum standard of hot water control required to comply with Document L of the Building Regulations.

Thermo-mechanical thermostats work on the principle of thermal expansion of a liquid or gas, in much the same way as thermostatic radiator valves except, with this valve, it is the temperature of the water that is sensed by a remote sensor. The sensor should be placed about a third of the way up from the bottom of the cylinder.

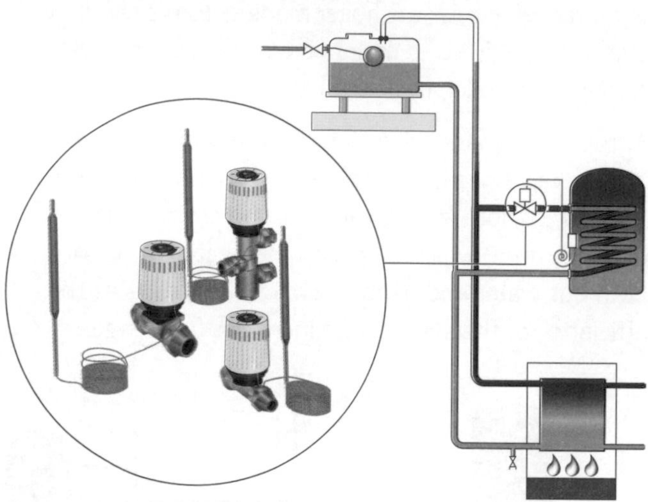

▲ Figure 7.48 Thermo-mechanical thermostat

Anti-gravity valves

Anti-gravity valves prevent unwanted gravity circulation to the upstairs radiators on semi-gravity systems when only the hot water is being heated. They are essential on all semi-gravity systems and especially in those systems fuelled by solid fuel. Anti-gravity valves should be positioned on the vertical flow to the upstairs heating circuit.

Anti-gravity valves are very similar in design to the single check valves mentioned in earlier chapters. They allow water to flow in only one direction and, when the heating system is off, they are in the closed position. In this position, gravity circulation cannot take place. As soon as the central heating circulating pump switches on, the flow of the water opens the valve to allow heating circulation.

Drain valves

Drain valves should be fitted at the lowest points in the heating installation to allow complete draining of the water in the system, and this includes all radiators, especially if the flow and returns to the radiators are on vertical drops from above. For this purpose, radiator valves with built-in drain valves are available.

▲ Figure 7.49 Expansion vessel

The expansion vessel

The expansion vessel is a key component of the system. It replaces the feed and expansion cistern on the vented system and allows the expansion of water to take place safely. It comprises a steel cylinder that is divided in two by a neoprene rubber diaphragm.

On one end of the expansion vessel is a Schrader air pressure valve where air is pumped into the vessel to 1 bar pressure; this forces the neoprene diaphragm to virtually fill the whole of the vessel.

On the other end is a ½-inch male BSP thread and this is the connection point to the system. When mains-pressure cold water enters the heating system via the filling loop and the system is filled to a pressure of around 1 bar, the water forces the diaphragm backwards away from the vessel walls, compressing the air slightly as the water enters the vessel. At this point, the pressure on both sides of the diaphragm is 1 bar.

As the water is heated, expansion takes place. The expanded water forces the diaphragm backwards, compressing the air behind it still further and, since water cannot be compressed, the system pressure increases.

On cooling, the water contracts, the air in the expansion vessel forces the water back into the system and the pressure reduces to its original pressure of 1 bar.

Periodically, the pressure in the vessel may require topping up. This can be done by removing the cap on the Schrader valve and pumping the vessel up to its original pressure with a foot pump.

IMPROVE YOUR MATHS

How much expansion takes place?

The amount of expansion that takes place will depend on how many litres of water the heating system contains. As we have found in previous chapters, water expands at atmospheric pressure by 4 per cent when it is heated but, in this case, the water is under pressure, so by how much does pressurised water expand?

To answer this question, we must first calculate the expansion factor, which can be used to calculate water expansion for a given volume and pressure. If the density of the cold water and the density of the water at maximum operating temperature are known, this is a fairly simple exercise. The calculation is as follows:

$$\frac{d1 - d2}{d2}$$

Where:

*d*1 = density of water at filling temperature (kg/m^3)

*d*2 = density of water at maximum operating temperature (kg/m^3)

If the system has 250 litres of water, and the system is filled with water at 4°C and the maximum temperature is 85°C, what is the expansion factor?

Water @ 4°C has a density of 1000 kg/m^3

Water @ 85°C has a density of 968 kg/m^3

The equation therefore is:

$$\frac{1000 - 968}{968} = 0.0330$$

So, the expansion factor (*e*) = 0.0330

Now, we must use this in another equation.

To find the amount of expansion of water in a system containing 250 litres of water operating at a maximum temperature of 85°C, the equation is:

$$V = \frac{eC}{1 - \frac{p1}{p2}}$$

Where:

V = The total volume of the expansion vessel

C = The total volume of water in the system in litres (250 litres)

*p*1 = The fill pressure in bar pressure (1 bar)

*p*2 = The setting of the pressure relief valve in bar pressure (3 bar)

e = The expansion factor (0.0330)

If these are entered into the equation, the equation becomes:

$$\frac{0.0330 \times 250}{1 - \frac{1}{3}} = 12.36$$

As a percentage of 250, 12.36 is:

$$\frac{12.36 \times 100}{250} = 4.94\%$$

Therefore:

Water under a pressure of 1 bar when cold expands by 4.94 per cent when heated to 85°C.

The pressure relief valve

The pressure relief valve (also known as the expansion valve) is installed onto the system to protect against over-pressurisation of the water. Pressure relief valves are usually set to 3 bar pressure. If the water pressure rises above the maximum pressure that the valve is set to, the valve opens and discharges the excess water pressure safely to the outside of the property through the discharge pipework. The pressure relief valve will open if the expansion vessel fails and the system pressure rises. When the valve opens it will drip hot system water to a safe location.

INDUSTRY TIP

Pressure relief valves are most likely to open because of lack of room in the system for expansion due to a malfunction with the expansion vessel. This can be caused by:

- the diaphragm in the expansion vessel rupturing, allowing water both sides of the diaphragm
- the vessel having lost its charge of air.

▲ Figure 7.50 A pressure relief valve

The filling loop

The filling loop is an essential part of any sealed system, and should contain an isolation valve at either end of the filling loop and a double check valve on the mains cold water supply side of the loop. The filling loop is the means by which sealed central heating systems are filled with water. Unlike open vented systems, sealed systems are filled directly from the mains cold water via a filling loop. The connection of a heating system to the mains cold water supply constitutes a cross-connection between the cold main (fluid category 1) and the heating system (fluid category 3), which is not allowed under the Water Supply (Water Fittings) Regulations. The filling loop must protect the cold water main from backflow and this is done in two ways:

1 a filling loop has a type EC verifiable double check valve included in the filling loop arrangement
2 the filling loop must be disconnected after filling, creating a type AUK3 air gap for protection against backflow.

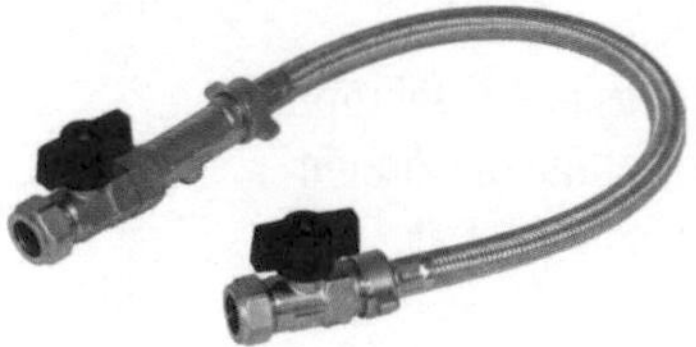

▲ Figure 7.51 The filling loop

Permanent filling connections to sealed heating systems

It is possible to permanently connect sealed heating systems to the mains cold water supply by using a type CA backflow prevention device. The type CA backflow prevention device, when used with a pressure reducing valve, can be used instead of a removable filling loop to connect a domestic heating system direct to the water undertaker's cold water supply. This is possible because the water in a domestic heating system is classified as fluid category 3 risk. A CA device can also be installed on a commercial heating system but only when the boiler is rated up to 45 kW. Over 45 kW, the water in the system is classified as fluid category 4 risk and so any permanent connection would require a type BA RPZ valve. An example of a CA backflow prevention device can be seen in Chapter 5, Cold water systems.

KEY POINT

An RPZ valve, or BA backflow prevention device, is used to protect fluid category 1 water from fluid category 4 water. They were described in detail in Chapter 5, Cold water systems.

The pressure gauge

This is to allow the correct water pressure to be set within the system. It also acts as a warning of component failure or an undetected leak should the pressure begin to inexplicably rise or fall.

The circulating pump

Circulating pumps were discussed within the fully pumped section (page 367).

Low loss headers

For a boiler to work at its maximum efficiency, the water velocity passing through the heat exchanger needs to remain constant, with little fluctuation. This is especially important for condensing boilers as they rely on a defined temperature drop across the flow and return before the condensing mode begins to work effectively. Installation of a low loss header allows the creation of two separate circuits. These are shown in Figure 7.53.

1 **The primary circuit:** the flow within the primary circuit can be maintained at the correct rate for the boilers so that the maximum efficiency of the boilers is maintained regardless of the demand placed on the secondary circuit. Each boiler has its own shunt pump so that equal velocity through the boilers is maintained.

Pressure gauge
Type CA device
Pressure relief valve and discharge pipework
Expansion vessel
Cold water supply

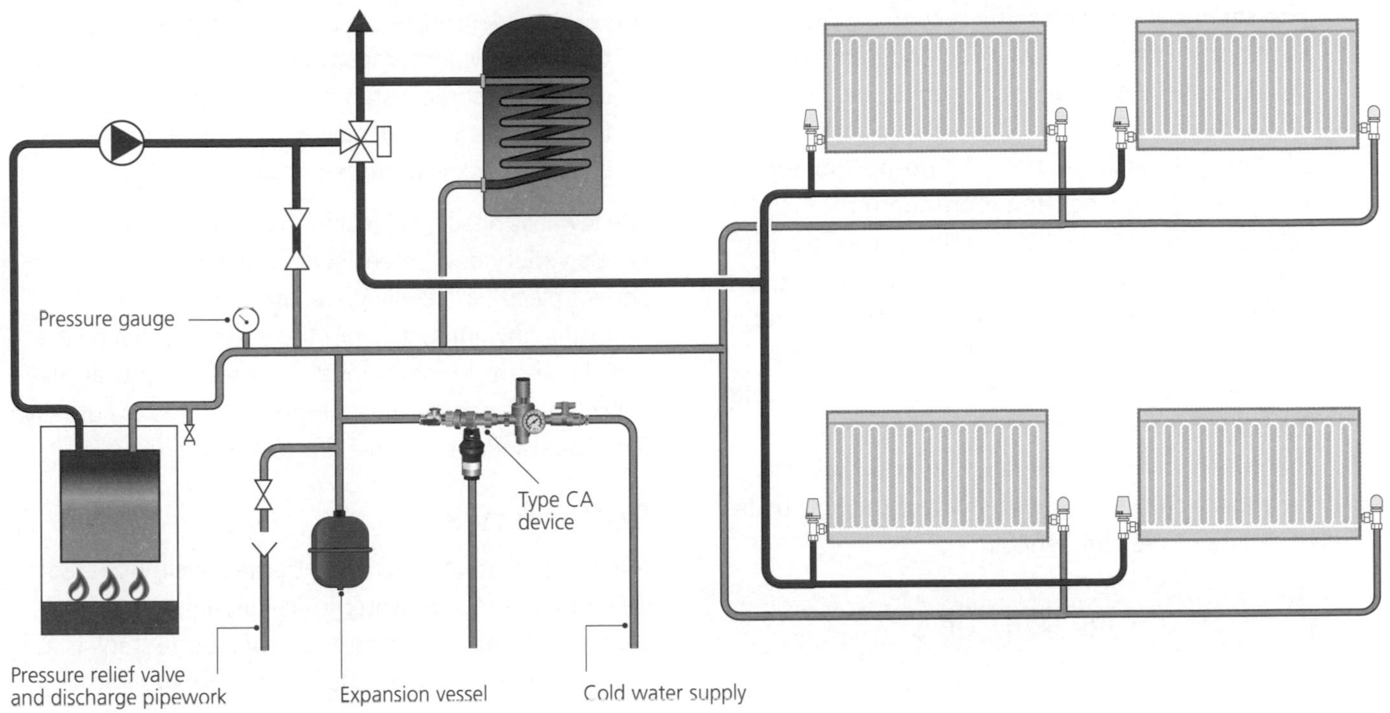

▲ Figure 7.52 A sealed system with CA backflow prevention device

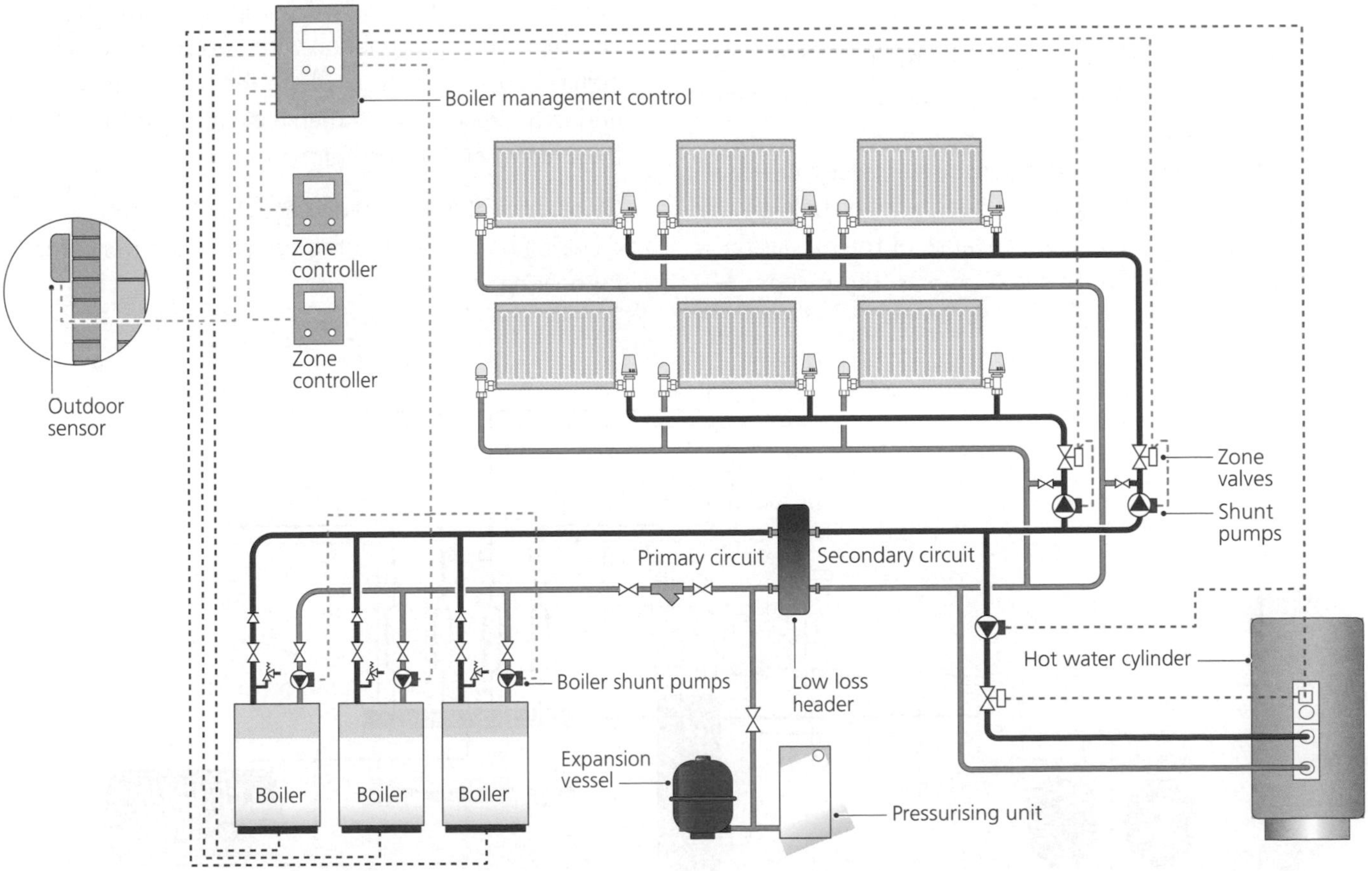

Boiler management control
Zone controller
Zone controller
Outdoor sensor
Zone valves
Shunt pumps
Primary circuit
Secondary circuit
Boiler shunt pumps
Low loss header
Hot water cylinder
Expansion vessel
Boiler
Boiler
Boiler
Pressurising unit

▲ Figure 7.53 A multiple boiler installation with a low loss header

2 **The secondary circuits:** the secondary circuits allow for varying flow rates demanded by the individual balanced zones or circuits. Each zone would be controlled by a shunt pump set to the flow rate for that particular zone. A two-port motorised zone valve, time clock and room thermostat control each zone independently, and these are often fitted in conjunction with other controls such as outdoor temperature sensors. In some cases, the flow rates through each secondary circuit will exceed that required by the boilers. In other cases, the opposite is true and the boiler flow rate will be greater than the maximum flow rate demanded by the secondary circuits, especially where multiple boiler installations are concerned.

INDUSTRY TIP

In essence, the low loss header can be used on multi-boiler or complex systems to ensure correct flow rates throughout the system.

Water velocity is just part of the problem. Water temperature is also important. There are two potential problems here:

1 If the difference in temperature between the flow and return is too great, it puts a huge strain on the boiler heat exchangers because of the expansion and contraction. This is known as 'thermal shock'.

2 For a condensing boiler to go into condensing mode, the return water temperature must be in the region of 55°C. In some instances, temperature sensors are fitted to the low loss header to allow temperature control over the primary circuit.

The low loss header is ideal for use with systems that have a variety of different heat emitters. It is the perfect place for installing an automatic air valve for removing unwanted air from the system. Drain points can also be fitted for removing any sediment that may collect in the header. Both of these features are usually fitted as standard on most low loss headers.

Buffer tanks

A buffer tank is basically an extremely well-insulated vessel that holds hot water for circulation around the heating system. The primary role of a buffer tank is to maintain a minimum volume of hot water in the heating circuit when demand for the heating is low. They are usually used with renewable heat sources such as ground or air source heat pumps. However, they can also be used in conjunction with gas/oil/wood pellet boilers and solar heating where the Sun heats the water during the day, which is then used at night to heat the dwelling when it is required.

Once the water is heated, it acts like a battery, releasing hot water into the system when the demand for heating increases.

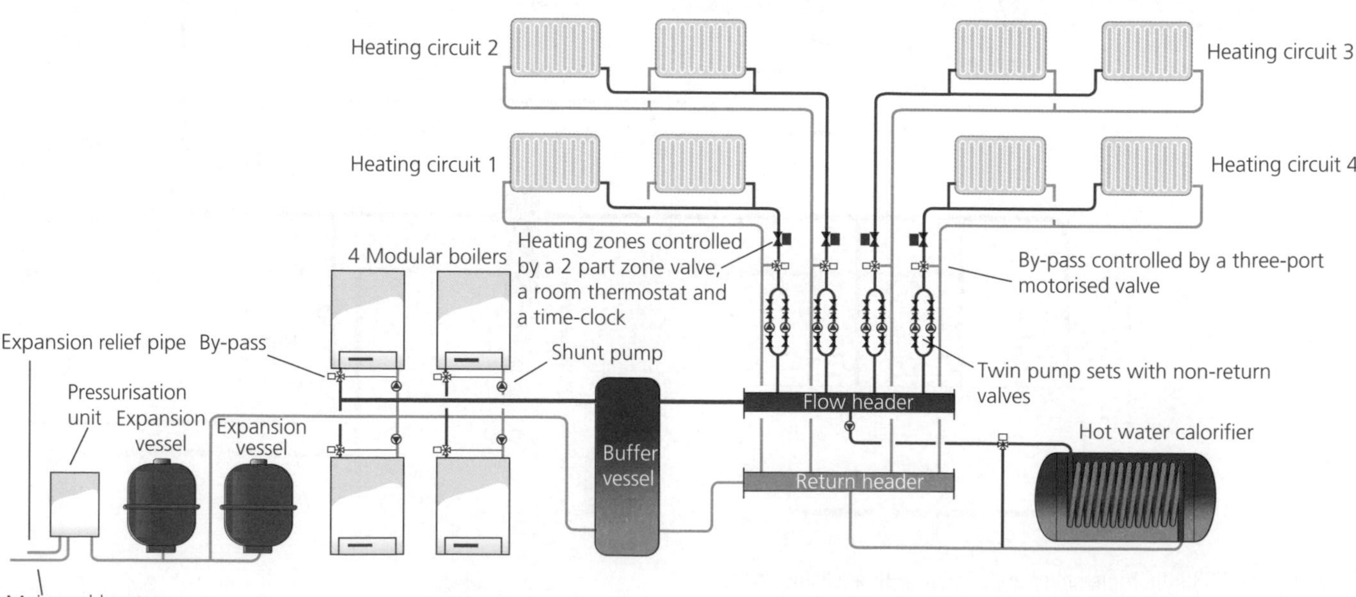

▲ Figure 7.54 Installation of a buffer tank

Expansion joints

Expansion joints are used in long runs of straight pipework where excessive expansion would damage the pipework. The expansion is taken up within the joint, thereby protecting the pipework from distortion and damage. They can either be prefabricated loops of pipework or manufactured bellows that expand and contract as the pipes heat up and cool down.

Corrosion protection

Corrosion is probably the biggest problem that takes place within central heating systems and it commonly occurs in two forms:

1. the formation of red oxide sludge (rust) because of constant **air infiltration**
2. the formation of black oxide sludge and sediment because of electrolytic corrosion; black oxide sludge can be prevented by the use of chemicals called 'inhibitors', which stop the sludge forming, or by the use of magnetic filters that use magnetism to attract the metallic black oxide sludge.

> **KEY TERM**
>
> **Air infiltration:** a process where air can get into a system and cause air locks and corrosion.

Corrosion can attack a system very quickly. As soon as the system is filled with water, corrosion begins to work to break down certain elements within it.

Air infiltration is a constant problem with some systems, especially those that are open vented. Central heating systems last longer once the water in the system has lost all of its oxygen. Without oxygen, rust cannot occur. Air infiltration happens for a number of reasons:

- micro leaks that let air in but do not show as a water leak; these are extremely hard to trace and usually occur around the packing glands of lockshield radiator valves and air release valves; they always occur on the negative pressure side of any system
- air being sucked down the vent pipe due to poor system design
- the constant see-sawing of water within the F and E cistern aerates the water
- small leaks introduce fresh aerated water into the system.

Electrolytic corrosion

Within central heating systems, there are a number of metals; steel radiators, brass valves (brass contains zinc), copper tubes and stainless steel heat exchangers. On older systems there may also be cast iron boilers or parts containing aluminium. All of these metals lie at different points on the electromotive series of metals (see Chapter 3, Scientific principles, page 136) and, once they are connected via water (an electrolyte), corrosion begins immediately. This problem is accelerated when the water becomes hot. The net result of this reaction is that the steel of the radiators begins to be eaten away, with the fine particles of steel falling to the bottom of the radiator as a sediment, which forms a magnetic black sludge. As a by-product, the radiator may also fill with hydrogen that requires constant venting. The sludge not only blocks pipework and finds its way into all of the low points of the system, but also causes boiler noise and creates pitting corrosion in the radiators. Figure 7.55 shows some of the problems that can result from system sludging.

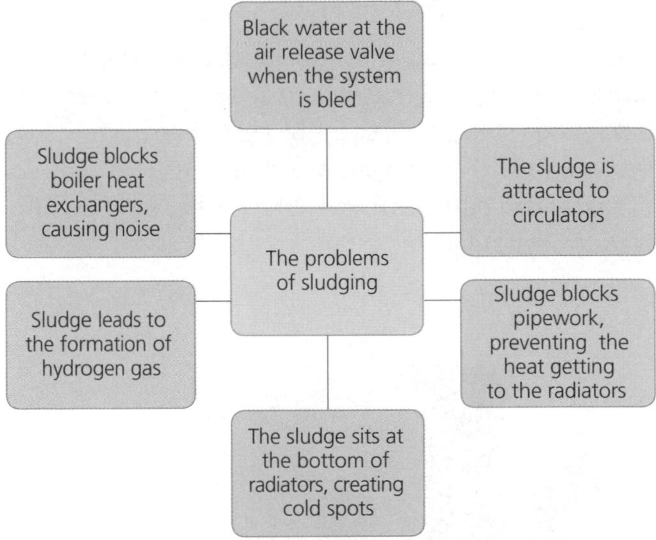

▲ Figure 7.55 The problems of sludging

Corrosion inhibitor

Corrosion inhibitor must be added to the system to comply with the manufacturer's warranty. Corrosion inhibitor stops corrosion from occurring and black sludge from forming, and helps to lubricate pump bearings and valves. Once added, corrosion inhibitor does not need to be replaced except when the system is drained down.

Corrosion inhibitor:

- stops a build-up of 'black oxide sludge', the major cause of central heating problems
- helps to reduce fuel costs
- helps prevent the formation of hydrogen gas
- has a non-acidic neutral formation and so is harmless to the environment
- prevents pin-holing of radiators and pipework
- prevents scale formation.

It should be remembered that corrosion inhibitor must not be added to systems that contain a single feed self-venting cylinder as these use air entrapment to separate the primary and secondary systems. Should the air bubbles within the cylinder break, this would lead to the inhibitor chemicals mixing with the domestic hot water supply, causing contamination.

The use of magnetic filters

As we have already seen, black oxide sludge is made up of minute particles of steel that have been 'robbed' by electrolytic corrosion and this is attracted to components such as circulating pumps, causing pump failure and damage to the system.

Magnetic filters protect central heating systems by using very powerful magnets to attract the suspended black oxide steel particles in the central heating system water. This can remove almost 100 per cent of suspended particles, preventing further build-up of black oxide sludge.

▲ Figure 7.56 A magnetic filter

Electrical central heating controls

Modern central heating systems cannot function without electrical controls. They are required at every stage of operation, from switching the system on to shutting it down when the temperature required has been reached. They provide both functional operation and safety, and are a requirement of Building Regulations Document L: Conservation of fuel and power.

No matter how good the central heating design, or how accurate the calculations, the system requires proper control to be effective, efficient and economical to run. The types of control that are added to a system can greatly improve its performance. Even older systems can benefit from the addition of modern and effective controls.

INDUSTRY TIP

Note that, under 'zoning', it is recommended that one zone valve is used for the heating system. However, the notes to the table state that: 'The SAP notional dwelling assumes at least two space heating zones...' (which was best practice in CHeSS). This means that heating systems should be divided between living and sleeping areas.

In this part of the chapter, we will look at the various controls for central heating systems, their function and how they 'fit' into modern systems.

Before we look at the various controls, we must first consider the implications of Document L, which was amended in 2016 and 2018. This document is now under consultation and is due to be re-published in 2022. This reflects the continual environmental considerations that governments have been striving to achieve in the light of global warming. This new publication of Building Regulations Part L is intended to consolidate CHeSS, the Domestic Heating Compliance Guide, Boiler Plus and the Building Services Compliance Guide. The most up-to-date requirements are to be found in the Building Services Compliance Guide on pages 14–20. The main points are listed in the tables below. Make sure you read the additional note 3 with Table 7.16a, concerning 'zoning'. Although it is not required under Part L, it is **good practice** to create two zones for space heating, usually upstairs and downstairs, in line with the SAP assumption.

▼ Table 7.16a Recommended minimum controls for new gas-fired wet central heating systems

Control type	Minimum standard
Boiler interlock	System controls should be wired so that when there is no demand for space heating or hot water, the boiler and pump are switched off.
Zoning	Dwellings with a total floor area > 150 m^2 should have at least two space heating zones, each with an independently controlled heating circuit[1]. Dwellings with a total floor area[2] ≤ 150 m^2 may have a single space heating zone[3].
Control of space heating	Each space heating circuit should be provided with: ● independent time control, and either: ● a room thermostat or programmable room thermostat located in a reference room[4] served by the heating circuit, together with individual radiator controls such as thermostatic radiator valves (TRVs) on all radiators outside the reference rooms, or ● individual networked radiator controls in each room on the circuit.
Control of hot water	Domestic hot water circuits supplied from a hot water store (i.e. not produced instantaneously as by a combination boiler) should be provided with: ● independent time control, and ● electric temperature control using, for example, a cylinder thermostat and a zone valve or three-port valve. (If the use of a zone valve is not appropriate, as with thermal stores, a second pump could be substituted for the zone valve.)

The standards in this table apply to new gas-fired wet central heating systems. In existing dwellings, the standards set out in Table 4 will apply in addition.

Always also follow manufacturers' instructions.

[1] A heating circuit refers to a pipework run serving a number of radiators that is controlled by its own zone valve.

[2] The relevant floor area is the area within the insulated envelope of the dwelling, including internal cupboards and stairwells.

[3] The SAP notional dwelling assumes at least two space heating zones for all floor areas, unless the dwelling is single storey, open plan with a living area > 70% of the total floor area.

[4] A reference room is a room that will act as the main temperature control for the whole circuit and where no other form of system temperature control is present.

▼ Table 7.16b Recommended minimum standards when replacing components of gas-fired wet central heating systems

Component	Reason	Minimum standard	Good practice[1]
1. Hot water cylinder	Emergency	For copper vented cylinders and combination units, the standard losses should not exceed $Q = 1.28 \times (0.2 + 0.0521V^{2/3})$ kWh/day, where V is the volume of the cylinder in litres. Install an electric temperature control, such as a cylinder thermostat. Where the cylinder or installation is of a type that precludes the fitting of wired controls, install either a wireless or thermo-mechanical hot water cylinder thermostat or electric temperature control. If separate time control for the heating circuit is not present, use of single time control for space heating and hot water is acceptable.	Upgrade gravity-fed systems to fully pumped. Install a boiler interlock and separate timing for space heating and hot water.
	Planned	Install a boiler interlock and separate timing for space heating and hot water.	Upgrade gravity-fed systems to fully pumped.

➜

2. Boiler	Emergency/ planned	**All boiler types except heating boilers that are combined with range cookers** The ErP[2] seasonal efficiency of the boiler should be a minimum of 92% and not significantly less than the efficiency of the appliance being replaced. In the exceptional circumstances defined in the *Guide to the condensing boiler installation assessment procedure for dwellings* (ODPM, 2005), the boiler SEDBUK 2009 efficiency should not be less than 78% if natural gas-fired, or not less than 80% if LPG-fired. In these circumstances the additional requirements for combination boilers would not apply. Install a boiler interlock as defined for new systems. Time and temperature control should be installed for the heating system. **Combination boilers** In addition to the above, at least one of the following energy efficiency measures should be installed. The measure(s) chosen should be appropriate to the system in which it is installed: ● Flue gas heat recovery ● Weather compensation ● Load compensation ● Smart thermostat with automation and optimisation.	Upgrade gravity-fed systems to fully pumped. Fit individual radiator controls such as thermostatic radiator valves (TRVs) on all radiators except those in the reference room.
3. Radiator	Emergency		Fit a TRV to the replacement radiator if in a room without a room thermostat.
	Planned		Fit TRVs to all radiators in rooms without a room thermostat.
4. New heating system – existing pipework retained	Planned	The new boiler and its controls should meet the standards in section 2 of this table. Fit individual radiator controls such as TRVs on all radiators except those in the reference room.	In dwellings with a total floor area > 150 m^2, install at least two heating circuits, each with independent time and temperature control, together with individual radiator controls such as TRVs on all radiators except those in the reference rooms.

Always also follow manufacturers' instructions.

[1] Best practice would be as for a new system.

[2] Refers to the efficiency methodology set out in Directive 2009/125/EC for energy performance related products.

Source: BEAMA Domestic Building Services Compliance Guide 2021, Tables 3 and 4

To comply with the requirements, the correct electrical controls must be fitted.

KEY POINT

Look up the Domestic Building Services Compliance Guide and print pages 18–19, but watch out for the new Part L to be published. https://assets.publishing.service.gov.uk/government/uploads/system/uploads/attachment_data/file/697525/DBSCG_secure.pdf

KEY POINT

Document L was implemented to save energy and power. A great deal of heat is lost from a building through not insulating properly and not having the necessary controls on a heating system to prevent the wastage of fuel. Document L prevents this by ensuring that even existing systems are brought up to date. It is vital for energy conservation that we follow the rules it lays out.

Time clocks and programmers

Time clocks are the simplest of all central heating timing devices. They are suitable for switching on only one circuit, such as the heating circuit, and so are ideally suited for combination boiler installations. Both mechanical and digital time clocks are available.

Programmers are two-way time clocks, being able to switch on both heating and hot water at various times throughout the day. There are three basic types:

1 A mini-programmer, which allows the heating and hot water circuits to be on together, or hot water alone, but not heating alone. Ideally suited to C-plan and C-plan plus systems.
2 A standard programmer: these use the same time settings for space heating and hot water.
3 A full programmer: allows the time settings for space heating and hot water to be fully independent. Some will allow seven-day programming of both heating and hot water so that the two circuits can be used individually or both together.

Programmers are often fitted to the front fascia of the boiler and integrated into the boiler design. This, however, is not always convenient, especially if the boiler is sited in a garage or roof space.

Room thermostats

A room thermostat senses air temperature. It is simply a temperature-controlled switch that connects or breaks an electrical circuit when either calling for heat or shutting the circuit down when the correct temperature has been reached. Most room thermostats contain a very small heater element called an accelerator, which 'tops up' the heat to the room thermostat by 1°C or 2°C, smoothing out the temperature cycle, preventing the boiler from 'cycling' when it isn't required.

Programmable room thermostats allow different temperatures to be set for different days of the week. It also provides a 'night set-back feature' where a minimum temperature can be maintained at night. Some units also allow the time control of the hot water cycle.

▲ Figure 7.57 A room thermostat

Cylinder thermostats

A simple control of stored hot water temperature, usually strapped to the side of the hot water cylinder about a quarter of the way up from the bottom. It is used with a motorised valve to provide close control of water temperature and should be set to 55°C.

Frost thermostats and pipe thermostats

The purpose of the frost thermostat is to stop the boiler and any other vulnerable parts of the system from freezing in extremely cold weather. It is wired in to the system to override all other programmers and thermostats. It should be set to between 3°C and 5°C, and should be placed close to the vulnerable parts of the system, especially if they are fitted in unheated garages and roof spaces.

Frost thermostats are much more effective when installed alongside a pipe thermostat.

A pipe thermostat is strapped to vulnerable pipework and senses water temperature. It is designed to override all other controls when the temperature of the water is close to 0°C and works in conjunction with the frost thermostat. The pipe thermostat and frost thermostat should be wired in series (see Chapter 11, Working with electricity, page 536).

Motorised valves

We have already seen that both the two-port zone valve and the three-port mid-position valve are key controls for the S-plan and Y-plan fully pumped systems, and the C-plan plus semi-gravity system. To recap the key points of these valves:

- **Three-port diverter valve:** very similar in appearance to the three-port mid-position valve, this valve is designed to control the flow of water on fully pumped central heating/hot water systems, where hot water priority is required.
- **Three-port mid-position valve:** used on fully pumped central heating/hot water systems to provide full temperature control of both the hot water and heating circuits when linked to cylinder and room thermostats. The circuits can operate together or independently of each other.
- **Two-port motorised zone valve:** these can be found on both C-plan plus systems, where a single valve linked to a cylinder thermostat controls the hot water temperature, and S-plan fully pumped systems where two two-port zone valves control the heating and hot water circuits via room and cylinder thermostats. They can also be used to zone different parts of the heating circuit.

Advanced controls: weather compensation, optimum start and delayed start

Domestic central heating systems can benefit from more advanced controls, which are often digital, especially when a condensing boiler is fitted. Condensing boilers respond better to lower flow and return temperatures than non-condensing appliances. Advanced controls enhance system efficiency and include:

- weather compensation controls
- delayed start controls
- optimum start controls.

Weather compensation controls

This type of control uses an externally fitted temperature sensor fitted on a north- or north-east-facing wall so as not to be in the direct path of solar radiation. As the external temperature rises, the weather compensator reduces the circulation temperature of the flow from the boiler to compensate for the warmer outside temperature. Similarly, the reverse occurs if the weather gets colder.

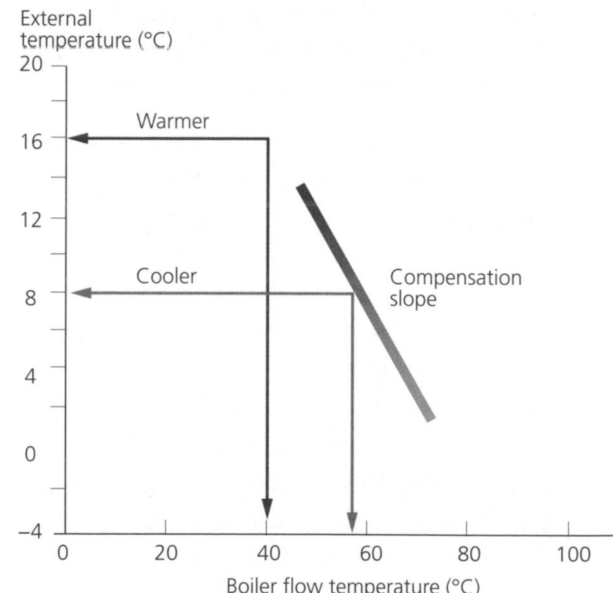

▲ Figure 7.58 A weather compensation graph

Delayed start

Here, the end user sets the time to switch on the heating, taking into account the time it would normally take to warm the dwelling – for example, most people would set the heat to come on at 5 pm if they were due to arrive home from work at 6 pm. A delayed start unit will, at the time the heat is due to come on, compare the current indoor temperature to that required by the room thermostat. It will then delay the start of the boiler firing if required. The benefits are that during milder weather, when the heat requirement is less, energy will be saved. Room thermostats with a delayed start function are now available.

Optimum start

With an optimum start system, the end user sets the required occupancy times and the required room temperature, and the controller calculates the necessary heat-up time so that the rooms are at the required temperature irrespective of the outside temperature. The idea is based around comfort rather than energy savings.

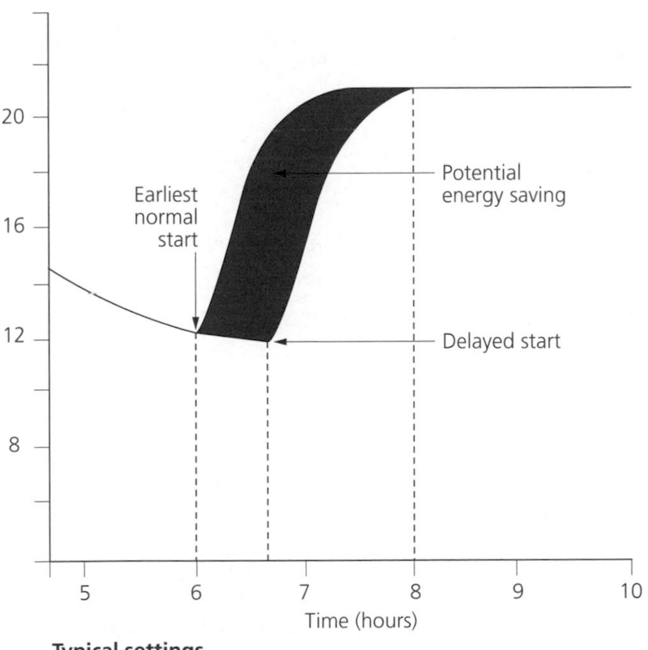

▲ Figure 7.59 Delayed/optimum start function

Domestic boiler management systems (home automation systems)

A boiler management system (BMS) is an electronic controller that provides bespoke control solutions for domestic central heating systems.

Standard functions of BMS control include real-time temperature and boiler/controls monitoring, room temperatures (known as set points) and time schedule adjustment, optimisation, and night set-back control.

The system remembers key points, such as how quickly the building heats up or cools down, and makes its own adjustments so that energy savings can be made. If it is very cold outside at, say, 2 am, the BMS will switch the heating on at 4.15 am to allow the building to be at the correct temperature by the time the user has set the heating to come on – say, 7 am – irrespective of the time that the user has set for the heating to activate. On milder nights, the heating may not come on until 6.15 am but it will still reach its set point by 7 am.

It will also learn how well your house retains heat and may shut down early if it calculates that your set point will still be maintained at your 'off' time of, say, 10 pm.

These systems provide a cost-effective means of monitoring system efficiency and can reduce heating costs by up to 30 per cent.

Electronic sensors are fitted to the flow and return pipework, and an external temperature sensor is fitted for weather compensation. The information is used to accurately vary the system output according to demand. This helps to significantly reduce fuel wastage caused by temperature overshoot, heat saturation of the heat exchanger, unnecessary boiler cycling and flue gas losses, while maintaining internal comfort levels and reducing CO_2 emissions.

System design and control

Now that we have seen the controls and the system layouts, we must look at how the controls work together to ensure efficient operation of the systems. We will concentrate on fully pumped systems as these are the systems that we must install on new installations.

▼ Table 7.17 How the Y-plan system works

The three-port valve	The flow from the boiler must be connected to the AB port, which is marked on the valve. The A port must be connected to the heating circuit. The B port must be connected to the hot water circuit. The valve must not be installed upside down as leakage of water could penetrate the electric actuator.
Time control	This must be provided by a programmer that allows individual use of hot water and heating circuits.
Heating circuit	Must have a room thermostat positioned in the coolest room, away from heat sources and cold draughts. It should be wall mounted at 1.5 m from floor level. The room thermostat controls the three-port valve. All radiators must have thermostatic radiator valves fitted.
Hot water circuit	The hot water temperature must be controlled by a cylinder thermostat placed a third of the way up from the base of the cylinder. The cylinder thermostat controls the three-port valve.
Bypass	An automatic bypass valve is required.
Frost/pipe thermostat	Must be provided where parts of the system are in vulnerable positions.

▼ Table 7.18 How the S-plan system works

The two-port zone valves	A single zone valve must be installed on the hot water circuit controlled by a cylinder thermostat. The heating circuit must contain one or more (if the system is to be zoned) two-port zone valves. These are controlled by individual room thermostats.
Time control	This must be provided by a programmer that allows individual use of hot water and heating circuits. A second time clock may be required if the system is zoned.
Heating circuit	One or more room thermostats controlling downstairs and upstairs heating circuits. These should be installed at 1.5 m from floor level.
Hot water circuit	The hot water temperature must be controlled by a cylinder thermostat placed a third of the way up from the base of the cylinder.
Bypass	An automatic bypass valve is required.
Frost/pipe thermostat	Must be provided where parts of the system are in vulnerable positions.

The S-plan gives better overall control of the system and this improves system efficiency.

Boiler interlock

The boiler interlock is not a single control device but the interconnection of all of the controls on the system, such as room thermostats, cylinder thermostats and motorised valves. The idea behind the boiler interlock is to prevent the boiler firing up when it is not required, a problem with older systems. A boiler interlock can also be achieved by the use of advanced controls, such as a BMS, usually reserved for larger systems but now available for domestic properties.

The selection of system and control types for single-family dwellings

The installation of an effective system of central heating controls has a major effect on the consumption of energy and the effectiveness of the system. Choosing the right controls will lead to:

- improved energy efficiency
- reduced fuel bills
- lower CO_2 emissions.

The establishment of a minimum standard of heating controls is vital if the heating system is to achieve satisfactory efficiencies when the system is in use. The efficiency of the boiler is only part of the story. For the boiler to achieve these efficiencies, at least a minimum standard of controls must be installed.

So, what is a good system of controls?

A good system of controls must:

- ensure that the boiler does not operate unless there is demand; this is known as 'boiler interlock'
- provide heat only when it is required to achieve the minimum temperatures.

There are two levels of controls for domestic properties and these are set out in Central Heating System Specification (CHeSS) CE51 2008:

1. **Good practice:** This set of controls achieves good energy efficiency in line with Approved Document L 2010. This is described in detail in the CHeSS document:
   a. **HR7** – Good practice for systems with a regular boiler and a separate hot water store:
      i. Full programmer
      ii. Room thermostat
      iii. Cylinder thermostat
      iv. Boiler interlock (see note 1)
      v. TRVs on all radiators, except in rooms with a room thermostat
      vi. Automatic bypass valve (see note 2).
   b. **HC7** – Good practice for systems using a combination boiler or Combined Primary Storage Unit boiler:
      i. Time switch
      ii. Room thermostat
      iii. Boiler interlock (see note 1)
      iv. TRVs on all radiators, except in rooms with a room thermostat
      v. Automatic bypass valve (see note 2).
2. **Best practice:** This standard uses enhanced controls to further enhance energy efficiency in line with Approved Document L1a/b 2010. This is described in detail in the CHeSS document:
   a. **HR8** – Best practice for systems with a regular boiler and a separate hot water store:
      i. Programmable room thermostat, with additional timing capability for hot water
      ii. Cylinder thermostat
      iii. Boiler interlock (see note 1)
      iv. TRVs on all radiators, except in rooms with a room thermostat
      v. Automatic bypass valve (see note 2)
      vi. More advanced controls, such as weather compensation, may be considered.
   b. **HC8** – Best practice for systems using a combination boiler or Combined Primary Storage Unit:
      i. Programmable room thermostat
      ii. Boiler interlock
      iii. TRVs on all radiators, except in rooms with a room thermostat
      iv. Automatic bypass valve (see note 2)
      v. More advanced controls, such as weather compensation, may be considered.

Note 1 (from CHeSS): Boiler interlock is not a physical device but an arrangement of the system controls (room thermostats, programmable room thermostats, cylinder thermostats, programmers and time switches) so as to ensure that the boiler does not fire when there is no demand for heat. In a system with a combi boiler this can be achieved by fitting a room thermostat. In a system with a regular boiler this can be achieved by correct wiring interconnection of the room thermostat, cylinder thermostat, and motorised valve(s). It may also be achieved by more advanced controls, such as a boiler energy manager. TRVs alone are not sufficient for boiler interlock.

Note 2 (from CHeSS): An automatic bypass valve controls water flow in accordance with the water pressure across it, and is used to maintain a minimum flow rate through the boiler and to limit circulation pressure when alternative water paths are closed. A bypass circuit must be installed if the boiler manufacturer requires one, or specifies that a minimum flow rate has to be maintained while the boiler is firing. The installed bypass circuit must then include an automatic bypass valve (not a fixed position valve).

Care must be taken to set up the automatic bypass valve correctly, in order to achieve the minimum flow rate required (but not more) when alternative water paths are closed.

Source: Energy Saving Trust (2008) *Central heating system specifications (CHeSS)*

Underfloor heating

Underfloor heating has been around for many years. The Romans used a warm air system 1500 years ago, to good effect. It is only fairly recently that its benefits have been rediscovered. With the arrival of new technologies such as air and ground source heat pumps and solar heating, underfloor heating becomes not only a viable option for the domestic dwelling but one that will also save money and energy, reduce CO_2 emissions and, as a consequence, help save the fragile planet on which we live.

The design principles of underfloor central heating systems

An underfloor heating system provides invisible warmth and creates a uniform heat, eliminating cold spots and hot areas. The temperature of the floor needs to be high enough to warm the room without being uncomfortable underfoot. There is no need for unsightly radiators/convectors because the heat literally comes from the ground up. Underfloor heating creates a low temperature heat source that is spread over the entire floor surface area. The key phrase here is low temperature. Whereas most wet central heating systems containing radiators and convectors operate at around 70°C to 80°C, underfloor heating operates at a much lower temperature, making it an ideal system for air and ground source heat pump fuel sources. Typical temperatures are:

- 40–45°C for concrete (screeded) floors
- 50–60°C for timber floor constructions.

Traditional wet central heating systems generate convection currents and radiated heat. Around 20 per cent of the heat is radiated from the hot surface of the radiators and, if furniture is placed in front of the radiator, the radiation emission is reduced. A total of 80 per cent of the heat is convection currents, which makes the hot air rise. This adds up to a very warm ceiling! Underfloor heating systems, however, rely on both conduction and radiation. The heat from the underfloor heating system conducts through the floor, warming the floor structure, making the floor surface a large storage heater; the heat is then released into the room as radiated heat. Around 50 to 60 per cent of the heat emission is in the form of radiation, providing a much more comfortable temperature at low room levels when compared to a traditional wet system with radiators and, with the whole floor being heated, furniture positioning no longer becomes a problem because as the furniture gains heat, it too emits warmth.

During the design stage, the pipe coils are fixed at specific centres depending on the heat requirement of the room and the heat emission (in watts) per metre of pipe. The whole floor is then covered with a screed to a specific depth, creating a large thermal storage heat emitter. The water in the pipework circulates from and to a central manifold and heats the floor. The heat is then released into the room at a steady rate. Once the room has reached the desired temperature, a room thermostat actuates a motorised head on the return manifold and closes the circuit to the room.

Such is the nature of underfloor heating that many fuel types can be used, some utilising environmentally friendly technology. Gas- and oil-fired boilers are common, but also biomass fuels, solar panels and heat pumps.

Floor coverings are an important aspect of underfloor heating. Some create a high thermal resistivity, making it difficult for the heat to permeate them. Carpet underlay and some carpets have particularly poor thermal transmittance, which means the heat is kept in and not released. The thermal resistivity of carpets and floor coverings is known as their TOG rating. The higher the TOG rating, the less heat will get through. Floor coverings used with underfloor heating should have a TOG rating of less than 1 and must never exceed 2.5.

Quite often, underfloor heating is used in conjunction with traditional wet radiators, especially in properties such as barn conversions. The higher temperatures required for radiators do not present a problem because the flow water for the underfloor system is blended with the return water via a thermostatic blending valve to maintain the steady temperature required for the underfloor system. Zoning the upstairs and downstairs circuits with two-port motorised zone valves and independent time control for the heat emitters also helps in this regard.

INDUSTRY TIP

A free copy of the CHeSS specifications is available at: http://bpec.org.uk/downloads/CE51%20CHeSS%20WEB%20FINAL%20JULY%2008.pdf

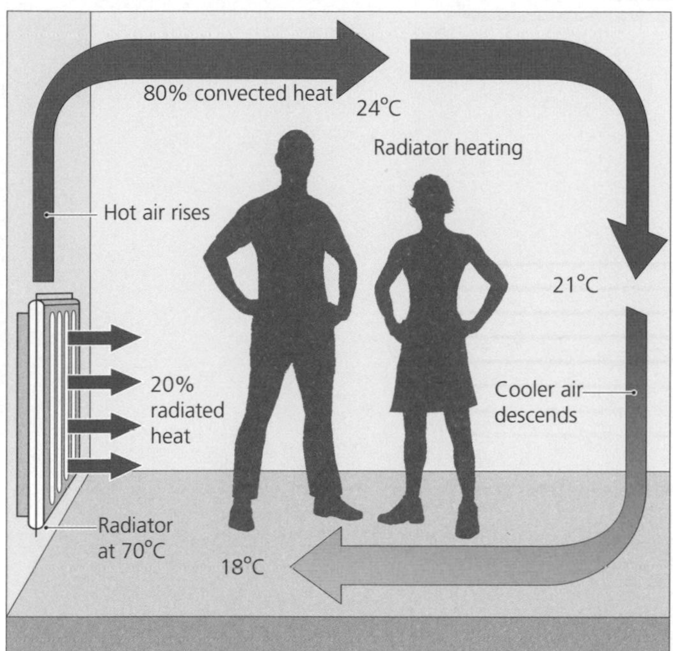

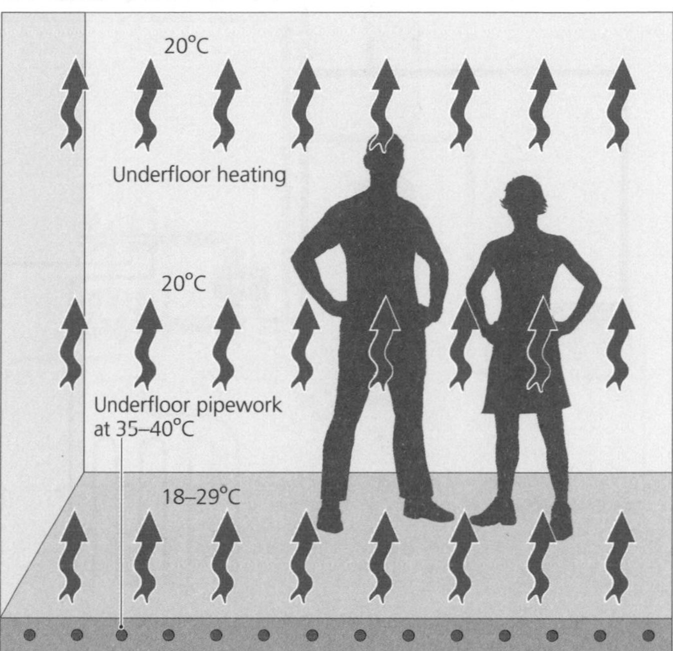

▲ Figure 7.60 The principle of underfloor heating

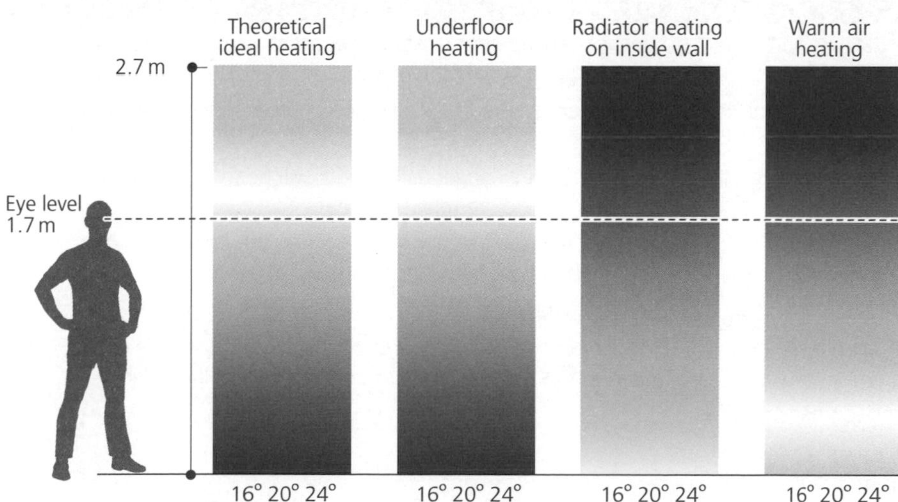

▲ Figure 7.61 Heating theory

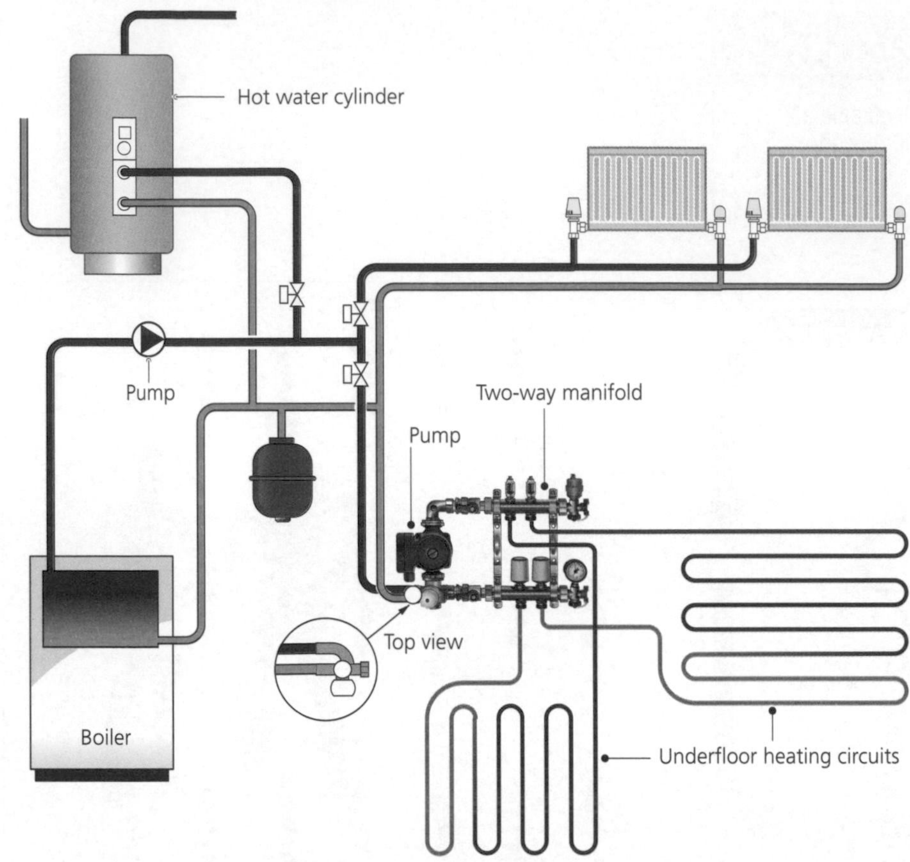

▲ Figure 7.62 Typical underfloor heating system combined with wet radiators

▼ Table 7.19 The advantages and disadvantages of underfloor heating

Advantages	Disadvantages
The pipework is hidden under the floor. This allows better positioning of furniture and interior design.	Not very suitable for existing properties unless a full renovation means the removal of floor surfaces.
The heat is uniform, giving a much better heat distribution than traditional systems.	Can be expensive to install when compared to more traditional systems.
These systems are very energy efficient, with low running costs.	Heat-up time is longer as the floor will need to get to full temperature before releasing heat.
Environmentally friendly fuels can be used.	Slower cool-down temperatures mean the floors may still be warm when heat is not required.
Underfloor heating is almost silent, with low noise levels when compared to other systems.	Longer installation time.
Cleaner operating, with little dust carried on convector currents. This can help those people who suffer from allergies, asthma and other breathing problems.	More electrical installation of controls is required, as each room will need its own room thermostat and associated wiring.
System maintenance is low, and decorating becomes easier as there are no radiators to drain and remove.	
Individual and accurate room temperatures, as every room has its own room thermostat that senses air temperature.	
Reduced possibility of leaks.	
Greater safety, as there are no hot surfaces that can burn the elderly, infirm or the very young.	
Better zone control as each room is, in effect, a separate zone.	

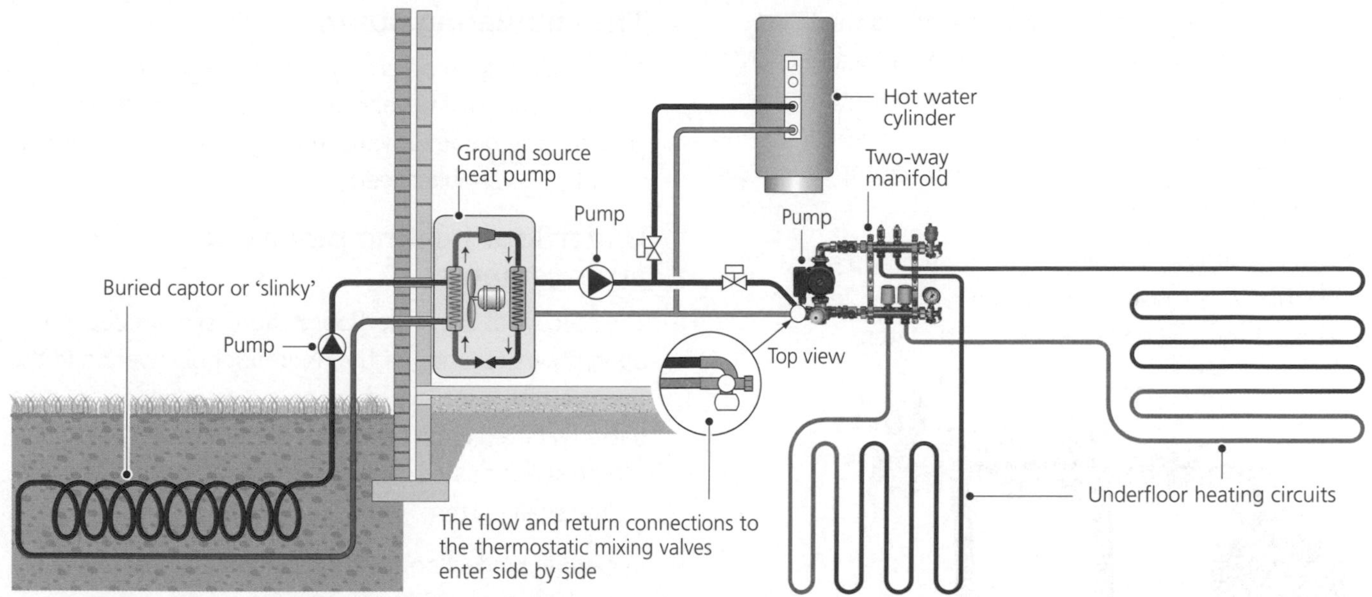

▲ Figure 7.63 Typical underfloor heating system using a ground source heat pump

The layout features of underfloor heating

Underfloor heating uses a system of continuous pipework, laid under a concrete or timber floor in a particular pattern and at set centre-to-centre pipe distances. Each room served by an underfloor heating system is connected at a central location to a flow and return manifold, which regulates the flow through each circuit. The manifold is connected to flow and return pipework from a central heat source, such as a boiler or heat pump.

The manifold arrangement also contains a thermostatic mixing valve to control the water to the low temperatures required by the system, and an independent pump to circulate the water through every circuit.

Each underfloor heating circuit is individually controlled by a room thermostat, which activates a motorised head on the return manifold to precisely control the heat to the room to suit the needs of the individual.

The working principles of underfloor central heating system pipework and components

Underfloor heating is becoming more commonly installed in new build properties where environmental technologies have been installed. This is due to its suitability to utilise the lower flow temperatures from environmental technologies. See Figure 7.63 and page 410. As we have already seen, underfloor heating works by distributing heat in a series of pipes laid under the floor of a room. To do this, certain components are required to distribute the flow of heat to ensure that the system warms the room. However, the components must be controlled in such a way as to maintain a steady flow of heat while ensuring that the floor does not become too hot to walk on. This is achieved by the use of:

- manifolds
- a thermostatic blending valve
- a circulating pump
- various pipework arrangements to suit the floor and its coverings
- the application of system controls – time and temperature to space heating zones.

The use of manifolds

In technical terms, the manifold is designed to minimise the amount of uncontrolled heat energy from the underfloor pipework. The manifold is at the centre of an underfloor heating system. It is the distribution point where water from the heat source is distributed to all of the individual room circuits and, as such, should be positioned as centrally as possible in the property. Room temperature is maintained via thermostatic motorised actuators on the return manifold, while the correct flow rate through each coil is balanced via the flow meters on the flow manifold. Both the flow and return manifolds contain isolation valves for maintenance activities, an automatic air valve to prevent air locks and a temperature gauge so that the return temperature can be monitored.

Most manifolds contain a circulating pump and a thermostatic mixing valve, often called a blending valve. These will be discussed a little later.

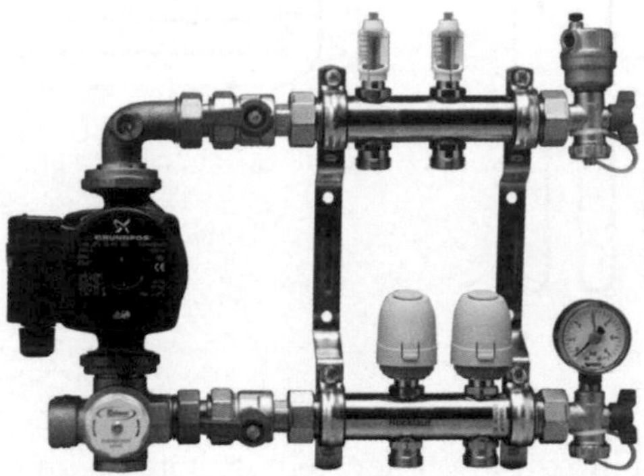

▲ Figure 7.64 Typical underfloor heating manifold

The thermostatic mixing (blending) valve

Thermostatic mixing or blending valves are designed to mix the flow and return water from the heat source to the required temperature for the underfloor heating circuits. They are available in many different formats, the most common being as part of the circulating pump module, as shown in Figure 7.65. The temperature of the water is variable by the use of an adjustable thermostatic cartridge inside the valve.

▲ Figure 7.65 Underfloor heating circulating pump/blending valve module

The circulating pump

The circulating pump is situated between the thermostatic mixing valve and the flow manifold to circulate the blended water through every circuit. Most models are variable speed.

Underfloor heating pipework arrangements

The success of the underfloor heating system depends upon the installation of the underfloor pipework and the floor pattern installed. There are many variations of pipe patterns based upon two main pattern types. These are:

1 the series pattern
2 the snail pattern.

In general, underfloor heating pipes should not be laid under kitchen or utility room units.

The series pattern

The series pattern (also known as the meander pattern) is designed to ensure an even temperature across the floor, especially in systems incorporating long pipework runs. It is often used in areas of high heat loss.

The flow pipe must be directed towards any windows or the coldest part of the room before returning backwards and forwards across the room at the defined pipe spacing centres.

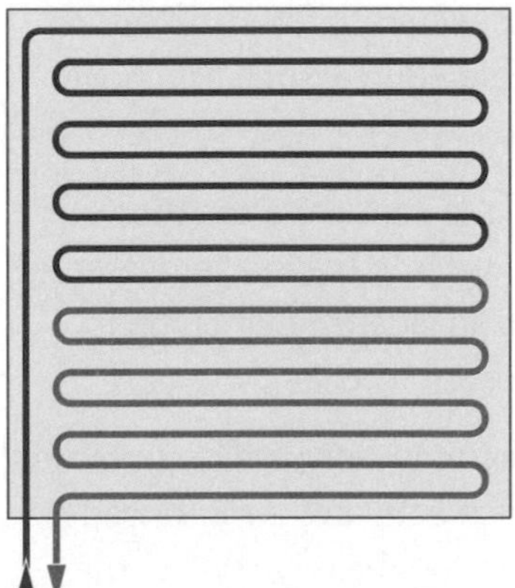

▲ Figure 7.66 The series pattern

The snail pattern

The snail pattern (also known as the bifilar pattern) is used where an even uniform temperature is required, such as under hardwood floors and vinyl floor tiles.

The flow pipe runs in ever decreasing circles until the centre of the room is reached; it then reverses direction and returns with parallel runs back to the starting point.

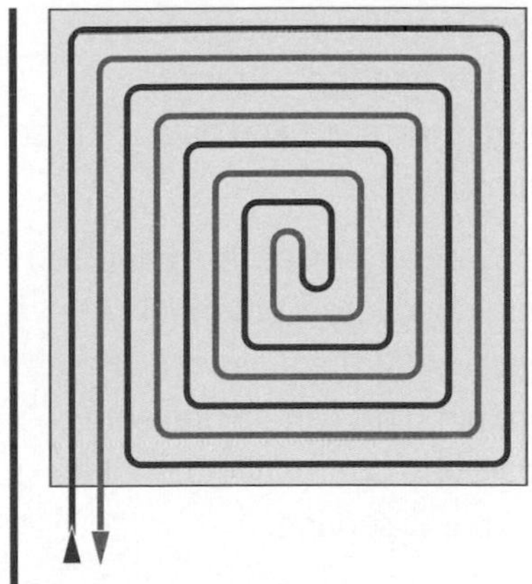

▲ Figure 7.67 The snail pattern

The application of system controls: time and temperature to space heating zones

The number of homes that require both time and temperature zone control has increased in recent years. In 2006, a survey showed that the average floor area of a domestic property with four bedrooms was around 157 m^2 and more than 200 m^2 for a five-bedroom domestic property. With properties of this size, zoning becomes a necessity and, in 2006, Document L1A/B of the Building Regulations requested that zoning of the heating system must be installed in all properties of 150 m^2 or more. This was updated in 2010 to include any property.

In most instances, zoning requires the separating of the upstairs circuit from the downstairs or, in the case of single-storey dwellings, separating the living space from the rest of the property. Separate time and temperature control of the individual circuits is a necessity.

Zoning with separate temperature control

Separate temperature-controlled zones provides a much better living environment because different parts of the dwelling can be maintained at different temperatures without relying on a single room to dictate the temperature across the whole system. Lower temperatures can be maintained in those rooms within the dwelling that are not occupied, allowing the dwelling to take full advantage of any solar gains, especially in rooms that face south, south-east or south-west. This can be quite pronounced, even in the winter sun. Significant savings on both energy usage and fuel costs can be made by simply taking advantage of the free heat that the Sun can provide. Outside sensors linked to weather compensators, and delayed start and optimum start controls, further help to reduce energy usage and cost.

Zoning with separate time control

Zoning with separate time control offers another dimension to the concept of zoning by allowing the heating to be controlled at different times of the day in different zones. The heat can be focused in those rooms that are occupied throughout the day, with the heating to other parts of the dwelling timed to come on in the early morning and evening. Separate zones reduce energy usage and costs while maintaining improved comfort levels throughout the property.

Zoning in practice

Zoning is required by Approved Document L1A/B of the Building Regulations 2010 and the installer must make decisions on the best way to arrange those zones to take the best advantage of energy savings while complying with the wishes of the customer/end user as well as the regulations. The only way this can be achieved is by talking to the customer and finding out their usage patterns. The main aim of zoning is to avoid overheating areas that require less heat to maintain the warmth or because the set point could be lower than in other areas. The point here is that the number of zones laid down by Document L is the minimum and there are real benefits to adding additional zones in key areas of the property.

An underfloor system lends itself naturally to zoning as each room is individually controlled by a room thermostat, which activates actuators on the individual circuits at the manifold. Further controls can be added where heat emitters and underfloor heating circuits are installed on the same system. In this case, the zones are both individually temperature controlled and timed.

The system can also be linked to other advanced controls such as night set-back and delayed start.

The choice of controls for the zones should be decided by the predicted activity in those zones. There are many options that can be used individually or collectively to achieve good system control:

- using individual temperature and timing controls in every zone
- multi-channel programmers allow the timing of individual rooms or multiple zones to be set from a single point; this is often more desirable than many individual programmers at different locations within the dwelling
- TRVs vary the heat output of individual heat emitters; this can be beneficial where solar gain adds to the room temperature as they are very fast reacting in most circumstances; some TRVs also have electronically timed thermostatic heads, which can be linked to a wireless programmer.

Zoning can help make significant energy savings. It allows the optimisation of the heating system while maintaining the dwelling at a comfortable temperature and saving money at the same time.

Positioning components in underfloor central heating systems

For an underfloor heating system to work effectively, the components require careful positioning to ensure that the efficiency of the system is maintained. All too often, systems fail to live up to their potential because of poor positioning of key components. Key areas include:

- manifolds
- pipework arrangements (cabling)
- pipework installation techniques.

Manifolds

The longer the circuit, the more energy is needed to push the water around it. Water will always take the line of least resistance and shorter circuits will always be served first. In many instances, balancing the system will help even out the circulation times so that all circuits receive the heat at the same time, but the system will only be as good as the slowest circuit. If the longest circuit is slow, once the system is balanced, then all circuits will be slow. In this regard, the positioning of the manifold is of great importance. By positioning the manifold centrally within the dwelling, the length of each circuit is balanced so that long circuits become shorter. Even if the short circuits become longer, the time for the heating system to reach full temperature will be shortened and balancing the system will become much easier.

A potential problem that may occur where the manifold is located is that the area may become a potential 'hot spot' on the system because of the pipework congestion around the manifold. This can be prevented by insulating the pipework around the manifold until the pipework enters the room it is serving.

Pipework arrangements (cabling)

There are many variations to the two basic layouts. The pattern should be set out in accordance with the orientation and the shape of the room. Window areas may be colder and may require the bulk of the heat in that area. Other considerations include the type of floor construction and the floor coverings. The pipework should be laid in one continuous length without joints. In some instances, the pipe is delivered on a continuous drum of up to 100 m to enable large areas to be covered without the need for joints. Large rooms may require more than one zone and the manufacturer's instructions should be checked for maximum floor coverage per zone.

▲ Figure 7.68 The series pattern laid out

▲ Figure 7.69 The snail pattern laid out

Pipework installation techniques

Solid floor

There are many types of underfloor heating installation techniques for a solid floor. Figure 7.70 shows one of the more common types using a plastic grid where the underfloor heating pipe is simply walked into the pre-made castellated grooves for a precise centre-to-centre guide for the pipework using a minimum radius bend.

The panels are laid on to pre-installed sheets of insulation to ensure a good performance and minimal heat loss downwards. Edge insulation is required to allow for expansion of the panels.

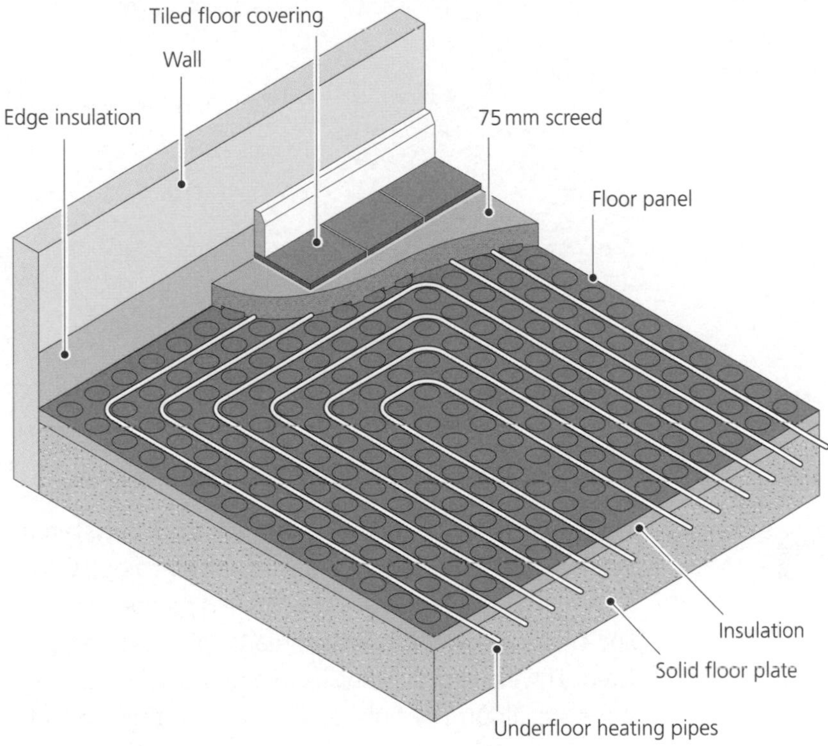

▲ Figure 7.70 Solid floor underfloor heating installation

▼ Table 7.20 Key design and installation information: solid floor

Maximum heat output	Approx. 100 W/m^2
Recommended design flow temp.	50°C
Maximum circuit length	100 m (15 mm pipe) 120 m (18 mm pipe)
Maximum coverage per circuit	12 m^2 @ 100 mm centres 22 m^2 @ 200 mm centres 30 m^2 @ 300 mm centres (18 mm pipe only)
Material requirements:	
Pipe	8.2 m/m^2 @ 100 mm centres 4.5 m/m^2 @ 200 mm centres 3.3 m/m^2 @ 300 mm centres (18 mm pipe only)
Floor plate usage	1 plate/m^2 allowing for cutting
Edging insulation strip	1.1 m/m^2
Conduit pipe	2 m/circuit

Suspended timber floor

This system is designed for use under timber suspended floors. It uses aluminium double heat spreader plates to transmit heat evenly across the finished floor surface.

This system is suitable for any timber suspended floor with joist widths up to 450 mm. The heat plates are simply fixed to the joists using small flat-headed nails or staples. A layer of insulation must be placed below the plates to prevent the heat penetrating downwards.

Where the pipework must cross the joists, the joists must be drilled in accordance with the building regulations.

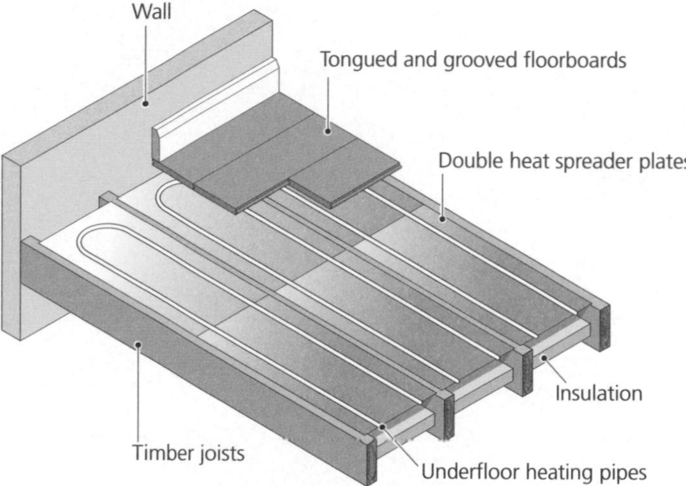

▲ Figure 7.71 Suspended floor underfloor heating installation

▼ Table 7.21 Key design and installation information: suspended floor

Maximum heat output	Approx. 70 W/m^2
Recommended design flow temp.	60°C
Maximum circuit length	80 m (15 mm pipe)
Maximum coverage per circuit using a double spreader plate	17 m^2 @ 225 mm centres
Material requirements:	
Pipe	4.5 m/m^2 @ 100 mm centres
Heat spreader plates	2 plate/m^2

Floating floor

This system is designed for use where a solid floor installation is not suitable due to structural limitations. It can be installed directly onto finished concrete or timber floors.

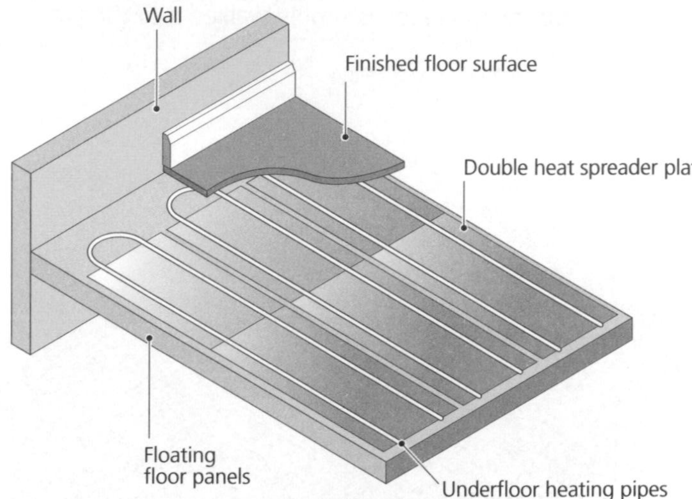

▲ Figure 7.72 Floating floor underfloor heating installation

The pipework is laid on top of 50 mm-thick polystyrene panels, each having a thermal transmittance of 0.036 W/m^{2}K. The insulation has pre-formed grooves that the pipe clips into after the heat spreader plates have been fitted. The insulation is not fixed and 'floats' on the top of the sub-floor. The finished flooring can then be laid directly onto the top of the pipework, completing the 'floating' structure.

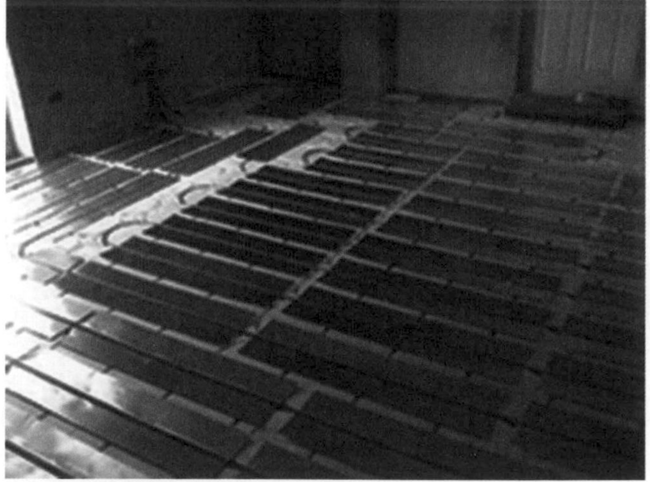

▲ Figure 7.73 Floating floor underfloor heating installation method

▼ Table 7.22 Key design and installation information: floating floor

Maximum heat output	Approx. 70 W/m^2
Recommended design flow temp.	60°C
Maximum circuit length	80 m (15 mm pipe) 100 m (18 mm pipe)
Maximum coverage per circuit using a double spreader plate	28.5 m^2 @ 300 mm centres (15 mm pipe) 30 m^2 @ 300 mm centres (18 mm pipe)
Material requirements:	
Pipe	3.1 m/m^2 @ 300 mm centres
Single heat spreader plates	3 plate/m^2
Floating floor panel	1 panel/1.4 m^2

Environmental technologies

Environmental issues are becoming more important to the plumbing industry. Environmental technologies are being installed and used in domestic properties. Government and house builders are under pressure to include a variety of systems to help cut back carbon emissions and reduce carbon footprints that are contributing to climate change.

As plumbers, we look to install systems that offer practical, affordable and sustainable alternatives to the use of fossil fuels that can heat and power the property.

We need to have a good working understanding of the systems that are being installed. This area of 'green technologies' is developed further in Book 2, but this section will introduce you to the following systems:

- solar thermal
- solar photovoltaic
- ground source heat pump
- air source heat pump
- biomass
- micro combined heat and power
- rainwater harvesting
- grey water harvesting.

With all systems being installed, environmental technologies come under a number of Acts, Building Regulations (see Chapter 1) and British Standards, which have to be followed along with the manufacturer's instructions.

Solar thermal

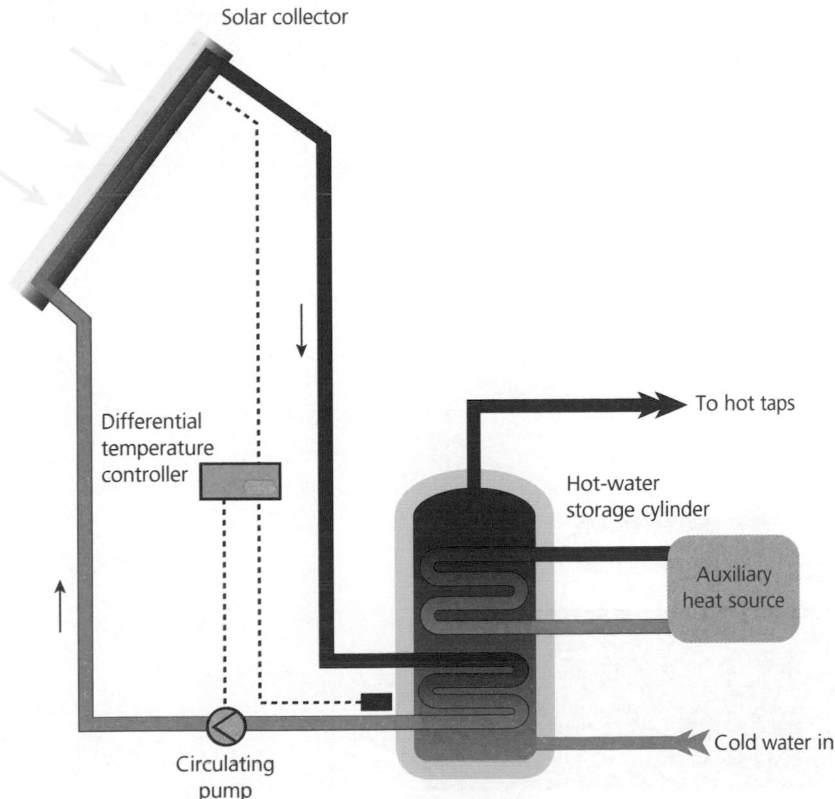

▲ Figure 7.74 A solar thermal system

This system is designed to collect heat by absorbing heat radiation from the Sun in a collector panel often located on the roof of a property. This heat energy is transferred into a fluid and transported down to the heat exchanger coil in a twin coil hot water cylinder. This transfer of heat is carefully controlled by the differential temperature controller. As some of the pipework in this system is vulnerable to freezing, being on the roof line of the property, the water transferring the heat would normally contain glycol (anti-freeze). In the UK, there will be insufficient solar energy to provide hot water to the property, so an auxiliary heat source, generally a boiler, would be linked to the system by a second coil heat exchanger or an immersion would be used.

There are several considerations that need to be made when locating the solar collectors: orientation of the panel; the angle of the panel; any shading of the panel during the day; suitability of the building to install the panel.

This system will reduce the use of fossil fuels for the property while heating the domestic hot water.

Solar photovoltaic

This system converts the sunlight into electricity. Collector panels (PV cells) are installed on the roof line which produce a charge which can either be stored in batteries for future use or can be connected to the grid to be used by the property straight away or sold 'back to the grid'. This means a household can generate extra income by selling surplus energy back to the energy provider via a government scheme.

▲ Figure 7.75 A solar photovoltaic system

The charge is carefully controlled by a charge controller.

This system offers the property electricity from the Sun, rather than using electricity produced by a power station.

Ground source heat pump

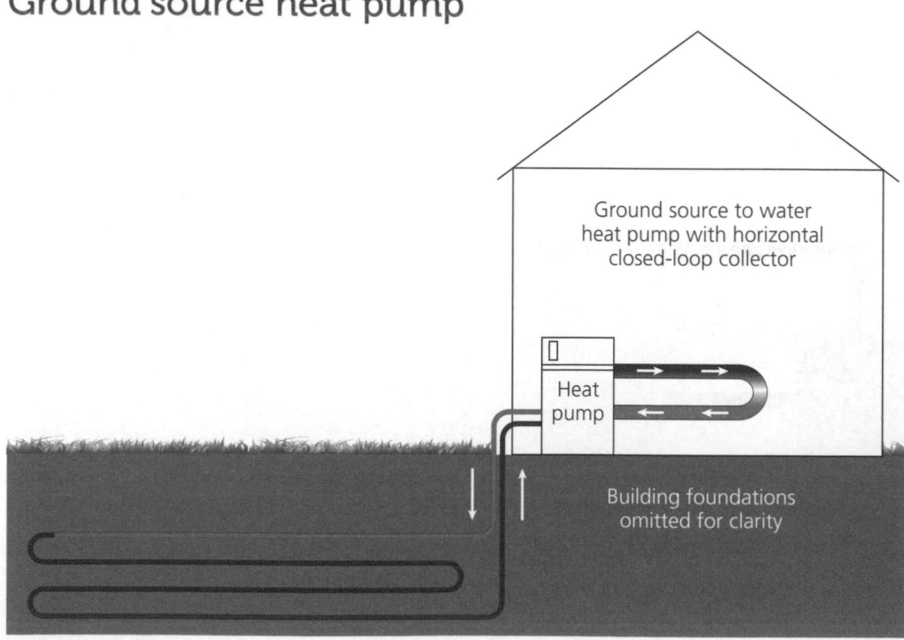

▲ Figure 7.76 A ground source heat pump

These extract low temperature free heat from the ground, magnifying it to a higher temperature and then releasing it when required for space and water heating. Heat is collected from the ground by means of pipe runs containing a mixture of water and glycol (anti-freeze) which are buried in the ground in one of three different ways. Figure 7.76 shows the pipe work for a horizontal loop. They can be vertical or slinkies.

The low temperature is passed through a heat exchanger, where the heat is transferred to refrigerant gas which is then compressed. It then passes through a second heat exchanger and releases the increased temperature to the domestic system.

The earth stores heat and maintains a constant low temperature of around 10°C, which the ground source heat pump utilises. The low temperature is constantly being replenished by nature and therefore is not a cost to the customer.

Air source heat pump

▲ Figure 7.77 An air source heat pump

These units extract free heat from low temperature air and release it where required for space and water heating. It works in a similar way to a refrigerator, with the cooled area on the outside of the property and the heat being released inside the property.

Like the ground source pump, the system contains refrigerants that pass the extracted heat on to the property.

Biomass

This is material that can be burnt to create heat, but biomass comes from animal matter or plants/trees that have recently been sourced, whereas fossil fuels have taken millions of years to source.

▲ Figure 7.78 Biomass fuels

Biomass fuels have to be burnt to produce the heat which is used within the system, but they produce carbon dioxide (a greenhouse gas) when burnt. The importance of biomass material is that the material absorbs carbon dioxide when it grows, reducing carbon dioxide levels in the atmosphere. When burnt, the carbon dioxide is released, the net result being no overall increase, so it is carbon neutral.

Biomass fuel does not have the same heat output as fossil fuels, which are more dense. However, if managed correctly, biomass is sustainable.

Micro combined heat and power

This is where the fuel source used to supply heat and hot water to the property is also used to supply electricity at the same time. The mCHP will achieve similar efficiencies as a condensing boiler but in different ways. 80 per cent of the heat is used to provide heat and 15 per cent is used to generate power. This system is known as 'heat led' and is carbon reducing rather than carbon free.

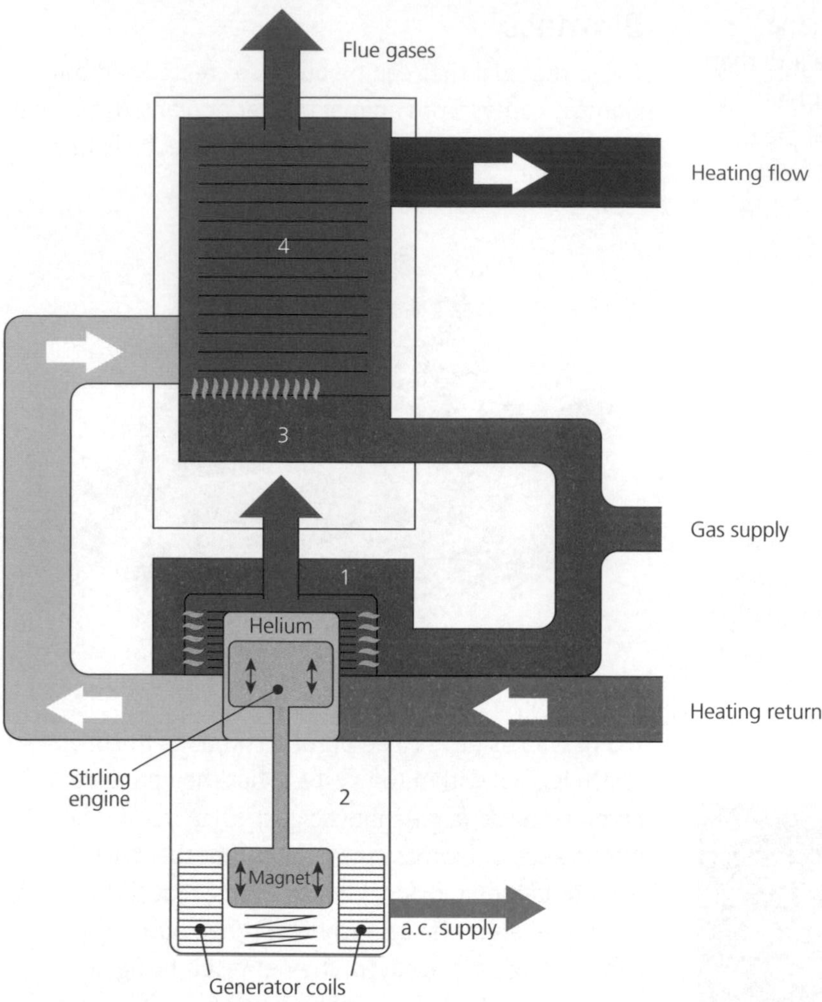

▲ Figure 7.79 Micro combined heat and power

These domestic systems contain a Stirling engine generator which uses the expansion and contraction of internal gases to operate a piston. When the engine fires, the gas (helium) expands and when the return water cools the engine down it contracts, moving the piston up and down, which creates power.

Rainwater harvesting

A further key area is water conservation due to expanding demands on the water systems. The water supplied to UK properties is potable drinking water, but this water is also used for washing, bathing and watering gardens. Water conservation is one way to make sure demand does not exceed supply.

Rainwater is captured off the roof line of a property and stored rather than letting it run off into the drains. Instead of flushing the WC with potable water, it could be re-filled with rainwater. The same goes for car washing, garden watering or even clothes washing, but rainwater must **never** be used for drinking because it is classified as Category 5 water.

There are two types of harvesting system: direct, which pumps the rainwater directly to the outlets, and indirect, which uses a header tank in the property's loft area to supply the outlets.

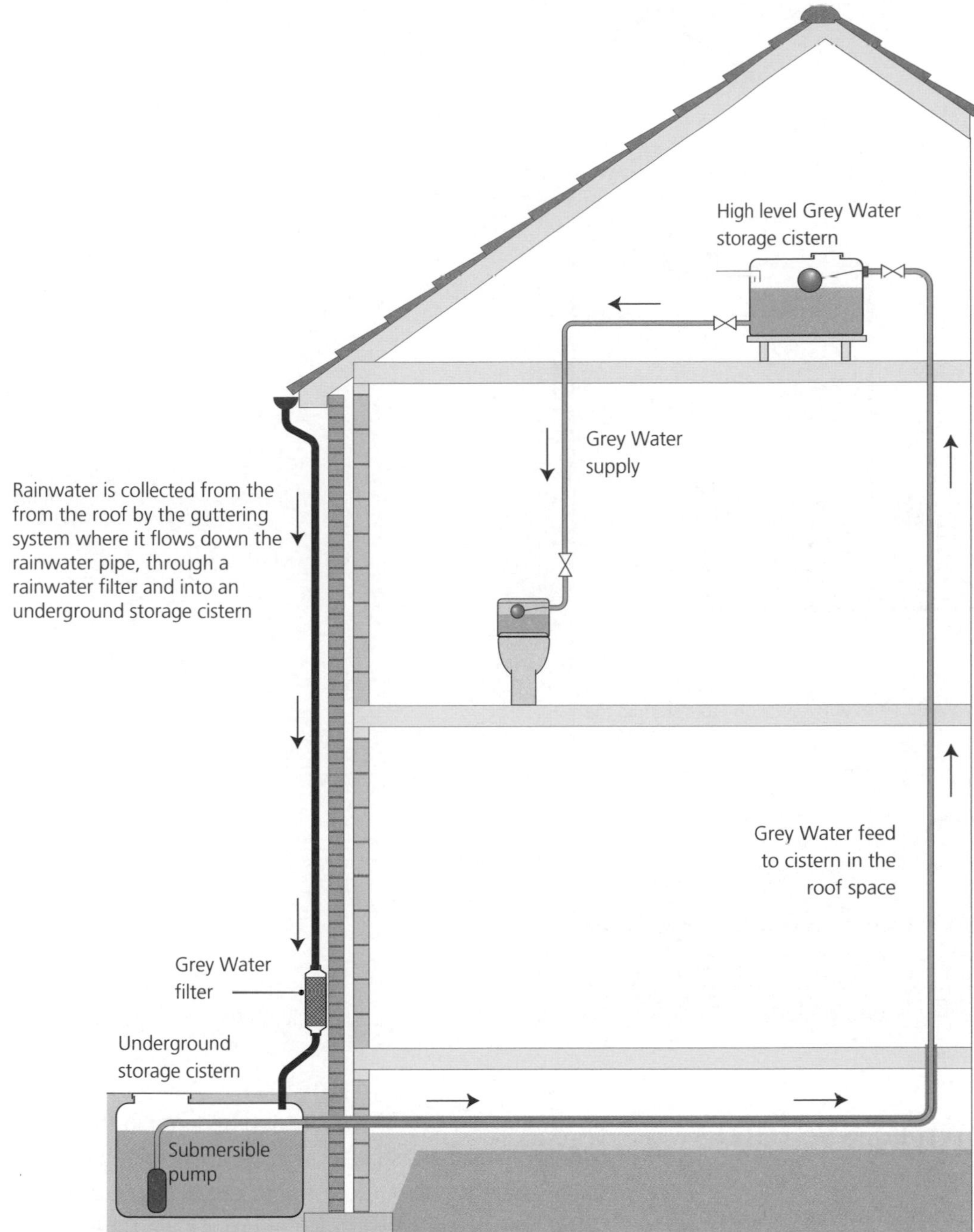

▲ Figure 7.80 Rainwater harvesting

Grey water harvesting

This is water from baths, showers, basins and washing machines and gets its name from the 'grey', cloudy appearance of the water. Instead of this water entering the sewers, it is captured and stored. This can be for a short while only, as bacteria will grow.

Again, grey water is classified as Category 5 so it cannot be used for drinking but can be used to flush toilets, wash cars, water gardens and wash clothes.

Both these systems will reduce demand on the main system and reduce water bills for the customers.

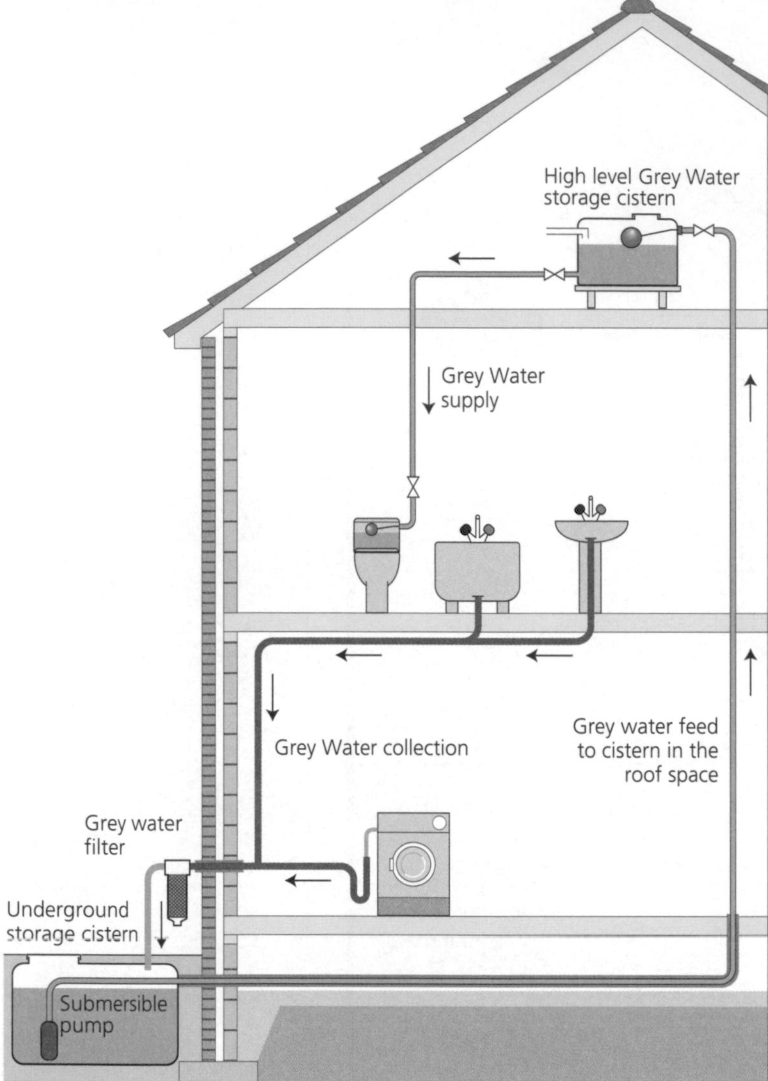

▲ Figure 7.81 Grey water harvesting

Filling and venting systems

The filling and venting of systems along with positioning of the circulator pump has been covered on page 367.

Sealed systems: positioning the expansion vessel, pressure gauge and filling loop

The expansion vessel is installed onto the return because the return water is generally 20°C cooler than the flow water and this does not place as much temperature stress on the expansion vessel's internal diaphragm as the hotter flow water. If installing the vessel on the flow is unavoidable, it should be placed on the suction side of the circulating pump in the same way as the cold feed and open vent pipe on the open vented system. Close to the expansion vessel is the pressure relief valve (sometimes called the expansion valve) and a pressure gauge.

The filling loop is generally fitted to the return pipe close to the expansion vessel and may even be supplied as part of the expansion vessel assembly.

Methods of releasing air from heating systems

When filling the system, air must be released to allow components such as heat emitters to work properly. Air is released by several methods:

- The open vent (on vented systems) allows air to escape from the pipework and the boiler as the system fills with water.
- Air release valves are sited on every heat emitter. These must be manually opened with a suitable radiator air release key to enable the air to escape as the system fills. Once water is detected, the air release valve is then closed.
- Automatic air release valves may be fitted in those places on the system where air is expected to collect, such as high points in the pipework and the coil heat exchanger on the cylinder. These were mentioned earlier in this chapter.
- Sealed systems must be filled, via the filling loop, taking care to avoid over-pressurising the system. The boiler manufacturer's recommendations on filling, venting and final system pressure must be followed.

Selection of fuels for heat-producing appliances

Types of fuels used for heat-producing appliances are:

- gas (both natural and LPG)
- oil
- solid fuel.

Gas

Natural gas is the most popular fuel for central heating in the UK. It comes from a variety of sources, including the North Sea, the Middle East and Russia. It is a naturally occurring gas that consists of a number of other flammable and inert gases:

- methane
- ethane
- propane
- butane
- hydrogen
- carbon dioxide
- nitrogen.

Oil

Oil-fired appliances are popular where access to mains gas is difficult. They offer a viable alternative to gas appliances. Most oil-fired appliances use C2 grade 28 second viscosity oil (kerosene), although other types of oil, such as heavy heating oil, are available.

Solid fuel

Solid fuel is still used in rural areas of the UK where access to piped fuel supply is difficult. Solid fuel is available in many different forms, including:

- coal
- coke
- anthracite
- biomass wood pellets (carbon neutral).

Warm air systems

Warm air systems are not often fitted in modern domestic properties in the UK, although they remain popular in parts of the USA. They work by blowing warm air through duct work into the rooms of the property via grilles, which can be located high on the wall or in the floor. The air is heated usually by a gas boiler and the heated air is distributed by an electric fan. Some boilers also contain a small water heat exchanger to heat the hot water for the property.

Electric storage heaters

Electric storage heaters were very popular in the 1980s and 1990s. They contain a series of ceramic bricks, which are heated by electric elements overnight using cheap rate electricity. The bricks store the heat and release it slowly during the day.

The heaters have two settings that can be operated either automatically or manually:

1. Charge – this controls the amount of heat that is stored
2. Draught – this controls the rate at which the heat is released.

They also have a day 'top-up' should the heaters require to be reheated for short periods during the day.

This type of heater is expensive to install and run. Because of this, their use today is very limited.

District heating

District heating utilises a very large central heat generating plant to heat hundreds of homes from the same system. They are very popular in Europe and some parts of the USA.

Pipes laid beneath the street distribute hot water (sometimes steam) from the central generating plant to a local substation. From here the heat is controlled to around 120°C, where it is delivered to all properties via a heating interface. The heating interface replaces the property's boiler as the heat source. From here, normal heating systems such as the S-plan and S-plan plus can be installed and controlled in the normal way. District heating systems are often installed as part of a combined heat and power system, as discussed in Chapter 10, Domestic fuel systems.

2 INSTALL CENTRAL HEATING SYSTEMS AND COMPONENTS

In this part of the chapter, we will consider the materials we can use to install domestic central heating systems and the installation methods for both new-build properties and existing installations. On any installation it is advisable to carry out a site survey prior to the installation, which will highlight any considerations that may be present. On larger installations and new builds you may be working to a works programme which will need to be regularly updated so that other trades are informed. This can also be uploaded onto information software so that it can be sent to people off site, which will enable remote management.

KEY POINT

The installation of pipework within domestic dwellings is covered in detail in Chapter 5, Cold water systems.

Effects of expansion and contraction

- **Pipework expansion:** if pipework expansion is not catered for during the installation phase, it can create ticking and creaking noises. To prevent this, any notches in joists should be deep enough to allow free movement of the pipework and any drilled holes should be large enough so that the pipework does not rub against them. Notches should be lined with hair felt to act as noise suppression.
- **Expansion in open vented systems (cistern):** the water in an open vented system expands into the feed and expansion cistern and up the vent pipe. When the water in the system gets hot, the expansion of water raises the water level in the cistern. It is therefore important that the feed and expansion cistern is large enough to accommodate the expanded water.
- **Expansion in sealed systems (expansion vessel):** the water in a sealed system expands into the vessel. It is therefore important that the expansion vessel is large enough to accommodate the expanded water without opening the pressure relief/expansion valve, as this would release the pressure to 0 bar and prevent the system from working correctly.

INDUSTRY TIP

A radiator bleed key is a useful tool for bleeding air from radiator air valves.

Connecting to existing systems

- **One pipe:** one-pipe circuits are a continuous loop with the heat emitters/radiators taken from the loop (see the section on the one-pipe circuit at the beginning of this chapter, page 360). Any extensions or new circuits should follow the one-pipe layout to prevent the new circuit from taking most of the heat flow. Swept tees should be used to encourage water flow around the new circuit.
- **Two pipe:** two-pipe systems are the easiest to extend. New circuits can be taken from existing circuits provided there are no more than three heat emitters on the run of pipework. Ideally, new circuits should be taken from the main heating flow and return pipes. If the system is a modern installation that has been zoned with two-port motorised zone valves, care should be taken to ensure that the new

heat emitter is taken from the correct circuit, i.e. living space heat emitters should be taken from the lounge/dining circuit and sleeping space heat emitters should be taken from the bedroom circuit.

- **Manifold (microbore):** when connecting extra circuits to a microbore system, the circuit must be connected to the flow and return manifolds. On no account must the circuit be connected to an existing circuit via tee pieces. Microbore pipework is only capable of carrying a heat load up to 3 kW per circuit.
- **Underfloor heating:** connecting extra pipework loops to an existing underfloor system is possible only if the manifold is replaced, allowing connection of the new circuit. The heat source/boiler and the circulating pump should be checked to ensure that they can cope with the extra heat load.

Soundness test requirements for pipework

Most domestic systems use three types of pipe materials:

1. **Copper tubes and fittings:** grades R220 and R250 are generally used for domestic central heating installations. Grade R250 in sizes 15 mm, 22 mm and 28 mm are used for minibore installations, while R220 is used for microbore systems, usually 10 mm.
2. **Low carbon steel pipes and fittings:** very rarely used for domestic installations but used extensively on commercial and industrial systems.
3. **Polybutylene pipes and fittings:** fast becoming the material of choice for new-build installations because of its ease of installation. It should be remembered, however, that the connections to any heat-producing appliances must be made using copper for the first metre away from the appliance.

> **KEY POINT**
>
> Testing of pipework, both metallic and plastic, is covered in Chapter 5, Cold water systems.

Installing and testing components

Types of information to be referred to for installation work:

- manufacturer instructions should always be read before any installation takes place
- specification drawings should be followed wherever possible; alterations to the original drawings should be done with care
- verbal instructions from the customer must be considered.

The installation of tubes and fittings has been covered extensively in earlier chapters of this book, but central heating systems demand careful consideration because of the temperature that the systems run at. With water at 80°C for the flow and 60°C for the return, the pipework, regardless of the material used, will expand and contract as the pipe heats up and cools down. Obviously, not all of the materials expand at the same rate, but provision should be made at the installation stage to allow for expansion and contraction if problems with noise are to be avoided. Here are some points to consider:

- Polybutylene pipe expands more than copper tube, but copper is much more rigid than polybutylene. When installing pipes in wooden floors, enough room should be allowed in any notches made. If the pipes are too tight in the joist, they will 'tick' as they expand and contract. This is very pronounced with central heating systems installed using copper tubes because the water reaches a higher temperature.
- Clipping and securing pipework becomes very important. The clipping distances for the various pipes and tubes we use are known from previous learning, but become critical where polybutylene pipe is concerned, especially when used with central heating installations, because as the pipe becomes hot, it starts to soften and this leads to the pipe 'drooping' between joists and between clips. This not only looks unsightly but can put excess strain on the joints.
- On new-build installations, it is common practice to install microbore pipework behind the dry lining plasterboard. In this instance, if the pipework is made from copper, it should be clipped well and wrapped to avoid noise and corrosion. Polybutylene pipe should be wrapped too, but because the expansion of pipe on a hard surface could cause undue abrasion on the soft plastic. A metallic tape should be placed at the back of

the polybutylene to allow the pipe to be found by metal-detecting tools when it is covered.
- Pipes placed in chases should be wrapped against corrosion and insulated where required.

General installation requirements are as follows:
- Feed and expansion cisterns must be fitted in accordance with the Water Supply (Water Fittings) Regulations.
- Filling loops, expansion vessels and associated equipment should be installed where they do not create an eyesore but are accessible. The installation of expansion vessels should always be in accordance with the manufacturer's installation instructions. With system boilers and combination boilers this does not present a problem as they are an integral part of the appliance.
- Radiator position should be considered with care. It is generally accepted that radiators be placed under windows, but this is not always the best position if an even circulation of warm air is to be achieved. On new builds and refurbishments, the radiator positions are usually marked on the detailed building plans.
- Pipework must be insulated in places where there is a risk of freezing, such as under a suspended timber floor and unheated garages. The Building Regulations also advise that pipework in airing cupboards must be insulated to prevent unwanted heat loss.
- All pipework and metal parts within the system must be electrically bonded to earth.

> **KEY POINT**
>
> Cistern requirements are mentioned in Chapter 5, Cold water systems.

Pressure testing and filling

Testing

The testing procedure is very similar to both hot and cold water installations, but the test pressure will depend on the type of system installed. As with other systems, the test pressure is 1.5 times normal operating pressure and that pressure will vary depending on the type of system installed. For instance:

- for sealed (pressurised) systems working at 1 bar pressure, the test pressure is 1.5 bar
- for open vented systems, where the head of pressure is, say, 8 m then the test pressure is 12 m, or 1.2 bar
- test timing should be in accordance with the Water Supply (Water Fittings) Regulations, **BS 8558** and **BS EN 806** and will depend on the material used in the installation. Although systems are commonly pressure tested to the standards outlined in **BS EN 8558** and **BS EN 806**, central heating systems have their own pressure testing criteria outlined in **BS EN 12828** (Design for water based heating systems) and **BS EN 14336** (Installation and commissioning of water based heating systems).

Testing should be conducted using a hydraulic test pump.

> **KEY POINT**
>
> - Before initial testing takes place, the system should be visually checked to make sure that it is correct, that all visible joints are tight and that all clipping is in accordance with the British Standard distances.
> - Pipework testing is covered extensively in Chapter 2, Common processes and techniques.

Filling

The procedure for filling central heating systems will again depend on the type of system that is installed. We will look at two separate procedures here.

Open vented systems

Filling open vented systems is a fairly simple procedure. Having conducted a pressure test at the installation stage, there should be no surprises when it comes to system filling:
- Ensure that all radiator valves and radiator air release points are closed.
- Check the F and E cistern to ensure that all joints are tight.
- Temporarily replace the pump with a short piece of tubing. This will ensure that no debris enters the pump.
- Ensure that all motorised valves are manually set to the open position for initial system filling.
- Turn on the service valve to the F and E cistern and allow the system to fill.

- Starting with the furthest-away radiator on the downstairs circuit, open the radiator valves and fill and bleed the air from each radiator. Work backwards towards the boiler, downstairs circuit first, then the upstairs circuit. This will ensure that air is not trapped in pockets around the system.
- Once the system is full, allow it to stand for a short while. Visually check for leaks at each radiator and all exposed pipework and controls/valves, etc.
- Check the water level in the F and E cistern.
- Drain down the system. This will flush the system through, removing any flux residues, steel wool, etc.
- Refit the pump and turn on the pump valves.
- Refill the system as before.

Sealed systems

The main difference when compared to the open vented system is that there is no F and E cistern, so the system will have to be filled in stages or short bursts via the filling loop. In other words, turn on the filling loop, fill the system up to operating pressure, turn off the filling loop, bleed the air from the radiators until the pressure has depleted and then restart the process until the system is full. All other points remain the same as for open vented systems, above.

Commissioning

Full commissioning details will be found in the manufacturer's instructions along with Domestic Building Service Compliance Guide; Domestic Heating Design Guide; **BS EN 12828** and **BS EN 14336**. All commissioning paperwork needs to be completed with a copy being handed to the customer and a copy kept for your records. This includes the Benchmark certificates.

Replace defective components

Maintenance of central heating systems takes many forms, from replacing valves to replacing boilers. It can also include adding to or altering an existing system.

In this part of the chapter, we will look at some of the more common maintenance activities and the processes involved.

Some of the general maintenance activities include:

- pump replacement
- radiator replacement
- radiator valve replacement
- tasks that may require system drain down
- power flushing a system
- routine maintenance tasks
- dealing with simple system faults.

Whatever maintenance activity is being undertaken, safe isolation of the system is paramount.

Replacing a central heating circulating pump, step by step

You should attempt this task only under supervision. The system should not require draining when replacing a pump.

Before attempting to remove the pump, the electricity should be isolated at the switched fused spur and the fuse retained to prevent accidental switching on of the circuit.

1. Check that the electrical circuit is dead using a GS38 proving unit or some other effective electrical testing device.
2. Make a simple drawing of the live/neutral/earth connections on the pump and disconnect the cable.
3. Turn off the isolating valves either side of the pump.
4. Carefully loosen the unions on the pump by turning them anti-clockwise using water pump pliers. It may be a good idea to have some old towels handy to catch any water.
5. Once both unions have been disconnected, remove the pump. The pump unions should have the old washers removed and the union faces cleaned. The new pump will include flat rubber washers in the box.
6. Position the new pump, with the sealing washers in place between the valves, and hand tighten the unions. Take care to ensure that the pump is facing the right direction for the system.
7. Fully tighten the unions with the water pump pliers. If the pump is installed horizontally, make sure that the bleed point is slightly above horizontal as this will help to remove any air in the pump.
8. Turn on the pump valves and check for leaks.
9. Carefully reconnect the electrics to the pump: live to the L point, neutral to the N point and earth to the E point. Make sure that all electrical connections are tight.
10. Remove the centre bleed point on the pump and release any air.

11 Reinstate the fuse in the consumer unit. Switch on and test for correct operation.
12 With open vented systems, check the F and E cistern in the roof space to ensure that the pump is not pumping water over the cistern through the vent pipe.

Replacing a radiator, step by step

If the new radiator is the same size as the one being replaced, the pipework should fit without too many problems. If the new radiator is either larger or smaller, then the pipework will either have to be altered or a radiator valve extension will need to be fitted. It is desirable, when replacing a radiator, to replace the valves as well, as they will probably be as old as the radiator you are replacing. If this is the case, then all or part of the system will need to be drained. We will assume that the radiator is downstairs, requiring complete system drain down.

Before attempting to remove the radiator, the electricity should be isolated at the switched fuse spur to the system and the fuse retained to prevent accidental switching on of the circuit. The system should also be cold. It may be a good idea to ask the customer to turn the central heating off before you get to the job. Before you begin, make sure you have protected carpets and furnishings with lots of dust sheets.

1 Isolate the F and E cistern at the service valve. If it is a sealed system, this will not be necessary.
2 Locate a suitable drain valve, attach a hose and drain the system. Take care that the system contents are disposed safely to a drain as they will probably be very dirty, especially if the system is an old one. The black water will stain all it comes into contact with.
3 As the system drains, open the air release valves on all radiators, starting upstairs, then working to the downstairs.
4 When the system is drained, carefully loosen the two radiator valve compression nuts and remove the radiator. It is a good idea to leave the valves on the radiator and to turn them off before removal. This will help in preventing any residual dirty water from leaking from the radiator. If possible, turn the radiator upside down (turn the air release valve off first!) as this will further prevent accidental spillage.
5 The new radiator should be dressed and hung as previously described.
6 Reconnect the pipework, ensuring that the old compression nuts and olives are removed first. If the old olives have crushed the pipe too much, then the pipe may have to be replaced.
7 Ensure all radiator unions and compression nuts are fully tight.
8 Turn off the drain valve. It may be a good idea to replace the drain-off valve washer at this stage. Drain valve washers quite often go stiff and brittle with the heat from the water.
9 Turn off all air release valves.
10 Turn on the service valve to the F and E cistern or (if applicable) reconnect the filling loop and refill the system.
11 Bleed the air from all the radiators, starting downstairs then upstairs. Leave the new radiator isolated at this stage. This will be the last radiator filled.
12 Open the valves to the new radiator and bleed the air from it. Check for leaks.
13 Replace the fuse in the fuse spur and run the system to full temperature to ensure that the new radiator is working perfectly.
14 If corrosion inhibitor had been added to the system in the past, this will need to be replaced. It must be replaced like for like. If this is not possible, the system should be flushed several times to ensure removal of all previous inhibitors.

Replacing faulty radiator valves, step by step

Faulty radiator valves are easy to replace provided that the new valve is of the same body size as the valve being replaced. Over the years, there have been many different styles and sizes of valve body and, sometimes, older valves are bigger than their modern equivalents. In this case, pipe alteration may be needed, which should be conducted with care to ensure that the customer's decorations and floor coverings are not damaged. To replace a like-for-like valve (assuming the radiator valves old and new are the same size), go through the following steps.

1 Isolate the F and E cistern at the service valve. If it is a sealed system, this will not be necessary.

2 Locate a suitable drain valve, attach a hose and drain the system. Take care that the system contents are disposed safely to a drain as they will probably be very dirty, especially if the system is an old one. The black water will stain all it comes into contact with.
3 As the system drains, open the air release valves on all radiators, starting upstairs, then working to the downstairs.
4 When the system is drained, carefully loosen the radiator valve compression nut and union and remove the valve.
5 Remove the old radiator union from the radiator and compression nut and olive from the pipework.
6 Wrap PTFE tape clockwise around the new valve union and screw it into the radiator using a radiator valve Allen key.
7 Slip the new compression nut and olive onto the pipework. If the old olives have crushed the pipe too much, then the pipe may have to be replaced.
8 Put the new valve onto the pipe, then hand tighten both the compression nut and the radiator union.
9 Using an adjustable spanner, and taking care not to damage the chrome plating, tighten both the compression nut and the radiator union. You may need to hold against excessive valve movement by using water pump pliers. Take care with the chrome plating.
10 Ensure all radiator unions and compression nuts are fully tight.
11 Turn off the drain valve. It may be a good idea to replace the drain-off valve washer at this stage. Drain valve washers quite often go stiff and brittle with the heat from the water.
12 Turn off all air release valves.
13 Turn on the service valve to the F and E cistern or (if applicable) reconnect the filling loop and refill the system.
14 Bleed the air from all the radiators starting downstairs then upstairs. Leave the radiator with the new valve isolated at this stage. This will be the last radiator filled.
15 Open the new valve to the radiator and bleed the air from it. Check for leaks.
16 Replace the fuse in the fuse spur and run the system to full temperature to ensure that the new radiator is working perfectly.
17 If corrosion inhibitor had been added to the system in the past, this will need to be replaced. It must be replaced like for like. If this is not possible, the system should be flushed several times to ensure removal of all previous inhibitors.

Tasks that may require system drain down

There are many situations where draining of the system is needed, such as:

- replacing the hot water storage cylinder
- boiler replacement
- decommissioning of components such as radiators; here, the radiator, brackets and pipework should be removed and the pipes capped off at the branch to the flow and return pipes
- replacement of motorised valves
- cutting into an existing system to alter or extend it; drain down should be conducted when all other installation work has been carried out
- power flushing.

Power flushing a system

During the last task we looked at, it may become apparent that the system contains a lot of black water and even sludge. If this is the case, the system may be in need of a power flush. When replacing boilers, a power flush is required to remove any sludge within the system as part of the warranty. Manufacturers' warranties are void if this is not carried out.

Power flushing involves using a special high-powered pump to circulate cleaning chemicals and de-sludging agents through the system. These powerful chemicals strip the old corrosion residue from the system, ensuring that the system does not contain sediment that may be harmful to new boilers, controls and valves.

Once the power flushing is complete, the system may have an inhibitor added to the system water to keep the system free from corrosion.

Routine maintenance tasks

Whenever carrying out any maintenance task, always refer to the manufacturer's instructions as they will outline what is required. It is also good practice to check with the customer to see if they have noticed any problems or changes in the performance of the system before starting work. At this point, look out for

- any potential causes of delay
- time to complete work
- delivery of parts
- need for additional labour or other trades.

After work has been completed, remember to fill in the job sheet so that a permanent record is kept of the work. It is worth looking at the parts that require maintenance in case they are modularised (the manufacturer in their parts list has a 'unit' that could be replaced). If pipework needs to be replaced, it is important to replace as little as possible, undoing as few watertight seals as possible to correct the situation. Routine maintenance includes:

- checking the pressure charge in expansion vessels on sealed systems, system boilers and combination boilers
- checking the operation of pressure relief valves on sealed systems, system boilers and combination boilers
- checking and topping up (if required) the pressure on system boilers and combination boilers
- visually checking for any signs of leakage on pipework, controls and appliances
- boiler servicing
- checking the correct operation of thermostats, motorised valves and thermostatic radiator valves
- checking the water level in F and E cisterns, and adjusting as necessary
- ensuring that the system is reaching full temperature.
- ensuring that there is no vibration from any components or appliances that could interfere with or affect the operation of the system.

Dealing with simple system faults

It is impossible to cover all scenarios when dealing with system faults. Often, the reason for a fault developing is clear and stems from poor design when the system was installed; others take rather more investigative work. Sometimes the system itself will be leading you to the problem by the way it is behaving or the noises it makes, and so diagnosis becomes an easy task. Long-term industrial experience will give you greater knowledge about system faults and how to rectify them. It is also good practice to keep up to date with courses that manufacturers may be offering as these will also add to your knowledge.

Here, we will look at some of the more common, simple system faults only, and the signs to watch out for. We will not be dealing with appliance faults. This is done at Level 3.

▼ Table 7.23 Common system faults and how to rectify them

Symptom	Fault	Rectification
Discoloured water appearing at hot water taps. System has a double feed indirect cylinder fitted.	The cylinder heat exchanger coil has pin-hole corrosion, allowing water to either pass from the F and E cistern to the hot water, or vice versa.	Drain down both hot water and heating systems and replace the hot water storage cylinder.
The overflow of the F and E cistern runs constantly, even when the heating system is off but the float-operated valve is working correctly.		
A radiator is cold at the top but works once the air has been bled. It then works for about four weeks before filling with air again.	It is not air that is filling the radiator. It is hydrogen and a clear sign of electrolytic corrosion at the radiator.	A very common occurrence with systems that contain single feed (Primatic) cylinders. Because inhibitor cannot be used here, the only action is to replace the cylinder with a double feed type, power flush the system and add corrosion inhibitor.
The hot water via the primary circulation pipes on a semi-gravity system is working correctly. However, the radiators on the system are lukewarm upstairs and cold downstairs.	Pump failure.	Replace the pump.

→

The radiators on a semi-gravity system work correctly but there is no hot water. The gravity primary circulation pipes are cold.	This is unlikely to be an air lock. The biggest cause of this problem is evaporation of water in the F and E cistern linked to the float-operated valve in the cistern sticking in the 'up' position.	Re-washer or replace the float-operated valve, and refill the F and E cistern.
A radiator is cold in the middle.	Black oxide sludge is blocking some of the radiator's water sections.	A temporary solution would be to take the radiator off and flush it out with cold water, but unless the problem is identified, it will reoccur. The system requires a power flush and corrosion inhibitor adding to the system water.
A number of radiators on a downstairs heating circuit only reach lukewarm temperature. All other radiators are working correctly.	Black oxide sludge is blocking the circuit pipework, leading to poor water circulation.	See above.
A boiler is noisy when the water begins to reach temperature.	This is known as 'kettling' because the noise resembles that a kettle makes just before it boils. Its correct name is 'localised boiling'; it occurs because the waterways of the boiler are partially blocked with either black oxide sludge or calcium deposits (limescale). As the water heats up, it momentarily boils before being moved away by the pump.	The system requires a power flush with sludge remover and descaler before corrosion inhibitor is added to the system. It is also a good idea to do a litmus paper test to see if the water is acid or alkali. Alkali water tells us that the likely cause is calcium deposits and a scale preventer can then be added to the system to stop the problem recurring.

3 UNDERSTAND THE DECOMMISSIONING REQUIREMENTS OF CENTRAL HEATING SYSTEMS AND THEIR COMPONENTS

The decommissioning of central heating systems follows much the same process as with other systems we have looked at. There are a number of scenarios where systems would need to be decommissioned:

- where the system is being completely stripped out prior to a new system installation or where the building is being demolished
- where the boiler is being replaced and the F and E cistern is being taken out
- where the system is being added to or altered
- where system components such as radiators are being permanently taken out
- general maintenance activities, such as:
  - a pump replacement
  - a radiator replacement
  - replacement of valves and other controls.

Decommissioning systems

There are two types of decommissioning of heating system:

1 **permanent decommissioning** is when a system is being taken out of service, completely dismantled and stripped out of the property
2 **temporary decommissioning** takes place when the system is being worked on for a short period of time, such as replacing the boiler, the pump or a radiator.

Preparing for decommissioning

When preparing to decommission central heating systems, always remember to:

- keep the customer and/or other trades informed of the work being carried out, i.e. when the system is being isolated and the expected length of time it will be out of service

- ensure that any services, such as electricity, gas, etc., are safely isolated and pipework capped
- use warning notices, such as 'do not use' or 'system drained' on any taps, valves, appliances, electrical components, etc.
- if possible, make alternative heating methods, such as warm air heaters and fans, available to the customer
- on larger systems or insurance repairs you may well have to undertake some calculations to put a value on the downtime while work is carried out
- if required, complete the decommissioning records so that a permanent record of the work is kept.

Decommissioning central heating systems

When decommissioning central heating systems, there are a number of procedures to be observed:

- **Isolation of services:** ensure that all relevant services, such as gas, water and electricity, are isolated before commencing the decommissioning process. If possible, localised isolation, such as removing fuses or isolating water at isolation valves, etc., is preferable so that the customer is not left without services for too long. You must seek the customer's permission before isolating any services.
- **Warning notices and signs:** warning signs saying that the system is isolated and must not be reinstated should be placed at the point of isolation, so that other users/customers know that the system is being worked on. If the isolation point is a long distance from your point of work, leave a contact or mobile number so that the customer can contact you with any queries.
- **Temporary continuity bonding:** temporary continuity bonding must be carried out when removing electrical components, such as circulating pumps and motorised valves. Temporary continuity bonding is absolutely essential when making new connections involves cutting into existing pipework. It is here that the risk from electrocution is at its greatest, as earth leakage faults on electrical systems are not always noticeable.
- **Drainage and disposal of systems' contents and components:** the water from central heating systems often contains chemicals that are very mildly toxic, such as inhibitors, scale preventers and sludge removers. These must be disposed of carefully down a foul water sewer and not a top water or rainwater drain. Top water drains often dispose of water straight into a watercourse, river or stream, and the chemicals may be harmful to aquatic life.
- **Capping of pipework:** no matter for how long the system is to be decommissioned, the capping of pipework is essential in case of the system being accidentally turned on. In the case of gas systems, the cap should be installed and the system fully tested according to the Gas Safety (Installation and Use) Regulations 1998 to ensure that, should the system be accidentally turned on, it is completely gas tight and safe.
- **A system may need to be re-instated after it has been decommissioned:** this is likely to be after a temporary decommission where a repair or replacement has taken place. This will require some or all of the commissioning procedure to be followed.
- **Make sure that all items and fluids are disposed of correctly and any items that can be recycled are:** metals like copper and brass hold a good value when recycled.

SUMMARY

This has been the most challenging chapter in the book for us so far. The myriad of systems, layouts, appliances, components and fuels are confusing, but each one has its tell-tale signs that make it unique. The art to good system recognition is looking – just as the key to good system fault diagnosis is listening. This chapter gives us the foundation to do both.

A good central heating system is one that is efficient in use, warms the home to the right temperature, is quiet in operation and is installed to the highest possible standards. This can be achieved only with the knowledge that allows us to recognise the possibilities of efficiency, design with the customer in mind and installation to the best of our ability.

Test your knowledge

1 In a modern sealed CH system incorporating a condensing type boiler, what is the expected temperature difference between the flow and return if designed correctly?

a 12°C

b 20°C

c 22°C

d 30°C

2 Which pipe within an open vented central heating system allows the system to remain at atmospheric pressure?

a Cold feed and expansion pipe

b Open vent pipe

c Warning pipe

d Discharge pipe

3 Which heating system incorporates 2 × two-port valves?

a The C-plan

b The Y-plan

c The W-plan

d The S-plan

4 The image below shows a three-port mid-position valve. How will the system function in its current position?

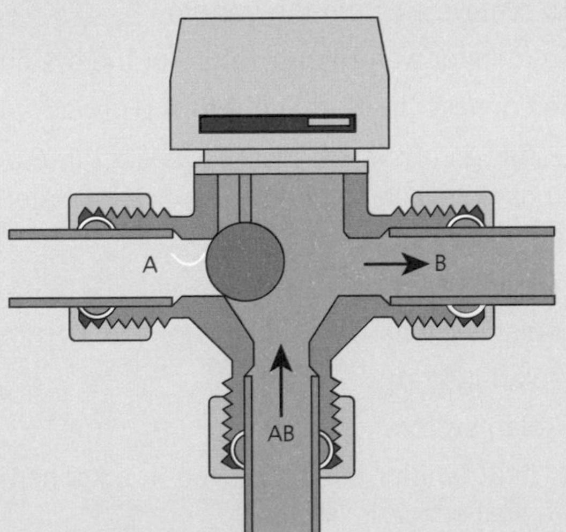

a Hot water only

b Heating only

c Heating and hot water

5 Which one of the three images below shows the best relative positions of the pump, cold feed and open vent to allow for a positive system pressure while minimising pumping over?

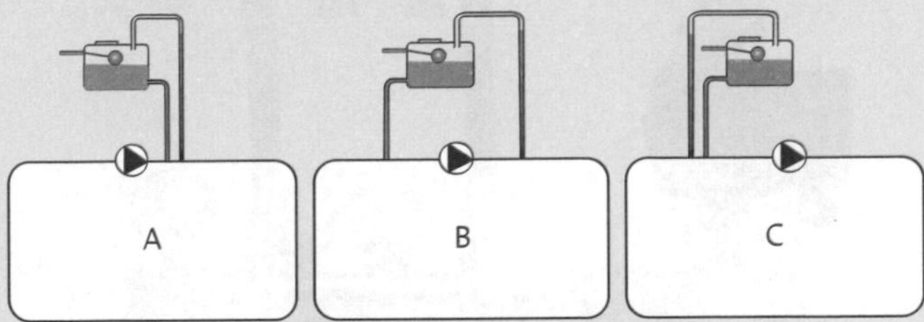

a A

b B

c C

6 Which type of boiler utilises increased efficiency by recovering latent heat from the flue gases?

a Combination boiler

b Traditional boiler

c Condensing boiler

d System boiler

7 What is the recommended installation height from the floor for a radiator?

a 100 mm

b 150 mm

c 200 mm

d 250 mm

8 Within a sealed heating system, which component accommodates the expansion of water during heating?

a Feed and expansion cistern

b Cold feed and expansion pipe

c Expansion bellows

d Expansion vessel

9 Which of the following should be provided to a central heating system within a dwelling that has a useable floor space greater than 150 m^2?

a Independent time-controlled zones

b A gravity hot water circuit

c Modular boiler arrangement

d A low loss header

10 What component is shown in the image below?

a Underfloor heating manifold
b Expansion relief valve
c Low loss header
d Boiler cascade system

11 Which Building Regulation does the Domestic Heating Compliance Guide 2021 follow?

a Approved Document H
b Approved Document L
c Approved Document P
d Approved Document J

12 What is the most common type of copper used in domestic central heating systems in the UK?

a R290 Hard copper
b R220 Soft copper
c R250 Half hard copper
d R320 Straight copper

13 Where would the automatic by-pass be installed on a central heating system?

a On the primary flow pipe immediately above the boiler
b On the primary flow pipe after the circulator
c Between the flow and return pipe near to the hot water cylinder
d Between the flow and return pipe of the index radiator

14 Where does the water come from on the 'AB' connection of a three-port mid-position valve?

a From the radiators
b From the cylinder
c From the return pipe
d From the boiler

15 What is the main purpose of a domestic central heating system?

a Thermal comfort
b Heat one room
c Thermal conductivity
d Heat the hot water

16 What is the primary purpose of the item circled in green, in this diagram?

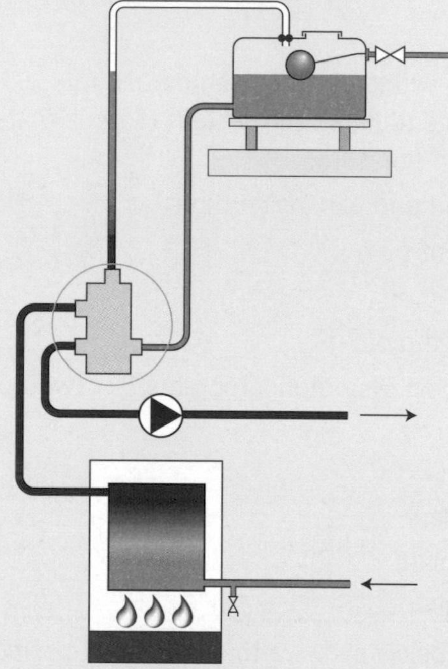

a To regulate the flow of system water
b To remove air from the system
c To monitor water temperature in the system
d To connect the flow and return pipework

17 What design of system incorporates a three-port mid-position valve to control the flow of water around the system?

a W-plan system
b C-plan system
c Y-plan system
d S-plan system

18 What heat emitter could be used in a kitchen where wall space is limited?

a Skirting board convector
b Single panel radiator
c Double panel convector
d Kick-space heater

19 What is the main difference between a condensing boiler and a non-condensing boiler?
   a The flow of water through the boiler
   b The type of gas valve
   c The heat exchanger
   d The fan

20 What is the principle method that a standard radiator uses to heat a room?
   a Convection
   b Conduction
   c Radiation
   d Infusion

21 You are called to a customer's property where a radiator is cold at the top and hot at the bottom. What course of action would you take?
   a Drain and re-fill the system
   b Replace the radiator
   c Remove the sludge from the radiator
   d Bleed the radiator

22 What is required to be installed either side of the circulator pump?
   a Air release valves
   b Isolators
   c Lock shield valves
   d Zone valves

23 When replacing a syncron motor from a two-port valve, what should be your first course of action?
   a Release the motor from the housing and pull the tab connectors off
   b Disconnect the wires making a note of the connections
   c Isolate the flow of water in the system
   d Isolate the electrics and test the supply is dead

24 Which of the following is an advantage of underfloor heating?
   a No visible heat emitters
   b Small heat emitters used
   c High surface temperature
   d Instant heat

25 Which of the following systems is approved by Part L?
   a C-plan system
   b One-pipe system
   c Fully pumped S-plan system
   d Semi-gravity two-pipe system

26 Describe a boiler interlock and its purpose.

27 What are the names of the two types of fully pumped vented system that can be installed?

28 What is the purpose of inhibitor within a CH system?

29 Explain why a filling loop should be disconnected after initial filling of the system.

30 Where should drain valves be installed within a central heating system?

31 Explain how a thermostatic radiator valve (TRV) works.

32 A customer asks why an S-plan plus system is better than an S-plan system. Describe why.

33 Outline the advantages of an underfloor heating system against a standard central heating system.

34 Describe where a district heating system would work best.

Answers can be found online at www.hoddereducation.co.uk/construction.

Practical activity

As a trainee or apprentice plumber it is important that you are able to position and fix (hang) a radiator. This may be a typical task an apprentice is asked to carry out on a regular basis. Develop your confidence by marking out for the installation of a given radiator.

You will need to ask your supervisor or tutor for a radiator to work from, and a space to mark and measure. Using the theory you have discussed at your training centre, measure the radiator/brackets and transfer the relevant marks onto a surface for fixing your brackets.

Ask your supervisor or tutor to check this work. If possible and convenient, perhaps you could select suitable fixings and continue to hang the radiator. Again, ask your supervisor or tutor to check once complete.

CHAPTER 8

RAINWATER SYSTEMS

INTRODUCTION

The UK has more than its fair share of rain. Rainfall varies greatly in the different regions. On average, the south-east of the UK has around 500 mm of rainfall a year compared to around 1.8 m for the north-west. Rain penetrating a building can do a vast amount of damage. Without guttering systems, the rainfall will run off a roof and erode the ground around a dwelling, it will penetrate the structure and may even affect a building's foundations.

In this chapter, we investigate the need for guttering systems, their function and design. We will also look at the various types of guttering system, the materials they are made from and their methods of jointing and installation.

Guttering installation invariably involves working at height and this brings with it immediate danger. During the course of the chapter, we will also review previous learning on working safely at height.

By the end of this chapter, you will have knowledge and understanding of the following:

- layouts of gravity rainwater systems
- installation of gravity rainwater systems
- the maintenance and service requirements of gravity rainwater systems
- the decommissioning of rainwater and gutter systems and components
- how to perform a soundness test, and commission rainwater, gutter systems and components.

1 UNDERSTAND LAYOUTS OF GRAVITY RAINWATER SYSTEMS

All dwellings have some form of rainfall collection system to take the rainfall that falls onto the building structure away from the building. This is achieved by the use of an eaves-level, usually fascia board-mounted, guttering system, which collects the water that runs off the roof and discharges it away harmlessly. The main purposes of a guttering system are:

- to protect the building's foundations
- to reduce ground erosion
- to prevent water penetration and damp in the building structure
- to provide a means for collecting rainwater for later use, i.e. rainwater harvesting.

Systems and materials used in gravity rainwater systems

The principle of any guttering and rainwater system is to remove the rainfall in such a way that it does not:

- constitute a nuisance for the occupiers of the dwelling, or
- damage the building structure or the building foundations or those of any adjacent building.

Domestic gutter and rainwater systems work by removing the rainwater that runs off roofs, in channels known as gutters, and discharges the water, via rainwater pipework, safely away from the building structure by gravity. The water may be discharged into:

- a surface (rain) water drain, used where the dwelling has a separate system of drainage for both foul water and surface water
- a combined sewer – a combined system of drainage where both foul and surface water discharge into a common drainage system
- a watercourse (stream, river, etc.), where the water discharges direct into a flowing, nearby water source
- a soakaway drain – a specifically designed and located pit, sited away from the dwelling, which allows the water to soak away naturally to the water table
- a rainwater harvesting system for further use within the dwelling; these are specifically designed to serve WCs.

The types of materials used for rainwater systems include:

- PVCu
- extruded aluminium
- cast iron
- copper.

Gutter systems and components

Over the years, gutters have been manufactured from many different materials and in many different profile shapes. In the past, the gutter profile was designed in line with the housing styles of the time. For example, the ornamental gutter profile (Ogee or OG) was designed during the Victorian era in the mid to late 1800s. As we shall see, a modern Ogee profile is still available today to give a dwelling a 'period' feel to its exterior.

In this part of the chapter, we will look at modern materials and profiles, as well as the different types of fittings for guttering and rainwater pipework and the typical methods of jointing:

- PVCu guttering systems
- cast iron guttering systems
- extruded aluminium guttering systems
- jointing guttering of different materials.

PVCu guttering systems

Unplasticised polyvinyl chloride (PVCu) guttering systems are manufactured to the following British Standards:

- **BS EN 607:2004 Eaves gutters and fittings made of PVCu**
- **BS EN 122001:2016 Plastic rainwater piping systems for above ground external use**.

Most of the guttering systems used on domestic dwellings today are made from PVCu, the characteristics of which are studied in Chapter 2, Common processes and techniques.

▼ Table 8.1 The advantages and disadvantages of PVCu as a material for guttering systems

Advantages	Disadvantages
It is easy to install It is lightweight and easy to handle Minimal maintenance is required It requires no painting It does not support combustion It is economical It is corrosion free It has a smooth internal bore It has a life expectancy of 50 years	It is adversely affected by wood preservatives It has a greater coefficient of thermal expansion (0.06 mm/m/°C) compared to other materials (see Chapter 3, Scientific principles) It goes brittle in cold temperatures and softens at a relatively low temperature

PVCu gutter profiles

There are four main gutter profiles manufactured from PVCu:

1 **Half round:** the standard gutter profile, used on many domestic properties throughout the UK.

2 **High capacity (often called deep half round or storm flow):** a deeper version of the half round profile. It is slightly elliptical in shape and generally used on larger or steeper-angled roofs where the velocity and volume of the water entering the gutter is high.

3 **Square section:** very popular in the 1980s and 1990s. Used with square section rainwater pipes.

4 **Ogee (or OG, ornamental gutter):** a modern redesign of a Victorian gutter profile. It is used where a 'period' look is important on new builds and on many Victorian refurbishments.

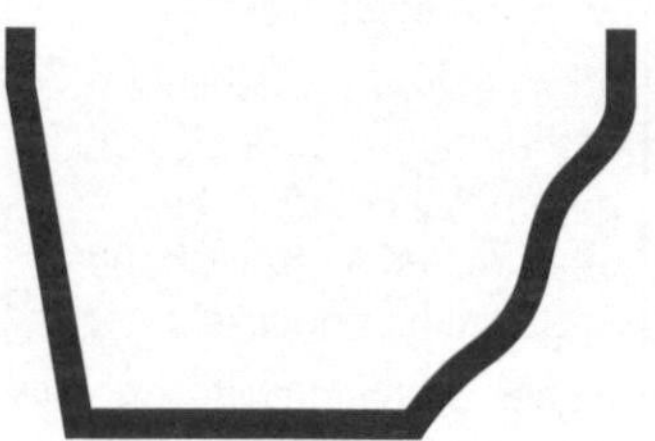

PVCu gutter fittings and jointing method

▲ Figure 8.1 A typical PVCu gutter system

Figure 8.1 shows the fittings in a typical 112 mm half round guttering system. The common fittings are shown in Table 8.2 in all profile styles.

▼ Table 8.2 Styles of common PVCu gutter fittings

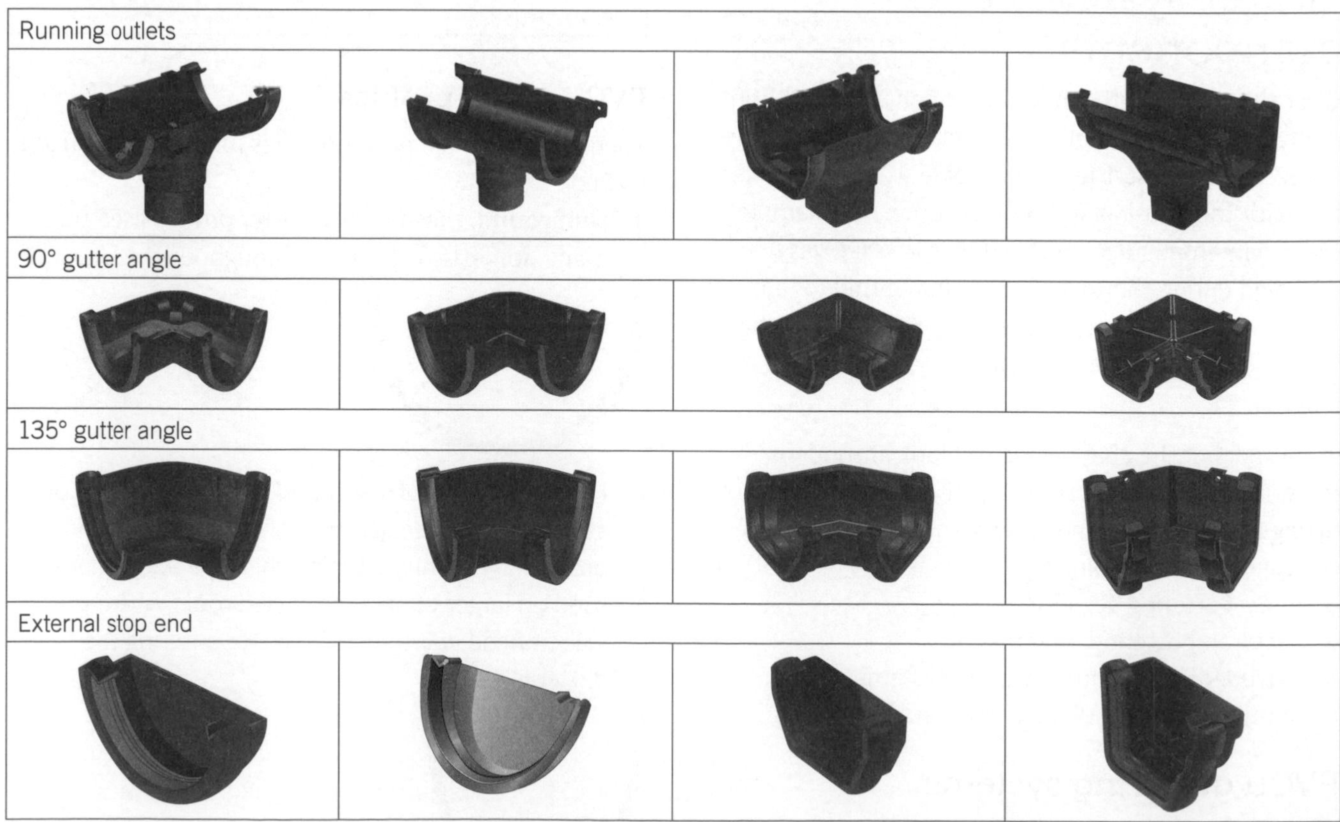

Running outlets			
90° gutter angle			
135° gutter angle			
External stop end			

→

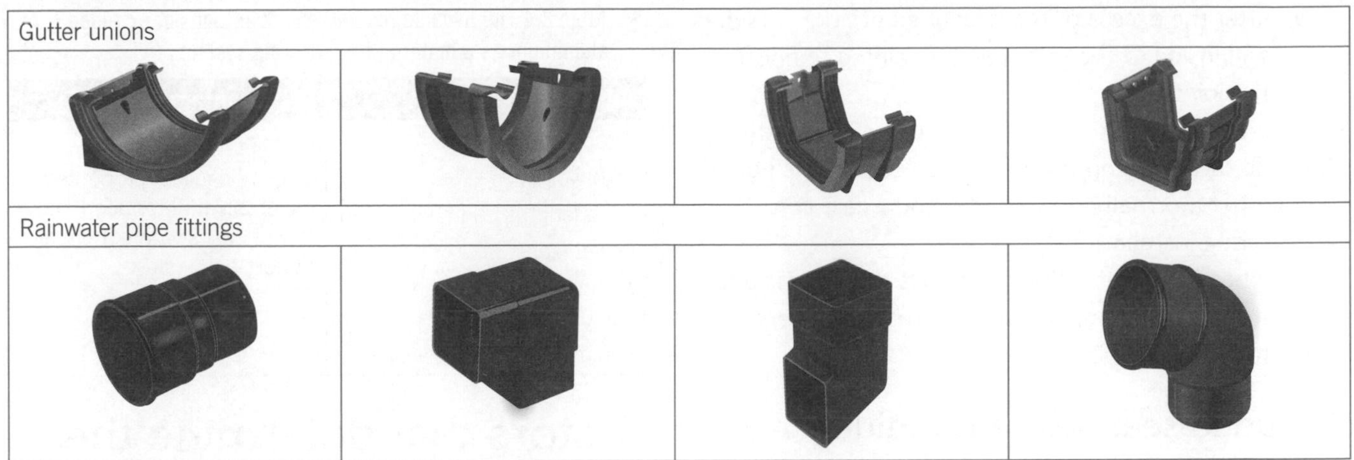

Cast iron guttering systems to BS 460:2002+A2:2007 Cast iron rainwater goods

Before PVCu guttering, cast iron was probably the most common material for gutters and rainwater pipework. It can still be seen on many older houses. It is strong and durable but can be difficult to maintain as it requires regular painting to prevent corrosion.

Cast iron may still be specified by the local authority, English Heritage or the National Trust if a building is listed or in a conservation area. The most common profiles for cast iron are:

- half round section – visually very similar in shape to PVCu half round profile
- Ogee section – there are several variations of the Ogee profile manufactured in cast iron, some that are specific to a particular area, such as Notts Ogee, which can be found only in the Nottinghamshire area
- deep half round – found on larger buildings.

Cast iron-type fittings and guttering are also available in cast aluminium.

▼ Table 8.3 The advantages and disadvantages of cast iron as a material for guttering systems

Advantages	Disadvantages
Strong and durable	Installation is expensive and time consuming Cast iron guttering is expensive Requires regular painting and maintenance to prevent corrosion Heavy Jointing is time consuming and messy

Jointing cast iron guttering systems

▲ Figure 8.2 Cast iron gutter

As you can see from Figure 8.2, cast iron guttering has a socket on one end. The other end is a plain gutter. A successful joint involves fitting the end of one length of gutter into the socket of another with a jointing material in between. The two lengths of gutter are then bolted together using special zinc-plated gutter bolts. The jointing material can be either:

- paint and putty joint – the traditional method of jointing cast iron guttering systems; the method of jointing is as follows
  - the inside of the socket and outside of the spigot are first painted with black bitumen paint
  - linseed oil putty is then placed into the socket before mating the socket and spigot together
  - a zinc gutter bolt is inserted through the holes on the socket and spigot, and the two sections bolted together; care should be taken not to over-tighten the bolt or the gutter will crack

  - after the excess putty is cleaned off, the outside and inside of the joint can be painted to finish the joint
- a special silicone sealant – the silicone is placed inside the joint and then the two sections are bolted together (normally only used on new cast iron guttering installations)
- a rubber grommet – this method is not generic and usually available only on specific manufacturers' gutter and fittings.

Extruded seamless aluminium guttering systems

This type of guttering system is usually installed by specialist companies. Extruded seamless aluminium guttering systems are a modern innovation that are light in weight and corrosion resistant. It is manufactured 'on-site' from a roll of coloured aluminium sheet by a special machine that is carried in the back of a van. The aluminium sheet is passed through the machine and this presses the sheet into the shape required. As the gutter exits the former, strengtheners are fitted at regular intervals to give the gutter added rigidity.

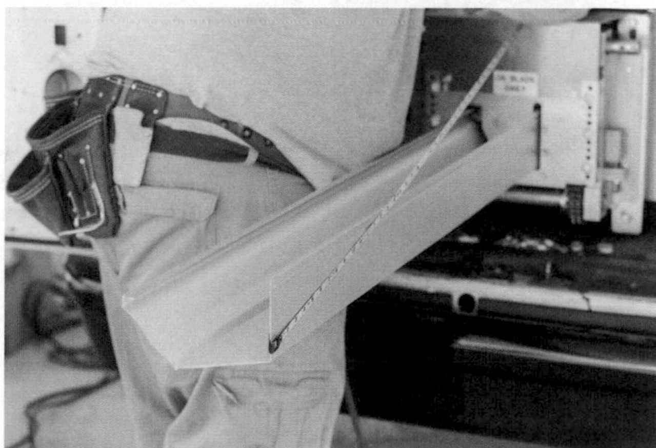

▲ Figure 8.3 How extruded seamless aluminium gutters are made

It can be manufactured in one continuous length of up to 30 m without the need for an expansion joint, reducing the amount of joints and, therefore, potential leaks. The gutter is installed with internal brackets spaced at 400 mm and this means it is able to withstand shock-load from ladders, etc.

Most companies offer a variety of profiles, including half round and Ogee, in a variety of colours.

▼ Table 8.4 The advantages and disadvantages of extruded aluminium as a material for guttering systems

Advantages	Disadvantages
Strong and durable Lightweight Long lengths can be installed Fewer leaks A variety of profiles and colours Minimal thermal expansion	An expensive system Does not suit all properties, especially mid-terraced and town houses where there are gutters either side

Factors that determine the type and size of guttering system

A guttering system should have sufficient capacity to carry the expected flow of water at any point on the system. When designing a guttering system for a dwelling, there are factors that must be considered if the system is to cope comfortably with the rain that falls on the roof surface. The actual flow in the system depends on the area to be drained, the rainfall intensity and the position of the rainwater outlets.

In this section, we will look at the design factors that enable us to install effective guttering systems. These are:

- rainfall intensity
- roof area
- running outlet position
- the fall of the gutter
- changes of direction in the gutter run.

Rainfall intensity

In the introduction to this chapter, it was mentioned that the amount of rainfall throughout the UK differs greatly, with the south-east being considerably drier than the north-west. In England, the county of Cumbria has the greatest total rainfall, at around 1.8 m per year, with Essex and Kent having considerably less at around 500 mm.

Average rainfall, however, is only half the story. While it may rain much more in Cumbria than in Essex over a 12-month period, the number of litres discharged in a single two-minute rainstorm is greater in Essex at 0.022 $l/s/m^2$ (litres per second per square metre) compared with Cumbria at 0.014 $l/s/m^2$. This is called rainfall intensity and must be factored into any guttering system design because the guttering system must be able to cope with the sudden, intense downpour.

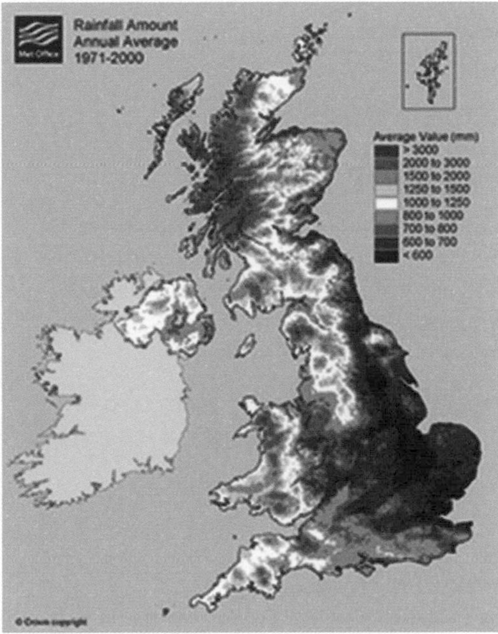

▲ Figure 8.4 Average rainfall in the UK

BS EN 12056–3:2000 gives rainfall intensity in litres per second per square metre (l/s/m²) for a two-minute storm event. The maps in the British Standard show the intensity for various periods from one year to 500 years. Rainfall intensity is divided into four categories (Table 8.5); the different categories are used depending on the type of building. Domestic dwellings are category 1.

INDUSTRY TIP

The Met Office website provides useful maps of rainfall in the UK, accessed at: www.metoffice.gov.uk/research/climate/maps-and-data/uk-actual-and-anomaly-maps

▼ Table 8.5 Categories of rainfall intensity

Cat. 1	Return period of 1 year	Eaves gutters and flat roofs
Cat. 2	Return period of 1.5 × design life of the building	Valley and parapet gutters for normal buildings
Cat. 3	Return period of 4.5 × design life of the building	Valley and parapet gutters for higher-risk buildings
Cat. 4	Maximum probable rainfall	Highest-risk buildings

Roof area

The angle and area of the roof is a key part of any guttering system design. Take a look at the diagram in Figure 8.5.

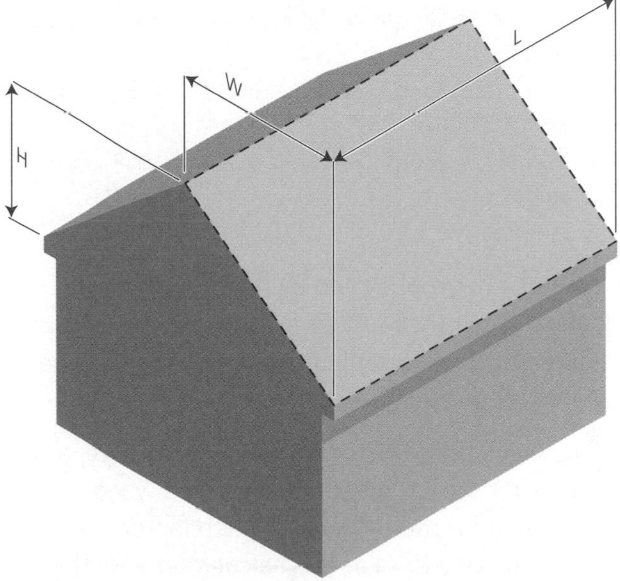

▲ Figure 8.5 Roof angle and area

The drawing shows the roof of a dwelling. If the area of the roof increases, the amount of water collected and discharged from it also increases. Similarly, if the angle of the roof increases then the area will increase, the amount of water will increase and the velocity at which the water enters the gutter will increase also.

IMPROVE YOUR MATHS

The area of a roof can be calculated by using the following formula in accordance with **BS EN 12056–3:2000**:

Effective maximum roof area (allowance for wind)

$$\left(W + \frac{H}{2}\right) \times L = \text{area in m}^2$$

Where:

W = horizontal span of slope

H = height of roof pitch

L = length of roof

Example 1

A roof has a length of 10 m, a width of 6 m and a height of 3 m. Calculate the effective area of the roof:

$$\left(6 + \frac{3}{2}\right) \times 10 = 75\,\text{m}^2$$

ACTIVITY

Calculation of effective roof area

Using the formula given above, calculate the following effective roof areas.

1 A roof has a length of 12 m, a width of 7 m and a height of 3 m.
2 A roof has a length of 8 m, a width of 8 m and a height of 4 m.
3 A roof has a length of 10 m, a width of 8 m and a height of 4 m.

The area of a flat roof should be regarded as the total plan area. If the roof has a complex layout, with different spans and pitches, each area should be calculated separately.

Building Regulations Document H3 gives an acceptable alternative for the calculation of roof area where the area of the roof is multiplied by a pitch factor. This is detailed in Table 8.6. For this calculation, only the length of the roof and the span are required.

INDUSTRY TIP

The Building Regulations 2010 Document H3 can be accessed at: www.gov.uk/government/uploads/system/uploads/attachment_data/file/442889/BR_PDF_AD_H_2015.pdf

▼ Table 8.6

Type of surface	Design area (m^2)
Flat roof	Plan area of relevant portion
Pitched roof at 30° Pitched roof at 45° Pitched roof at 60°	Plan area of portion × 1.29 Plan area of portion × 1.50 Plan area of portion × 1.87
Pitched roof over 70° or any wall	Elevational area × 0.5
To calculate flow in litres/second for 75 mm/hour intensity, multiply effective roof area m^2 by 0.0208	

Source: The Building Regulations 2010 Approved Document H

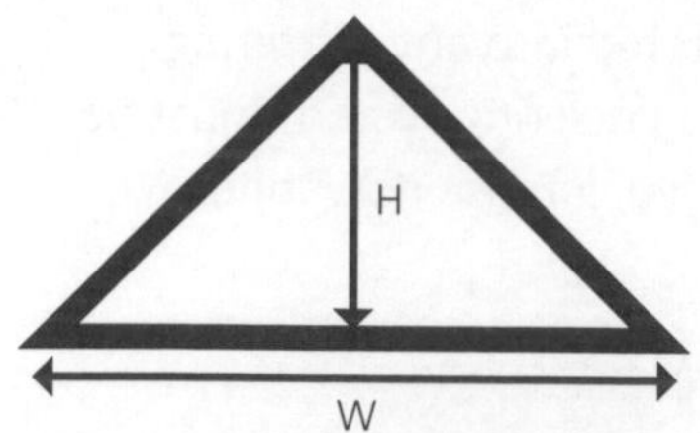

▲ Figure 8.6 Elevational area

IMPROVE YOUR MATHS

In this instance, if the angle of the pitch of the roof is known, the calculation is simplified. For example, if we use the data from the previous example, we arrive at the following.

Example 2

A roof has a length of 10 m and a width of 6 m. Calculate the effective area of the roof if the pitch of the roof is 30°.

Length of roof = 10 m

Width of roof = 6 m

The pitch factor from the table = 1.29

Therefore:

$$10 \times 6 \times 1.29 = 77.4 \text{ m}^2$$

ACTIVITY

Calculation of effective roof area using pitch factors

Using the pitch factors given in Table 8.6, calculate the following effective roof areas.

1 A roof has a length of 12 m, a width of 7 m and pitch of 45°.
2 A roof has a length of 8 m, a width of 8 m and pitch of 60°.
3 A roof has a length of 10 m, a width of 8 m and pitch of 30°.

IMPROVE YOUR MATHS

We can now calculate the amount of rainwater to be expected on any given roof area in a sudden storm deluge of 75 mm of rainfall per hour. To convert the area to litres per second (l/s), multiply the roof area (m^2) by 0.0208.

Example 3

The area of the roof in Example 1 is 75 m^2. What is the expected rainfall in l/s?

$$75 \times 0.0208 = 1.56 \text{ l/s}$$

ACTIVITY

Calculation of rainfall rate in litres per second

Using the effective areas calculated in the above activity box, calculate the rainfall rate in litres per second.

Running outlet position

Figure 8.7 shows a running outlet. It is the connection between the guttering and the rainwater pipe.

▲ Figure 8.7 A running outlet

The position of the running outlets is usually based upon the position of the gullies for the surface water sewer/drain to the property. These can be found on the building layout drawing (Figure 8.8).

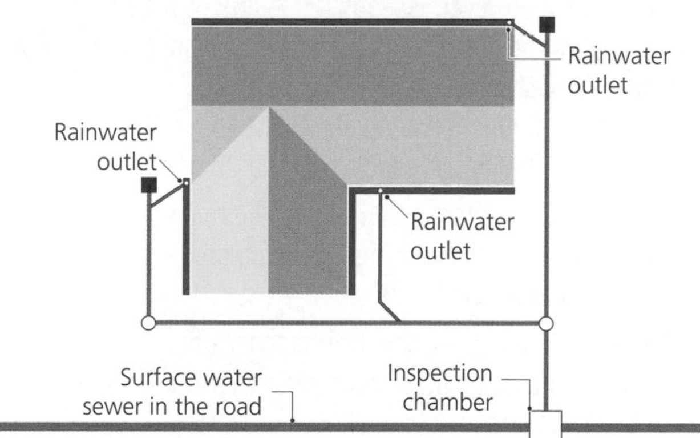

▲ Figure 8.8 Building layout drawing

The more outlets there are on a gutter system, the shorter the distance the water has to travel and the more effective the system is at discharging the rainwater. Consider Figure 8.9.

In drawing 1, the outlet has to be able to cope with the total rainwater run-off from the whole roof area. The outlet in this situation could be positioned at either end of the roof, but the total flow rate would be the same. Running outlets are designed to cope with rainwater from two directions, so the outlet at either end can cope with only half the flow rate. Only half the capacity of the outlet can effectively be used. Placing the outlet centrally would increase the total area of roof that the gutter can serve.

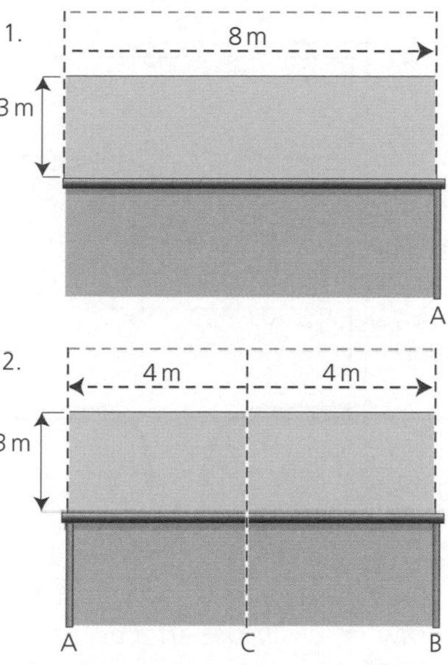

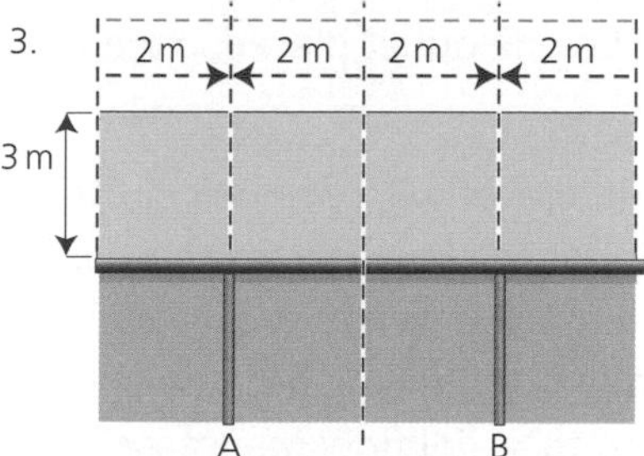

▲ Figure 8.9 Outlet positions

Outlet design

This is the least effective outlet design. The corners are sharp edged, which restricts the flow of water down the outlet by causing a clash of water streams at the shaded area. This creates turbulent water flow. Some water will travel across the outlet and against the flow on the opposite side of the outlet.	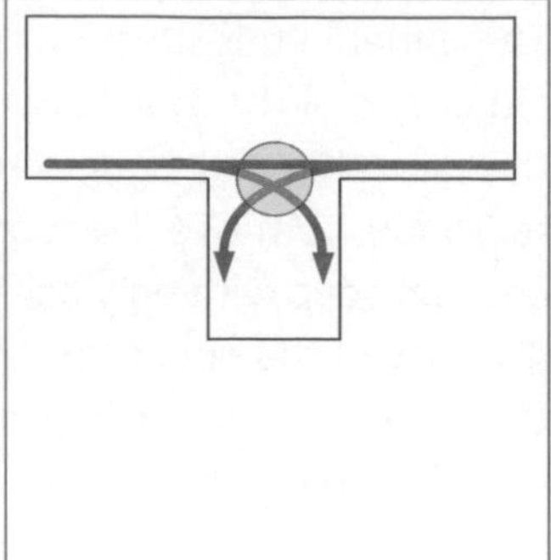
Here the corners are slightly rounded, which assists the flow of water down the outlet. However, the two water streams are likely to clash, creating some turbulence.	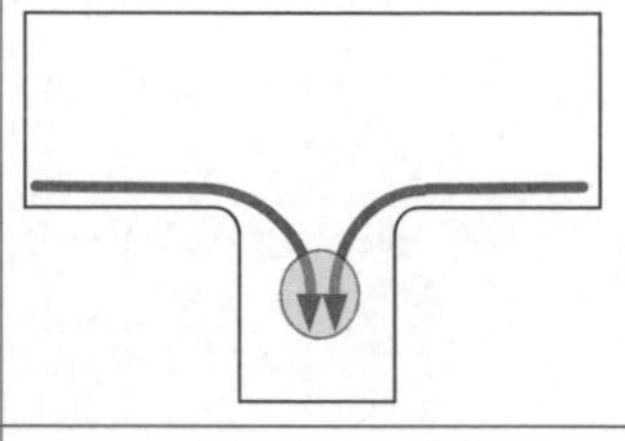
Fully rounded corners give a much better flow of water down the outlet. The two streams are kept more or less separate, which assists gravity flow down the rainwater pipe. This is known as hydraulic efficiency.	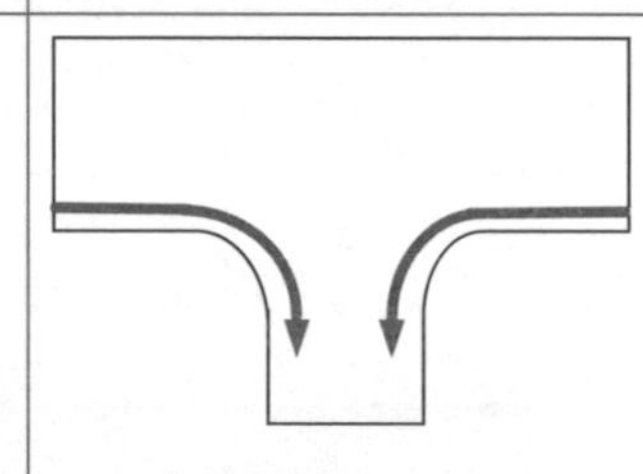

The outlet position in drawing 2 is more effective than drawing 1 simply because there are now two outlets and each outlet is coping with half the expected rainwater run-off. Again, an alternative, but equally effective, layout would be one outlet placed in the centre of the gutter run.

With outlets placed as in drawing 3, each half of the outlet has only a quarter of the flow rate to cope with and so layout 3 is much more effective at discharging the rainfall without the risk of flooding because both outlets are being used to their full flow rate capacity.

Each manufacturer will have different rainwater flow rates for its own running outlet designs. It should not be assumed that all manufacturers' flow rates will be equal. Therefore, manufacturers' data should be considered before the installation begins.

To find out how many outlets are required on a rainwater system design, simply divide the expected flow rate of the roof area by the flow rate for the outlet given in the manufacturer's technical literature.

The fall of the gutter

BS EN 12056–3:2000, Section 7.2.1 and NE.2.1 states that:

1 Gutters should be laid to a nominal gradient of between 1 mm/m and 3 mm/m where practicable.
2 The gradient of an eaves gutter shall not be so steep that the gutter drops below the level of the roof to such an extent that water discharging from the roof will pass over the front edge of the gutter.

In most cases, manufacturers interpret these two points as a slight fall of 1:600 (25 mm in 15 m). Laying a gutter with a fall greatly increases the flow capacity and, therefore, the area of roof that can be drained. It also ensures that silting of the gutter does not occur. However, manufacturers design guttering systems in such a way that the performance of the gutter is not compromised if it is laid level, with little or no fall. A fall of 1:600 ensures that the gutter will not fall so low as to be below the discharge point of the roof.

INDUSTRY TIP

Gutter falls

Not all manufacturers recommend a fall of 1:600. Some manufacturers advocate a fall of 1:350. This increases the amount of fall, thereby increasing the flow rate of the gutter. It also, however, lessens the length of the run of gutter before the gutter will dip below the discharge point of the roof. For instance:

- a fall of 1:600 is the equivalent of a 25 mm fall in 15 m
- a fall of 1:350 is the equivalent of a 22 mm fall in 8 m.

A fall of 1:600 therefore ensures that the rainwater will clear the gutter effectively and cause no problems with discharge from the roof.

Before installing a guttering system, check the manufacturer's installation instructions for the fall gradient that is recommended.

Changes of direction in the gutter run

In most domestic gutter systems, changes of direction cannot be avoided. Where changes in direction greater than 10° occur within a guttering system, they restrict the flow of water through the system. A 90° gutter angle reduces the effectiveness of the run of gutter where the angle is situated by 15 per cent, effectively reducing the roof area that the gutter can usefully serve. Each subsequent change of direction reduces the gutter's effectiveness still further. A gutter angle that is placed near an outlet will also reduce the effectiveness of the outlet.

Thermal expansion of PVCu gutters and fittings

One of the problems with PVCu gutters is the large expansion rate. This can cause the gutters to creak as they are warmed by the Sun and, in extreme cases, it can cause joint failure.

ACTIVITY

Calculation of thermal expansion

Using the method shown in the worked example above, calculate the following:

1 A south-facing gutter 10 m long is subjected to a 15°C temperature rise. What is the expansion of the gutter when the coefficient of linear expansion of the gutter is 0.06 mm/m/°C?
2 A south-facing gutter 20 m long is subjected to a 30°C temperature rise. What is the expansion of the gutter when the coefficient of linear expansion of the gutter is 0.06 mm/m/°C?
3 A south-facing gutter 5 m long is subjected to a 20°C temperature rise. What is the expansion of the gutter when the coefficient of linear expansion of the gutter is 0.06 mm/m/°C?

IMPROVE YOUR MATHS

PVCu has a coefficient of linear expansion of 0.06 mm/m/°C. This means that for every metre (1 m) of gutter, PVCu expands by 0.06 mm for every degree rise in temperature. For example:

If a 1 m length of gutter is subjected to a rise in temperature of 10°C, it will expand by the following amount:

$$1 \times 0.06 \times 10 = 0.6 \text{ mm}$$

This might not seem a lot, but let's look at this in more detail.

Example 4

A south-facing gutter 15 m long is subjected to a 25°C temperature rise. What is the expansion of the gutter when the coefficient of linear expansion of the gutter is 0.06 mm/m/°C?

All the information we need to be able to calculate this is in the question:

Length of gutter = 15 m

Temperature diff. (Δt) = 25°C

Coefficient of linear expansion = 0.06 mm/m/°C

Therefore:

$$15 \times 25 \times 0.06 = 22.5 \text{ mm}$$

KEY POINT

To counteract the expansion, all manufacturers build in to their fittings a 10 mm expansion gap. This must be observed when installing PVCu gutters if problems with thermal expansion are to be avoided.

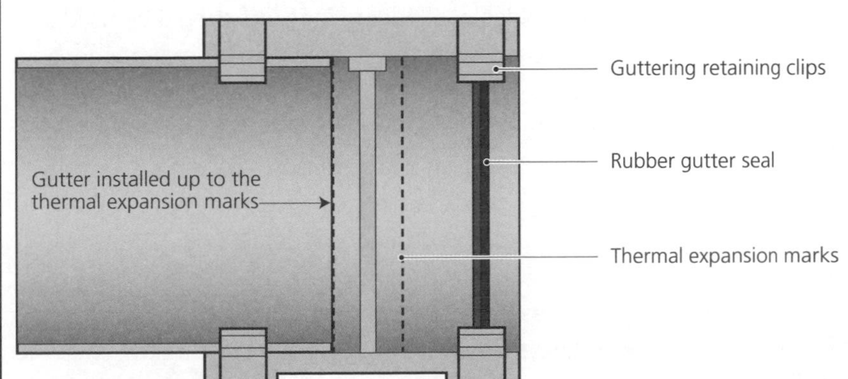

▲ Figure 8.10 Expansion gap on PVCu gutter fittings 1

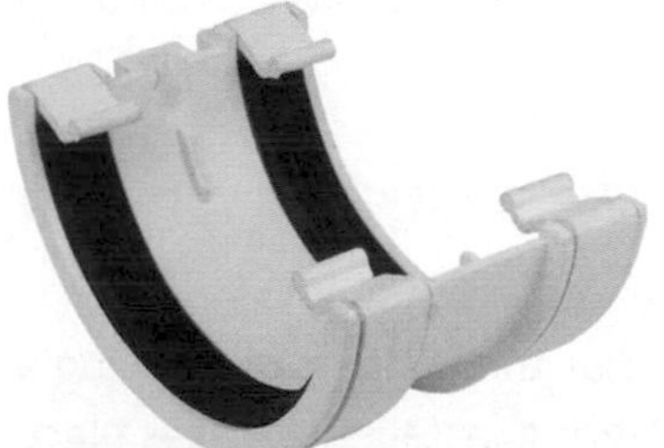

▲ Figure 8.11 Expansion gap on PVCu gutter fittings 2

2 INSTALLATION OF GRAVITY RAINWATER SYSTEMS

The sources of information required when carrying out work on gravity rainwater systems

There are a number of documents we must consult when designing and installing rainwater systems. Like all other aspects of the building process, gutters and rainwater systems are subject to various legislative restrictions to ensure that the systems we design and install collect the rainwater from the roof structure and dispose of it safely. To ensure the correct design and installation of rainwater systems, we must, therefore, refer to the following resources.

- **Building Regulations Approved Document, Section H3:** Rainwater drainage: this section states that adequate provision shall be made for rainwater to be carried from the roof of a building. It contains important information regarding design and installation of rainwater systems. It makes reference to **BS EN 12056–3:2000**.
- **BS EN 12056–3:2000 Gravity drainage systems inside buildings. Roof drainage, layout and calculation.** Like all British Standards, this document takes the form of recommendations. It relays the more technical aspects of rainwater system design, such as rainfall intensity calculations and outlet provision. It should be used in conjunction with the Building Regulations.
- **Manufacturers' instructions:** the manufacturers of gutters and rainwater pipework will have designed their systems to accommodate both the Building Regulations and British Standards. Wherever possible, manufacturers' recommendations must be followed.

Safe working practices

In Chapter 1, Health and safety practices and systems, we looked at the dangers of working at height and, since guttering installation takes place at heights above head level, it is relevant here that we look again at some of the more important aspects of these procedures.

In this part of the chapter, we will be revisiting some past learning with regard to working at height and also investigating how we can protect the customer's property while we are working above ground level.

Working at height

The safest way to install gutters and rainwater pipes is from a correctly erected and secured scaffold and on new-build housing, and this is usually the case. Unfortunately, erecting a scaffold for the purpose of replacing existing gutters and rainwater pipes is uneconomical because of the cost and so most of this type of work is performed using ladders. It should always be remembered that a ladder is not a safe working platform and extreme care should be taken when working from a ladder. Here are some points to remember:

- Always assess the work before any working at height is performed. A risk assessment should be performed.
- Never attempt the job alone. PVCu gutter is very light but it can catch the wind.
- There is no height threshold but if you are high enough to become injured from a fall, you must adhere to the Work at Height Regulations 2005.
- Always select the most appropriate equipment for the task, such as mobile scaffold towers or elevated working platforms. If working from a ladder is unavoidable, a ladder stand-off should be used, especially when performing gutter maintenance tasks.
- Ensure that you are properly trained in the use of ladders and mobile scaffolds.
- Always check ladders to ensure that they are in good order and free from defects.
- Always use the appropriate fall restraints and harnesses when working at height.
- Always be aware of what or who is below you when working at height. Never drop tools, equipment or materials.

- Always make sure that the ladder is secure before attempting the work. If securing the ladder is not possible, then a second person should 'foot' the ladder.

▲ Figure 8.12 Using a ladder stand-off

HEALTH AND SAFETY

A ladder is not a safe working platform. Take extreme care and have proper supervision at all times.

Be safe when working at height – don't take risks!

More information about working at height can be found in Chapter 1, Health and safety practices and systems, and on the Heath and Safety Executive website at: www.hse.gov.uk

Protecting the customer's property

In previous chapters, we have seen how we should protect the customer's property when working inside the dwelling. The same care and attention should extend to outside the property.

It is important that the outside of the property is checked for any existing damage before work begins and this should be pointed out to the customer.

VALUES AND BEHAVIOURS

Precautions that can be taken to protect the customer's property are:

- when using a ladder, a ladder stand-off should be fitted to prevent the scraping damage that can be caused by ladders to brickwork or masonry
- if a ladder is to be erected on a lawn, first cover the lawn with a plywood sheet to prevent damage to grass and flower beds
- lawns should have walk boards placed on them to prevent lawn damage
- take care where vehicles are parked on the customer's drive; to prevent possible damage, ask the customer to move them
- take care not to erect ladders on soft ground as they could sink, causing slippage; if this is unavoidable, ensure the ground is supported beforehand
- place barriers around where work is being carried out, to prevent people from being injured when walking near by.

Preparatory work to be carried out on building fabric

Before starting the installation, fascia boards must be checked to ensure that they are straight and level, and that they do not need replacing. Fascia boards that are not level or straight can give the gutter a crooked or wavy appearance, and rotted fascias will not hold the gutters properly. Occasionally, it may be necessary for fascia boards to be painted before the gutter is installed. The underfelt drip, which is a strip of felt positioned under the front row of tiles, should also be checked to ensure that it has not ripped or rotted. The felt stops rainwater from leaking behind the gutter and should be replaced if it is found to be defective.

Existing rainwater systems should be removed with care, to avoid damage to the outside wall surfaces and existing fascia boards.

On new-build properties, it is likely that the gutters will be installed before the roof is laid.

HEALTH AND SAFETY

All materials that are removed should be disposed of safely and responsibly.

Occasionally, asbestos gutters may be found. Where these are to be removed, this should be carried out by specialist asbestos removal companies.

Installing PVCu gutters

The hand and power tools that will be required when installing PVCu gutters and rainwater pipes are listed in Table 8.7.

▼ Table 8.7 The hand and power tools required when installing PVCu gutters and rainwater pipes

Hand tools	Power tools
Pozidriv screwdrivers Hacksaw Claw hammer String/plumb line Spirit level/line level Bradawl File/rasp	110 V SDS power drill 24 V battery-powered cordless drill

When installing a rainwater system, a survey of the property must be carried out, and there must be a discussion with the customer about their requirements and choices. It could transpire that an existing installation may not be installed correctly. Therefore, it is beneficial to ask the customer about how the system has performed in the past.

IMPROVE YOUR ENGLISH

Remember: communication skills are key. When working on a new build or alteration, you need to understand the customer's needs and requirements. You could use visual aids, such as manufacturers' brochures to show the customer, so they can select an aesthetically pleasing style. This can also help a plumber specify a suitable style as the discharge rates will be provided.

It is wise to establish the type of drainage for the premises, and plan your system around whether or not it will be combined, separate or partially separate. If it is the latter, check that any soakaway is fit for purpose.

Sometimes, a new building extension will require a rainwater system. This could be connected to a functioning system already installed on the premises. In this situation, a recalculation of the existing system may be required to estimate whether the gutter size and outlets are sufficient to accommodate the additional flow rates caused by the new extension.

There are many documents that need to be consulted when designing or installing rainwater systems. This is because there are a range of restrictions in the legislation to ensure that the water is collected efficiently and safely discharged from a building.

Installation of PVCu gutters, step by step

1 Establish the position of the outlets.
2 Establish the high point of the gutter and fix a fascia bracket at this high point onto the fascia board using 25 mm × no. 10 zinc-plated roundhead screws.
3 Using a plumb line, centre the outlet over the gully or drain.
4 The distance between the high point and the outlet should be measured and a fall of 1:600 determined. Using this fall, fix the outlet at the low point on the fascia board.
5 A line can now be strung between the high fascia bracket and the outlet. For gutters that are to be fixed level, a spirit level should be used against the string. (See Figure 8.13.)
6 Screw further fascia brackets onto the fascia board, working away from the running outlet. The brackets should just touch the line but not distort it. Most manufacturers recommend a distance between the fascia brackets of 1 m (750 mm in areas that suffer heavy snowfall), but the manufacturer's instructions should be checked beforehand. (See Figure 8.14.) There is no need to fix a bracket close to the running outlet as it is secured using screws and therefore acts as a bracket.

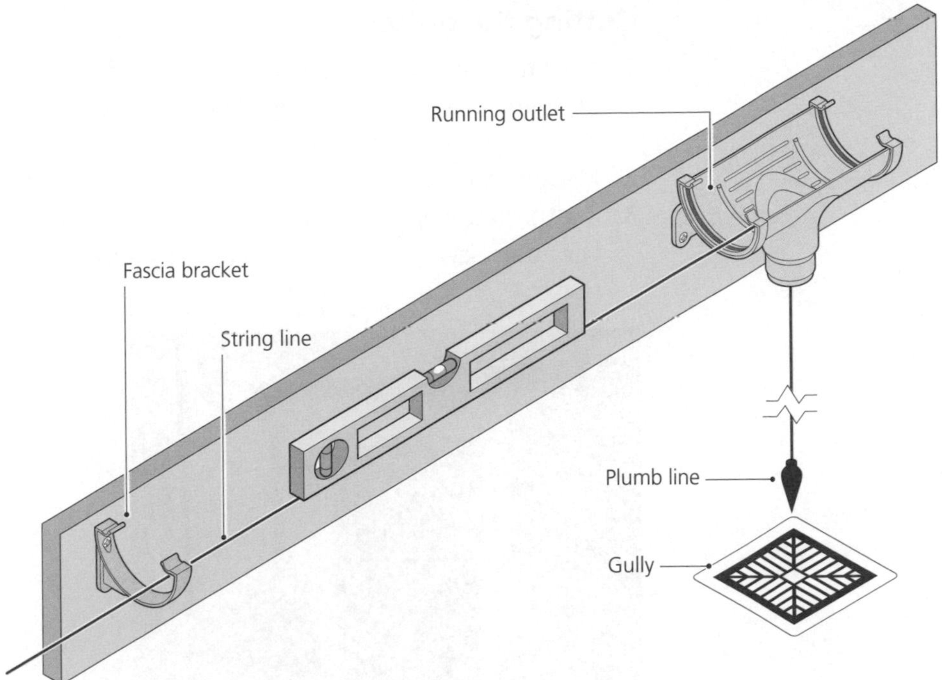

▲ Figure 8.13 Setting the gutter fall

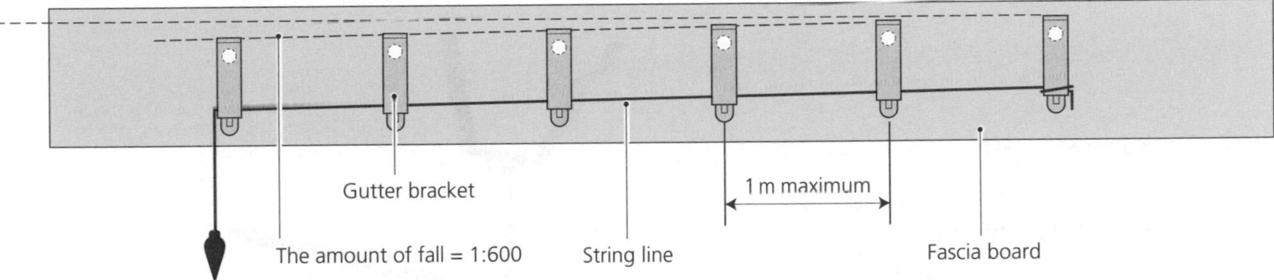

▲ Figure 8.14 Installing fascia brackets

For buildings without fascia boards

There are two methods of fixing to dwellings without fascia boards. These are as follows.

1 **The use of top-or side-fitted rafter brackets:** these are galvanised steel brackets that are screwed to the top or the side of the roof rafters. The fascia brackets are then bolted to the rafter brackets. It is often necessary to replace sections of rafters that have been exposed to the elements. The rafters should be checked before installation.
2 **The use of drive-in rise and fall brackets (also known as rise and fall irons):** these are flat pointed strips of galvanised steel that are built into the brickwork joints. Threaded rod is then fitted with a gutter bracket attached, which can be adjusted up or down to give a fall.

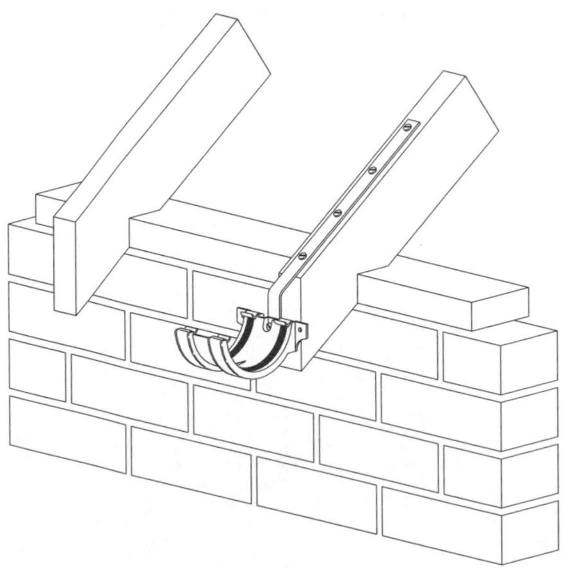

▲ Figure 8.15 Top-fitted rafter brackets

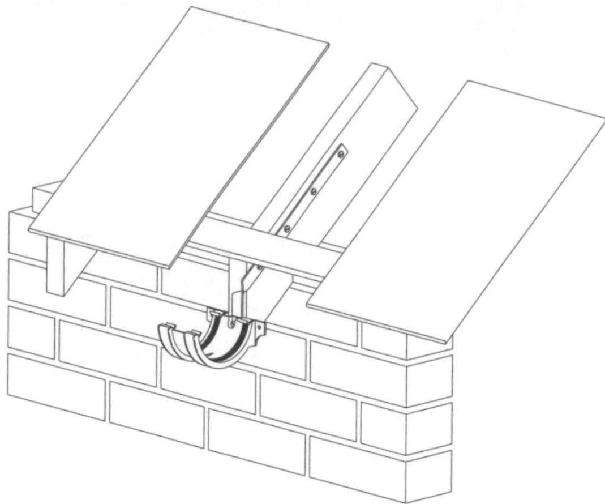

▲ Figure 8.16 Side-fitted rafter brackets

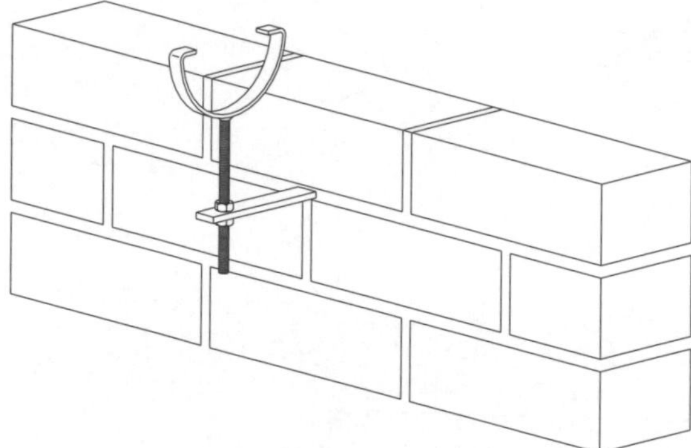

▲ Figure 8.17 Rise and fall brackets

Installation of the guttering

When installing gutter angle fittings, stop ends and gutter unions that are unsupported, fascia brackets should be fitted no more than 150 mm away from either side of the fitting or end of the gutter.

Once all the fascia brackets have been fixed, the gutter can be fitted. It is advisable to work away from the outlet towards the high point – this will save time on installation as fewer cuts will be needed.

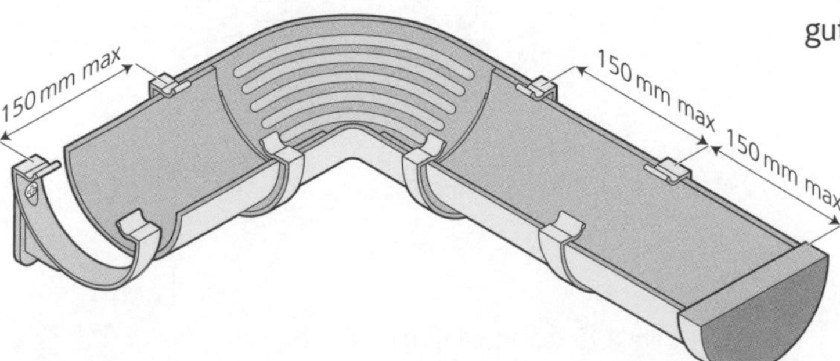

▲ Figure 8.18 Installing gutter angles

Cutting the gutter

1 Manufacturers recommend that gutter and rainwater pipes are cut using a fine-tooth saw or a hacksaw with a 24-teeth/inch blade.
2 Measure and mark the gutter to the required length.
3 Cut the gutter or rainwater pipe carefully using a fine-tooth saw and de-burr the cut using a file or rasp.

▲ Figure 8.19 Cast iron gutter fitted to rafter brackets

▲ Figure 8.20 Rise and fall iron

Making the guttering into the fittings

PVCu gutter systems use a snap-fit jointing system. To make a watertight joint, simply insert the gutter into the fitting up to the expansion mark. Push the gutter up into the back of the gutter fitting clip. Pull the front of the gutter down and clip the gutter in with the front gutter fitting clip using the thumb.

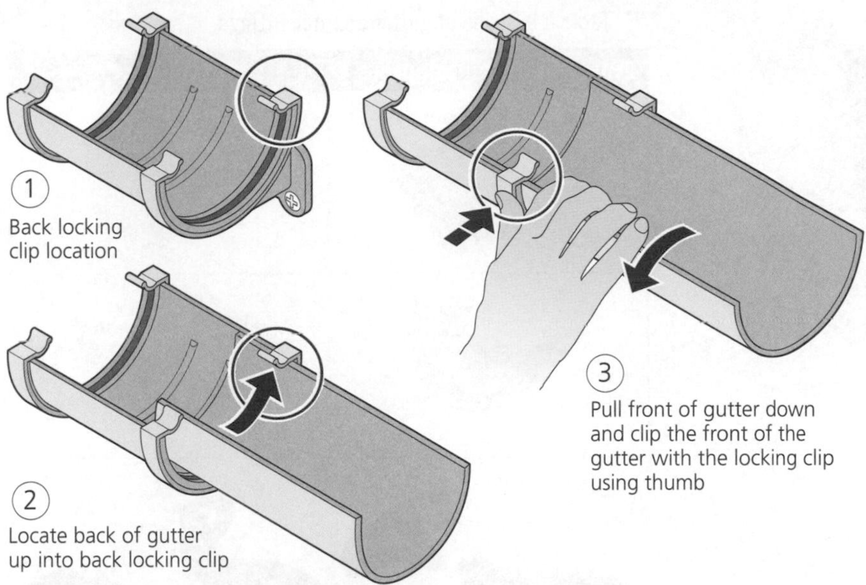
1
Back locking clip location
2
Locate back of gutter up into back locking clip
3
Pull front of gutter down and clip the front of the gutter with the locking clip using thumb

▲ Figure 8.21 Installing the gutter

Installation of the rainwater pipe

1. Before installing the rainwater pipe, it is advisable to fabricate the swan neck bend at the top of the pipe where it connects to the running outlet:
   - measure the distance between the two 112.5° bends marked 'L' on Figure 8.22
   - cut the length of rainwater pipe, de-burr the pipe and, using solvent weld adhesive, glue the swan neck bend together; this should be left for 5 minutes to set.
2. Install the swan neck onto the outlet and measure the distance to the shoe at the base of the rainwater pipe (if the pipe is to be fitted directly to the drain, measure the distance to the drain connection).
3. Cut the length of pipe required and de-burr. Install the pipe onto the bottom of the swan neck and, using a level, mark and drill the bottom rainwater pipe clip, and screw the clip and pipe against the wall using wall plugs and 50 mm × 10 alloy or stainless steel screws.
4. There is no restriction to the number of bends that can be installed on rainwater pipes.
5. Where two rainwater pipes converge, it is possible to take both into a hopper head (Figure 8.23).
6. The distance between the rainwater pipe clips is shown in Table 8.8. Before installing the rainwater pipe clips, always check the clip distances in the manufacturer's instructions.

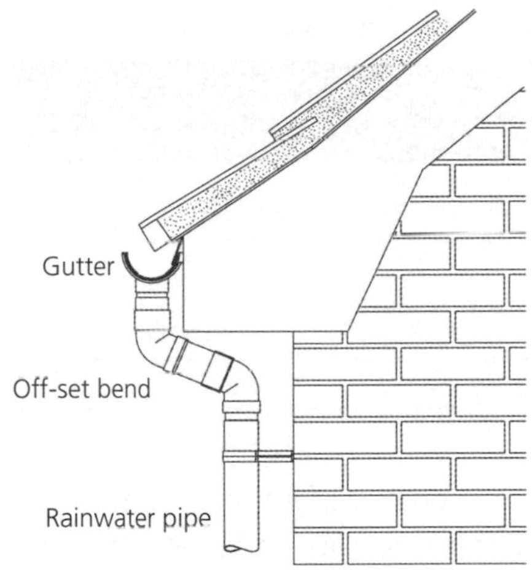
Gutter
Off-set bend
Rainwater pipe

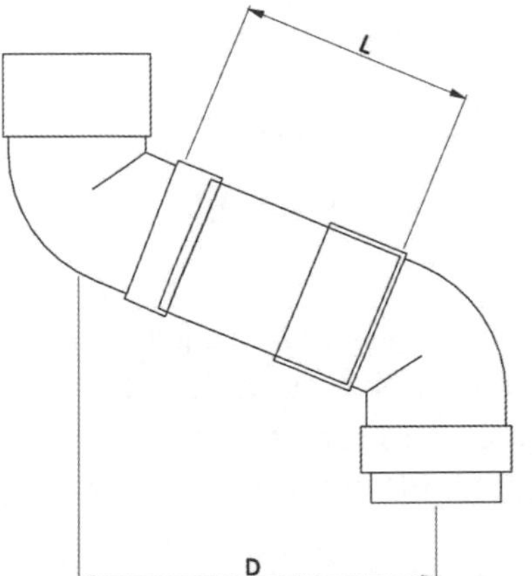
L
D

▲ Figure 8.22 Making the swan neck bend

▲ Figure 8.23 A hopper head

▼ Table 8.8 Distance between the rainwater pipe clips

Rainwater clip support centres		
Pipe size	**Vertical (m)**	**Horizontal (m)**
55 mm	1.2	0.6
62 mm	2.0	1.2
68 mm	2.0	1.2
70 mm	2.0	1.2
82 mm	2.0	1.2
110 mm	2.0	1.2

7 Measure the required distance for the clips, mark and drill the rainwater pipe clips, and screw the clips and pipe against the wall using wall plugs and 50 mm × 10 alloy or stainless steel screws.

Jointing guttering of different materials and profiles

Occasionally, it may be necessary to make joints between systems of guttering that use different materials or profile shapes. This can be done easily using specific adapter fittings. Gutter adapters include:

- half round PVCu to half round cast iron
- half round PVCu to Ogee PVCu
- half round PVCu to Ogee cast iron
- half round PVCu to square section PVCu.

▼ Table 8.9 Styles of gutter adapter fittings

Gutter-to-gutter adapters
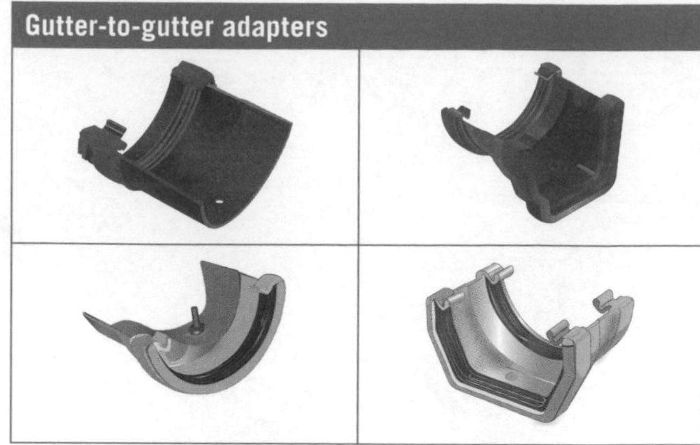

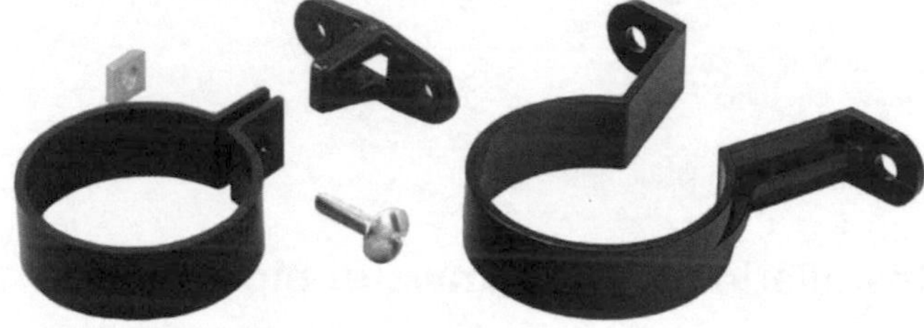

▲ Figure 8.24 Rainwater pipe clips

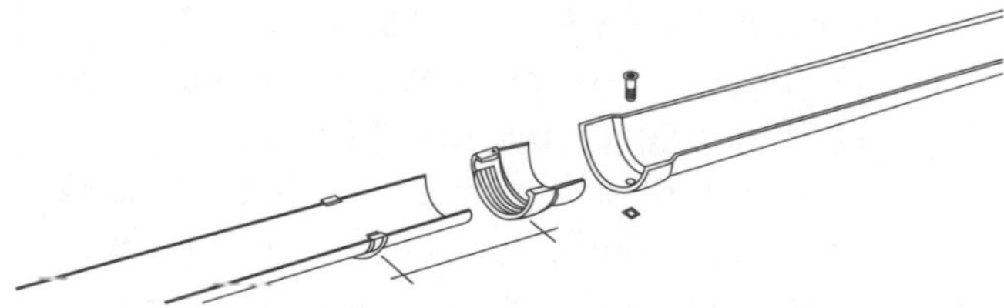

▲ Figure 8.25 Connections to existing cast iron gutters

Handling and storage of materials

Care should be taken when handling PVCu gutter and rainwater pipes. Excessive scratching can ruin the aesthetic appearance of the gutter and affect joint sealing. Cold weather reduces the impact strength of PVCu and extra care is needed in wintery conditions.

When pipe is delivered to site, it is recommended that loading and unloading of pipe and gutter lengths is performed by hand without the use of mechanical lifting aids.

Always store pipes and gutters on flat surfaces, ensuring that the surface is free from sharp protrusions. Bundles of pipes and gutters can be stored up to 3 m high without support. Loose gutter and pipe requires supports every 2 m. Fittings should remain in their packaging until needed to reduce damage by scratching.

Testing completed rainwater systems

Once the system installation is complete, testing can be carried out by discharging water, from a hosepipe, at all high points in the system and checking to make sure that the water discharges down the outlets and through the rainwater pipes without leakage or pooling of water in the gutter.

3 THE MAINTENANCE AND SERVICE REQUIREMENTS OF GRAVITY RAINWATER SYSTEMS

Maintenance of guttering systems is an essential activity to keep systems working correctly. In this next section of the chapter, we will look at those essential items of maintenance that are carried out during planned preventative maintenance or fault rectification.

These include:

- visual inspections and fault finding
- leakage repairs
- replacement of defective gutters and fittings
- cleaning and clearing blockages.

Different gutter materials require different methods of working and repair, and it is important that we have knowledge of the basic repair techniques required.

Visual inspections and fault finding

Visual inspections are the first part of the maintenance and repairing activity. Visual inspections help in establishing the overall condition of the gutter and rainwater pipe installation, joints and fittings, and in pinpointing specific problems, such as those listed in Table 8.10.

▼ Table 8.10

Fault	Remedy
Leaking joints	Carry out rectification operations (see below)
Cracked and broken gutters and rainwater pipes	Carry out rectification operations (see below)
Bad falls and bowing gutters	These will require realigning with the correct fall or the installation of extra fascia brackets
Blocked gutters and rainwater pipes causing water to overflow at the outlets	These require cleaning and clearing (see below)
Incorrectly spaced fascia and rainwater pipe brackets	Fascia brackets at 1 m distance, vertical pipework brackets at 2 m and horizontal at 1.2 m
Water overflowing from the gutter during periods of heavy rain after a major extension to the gutter system	Generally, a sign of too big a roof area – install more rainwater pipe outlets, or replace the gutter with high-capacity gutter

Leakage repairs

There are different visual signs for leaking joints depending on the material that the gutter is made from.

Leaking PVCu gutters

The leak may not be obvious until water is discharged down the gutter, especially if the gutter is black in colour. In some instances, leaks may show on the surface of the gutter as a black/green moss growth. A joint that is leaking, usually because the rubber seal has either shrunk or become misaligned, is generally an easy problem to fix by replacing the defective fitting.

INDUSTRY TIP

It should be remembered, however, that some gutter manufacturers use different fitting dimensions and one type of gutter may not fit another. Most manufacturers produce 'compatibility charts' showing which gutter fits another.

With leaking PVCu fittings, remember:

- always try to replace like for like; this is sometimes not possible as there have been many manufacturers in the past that no longer exist or the company has changed its specifications and fittings have been improved/updated

- do not be tempted to repair leaking joints with silicone sealant; while the joint may be sealed initially, as soon as the gutter expands and contracts, it will break again and begin to leak.

Leaking cast iron fittings

These are generally visible from the ground without the need to pour water down the gutter. Leaking cast iron joints have visual tell-tale signs, such as:

- rust staining on the mouth of the joint
- moss and lichen growth on the mouth of the joint
- water staining in the joint area
- rust around the gutter bolt.

Repairing a leaking cast iron joint is a reasonably easy task that involves removing the gutter bolt, breaking (parting) the joint, cleaning out the old jointing medium (usually paint and putty), repainting and re-puttying the joint before remaking the joint with a fresh gutter bolt. Care should be taken, however, as movement of the gutter can break further joints down the gutter run. Again, silicone sealant is not a satisfactory jointing medium in this situation and the joint must be dry before jointing is attempted.

Replacement of defective gutters and fittings

Perhaps the most obvious of all gutter defects are cracked and broken gutters and rainwater pipes.

PVCu gutters and rainwater pipes are at constant risk from ultraviolet (UV) rays from sunlight. This can often lead to gutters becoming brittle, causing them to shatter or crack. Placing ladders directly against PVCu gutters, when undertaking maintenance and cleaning, can also damage them further.

Look back at Chapter 2, Common processes and techniques, for more information on the effects of UV light on plastics.

The main problem here, especially where replacement is necessary, is compatibility. Most manufacturers now use generic gutter and rainwater pipe sizes, but older guttering systems are often smaller in size with no adapters available. In this case, replacement of the entire system is the only option.

Where the gutter is compatible with other systems, the replacement of gutter is a fairly simple process, as described below.

1. Visually inspect the job and assess the risks. A risk assessment should be carried out. Guttering is a two-man job if working from a ladder.
2. The correct PPE should be worn when attempting this task. Eye protection is essential.
3. If it is possible to remove the cracked section between two fittings, this will be the simpler option. It is advisable to replace the fittings either side as well as the length of gutter as the rubber seals may not create a seal when the new gutter is installed.
4. Unclip the gutter from the fittings and begin to remove the gutter from the fascia brackets by pulling the gutter and bracket towards you and down. Unclip the gutter by lifting the front edge of the fascia bracket and clicking it over the gutter. Be careful here. The brackets may be as brittle as the gutter itself.
5. Once all the brackets and fittings have been unclipped, carefully lift out the gutter by twisting the front face of the gutter upwards and out of the brackets.
6. Replace the fittings (gutter unions, angles, etc.) as necessary, taking care not to alter the fall of the gutter.
7. Measure the distance between the expansion marks of the fittings, and cut and de-burr the new length of gutter.
8. Install the new gutter by inserting the back edge first, and twisting down and away from you.
9. Carefully re-clip the gutter into the first fitting and, working towards the second fitting, re-clip the gutter into the fascia brackets.
10. When the gutter has been clipped into the last fitting, testing the gutter with a hosepipe can begin.
11. Check for leaks and clearance of the water from the gutter.

HEALTH AND SAFETY

Take care when working in the same space as cast iron gutters as, occasionally, the gutter may fall without warning.

Problems with cast iron gutters

Cast iron gutters present very different problems to PVCu. Cast iron gutters, if not regularly painted, rust from the back edge towards the front, causing weakness of the metal. The rust also attacks the rafter brackets so they too become weak. When this happens, the weight of the gutter will cause the gutter to drop and become unstable.

The procedure is as follows.

1 Visually inspect the job and assess the risks. A risk assessment should be carried out. Removing sections of cast iron guttering is a two-man job if working from ladders. The guttering is very heavy and this task should not be handled alone.
2 It may be beneficial to clean the gutter out beforehand as this often reduces the weight.
3 Carefully cut through the gutter bolts above the nut with a junior hacksaw.
4 Using a nail punch, punch the cut bolts upwards from the cut end.

INDUSTRY TIP

Do not be tempted to punch downwards as gutter bolts are either large dome-headed or countersunk-style bolts and you risk breaking further lengths of gutter.

5 Once the bolts are removed, carefully break the joints at either end. Be careful as cast iron gutter often has only one rafter bracket in the centre of the gutter length and the gutter may drop suddenly.
6 Carefully lift out the gutter by twisting towards you and upwards.
7 With the section of gutter removed, clean the socket and spigot of the gutter either side of the removed length to remove the old jointing material, and paint the inside of the socket and the outside of the spigot using black bitumen paint.

VALUES AND BEHAVIOURS

Always try to maintain the customer's property as it was – the original colour paint of the gutter can be used if the customer requests it.

8 There should be no need to cut the gutter if it is a full length being replaced as cast iron gutter is supplied in 6 ft (imperial) lengths to be compatible with existing systems. Should cutting be required, a hacksaw or angle grinder with an appropriate metal-cutting blade can be used. Eye protection is essential.
9 Mark and re-drill the bolt hole (if required after cutting).
10 Paint the inside of the socket and outside of the spigot of the new length, and place a 20 mm-thick bead of soft linseed oil putty in the socket.
11 Place another bead of putty in the existing gutter socket.
12 Carefully lift the new section of gutter to roof height and, ensuring spigot is to socket, lift the new section of gutter into place by inserting the back edge first, and twisting down and away from you.
13 Gently press the joints together and insert the gutter bolts at both joints. Re-tighten the gutter bolts. Do not over-tighten as the gutter may crack.
14 Remove any excess putty from inside and outside the joints, and paint the joint both internally and externally.
15 Test the gutter by discharging water from a hosepipe down the guttering and check for leaks.

HEALTH AND SAFETY

The correct PPE should be worn when attempting this task. Eye protection is essential.

INDUSTRY TIP

Replacement of broken or rusted cast iron gutter sections is often difficult and time consuming, and should be attempted only with an experienced plumber to supervise the activity.

Replacing cast iron with PVCu

Replacing cast iron gutters with PVCu is possible with special adapters that convert from cast iron to PVCu (see page 444 of this chapter). When replacing cast iron gutters, do not be tempted to reuse the rafter brackets as these are not secure enough for PVCu and the new gutter may flap in the wind. Any existing rafter brackets should be removed beforehand and a string line put up between the sockets of the cast iron gutter. This can be done by installing the line between the bolt holes of the existing cast iron to maintain the correct fall. The new fascia can then be installed to the line as previously described.

Cast iron rainwater pipes are easily replaced with PVCu equivalents. The cast iron rainwater pipe should be replaced to the nearest downstream joint or, better still, replace the whole length of cast iron with PVCu pipe.

Cleaning and clearing blockages

Probably the most common of all maintenance procedures is the cleaning and painting (cast iron only) of gutters. Over a period of time, silt can build up in gutters, especially when the roof tiles are made from concrete. Silting can lead to moss growth and eventual blockage, causing gutters to overflow, and this could possibly cause fascia boards and roof joists to rot away and walls to become damp. Cleaning (and painting both inside and out on cast iron gutters) should be carried out during the scheduled preventative maintenance programme on a yearly basis.

▲ Figure 8.26 Blocked gutters

HEALTH AND SAFETY

Where the gutter is found to contain bird droppings, this should be handled with extreme care as these carry disease and should not be ingested into the body by breathing in. A face mask and waterproof gloves should be worn at all times.

4 DECOMMISSION RAINWATER AND GUTTER SYSTEMS AND COMPONENTS

Decommissioning rainwater systems can be messy and often dangerous as most of the work is done at height. The following points must be considered.

- **Notify relevant person:** inform the customer that you are going to start removing the old guttering and rainwater pipework.
- **Apply warning notices and signs:** position warning signs and notices that there are operatives working overhead.
- **Wear the correct PPE:** gutters are often full of sludge and plant life and, occasionally, dead birds and small mammals. These can present a health hazard. It is therefore important to use the correct PPE, including goggles, face mask and rubber gloves.
- **Check for hazardous materials:** asbestos was used for both gutters and rainwater pipes. This must not be touched and must be removed by specialist asbestos removal contractors.
- **Appropriate access equipment:** gutters should be removed and installed from a properly constructed and erected scaffold. Gutters should not be installed or removed from a ladder.
- **Removal of components:** remember, some gutter components are heavy, such as cast iron. Gutters and rainwater pipes should be removed with care.
- **Dispose of materials appropriately:** old gutter systems should be disposed of responsibly at a recognised disposal point or recycling centre. Alternatively, the hire of a mini-skip would prove useful.

5 PERFORM A SOUNDNESS TEST, AND COMMISSION RAINWATER AND GUTTER SYSTEMS AND COMPONENTS

Visual inspection

On completion of the gutter and rainwater pipe installation, a visual inspection should be conducted. Check that:

- the gutters are clear of any debris that may hinder the free flow of water
- the gutters and rainwater pipes are adequately and correctly supported to the manufacturer's instructions
- the fall of the gutter complies with the manufacturer's installation data.

Before testing takes place

Before undertaking testing procedures, notify the customer or the responsible person of what you are about to do, and give a rough estimate of how long this will take. Ask the customer to move any obstructions or property that may hinder the testing process.

Testing the gutters and rainwater pipes

All gutter and rainwater pipes should be wet tested after completion. This can be done either using a hosepipe or a bucket full of water. Do not discharge the water directly into the gutter as this often causes spillage, which can be mistaken for a leak. Instead, spread the water over the roof to simulate how rainfall would enter the gutter naturally. While the gutters are full of water, check for:

- leaks at the gutter and rainwater joints
- clearance of the water from the gutter without pooling in the gutter bottom
- free flowing of the water down the rainwater pipework without any backing up and spilling over the top of the gutter
- a gentle flow of water without the water flowing too fast, which is an indicator of an incorrect fall
- any signs of dampness on the gutter or the building structure.

On completion

Complete any commissioning documentation required. This is often a requirement on new housing construction sites.

SUMMARY

This chapter has shown the importance of correctly designed and installed gutter and rainwater systems. But this is only half the story. All too often, good, well-installed gutter systems are neglected and left to depreciate in the elements. The important points of this comprehensive insight into rainwater management are:

- think about the design and comply with Building Regulations Approved Document H3 and the recommendations of British Standard **BS EN 12056-3**
- use manufacturers' installation instructions for fall ratios, clipping distances and rainwater pipe positioning
- protect the customer's property during installation operations
- be aware of health and safety at all times.

Test your knowledge

1 What are the most important factors to be considered when sizing and selecting a guttering system?

a Drainage connection, rainfall direction, durability
b Rainfall intensity, roof area, gutter fall
c Material type, height, drainage system type
d Environmental factors, weight, labour availability

2 On completion of the installation of a rainwater gutter system, how would you check that the system does not leak?

a Return on a day when it is raining to visually check
b Carry out a visual inspection
c Discharge water on the roof with a hosepipe
d Use a manometer and weir cup

3 What type of gutter profile is shown in the image below?

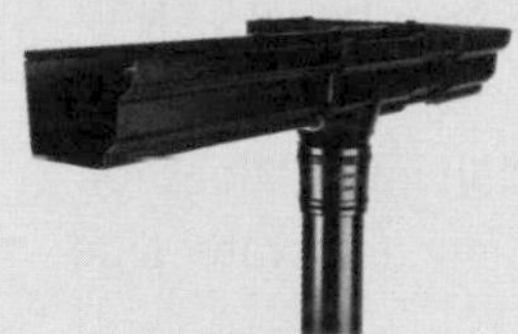

a Half round
b Deep flow
c Ogee
d Square

4 Which is the most suitable gradient of fall for a standard gutter installation?

a 2 mm/m
b 8 mm/m
c 10 mm/m
d 60 mm/m

5 A south-facing gutter 20 m long is subjected to a 30°C temperature rise. What is the expansion of the gutter when the coefficient of linear expansion of the gutter is 0.06 mm/m/°C?

a 12 mm
b 18 mm
c 36 mm
d 42 mm

6 Which part of the Building Regulations specifically covers rainwater systems?

a Document G
b Document H
c Document J
d Document P

7 What is the recommended clipping distance for PVCu gutters to fascia boards?

a 650 mm
b 750 mm
c 800 mm
d 1000 mm

8 What is the component marked 'X' in the image below?

a Rainwater shoe
b Expansion joint
c Gulley
d Hopper

9 Which of the following is not a common material used in the manufacture of rainwater systems?

a PVCu
b MDPE
c Copper
d Extruded aluminium

10 Which of the following British Standards gives specific information on the intensity of rainfall in the UK?

a BS EN 12056–3:2000
b BS EN 806
c BS EN 1057
d BS 6565–4

11 Which item would be used to inspect a guttering system located on the first floor for a blockage?

a Trestle
b Step-up
c Step ladder
d Extension ladder

12 What is another common term used to describe the offset from a running outlet to the vertical downpipe?

- **a** Swan neck
- **b** Cranked turn
- **c** Goose neck
- **d** Direct change

13 What is a common diameter for round domestic downpipe?

- **a** 32 mm
- **b** 50 mm
- **c** 68 mm
- **d** 75 mm

14 Which of the following factors needs to be considered before the installation of a guttering system?

- **a** How tall the building is
- **b** What type of tiles are on the roof line
- **c** The roof area
- **d** What type of brickwork the building is made from

15 What is the recognised ratio of fall for the gutter length?

- **a** 1:200
- **b** 1:300
- **c** 1:600
- **d** 1:800

16 A roof has a length of 12 m and a width of 7 m. If the pitch of the roof is 45° then what is the effective area of the roof?

17 How is expansion accommodated within PVCu guttering?

18 Calculate the amount of rainwater expected at any one time from a roof with an effective area of 104 m^2 in an area where the number of litres discharged in a single two-minute rainstorm is 0.022 $l/s/m^2$.

19 Give at least three advantages to the use of extruded aluminium rainwater systems.

20 What tools are required to install a PVCu gutter including running outlet to a wooden fascia?

21 Show your full calculations to find the drop or fall of a 5.25 m length of guttering.

22 A customer complains that the same section of guttering is overflowing every time it rains. Outline what could be wrong with the installation.

Answers can be found online at www.hoddereducation.co.uk/construction.

Practical activity

Practise your cutting and measuring by producing the gutter arrangement shown in the image below. Ensure cuts are straight and smooth, and brackets are installed at the correct distance to allow for support of the bend. This does not need to be fixed but if your training centre allows it or you have the facility and time on-site, then try positioning and fixing using suitable fixings. The short length shown will have very little fall but if possible (and, again, if time permits) a 3 m length of gutter could be installed, ensuring correct fall is provided.

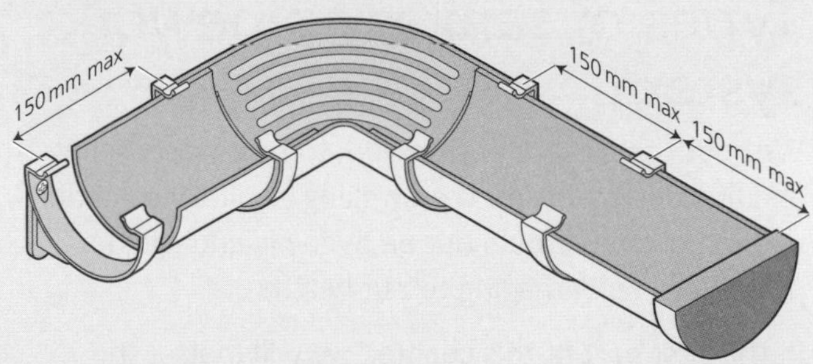

CHAPTER 9

SANITATION SYSTEMS

INTRODUCTION

Some 200 years ago, waste water and sewage simply ran down the centre of streets and alleys. These were open sewers breeding disease that, on many occasions, caused severe illness and death. Today, the effluent we produce is directed safely away from our homes by a network of pipes called sanitation systems.

In this chapter, we will investigate domestic sanitation systems. We will look at the many different sanitary appliances available and the systems of above-ground sanitation pipework they are connected to, which ensure hygienic living conditions in our homes and in the surrounding environment.

By the end of this chapter, you will have knowledge and understanding of the following:

- sanitary pipework and appliances used in dwellings
- installing sanitary appliances and connecting pipework systems
- service and maintenance requirements for sanitary appliances and connecting pipework systems
- the principles of grey water recycling.

1 SANITARY PIPEWORK AND APPLIANCES USED IN DWELLINGS

Types of sanitary pipework system

Without the system of pipework to take waste solids and liquids away from the dwelling, sanitary conditions within buildings would not be hygienic and could potentially be damaging to our health.

In this first part of this chapter, we will look at the various systems of sanitary pipework, often called above-ground discharge systems (AGDS), and investigate where these systems should be installed. The systems are:

- primary ventilated stack system
- ventilated branch discharge system
- secondary ventilated stack system
- stub stack.

All sanitary systems contain two sections:

1. **The soil pipe:** also known as the soil stack, this is the lower, wet part of the system, which takes the effluent away from the building.
2. **The vent pipe:** also known as the vent stack, this is the upper part of the system that introduces air into the system to help prevent loss of trap seal. Ventilation of a soil and waste system is necessary to prevent water seals in traps being broken due to negative pressure or pressure fluctuations within the system. Broken seals allow foul air and smells to enter the building. The vent pipe is the dry part of the system.

KEY POINT

Together, the two sections are referred to as the soil and vent pipe.

Before we look at sanitary systems, we must remember that all sanitary pipework and drainage systems need to comply with Approved Document H of the Building Regulations. These requirements will be met if the recommendations of **BS EN 12056:2000** – which contains recommendations for design, testing, installation, and maintenance for all above-ground non-pressure pipework systems – are followed.

To comply with Document H, all appliances must be fitted with a water trap seal to prevent foul air from entering the building. Also, the waste pipe diameter and gradient must maintain a water seal in the trap of at least 25 mm after the appliance has been used.

INDUSTRY TIP

Access Building Regulations 2010 Approved Document H at: www.gov.uk/government/uploads/system/uploads/attachment_data/file/442889/BR_PDF_AD_H_2015.pdf

Primary ventilated stack system

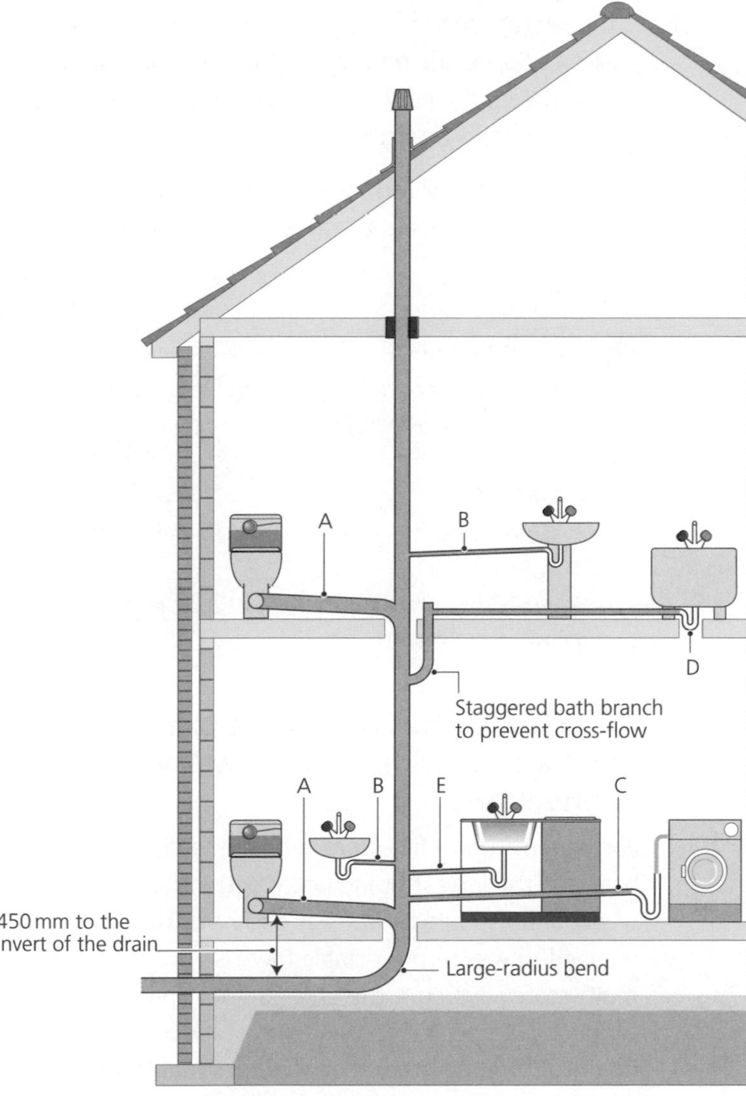

▲ Figure 9.1 The primary ventilated stack system

▼ Table 9.1 Branch and waste pipe sizes, gradients and trap seal depths

Appliance		Pipe size (mm)	Max. length (m)	Gradient (mm/m)	Trap seal depth (mm)
A	WC branch	75–100	6	18	50
B	Washbasin and bidet	32	1.7	18–22 (see Figure 9.2)	75
C	Washing machine/dishwasher	40	3	18–90	75
D	Bath	40	3	18–90	50
E	Kitchen/utility sink	40	3	18–90	75
Where these lengths are exceeded, then the next pipe size up should be used; 40 mm appliances will need to increase to 50 mm pipe, the length and gradient of which are listed below.					
Appliances with 50 mm waste pipe			4	18–90	75

The primary ventilated stack is probably the most common system installed in domestic dwellings. It relies on all the appliances being closely grouped around the stack and therefore does not need an extra ventilating stack like other systems. It is used in situations where the discharge stack is large enough to limit pressure fluctuations without the need for a separate ventilating stack.

Waste pipe sizes and lengths

Waste pipes need to fall away from the appliances with enough of a fall for the water to reach what is known as a 'self-cleansing velocity'. The fall is known as the gradient.

Table 9.1 shows the size of waste pipe for a given appliance installed on a primary ventilating stack, and its maximum length and gradient.

The rules regarding the gradient for washbasins are slightly different to those for other appliances. If the maximum length of 1.7 m is used, then the gradient is 18–22 mm/m. For shorter lengths than this, the gradient can increase and a gradient graph, like that shown in Figure 9.2, can be used to calculate the gradient needed.

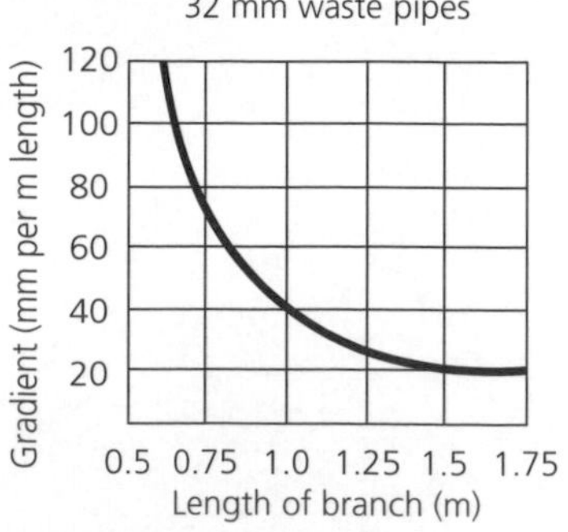

▲ Figure 9.2 Gradient graph

Reading the graph is a simple task. The horizontal line is the length of the waste pipe. The vertical line is the gradient. So, decide on the length, trace the line up until it meets the curve, then follow it across to the left side to read the gradient.

For example, if a 32 mm waste pipe is to be installed that is 1 m in length, then the gradient will be 40 mm/m.

ACTIVITY

The gradient curve

Try the gradient curve for yourself. Determine the answers to the following questions:

1 A waste pipe has a length of 1.5 m. What is its gradient?
2 A waste pipe has a length of 750 mm. What is its gradient?
3 A waste pipe has a gradient of 120 mm/m. What is its length?

Branches at the base of the primary ventilated stack system: low-level connections

For systems up to five storeys high, the distance between the lowest branch connections and the invert of the drain should be at least 750 mm. This can be reduced to 450 mm for single low-rise dwellings. For multi-storey systems, the ground floor appliances should be connected to their own stack or drain but not into the main stack. For buildings that have more than 20 storeys, the ground and first floors should be connected in this way.

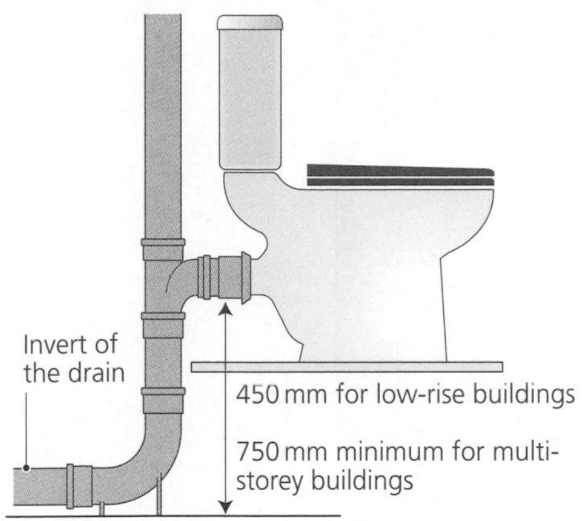

▲ Figure 9.3 Branch connections at the invert of the drain

Bends and off-sets

Bends at the base of discharge stacks should be large radius, the minimum radius being 200 mm. Two 45° bends can be used as an alternative. This ensures the smooth flow of water and solid waste into the drainage system. Tight bends can cause a problem called **compression**, where the water hitting the bend forces a shock wave of air upwards, which can blow the water out of waste pipe traps, causing them to lose their seal and let obnoxious smells into the dwelling.

> **KEY TERM**
>
> **Compression:** the process of water hitting a bend at forces that cause a shock wave of air upwards.

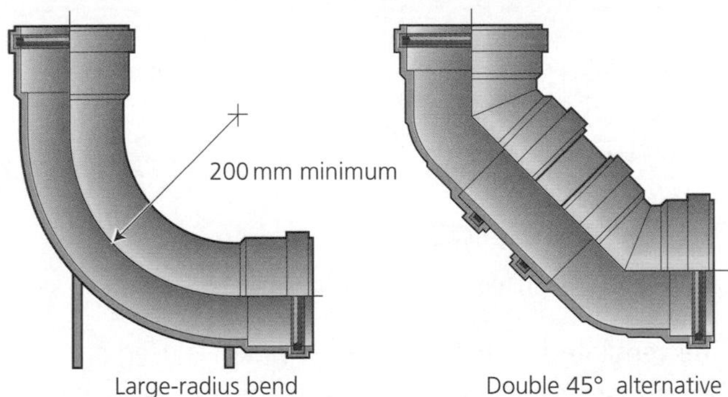

▲ Figure 9.4 Large-radius bends at the base of the stack

Off-sets in the wet part of the stack should be avoided if possible. Where there is no option, again large-radius bends should be used, with no branch connections within 750 mm of the off-set. If an off-set is to be placed in the wet part of a soil stack, in a building of up to five storeys, then the stack must be ventilated both above and below the off-set.

Branch connections for waste pipes

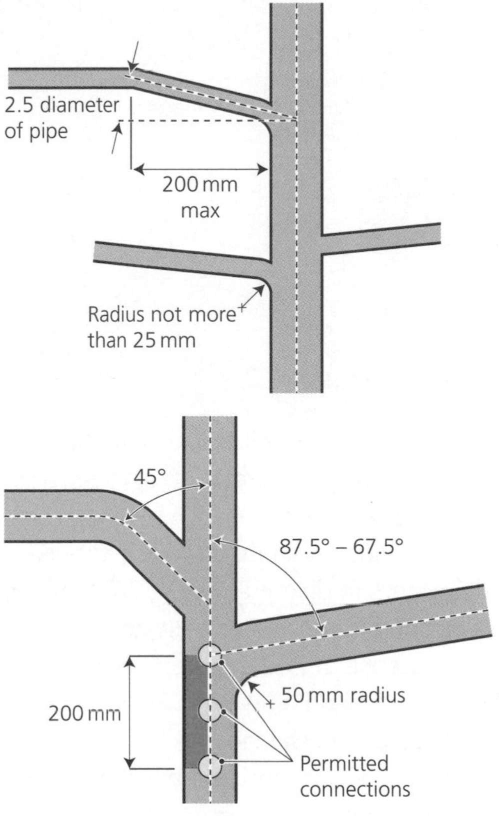

▲ Figure 9.5 Branch connections

In Figure 9.5, the upper drawing shows that junctions, including branch pipe connections of less than 75 mm, should be made at a 45° angle or with a 25 mm bend radius. The prohibited zone in the lower drawing shows the area (opposite the WC connection) in which a branch pipe may not be connected to a distance of 200 mm. Branch connection pipes of over 75 mm diameter must either connect to the stack at a 45° angle or with a minimum bend radius of 50 mm.

Prevention of cross-flow

A branch pipe should not discharge into a stack in such a way that it could cause cross-flow into any other branch pipe. This can cause loss of trap seal by effluent back-flowing up the opposite connection.

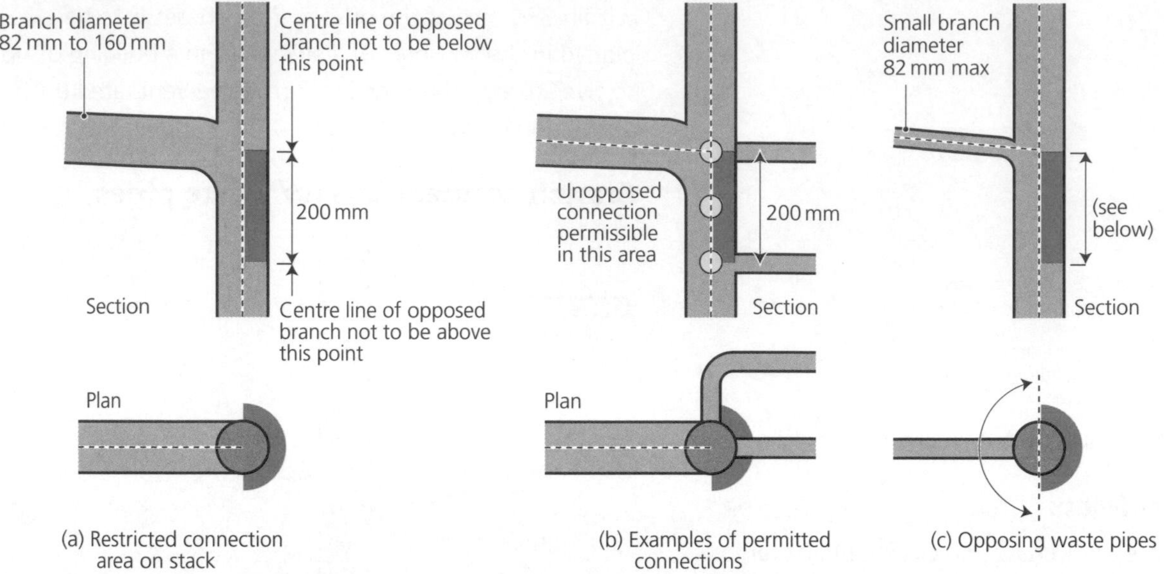

▲ Figure 9.6 Preventing cross-flow

Figure 9.6 shows the areas of a soil stack where branch connections directly opposite are restricted. In general, there are several rules, as follows.

- Where a branch connection into a stack is between 82 mm and 160 mm in diameter (e.g. a WC branch), no other connection is allowed to be installed opposite for a distance of 200 mm vertically downwards. The left-hand drawing in Figure 9.6 shows the restricted area shaded.
- The middle drawing shows that side connections at 90° to the branch are allowed.
- Where the branches are of similar size – say, two 40 mm connections – then the restricted distance will depend upon the size of the main stack:
  - on a stack up to 82 mm in diameter, no connection is allowed for a distance of 90 mm
  - on a stack up to 110 mm in diameter, no connection is allowed for a distance of 110 mm
  - on a stack up to 160 mm in diameter, no connection is allowed for a distance of 250 mm.

Where it is not possible to meet the requirements of the primary ventilated stack (e.g. excessive waste pipe lengths), then extra ventilation to the system will need to be added to safeguard the trap seal. This can be done by installing either of the following:

- a ventilated branch discharge system, where each waste pipe branch is separately ventilated
- a secondary ventilated stack system, where the waste stack is directly ventilated.

These are described below.

The ventilated branch discharge system

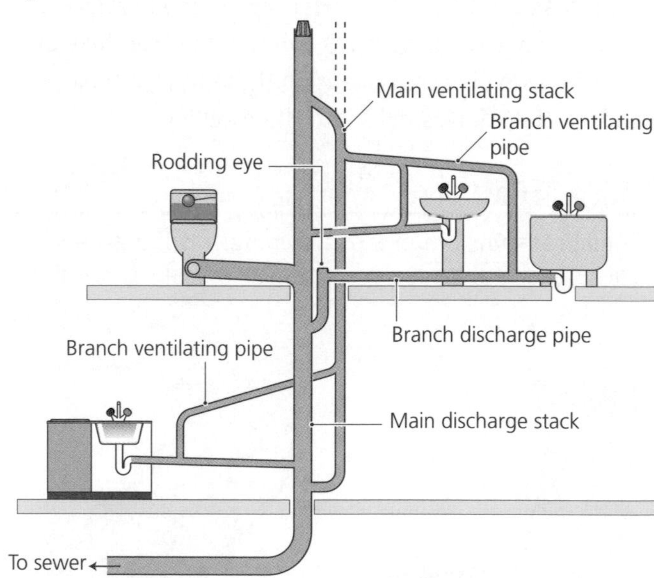

▲ Figure 9.7 Ventilated branch discharge system

The ventilated branch discharge system is used on larger systems where there is a risk of trap seal loss because the waste pipe lengths are excessive. Control of the pressure in the waste pipe (the discharge branch) is achieved by ventilating it no further than 750 mm from the appliance as this safeguards against trap seal loss by induced or self-siphonage. Alternatively, small air admittance valves may be used at each appliance. These allow air into the system when the appliance is in operation.

Secondary ventilated stack system

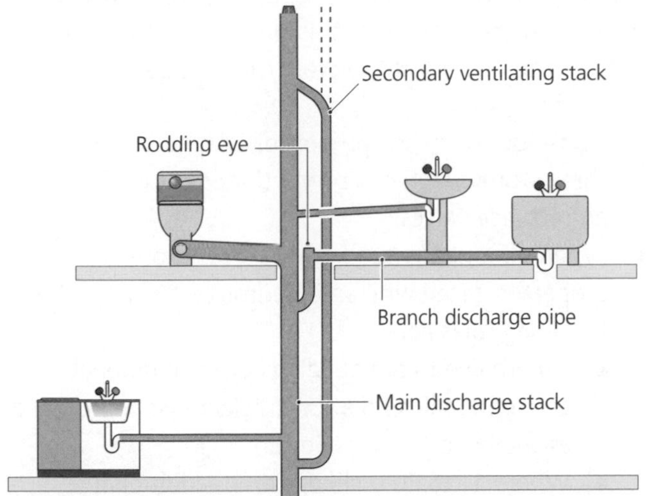

▲ Figure 9.8 Secondary ventilated stack system

With a secondary ventilated stack system, only the main discharge stack is ventilated. This system arrangement safeguards against positive and negative pressure fluctuations.

The rules regarding branch ventilating pipes

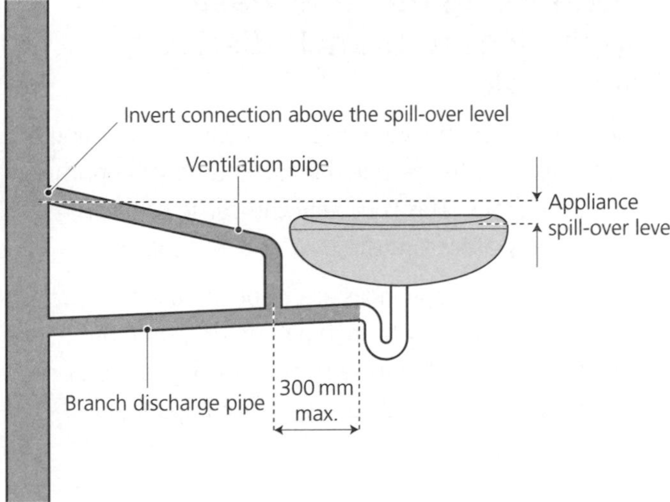

▲ Figure 9.9 Branch ventilating pipe rules

Where branch ventilating pipes must be installed, the following rules apply.

- Any branch ventilating pipe must be connected to the discharge stack above the spill-over level of the highest appliance fitted to the stack. The ventilating pipe must also rise away from the appliance.
- The minimum size of any ventilating pipe to a single appliance is 25 mm. However, if it is longer than 15 m, or the ventilating pipe serves more than one appliance, then the size must be 32 mm.
- The main ventilation stack must be a minimum of 75 mm. This also applies to the dry part of the primary ventilating stack.

Stub stack system: low-level WC connections to the drain

When a group of appliances are connected direct to the drain, under certain circumstances a 110 mm stub stack may be used. Figure 9.10 shows a typical ground-floor stub stack. Ventilation is required when the connection from the invert of the drain to the highest connection of an appliance to the stack exceeds 2 m, or the WC crown connection to the invert of the drain exceeds 1.3 m. Ventilation of a stub stack is via an air admittance valve.

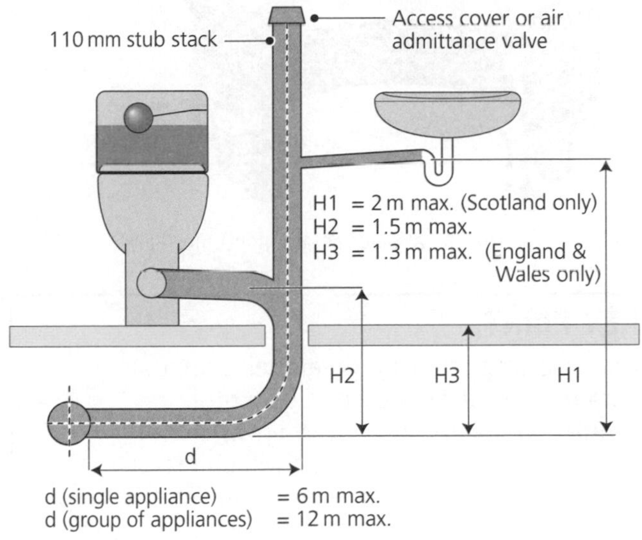

▲ Figure 9.10 The stub stack

Air admittance valves

An air admittance valve allows air into a stub stack to prevent the loss of trap seals. The subsequent suction action, when an appliance is used, opens the valve. This stabilises the air condition in the stack because air is sucked into the stack through the valve, also preventing smells and foul air escaping out. When the appliance has finished its operation, the valve closes, preventing smells escaping into the space where the valve is installed.

Air admittance valves should be fitted in an uninhabited space such as a roof space. This minimises the risk of freezing while keeping the valve accessible.

On no account should they be fitted outside because of the risk of the valves freezing up in the closed position during cold weather. If air admittance valves are installed within a boxing, the boxing must be ventilated. In all cases, the valve must be accessible for repair or replacement.

▲ Figure 9.11 The operation of an air admittance valve

KEY POINT

An important point to remember is that air admittance valves are not a substitute for ventilation stacks and any drain where an air admittance valve is fitted will still require conventional venting at some point. This is simply to minimise the effects of back pressure, which could occur if the underground drainage system becomes blocked.

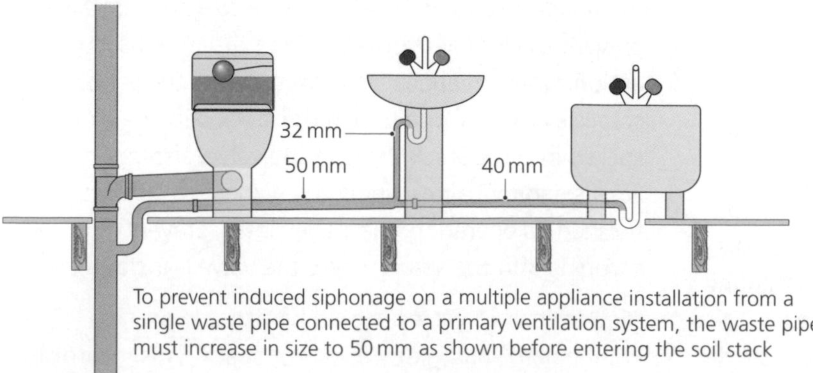
32 mm
50 mm
40 mm
To prevent induced siphonage on a multiple appliance installation from a single waste pipe connected to a primary ventilation system, the waste pipe must increase in size to 50 mm as shown before entering the soil stack

▲ Figure 9.12 Multiple appliance installations

The requirements are that one stack in five must be ventilated to the outside air using a conventional ventilation stack, and that this should usually be done at the head or start of the drain run. The general rules are as follows:

- Up to four domestic properties of no more than three storeys high can be ventilated using air admittance valves.
- Where an underground drain serves more than four properties fitted with an air admittance valve, the following rules apply.
  - Where five to ten buildings exist, additional conventional ventilation stacks must be installed at the head of the drain run.
  - Where 11 to 20 buildings exist, additional conventional ventilation stacks must be installed at the head of the drain and at the mid-point in the run of the drain.
  - All multi-storey domestic properties will require additional conventional ventilation if more than one property is fitted with an air admittance valve and is connected to a common drain that is not ventilated by a conventional ventilation stack.

Connecting multiple waste appliances to branch discharge pipework

The connection of two or more appliances on a single waste pipe is often installed incorrectly on the primary ventilated stack. This is usually the cause of baths pulling the water from the trap of a washbasin.

Where multiple appliances are to be installed, then the use of ventilating branch pipework should be considered to avoid trap seal loss (see the section on the ventilated branch discharge system, page 456).

General sanitary pipework requirements

As well as the requirements we have already looked at, sanitary pipework systems should follow the general rules listed below.

- Where a ventilation stack is installed within 3 m of an opening window, the stack should be installed at least 900 mm above the window.
- A cage should be fitted to the top of the vent pipe to prevent birds nesting at the top of the stack. Birds' nests have the effect of blocking off the air supply to the stack, causing waste pipes to lose their trap seal.
- A vent cowl should be fitted in exposed or windy positions to prevent 'wavering out', where the wind blowing across the top of the stack causes the trap water to move from side to side, potentially resulting in trap seal loss by the momentum of the water.
- Access should be provided above the spill-over level of the highest appliance, to allow for clearing blockages.
- When installing a soil stack for waste pipes only, the size of the stack must be at least the same size as the largest trap or branch connection to it.

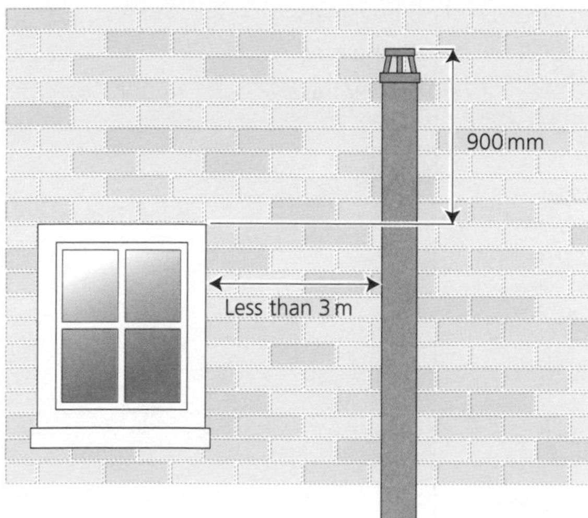

▲ Figure 9.13 The position of a vent stack next to an opening window

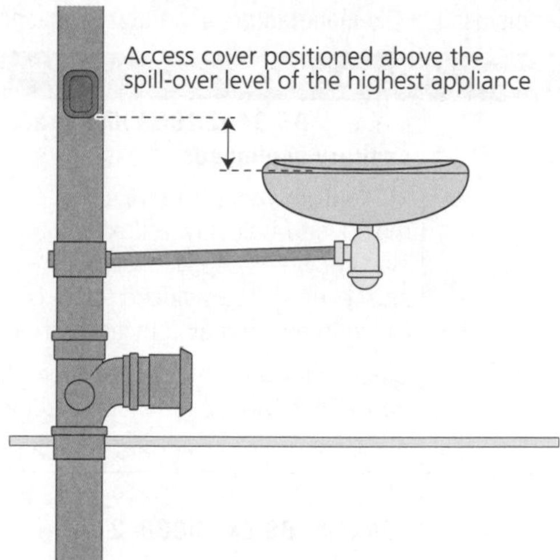

▲ Figure 9.14 The position of access

- Waste pipes may discharge over a gulley provided that:
  - the gulley is capable of accepting discharge from a waste pipe and is not connected to a rainwater drain
  - the waste pipe discharges below the gulley grate but above the water level in the trap
  - appliances connected to the gulley may use a trap with a 38 mm trap seal.

Sanitary appliances

There are two purposes of sanitary appliances: to maintain personal hygiene by washing, bathing or showering, and the removal of solid and fluid human waste. In this part of the chapter we will look at the types of sanitary appliances used in dwellings and their working principles, including:

- materials used for sanitary appliances
- conventional WCs
- washbasins
- bidets
- baths
- shower trays and cubicles
- sinks
- urinals.

Materials used for sanitary appliances

The materials used in the manufacture of sanitary appliances are listed in Table 9.2. They must be robust, hygienic and easy to clean.

▼ Table 9.2 Materials used in the manufacture of sanitary appliances

Material	Description	Appliance
Vitreous china	Made to **BS 3402:1969 High grade ceramic ware used for sanitary appliances**. Made from white burning clays and finely grained material mixed with ball clay, a fluxing agent and water, into casting clay known as slip. The slip is fired to a high temperature and, even in its unglazed state, cannot be contaminated by bacteria and remains hygienic in all situations. Glazed vitreous china is stain-proof, burn-proof, rot-proof and non-fading, and is resistant to acids and alkalis. Available in many colours and shades.	WC pans and cisterns Washbasins Bidets Urinals
Stainless steel	Made from 304- or 316-grade stainless steel to European Standard **BS EN 10088–2**. Usually fitted where the general public has access and highly resistant to vandalism. All stainless steel sanitary ware conforms to the Department of Health specification.	WCs and cisterns Washbasins Kitchen sinks Urinals
Fireclay	Made from buff-coloured ball clays from Devon and Dorset in the UK. Fireclay is very robust to withstand rough treatment but, unlike vitreous china, it is porous. Because of this, it requires 'firing' with a ceramic undercoat to seal the clay before being coated with two coats of white glaze and then re-fired.	Belfast sinks London sinks Butler's sinks Urinals Heavy-duty WC pans and washbasins for hospitals
High-impact plastic	Usually manufactured by injection moulding techniques.	WC seats WC cisterns Bath panels
Acrylic	Comes in varying thicknesses, between 3 mm and 8 mm. Heated until it becomes soft and pliable, and then placed over an aluminium mould, where it is sucked into place (known as vacuum forming). Warm to the touch. Can be moulded into many shapes. However, easily damaged by scratching and abrasive cleaners. Acrylic baths are often strengthened by a base board made from chipboard and glass-reinforced polyester (GRP). Very lightweight; appliances are usually aimed at the domestic market. Acrylic baths require a supporting cradle.	Baths Bath panels Washbasins Shower trays
Enamelled cast iron	Extremely robust but is very heavy and very cold to the touch. Because of the nature of cast iron, bath designs tend to be very traditional.	Baths
Porcelain enamelled pressed steel	The steel sheet used in the manufacture of sanitary ware must be of the highest grade low-carbon steel. The enamel is sprayed on and then kiln fired. It is rigid but light, very robust but the enamel is easily damaged.	Baths Washbasins

Conventional WCs

INDUSTRY TIP

Smaller versions of WC pans are available for use in infant schools.

WC stands for water closet. It consists of a WC pan and a flushing cistern. There are different types of WC pan, as follows:

- **The wash down type:** the most common type of WC fitted in the UK. The pan is cleared by a carefully designed water distribution system, which uses the force of the water flush and volume of water delivered to the bowl to clear the contents. Wash down-type WC pans are usually around 400 mm high, depending on the manufacturer, and have 50 mm of water seal in the trap. The bowl is shaped to provide efficient effluent clearance while maintaining easy cleaning.

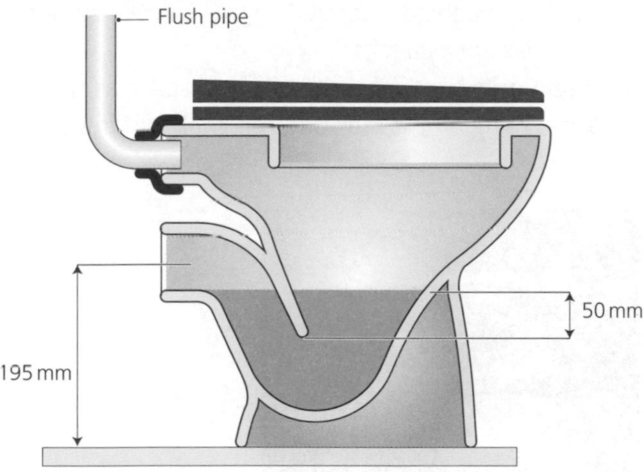

▲ Figure 9.15 A wash down WC pan

- **The siphonic type:** the flushing operation creates a vacuum, which contributes to clearing the pan. There are two pan types:
  1. **The single trap siphonic WC pan (or 'Malvern type'):** this pan has a lower outlet than other pan designs. It is usually installed only on replacements as the design tends to look very outdated. They work by restricting the flow of water from the cistern, which allows a build-up of water in the pan, which is then forced through the restricted neck of the trap creating a vacuum behind it and clearing the pan contents completely.

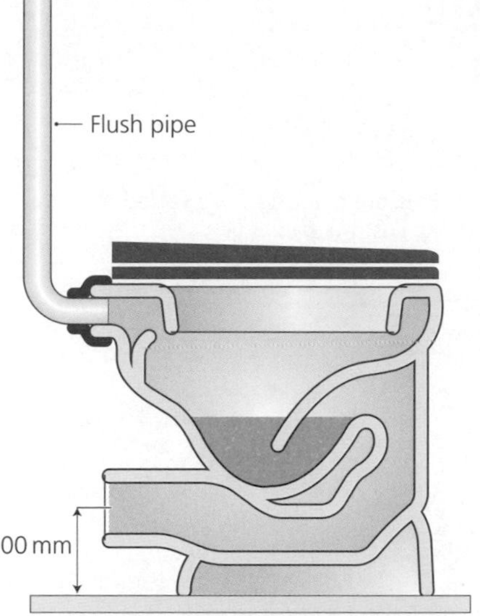

▲ Figure 9.16 A single trap siphonic WC pan

  2. **The double trap siphonic WC pan:** very rarely sold in the UK since the flushing volume of WC cisterns was reduced to six litres by the Water Supply (Water Fittings) Regulations 1999. This kind of WC pan is very quiet and extremely efficient at removing the pan contents. Unlike the single trap siphonic pan, the double trap siphonic has an unrestricted outlet and two water traps. A special pressure reducing valve, called an aspirator (or bomb), is fitted to the bottom of the siphon.
     When the cistern is flushed, a negative pressure is caused in the chamber between the two traps by the aspirator. The aspirator follows Bernoulli's principle (see page 317). It sucks out the air from the chamber as the water from the flush passes through it, which causes the contents of the bowl to be sucked through the two traps. The aspirator holds a little water back to refill the second trap after the flush is complete.
     Double trap siphonic WCs tend to be longer than wash down types because of the extra water trap.

INDUSTRY TIP

Access the Water Supply (Water Fittings) Regulations 1999 at: www.legislation.gov.uk/uksi/1999/1148/contents/made

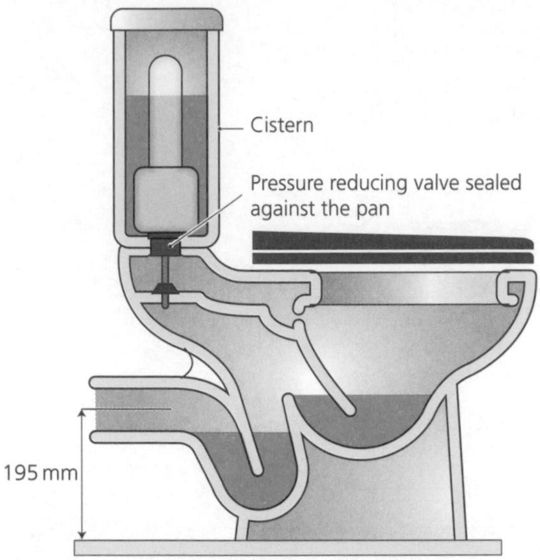

▲ Figure 9.17 A double trap siphonic WC pan

WC styles

WCs can be manufactured in five main styles, as described below.

1 **Close coupled:** the WC pan is designed to have the cistern bolted to the back of the pan to form one unit.
2 **Low level:** the cistern is connected to the WC pan by a short flush pipe to convey the water from the cistern to the WC pan.
3 **High level:** similar to the low level but the flush pipe is much longer and the cistern is at high level. Usually used when designing period bathroom suites.

▲ Figure 9.18 A close coupled WC suite

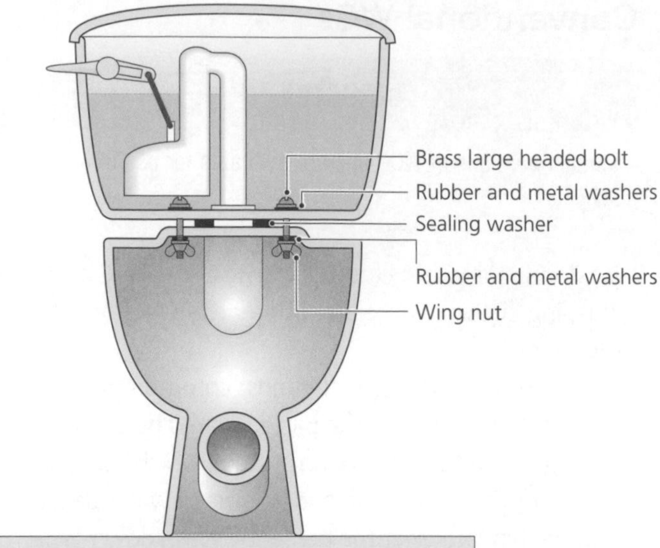

▲ Figure 9.19 A diagram showing how the cistern is fixed to the WC pan

4 **Back to wall/concealed:** becoming more popular due to the fact that the cistern is concealed in a cabinet or behind a panel. The WC pan sits close to the cabinet or panel.
5 **Wall hung:** these give the effect of space as the WC pan is hung on the wall and is completely free of the floor.

▲ Figure 9.20 A back to wall WC suite

▲ Figure 9.21 A modern wall-mounted WC pan

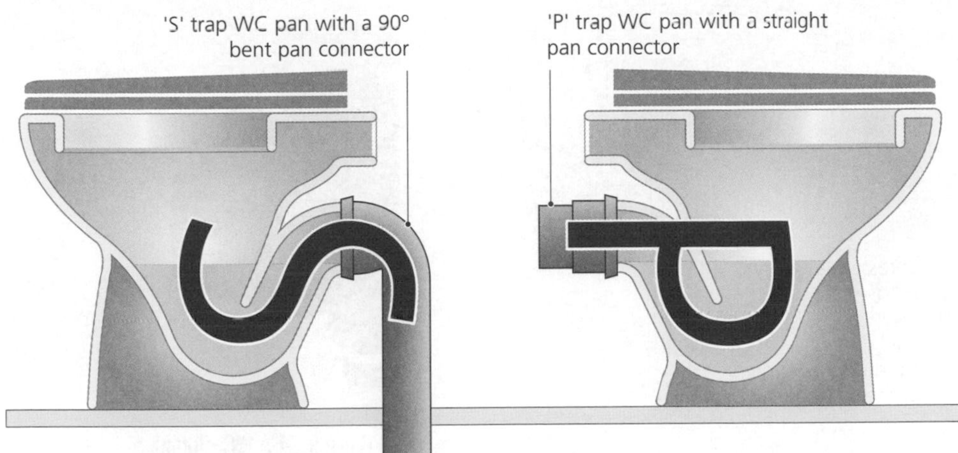

▲ Figure 9.22 'P' trap and 'S' trap WC suite

In the past, WC pans were manufactured with a variety of 'P' trap and 'S' trap configurations formed as part of the pan casting, but this proved expensive. Today, most WC pans are manufactured with the 'P' trap configuration. However, with the use of an angled WC pan connector, they can be made into an 'S' trap or left or right outlet depending on the installation requirements.

The WC cistern

Prior to 1986 regulations, the flush volume was 9 litres. This was lowered in the Model Water Bylaws of 1986 to 7.5 litres. The WC cistern is the method by which the water is discharged into the WC pan. Today, the Water Supply (Water Fittings) Regulations 1999 restrict the flushing volumes of new WC cisterns to 6 litres for a long flush and 4 litres for a short flush. The water can be delivered to the WC pan in several different ways, depending on the cistern design:

- **By the use of a siphon:** the traditional way to flush a WC cistern. The cistern is flushed using siphonic action (see Chapter 3, Scientific principles, page 153). The WC flushing handle is connected to the siphon by a link pin. When the WC cistern handle is depressed, the link pin lifts a plunger in the siphon bell, which has a large thin plastic or thin rubber diaphragm at the end of it. The diaphragm lifts a column of water up and over the top of the siphon to begin the siphonic action.

  There are many different styles and sizes of WC siphon available and the correct one must be chosen depending on the cistern size. Some siphons allow different flushing volumes to be set by adjusting the height at which air is let into the siphon bell to stop the siphonic action.
- **By the use of a dual flush valve (also known as a drop valve):** these can be operated by pressing a button on the top of the WC cistern, or remotely by air, which is blown through a tube when the button is depressed. They work by simply opening up a valve when the button is activated and this allows water to flow by gravity to the cistern. Siphonic action is not needed. Flush valves have a 6-litre and 4-litre flush action. Flush valves have an integrated overflow that allows water to flow straight to the WC pan should the float-operated valve begin to overflow, so a separate overflow pipe is not required.

INDUSTRY TIP

Older WC pans will not flush with such a low water volume, so 9- and 7.5-litre cisterns are still available for the replacement market.

ACTIVITY

Refresh your knowledge of service valves and float-operated valves; these were covered in detail in Chapter 5, Cold water systems.

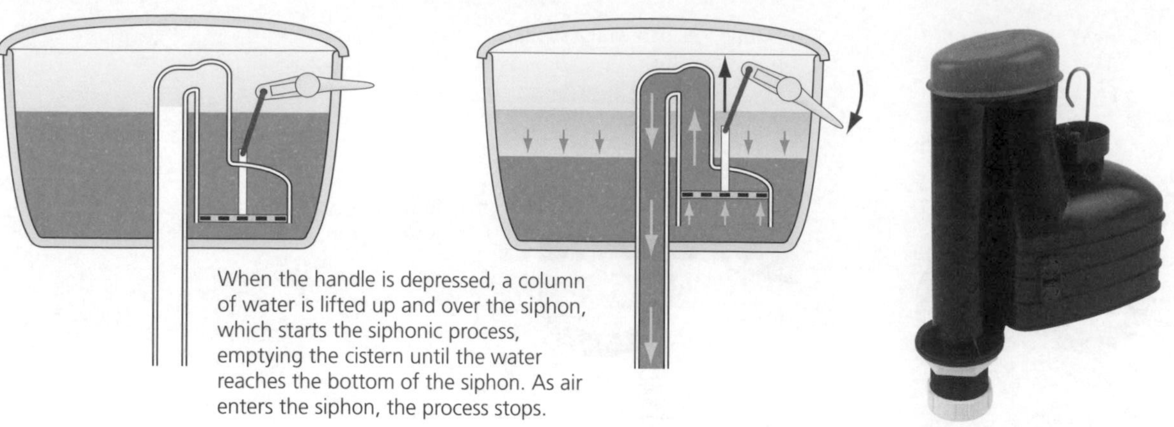

When the handle is depressed, a column
of water is lifted up and over the siphon,
which starts the siphonic process,
emptying the cistern until the water
reaches the bottom of the siphon. As air
enters the siphon, the process stops.

▲ Figure 9.23 How a WC siphon works

▲ Figure 9.24 A WC siphon

▲ Figure 9.25 A dual flush valve

▲ Figure 9.26 A flapper valve

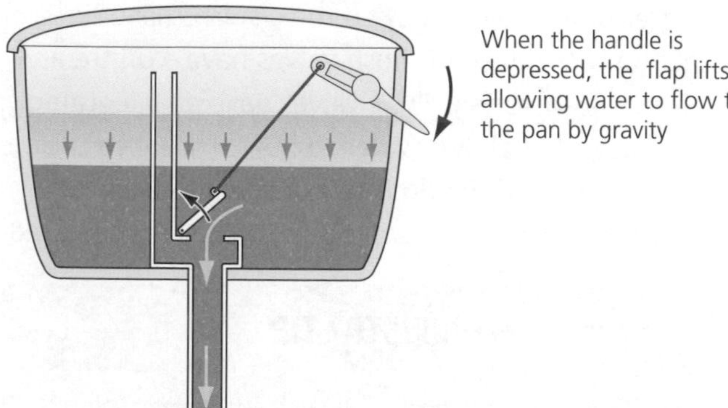

▲ Figure 9.27 The operation of a flapper valve

- **By the use of a flapper valve:** a very simple valve that allows water to flow by gravity to the cistern. In the closed position, it is the weight of the water that makes a watertight seal. When the WC handle is depressed, a link pin simply lifts the valve up. These are not dual flow and will flush only as long as the handle is pressed down. Most flap valves have an integral overflow.

The water in the cistern is controlled using a float-operated valve conforming to **BS 1212 Parts 2, 3 and 4**. The cistern must also have a service valve fitted as close to the cistern as possible. A separate overflow must be installed with WC cisterns not having an integral overflow, and this must discharge safely in a conspicuous position, usually outside the building.

INDUSTRY TIP

Float-operated valves and service valves are covered in detail in Chapter 2, Common processes and techniques, and Chapter 5, Cold water systems.

WC cisterns can be made from a variety of materials, including vitreous china, plastic and hard rubber, but other materials such as cast iron and lead-lined wood have also been used in the past.

Washbasins

There is a huge variety of different styles of wash hand basin and many of these also come in various sizes and tap arrangements. Corner washbasins are also available. Washbasins should be installed approximately 800 mm from the floor to the front lip of the basin. Washbasins can be divided into three basic types:

1 **Wall-hung washbasins:** this type of washbasin is mounted on wall-fixed brackets or bolted directly to the wall. There are several different types of mounting bracket, including towel rail type or concealed, depending on the washbasin style. The mounting wall must be able to take the weight of the washbasin. If there is any doubt, either a centre leg or a pair of legs should be used.

▲ Figure 9.28 A wall-hung washbasin

2 **Pedestal washbasins:** there are two different types of these:
- **pedestal washbasins** are fixed to the wall but rely on the pedestal for their main support; the pedestal is designed to hide the pipework
- **semi-pedestal washbasins** are becoming increasingly popular; the pedestal does not carry the weight of the basin as it does not reach the floor, and is designed to hide the associated pipework.

▲ Figure 9.29 A pedestal washbasin

▲ Figure 9.30 A semi-pedestal washbasin

3 **Countertop washbasins:** there are several different types of countertop basins:
- **countertop style washbasins** are also known as inset washbasins; they sit snugly into a worktop surface
- **semi-countertop style**, also known as the semi-recessed basin, this basin style sits half on and half off a work surface
- **under-countertop style**, as its name suggests, is mounted under a work surface; the work surface is usually marble, agglomerate marble or granite
- **vessel washbasins** are designed to be supported by a mounting surface such as a worktop or cabinet.

▲ Figure 9.31 A countertop washbasin

▲ Figure 9.32 A semi-countertop washbasin

▲ Figure 9.33 An under-countertop washbasin

▲ Figure 9.34 A typical vessel washbasin

Washbasins can be made from a variety of materials, including vitreous china, stainless steel and porcelain enamelled pressed steel (refer back to Table 9.2).

Tap hole and waste arrangements for washbasins

There are four main tap hole arrangements for washbasins; these are shown in Table 9.3.

▼ Table 9.3 The main tap hole arrangements for washbasins

One tap hole basin with monobloc mixer tap 	Specifically designed for use with a monobloc mixer tap.
Two tap hole basin with hot and cold taps 	The traditional tap hole arrangement, for use with hot and cold ½-inch BSP pillar taps.
Three tap hole basin with remote mixer tap 	This is a little used tap arrangement where the tap bodies are fitted below the basin with just the wheel heads showing. The spout and the tap bodies are connected secretly below the washbasin.
No tap hole basin with wall-mounted taps 	Becoming more popular for bespoke bathrooms. These use wall-mounted bib taps with concealed pipework.

Washbasins are manufactured with an integral overflow for use with a 1¼-inch slotted waste for connection to a 32 mm waste trap. There are two basic waste types available, as follows.

1 **Slotted waste, plug and chain:** the old-fashioned method of providing a waste stopper. The slots in the waste are to allow water that has flowed down the integral overflow to find its way safely down to the trap. These are usually 'made in' to the basin with silicone sealant, with a plastic poly-washer inserted between the securing nut and the basin. Care should be taken when using gold-plated fittings and silicone sealant as some sealants can discolour the gold plating.
2 **Pop-up waste:** these provide a handle, typically designed as part of the tap, which, when pushed down, pops the waste plug up. They tend to have specific sealing washers to seal the waste into the basin.

▲ Figure 9.35 Waste, plug and chain arrangement

▲ Figure 9.36 Pop-up waste arrangement

Bidets

Very similar in design to a WC pan, the bidet is often called a 'sit-on washbasin'. It is a hygienic method of ensuring personal cleansing, especially after using the WC. It is often also used as a footbath. There are two distinctly separate types, as described below.

1 **Over-rim bidet:** the over-rim type is the most common bidet. It is installed in the same way we would install a washbasin. It is available with one or two tap holes, depending on the bidet design, and can be fitted with a variety of taps, including monobloc mixers, pillar taps and hand spray-type mixers with a hose connection.

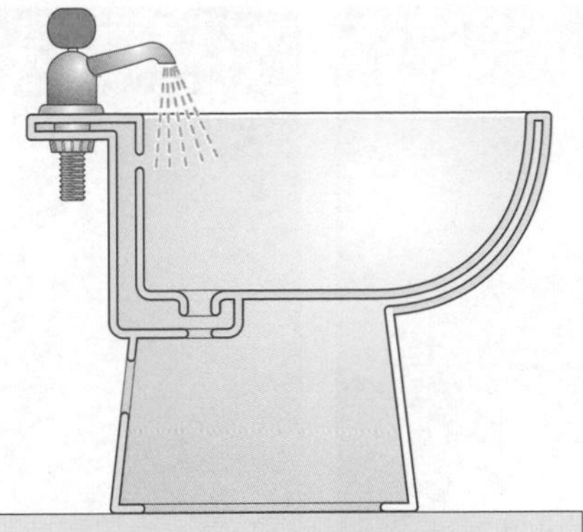

▲ Figure 9.37 The over-rim bidet

2 **Ascending spray bidet:** very rarely seen in the UK, the ascending spray bidet uses a special tap arrangement to discharge water upwards from inside the bowl of the bidet in a spray similar to a small showerhead.

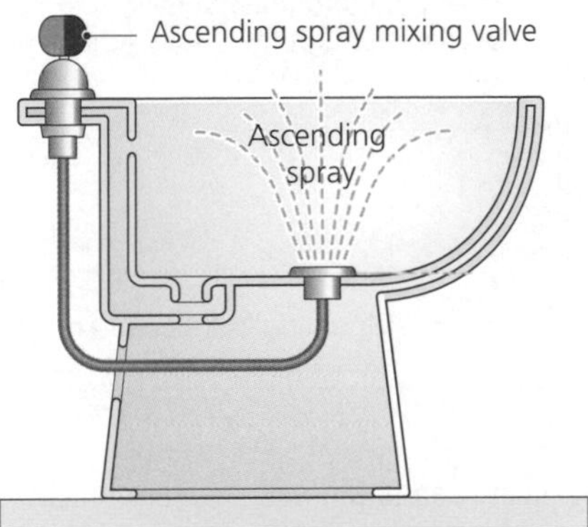

▲ Figure 9.38 The ascending spray bidet

HEALTH AND SAFETY

Special installation arrangements exist for the ascending spray bidet because of the risk of contamination of water by backflow through the spray head. It must not be installed on mains pressure systems, and the Water Regulations should be consulted for all installations of this type of appliance.

INDUSTRY TIP

The installation of ascending spray bidets will be covered in later phases of your qualification.

Bidets are usually made from vitreous china. Styles include floor mounted, back-to-wall and wall-hung types (Figures 9.39–9.41).

▲ Figure 9.39 Floor-mounted bidet

▲ Figure 9.40 Back-to-wall bidet

▲ Figure 9.41 Wall-hung bidet

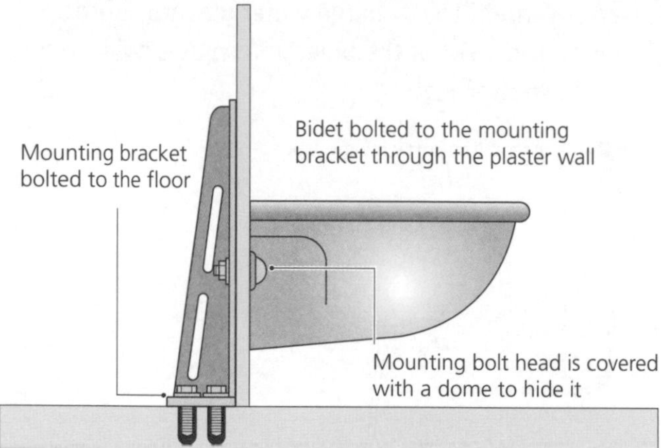
Mounting bracket bolted to the floor
Bidet bolted to the mounting bracket through the plaster wall
Mounting bolt head is covered with a dome to hide it

▲ Figure 9.42 Wall-hung bidet fixed to the wall mounting bracket

Baths

Baths are manufactured to **BS EN 198:2008** and can be supplied manufactured from the following materials (refer back to Table 9.2).

- **Reinforced cast acrylic sheet:** this is the most common material for baths. Some acrylic baths require reinforcement in the form of glass-reinforced polyester (GRP), and all types require a steel tubular cradle, a top frame and a base board.
- **Porcelain enamelled steel:** these tend to be used in commercial situations such as hotels, hostels, etc., or in housing association and local authority housing, where durability is important.
- **Porcelain enamelled cast iron:** these are much less common and tend to be used for the more traditional designs, such as roll top freestanding and rectangular shapes.

INDUSTRY TIP

In Europe, pressed steel baths are more common than acrylic.

Other materials are also available, such as:

- a resin-bonded cementitious slurry, known as resin stone
- gel-coated reinforced polyester but this is generally used in specialist and niche markets.

Each material has its own unique characteristics that influence the bath design. Baths can be manufactured in a wide variety of styles and designs, including those described below.

- **Standard baths:** rectangular shaped with many size and design options. They are usually fitted with a front panel and/or end panels as required.
- **Corner baths:** these fit into the corner of the bathroom. They require a curved bath panel, which is easily cut and trimmed to specific installation requirements.
- **Off-set corner baths:** similar to a standard corner bath but they have sides of unequal length. This design utilises the space available while optimising the bathing space. They are available left or right handed, depending on the installation requirements.
- **Freestanding baths:** these are designed to stand on their own feet and are not usually fitted against a supporting wall. A range of styles are available, from traditional roll top and ball-and-claw styles to more contemporary designs.

▲ Figure 9.43 Off-set corner bath

▲ Figure 9.44 Freestanding ball-and-claw feet bath

▲ Figure 9.45 Double-ended bath

▲ Figure 9.46 Shower bath

- **Double-ended baths:** usually rectangular in shape but they have two ends and are designed with two people in mind.
- **Tapered baths:** designed for situations where space is limited. They are wider at one end than the other, and are usually fitted with a shower at the wider end.
- **Shower/baths:** again, usually bulged at one end to maximise the space available for showering.

The most common sizes range from 1600 mm to 1800 mm in length and 700 mm to 800 mm wide. The most popular shapes require a front bath panel and, very often, end panels to hide the frame, the cradle and the plumbing.

Tap hole and waste arrangements for baths

Tap holes for baths come supplied in three ways:

1. no tap hole – this type of bath needs to be drilled so the taps can be installed in the position of the customer's choice
2. two tap hole – the standard arrangement; they can be either on the end of the bath or the side
3. three tap hole – for remote-type taps (one hole for the spout and two holes for the taps).

Waste connections for baths can be made by the following methods.

- **'Banjo'-type bath waste fitting:** this uses a long threaded waste fitting with slots on opposite sides near the top. The waste from the overflow comes via a flexible pipe connected by a 'banjo' connection. This is assembled over the waste fitting and is held in place on the underside of the bath by a large 1½-inch BSP nut fitted to the bottom of the waste fitting and tightened against the banjo. Silicone sealant should be used at the joint between the bath and the banjo, and the banjo and the nut. This type of bath waste connection is very prone to leakage.

▲ Figure 9.47 The 'banjo'-type bath waste fitting

- **Bath waste and overflow kit:** where there is sufficient space underneath the bath, this is the easiest bath waste connection to fit. It uses a one-piece bath waste connection, which is held in position by a long bolt placed through the centre of the bath waste grille. Both the waste connector and the grille have sealing washers. The bath waste connection incorporates the overflow connection. The bolt pulls the waste connector and the bath waste grille together, and this compresses the washers to make a watertight seal.
- **Bath pop-up waste and overflow fitting:** bath pop-up waste systems are becoming increasingly popular. They are fitted in the same way as a bath waste and overflow kit, but feature a 'twist action' chrome or gold plate overflow, which operates a lever to raise or lower the bath waste plug.
- **Combined waste and trap:** this is a fitting that combines the bath waste and overflow with the bath trap.

▲ Figure 9.48 The bath waste and overflow fitting

▲ Figure 9.49 The bath waste and trap combination fitting

Whirlpool and air spa baths

The whirlpool and air spa bath is a recent addition to the bath range.

INDUSTRY TIP

The idea of the whirlpool bath to relax the body is not a new one. It was used by the Ancient Romans. In the 1950s, the idea resurfaced when the Jacuzzi brothers developed the whirlpool bath for domestic use.

Whirlpool and air spa baths are considered luxury fittings and can take many forms, such as jetted baths, hydro-pools, hydro-spa and air spa types, all of which use the pumping of air and water through nozzles installed into the side or floor of the bath. They can also be retro-fitted to any acrylic or pressed steel bath. The pump is usually situated at one end of the bath.

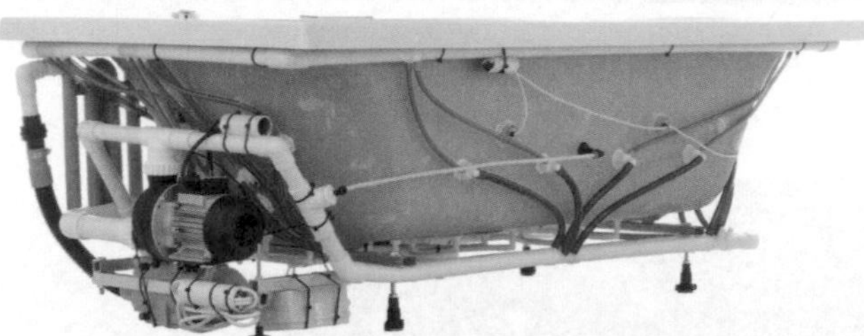

▲ Figure 9.50 The workings of a whirlpool bath

All whirlpool baths require regular cleaning to remove any build-up of soap and other impurities. Circulation cleansers should be run through the system every month to six weeks depending on use. Additionally, sanitiser tablets can be used after each bath to sanitise the system ready for the next user (particularly important in hotels and guesthouses).

All baths of this type incorporate a safety cut-out to suspend the pump or suction if anything blocks the water suction pipe.

Shower trays and cubicles

Shower trays (also known as shower bases) vary in size from compact square shapes to large rectangular, quadrant and five-sided models. They are made from a variety of materials, such as heavy-duty reinforced acrylic sheet, fireclay and resin bonded. The choice of shower tray depends largely on the space and budget available.

Many shower trays have a raised lip that, when placed against the wall, allows tiles to be placed over it to help the sealing of the tray. Some trays have adjustable feet to assist in levelling the tray. Resin-bonded and fireclay trays are bedded on a weak bed of sand and cement.

▲ Figure 9.51 Square shower tray

▲ Figure 9.52 Rectangular shower tray

▲ Figure 9.53 Five-sided shower tray

Waste arrangements for shower trays

The most common waste arrangement for a shower tray is by use of a combined shower waste and trap. Most modern trays are bedded to the floor and, because of the position of the waste on the tray, the trap is often inaccessible, making cleaning and clearing of blockages almost impossible. The combined shower waste and trap allows the trap to be cleaned of potential blockages, such as hair, from the top of the waste on the shower tray. The inside of the trap is removable from above.

Shower cubicles and enclosures

Shower enclosures are available in three distinct forms, as described below.

1 **The freestanding shower cubicle:** as the name suggests, a freestanding shower cubicle is one that does not use any of the walls of the building in its construction. However, the cubicle may be fixed to the wall for support.
2 **The shower enclosure that uses one or more walls:** where a shower enclosure uses either one or two walls to form part of the showering area. This is the most common of all shower enclosures installed.
3 **The shower door:** a single shower door is fixed between two opposing walls. This uses three walls of the building to form the enclosure.

Sinks

Sinks are appliances typically fitted in a kitchen or utility room. The ideal sink has to be hardwearing and robust enough to be able to withstand the abuse it is likely to receive. There are several different types, as described here.

- **Kitchen sinks:** these come in a variety of different shapes and sizes. Common arrangements are single bowl and drainer, bowl and a half, and double bowl. They are usually set into the work surface and can be made from a variety of materials, such as stainless steel, granite, astro-cast and polycarbonate materials. Vitreous china sinks are also available, but these tend to chip easily and will shatter if heavy pans are dropped into them.

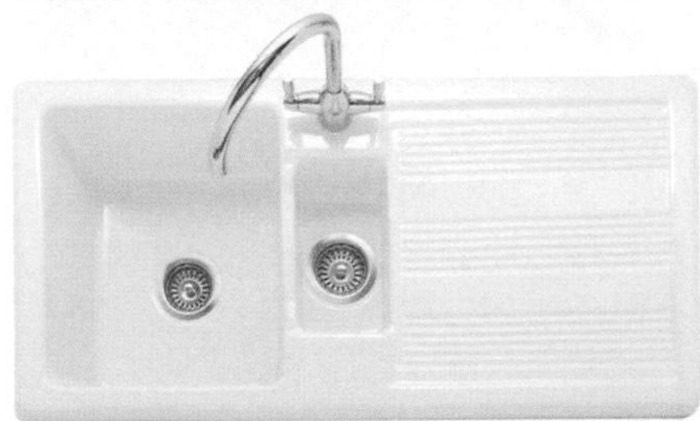

▲ Figure 9.54 The styles of kitchen sink

INDUSTRY TIP

Stainless steel sinks will require bonding to the electrical earthing in the property.

The bonding of metalwork and pipes is covered in detail in Chapter 3, Scientific principles.

- **Butler's sinks:** similar to the London sink with two main differences: the sink has a high splash-back, and also has a bucket grille.
- **Cleaners' sinks:** there are three types of cleaners' sink, and all are large, deep, rectangular sinks made of very thick white-glazed fireclay. They are usually mounted on cast iron cantilever brackets, but modern installations allow them to be fitted into kitchen units.
- **The Belfast sink:** originates from the early 18th century when they were fitted into the servants' quarters and the butler's area. Today, they are primarily used in utility and cleaners' rooms, although they can also be used in period-style kitchens. Recognisable from their integral weir-type overflow. The taps are usually bib type, fixed to the wall above the sink.
- **The London sink:** visually very similar to the Belfast sink, but does not have a weir overflow.

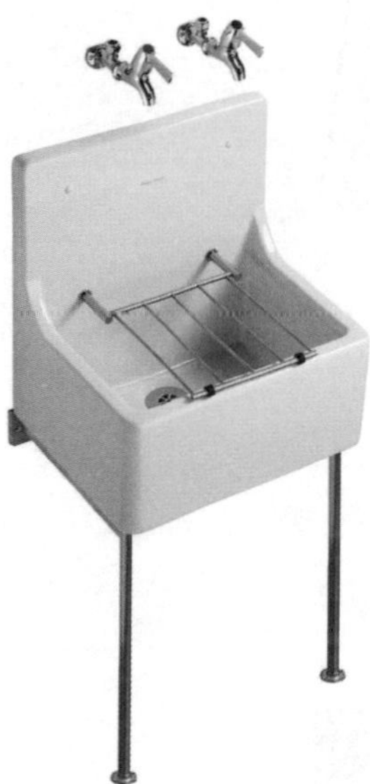

▲ Figure 9.55 The butler's sink

Urinals

Urinals are fitted in non-domestic buildings and there are three different styles:

1 **Bowl urinals:** usually made of vitreous china and stainless steel, these are the most commonly used urinal type and are the easiest to install. Dividers may be placed between the urinal bowls to give a little privacy. The bowl should be fixed at around 600 mm from the floor to the front lip. This can be reduced for urinals installed in schools.

2 **Trough urinals:** generally made from stainless steel and installed where the risk of vandalism is high – for example, in public conveniences. The trough is available in different lengths according to the number of people that are expected to use it. The trough has a waste connection and the trough floor has a built-in slight fall to allow the urinal to be installed level.

3 **Slab urinals:** manufactured from fireclay and assembled on-site. The channel in the base of the urinal is laid to a slight fall and the waste connection is made directly to the drain via the channel into a trapped gulley.

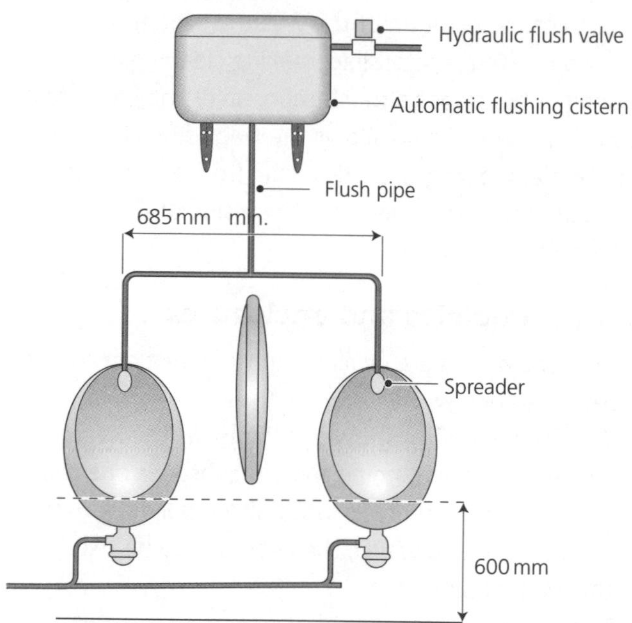

▲ Figure 9.56 The bowl urinal layout

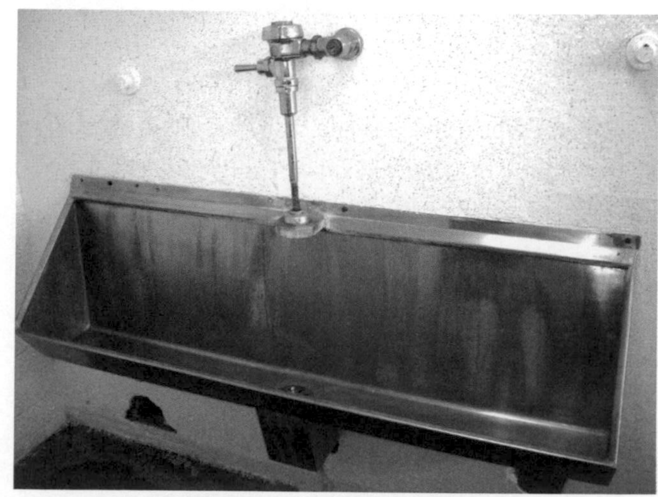

▲ Figure 9.57 The stainless steel trough urinal

▲ Figure 9.58 The slab urinal

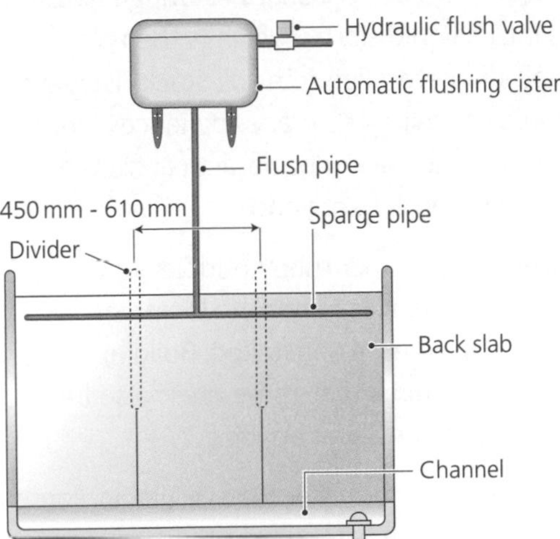

▲ Figure 9.59 The slab urinal layout

Flushing the urinal

Defra's guidance to the Water Supply (Water Fittings) Regulations 1999 states that urinals may be flushed with either:

a manual or automatically operated cistern;

or

b a pressure flushing valve directly connected to a supply or distributing pipe which is designed to flush the urinal, either directly manually or automatically, provided that the flushing arrangement incorporates a backflow arrangement or device appropriate to fluid category 5.

Clause G25.13 states:

> Where manually or automatically operated pressure flushing valves are used for flushing urinals, the flushing valve should deliver a flush volume not exceeding 1.5 litres per bowl per position each time the device is operated.

There are two ways that urinals can be flushed:

1 by the use of an automatic flushing cistern
2 by the use of a flushing valve.

KEY POINT

How an automatic flushing siphon works

Refer to Figure 9.60. When the level of the water reaches the top of the dome, the head of water at point A becomes greater than the pressure at point B. The water pressure in the trap (point C) overcomes the air pressure inside the siphon and this initiates siphonic action, emptying the cistern.

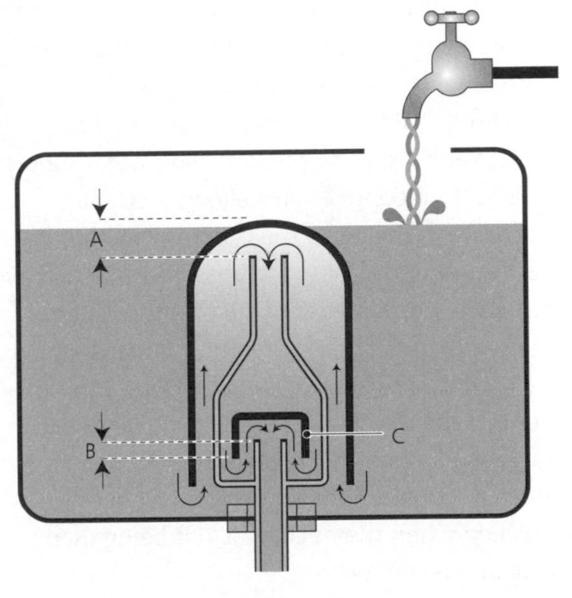

▲ Figure 9.60 Automatic flushing siphon

▲ Figure 9.61 Bowl urinals can be fed by an automatic flushing siphon

The automatic flushing cistern

As the name suggests, automatic flushing cisterns use an automatic flushing siphon to flush the urinals automatically when the water reaches a predetermined level in the cistern. The Water Regulations state that any auto-flushing cistern must not exceed the following water volumes:

- 10 litres per hour for a single bowl or stall
- 7.5 litres per hour per urinal position for a cistern serving two or more urinal bowls or 700 mm of slab.

The maximum flow rate from any automatic flushing cistern must be regulated by the inflow of water from the cold supply. This can be done quite easily by the use of urinal flush control valves such as a hydraulic flush control valve fitted to the incoming water supply. The hydraulic flush control valve allows a certain amount of water through to the cistern when other appliances like taps and WCs are used, rather than have a constant supply of water dripping into the cistern. The sudden reduction in pressure on the mains supply opens the hydraulic flushing valve to allow a certain amount of water through. The amount of water can be varied depending on the installation requirements and number of urinals. The idea here is to prevent the urinals flushing when the building is not being used, thus saving on wasted water.

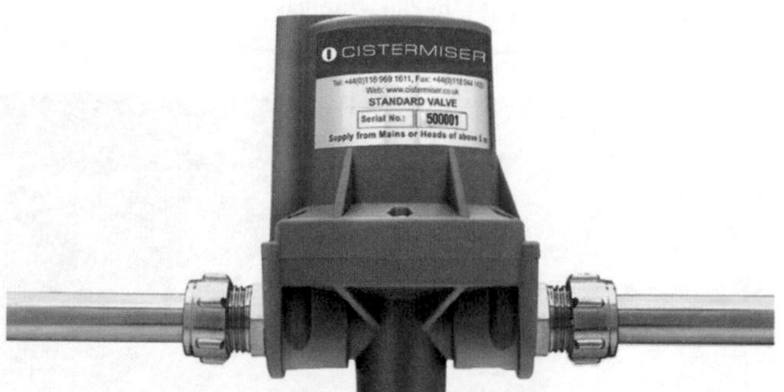

▲ Figure 9.62 The hydraulic flush control valve

The flushing valve

This is a new method of flushing a urinal that involves the use of either a manual or automatic valve, which delivers a short 1.5-litre flush to an individual urinal bowl. The water can be supplied either direct from the water main, from a boosted cold water system, or by low pressure from a cistern supplied by a distribution pipe.

Manual valves are lever operated and are located just above the urinal bowl. Automatic valves are activated via an infrared sensor. The sensor should sense a person for at least ten seconds to prevent accidental activation by someone walking by. The sensor activates a solenoid valve and this allows the minimum short flush.

Automatic flushing valves require a backflow prevention device to be included, which prevents backflow of fluid category 5 contaminated water.

WC macerators

Macerators use a series of very sharp rotating blades to turn solids into a liquid slurry, which is then pumped through a small-diameter pipe to a soil stack. They also offer a solution to installing sanitary appliances where access to the main soil stack is not practical from a conventional gravity outlet appliance.

Macerators offer the plumber many options when installing sanitary appliances in remote locations. However, if a WC macerator is installed, Building Regulations Part G requires that there must also be a gravity WC located in the same building.

There are many versions of macerators available, some purely for pumping from a WC, while others may be used to install entire bathrooms in difficult locations such as a basement or cellar.

▲ Figure 9.63 WC with a macerator

Pump stations used in domestic dwellings

Compact pump systems for small domestic waste water applications are suitable in situations where foul drainage by gravity is not possible. Larger domestic pumping stations are recommended for 8 to 13 people, for the removal of sewage and effluent. They are fitted with an alarm in the event of high fluid levels.

Waste water lifters used in domestic dwellings

Waste water lifters are used for pumping waste water and sewage from a low level to a higher level. They are used when it is not possible to remove waste and sewage from a normal gravity drainage system. They are usually factory-manufactured units. Key components of waste water lifters include a waste water treatment receiving well called a 'wet well', equipped with lift pumps and piping with valves, a junction box, and an equipment control panel with an alarm system.

The installation of a waste water lifter can be below or on the same finished floor level of a dwelling or premises. The discharge pipework enters the soil stack and forms a backflow loop as shown in Figure 9.64. The vent pipe must discharge in accordance with **BS EN 12050–1** for faecal lifting plants to above roof level, to avoid foul smells from entering the dwelling.

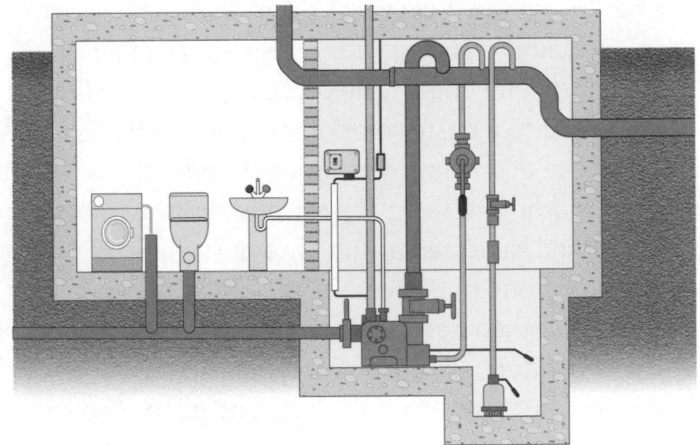

▲ Figure 9.64 A waste water lifter

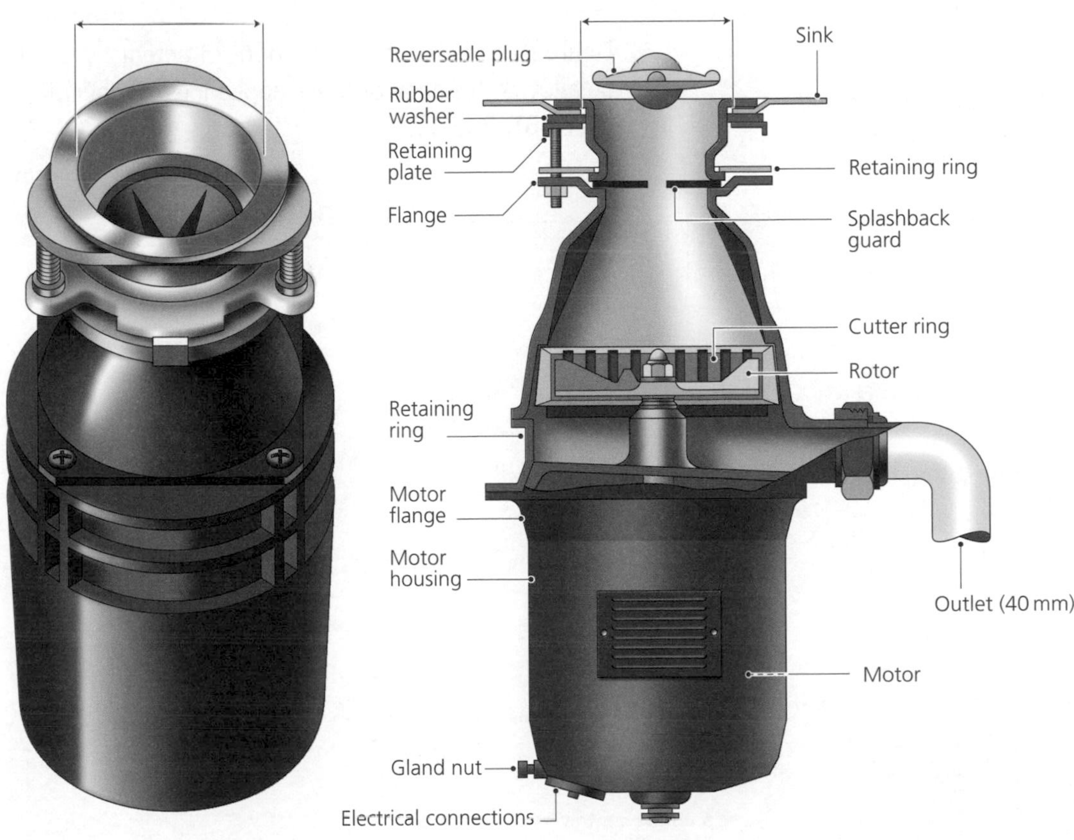

▲ Figure 9.65 A waste disposal unit

Sink waste disposal units

These units are installed in kitchen sinks and need a pre-made hole, 89–90 mm in diameter, in the sink to fit the unit. A standard 40 mm trap will fit on the outlet of the waste disposal unit.

These are installed under sinks to dispose of waste food and cooking products from a kitchen, and then discharge into the drainage system. The cutting or grinding blades can deal with a large range of food matter, including bones. The process turns anything in the unit into a paste solution, then water flushes this into the drain via a 40 mm waste outlet. The electric motor that turns the rotor where the blades are attached is located at the base of the unit. A sink housing one of these units requires a larger waste outlet than normal, approximately 89 mm, and manufacturers usually supply this on the cutlery bowl. The motor on the unit should be connected to an electrical supply via the correctly sized fused spur outlet, with a fuse appropriately sized in relation to the load (typically 10 amp).

Bathroom layout specifications

Sanitary appliances within a dwelling should be installed so that the minimum amount of space is provided for each appliance for both personal use, and for an adult to supervise the bathing and washing of children. British Standard **BS 6465-2 2017 Code of practice for space requirements for sanitary appliances**, recommends the minimum space required by each appliance for adequate usage.

British Standard **BS 6465–1:2006 Code of practice for the design of sanitary facilities** informs us that there must be a minimum number of appliances within a dwelling based on the number of people occupying the property.

▼ Table 9.4 Minimum number of appliances within a dwelling based on the number of people occupying the property

Sanitary appliance	Number per dwelling	Notes
WC	1 for up to 4 people 2 for 5 or more people	There should be a washbasin in or adjacent to every WC in the property
Washbasin	1	
Bath or shower	1 for every 4 people	
Kitchen sink	1	

Figure 9.66 illustrates the amount of activity space to be allowed for each of these appliances, as specified in **BS 6465-2 2017**.

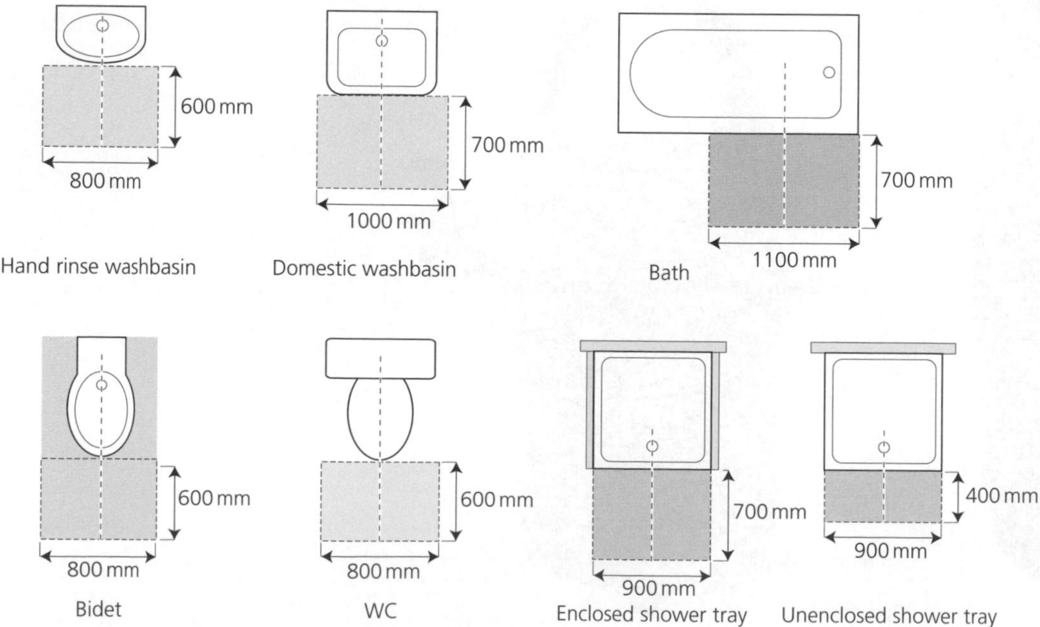

▲ Figure 9.66 Space provision for sanitary appliances

In some cases, it is not possible to maintain these distances, especially when the bathroom is small. In these situations, the British Standard allows overlap of the appliance space.

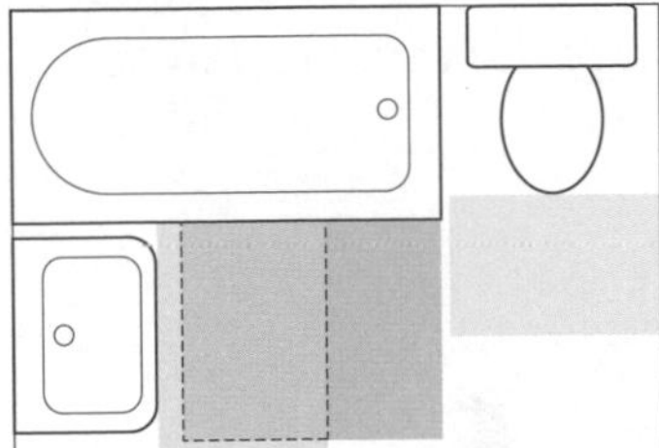

In this layout, the activity spaces of the bath and the washbasin overlap.
The space for the WC usage is not affected.

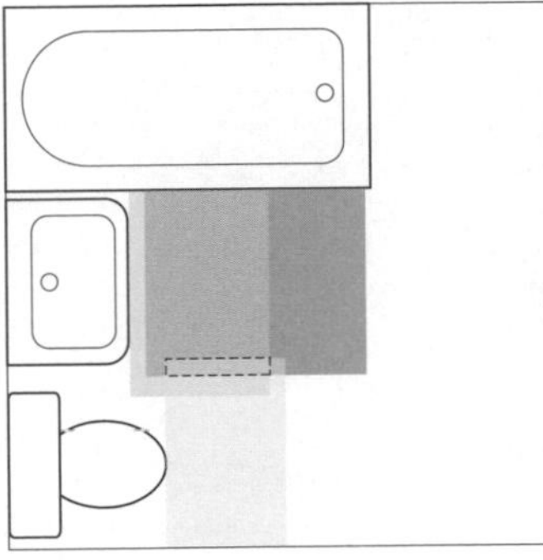

In this layout, the activity space of the bath, wash hand basin and WC all overlap. The overlap is shown by the dotted line rectangle on the drawing.

This one is the most common of all bathroom layouts.

▲ Figure 9.67 Overlap of the appliance space

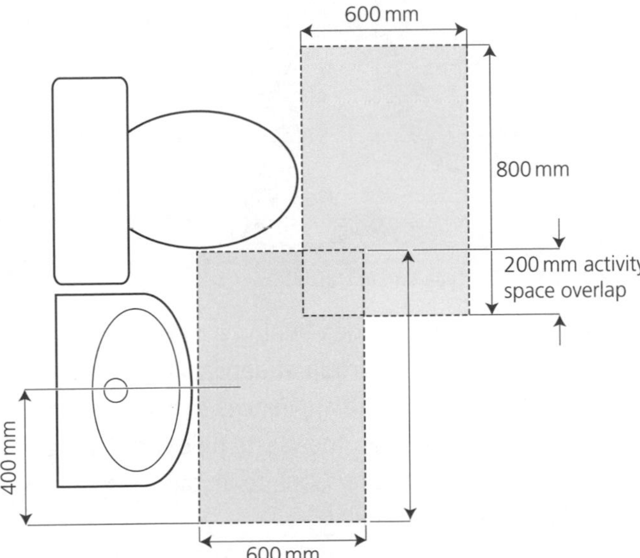

▲ Figure 9.68 Overlap in a downstairs WC

INDUSTRY TIP

Overlaps often occur in cloakrooms and downstairs WCs.

Traps

The purpose of a trap is to stop obnoxious smells from entering the dwelling. There are many different types of trap, to suit numerous appliances and applications. Traps are generally manufactured from polypropylene to **BS EN274 PT1-3** for domestic applications but can also be made from copper or brass. Jointing methods include push-fit type joints and compression type with a rubber compression ring.

INDUSTRY TIP

Copper- and brass-made traps can be chrome plated for a luxury finish and are usually used with chrome-plated copper waste pipe.

Trap depths and sizes

To recap pipe sizes and trap depths, remember where a trap diameter is 50 mm and above, the trap seal must be 50 mm, such as the traps in WC pans. There are two reasons for this, both of which are reliant upon the cohesive quality of water:

1 A trap with a diameter of 50 mm and over contains more water than, say, a 32 mm or 40 mm diameter trap. This makes the water much more difficult to move by induced siphonage, wavering out or compression.
2 Because of the pipe size, it is unlikely that an appliance will discharge at full bore. If a pipe runs at full bore it will try to pull air along with it. If there is no air to pull, then the water in the trap is pulled instead until the trap is empty and the pipe can pull air, thus breaking the siphonic action.

INDUSTRY TIP

Water has both cohesive and adhesive qualities and these were explored in Chapter 3, Scientific principles, page 141.

Where a waste pipe runs into a hopper head or a gulley, the trap depth can be reduced to 50 mm for washbasins, kitchen sinks and electrical appliances such as washing machines and dishwashers, and 38 mm for baths and shower trays. The reason for the bath trap difference is that baths and shower trays are large, flat-bottomed appliances, which by their nature discharge water more slowly than, say, a washbasin. The flat bottom of the bath means that the last drops of water run away more slowly than the water from a washbasin and so trap seal is retained.

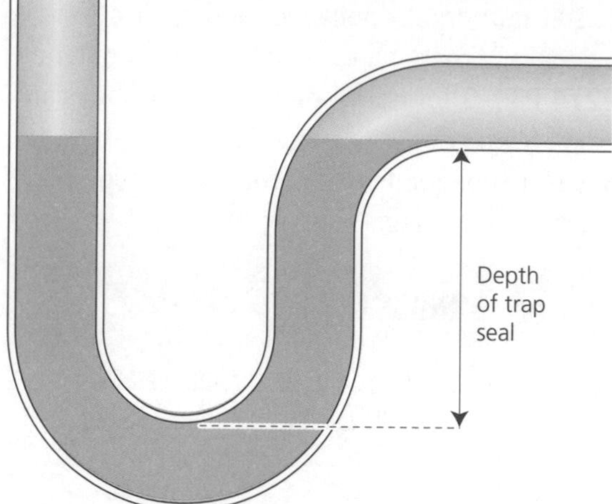

▲ Figure 9.69 Trap seal depth

▼ Table 9.5 Appliance waste pipe size and trap seal depth

Appliance	Waste fitting size (inches)	Diameter of trap (mm)	Trap seal depth when fitted to a primary ventilated system (BS EN 12056–2) (mm)
Washbasin	1¼	32	75
Bidet			
Bath	1½	40	50
Shower			
Bowl urinal		40	75
Washing machine		40	75
Dishwasher			
WC pan	N/A	75	50
		100	50

The way that trap seal depth is measured is shown in Figure 9.69.

Tubular traps

Tubular traps can take several different forms.

INDUSTRY TIP

'P' traps and 'S' traps are named after their shape. A 'P' trap is used where the waste pipe is installed from the appliance horizontally, directly through the wall and into a gulley or stack. The 'S' trap, because of its shape, will allow pipework to be installed vertically downwards from the trap into a waste pipe serving a number of appliances or into an underfloor waste pipe.

- **Swivel traps:** often used on new work and appliance replacements, they have a union connection in the centre that allows the trap to swivel 360°. This facilitates multi-positioning, allowing many different pipe connection options. They can be either 'P' trap, 'S' trap or running trap types.

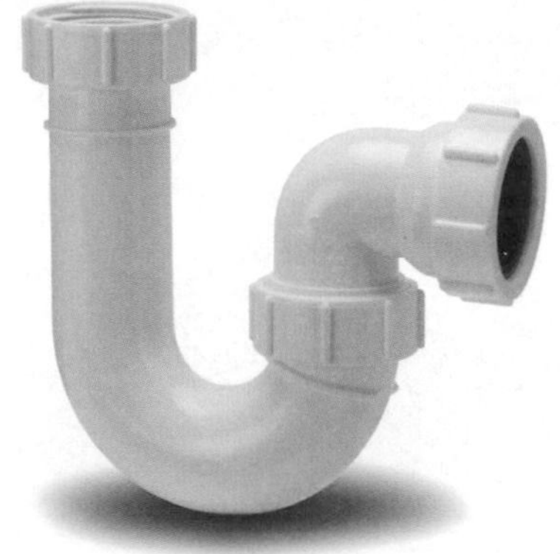

▲ Figure 9.70 'P'-type swivel traps

▲ Figure 9.71 'S'-type swivel traps

- **Running traps:** the idea behind a running trap is that an appliance or group of appliances can be trapped away from the appliances themselves, the trap being installed on the waste pipe run. They are sometimes used where space to install a trap at the appliance is limited.

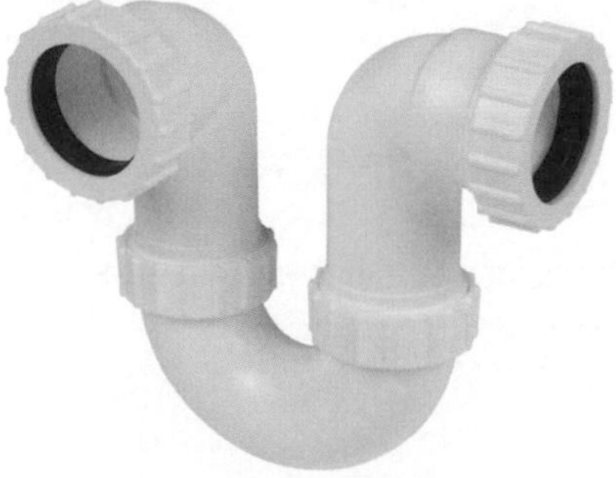

▲ Figure 9.72 Running trap

- **In-line traps:** specifically designed with washbasins in mind, the in-line trap is essentially an 'S' trap where both inlet and outlet are in line. They allow washbasin wastes to be completely hidden behind a pedestal, but can be restrictive and tend to block easily.

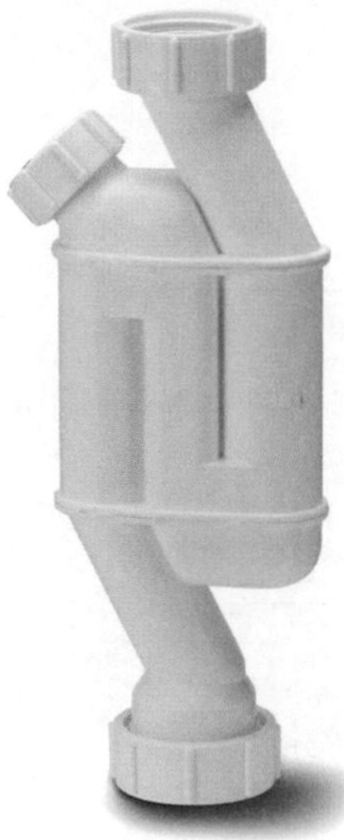

▲ Figure 9.73 In-line trap

- **Washing machine traps:** generally used for appliances such as washing machines and dishwashers with a 'P' trap configuration. They have an extended neck to facilitate a washing machine/dishwasher outlet hose.

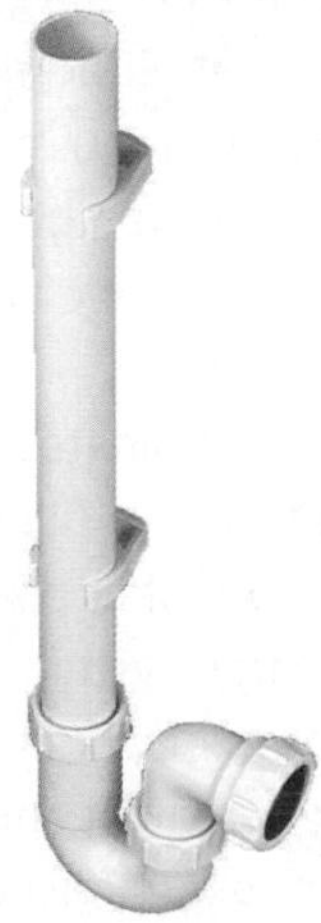

▲ Figure 9.74 Washing machine trap

- **Bath traps:** two different types are available. One is a swivel type with a 50 mm trap seal and the other has a 38 mm trap seal. They are made specifically to be fitted in the restricted space under a bath or shower tray.

▲ Figure 9.75 Swivel-type 50 mm seal bath trap

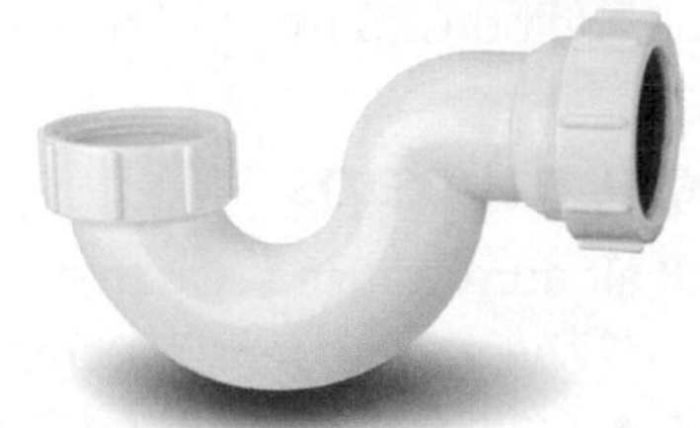

▲ Figure 9.76 38 mm seal bath trap

Bottle traps

Bottle traps are used on washbasins because of their neat appearance. However, they can be very restrictive to the flow of water. There are certain appliances where a bottle trap is not suited, such as on a kitchen sink where they could become blocked with foodstuff. Regular trap cleaning is important to maintain an adequate water flow. There are two different types:

1. **Bottle traps:** used with washbasins and bidets. Access for cleaning is via the bottom of the trap, which unscrews to facilitate the removal of blockages.

▲ Figure 9.77 Bottle trap

2. **Shower traps:** although not strictly a bottle trap, the operating principle of the shower trap is exactly the same as that of a bottle trap. The main difference here is that the trap seal depth is much less than 75 mm and access for cleaning is through the grille on the top of the trap rather than underneath.

Anti-vac and resealing (anti-siphon) traps

There is no substitute for a well-planned, well-designed system of sanitary pipework. If the system is designed in accordance with **BS EN 12056**, then problems with trap seal loss should not occur. Anti-vac (anti-vacuum) and resealing traps are not an alternative to a good sanitary pipework system, but they can reduce the problems that occur with existing systems due to upgrades to appliances and additions to the system.

Anti-vac (anti-vacuum) traps

Anti-vac traps use a small air admittance valve that is located after the water seal. The valve opens if a drop in pressure occurs when the appliance is used, and this allows air into the system to break any siphonic action that may occur.

INDUSTRY TIP

Fitting an anti-vac trap is not recommended on new systems as they have a habit of not holding an air test at the installation and testing stage.

Resealing traps

Externally, resealing traps are identical to a normal bottle trap. Internally, however, they are quite different: the trap has a bypass within the body of the bottle trap, allowing air to enter the trap, via a dip pipe, in the event of siphonic action occurring. They are available only as a bottle trap and so are not suitable for all installation situations.

Self-sealing traps

The self-sealing trap is a waterless valve that uses a thin neoprene rubber membrane to create an airtight seal preventing foul air from entering the dwelling, while maintaining equal pressure within the soil and vent system.

The membrane opens under the pressure of water from an appliance, closing again when the water discharge has finished.

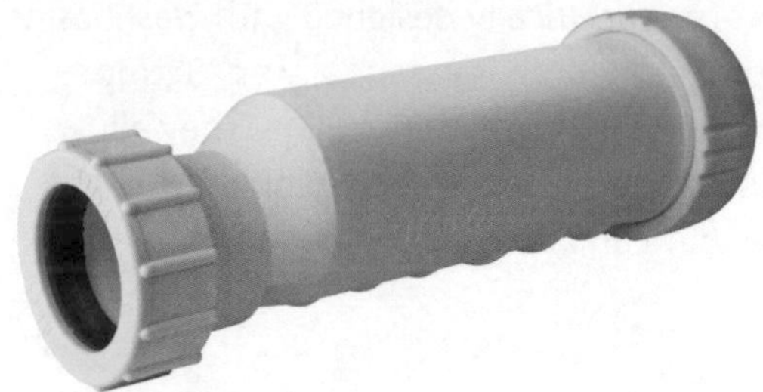

▲ Figure 9.78 The self-sealing trap

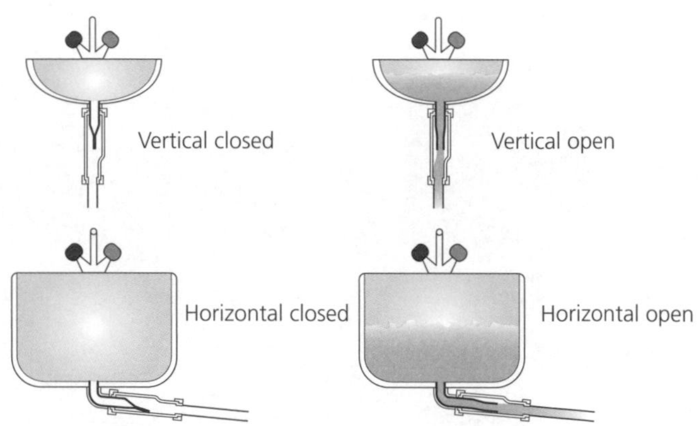

▲ Figure 9.79 The operation of the self-sealing trap

The self-sealing trap has certain advantages over conventional traps:

- The valve removes the problems associated with negative pressure within a system by opening to allow air in, in much the same way as an air admittance valve. This creates a state of equilibrium within the system, and means that air admittance valves and extra vent pipes are not required.
- Because there is no water in the valve, the problems of self-siphonage and induced siphonage are eliminated.
- The valve operates silently. This eliminates the noises generally associated with water-filled traps.
- The valve allows a greater number of appliances to be installed on the same discharge system without the risk of compromising system efficiency.
- The valve can withstand back pressures equivalent to ten times greater than those experienced in a typical sanitary pipework system.

KEY POINT

The self-sealing trap is so effective that it can be used safely on primary ventilated stack systems and ventilated discharge branch systems.

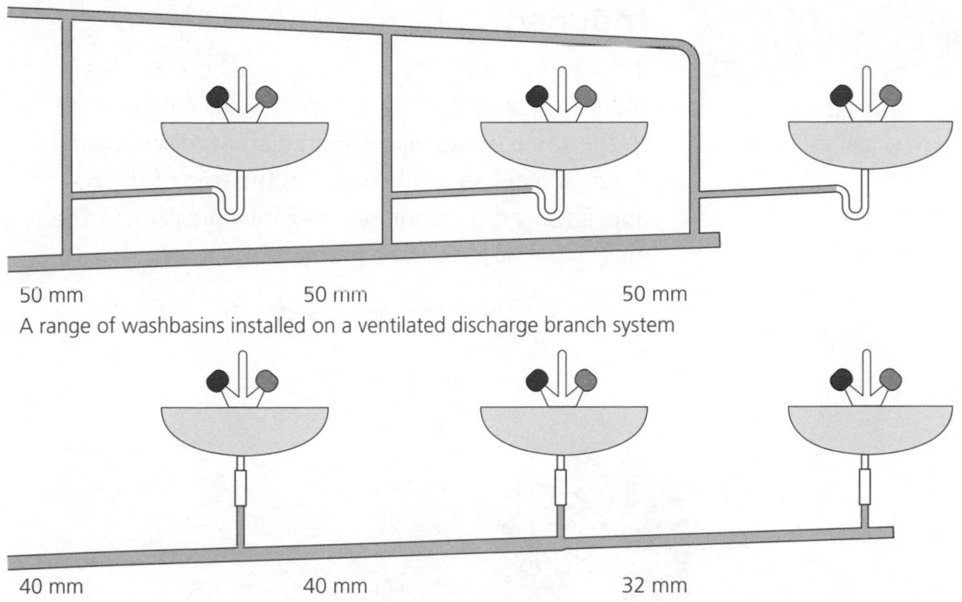

▲ Figure 9.80 Multiple installations of the self-sealing trap

Loss of trap seal

Provided the recommendations in **BS EN 12056:2000** are followed, problems of trap seal loss should be avoided. Most trap seal problems occur even before water has been let down the trap, simply because they can be attributed to design and installation issues with the sanitary pipework system. When loss of trap seal occurs, obnoxious smells will permeate the dwelling. Most trap seal problems can be traced back to the following faults:

- waste pipes that are too long
- waste pipes that are too small for the appliance
- waste pipes that are laid to an incorrect fall
- incorrect bends at the foot of the soil stack
- too many appliances on the same waste branch
- too many changes of direction.

Reasons for loss of tap/trap seal

There are various ways in which tap/trap seal loss occurs:

- self-siphonage
- induced siphonage
- compression
- wavering out
- evaporation
- capillary attraction
- momentum
- foaming.

Self-siphonage

Self-siphonage occurs when water is discharged from an appliance. The water forms a plug, which, as it disappears down the appliance waste, creates a partial vacuum in the waste pipe between the plug of water and the water in the trap. This then pulls the water from the trap.

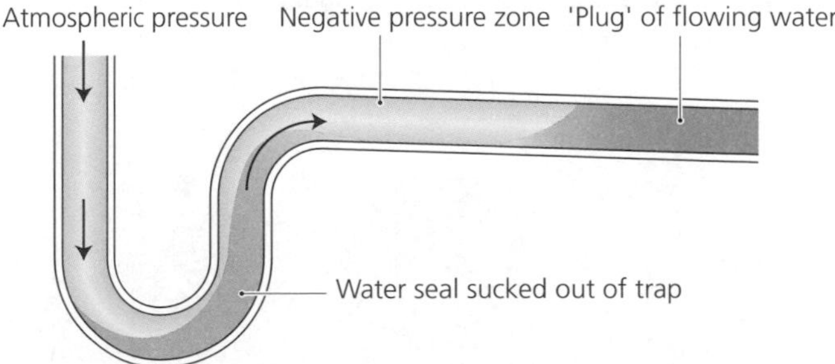

▲ Figure 9.81 Self-siphonage

In most cases, self-siphonage will not occur if the waste pipe length is kept within the recommended lengths of **BS EN 12056**. If it does occur, the installation of a vent on the waste pipe branch may be necessary or an anti-siphon trap could be fitted.

INDUSTRY TIP

Self-siphonage is most common on washbasins due to the rapid evacuation of the water from the bowl and the small size of the waste pipe.

Induced siphonage

Induced siphonage can occur by one appliance causing the loss of trap seal of another appliance connected to the same waste pipe. When water is discharged down an appliance, the water in the trap of the next appliance is drawn out by a negative pressure as the plug of water passes the branch connection.

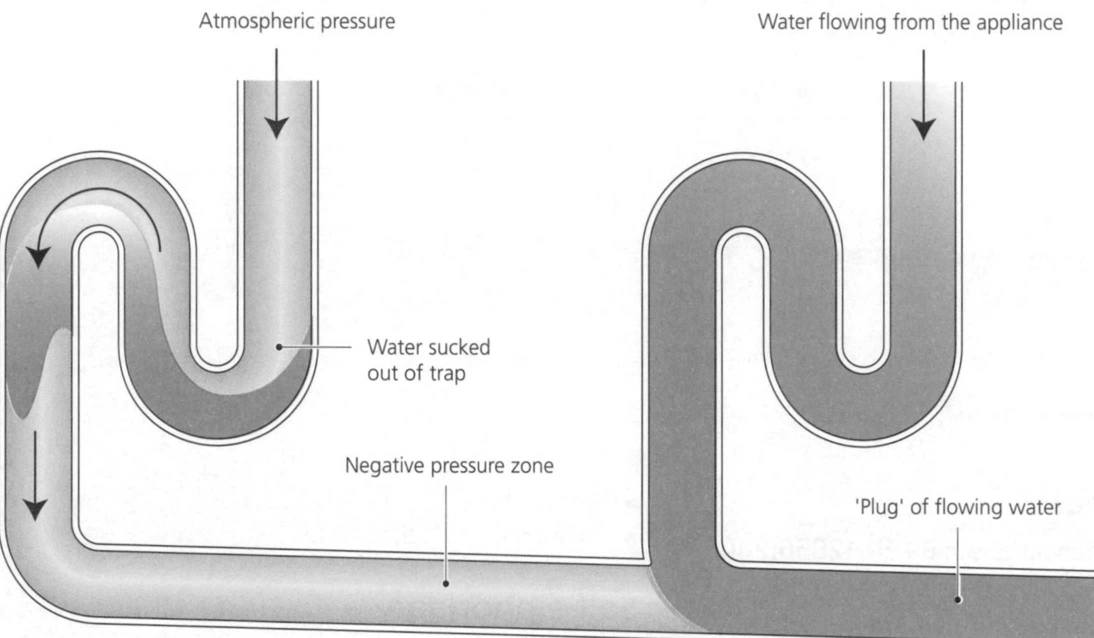
Atmospheric pressure
Water flowing from the appliance
Water sucked out of trap
Negative pressure zone
'Plug' of flowing water

▲ Figure 9.82 Induced siphonage

INDUSTRY TIP

Connecting two appliances on a single waste pipe often leads to incorrect installation on the primary ventilated stack, and is regularly the reason that baths pull the trap of the washbasin in many bathroom installations. This will always become a problem unless the pipe size leading to the final branch connection is increased to 50 mm (see Figure 9.83).

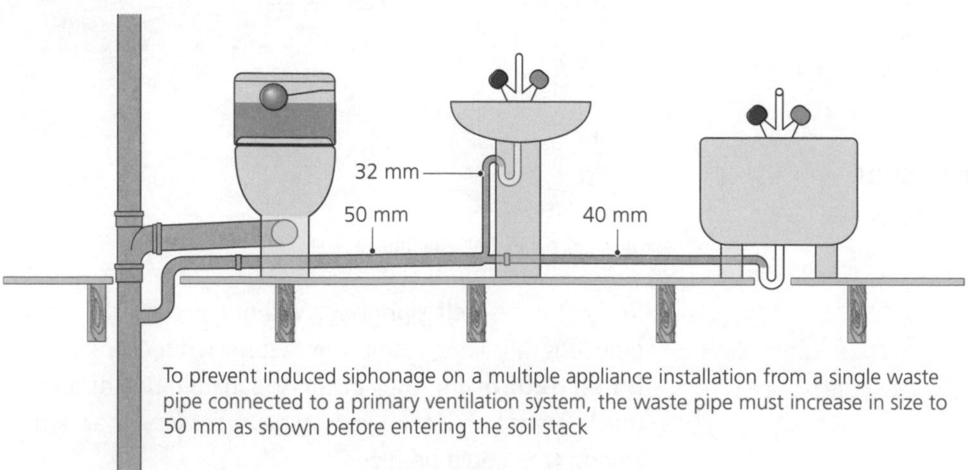
32 mm
50 mm
40 mm
To prevent induced siphonage on a multiple appliance installation from a single waste pipe connected to a primary ventilation system, the waste pipe must increase in size to 50 mm as shown before entering the soil stack

▲ Figure 9.83 Multiple appliance installations

Compression

When water is discharged from a WC at first-floor level, it falls rapidly to the base of the stack. If the bend at the base of the stack has a tight radius, the water momentarily stops flowing, causing the water to back up, which creates a back pressure of air. The back pressure travels up the stack and moves through ground-floor waste pipes, eventually blowing the water out of the traps.

The use of large-radius bends, or two 45° bends, at the base of the stack (see page 455) prevents this from happening by allowing the easy flow of water from the soil stack to the drain, allowing the water to maintain its forward motion and velocity.

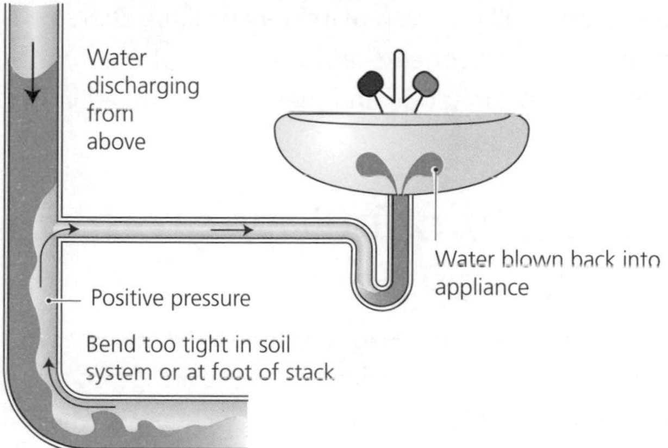

▲ Figure 9.84 Compression

Wavering out

Wavering out is caused naturally by the wind. In high winds or exposed positions, the effect of the wind blowing across the top of the vent pipe will cause the water in the traps of appliances to move with a wave-like motion because of pressure fluctuations. This momentum can often cause water to disappear over the top of the trap, resulting in trap seal loss. It can be prevented by fitting a wind cowl onto the top of the vent pipe.

KEY TERM

Wavering out: the process of water in traps in appliances moving with a wave-like motion because of pressure fluctuations due to exposure to winds.

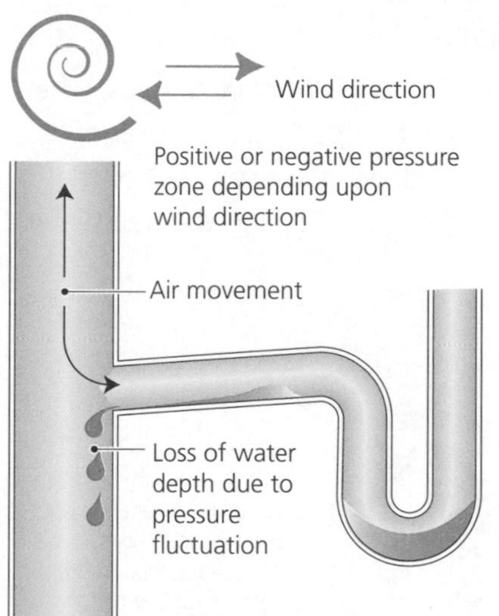

▲ Figure 9.85 Wavering out

Evaporation

Evaporation is a natural form of trap seal loss caused by lack of use of the appliance. Traps, to some extent, rely on the appliance being used regularly to keep the trap 'topped up' with water. When the appliance is not used, the water in the trap will begin to evaporate away until all the water is gone.

INDUSTRY TIP

The rate of evaporation can vary but, on average, the rate of evaporation is about 2.5 mm of trap seal per week, increasing when the weather is hot and dry.

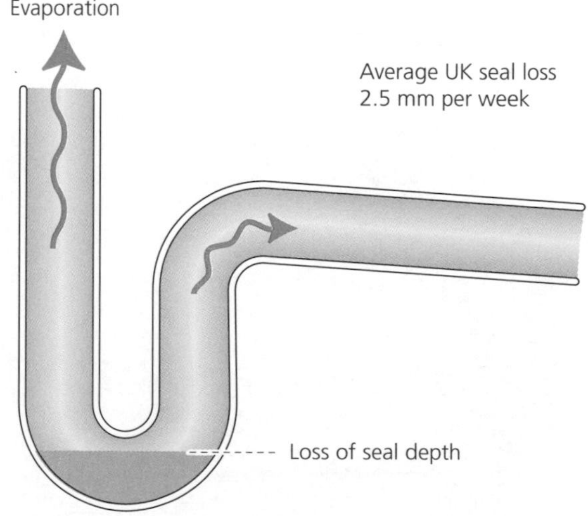

▲ Figure 9.86 Evaporation

Capillary action

Capillary action generally occurs only in 'S' traps, when long fibres or long hairs get lodged over the weir of the trap. Capillary action draws the water out of the trap and down the lodged material, and the trap water slowly drips away.

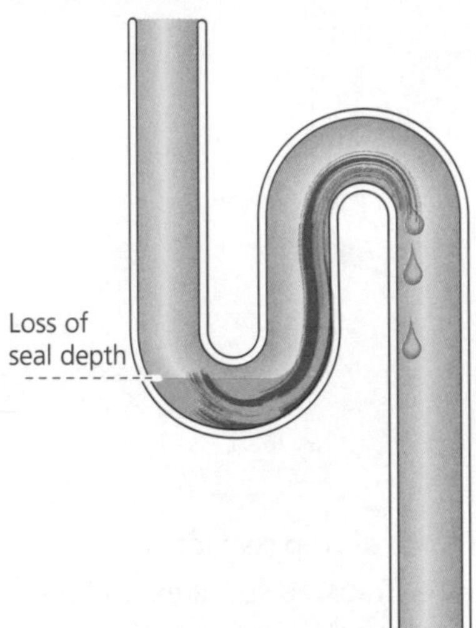
Strand of material or hair hanging over the trap weir draws water out of the trap by capillary action
Loss of seal depth

▲ Figure 9.87 Capillary action

INDUSTRY TIP

Capillary action is covered in Chapter 3, Scientific principles, page 140.

Momentum

Loss of trap seal by momentum occurs only when a large amount of water is suddenly discharged down the trap of an appliance. The force of the water will dislodge most of the water in the trap in a similar way to self-siphonage.

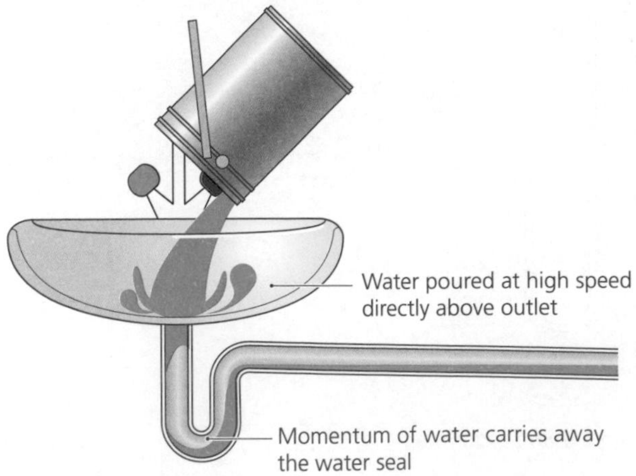
Water poured at high speed directly above outlet
Momentum of water carries away the water seal

▲ Figure 9.88 Momentum

Foaming

Foaming is a direct result of too much detergent (soap) being used. The excessive foam can back up waste pipes and soak away the water in the trap. It can usually be detected by the appearance of foam emerging from traps in appliances.

Sanitary pipework systems, positioning fixing, connection and operation of components

This subject was covered in detail in Chapter 2, Common processes and techniques, which covers the installation of all pipework within a dwelling. There are, however, components and systems that are specific within sanitation and these are discussed in the sections that follow.

Suitability of below-ground drainage systems

A below-ground drainage system takes the soiled water away from the dwelling and deposits it into a main sewer in the road. From here it will flow to the sewerage plant. The systems of below-ground drainage are:

- the separate system
- the combined system
- the partially separate system (sometimes called the partially combined system)
- soakaways
- cesspits
- septic tanks.

The separate system

This is the system favoured by local authorities. With this system, foul water and rainwater flow into separate drainage systems. These are then connected to a separate foul water sewer and surface water drain in the road. The foul water from WCs, baths, washbasins and kitchen sinks is conveyed to the sewage treatment plant and the rainwater flows to the nearest watercourse. The layout of the design is shown in Figure 9.89.

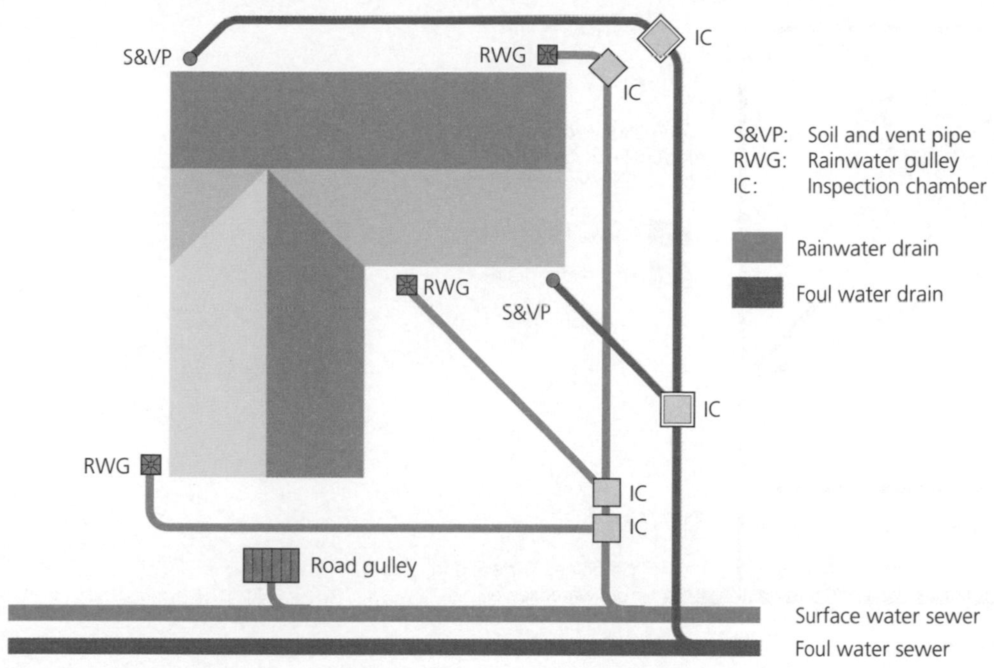

▲ Figure 9.89 Separate drainage system

▼ Table 9.6 The advantages and disadvantages of the separate system

Advantages	Disadvantages
Because the drains are separate, the sewerage plant does not get inundated with water when it rains heavily Trapped gulleys are not required for the rainwater connections. This helps to identify the drainage system in use	It is an expensive system to install because two drains are required The foul water drain does not get flushed and cleaned out by the rain There is a risk of making incorrect connections onto the rainwater drain The number of inspection chambers required is excessive

The combined system

With the combined system, both foul and rainwater drains discharge into a common sewer. This makes connections to the drains much simpler. It is a simple and economic system to install.

KEY POINT

The combined system is no longer recognised by the Building Regulations as a viable system on new installations.

▼ Table 9.7 The advantages and disadvantages of the combined system

Advantages	Disadvantages
Maintenance of the drains is much easier It is a cheaper system to install It is impossible to connect to the wrong drain All drains are flushed out when it rains	All discharge must pass through the sewage treatment plant, which is expensive and difficult to handle during heavy rainfall

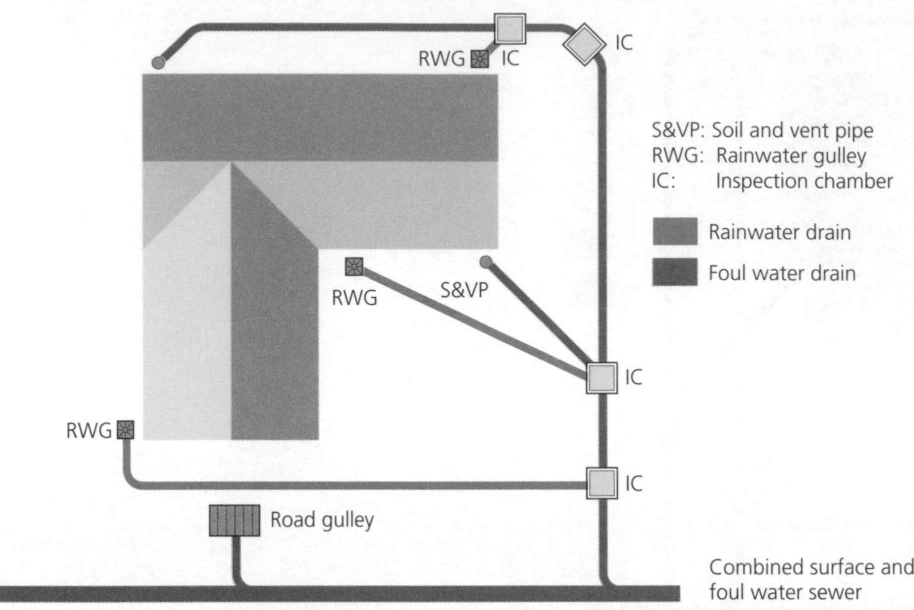

▲ Figure 9.90 Combined drainage system

The partially separate system

The partially separate system is a compromise between the separate and the combined systems. Two drainage systems are used: one that carries part of the rainwater discharge from the roof, and one that carries foul water and part of the rainwater discharge.

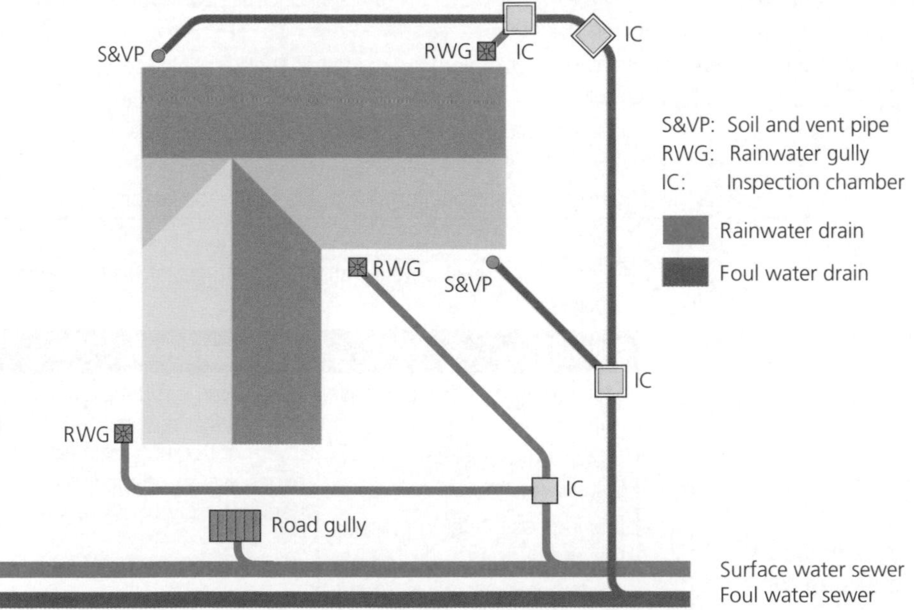

▲ Figure 9.91 Partially separate drainage system

▼ Table 9.8 The advantages and disadvantages of the partially separate system

Advantages	Disadvantages
It can reduce costs by allowing isolated rainwater connections to the foul water drain Rodding eyes can be used at strategic points, instead of costly inspection chambers	Care must be taken when installing foul water outlets to ensure the correct system is used

Soakaways

Soakaways are also used with partially separate systems to collect water from a roof not connected to the surface water drain. This is also known as the partially combined system.

KEY TERM

Soakaway: a pit, usually 1 m × 1 m × 1 m, dug into the ground and filled with gravel, into which the rainwater pipe discharges. It allows rainwater to soak naturally away to the water table. A soakaway should be situated at least 5 m away from the property.

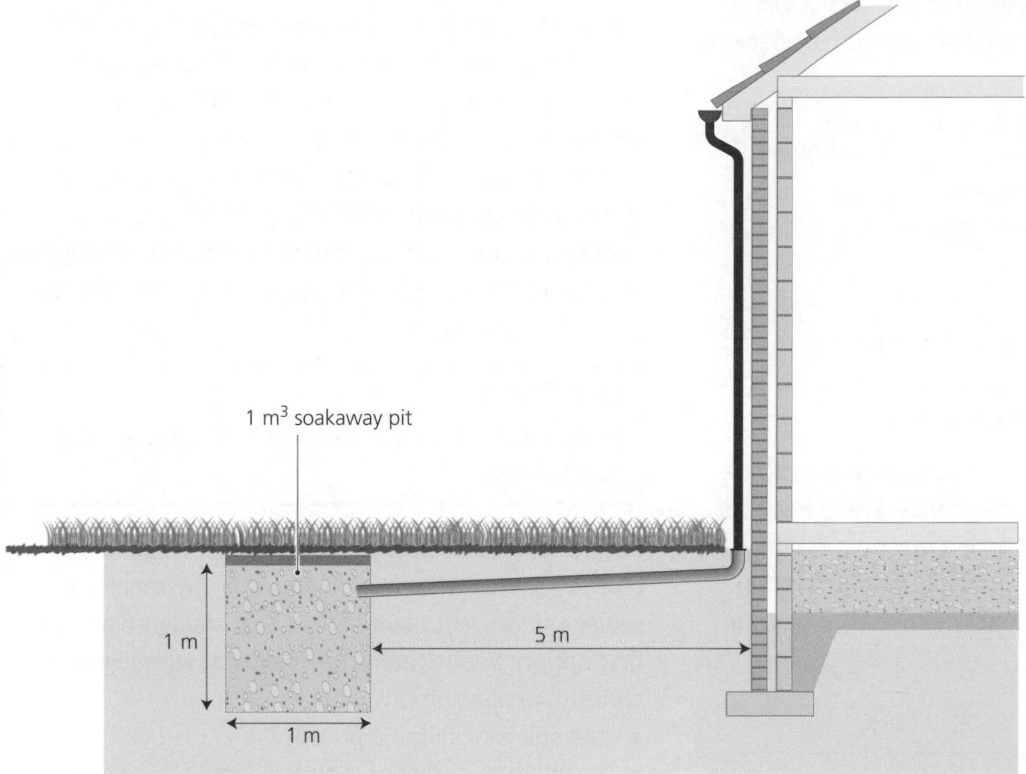

▲ Figure 9.92 Soakaway

Appliance connections to existing below-ground drainage systems

The method we use for connecting appliances to below-ground drainage systems will depend on the appliance and the material that the below-ground drainage pipework is made from.

Waste pipes up to 50 mm diameter

Appliances such as kitchen sinks, cleaners' sinks and washbasins may discharge direct into the back-inlet gulley of a below-ground drainage system. The waste pipe must discharge below the grating but above the water line in the gulley. This ensures an air break is maintained and that no smells can enter the building.

WC connections to ground-floor drains

Where a WC is to be connected to a ground-floor drain, this can be simply done by the use of a WC pan connector. Pan connectors are available in a number of lengths and outlet sizes to suit 75 mm to 110 mm drainage systems, and to fit both modern and existing WC pans.

▲ Figure 9.93 WC pan connectors

Cesspits

In rural areas, many homes and villages are self-contained, and the combined waste ends up in a local cesspit, septic tank or treatment plant. This has no connection to the public sewer system and is known as off-mains.

A cesspit is an underground tank that stores sewage until the time of its disposal. The design of the cesspit will incorporate an inlet pipe but will have no outlet pipework.

INDUSTRY TIP

Older cesspits were usually constructed of brick, but modern ones are made from glass-reinforced plastic (GRP) (also referred to as 'fibreglass').

Cesspits must be constructed so that they are watertight, to prevent the leakage of any foul water or the ingress of surrounding groundwater.

There are problems that may arise, such as overflow of effluent, so cesspits must be emptied on a frequent basis. This process must be carried out by a drainage contractor, who will use the principle of **mechanical suction** to draw up the contents of the cesspit into a tanker vehicle.

KEY TERM

Mechanical suction: suction that is created by an electrical/mechanical pump installed onto a tanker truck.

KEY POINT

In the past, older cesspits had overflow pipes that no longer conform to the recommended current design requirements.

Leakage is another problem that is more common with brick-built designs because the fabric of the structure can break down, leading to the ingress of groundwater and leakage of foul effluent, and resulting in foul smells and pollution of the surrounding area. As a result, the use of cesspits is no longer an option in most instances.

INDUSTRY TIP

Sometimes even the inlet connections to cesspits can leak.

Septic tanks

A septic tank is a multi-chamber storage tank allowing liquid and solid waste to separate. The liquid is then allowed to flow out of the tank and be disposed of separately. First, the sewage enters a settlement chamber, allowing solid waste (sludge) to sink and the liquid to rise to the surface. The surface liquid makes contact with oxygen and the organic matter starts to break down biologically. This liquid still contains sewage but the particles are small enough to be carried through the discharge outlet and into the ground (soakaway).

KEY POINT

Many areas of the UK prohibit the installation of septic tanks.

Basic septic tanks only partially treat sewage and discharge effluent of low quality. In all instances, a sewage treatment plant should be considered as the first option. Septic tanks may be installed, subject to consent, in applications where:

- the soil is of suitable porosity
- installation complies with Building Regulations (Approved Document H)
- the installation will not contaminate any ditch, stream or other watercourse.

Septic tanks must be at least 7 m minimum from buildings and within 30 m of access for a vehicle if being emptied by pumping truck.

Suitable termination of condensing boiler condensate drain connections

Condensing boilers, during the course of their operation, make **condensate**.

KEY TERM

Condensate: the water vapour that is present in the CO_2 emissions resulting from burning gas. As the flue gases cool down, the water vapour condenses back into water droplets that are collected in the boiler and discharged via the condense pipework.

Condensate is very acidic and needs to be dealt with correctly. The important points are as follows.

- The condensate outlet on boilers will accept 21.5 mm overflow pipe and it is strongly recommended that this discharges into the building drainage system. Where this is not possible, it may discharge into an outside drain, provided that precautions have been taken to prevent freezing.
- The condense pipe should be made of a proprietary drainpipe material, such as PVCu, ABS, PP or PVC.
- Metal pipework of any kind is not acceptable due to the acidic nature of the condensate.
- The pipework must be at least 21.5 mm diameter and fully supported with suitably spaced clips to prevent sagging.
- The pipe must be installed to a minimum angle of fall of at least 2.5° or 50 mm in every 1 m.
- The length of pipe should not exceed 3 m.
- External runs should be as short as possible, and a minimum of 32 mm diameter waste pipe.
- External pipework must be protected against freezing by insulation of an appropriate thickness along its entire run.
- Connections to rainwater drains are not permitted unless the rainwater drain connects to a foul drain.
- Since most combination boilers contain a 75 mm condensate trap, external traps are not required unless the boiler manufacturer states otherwise. Manufacturers' data should be checked before the installation of condense pipework.

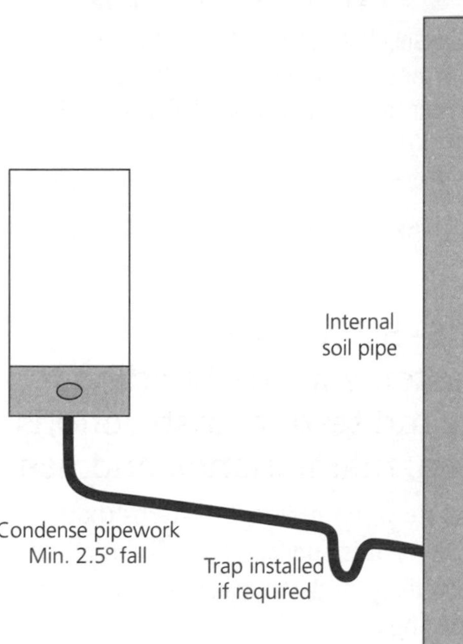

▲ Figure 9.94 Termination to a soil stack

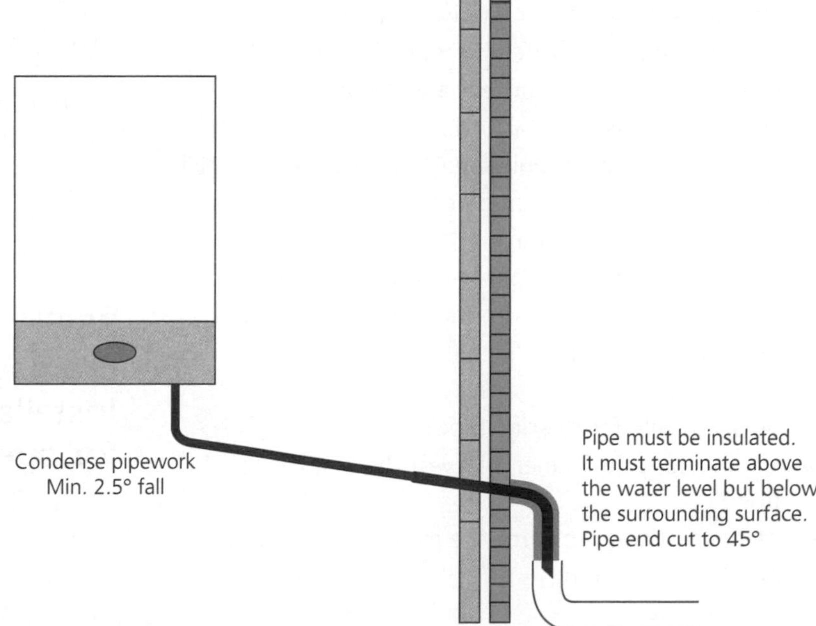

▲ Figure 9.95 Termination to a drain or gulley

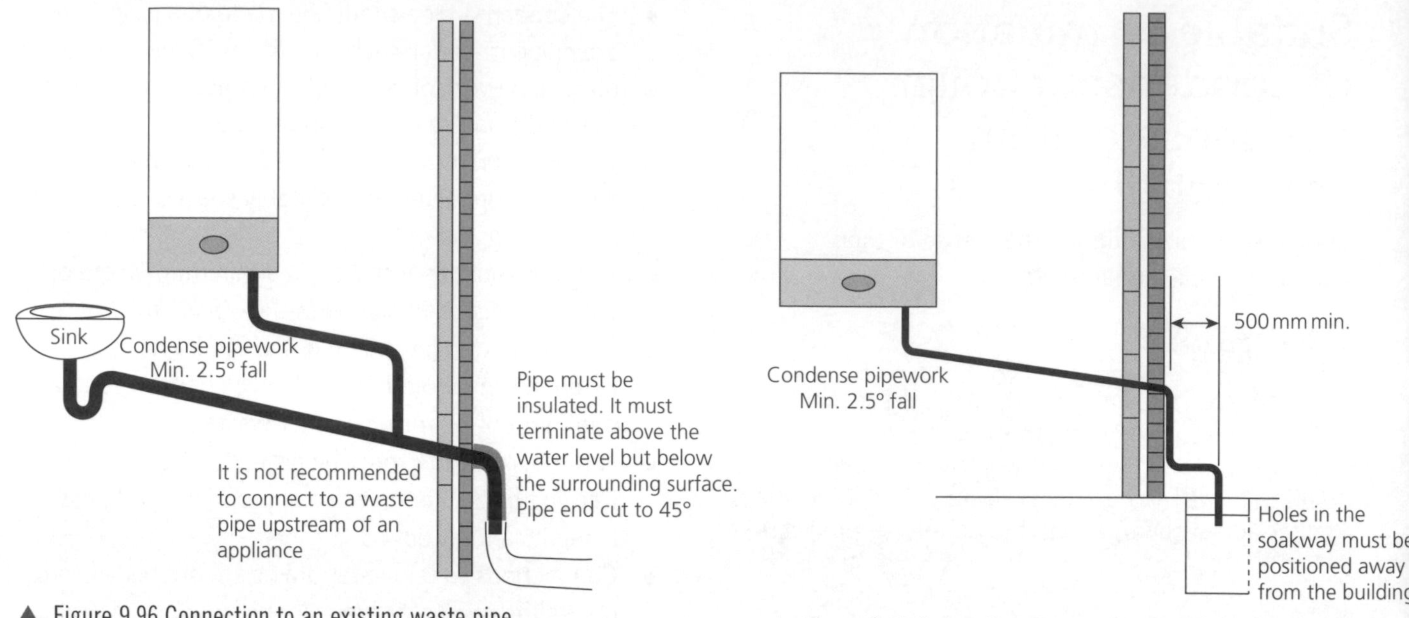

▲ Figure 9.96 Connection to an existing waste pipe

▲ Figure 9.97 Connections to a purpose-made soakaway drain

2 INSTALL SANITARY APPLIANCES AND CONNECTING PIPEWORK SYSTEMS

Sources of information

Sanitation systems must comply with Building Regulations Approved Document H1. The general requirements of this document are that a foul water system must:

- convey the flow of foul water to a foul water outfall; this can be a foul or combined foul/rainwater sewer, a cesspool or septic tank
- minimise the risk of blockage and/or leakage
- prevent foul air from entering the building under working conditions
- be ventilated
- be accessible for clearing blockages
- not increase the vulnerability of the building to flooding
- be large enough to carry the expected flow at any point in the system.

INDUSTRY TIP

You can access Building Regulations 2010 Approved Document H1 at: www.gov.uk/government/uploads/system/uploads/attachment_data/file/442889/BR_PDF_AD_H_2015.pdf

To successfully achieve this, we must consult several documents.

BS EN 12056–5:2000: Gravity drainage systems inside buildings. Installation and testing, instructions for operation, maintenance and use

This applies to waste water drainage systems that operate under gravity. It is applicable for drainage systems inside dwellings and commercial, institutional and industrial buildings.

Part 5 of this standard gives information that should be followed when installing and maintaining waste water gravity drainage systems, as well as the materials that can be used.

BS 8000 Part 13: 1989: Workmanship on building sites. Code of practice for above ground drainage and sanitary appliances

This provides recommendations on basic workmanship, and covers tasks that are carried out in relation to above-ground drainage and sanitary appliance installation.

The Water Supply (Water Fittings) Regulations 1999

This is not strictly relevant with regard to sanitation, but sanitary appliances require hot and cold water supplies, and appliances are used in connection with foul and waste water. Consequently, the Water Regulations must be consulted to guard against backflow of contaminated water.

▼ Table 9.9 Key British and European standards for system installation and materials

BS EN and BS Standards for the installation of sanitation systems	
BS EN 12056–2:2000	Gravity drainage systems inside buildings. Sanitary pipework, layout and calculation
BS EN 12056–5:2000	Gravity drainage systems inside buildings. Installation and testing, instructions for operation, maintenance and use
BS 8000–13	Workmanship on building sites Part 13 covers the installation of drainage and sanitation systems
BS Standards for materials	
BS 5627	Plastics connectors for use with horizontal outlet vitreous china WC pans
BS EN Standards for materials	
BS EN 1329	Plastics piping systems for soil and waste discharge (low and high temperature) within the building structure – Unplasticised polyvinyl chloride (PVCu)
BS EN 1451	Plastics piping systems for soil and waste discharge (low and high temperature) within the building structure – Polypropylene (PP) requirements and test methods
BS EN 1453	Plastics piping systems with structured-wall pipes for soil and waste discharge (low and high temperature) inside buildings – Unplasticised polyvinyl chloride (PVCu)
BS EN 1455	Plastics piping systems for soil and waste discharge (low and high temperature) within the building structure – Acrylonitrile butadiene styrene (ABS)
BS EN 1519	Plastics piping systems for soil and waste discharge (low and high temperature) within the building structure – Polyethylene (PE)
BS EN 1566	Plastics piping systems for soil and waste discharge (low and high temperature) within the building structure – Chlorinated polyvinyl chloride (PVCc)
BS EN 12380	Air admittance valves – for use in drainage systems
BS EN 274	Waste fittings for sanitary appliances
BS EN 14680:2015	Solvent cement for non-pressure pipe systems

Manufacturer technical instructions

Manufacturers' instructions should be used when assembling, installing, repairing and maintaining sanitary equipment, components and appliances. Installing to the written instructions provided is the best way to ensure compliance with the Regulations in force and the recommendations of the British Standards.

> **INDUSTRY TIP**
>
> Following the manufacturer's installation and servicing instructions is a requirement of any guarantee or warranty given with the appliance or component.

Design requirements

The design of the system will give vital information with regard to the position and number of appliances installed, and the pipework size and material type. It is important that the design is followed.

Storage and protection of sanitary appliances

Sanitary appliances are expensive, and require great care when handling, transporting, storing and installing them. To protect sanitary appliances:

- store in a clean, dry place away from cements, mortars and plaster
- always leave the protective cover on for as long as possible; the protective wrappers should be removed before installation so that a check can be made for damage and defects
- always store the items on timber battens and not directly on the ground
- stack items very carefully
- leave plenty of space for removing and replacing items from storage.

We must remember that the installation of AGDS is covered both by Building Regulations Document H3 (Rainwater Drainage) and **BS EN 12056**. These important documents restrict the systems we install and the materials we use, to ensure that hygienic conditions are maintained in dwellings and buildings at all times. The way we install them is also an important issue and is subject to a **code of practice**. This is **BS 8000 Part 13: Workmanship on building sites. Code of Practice for above ground drainage and sanitary appliances: 1989**.

> **KEY TERM**
>
> **Code of practice:** similar to a British Standard, this is a set of rules that explains how people should behave in their chosen profession.

> **INDUSTRY TIP**
>
> You can access Building Regulations 2010 Approved Document H3 (Rainwater Drainage) at: www.gov.uk/government/uploads/system/uploads/attachment_data/file/442889/BR_PDF_AD_H_2015.pdf

Preparation before installation

On new-build installations such as multi-dwelling housing developments, the position of the soil and vent pipework will be determined by the drain connection installed to the architect's drawings. Any preparation work required to allow the installation of the sanitary pipework should be agreed with the relevant trades beforehand. For instance, on some sites, any holes required in brickwork or timberwork is undertaken by the building or joinery contractors. This must be completed prior to our installation to avoid unnecessary and costly delays.

> **INDUSTRY TIP**
>
> It is a good idea to check the preparation work to ensure that pipe and fitting clearances are adequate.

Types of materials

Generally speaking, the materials used on modern AGDS are made from plastic. The range of plastics used are covered in Chapter 2, Common processes and techniques, but are briefly summarised here in Table 9.10.

▼ Table 9.10 Plastics used on modern AGDS

Material	BS number	Characteristics
PVCu	**BS EN 1329–1**	These three materials can be either solvent welded using solvent cement to **BS EN 14680:2015** for waste pipes from 32 mm to 50 mm diameter, or push-fit and solvent welded for soil and vent pipes from 82 mm to 160 mm diameter. Push-fit soil and waste fittings should be to **BS 4514**. Pipe is available in lengths of 2.5 m, 3 m or 4 m, either plain ended or socket and spigot ended.
MuPVC	**BS EN 1566–1**	
ABS	**BS EN 1455–1**	
Polypropylene	**BS EN 1451–1**	Polypropylene is a push-fit waste system with sizes ranging from 32 mm to 50 mm diameter. It cannot be solvent welded and is identifiable by a warm but slightly greasy feel to the pipe. It is more flexible than PVCu or ABS, and does not break or shatter.

The choice between push-fit and solvent weld waste pipes and fittings is down to personal preference, although on some housing contracts, solvent weld will be specified. Each system has its benefits and drawbacks, as listed in Table 9.11.

▼ Table 9.11 The advantages and disadvantages of different waste pipe types

Waste pipe type	Advantages	Disadvantages
Push-fit	Easy to make a watertight joint Pipe is light and easy to install Easy to take apart for unblocking and maintenance Joints allow movement for thermal expansion and multi-positioning Can be tested immediately after jointing	Pipe tends to sag if not clipped correctly Joints can pull apart easily, causing unsuspected leaks Suffers from UV light degradation so may require painting if installed outside
Solvent weld	The pipe is much more rigid than polypropylene pipe and does not suffer as much from sagging Neater appearance Joints will not push apart Will resist most acids, alkalis and chemicals	Joints are permanent and will not allow for repositioning Joints must be left for a period of time before testing can begin Fumes from the solvent cement can be damaging to health Suffers from UV light degradation so may require painting if installed outside

Other materials: cast iron

Cast iron was used for many years in both domestic and industrial installations. Cast iron has the advantage of being very robust, but it is also very heavy and difficult to work with. The jointing system is much easier than it used to be. Today, cast iron is jointed using a special jointing system called 'timesaver', which is simply bolted together using special torque wrenches so that the joints are not over-tightened.

INDUSTRY TIP

Today, cast iron is restricted to large installations and public buildings such as hospitals. You still may be required to work on cast iron, especially when refurbishing existing dwellings.

Waste pipe connections to the soil stack

Waste pipe connections to the soil stack can be made in two ways:

1 **By the use of a boss pipe:** these can be push-fit or solvent cement type. Each connection for the waste pipe will need to be drilled out using an appropriately sized hole saw beforehand, and the correct insert for the waste pipe size used.

▼ Figure 9.98 Boss pipe adapter

2 **By the use of a strap boss:** these are solvent welded onto the soil pipe.

Both methods were featured in Chapter 2, Common processes and techniques. Care should be taken to ensure we install them the right way up, as both boss pipes and strap bosses have a slight gradient in the moulding to ensure the correct fall for the waste pipe.

Installing a strap boss, step by step

1 Determine where the strap boss is to be installed and mark the centre of the hole.
2 Using the correct size of hole saw and a cordless drill, drill the hole for the strap boss, ensuring that the lip on the inside face of the strap boss fits snugly inside the hole. It is important not to have too much play in the hole as this may result in leakage once the solvent cement has set.
3 Clean around the hole and the surface of the strap boss with cleaning fluid.
4 Apply solvent cement to the strap boss first and then around the hole on the soil pipe.
5 Place the strap boss into position, insert the nut and bolt at the back of the boss, and tighten gently.
6 Clean away any excess solvent cement with a clean, dry cloth.
7 The boss must be left for at least five minutes for the solvent cement to cure enough for testing to be carried out.

KEY POINT

Not all strap bosses have nuts and bolts to keep them in place. Some just clip together to make a watertight seal. All strap bosses, however, must be solvent welded.

Waste pipes that are to be installed on an internal soil stack can use a waste pipe manifold. This is an adapter that allows multiple waste pipe connections and avoids problems with cross-flow exclusion zones.

▲ Figure 9.99 A waste pipe manifold

Access to pipework

Access to AGDS pipework for cleaning, clearing blockages and maintenance is a requirement of the Building Regulations. There are various ways that we can fulfil this requirement:

- by the use of access plugs inserted into soil pipe junctions and waste pipe tees
- by the use of purpose-designed access covers and fittings.

▲ Figure 9.100 Access plug in soil junctions

▲ Figure 9.101 Purpose-made access fitting

PVCu	HepSleve	Salt-glazed earthenware	Cast iron
These two sockets are simple push-fit types. The soil pipe should be chamfered and silicone lubricant applied before inserting into the socket.		These sockets require jointing with a strong 2:1 ratio sand and cement mortar. They should be left for 24 hours before testing is carried out.	

Soil stack connection to the drain

The connection to the drain could be one of several materials depending on the age of the building and its use. Older properties tend to have salt-glazed earthenware drains, and public buildings often use cast iron drains. Connection to these materials is usually by a collar, which is sealed with a sand and cement mortar joint. Modern houses use either PVCu or HepSleve clay piping. The jointing methods to these materials are shown below.

Multi-fit pipe adapters are also available for connecting differing pipe materials below ground. The drain connection, as we have already seen, is made to a large-radius bend.

If the soil stack is external (outside the building), an access pipe can be used as the drain exits the ground. On internal soil stacks, access must be above the spill-over level of the highest appliance.

INDUSTRY TIP

On new dwellings, the position of the drain connections will be marked on the building plan. On older buildings, we will have to use what is already in place, so careful consideration should be made as to the method of jointing we are going to use.

Installation requirements of appliances and systems

Good preparation for the installation of sanitary appliances is essential as it is probably the most visual of all the installations that we undertake. Customers can invest a great deal of money replacing their bathroom suites and it is vital that we get it right.

The preparations we make before we install sanitary appliances need very careful consideration. Good planning includes:

- making sure that the hot and cold pipework has been installed in accordance with drawings
- making sure that any chases and holes necessary have been prepared
- checking that sanitary ware has been delivered on time, is correctly stored and is free from damage; it is a fact that one in four bathroom suites delivered to site are either damaged, incorrect or have parts missing; these hold-ups can be costly in terms of time and repeat customer business.

Remember, when ordering and receiving delivery of materials:

- contact the merchant before you start the job to ensure that the bathroom suite is going to be delivered to the correct address and on the correct day

VALUES AND BEHAVIOURS

British Standard **BS 8000 Part 13** gives essential guidance on the class of workmanship that is expected of us when installing above-ground sanitation systems and sanitary appliances.

- always check the delivery note to ensure that the equipment on the sheet is the same as that being delivered
- always handle sanitary ware with care – most appliances are easily scuffed or damaged
- when storing materials, ensure that the store is secure and the materials have been stacked correctly.

Before the job commences, you should ensure that:

- the work area is completely clear of all debris
- the customer's carpets and furniture are protected
- you have all the manufacturers' instructions to hand; these will need to be left with the customer at the end of the job
- you consult with the customer so that there are no last-minute changes that may need the intervention of your supervisor.

Dressing the appliances

Dressing of sanitary ware includes the following stages.

KEY TERM

Dressing: the term used by plumbers to describe the preparation of the appliances ready for installation.

- Installing the taps and wastes to the bath, washbasin and bidet:
  - Taps should always be fitted in accordance with the manufacturer's instructions; the washers provided for sealing the taps to the appliance should always be used, and care should be taken to ensure that they are not over-tightened in the appliance or we risk cracking the appliance itself.
  - Wastes will either be slotted for appliances with integral overflows, unslotted or a pop-up waste system; wastes should be made into the appliance with silicone sealant or specific washers if the manufacturer provides them.

INDUSTRY TIP

Never use bare grips to hold a tap as this will mark the chrome/gold plate.

INDUSTRY TIP

If silicone sealant is used with the wastes, try not to use too much as it is difficult to remove from the glaze of the appliance.

VALUES AND BEHAVIOURS

Taps are a personal choice and will have been chosen by the customer with a lot of thought. We must treat them with care to ensure that they are not damaged during the installation.

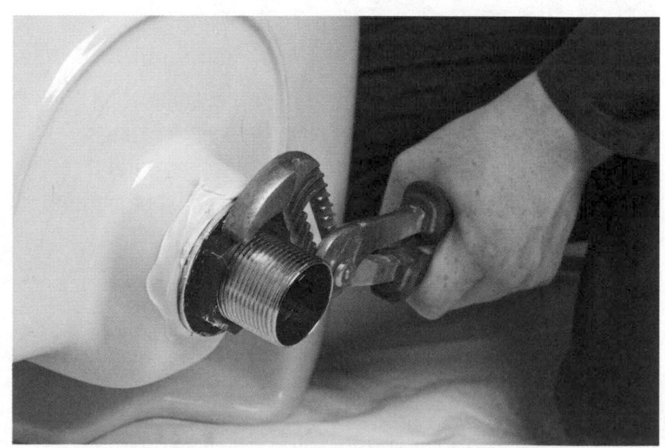

▲ Figure 9.102 Taps and waste being fitted to a bath and washbasin

- Assembling the WC cistern: this means installing the siphon, float-operated valve, overflow (if applicable), flushing handle and close coupling bracket (if the WC is a close coupled model).
- Carefully fixing the bath cradle and feet to the bath: the bath should be carefully turned upside down on a clean dust sheet for this operation. The bath feet are adjustable to enable the bath to be fitted level and to the correct height. Great care must be taken here as the cradle comes with specific screws for different positions. If we use the wrong screws, we could pierce the bath itself. Always read the manufacturer's instructions beforehand.

INDUSTRY TIP

Specialist tools you will need to install sanitary appliances will include: box spanner, tap spanner, basin wrench, soft jaw pliers, toilet seat installation tool, tap reseating tool, magnetic telescopic mirror.

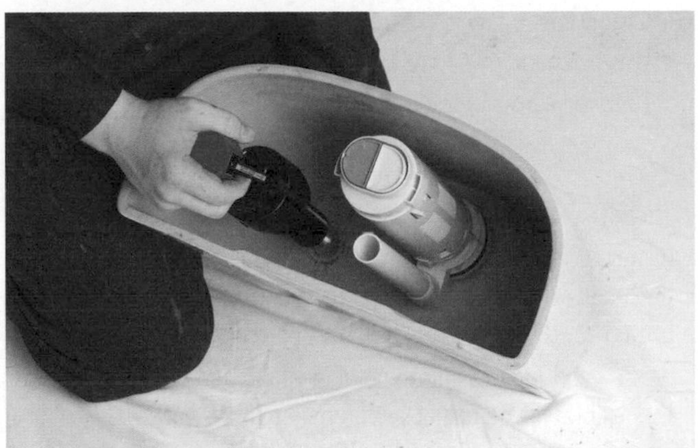

▲ Figure 9.103 A WC cistern being assembled

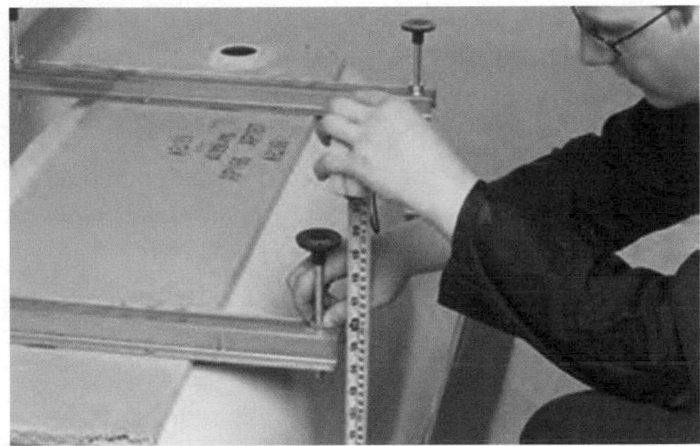

▲ Figure 9.104 A bath being assembled

Install and test systems and appliances

The installation process for bathrooms for refurbishments is quite different from that for new-build installations. Here, we will deal only with the installation of the three most common appliances:

1 the bath
2 the washbasin
3 the WC suite.

VALUES AND BEHAVIOURS

Before work begins, make sure the customer is aware that appliances will be taken out of commission, and that warning notices have been placed in strategic areas to prevent the accidental turning on of water supplies and unintentional use of partially fitted appliances such as WCs.

INDUSTRY TIP

The protective tape around the sanitary ware should be removed before installation takes place so that the appliances can be checked for damage prior to being fitted.

New-build installations

On new-build installations, the choice of bathroom suite is often not as varied as it is for a private customer, especially on housing contracts where there are only two or three house styles being constructed. The work can become very repetitive, with the same suite types being installed time and again, and always in the same positions. The appliance positions are set by the architect and it is often difficult to **deviate** from these plans. It is usual for the first fix to have been installed beforehand, with hot, cold and waste pipework tails visible.

KEY TERM

Deviate: change; do differently from the original plan.

INDUSTRY TIP

The hot and cold will have been hydraulically pressure tested previously.

Installing the bath, step by step

Although there are no set rules for the order that appliances are installed, it is common practice to install the bath first as this is the largest of all the appliances and is much easier to manoeuvre into position in an empty bathroom.

1 The bath should be placed into position and the feet adjusted until it is level on all sides at the correct height to suit the bath panel (if one is being fitted).

2 When you are sure that it is ready to be fixed to the wall, mark the brackets that hold the bath to the wall. Remove the bath temporarily, and drill the fixing holes. The type of fixings you use will depend on the type of wall it is:
- for masonry, concrete block and thermalite block walls, wall plugs and brass screws may be used
- plasterboard studded walls will require plasterboard fixings unless wooden noggins have been placed in the wall previously.

3 Fix the bath in its permanent position and, after checking once more to ensure correct level and height, screw the feet to the floor. Make sure that all the feet are screwed down as this is often missed and can cause bath movement later if not done correctly.

▲ Figure 9.105 A bath being levelled

4 Once the bath has been fixed into place, it can be connected to the hot and cold pipework. How this is done will depend on the first-fix pipework material:
- polybutylene is by far the easiest material to work with
- copper adds rigidity to the installation.

5 Make sure that both tap connectors are fully tightened. The waste pipe to the bath can also be installed at this stage while all other appliances are out of position. It is often difficult to work under a bath, especially if the washbasin pedestal or WC pan is in the way. Ensure that the service valves are in the off position prior to commissioning.

6 Once the bath is fixed, it is normal practice for the bathroom to be tiled and grouted before any further appliances are installed.

INDUSTRY TIPS

It may be a good idea to fix timber to the floor where the bath feet will sit as this helps to spread the weight of the bath, the water and the person using it.

It is a good idea to install service valves to both taps, as this will facilitate easier maintenance in the future.

Installing the washbasin, step by step

The washbasin often comes next. This can be a tricky installation.

1 The centre line of the basin should be marked lightly in pencil on the tiles. This is usually the centre between the hot and cold pipework. Also, mark the centre on the washbasin itself. This will allow both centre lines to be lined up, ensuring that the basin is in the correct position for the pipework and the drawing specification.

2 Place the pedestal into position and gently lower the washbasin, complete with the waste trap fitted, onto the pedestal, ensuring that the centre lines match. Do not use any silicone on the pedestal face at this time. It is important first to ensure that the washbasin's position is correct, that the basin and pedestal match properly, and that the basin is level. Place a level on the top of the washbasin and, once the appliance has been adjusted level, mark the fixing holes underneath the basin and also mark around the pedestal at floor level. This will ensure that both pedestal and basin go back into the same position once the wall has been drilled.

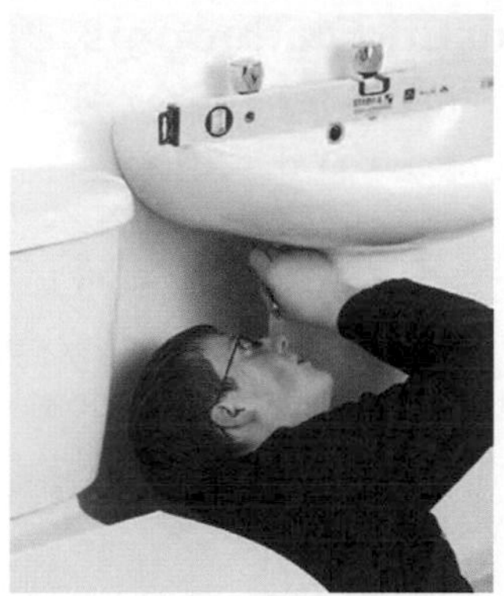

▲ Figure 9.106 Washbasin fixing holes being marked

3 Carefully drill the tiles and the wall, ensuring that the fixing holes are deep enough to allow the wall plugs to be inserted below the tile surface. Reposition the pedestal in line with the previous floor mark.

INDUSTRY TIP

It is a good idea at this stage to put a thin bead of silicone sealant around the face (lip) of the pedestal where the washbasin sits. This will ensure that both washbasin and pedestal are fixed together once it has cured.

4 Again, carefully reposition the washbasin and carefully screw back to the wall using brass or stainless steel screws. Do not over-tighten the screws or the fixing holes will break. Once again, check for level and clean any surplus silicone sealant from the pedestal.
5 The hot and cold pipework is placed behind the pedestal to hide it as much as possible. It is often difficult to install the pipework inside the pedestal itself. Any bends in the pipework need to be as high as possible so that they cannot be seen when a person is standing up. Do not be tempted to solder pipework joints near to the pedestal or cracking of the pedestal (and washbasin too) may occur. Again, it is considered good practice to install service valves on the pipework. Ensure that both the tap connectors (or compression joints if a monobloc mixer tap is being used) are fully tightened. The waste pipe can now be finished onto the previously fitted trap. Some pedestals have fixing holes at floor level and, if these are present, carefully screw the pedestal to the floor. Ensure that the service valves are in the off position prior to commissioning.

Installing the WC suite, step by step

Most WC suites today are of the close coupled style. The following procedure is based on this type of installation.

1 When installing the WC suite, the distance between the WC pan outlet and the wall should be measured so that the soil pipe can be trimmed to the correct length. The distance from the wall can also be obtained from the manufacturer's instructions. Remember to put the pan connector on the pan outlet first so that an accurate measurement can be taken.
2 Once the soil pipe has been cut to length, insert the pan connector into the soil pipework and carefully place the WC pan into position. Now place the cistern onto the pan, and fix it using the nuts, bolts and washers provided. At this stage, it is better to step back from the pan and look to make sure that the pan and cistern sit correctly. Place a spirit level across the back of the cistern to ensure that it is level and mark the cistern fixing holes with a pencil. If an overflow pipe is required this can also be marked. Remove the cistern and carefully drill the fixing holes, again ensuring that the holes are deep enough for the wall plugs to be pushed below the surface of the tiles. Carefully drill the hole for the overflow (if required).
3 Refix the cistern to the WC pan, ensuring that the large foam or rubber sealing washer, often referred to as a 'doughnut', that seals the cistern to the pan is in place around the WC siphon tail on the bottom of the cistern.
4 You can then proceed to screw the WC pan down, again using the correct gauge and length of screw. Brass or stainless steel screws are best used in this situation to prevent the screw from corroding.
5 The water connection should be installed as neatly as possible as it will be on view all the time. It is a requirement of the Water Supply (Water Fittings) Regulations that the cold water supply to WC cisterns contains a service valve. Be careful when connecting the tap connector as it is very easy to strip the thread of the plastic float-operated valve. Do not over-tighten the connector as this may also strip the thread. Ensure that the service valve is in the off position prior to commissioning.

▲ Figure 9.107 A WC pan and cistern being assembled

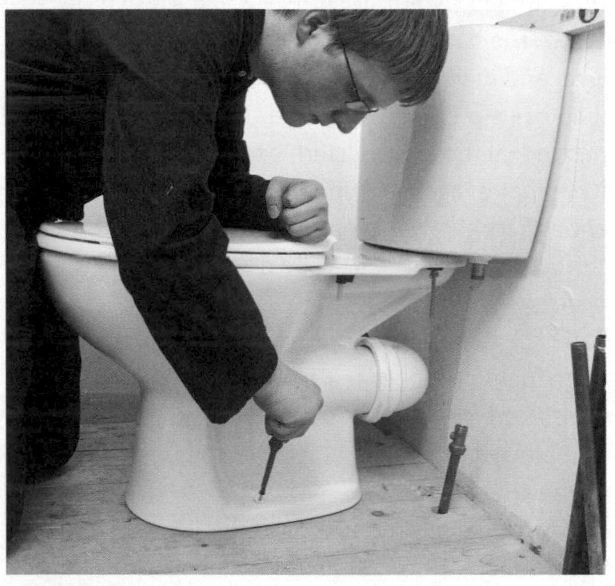

▲ Figure 9.108 A WC pan being screwed to the floor and cistern to the wall

INDUSTRY TIP

Before screwing the cistern to the wall, it is worth considering putting spacing washers (tap washers will work for this) between the cistern and the wall. This helps to prevent the build-up of condensation at the back of the cistern by allowing air movement, which in turn prevents the build-up of black mould on the tiles where the cistern is fitted. Also, to prevent breaking the cistern, place a tap washer over both brass screws before the cistern is screwed back to the wall.

Refurbishments of existing bathrooms

This is where plumbers can show their creativity by designing bespoke installations to suit the customer's requirements. The customer may already have an idea of how they want their bathroom to look, so it is important that we consider the ideas that they may put forward. It may also mean that the original bathroom layout will be altered, with appliances occupying different positions than they did originally.

IMPROVE YOUR ENGLISH

As noted, customers often invest a lot of thought, planning and money into achieving their dream bathroom design. It is your job to listen carefully to their wishes and offer them all feasible options that can be achieved for their space and budget.

We will presume here that the appliances are returning to their original positions and that the first-fix pipework has been completed.

VALUES AND BEHAVIOURS

Remember: customers, particularly those in domestic properties, cannot be left without a WC.

Although the method of installing the appliances is identical to that for the new-build installation, the order in which they are fitted might not be. In this situation, the customer cannot be without a WC, especially if the one you are replacing is the only one in the property. There are two choices in this situation:

1 **Leave the existing WC in place until all other appliances have been fitted:** if the appliances are to be installed in the same positions as the original bathroom suite, this is probably the better option as the WC will eventually be removed and any damage done to it while installing the bath will not matter. If the new WC is fitted first, in this situation damage could be costly in terms of materials and labour charges.
2 **Replace the WC first:** if the bathroom layout is being altered, with the WC occupying a new position, this will obviously be the only option as the soil stack connection will need to be altered before

the bathroom installation can begin. It is pointless installing the original WC on a new soil stack as this wastes precious installation time.

In both cases, the soil pipe to the WC should be blanked off (temporarily capped) when there is no WC fitted. This will prevent obnoxious smells from entering the working area. This can be done by the temporary use of a drain plug or PVCu cap end.

Once the bathroom suite has been installed, we can think about testing the sanitary pipework.

Jointing methods used in sanitary appliances pipework systems

INDUSTRY TIP

A full range of fittings are available for both polypropylene and PVCu, and some of these are looked at in Chapter 2, Common processes and techniques. It may be a good idea to keep a fittings catalogue handy when working on-site so that you are aware of the full range of fittings available.

The fixing details for polypropylene and PVCu (ABS and MuPVC) are covered in Chapter 2, Common processes and techniques, pages 100–102.

ACTIVITY

Soil and waste fittings

For more information on soil and waste pipe and fittings, revisit Chapter 2, Common processes and techniques, or check out the manufacturers' literature. Most manufacturers produce fittings catalogues and these are available from your local plumbers' merchant or by downloading as pdf files from the manufacturers' websites. Check out the following websites:

- www.hunterplastics.co.uk
- www.wavin.com/en-gb/Catalog/Potable-Water/Plumbing-Systems/Push-fit-Hep2O
- www.osma.co.uk
- www.marley.co.uk
- www.polypipe.com

Commissioning process for appliances and systems

By using the job specification and manufacturers' instructions during a visual inspection of a sanitary system, a plumber can verify that all complies with the original design and nothing has been changed.

Once the plumber is satisfied that the installation of a sanitary system is complete, it is important to check that all connections are properly fitted, such as any push-fit spigots are completely engaged in the fitting socket and any solvent welding is complete on waste and soil pipe joints.

It is essential that none of the joints or components leaks. Appliance pipework falls should be inspected and tested to confirm a smooth and efficient discharge. If a macerator is fitted, make sure there are no push-fit connections on the discharge pipework.

Any WC cistern mechanisms should be adjusted to discharge so that they comply with requirements for low water consumption. Mixing valves should be tested to make sure they are operating at a safe temperature, and the flow rates of showers and basins confirmed as being satisfactory.

Clips are often overlooked but they must be checked to confirm that they are properly anchored and spaced in accordance with the Standards, as their performance will be tested under load conditions.

If problems are identified during the commissioning process, then an apprentice plumber should consult with their supervisor and seek guidance as to a remedy. Quite often only an adjustment is required to a float-operated valve or temperature settings to a mixing valve. On some occasions, though, there could be a more serious problem, such as very slow discharge from an appliance that previously performed normally. It could be that, since the first fix, some debris has entered a trap or discharge pipe. It is possible that an improvised plug of compressed plastic has travelled down the soil stack and caused a major blockage, affecting all the appliances that are connected. The solution would be to remove it via an access point on the soil stack.

Once the installation procedures of all the sanitary appliances to a new soil stack have been completed, soundness testing can begin. In the case of multi-storey property installations, testing of appliances on a floor-to-floor basis is required. The installation needs to be checked in accordance with **BS 12056–2:2000** to ensure that there are no leaks as this will result in the ingression of foul odours into a property.

Soundness and performance testing of above-ground sanitation systems

The testing of above-ground sanitation systems is the final part of the installation process. When we test sanitary pipework, there are two elements we are looking at:

1 ensuring that the pipework is sound and does not have any leaks
2 ensuring that it performs to the recommendations of **BS 12056**.

Soundness testing, step by step

Before testing begins, make a visual inspection of the system to ensure that it conforms to the British Standards, that you are happy with the clipping distances and that all joints appear to be made correctly.

Testing should be completed in accordance with **BS 12056–2**.

> **INDUSTRY TIP**
>
> If the system is installed in a multi-storey property, it may need to be tested in stages or floor by floor.

1 First, seal the pipework at the top and the bottom by using either drain plugs or drain testing bags. The bottom drain plug can be inserted through the access cover at the base of the stack.
2 Fill all the traps on the system by letting a little water down each appliance and a little water down the WC to cover the bottom plug. This will ensure the plug's airtightness.
3 On the top plug, a rubber tube is fastened. The tube needs to have a tee piece inserted. On one side of the tee is a hand pump and an air inlet valve, on the other side a manometer (water gauge) is installed. The manometer is measured in mm.
4 The hand pump is pumped until a measurement of 38 mm is reached and the air inlet valve is turned off; 38 mm is the maximum pressure that should be pumped into the system so that the pressure does not push water out of any trap in the system (38 mm shallow bath trap being the minimum trap seal). The 38 mm test pressure must remain constant for a minimum of three minutes.

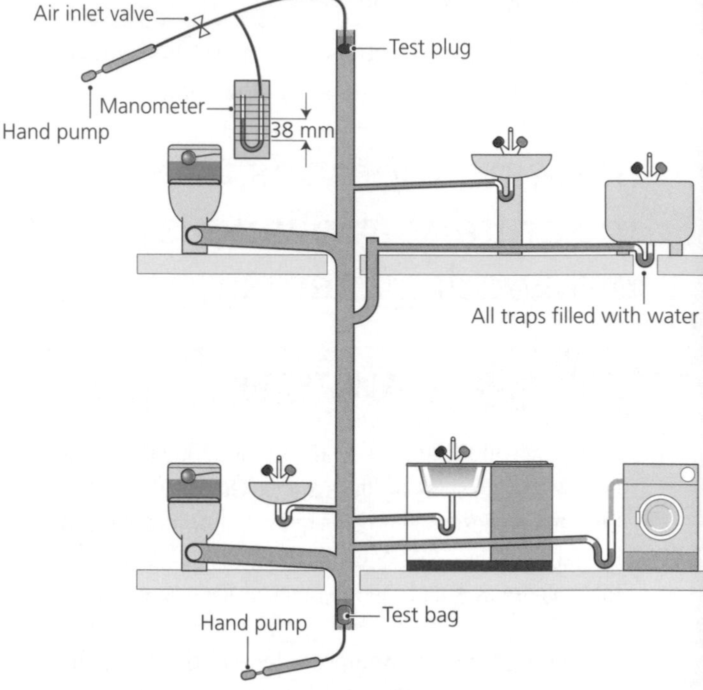

▲ Figure 9.109 The test equipment and procedure

Performance testing, step by step

With the soundness test complete and the test equipment removed, performance testing can begin. This is done to confirm that the system meets the recommendations of the British Standards and the Building Regulations.

1 Fill all of the appliances with water up to their overflow levels and release the water from the appliances simultaneously.
2 At the same time, flush the WCs.
3 When all of the appliances have emptied and the WC flushes have finished, the traps of all the appliances can be checked for water seal depth. The trap seal depth after all of the appliances have discharged their water must be at least 25 mm. This can be checked with a dipstick.

KEY POINT

Testing trap seal depth

The dipstick is a thin piece of wood painted matt black (wooden coffee stirrers painted matt black are sometimes used). It is inserted down the centre of the waste fitting until it reaches the bottom of the trap. When it is withdrawn, the wetness should be measured and the diameter of the trap deducted. What is left is the depth of trap seal.

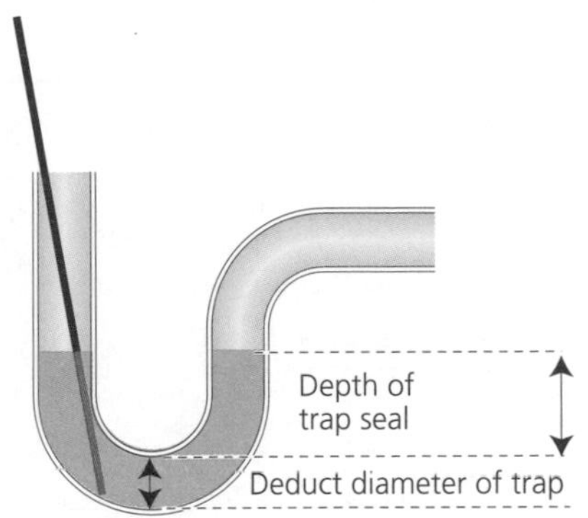

▲ Figure 9.110 Testing trap seal depth after a performance test

Expansion and contraction in sanitary appliances pipework systems and negative effects

One of the problems with PVCu soil and waste pipe is the large expansion rate. This can cause the joints to move as they get warmed by hot water discharging down them or the Sun and, in extreme cases, can cause joint failure.

INDUSTRY TIP

To counteract the expansion, push-fit soil and vent pipe should be withdrawn 10 mm. This will also prevent joint failure.

IMPROVE YOUR MATHS

PVCu has a coefficient of linear expansion of 0.06 mm/m/°C. This means that, for every metre (1 m) of pipe, PVCu expands by 0.06 mm for every degree rise in temperature.

For example:

If a 1 m length of pipe is subjected to a rise in temperature of 10°C, it will expand by the following amount:

$$1 \times 0.06 \times 10 = 0.6 \text{ mm}$$

This might not seem a lot, but let us look at this in more detail.

Example

A south-facing external soil and vent pipe 10 m high is subjected to a 25°C temperature rise. What is the expansion of the pipe when the coefficient of linear expansion of the pipe is 0.06 mm/m/°C?

All the information we need to be able to calculate this is in the question:

Length of pipe = 10 m

Temperature diff. (Δt) = 25°C

Coefficient of linear expansion = 0.06 mm/m/°C

Therefore:

$$10 \times 25 \times 0.06 = 15 \text{ mm}$$

ACTIVITY

Calculation of thermal expansion

Using the method shown in the worked example above, calculate the following:

1 A south-facing soil and vent pipe 15 m high is subjected to a 15°C temperature rise. What is the expansion of the pipe when the coefficient of linear expansion is 0.06 mm/m/°C?

2 A south-facing waste pipe 10 m long is subjected to a 30°C temperature rise. What is the expansion of the pipe when the coefficient of linear expansion is 0.06 mm/m/°C?

3 A south-facing soil and vent pipe 5 m long is subjected to a 20°C temperature rise. What is the expansion of the pipe when the coefficient of linear expansion is 0.06 mm/m/°C?

The procedure for notifying works carried out to the relevant authority

Once an installation is finished, the appropriate commissioning form should be completed and, if an installer is not part of a self-certification scheme, then the Local Authority Building Control (LABC) department should be informed.

However, if you are a member of a certification scheme and you have completed your commissioning paperwork, you have 30 days to submit your details. It is important to keep a record of all the tests that you have carried out as this will help in the event of a query at a later date, and will help you diagnose any problems that may have occurred since the installation. Invariably, you will also be installing sanitary fittings so the forms shown in Figures 9.111–9.113 may require completion.

Regulation 5.
Notification

5.-(1) Subject to paragraph (2), any person who proposes to install a water fitting in connection with any of the operations listed in the Table below-
(a) shall give notice to the water undertaker that he proposes to commence work;
(b) shall not begin that work without the consent of that undertaker which shall not be withheld unreasonably; and
(c) shall comply with any condition to which the undertaker's consent is subject.

TABLE

1. The erection of a building or other structure not being a pond or swimming pool.

2. The extension or alteration of a water system on any premises other than a house.

3. A material change of use of any premises.

4. The installation of-
- *(a) a bath having a capacity, as measured to the centre line of overflow, of more than 230 litres;*
- *(b) a bidet with an ascending spray or flexible hose;*
- *(c) a single shower unit (which may consist of one or more shower heads within a single unit), not being a drencher shower installed for reasons of safety or health, connected directly or indirectly to a supply pipe which is of a type specified by the regulator;*
- *(d) a pump or water booster drawing more than 12 litres per minute, connected directly or indirectly to a supply pipe;*
- *(e) a unit which incorporates reverse osmosis;*
- *(f) a water treatment unit which produces a waste water discharge or which requires the use of water for regeneration or cleaning;*
- *(g) a reduced pressure zone valve assembly or other mechanical device for protection against a fluid which is in fluid category 4 or 5;*
- *(h) a garden watering system unless designed to be operated by hand; or*
- *(i) any water system laid outside a building and either less than 750 mm or more than 1350 mm below ground level.*

5. the construction of a pond or swimming pool with a capacity greater than 10,000 litres which is designed to be replenished by automatic means and is to be filled with water supplied by a water undertaker.

(2) This regulation does not apply to the installation by an approved contractor of a water fitting falling within paragraph 2, 4(b) or 4 (g) in the Table.

(3) The notice required by paragraph (1) shall include or be accompanied by-
(a) the name and address of the person giving notice, and (if different) the name and address of the person on whom notice may be served under paragraph (4) below;
(b) a description of the proposed work or material change of use, and
(c) particulars of the location of the premises to which the proposal relates, and the use or intended use of those premises;
(d) except in the case of a fitting falling within paragraph 4(a), (c), (h) or 5 in the Table above,
(i) a plan of those parts of the premises to which the proposal relates; and
(ii) a diagram showing the pipework and fittings to be installed; and
(e) where the work is to be carried out by an approved contractor, the name of the contractor.

(4) The water undertaker may withhold consent under paragraph (1), or grant it subject to conditions, by a notice served before the expiry of the period of ten working days commencing with the day on which the notice under that paragraph was given.

(5) If no notice is given by the water undertaker within the period mentioned in paragraph (4), the consent required under paragraph (1) shall be deemed to have been granted unconditionally.

NOTICE OF INTENTION TO INSTALL WATER FITTINGS

I hereby give notice as required under Regulation 5 of the Water Supply (Water Fittings) Regulations 1999 that I intend to install water fittings as follows:

Intended installation date ☐

Location of premises where work is to be done...
....
Use of the buildings to which the notice refers...
....
Description of proposed work/fittings...
....
....

Is plan of proposed installation included? *Yes* *No*
Will there be a material change of use of the premises? *Yes* *No* *if yes give details*

Name of installer.............................. ***Approved Contractor Number***..............................
....
Company name and address...
....
Name of person on whom the notice may be served (if different to above)..............................
and address...
....
Signed.. ***Date***..............................

▲ Figure 9.111 Water undertaker's notification

To the customer: please keep this certificate safe, you may need to show it to an authorised water inspector.

Water Supply (Water Fittings) Regulations 1999

1: Installation of water fittings at: (insert name and address of premises where work has been undertaken)

Certificate of compliance

I certify that the work indicated below, carried out at the above premises complies with the requirements of the Water Supply (Water Fittings) Regulations 1999

2: Installation work carried out at the premises indicated by this notice	
the erection of a building or other structure, not being a pond or swimming pool	
the extension or alteration of a water system in premises other than a house	
a material change of use of premises	
the installation of:-	
a bath with a capacity, measured to the centre of the overflow, of over 230L	
a bidet with an ascending spray or flexible hose	
a single shower unit of a type specified by the regulator	
a pump or booster drawing more than 12 litres per minute from a supply pipe	
a unit that incorporates reverse osmosis	
a water treatment unit with waste water discharge or use of water for regeneration or cleaning	
a reduced pressure zone valve assembly or mechanical device for fluid category 4 or 5 protection	
an automatic garden watering system	
an outside water system laid less than 750 mm or more than 1350 mm below ground level an automatically filled pond or swimming pool with a capacity greater than 10,000L	

3: Name and address of contractor supplying this certificate

4:

Signature ....Date..

▲ Figure 9.112 Certificate of compliance

Regulation 6.

Contractor's certificate

6.-(1) *Where a water fitting is installed, altered, connected or disconnected by an approved contractor, the contractor shall upon completion of the work furnish a signed certificate stating whether the water fitting complies with the requirements of these Regulations to the person who commissioned the work.*

(2) In case of a fitting for which notice is required under Regulation 5(1) above, the contractor shall send a copy of their certificate to the water undertaker.

▲ Figure 9.113 Contractor's certificate

Handing over a completed system to the end user

Once all the tests have been carried out satisfactorily, the system operates as planned, and the work area, appliances and fittings are clean and ready for use, it is time to hand over to the customer. The customer will need to know how their newly installed system works, and will need to be given clear information on how appliances such as macerators, waste water lifters and showers operate. A full demonstration of how they operate is required, and advice on the limitations of an appliance's performance should be discussed in line with the manufacturer's instructions.

Decommissioning process for appliances and systems

When we remove old sanitary appliances and replace them with new ones, we are decommissioning the above-ground drainage system. In many cases, this will mean the removal of the soil and vent stack, and waste pipes too. These are procedures that need careful consideration. The following sections will assess the most effective way of decommissioning an existing system of above-ground drainage.

The old sanitary appliances

Removal of the old appliances should be carried out with care. We must use appropriate personal protective equipment (PPE) for this as we will be handling sanitary ware that has been used for personal ablutions and will most definitely be carrying disease.

HEALTH AND SAFETY

A risk assessment must be carried out as some of the appliances, such as cast iron baths, will be heavy and may require the assistance of a second person, especially if the bath is to be carried out of the property in one piece. Ensure the area and pathway to outside the property is clear and free from obstacles.

HEALTH AND SAFETY

Try not to damage vitreous china sanitary ware such as WCs and washbasins when you are removing them. Vitreous china is extremely sharp when broken. Always wear rubber gauntlets and eye protection during these operations.

Cast iron baths are often broken into four pieces before being carried outside. This is quite a dangerous task as the enamel on the bath is glass and will fly in all directions when it is hit. It is best to use a lump hammer for this. Start at the waste hole, as this is the weakest and thinnest point, and work down the spine of the bath, then work across the bath. Eventually, the bath will break into four almost identically sized pieces, which are much easier to carry. PPE required will be eye protection, gloves and ear defenders. It may be that the bath has more value in one piece as there is a market for second-hand cast iron baths. In this case, help will be needed to manoeuvre the bath outside safely.

VALUES AND BEHAVIOURS

You must be aware that this task is extremely noisy and can be alarming to the customer, so advise them of your intentions beforehand.

When all of the appliances have been removed from the property, they should be stripped of any scrap metal as this is recyclable.

The old sanitary pipework

Old sanitary pipework can be made from a variety of materials, including cast iron, lead and asbestos. Each of these materials has its own health and safety issues, which must be observed. We will look at the methods of removal individually.

HEALTH AND SAFETY

Remember: old sanitary pipework can be made of a variety of materials, and each one has particular health and safety considerations.

KEY POINT

A risk assessment must be carried out when dealing with the removal of all old sanitary pipework materials.

Cast iron: health and safety considerations

Cast iron is a heavy material. As you will be working at height when removing this kind of pipework, precautions must be taken so that the pathway around you is blocked off and signs posted warning of the dangers of falling debris. The most common type of fixing for cast iron was nails and bobbins fixed through lugs on the cast iron pipe sockets, known as ears. It is usual practice to break the ears of the pipe to free it from the wall. Care should be taken as these can fly off when being broken. Breaking the pipe in sections, working from the top, is the best way of removing this pipe but you must take care that pieces of broken pipe do not enter and block the drain. The correct PPE should be worn during this process, which includes hard hat, goggles, gauntlets and eye protection.

VALUES AND BEHAVIOURS

All metals, like cast iron and lead, have scrap value. They should be disposed of for recycling purposes so ensure you are working with the environment and sustainability in mind.

Asbestos: health and safety considerations

If you suspect that the soil and vent pipes are made from asbestos, you must seek advice from your supervisor. On no account must you break the pipe or you risk releasing potentially dangerous fibres into the atmosphere.

INDUSTRY TIP

If the material is asbestos, it is the law that this is removed by a specialist asbestos removal company (see Chapter 1, Health and safety practices and systems, page 26).

Lead: health and safety considerations

Traditionally, lead pipe was used for WC branches and waste pipes. Occasionally, soil and vent pipes made from lead can be found, though this is extremely rare. As with all lead, it should be handled with great care. Lead can sometimes corrode, leaving a fine white powder residue known as lead oxide. This material is extremely dangerous as it offers the quickest way of being ingested into the body through breathing in the powder. Always wear the correct PPE when handling lead, such as barrier cream on the hands, or wear gloves, a face mask and goggles. Lead is also a heavy material so, again, take care when lifting.

General points about decommissioning

- When working at height, place barriers and warning notices around where you are working.
- Large systems should be decommissioned in sections to minimise any disruption.
- If the system is being decommissioned for a short period, ensure that warning notices are placed at the appliances to prevent accidental usage while you are working on the system.
- Inform the customer of the length of time you expect the system to be out of action.
- If possible, arrange for decommissioning to be completed outside normal working hours to minimise disruption to users and residents.
- Arrange for alternative welfare facilities, such as portaloos and temporary showering accommodation, if the system is going to be out of commission for long periods.
- Always wear an appropriate level of PPE and conduct a risk assessment.

Permanent and temporary decommissioning

Decommissioning can be either temporary or permanent.

Temporary decommissioning

Temporary decommissioning is usually carried out when replacing bathroom suites or updating soil stacks and vent pipes.

General points to consider:

- Pipework should be properly capped or plugged with purpose-made fittings to prevent smells from infiltrating the building.
- Keep the customer informed of the probable length of time the system will be out of use.
- Keep mess and disruption to a minimum, and always clear away any waste or unwanted materials.

Permanent decommissioning

Permanent decommissioning is usually carried out when a soil and vent stack is being removed and not replaced. In this case, the stack and all appliances should be removed, and the drain properly capped at ground level.

Health hazards working with drainage systems

Weil's disease

Weil's disease is an acute human form of a bacterial infection with an array of different names – it is also known as mud fever, swamp fever, haemorrhagic jaundice, swineherd's disease and sewerman's flu.

Weil's disease is also known as leptospirosis and is caught through contact with infected animal urine, mainly from rodents which are found in drains, and typically enters the body through cuts, scrapes, or the lining of the nose, mouth, throat or eyes. This disease kills up to two or three people a year in Britain because the correct precautions and PPE have not been used or put in place.

Hepatitis

There are three known strains of hepatitis that you could contract. They are known as A, B and C. While not commonly contracted in Britain, the risks are higher when working in close contact with waste products. Symptoms are flu-like, which can continue for months and finally turn into liver failure. A 'healthy carrier' may not realise they have the disease until organ failure begins. Normal precautions in your day-to-day routine will limit your chances of contracting it.

Dermatitis

Dermatitis means 'skin inflammation', and in most cases the early stages are characterised by red, itchy skin, which can be mistaken for eczema. Acute attacks may result in crusty scales or blisters that results in pussy fluid discharges. Dermatitis is contracted by coming into contact with hazardous chemicals in the system. People who discard chemicals or other harmful substances down the drain network are putting other people's health at risk. PPE is the best way to prevent yourself from harm.

3 SERVICE AND MAINTENANCE REQUIREMENTS FOR SANITARY APPLIANCES AND CONNECTING PIPEWORK SYSTEMS

Like all the systems we have seen in this book, maintenance of AGDS should be carried out regularly to ensure problem-free operation. This is especially important with older systems, as some of the materials used in the past corrode over time, such as cast iron, and others, such as asbestos pipework, may bring health and safety issues. On larger systems or housing contracts, periodic maintenance will be carried out to a maintenance schedule that lists the properties and systems to be checked.

Simple maintenance tasks

The simple tasks that can be performed are described below.

Cleaning out traps

Traps, especially bath and shower traps, accumulate hair and soap residue that will eventually cause slow discharge of water, or even complete blockage. If left, this will eventually begin to smell.

Kitchen sink traps collect grease; this can be a constant source of problems as the grease clings to the waste pipework, making the smooth flow of water less likely in the future. These can be cleaned by disconnection from the appliance and thorough cleaning. Cleaning chemicals can be used but should be administered with caution as some can burn the skin on contact.

HEALTH AND SAFETY

- Always wear appropriate PPE when using chemicals or cleaning traps, such as rubber gloves and eye protection, and always read the dosage instructions.
- Never mix different cleaning chemicals as this could result in dangerous fumes developing and even explosive mixtures.

Cleaning out the overflows of the appliances

Belfast sink overflow are notorious for blocking. These can be cleaned with stiff wire and then thoroughly flushed out.

Checking access covers

These should be checked for leakage and tested to ensure that the bolts on the access door are free moving. A little silicone grease will prevent the bolts from rusting. Also, check the rubber seals to make sure that they are not showing signs of perishing.

Checking the pipework

Pipework is often neglected during periodic maintenance. Always check for signs of leakage and that the clips are in good order, especially if the soil and waste pipes are external as they can be affected by the weather. Direct sunlight is especially damaging to pipework and clips. Damaged or broken clips should be replaced. Also, check the cage on the top of the stack as these often blow off in high winds. These should be replaced as necessary.

Check for signs of overflowing WC cisterns

Adjust the water levels and check their correct operation.

KEY POINT

Make a note of all actions taken on the maintenance report.

Dealing with blockages

Unblocking drains and soil stack pipework is probably the most unpleasant of all the jobs a plumber undertakes, and can pose a real health risk.

HEALTH AND SAFETY

When dealing with blockages, always wear the correct PPE, including rubber gauntlets, eye protection, a face mask, full boiler suit and wellington boots.

Blocked soil pipes

There are a number of reasons why soil pipes and drains block. Often, it can be attributed to three possible causes:

1. **A broken drain:** if this is suspected, there is very little we, as plumbers, can do. The drain will probably need a camera inspection to accurately pinpoint the problem. Broken drains often occur because of ground compression or movement.
2. **A tree root growing through the drain:** again, if this is suspected, a camera inspection will be necessary.
3. **A physical blockage:** these are usually caused by something being flushed down the toilet and eventually becoming wedged in the drain. These can be moved by the use of drain rods (Figure 9.114), which come with various attachments to deal with a variety of blockage situations.

Sinks, washbasins and baths can often be cleared by a tool known as a force cup (Figure 9.115) (also known as a plunger). The blockage is cleared by filling the appliance with water and pressing down repeatedly on the handle of the force cup. This creates a positive pressure on the downward push and a negative pressure on the upward pull. This results in movement of water in the waste pipe, which is usually enough to dislodge the blockage. The force of the water when the force cup is removed will move the blockage down the wastepipe, breaking it up as the water flows.

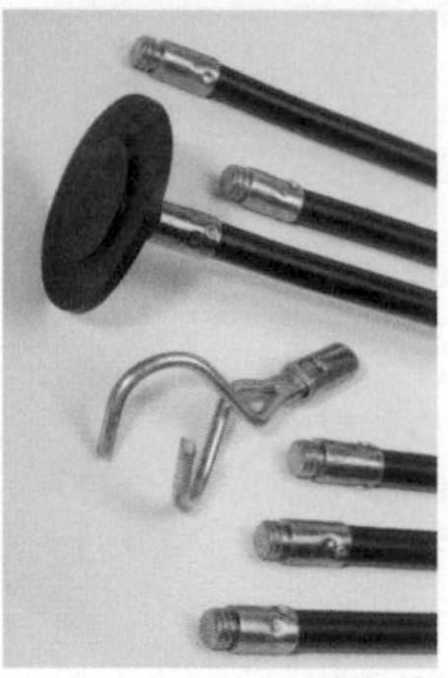

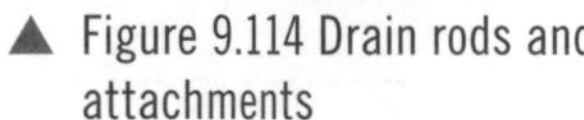

▲ Figure 9.114 Drain rods and attachments

▲ Figure 9.115 A force cup

Blocked WCs

Again, a blocked WC can often be cleared using a special kind of force cup, designed specifically to unblock WC pans and external gullies. It is known as a WC plunger (Figure 9.116).

Blocked waste pipes

Blocked waste pipes can often be cleared by the use of a hand spinner (Figure 9.117). An auger at the end of the hand spinner rotates as it enters the waste pipe, breaking up the blockage on contact. Care must be taken if this tool is being used with push-fit waste systems to ensure the joints are not being forced apart. Thorough testing should be conducted after use to make certain that leaks have not been created.

▲ Figure 9.116 A WC plunger

▲ Figure 9.117 A hand spinner

4 THE PRINCIPLES OF GREY WATER RECYCLING

Grey water

About a third of all water used in the average household is for WC flushing. The **grey water** used for bathing from baths, showers and washbasins can be collected, cleaned and reused for this purpose.

KEY TERM

Grey water: waste water from baths, showers, washing machines, dishwashers and sinks.

Grey water is usually clean enough for use in WCs with only minimal disinfection or micro-biological treatment. Problems can arise, however, when the warm grey water is stored because it quickly deteriorates and the bacteria it contains quickly multiply, making the water smell; this can be overcome by filtration and treatment with chemicals. There must also be a means of protecting the mains water against contamination by backflow from a grey water system in order to comply with the Water Supply (Water Fittings) Regulations 1999.

Rainwater harvesting

Rainwater harvesting has the potential to save a large volume of mains water and reduce pressure on resources because water that would otherwise be lost can be used to flush toilets, to water gardens and feed washing machines, instead of using water direct from the mains supply.

VALUES AND BEHAVIOURS

Water is a precious resource and rainwater harvesting is another means by which we can minimise our impact on the environment.

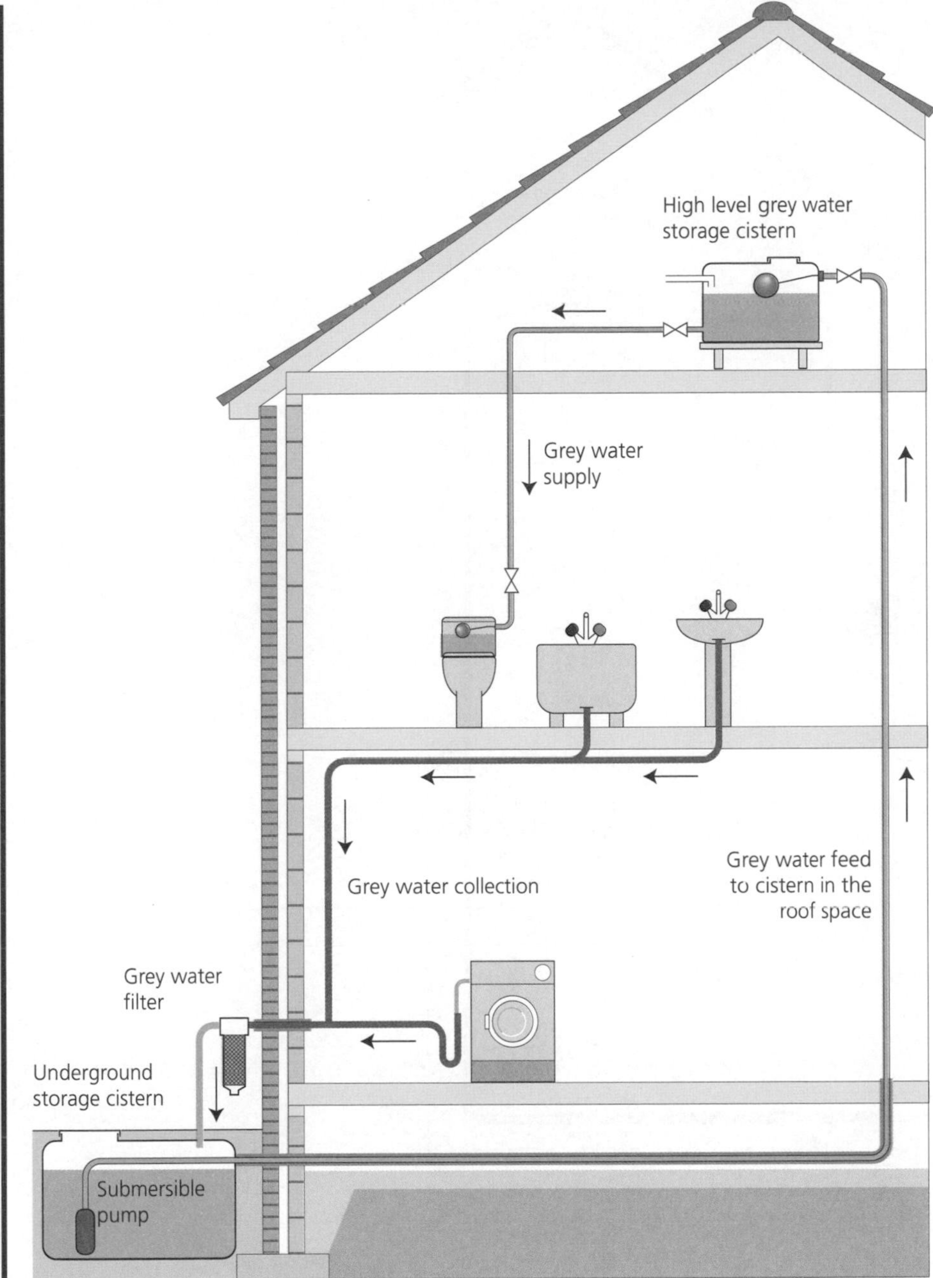

▲ Figure 9.118 Grey water system feeding a WC

Rainwater harvesters can be installed at domestic or commercial sites, and average households can expect to save up to 50 per cent of their water consumption by installing a rainwater harvesting system.

Harvesters are usually installed beneath the ground in an underground storage cistern, or on the roof of a flat-roofed building. A typical four-bedroom house will capture enough water to keep a 5000-litre cistern in use through most of the year.

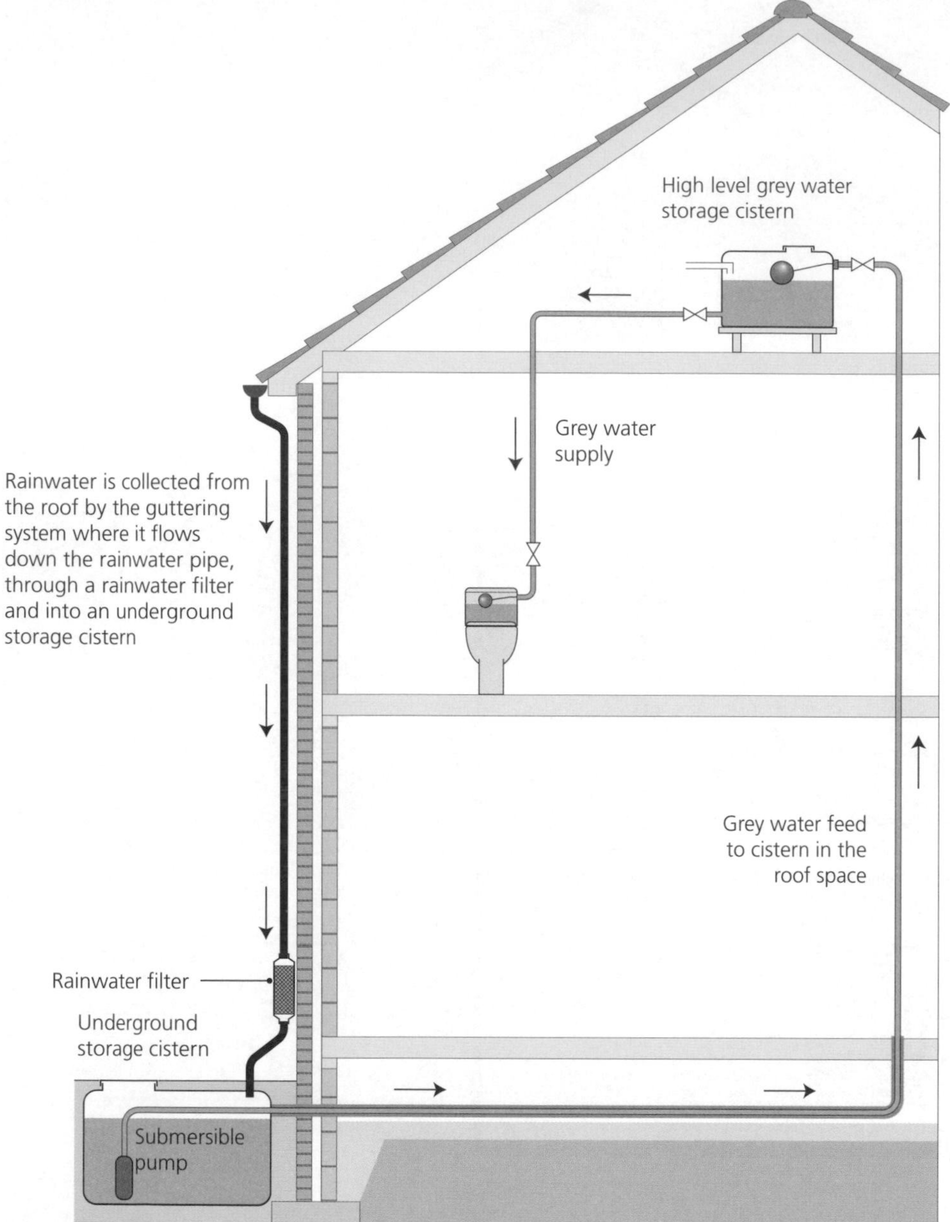

▲ Figure 9.119 Simple rainwater harvesting system

SUMMARY

We have seen as we have worked through this chapter just how important above-ground sanitation systems are with regard to both personal and environmental hygiene. Correctly installed and functioning sanitary appliances and pipework protect us from diseases that were rife in the UK just 200 years ago, and still continue to cause severe illness in other parts of the world to this day.

Properly installed sanitary appliances and pipework are a visual reminder of how well we can portray our plumbing skills while providing a necessary, hygienic environment for ourselves, the customer and the environment at large.

Test your knowledge

1 Which building regulation must sanitation systems comply with?

a Part G
c Part F
b Part H
d Part L

2 What is the correct pressure for the soundness testing of above-ground drainage systems?

a 21 mm
c 32 mm
b 28 mm
d 38 mm

3 What is the minimum trap seal depth for a bath?

a 25 mm
c 75 mm
b 50 mm
d 100 mm

4 What is the maximum length of waste pipe from a bath when connecting to a primary vented stack?

a 1.7 m
c 5 m
b 3 m
d 6 m

5 When a stack is installed within 3 m of an openable window, what minimum distance above that window should it terminate?

a 300 mm
c 900 mm
b 500 mm
d 1000 mm

6 What type of trap is shown in the image below?

a Bath trap
c Waterless trap
b In-line trap
d Washing machine trap

7 What problem can be caused by the use of short-radius bends at the bottom of discharge stacks?

a Compression
b Evaporation
c Excessive velocity
d Induced siphonage

8 Where a branch connection into a stack is between 82 mm and 160 mm in diameter (e.g. a WC branch), no other connection is allowed to be installed within which of the following distances vertically downwards?

a 90 mm
c 200 mm
b 110 mm
d 250 mm

9 Which of the following British Standards offers specific guidance on the space requirements for sanitary appliances?

a BS 6465–2
c BS 12056–2
b BS 3943
d BS 4305

10 What is the cause of loss of trap seal in the image below?

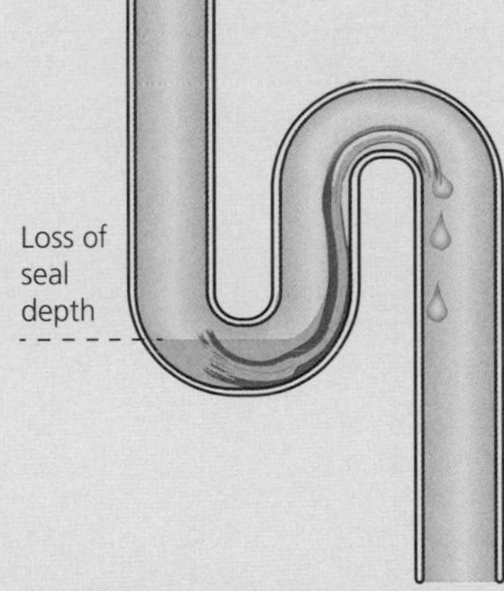

a Self-siphonage
b Compression
c Evaporation
d Capillary action

11 If a discharge branch is 4.5 m long and has a fall of 18 mm/m, what is the vertical drop?

a 250 mm
b 45 mm
c 22.5 mm
d 81 mm

12 What type of fitting is NOT used on a drainage system?

- **a** Compression
- **b** Push fit
- **c** Solder
- **d** Solvent weld

13 Which one of the following is an essential installation requirement for an air admittance valve located in boxing?

- **a** Coolness for operation
- **b** Access for maintenance
- **c** Consistent temperature for operation
- **d** Condensation for watertight seal

14 What equipment do you piece together (join) to help unblock drains?

- **a** Plunger
- **b** Drive in wedge
- **c** Boss strap
- **d** Drain rods

15 What type of soil stack is shown in this diagram?

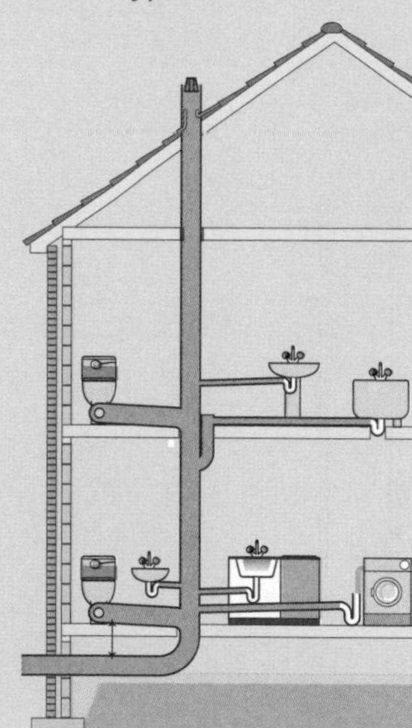

- **a** Ventilated discharge stack
- **b** Secondary ventilated stack
- **c** Primary ventilated stack
- **d** Combined discharge stack

16 What appliance is unsuitable for a bottle trap to be installed under?

- **a** Basin
- **b** Sink
- **c** Urinal
- **d** Bidet

17 At what pressure is a soil stack tested to prove soundness?

- **a** 1 bar
- **b** 1.2 times working pressure
- **c** 38 mm
- **d** 75 mm

18 What is the best way to prevent crossflow in a soil stack?

- **a** Ensure branches are not positioned directly opposite each other
- **b** Ensure only one strap boss is used on the system
- **c** Ensure branches are positioned leaving 50 mm vertical distance
- **d** Ensure the same size branches are not opposite each other

19 Which item is used if a soil stack needs to be temporarily decommissioned?

- **a** End cowl
- **b** Drain plug
- **c** Manifold plate
- **d** Access cap

20 Identify this style of trap:

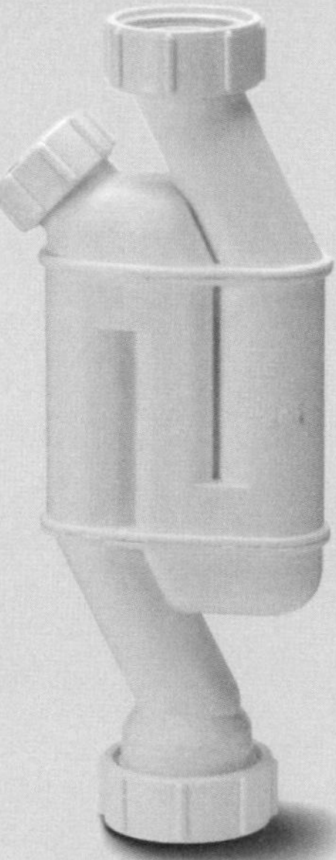

- **a** Running trap
- **b** Waterless valve
- **c** In line trap
- **d** P trap

21 What style of below-ground drainage system allows all the surface water and sewage to enter the foul water sewer in the road?

- **a** Separate
- **b** Combined
- **c** Partially separate
- **d** Integrated

22 What is the sole purpose of a trap below an appliance?

- **a** To aid the self-cleaning process of the appliance
- **b** To stop noxious smells entering the property
- **c** To prevent blockages in the discharge branch
- **d** To prevent the fast flow of liquids in the system

23 The vent termination must rise 900 mm above an opening window if the window is within what distance?

- **a** 6.0 m
- **b** 9.0 m
- **c** 10.0 m
- **d** 3.0 m

24 When would an anti-vac trap be used?

- **a** When there is a possibility of bad weather
- **b** When there is a possibility of trap seal loss
- **c** When there is a possibility of large fluid discharges
- **d** When there is a possibility of small fluid discharges

25 How many times must a performance test be carried out?

- **a** Once
- **b** Twice
- **c** Three times
- **d** Four times

26 What hazards are involved in the removal of a cast iron bath from an existing bathroom?

27 What are the maximum flush limits for toilet cisterns for both long and short flushes?

28 Explain the purpose and operation of the component in the image below.

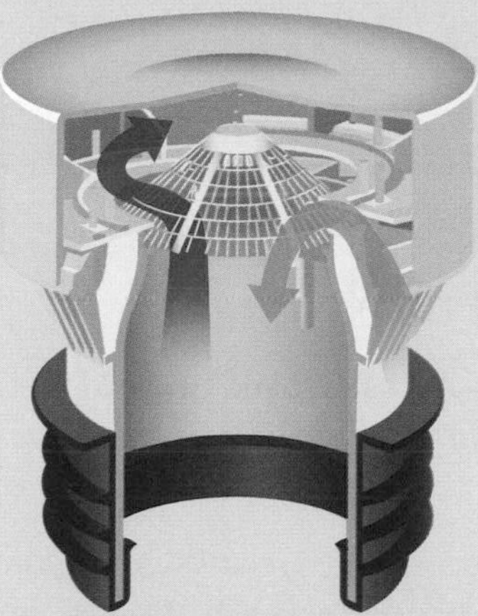

29 Give at least three advantages of the waterless (self-sealing) trap.

30 A 32 mm diameter waste pipe is to be installed from a wash hand basin. The length of the branch is 1.5 m, what is the recommended gradient in mm/m?

31 Describe what self-siphonage is and where it could occur.

32 What advantage does a secondary ventilated stack have over a primary ventilated stack in larger installations?

33 Explain why care must be taken when installing a bidet in a customer's property.

34 Outline how a self-sealing trap operates.

Answers can be found online at www.hoddereducation.co.uk/construction.

Practical activity

Ask your tutor, or perhaps your supervisor on-site, if you could try installing a trap to a basin (or other appliance). First, you will need to install a suitable waste to the basin to allow for the trap to be connected. Select a suitable trap for the appliance and connect the trap and waste. Be careful not to damage the appliance, which could easily be chipped or marked if not handled carefully.

CHAPTER 10

DOMESTIC FUEL SYSTEMS 

INTRODUCTION

For hundreds of years humans relied on solid fuel in the form of wood and coal to heat their homes. Then, in the 1850s, gas in the form of coal gas was used to heat and light dwellings and factories. This was followed soon after by oil.

These fuels – coal, gas and oil – are known as hydrocarbons and, because of the way they were formed millions of years ago, they are very carbon rich. When they are combusted, they produce copious amounts of carbon dioxide (CO_2), which has systematically altered the Earth's climate and this has led to the phenomenon known as global warming.

Now, less than 300 years later, fossil fuels are all but depleted and the damage to the climate they have caused is practically irreparable. With gas and oil reserves set to last only 50 years, and much of the coal left below the Earth's surface unreachable, we have to look for alternative forms of energy for our heat and light.

This chapter will investigate the types of fuels used in the appliances we install and identify the reasons that certain fuels are chosen. We will also take a look at how these fuels are supplied and stored.

By the end of this chapter, you will have knowledge and understanding of the following:

- types of fuels used in appliances
- factors that affect the selection of fuels
- sources of information for fuel supply installation
- the regulatory bodies that govern the installation of fuel systems
- storage requirements for fuels
- considerations that could affect the storage requirements of fuels.

1 IDENTIFY THE TYPES OF FUELS USED IN APPLIANCES

The heating appliances that we install are fuelled by a selection of energy sources, some of which have been around for many years and some that are relatively new technology. In this first section, we will investigate these fuels, both old and new. We will learn where they come from and the consequences of using them.

KEY POINT

The information in this chapter relates to the Plumbing and Domestic Heating Technician Apprenticeship only. It covers LO1 of Unit 11, Domestic fuel systems.

There are five categories of fuels:

1 natural gas
2 liquid petroleum gas (LPG)
3 oil
4 solid fuel
5 biomass fuel.

Natural gas

Natural gas is a combustible mixture of hydrocarbon gases and is probably the most widely used hydrocarbon fuel on Earth. It is colourless and odourless in its purest form and, when it is combusted, it releases a vast amount of energy with fewer emissions than many other common fossil fuels. Natural gas is naturally occurring and is usually found during the extraction of oil from deep below the Earth's surface, but it can also be found near coal formations and seams.

Natural gas is composed primarily of five combustible gases, two inert gases and water vapour (see Table 10.1).

▼ Table 10.1 The composition of natural gas

Gas	Chemical symbol	Percentage
Methane	CH_4	70–90%
Ethane	C_6H_6	0–5%
Propane	C_3H_8	0–20%
Butane	C_4H_{10}	0–5%
Hydrogen sulphide	H_2S	0–5%
Nitrogen	N_2	0–5%
Carbon dioxide	CO_2	0–8%
Water vapour	H_2O	

The distinctive 'rotten eggs' smell of natural gas is added to the gas when it is cleaned of the impurities and **naphtha** it contains at the refinery. The smell is a chemical called mercaptan, which is added to aid the detection of gas leaks.

Natural gas is lighter than air, having a specific gravity of 0.6–0.8. It is available in most cities, towns and villages through a national grid of underground pipes, with only the most isolated of places not connected to this.

The **calorific value** (CV) of gas is usually 37.8–43 MJ/m^3 depending on where the gas was extracted from.

Most natural gas used in the UK comes from the North Sea, but other sources include Russia and the Middle East.

KEY TERMS

Naphtha: a waxy oil deposit that is present in natural gas in its unrefined state. It is removed and later reused in other products such as cosmetics.

Calorific value: the amount of energy stored in the gas in its uncombusted state. It is the amount of energy released when the gas is combusted. It is measured in megajoules (MJ) per cubic metre or MJ/m^3.

Liquid petroleum gas (LPG)

Liquid petroleum gas, like natural gas, is a fossil hydrocarbon fuel that is closely linked to oil. About two-thirds of all LPG used is extracted direct from oil wells; the rest is extracted during the manufacture of petroleum from crude oil.

There are many types of LPG but, generally, only three of these are used commercially: propane, butane and iso-butane. These gases share common elements but in different quantities and these are reflected in their chemical symbols:

- propane – three atoms of carbon and eight atoms of hydrogen (C_3H_8)
- butane – four atoms of carbon and ten atoms of hydrogen (C_4H_{10})
- iso-butane is butane that has the same elements, but these are connected in a slightly different way.

The composition of LPG fuels is shown in Table 10.2.

▼ Table 10.2 The composition of LPG fuels

LPG attributes	Propane	Butane
Chemical formula	C_3H_8	C_4H_{10}
Energy content: MJ/m^3	95.8	111.4
Energy content: MJ/kg	49.58	47.39
Boiling temp.: °C	–42	–4
Pressure @ 21°C: kPa	858.7	215.1
Flame temp.: °C	1967	1970
Gas volume: m^3/kg	0.540	0.405
Relative density: H_2O	0.51	0.58
Relative density: air	1.53	2.00
L per kg	1.96	1.724
kg per L	0.51	0.58
Specific gravity @ 25°C	1.55	2.07
Density @ 15°C: kg/m^3	1.899	2.544
Combustion temp.:°C	482–540	410–470

▼ Table 10.3 The uses of LPG fuels

Compound	Uses
Butane C_4H_{10}	Used for portable supplies, such as camping equipment, boats and barbecues. Not much use for plumbing or heating installation as it boils (turns from a liquid to a gas) at –4°C.

Compound	Uses
Propane C_3H_8	Has a very low boiling point at –42°C. Can be used in domestic situations as an alternative to natural gas where the mains gas supply is not available. Many appliances are available for use with propane, including boilers, cookers, fires and water heaters.
Iso-butane	Used as a refrigerant in domestic refrigerators and fridge-freezers.

Both of these compounds are heavier than air in their gaseous form, with propane having a specific gravity of 1.5 and butane having a specific gravity of 2.0. In liquid form, both are thinner than water, butane having a relative density of 0.58 and propane 0.51.

When LPG gas is subjected to high pressure it turns into a liquid, but it also takes up less space than the gas; 1 litre of LPG in its liquid state makes 274 litres of LPG gas. This means that one cylinder of LPG liquid is equivalent to 274 cylinders of LPG gas.

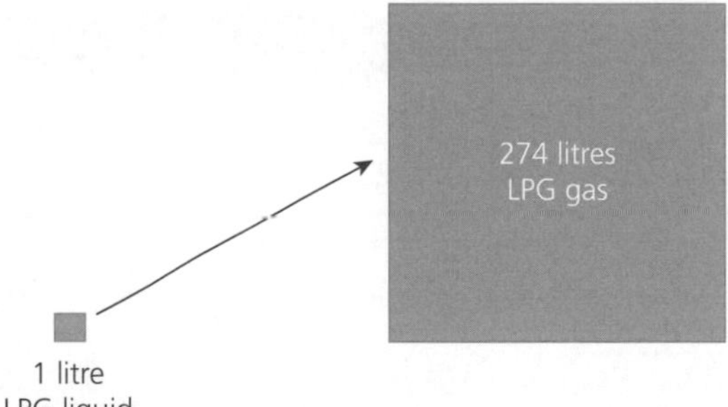

▲ Figure 10.1 Liquid to gas ratio

Environmentally, LPG is relatively clean when compared to other fuels such as coal or oil, creating far less air pollution in the form of soot and carbon particulates, sulphur and carbon dioxide, and therefore adds less to global warming than might be realised.

Cost, however, is an issue, since LPG is much more expensive than conventional natural gas.

Fuel oil (kerosene grade C2, 28 second viscosity oil to **BS 2869:2017**)

A simple definition for fuel oil is a liquid by-product of crude oil, which is produced during petroleum refining. There are two main categories under which it is classified:

1 distillate oils – such as diesel fuel
2 residual oils – includes heating kerosene, generally used for home heating.

Around 95 per cent of boilers burning fuel oil in domestic properties use kerosene, which is also known generically as C2 grade, 28 second viscosity oil. This is the preferred oil fuel grade for domestic heating, due to its clean combustion. Modern oil central heating boilers require only a single annual service if being used with an atomising pressure jet burner. It is the only oil grade that can be used with balanced or low-level flues.

Kerosene has very good cold-weather characteristics and remains fluid beyond minus 40°C, although it does tend to thicken slightly during extremely cold weather.

Kerosene is a high-carbon fuel and is clear or very pale yellow in colour. Newer boilers have a label inside the casing, with information on nozzle size and pump pressure, which show that the boiler has been set up to use kerosene. It may also reference the British Standard for kerosene **BS 2869** grade C2.

▲ Figure 10.2 A domestic kerosene oil tank for oil-fired heating

Solid fuel (coal, coke and peat)

There are three main types of solid fuel. These are:

1 coal
2 coke
3 peat.

▼ Table 10.4 The different types of commercially available coal

Coal type	Heat content kW/kg	Carbon content %	Description
Lignite	2.2–5.5	25–35	The lowest type of coal, lignite is crumbly and has high moisture content. Most lignite is used to produce electricity.
Sub-bituminous	5.5–8.3	35–45	Typically contains less heating value than bituminous coal, but more moisture.
Bituminous	7–10	45–86	Formed by added heat and pressure on lignite. Made of many tiny layers, bituminous coal looks smooth and sometimes shiny. It has two to three times the heating value of lignite. Bituminous coal is used to generate electricity, and is an important fuel for the steel and iron industries.
Anthracite	10	86–97	Created where additional pressure combined with very high temperature inside the Earth. It is deep black and looks almost metallic due to its glossy surface.

Coal

This is a fossil fuel created from the remains of plants that lived and died between 100 and 400 million years ago, when large areas of the Earth were covered with huge swamps and forest bogs.

The energy that we get from coal comes from the energy that the plants absorbed from the Sun millions of years ago. The process is called photosynthesis. When plants die, this stored energy is usually released during the decaying process, but when coal is formed the process is interrupted, preventing the release of the trapped solar energy.

As the Earth's climate evolved and the vegetation died, a thick layer of rotting vegetation built up that was covered with water, silt and mud, stopping the decaying process. The weight of the water and the top layer of mud compressed the partially decayed vegetation under heat and pressure, squeezing out the remaining oxygen and leaving rich hydrocarbon deposits. What once had been plants gradually fossilised into the combustible carbon-rich rock we call coal.

Types of coal

Coal is classified into four main types (see Table 10.4), depending on the amount of carbon, oxygen and hydrogen present. The higher the carbon content, the more energy the coal contains.

Coal is still used for central heating boilers, both domestic and industrial, and for steam and electricity generation.

▲ Figure 10.3 Open-cast coal mine

Coke

Coke is produced by heating coal in coke ovens to around 1000°C. During this process, the coal gives off methane gas and coal tar, both of which can be cleaned and reused. Coke burns clearly and without a flame, and gives out a lot of heat. However, it has to be mixed with coal as it will not burn by itself.

Coke is a smokeless fuel that is valued in industry because it has a calorific (heat) value higher than any form of natural coal. It is widely used in steel making and in certain chemical processes, but can also be used in some domestic boilers and room heaters.

Peat

Peat is an organic material that forms over hundreds of thousands of years from the decay of plant material in the absence of oxygen, in boggy, waterlogged ground. This encourages the growth of moss, which forms the basis of the peat. As the plants die, they do not decompose. Instead, the organic material slowly accumulates as peat because of the lack of oxygen in the bog. Peat is a poor-quality fossil fuel that is easily cut and dried.

Peat has a high carbon content but much less than coal, with large amounts of ash produced during combustion.

It is used in many domestic fires, room heaters and peat-burning stoves.

Sustainable, low-carbon fuels

Low carbon can be classified as fuels made from renewable sources like those described below.

- **Solar thermal:** solar thermal technology utilises the heat from the Sun to generate domestic hot water supply to off-set the water heating demand from other sources, such as electricity or gas.
- **Solid fuel (biomass):** the term biomass can be used to describe many different types of solid and liquid fuels. It is defined as any plant matter used directly as a fuel or that has been converted into other fuel types before combustion. When used as a heating fuel, it is generally solid biomass including wood pellets, vegetal waste (including wood waste and crops used for energy production), animal materials/wastes and other solid biomass.
- **Heat pumps:** a heat pump is an electrical device with reversible heating and cooling capability.
  It extracts heat from one medium at a low temperature (the source of heat) and transfers it to another at a high temperature (called the heat sink), cooling the first and warming the second. They work in the same way as a refrigerator, moving heat from one place to another. Heat pumps can provide space heating, cooling, water heating and air heat recovery. There are several different types:
  - ground source heat pumps
  - air source heat pumps
  - water source heat pumps
  - geo-thermal heat pumps.

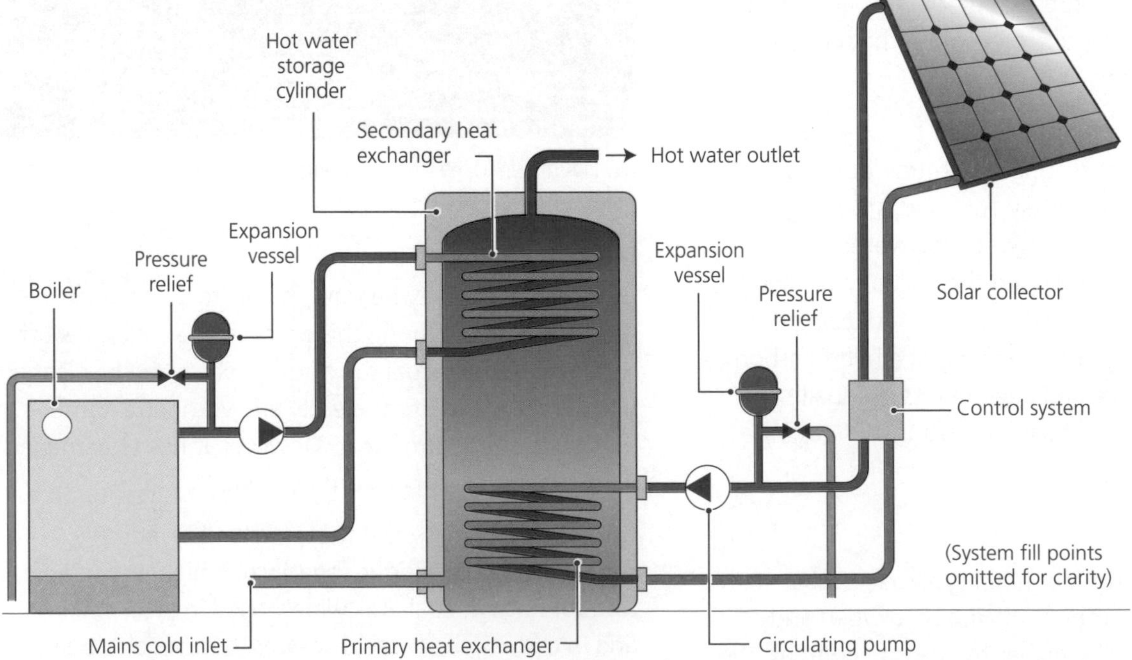

▲ Figure 10.4 Solar thermal system

▲ Figure 10.5 Biomass wood pellets

- **Combined heat and power (CHP):** combined heat and power is a plant where electricity is generated and the excess heat generated is used for heating. It is used primarily for district heating systems but micro-CHP has also been developed for domestic properties.
- **District heating:** This uses the CHP principle to supply low-carbon, high-efficiency heat to high-density homes, businesses and public buildings. This system is used extensively over Europe to good effect. In the UK there are now more than 17,000 district heating systems, mainly in city areas. District heating relies on high efficiency and areas of dense population. There is one highly efficient heat source and this heat is transferred through plate heat exchangers into the primary pipe network. The primary pipe network transports the heat to individual properties. Each property has its own heat interface unit (HIU). This interface becomes the 'boiler' for the property, which allows the heat to enter the systems within the property.

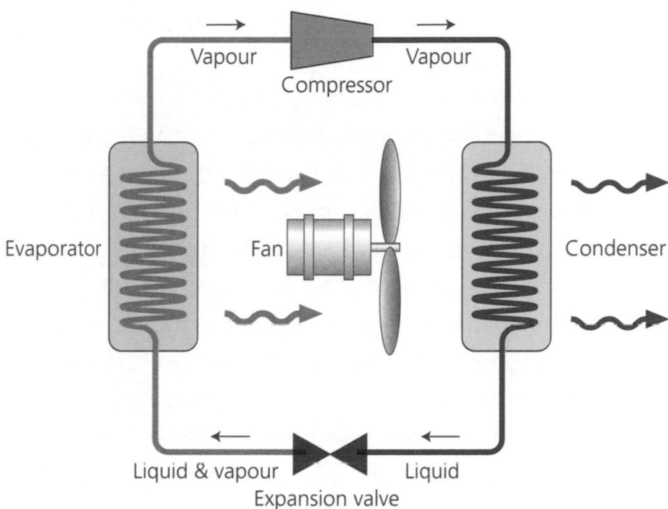

▲ Figure 10.6 Heat pump theory

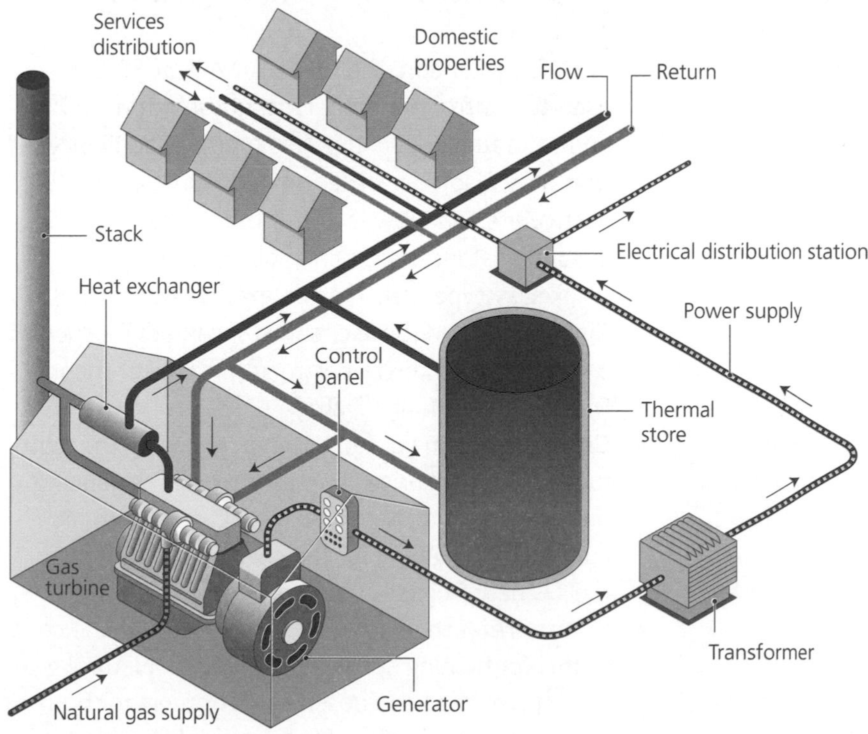

▲ Figure 10.7 A CHP system

- **Combined cooling, heat and power (CCHP):** very similar to CHP, combined cooling, heat and power uses the excess heat from electricity generation to achieve additional building heating or cooling.

INDUSTRY TIP

Watch this video clip: https://www.youtube.com/watch?v=MhJVsSkxg7s

2 FACTORS THAT AFFECT THE SELECTION OF FUELS

There are many reasons why the fuels used in domestic appliances are chosen:

- **Availability:** the availability of fuels plays a big part when choosing the right fuel for an installation. For example, for most of the UK, natural gas is available piped to the home with no interruption of supply. However, in many rural areas, the piped gas supply is many miles away. In this instance, like other fuels such as coal and oil, gas is delivered by suppliers and the customer is dependent on a regular fuel delivery. While, in most cases, this does not pose a problem, in times of inclement weather, deliveries may be interrupted or cancelled, leaving the customer with no means of heating or cooking.
- **Appliance type:** the type of fuel available will dictate the type of appliance used, and vice versa. Some appliances may be dual-fuel types, where two types of fuel may be used in the same appliance. It must be remembered that gas appliances must be purchased in line with gas type available. A natural gas appliance cannot be used on an LPG supply. Similarly, a coal-fired boiler will have solid fuels recommended by the manufacturer and must not be used with other types of solid fuel.
- **Fuel storage requirements:** with the exception of natural gas, all fuels require storage space within the boundary of the property. With solid fuels and biomass, these can take up considerable space. Biomass also needs to be kept reasonably dry. Where oil and LPG are concerned, space may not be the issue. However, safe distances from the property to store the fuel may be dictated either by legislation or manufacturers' instructions.
- **Environmental considerations:** most fuels used in hot water and heating systems directly or indirectly create waste products that are harmful to the atmosphere. These may be by direct pollution, such as soot and sulphur emissions from coal and oil combustion, or saturation of the atmosphere by CO_2. Some gases released by fossil fuel combustion, such as nitrogen dioxide (NO_2), are extremely toxic in large quantities.

  With solid fuels, there is an added environmental problem in the form of ash and clinker left over from the combustion process that requires careful consideration and disposal. Consultation and advice should be sought from the fuel supplier and local authority as to the recommended disposal methods.
- **Smoke control legislation:** under the Clean Air Act of 1993, local authorities may declare that a district is a smoke control area. It is an offence under this Act to emit smoke from a chimney from a boiler or furnace located within an area designated a smoke control area.

  In Greater London, the Clean Air Act of 1993 is used to control the emissions, not just from oil and solid fuel boilers but also biomass appliances that may also emit other noxious fumes, fly ash particulates and low levels of ozone gas.

  In some instances, certain appliances and smokeless-type fuels may be exempt from the Clean Air Act. In these circumstances, advice should be sought from the Department for Environment, Food & Rural Affairs (Defra).
- **Cost:** this is a major factor when choosing the right fuel. Heating oil prices fluctuate widely, depending on the price of crude oil, while LPG prices remain consistently high. By far the cheapest of the fossil fuels is natural gas.
- **Client preference:** in towns and cities, the choice of fuel for heating appliances is limited. Natural gas is the preferred fuel chosen by customers for both heating and cooking appliances, simply because it is readily available. Solid fuel, in the form of smokeless

fuel, is still used in some areas. In rural settings, heating oil is preferred. LPG is expensive, and this is often the reason that this fuel is rejected. Many new-build properties are actively seeking greener alternatives to fossil fuels, with heat pumps and electric boilers being chosen because of their very low carbon emissions.

INDUSTRY TIP

The Department for Environment, Food & Rural Affairs' website can be accessed at: www.gov.uk/government/organisations/department-for-environment-food-rural-affairs

3 SOURCES OF INFORMATION FOR FUEL SUPPLY INSTALLATION

Boilers, cookers, room heaters and fires require a supply of fuel, whether that fuel is piped direct to the door or delivered by a tanker. Each fuel has specific supply and storage requirements that must comply with certain documents:

- **Regulations:** there are certain regulations that fuel supply systems must comply with to maintain the safety of the property where the appliances are installed and the safety of the building occupants. Solid fuel and oil systems, such as coal, coke, biomass and heating oils, are simple to understand, since the fuel is readily visible. However, gaseous fuel systems such as those for natural gas and LPG tend to be much more stringent, since these fuels are at pressure and cannot be seen. Regulations include:
  - the Gas Safety (Installation and Use) Regulations 1998
  - Approved Document J of the Building Regulations – Combustion Appliances and Fuel Storage systems (2010 edition incorporating 2010 and 2013 amendments)
  - the 18th Edition of the IET Regulations.
- **British Standards:** there are many British Standards and European Standards that give recommendations when installing fuel systems.
- **Manufacturers' instructions:** manufacturers of appliances and components will often give advice about the installation of the fuel system to the appliance. These may sometimes conflict with the Regulations and British Standards. In this instance, the manufacturer's instructions must always be followed.
- **Guidance notes:** guidance notes are produced by regulatory bodies and professional associations to assist in compliance with the Regulations. Many guidance notes are produced by the HSE. They should be read in conjunction with the Regulations and manufacturers' instructions.

KEY POINT

There is a comprehensive list of British Standards in Approved Document J of the Building Regulations – Combustion Appliances and Fuel Storage Systems.

4 REGULATORY BODIES THAT GOVERN THE INSTALLATION OF FUEL SYSTEMS

Before we investigate the **regulatory bodies** concerned with the installation of fuel systems and appliances, we must first understand what a regulatory body is. Its aims are to protect consumers, and to educate and guide installers in the ways of good practice. Occasionally, it may be necessary for a regulatory body to prosecute, in the interests of public safety, those installers who refuse to comply with regulations. In the plumbing and heating industry, it is compulsory to belong to the regulatory bodies if you engage in the installation of either gas, oil or solid fuel appliances and fuel supply systems.

KEY TERM

Regulatory body: an organisation set up by the government to monitor, control and guide various sectors within industry.

In the plumbing and heating industry, there are three regulatory bodies related to fuels:

1 **Gas Safe:** this is the UK registration body for the installation, maintenance and repair of gas installations and appliances. By law, all operatives engaging in domestic natural gas and LPG installations must be registered with Gas Safe and must hold various qualifications within the gas industry.
2 **OFTEC:** this is the registration body for the installation and maintenance of oil-fired heating appliances and fuel systems. Registration is voluntary, but being a member is considered good practice. OFTEC registration means that installers are able to self-certify installations without the need for local authority intervention and inspection. OFTEC also administers recognised and authorised training courses for installers.
3 **HETAS:** this is the official body recognised by the UK Government for approving solid fuel and biomass domestic heating systems, fuels and appliances. HETAS also manages a register of approved, competent installers and servicing businesses, and oversees HETAS-registered training courses.

▲ Figure 10.8 OFTEC logo

▲ Figure 10.9 HETAS logo

INDUSTRY TIP

- More information can be found on the Gas Safe website at: www.gassaferegister.co.uk
- More information can be found on the OFTEC website at: www.oftec.co.uk/technicians/industry-news-and-training/training-courses-assessments
- More information can be found on the HETAS website at: www.hetas.co.uk/professionals/training-courses/

5 STORAGE REQUIREMENTS FOR FUELS

In this section, we will investigate the methods of safely storing:

- coal and smokeless fuels
- oil
- LPG
- biomass.

Storing coal and smokeless fuels

The Solid Fuel Association recommends that coal should be stored outside of any dwelling in a purpose-made bunker, to protect the fuel from damage. There are a number of recommendations as to how coal should be stored:

- Coal may be stored inside or outside the property.
- Coal should be covered to reduce the contaminants that can enter the fuel.

- A smooth, hard floor is important as it allows easy shovelling of the fuel.
- If the fuel is stored in a coal bunker, a slight slope on the base of a coal bunker prevents water from collecting inside it. Keeping the fuel dry makes it easier to combust.
- The area around the coal bunker should be well lit to ensure safe bagging and shovelling.
- Good ventilation of the bunker helps to prevent a build-up of moisture, allowing the fuel to stay dry.

INDUSTRY TIP

Unlike other fuel sources, there are no special rules, regulations or restrictions when it comes to storing coal and smokeless fuels, other than storing it away from the heating appliance or boiler.

Storing fuel oil

The following information is intended as a general guide as the Regulations regarding oil storage may vary slightly, depending on the location of the installation.

Oil storage tank specifications

Generally, oil storage tanks of up to 3500 litres capacity, supplying oil to a single domestic property, can be made of either plastic or steel. The actual size for any given installation will depend on the individual requirements. Any tank installed should conform to the following specifications:

- OFS T100 for plastic storage tanks
- OFS T200 for steel storage tanks
- **BS 799 PT5:1987**.

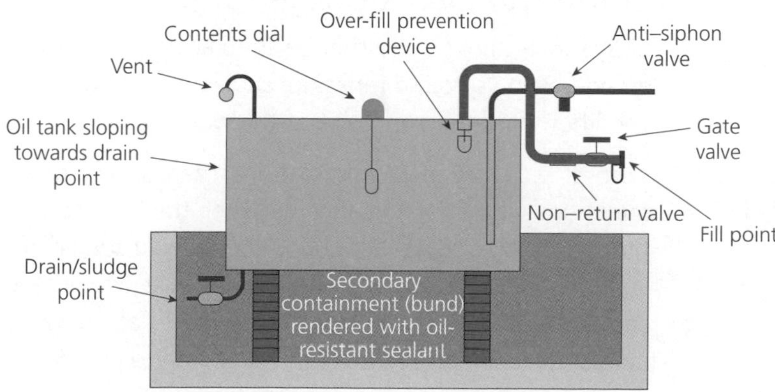

▲ Figure 10.11 An oil tank with an oil-proof bund wall

Oil tanks should be inspected annually as part of the heating system's regular servicing. Oil tanks have a useful working life of around 20 years and using a tank beyond this time carries the risk of failure.

Protection of the environment

Some tank installations require a secondary containment system, known as a bund, to counteract the risk of pollution from oil spillage. This may be achieved by using an integrally bunded oil tank with secondary oil containment built in, or building an oil-impermeable containment wall around the tank installation. These are generally required where the tank is close to a river or water source. The bund must be capable of holding 110 per cent of the oil tank's contents. Usually, a standard risk assessment is required by a registered oil installer to ascertain whether a bunded installation is required.

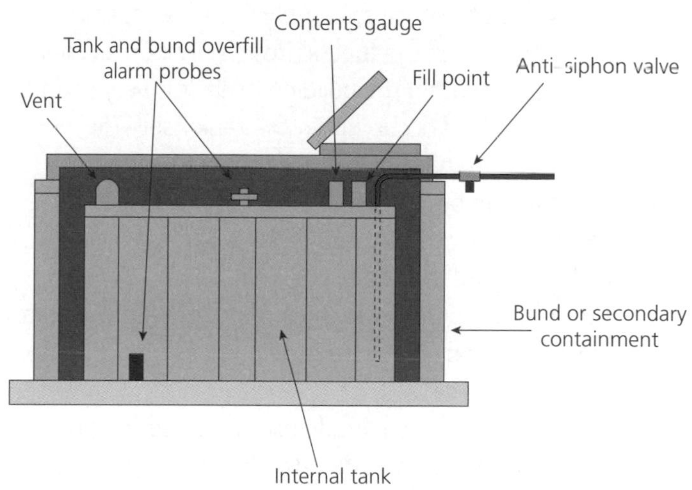

▲ Figure 10.10 Secondary containment tank

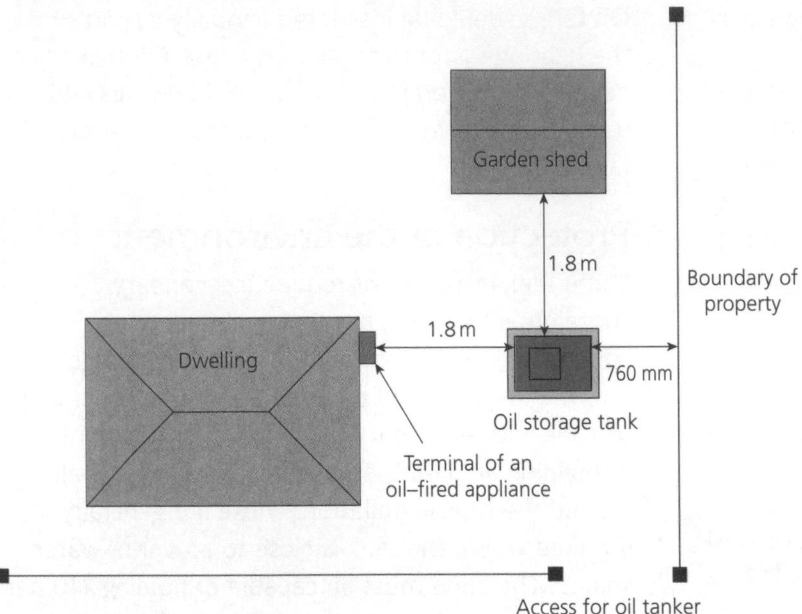

▲ Figure 10.12 Siting oil tanks

The location of fuel oil tanks

The siting of oil tanks must comply with fire separation distances, to protect the fuel oil from a fire or heat source that may occur within the building itself. It is very unlikely that any fire would occur within the tank itself. The regulations state that fuel oil tanks should be sited:

- 1.8 m from non-fire rated eaves of a building
- 1.8 m from a non-fire rated building or structure such as a garden shed
- 1.8 m from openings such as doors or windows in a fire rated building or structure such as a brick-built house or garage
- 1.8 m away from oil-fired appliance flue terminals
- 760 mm from a non-fire rated boundary such as a wooden boundary fence
- 600 mm from any trellis or foliage that does not form part of the boundary.

If any of these requirements cannot be met, then a fire protection barrier with at least a 30-minute fire rating must be provided. A minimum separation space of 100 mm is required between the tank and any fire-resistant barrier unless the tank manufacturer specifies a larger distance.

Storage of liquid petroleum gas (LPG)

It should be remembered that LPG is heavier than air and will 'search' for the lowest position if a leak occurs – and, although LPG has a distinctive smell, this will not be apparent until a person is at the same level as the low-lying gas.

KEY POINT

Above all else, LPG is extremely flammable and explosive. The siting of any LPG storage tanks must comply with certain recommendations and any gas installation is subject to the Gas Safety (Installation and Use) Regulations 1998.

Siting the LPG storage tank

LPG storage tanks can either be sited above or below ground. Below-ground tanks are subject to ground conditions and the proximity of the water table.

According to HSE recommendations, there should be a minimum separation distance between the LPG storage tank and any building, boundary line or fixed source of ignition. These distances are shown in Table 10.5. There should be no drains or gullies in the vicinity of the tank, unless these are protected by a water trap to prevent the gas from entering the drainage system.

▼ Table 10.5 The distances from buildings and structures for LPG storage tanks

Maximum LPG capacity of any single vessel in a group			Minimum separation distances of all vessels in a group		
LPG capacity (tonnes)	Typical water capacity (litres)	LPG capacity (tonnes)	From buildings, boundary, property line or fixed source of ignition		Between vessels (m)
			Without a fire wall (m)	With a fire wall (m)	
0.05 to 0.25	150 to 500	0.8	2.5	0.3	1
> 0.25 to 1.1	> 500 to 2500	3.5	3	1.5	1
> 1.1 to 4	> 2500 to 9000	12.5	7.5	4	1

Ventilation and conditions around the LPG storage tank

There should be plenty of room around the tank to allow good air circulation so that pockets of the heavier-than-air gas cannot build up around it should a leak occur. The area should also be kept free of rubbish and weeds, and any grass should be kept short.

Protection against impact

Tanks and their associated pipework should not be located in areas where motor traffic is likely. However, if this is unavoidable, then a suitable protective barrier should be installed in the form of either bollards or crash barriers. A security fence is not suitable since this is unlikely to offer the required protection.

> **KEY POINT**
>
> Further guidance on location and spacing for vessels and requirements concerning fire wall provision and for buried vessels can be found at: www.liquidgasuk.org

The LPG gas cylinder option

It is often a good idea to start using LPG with an LPG cylinder installation until the exact usage of the installation is known. Large bulk storage tank installations become viable only when usage exceeds 2000–2500 litres per year. The average bulk storage tank user uses around 2300 litres per year.

An LPG gas cylinder installation typically uses 47 kg propane gas cylinders located at the dwelling in a lockable cabinet. This type of cylinder installation usually uses either 2 × 47 kg cylinders or 4 × 47 kg cylinders.

▲ Figure 10.13 LPG gas cylinders

Storage requirements for biomass fuels

The storage requirements of the various types of biomass fuels can influence a client's decision because key points, such as site access, space requirements and even the aesthetics of the storage vessel itself, need careful consideration before the installation begins. Storage considerations for biomass fuels such as wood chips or pellets should be considered early on in any biomass system design.

There are many storage options for biomass, and all of them need to be watertight. Water ingress can severely affect biomass fuel quality and, as a consequence, the operation of the biomass boiler. Wood pellets, for example, that have a low moisture content will expand if they get wet and this can even damage the wood store itself.

- **Container or hook bin:** wood chips can be delivered in a container, often called a hook bin, where the container forms the fuel storage, which connects direct to the fuel extraction system. However, these are quite expensive because at least two bins are required.
- **Covered shed:** these are relatively cheap and easy to install. Fuel delivery is quite straightforward. For large stores, the use of manual handling equipment, such as a front-end loader or mechanical grab, is recommended.
- **Hoppers:** the hopper is a chute with extra storage capacity. They are relatively inexpensive to install. The hopper has a 'V'-shaped floor, sloped at approximately 40°. This allows the fuel to fall directly onto the boiler feed screw located at the base of the floor.

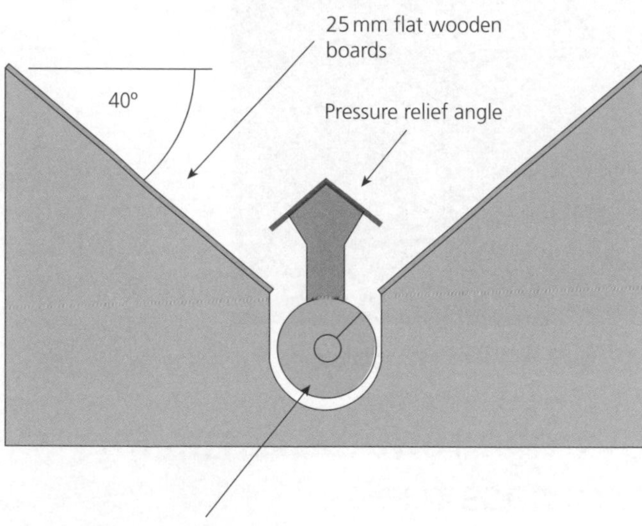

▲ Figure 10.14 A biomass hopper floor

- **Silos:** these are purpose-made rigid structures that are relatively inexpensive to install but may require special delivery equipment to maintain the biomass supply.

▲ Figure 10.15 A biomass silo

- **Flexible silos:** these are prefabricated, collapsible structures designed specifically for smaller installations where access may be restricted, such as in a confined space or a roof space. The fuel delivery system is usually where the fuel is blown into the hopper. This system uses two hoses: one to blow in the fuel and the other to extract any dust.
- **Underground bunker:** underground bunkers are ideal for larger installations with easy access for tipper-truck delivery. The feasibility of an underground bunker will depend upon such factors as ground type, water table and cost.

HEALTH AND SAFETY

Safety masks should be worn when moving wood chips or wood pellets as the dust can pose a health risk. Dust can also pose a significant explosion risk if the area is not ventilated sufficiently.

6 CONSIDERATIONS THAT COULD AFFECT THE STORAGE REQUIREMENTS OF FUELS

When considering the type of fuel system to be used in a dwelling, there are several factors that need special consideration:

- space for fuel storage
- delivery requirements
- safety
- weather conditions
- distribution
- proximity to dwelling.

Space for fuel storage

Space for fuel storage is a major factor when deciding which fuel system to use. Most fuels require specific distances in which to site storage vessels, tanks or silos. This may take the form of environmental concerns, as with heating oil, or explosion or fire risks, as with LPG. Where biomass is concerned, it may be the sheer mass of the fuel that is problematic.

Delivery requirements

The transportation and delivery requirements for domestic fuels differ according to the fuel, as described below.

- **Heating oil:** most oil tankers carrying domestic heating oil carry 45 m of hose. This is suitable for most installations. However, extra-long hoses can be requested. Consumers should remember to measure around any corners or obstacles when stipulating the oil tank distance from the access point.

▲ Figure 10.16 A typical small LPG tanker for domestic deliveries

- **LPG:** LPG bulk deliveries are usually delivered in mini-LPG tankers. These are 2.6 m wide and require access 2.75 m wide with a minimum access road width of at least 3 m. It is a requirement that a line of sight is maintained between the storage vessel and the tanker, with a maximum hose length of 40 m, to ensure the safe delivery of LPG to the bulk storage tank.
- **Coal/smokeless fuels:** solid fuels such as coal and smokeless fuels are delivered in sealed 25 kg bags. Deliveries are arranged as required.
- **Biomass:** in October 2015, rules to support sustainable fuels for the Domestic Renewable Heat Initiative (RHI) came into effect for all biomass heating systems. The Domestic RHI scheme aims to support homeowners and landlords who have invested in renewable heating technologies. This includes biomass, heat pumps and solar thermal panels. The idea behind the RHI scheme is to reward those people that stick to the RHI rules regarding sustainable supplies of fuel by paying them a tariff per kW/h. Payments are made every three months for a period of seven years.
- **Access for biomass fuel deliveries:** biomass pellets can be blown up to 30 m via hoses, but this distance often causes problems such as clogging of the hose and break-up of the fuel. It is recommended that deliveries of biomass should be within a 20 m limit of the fuel store. A lorry of around 2 m wide will need to be able to gain access to the property.

Safety

All fuels, by their very nature, are flammable and some are even explosive. With this in mind, the storage of fuels should be considered with care. Here are some points to consider:

- **Confined spaces:** solid fuels, such as biomass and coal, are kept in confined spaces. There are several problems with this:
  - **Fire:** although rare, bunker and fuel store fires can occur, especially where the store is directly connected to the boiler room. Generally, biomass wood chips are too wet to ignite but if they begin to decompose they may self-ignite.
    Liquid fuels, such as kerosene, do not usually combust unless they are either atomised or vaporised, but they can become dangerous near excess heat because the vaporisation process begins at a relatively low temperature of around 65°C.
  - **Explosion:** LPG, because it is heavier than air, settles at low level. In the event of a leak, the build-up of gas may not be noticed, despite the fact that a chemical called mercaptan is added to make the gas detectable by smell.
    Some fuels, such as coal and biomass, create dust. Excessive dust in the atmosphere can also be extremely explosive. A good air-extraction ventilation system is vital in confined spaces.
    The HSE recommends building in an 'explosion relief' into any storage space used for solid fuels that create dust. This can be a plywood panel

in a bunker or silo that creates a weak spot to release the explosive energy. The HSE produces a fact sheet, HSG103 Safe handling of combustible dust, which is available from its website at: www.hse.gov.uk

- **Carbon monoxide build-up:** for any confined space close to the place of combustion of a fossil fuel, combustion problems may lead to a build-up of carbon monoxide (CO), which is highly toxic. An audible CO alarm installation is recommended in fuel storage facilities.
- **Slips, trips and falls:** fuel stores of all kinds are dangerous places. Build-up or spillages of fuel create slip, trip and fall hazards. Some hazards may be limited by fuel store design. However, where solid fuel and biomass are concerned, the fuel storage space height may be high and so safety nets and harnesses should be considered.

- **Fuel delivery:** fuels are delivered to properties by either tanker (heating oil, LPG, biomass) or flat-bed truck (solid fuels – coal, coke, etc.). Care should be exercised while fuel deliveries are taking place. Follow the recommendations of the fuel delivery driver.
- **Personal hygiene:** there should be no reason for the fuel itself to be handled. However, in the event that contact with the fuel must be made, always wear appropriate PPE, such as overalls, gloves, hard hat, goggles and respirator (especially in dusty environments).

Weather conditions

The prevailing weather can have a severe effect on the storage of fuels. Bad weather, such as wind, rain, hail and snow, is often a cause for late deliveries and even cancellations of fuel deliveries, especially in rural areas. In almost all cases, fuel is delivered by large tanker or flat-bed vehicles that find it next to impossible to negotiate small, narrow roads when the weather conditions are poor. While the weather can be unpredictable in the UK, good planning of fuel deliveries can reduce the impacts of bad weather. Ordering more when severe weather is forecast can often mean the difference between running out of fuel and keeping the heating on.

Similarly, bad weather can render some fuels, such as wood chip and wood pellet biomass, almost unusable. Coal and coke too suffer from the negative effect of excessive rain, whereby the fuel can become too wet to burn effectively. Wood pellets swell from the effects of the rain and these then clog fuel delivery to the fuel bed of the boiler. Wood chip biomass can begin to de-compost if it gets too damp and this, paradoxically, can cause the fuel to heat internally and spontaneously combust.

KEY POINT

It is vital that fuels are kept dry and that they are delivered in good condition for optimum combustion efficiency to occur.

Distribution

The distribution of fuels becomes a vital consideration, especially the further outside a major town or city you live. Natural gas coverage in the UK through the national grid stands at around 7000 km of pipelines, but there are still many rural areas that are too far away from the grid for a supply to be economically viable. In these cases, other fuel supplies have to be considered.

By far the most viable fuel in rural areas is domestic heating oil, otherwise known as C2 grade, 28 second viscosity kerosene. Distribution of this still vital fuel is nationwide. However, kerosene poses an environmental risk if leakage occurs, especially where the installation lies close to a watercourse, river or stream or where the water table is high.

LPG distribution is also very comprehensive, with most areas in the UK reachable by tanker. However, there are certain restrictions with LPG that do not exist with heating oil, such as that the delivery driver must have line of sight to the LPG storage tank at all times during delivery of the liquid gas. LPG is also very expensive as a domestic heating fuel.

Coal and coke solid fuels continue to be readily available all over the UK, although many areas now forbid the use of these fuels because of the environmental pollution they release. If solid fuel is to be used, then local authority advice should be sought.

The use of biomass in rural and suburban areas is permitted under the Permitted Development legislation, which came into force in 2008. However, some areas, especially suburban districts, may put restrictions on its use if they lie within a smoke control zone. Outside of these zones, there are no major restrictions other than a requirement not to emit 'dark smoke'. In most cases, domestic biomass does not fall into this category.

INDUSTRY TIP

Access the Permitted Development legislation at: www.legislation.gov.uk/uksi/2015/596/contents

INDUSTRY TIP

Biomass fuel sources are available the length and breadth of the UK, with many companies supplying a wide variety of biomass pellets and chips, as well as liquid biofuels such as biodiesel. A list of local biomass suppliers is available at: www.hetas.co.uk

Proximity to dwelling

The installation of fuel storage and its requirements with regard to the proximity of the dwelling is covered elsewhere in this chapter (page 524).

SUMMARY

Fossil fuels, and more specifically coal and coal related products, have fallen out of favour over recent years because of the damage fossil fuel combustion and the resulting CO_2 is causing to the climate of Planet Earth. Yet, as far as the 'home' is concerned, natural gas continues to be the fuel of choice for home heating and cooking. Similarly, natural gas still has the largest fuel usage in the generation of electricity in the UK.

In both of these uses – electricity generation and home heating/cooking – natural gas looks set to be the leading fuel for many years to come until a viable renewable, cheaper and less polluting alternative becomes available.

Test your knowledge

1 What is the chemical formula for propane?
  a CH_4
  b C_3H_8
  c C_4H_{10}
  d CO_2

2 What is the approximate calorific value of natural gas?
  a 38 MJ/m^3
  b 38 KJ/m^3
  c 21 MJ/m^3
  d 21 KJ/m^3

3 Which of the following types of coal has the highest heat content?
  a Lignite
  b Anthracite
  c Bituminous
  d Sub-bituminous

4 Which Building Regulation document gives specific guidance on the requirements for fuel-burning appliances?
  a Part A
  b Part L
  c Part J
  d Part P

5 When a bund wall is required for an oil storage tank, what volume of oil must it be capable of holding?
  a 150 litres
  b 230 litres
  c 100% of the volume of the tank
  d 110% of the volume of the tank

6 Which regulatory body monitors, controls and guides on the use of solid fuel and biomass fuels?
  a Gas Safe
  b OFTEC
  c HETAS
  d Defra

7 What is the minimum recommended distance between an oil storage tank and the flue from an oil-fired appliance?
  a 600 mm
  b 760 mm
  c 1000 mm
  d 1800 mm

8 Which of the following toxic gases that can lead to dizziness, nausea and, in some cases, death is produced from a fossil fuel-burning appliance if not adjusted properly?
  a CO
  b H_2O
  c CH_4
  d CO_2

9 What grade of oil is domestic kerosene?
  a C
  b D
  c E
  d F

10 Which of the following fuels is deemed carbon neutral?
  a Natural gas
  b Oil
  c Coal
  d Biomass

11 Which phrase best describes 'calorific value'?
  a The amount of heat energy stored in a fuel
  b The amount of dynamic energy stored in a fuel
  c The amount of energy per kilogram of fuel
  d The amount of combustion within a fuel

12 What is the specific gravity (relative density) of natural gas?
  a 0.2–0.3
  b 0.6–0.8
  c 1.1–1.3
  d 1.8–1.9

13 Which of the following fuel systems is not a low-carbon fuel?

a Solar thermal

b Heat pumps

c Peat

d Biomass

14 Which regulatory body oversees solid fuels?

a Gas Safe

b HETAS

c OFTEC

d HSE

15 What is the minimum distance an oil storage tank must be from a property?

a 2.7 m

b 2.1 m

c 1.8 m

d 1.3 m

16 Which of the following is not a consideration for LPG cylinder locations?

a Minimum of 760 mm from a property boundary

b Should be kept clear of weeds and plants

c Must be well ventilated

d Should be clearly labelled

17 Which of the following fuels does not require any storage facilities?

a Biomass

b Oil

c Natural gas

d Liquid petroleum gas

18 How often should an oil storage tank be inspected?

a Every 3 months

b Every 6 months

c Annually

d Bi-annually

19 Natural gas is made up of eight different components. What is the main combustible gas in natural gas?

a Ethane

b Propane

c Methane

d Butane

20 What are the five categories of fuel?

21 What is added to natural gas to give it a smell that enables better detection?

22 List at least five factors that might be considered when a client is selecting a fuel type.

23 What is the role of Gas Safe?

24 What is combined heat and power (CHP)?

25 Explain the important criteria for the use of a district heating system.

Answers can be found online at www.hoddereducation.co.uk/construction.

CHAPTER 11

WORKING WITH ELECTRICITY

INTRODUCTION

When working on most plumbing systems such as heating, hot-water, or renewable technologies, you are likely to come into contact with electrical parts and components. Having a good understanding of basic electrical principles will reduce the risks that electricity can bring.

In this chapter, we will explore what electricity is and how it behaves in a circuit, causes of electric shock and how to protect against the risk of it occurring, common electrical components you will encounter, how to work safely around electricity, and basic electrical tasks.

By the end of this chapter, you will have knowledge and understanding of the following:

- electrical principles
- conductors and insulators
- resistors in series and in parallel
- protection against electric shock
- electrical supply systems
- protective devices
- working on electrical systems, including safe isolation procedures
- how to install wiring systems.

1 ELECTRICAL PRINCIPLES

Electricity is a flow of electrons in a material. You cannot see electricity flow, you cannot smell it, and you must not touch it.

To understand the basic principles of something you cannot see, it helps to liken it to something you can, so it is useful consider the flow of electricity as being similar to the flow of water in a pipe.

There are three main quantities that you need to focus on to understand electricity.

Current

Current measures the rate of flow of electrons in a circuit; it represents the amount of electricity that flows. If you push three litres of water through one end of a pipe, three litres of water will come out of the other end. Similarly, if three amperes of current flows into one end of a circuit, three amperes will flow out of the other end. Amperes (or amps for short, symbol A) is the unit used to measure current flow.

Voltage

Voltage is the pressure or electromotive force (emf) that pushes the current around a circuit. It is measured in volts (V). If you need to force water through a pipe you can use a pump to push it. Voltage is essentially the same. Like water pressure, voltage can drop through a circuit if there is resistance. This pressure drop in a circuit is called volt drop.

Resistance

Resistance is opposition to current flow. Like putting a kink in a pipe, it restricts the amount of current that can flow unless you increase the pressure. Different materials, for example electrical loads such as heater elements, offer different resistances. In this case, the resistance causes the current flowing through the element to release thermal energy.

Ohm's law

The relationship between current, voltage and resistance can be expressed using Ohm's law:

$$V = I \times R$$

where

V represents the electromotive force measured in volts (V)

I represents the current flow in the circuit measured in amperes (A)

R represents the resistance in the circuit measured in ohms (Ω).

Ohm's law states that if you know any two electrical values, you can always find a third since

$$V = I \times R \text{ or } I = \frac{V}{R} \text{ or } R = \frac{V}{I}$$

A good way to remember the Ohm's law relationship is by using a formula triangle.

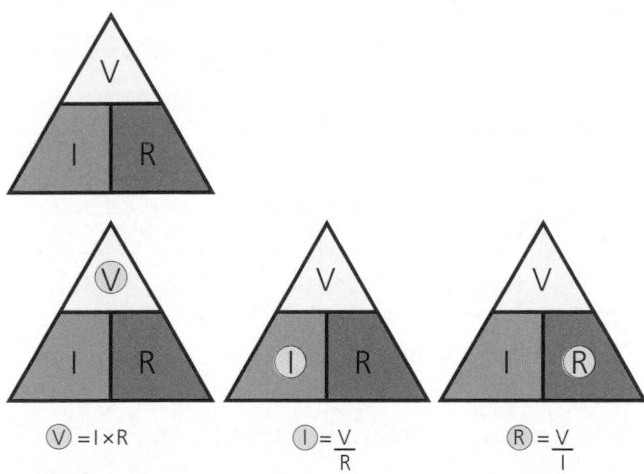

▲ Figure 11.1 Ohm's law triangles

Example

An electric heater element draws 10 A when connected to a 230 V supply. Calculate the resistance of the element.

Answer

Using

$$R = \frac{V}{I}$$

$$R = \frac{230}{10} = 23\,\Omega$$

ACTIVITY

A circuit has a total resistance of 1500 Ω when connected to a 12 V supply. Calculate the current drawn.

Answer

$$I = \frac{V}{R}$$

so

$$I = \frac{12\text{ V}}{1500\,\Omega}$$

$$= 0.008\text{ A or } 8\text{ mA (milliamperes)}$$

Electron flow

Current is a flow of electrons. So, what are electrons?

Electrons are parts of an atom. Every solid, liquid or gas is made of atoms. Atoms consist of a nucleus, containing neutrons and protons, and electrons which orbit the nucleus. This can be seen in Figure 11.2.

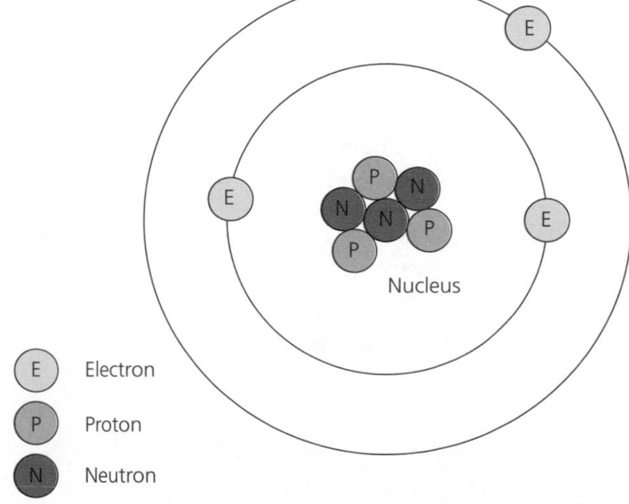

▲ Figure 11.2 Protons, neutrons and electrons making an atom

The number of electrons orbiting a nucleus is equal to the number of protons in the nucleus, and this number is the atomic number of the material. For example, copper has an atomic number of 29 and so has 29 electrons orbiting the nucleus and 29 protons in the nucleus.

Because the atoms in a metal are packed closely together, some of these electrons may get transferred from one atom to another so that they are no longer associated with a single atom. These delocalised electrons are free to move when the metal is subjected to an electromotive force, or voltage. This flow of electrons in a particular direction results in an electric current.

Electrons have a negative charge. Charge in a circuit is measured in coulombs (C); when one coulomb of charge flows in one second, it creates a current of one ampere (1 A).

ACTIVITY

Use the internet to find out how many electrons flowing is equivalent to one coulomb (1 C).

Answer

6.24×10^{18} electrons make up one coulomb. This is 6,240,000,000,000,000,000 electrons!

2 CONDUCTORS AND INSULATORS

Because different materials have different atomic structures, some are better than others at conducting electricity.

A good conductor is a material that has a low resistance to the flow of current. An example is copper.

A good insulator (or poor conductor) is a material that does not allow a current to flow at a given voltage. An example is PVC. These materials are useful for preventing contact with live parts of a circuit.

Materials of the same size have different values of resistance depending on their **resistivity**.

KEY TERM

Resistivity: the measure of the resistance of a 1 m^3 block of a material at 20°C; it is measured in ohm-metres (Ω m).

In a two or three-core electric cable, such as a flexible cable from a plug to an appliance, the conductor is the copper wire that carries the current, and the insulation, which prevents contact between conductors, is the coloured PVC surrounding the copper. The **colour coding** of the insulation indicates the function of the conductor. Around the insulation there is a further layer of PVC known as the sheath, which acts as mechanical protection.

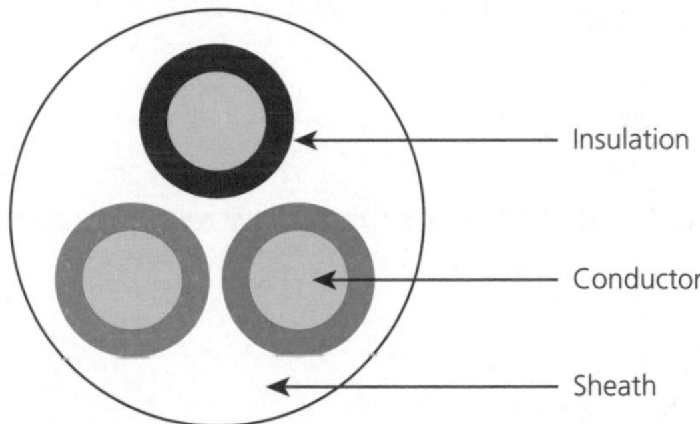

▲ Figure 11.3 A flexible cable having a conductor, insulation and sheath

KEY TERM

Colour coding: the colour coding of electric cable insulation is

Brown – line conductor (L)

Blue – neutral conductor (N)

Green/yellow – earth (E) or circuit protective conductor (cpc)

HEALTH AND SAFETY

The line and neutral conductors are both classed as **LIVE** conductors as they both carry a current under normal operating conditions.

Temperature effects

As previously stated, good conductors have a low resistance, so the opposition to current flow is minimal. Temperature also influences the ability of a conductor to carry a current. The warmer a conductor gets, the higher its resistance, so the greater the opposition to the current flow.

Temperature also affects common materials used as insulators. As some materials, like PVC, get warmer, they begin to melt, which reduces the resistance of the insulation and causes current to leak between conductors. This increases the risk of contact and therefore electric shock.

If cables are allowed to become **overloaded**, it will cause the conductor to heat up sufficiently for the insulation to break down and cause failure of the circuit.

Voltage drop

It is also worth considering at this point that the higher the resistance of a conductor, the greater the voltage drop across it. Several factors affect voltage drop. These include:

- Length – the longer a circuit or cable, the more resistance it has and therefore the more volts lost.
- Current flow – the more current that flows, the hotter the conductor gets and the greater is the resistance of the conductor. A higher resistance means a greater voltage drop.
- Conductor material – different materials have different resistivities and therefore different conductors have different resistances. For example, steel has a higher resistivity than copper, so its resistance will be higher than a copper cable of the same diameter and length.
- Cross-sectional area (csa) – the csa of the cable is the area of the end face. The bigger the csa, the thicker the cable, so it can take more current without significantly heating up. The bigger the csa, the better the **capacity** of the cable.

KEY TERMS

Overload: where more is drawn into a circuit than the circuit is designed for. This is usually caused by misuse such as plugging in too many appliances.

Capacity: the ability of a cable to carry current. This is directly linked to the cable csa.

INDUSTRY TIP

Typical cross-sectional areas (csa) for cables commonly used in domestic installations are

1 mm^2, 1.5 mm^2, 2.5 mm^2, 4.0 mm^2, 6.0 mm^2 and 10 mm^2

If there is too great a voltage drop in a cable, then the voltage at an appliance that it is connected to will be reduced. This will have a negative effect on the performance of the appliance. Remember that voltage drop in an electrical circuit is like pressure drop in a pipe. As a result, it is better to use conducting cables with a low resistance so that the voltage drop will be lower.

INDUSTRY TIP

BS 7671: The IET Wiring Regulations states that the maximum permitted voltage drop is:

- 3% of the supply voltage for lighting circuits
- 5% of the supply voltage for power circuits.

IMPROVE YOUR MATHS

What is the **maximum** permitted voltage drop for the following circuits connected to a 230V supply?

a Lighting circuit

b Socket-outlet (power) circuit

Answer

a Lighting is 3% of 230V which is 6.9V

b Socket-outlets is 5% of 230V which is 11.5V

3 RESISTANCES IN SERIES AND PARALLEL

How components and equipment are arranged has an effect on the behaviour of the current and voltage in a circuit.

Circuits can be arranged with resistances in:

- series – where each resistance follows on from another
- parallel – where each resistance is in line with another.

Series circuits

The resistances in a series circuit follow after one another, as shown in Figure 11.4.

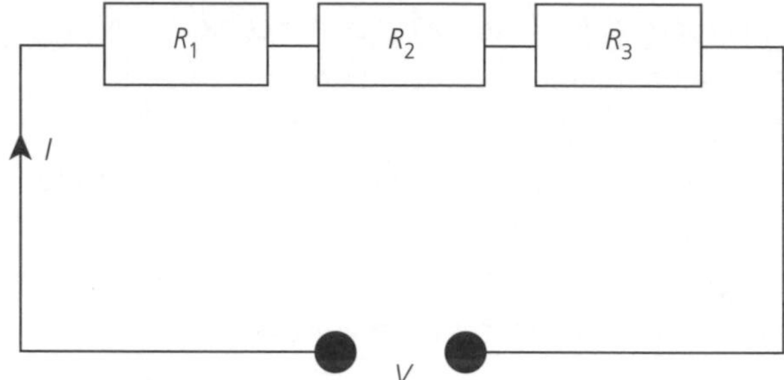

▲ Figure 11.4 Three resistances in series

It is worth noting, at this point, that a circuit resistance can be a current consuming item of equipment, such as a lamp or heater, or a conductor in a cable. Figure 11.5 shows a circuit supplied by a 36 V battery. The circuit contains one conductor supplying a heater element and one conductor returning back to the power supply. Each conductor has an overall resistance of 0.4 Ω. The heater element has a known resistance of 12 Ω.

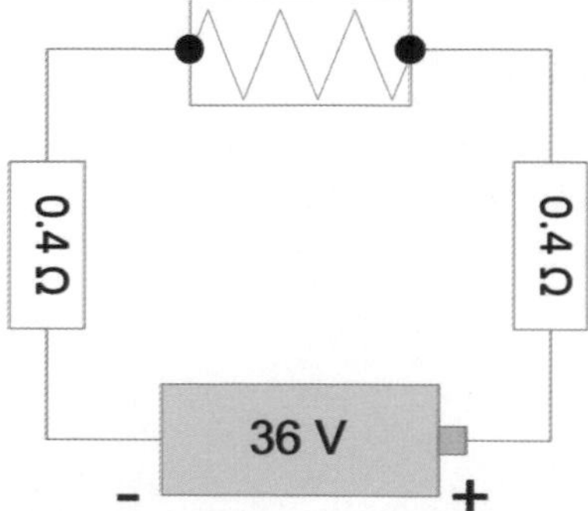

▲ Figure 11.5 A circuit supplied by a battery and containing three resistances: two conductor resistances of 0.4 Ω each and one heater element having 12 Ω resistance

Consider the three resistors shown in the circuit in Figure 11.4. Suppose:

- resistance R_1 is 0.4 Ω
- resistance R_2 is 12 Ω
- resistance R_3 is 0.4 Ω
- supply voltage (V) is 36 V
- circuit current (I) is not yet known.

In a series circuit, because there is only one path for the current and it flows through each resistor one after another, the total resistance is found by adding all the resistances together. So

$$R_{total} = R_1 + R_2 + R_3$$

For the circuit in Figure 11.4, the total resistance is

$$R_{total} = 0.4\ \Omega + 12\ \Omega + 0.4\ \Omega = 12.8\ \Omega$$

INDUSTRY TIP

Remember the Ohm's law triangle:

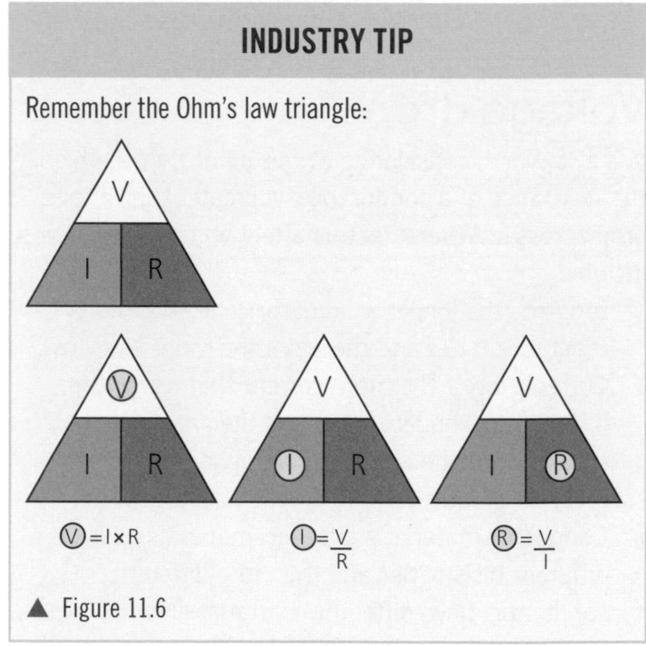

▲ Figure 11.6

In a series circuit, the current is the same through each component. To determine the total current drawn by the circuit, we apply Ohm's law:

$$I = \frac{V}{R}$$

so

$$I = \frac{36}{12.8} = 2.81\text{ A}$$

Therefore, because of the resistances in this circuit, the current drawn is 2.81 A.

IMPROVE YOUR MATHS

Using the series circuit in Figure 11.7, determine the total circuit current (I) and the voltage drop across each 3 Ω resistor. Assuming that the two 3 Ω resistances are the resistances of the conductors, what is the voltage across the 14 Ω resistor?

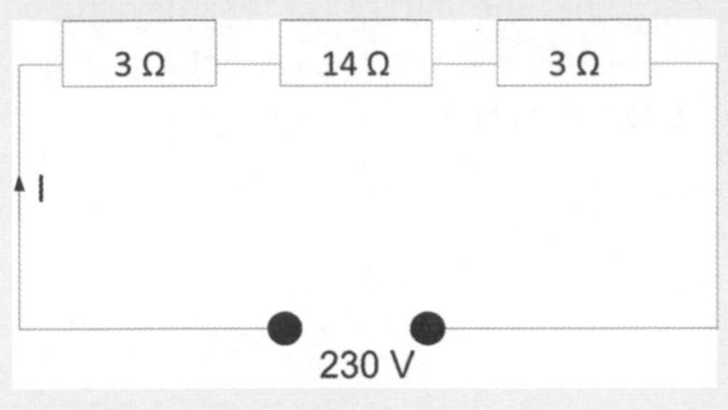

▲ Figure 11.7

Answer

$$\text{Total resistance} = 3 + 14 + 3 = 20\ \Omega$$

$$\text{Current } (I) = \frac{230}{20} = 11.5\ \text{A}$$

$$\text{Voltage lost} = 11.5 \times 3 = 34.5\ \text{V per conductor}$$

$$\text{Total voltage lost} = 34.5 + 34.5 = 69\ \text{V}$$

Voltage across the 14 Ω resistor is $230 - 69$ $= 161\ \text{V}$

This could also be shown using

$$11.5\ \text{A} \times 14\ \Omega = 161\ \text{V}$$

We could also work out the voltage lost due to the circuit conductors, using the fact that each has a resistance of 0.4 Ω. Again, this is done by applying Ohm's law, but this time we know the current flowing through each resistance or, in this case, cable. The voltage drop across each resistance is found using

$$V = I \times R$$

So, the voltage drop through resistor R_1 is

$$V = 2.81 \times 0.4 = 1.12\ \text{V}$$

The voltage drop through the second resistor or cable is the same, as the current is the same and the resistance is the same, so

$$\text{total voltage drop} = 1.12\ \text{V} + 1.12\ \text{V} = 2.24\ \text{V}$$

If 2.24 V is lost due to the resistance of the circuit conductors, the voltage across the heater is the remaining voltage from the 36 V supply, so

$$\text{voltage across heater} = 36 - 2.24 = 33.76\ V$$

INDUSTRY TIP

Remember: the longer the cable, the greater the voltage drop as resistance increases with length.

Parallel circuits

In a parallel circuit, the resistances, or loads, are in line so they have an equal voltage across them. In normal household 230 V circuits supplying lights and appliances, the loads are arranged in parallel to ensure that each receives the maximum voltage rather than having it shared between them. Figure 11.8 shows three resistors in parallel.

This is similar to radiators in a central heating system, which are also connected in parallel. All the radiators are supplied by the flow pipe and the water exits the radiators into the return pipe.

In a parallel circuit, the voltage is the same across each resistance or load and the current splits through each resistance depending on its value.

The total resistance is found by calculating the reciprocal value using

$$\frac{1}{R_{total}} = \frac{1}{R_1} + \frac{1}{R_2} + \frac{1}{R_3} \ldots$$

In general, when you increase the number of resistances in parallel, the total resistance decreases as this introduces more paths for the current to take.

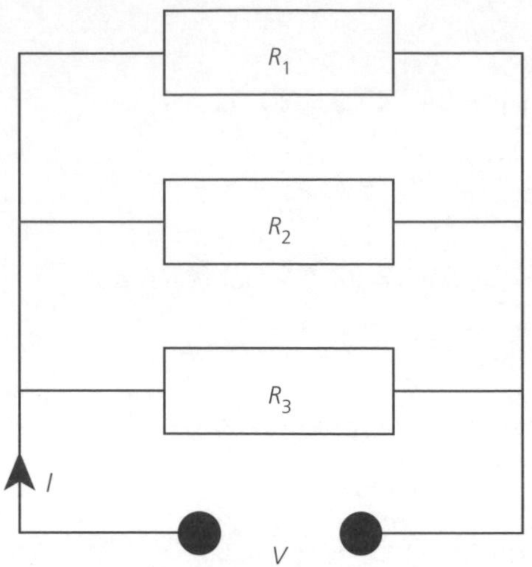

▲ Figure 11.8 A parallel circuit

For example, if the values of the components in Figure 11.8 are:

- $R_1 = 15\,\Omega$
- $R_2 = 15\,\Omega$
- $R_3 = 15\,\Omega$
- and the supply voltage = 100 V

we can determine the total resistance using

$$\frac{1}{R_{total}} = \frac{1}{15} + \frac{1}{15} + \frac{1}{15} = \frac{1}{0.2}$$

So the total resistance = 5 Ω

KEY POINT

If you have a calculator with a button marked x^{-1} you can calculate this by pressing the buttons in this order

[15][x^{-1}] [+] [15] [x^{-1}] [+] [15] [x^{-1}] [=] [x^{-1}] [=]

Note that after the first =, the value is 0.2; this is known as the reciprocal to one value, which must be divided into 1, so the [x^{-1}] key needs to be pressed again to get the correct answer 5 Ω.

Once the total resistance is found, the current in the main circuit can be found by using Ohm's law

$$I = \frac{V}{R}$$

So the current flowing from the power supply is

$$\frac{100}{5} = 20\text{ A}$$

The current through each resistor can be found by applying Ohm's law using the value of each resistance and remembering that the voltage is the same across each branch of a parallel circuit. The current through resistance R_1 is found using

$$I_{R1} = \frac{V}{R_1}$$

so

$$\frac{100}{15} = 6.667\text{ A}$$

In the circuit in Figure 11.8, both R_2 and R_3 have the same resistance as R_1, so the current through them is the same. As a check, when the three current values are added together, they equal the supply current of 20 A.

So

$$6.667 + 6.667 + 6.667 = 20\text{ A}$$

IMPROVE YOUR MATHS

For the circuit shown in Figure 11.8, let

- $R_1 = 5\,\Omega$
- $R_2 = 10\,\Omega$
- $R_3 = 15\,\Omega$
- and the supply voltage = 80 V

Determine the total resistance of the circuit, the current through the power supply and the current through each resistor.

Answer

$$\frac{1}{R_{total}} = \frac{1}{5}(0.2) + \frac{1}{10}(0.1) + \frac{1}{15}(0.0667)$$

$$= \frac{1}{0.2 + 0.1 + 0.0667} = 2.73\,\Omega$$

$$\text{Current through power supply} = \frac{80}{2.73} = 29.3\text{ A}$$

$$\text{Current through } 5\,\Omega \text{ resister} = \frac{80}{5} = 16\text{ A}$$

→

Current through 10 Ω resistor $= \frac{80}{10} = 8$ A

Current through 15 Ω resistor $= \frac{80}{15} = 5.3$ A

And, as a check,

$$16 + 8 + 5.3 = 29.3 \text{ A}$$

Power

Power (P), measured in watts, is the relationship between the circuit or load current and the voltage applied. This is expressed as:

$$P = V \times I$$

For example, a heater element connected to a 230 V power supply that draws a current of 13 A has a power rating of:

$$230 \times 13 = 2990 \text{ W}$$

More commonly, the current drawn by a load is calculated from the load's power rating, so that the correct size cable is selected.

Example

If a 4500 W load is connected to a 230 V supply, what is the current drawn by the load?

Answer

$$P = V \times I$$

so

$$I = \frac{P}{V}$$

and

$$\frac{4500}{230} = 19.57 \text{ A}$$

IMPROVE YOUR MATHS

Calculate the current drawn by an 80 W motor using a 12 V power supply.

Answer

$$\frac{80}{12} = 6.67 \text{ A}$$

Measurement of electrical quantities

The following instruments are used to measure electrical quantities and are connected as shown in Figure 11.9

Quantity to be measured	Instrument used	Connection	Notes
Voltage	Voltmeter	In parallel with the item being measured	The instrument measures the voltage difference from one side of the load to the other. This is called potential difference.
Current	Ammeter	In series with the load	Ammeters can measure small currents, but for much larger currents, a current transformer is needed such as those used in clamp-meters.
Resistance	Ohmmeter	In parallel with what is being measured	This will only work if the circuit or item being measured is disconnected from any power source.
Power	Wattmeter	Use in both parallel and series	This measures the voltage and current and calculates the resulting power.

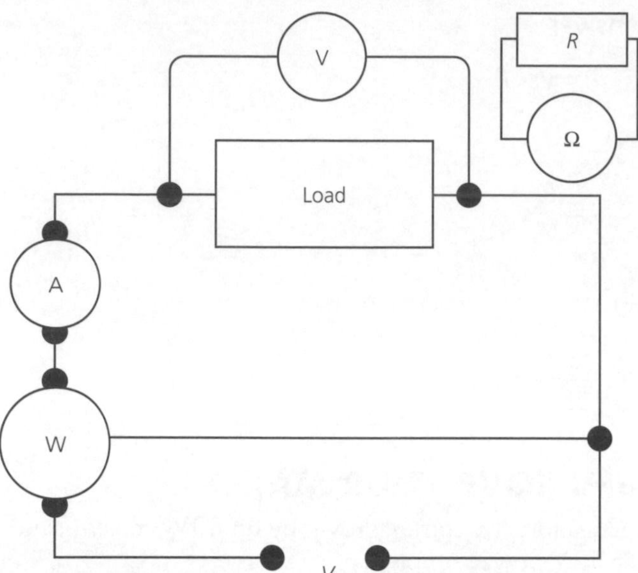

▲ Figure 11.9 How instruments are connected to measure circuit quantities

HEALTH AND SAFETY

Never expose yourself to live parts when measuring electrical quantities, as the risk of electrocution is great. Instruments and test leads need to comply with HSE guidance GS38. Instruments need to be correctly set or they could explode. Unless you have received suitable training, only measure circuit values if the circuit is completely and securely isolated, as described later in this chapter.

4 PROTECTION AGAINST ELECTRIC SHOCK

Electric shock is undoubtedly the greatest risk when working around electrical appliances. This does not only apply to those working on them but also, if the work is not carried out correctly, to other people who use the equipment afterwards.

Before we consider how to protect against electric shock, we need to consider how you can receive an electric shock.

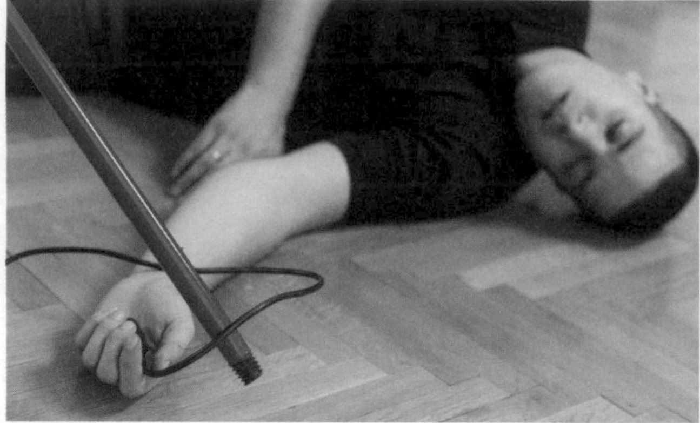

▲ Figure 11.10 A person who has just received an electric shock

An electric shock is defined as an accident or injury as a result of contact with electricity. It occurs when current flows through the body. This can cause muscles, such as the heart, to react.

Current as low as 15 mA (0.015 A) can cause involuntary reactions such as flinching of leg and arm muscles, and this can cause a person to fall. A current of about 40 mA (0.04 A) can cause the heart to react. If a person comes into contact with a current exceeding 40 mA, it can interrupt the body's normal electrical impulses which make the heart beat at regular intervals.

If a person touches a supply of alternating current (AC) with a frequency of 50 hertz (Hz), which changes direction 50 times a second in the UK, the muscles would react in time with this.

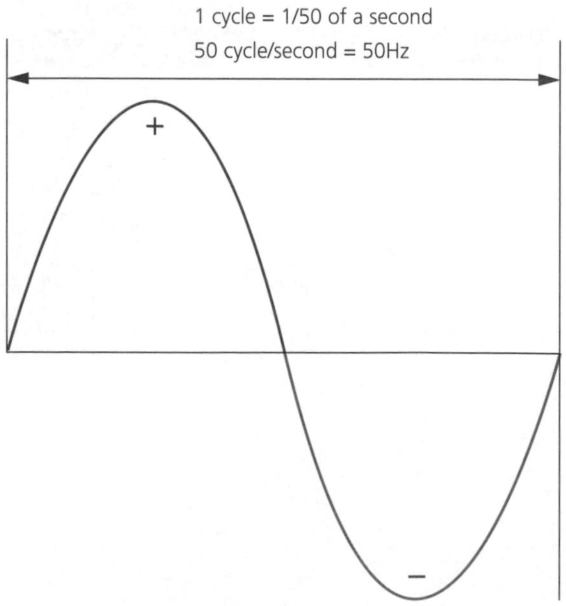

▲ Figure 11.11 The AC sine wave

So, in one cycle, the current turns on (+ flow) and off (– flow) twice; this repeats 50 times a second so, effectively, the current turns on and off 100 times a second.

If someone comes into contact with alternating current higher than 40 mA on a 50 Hz supply, it will cause the heart to attempt to beat 100 times a second. As a result, in a relatively short time, the heart would not be able to cope and would stop.

If a person comes into contact with a direct current (DC) supply, the current is on all the time and this would cause the muscles to seize.

HEALTH AND SAFETY

It is said that people stick to DC. This is because the arm muscles controlling a person's grip can seize at a particular level of current (15–25 mA) and this locks their grip. So, even though the person's brain is telling the hand to let go, it will not!

For current to flow, there needs to be a passage into the body and a passage out. The current needs to pass through the skin and overcome its resistance.

Remember: a voltage is required to overcome a resistance.

For the average person in **average conditions**, 50 V AC or 120 V DC is required to overcome the body's skin resistance. Is there such a thing as an average person? Some people have tough skin with high resistance, so higher voltages are needed to overcome this. Some people have very soft skin, so lower voltages can cause electric shock. An average person is somewhere in the middle.

If humidity and temperature increase beyond average conditions, the body starts to sweat and skin resistance falls, creating a higher risk of shock.

KEY TERM

Average conditions: somewhere with a comfortable indoor temperature of about 20°C and a low humidity.

If a person touches a live part of a circuit but is standing on a wooden or rubber, insulated floor, and is not in contact with any earthed parts, then the current passing through their body would be quite low and the person may survive the shock.

If the person touching a live part of a circuit is standing on the ground or is in contact with earthed parts of the circuit, their body would form a low resistance path and a current that is high enough to kill will flow through them.

A person can receive an electric shock in two ways:

- by directly touching parts that are intended to be live
- by contacting parts that have become live due to a fault.

BS 7671 gives several methods of protecting people against electric shock. These include:

- basic protection by insulation, barriers and enclosures to stop contact with parts intended to be live
- automatic disconnection of the supply (ADS) in the case of a fault.

Basic protection

Basic protection is provided by the following two methods:

- insulation of live parts by insulating material such as PVC around cables, which stops people coming into contact with the live conductors within
- barriers and enclosures housing electrical connections and live parts of equipment, which are designed so that:
    - all surfaces except the top have no hole allowing penetration by the finger (**IP2X**)
    - the horizontal top surface of a barrier or enclosure must have no hole greater than 1 mm in diameter (**IP4X**)
    - barriers and enclosures can only be opened and live parts accessed by using a tool or key to stop unintended contact.

KEY TERM

IP2X and IP4X: part of the international coding system for the level of Ingress Protection for enclosures. This is the level of protection against the ingress of foreign bodies, such as fingers (first number) and the level of protection against the ingress of water and moisture (second number).

Basic protection really is basic in the sense that it stops contact with all the parts that are meant to be live and must always be present whenever a system is live.

Automatic disconnection of supply (ADS)

Should any fault occur in a circuit, the device protecting the circuit needs to disconnect very quickly to remove the danger.

As the parts of an appliance that are likely to become live if a fault occurs are normally made of metal and are part of the electrical system, these parts must always be connected to the circuit protective conductor (cpc) which is the green/yellow coloured conductor in a circuit.

This ensures that all **exposed conductive parts** have a low resistance path to earth, which, in turn, will induce a high fault current. This fault current will be high enough to trip the device protecting the circuit, for example a circuit breaker or residual current circuit breaker with overcurrent protection (RCBO).

INDUSTRY TIP

Remember Ohm's law. If the voltage is fixed, e.g. 230 V, and the resistance decreases, the lower the resistance, the higher the current.

KEY TERM

Exposed conductive part: a metallic or conductive material that forms part of the electrical circuit and therefore is likely to become live if an electrical fault occurs. Examples include a metal consumer unit, a metal central heating pump or a metal-cased boiler.

As a current of 0.04 A can kill someone, a fault current of 400 A can too, so it is crucial that the protection device disconnects the circuit very quickly to prevent someone who touches an appliance receiving an electric shock or being electrocuted if the current becomes dangerously high. The higher the fault current, the quicker the device will break the circuit. This is why earthing is important for saving lives.

Earthing and bonding

Both of these safety measures use green and yellow cables, but they protect in different ways. Nowadays, there is less need for bonding than there used to be.

Earthing

Earthing provides a low resistance path for a fault current to flow to earth. The high current trips the circuit breaker or blows the fuse which rapidly switches off the current in the faulty circuit.

Faults can happen in both equipment and cables so even if an item of equipment does not have any exposed conductive parts, a cpc must still be installed to protect the cable and be there in case the non-metallic equipment is changed for metallic equipment at a later date.

INDUSTRY TIP

Some cables, such as flat profile twin and cpc cable, have no insulation on the cpc. It is up to the person connecting it to place green/yellow sleeving on the cpc when they terminate the cable.

Bonding

Like earthing, bonding uses green and yellow cable, but whereas an earth wire is normally a core within a three-core cable, called the cpc, bonding is likely to be a single-core green and yellow cable running on its own.

▲ Figure 11.12 A clamp used for bonding extraneous conductive parts such as pipework

There are three classifications of bonding conductor.

- **Main protective bonding conductor** This connects the **main earthing terminal (MET)** for the installation to any incoming metallic gas and water services, amongst other **extraneous conductive parts**. This conductor, in most cases, will be a 10 mm^2 green and yellow insulated single-core cable terminated on the pipe using a bonding clamp. It is important to note that if the utility such as water or gas enters through a plastic pipe or has an insulated section of pipe, known as an IJ, there is generally no requirement to bond the pipework. If in doubt, consult an electrician.

INDUSTRY TIP

Main protective bonding, if needed, must be connected within 600 mm at the point of entry of the pipe or within 600 mm of any meter or stop cock on the consumer's side and before any tee in the pipework.

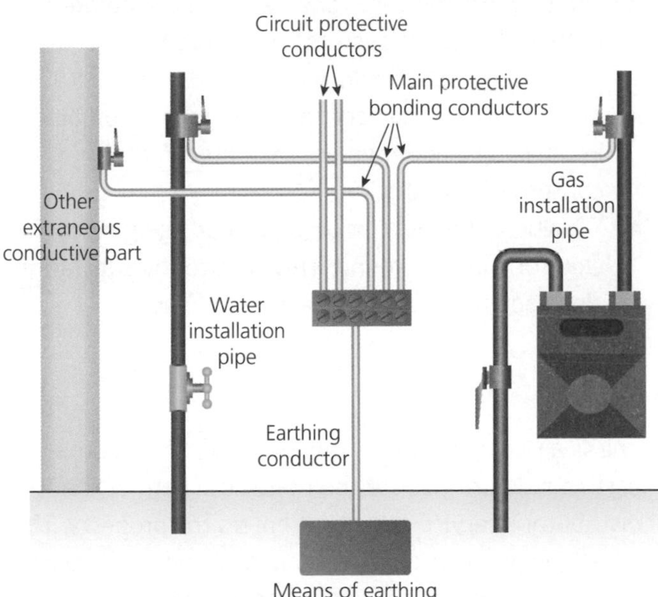

▲ Figure 11.13 Protective conductor arrangement for earthing and bonding

- **Supplementary equipotential bonding** This is now very rarely installed in new buildings where residual current devices (RCDs) are used to protect electrical circuits. Supplementary bonding, often called cross bonding, normally uses 4 mm^2 single-core cable and is often found linking hot and cold pipes, sinks or pipework in bathrooms. If you ever remove a bonding clamp and reuse or replace copper pipes, always put the bonding back unless it can be verified by an electrician that it is not required.
- **Temporary continuity bonding** This is used when cutting metal pipes in operational buildings. It consists of a simple short length of earth wire with two crocodile clamps on either end. If, due to faults or **diverted neutral currents**, a pipe is cut and parted without a temporary continuity bond, the person parting the pipes could then become part of the circuit and suffer an electric shock. The temporary bond offers a path of least resistance for the current rather than through the person.

Sometimes problems occur in the electricity supply, for example high resistance joints in the neutral side of the supply cable. This could cause the neutral current to try to find an alternative route back to the local electrical substation transformer. This alternative route could be through a metal pipe and bonding cable. If you disconnect a bond or cut a pipe, you could become part of that circuit and receive an electric shock. This situation is known as neutral current diversion.

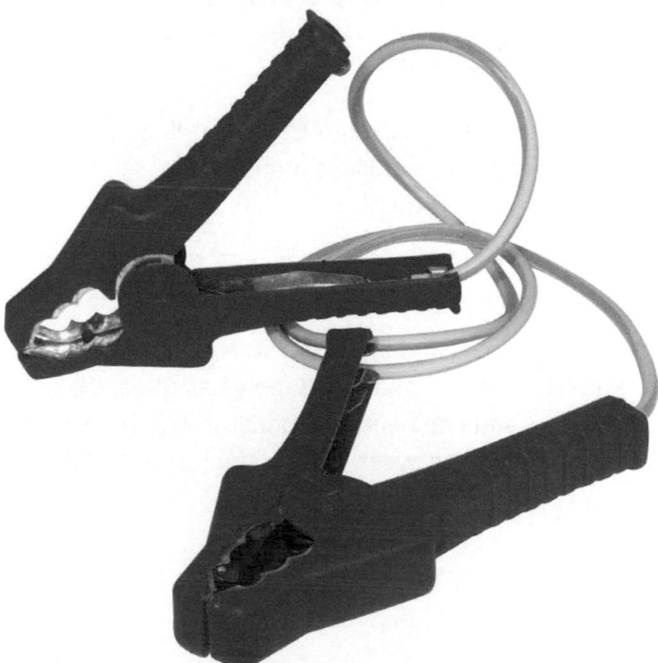

▲ Figure 11.14 Temporary continuity bonding set

KEY TERMS

Main earthing terminal (MET): every electrical system in a building will have a main earthing terminal (MET) which is either in the main consumer unit housing the fuses/circuit breakers or a separate metal block close to the electricity meter.

Extraneous conductive parts: conductive metal parts that are not part of the electrical system but provide a direct path to earth, for example steel framed buildings with steel structural parts embedded into the ground or metal gas pipes coming out of the ground.

Diverted neutral current: a high-risk situation in which a problem in the electricity supply causes neutral current to try and find alternative routes back to the local electrical substation transformer.

HEALTH AND SAFETY

Always use a temporary continuity bond when cutting metal pipes in an operational building as the risk of diverted neutral currents becomes greater as infrastructure ages.

5 ELECTRICAL SUPPLY SYSTEMS

Electrical supplies are categorised according to:

- number and type of live conductors
- earthing arrangement.

The number and type of live conductors depends on whether the supply is three-phase or single-phase.

Most commercial and industrial systems are three-phase, which means that the supply has three line conductors and one neutral conductor. The voltage, when measured, between line conductors is 400 V, and between any one line conductor and neutral or earth the voltage is 230 V.

Domestic and small commercial buildings are likely to have single-phase supplies with two live conductors which are a line and neutral and where the voltage between the line conductor and neutral or earth is 230 V.

There are three common earthing arrangements found in the UK. The arrangement does not depend on whether the supply is single- or three-phase.

The three commonly used systems use letters to signify the earth and neutral status from the local substation transformer to the installation and within the installation. The letters used are:

- T – this stands for *terre* meaning earth in French
- N – stands for neutral
- C – stands for combined
- S – stands for separated.

The three common systems are:

- TN-S, where the earth and neutral are separated both in the supply system and within the consumer's installation
- TN-C-S, where the earth and neutral are combined in the supply and then are separated in the consumer's installation
- TT, where the installation is earthed using an earth electrode in the ground. This links to the substation electrode using the general mass of earth.

TN-S

This is a very common system, which connects the earth conductor back to the metallic sheath of the distributor's service cable as it enters the property. This sheathing provides a separate neutral and earth route back to the substation supply transformer.

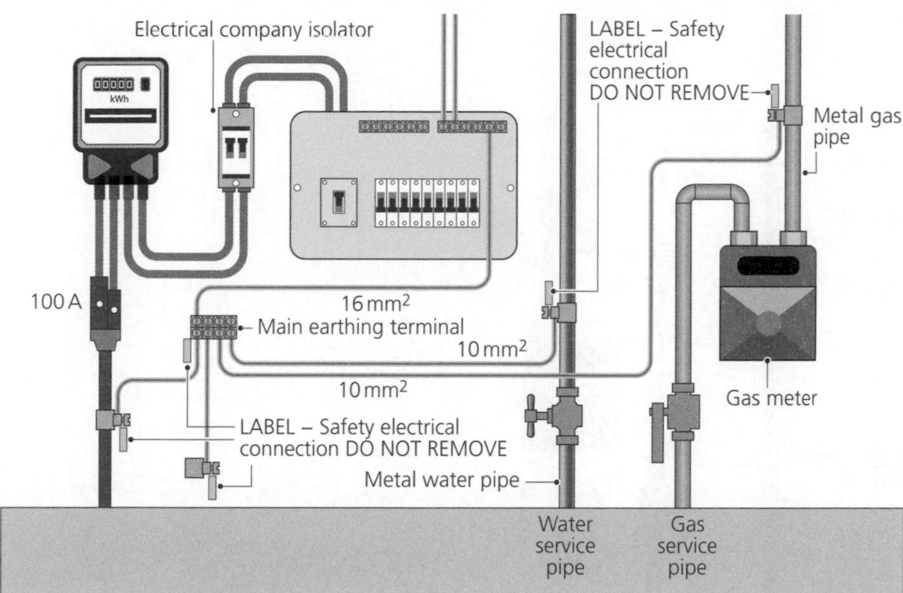

▲ Figure 11.15 TN-S system

TN-C-S

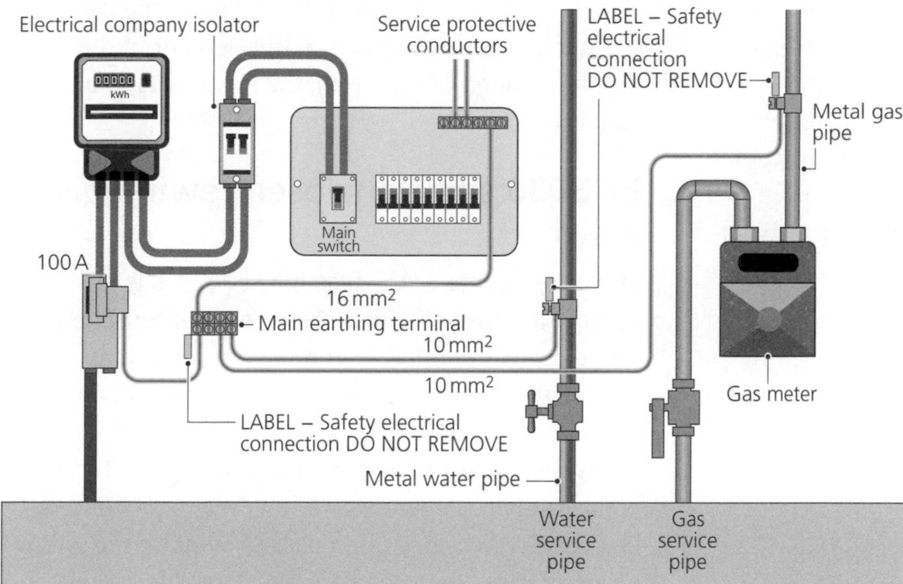

▲ Figure 11.16 TN-C-S system

This system is also known as a protective multiple earthing system, or PME. It relies on the neutral being earthed close to the source of supply and at points throughout the distribution system. There is also a neutral-to-earth connection at the intake of the installation. The 'C' means that the supplier uses a combined earth and neutral. The 'S' means that the earth and neutral are separate within the customer's property.

TT

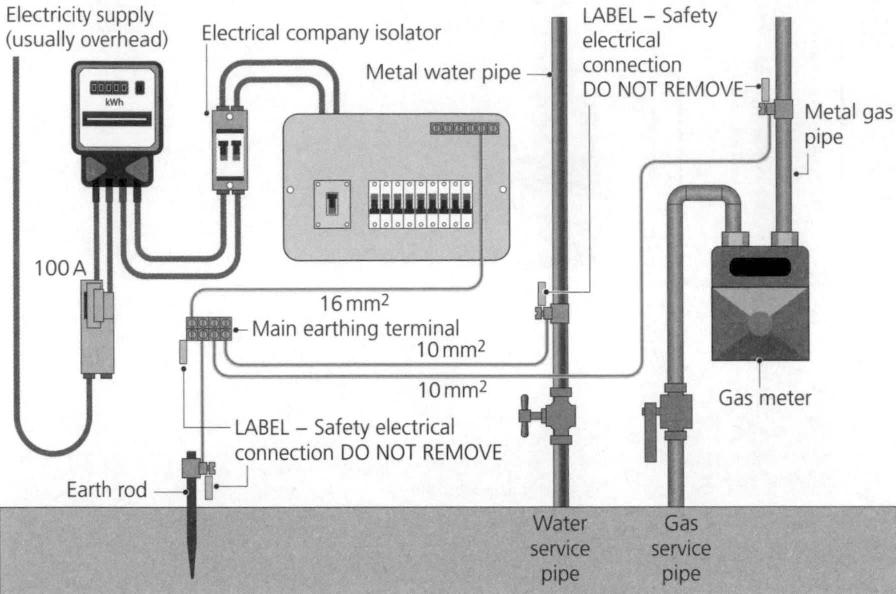

▲ Figure 11.17 TT system

The first 'T' stands for *terre*, meaning that all exposed metalwork is connected directly to earth via a large copper rod stuck directly into the ground outside. The second T means that the substation is also connected to earth in the same way. This system is commonly used in rural areas and the electrode/rod for the installation can be very hard to find as foliage may hide it. If you do find it, do not touch or disconnect the connections for any reason.

6 PROTECTIVE DEVICES

Within a consumer's electrical installation, there are several different types of protective device that can be used to protect circuits and persons against faults.

Fuses

Fuses have been a tried and tested method of circuit protection for many years. A fuse is a very basic protection device that is destroyed and breaks the circuit if the current exceeds the rating of the fuse. Once the fuse has 'blown' (i.e. the element of the fuse has melted or ruptured), the fuse needs to be replaced.

BS 3036 semi-enclosed rewireable fuses

In older equipment, the fuse may be just a length of appropriate fuse wire fixed between two terminals. These devices are now becoming uncommon as electrical installations are rewired or updated. One of the main problems associated with re-wirable fuses is the overall lack of protection, including insufficient breaking capacity ratings. Another major problem is unreliability, which is often caused by using the wrong gauge of wire when changing the fuse. The overall reliability of such fuses cannot be guaranteed because a wide range of factors can cause them to fail. Typical factors include wire being labelled with an incorrect current rating and the number of times and the length of time that a fuse wire has been subjected to overload.

▲ Figure 11.18 A range of **BS 3036** fuses. The dots on the front of the carrier signify the rating: red 32 A, yellow 20 A, blue 15 A and white 5 A

BS 88 fuses

These modern fuses are generally incorporated within sealed ceramic cylindrical bodies (or cartridges). The whole cartridge needs to be replaced if the fuse ruptures. It is not common to find these fuses in domestic properties protecting circuits but they are likely to be used in the service head just before the electricity meter. These fuses are the property of the energy supplier and should not be touched.

▲ Figure 11.19 A range of **BS 88** fuses; some may not have the fixing lugs

BS 1362 plug fuses

These are a type of cartridge fuse that fit inside plugs and fused connection units. The fuse is intended to protect the flexible cable that supplies the power to the appliance. They come in various ratings from 1 A to 13 A but the two main ratings are 3 A for any appliance up to 700 W and 13 A for any appliance over 700 W.

Circuit breakers

Circuit breakers are thermomagnetic devices capable of making, carrying and interrupting currents under normal and abnormal conditions. They fall into two categories: miniature circuit breakers (MCBs), which are common in most installations for the protection of final circuits, and moulded case circuit breakers (MCCBs), which are normally used for larger distribution circuits.

Circuit breaker nominal ratings (I_n) relate to the current rating of the device for continuous service under specified installation conditions.

There are three circuit breaker types: Type B, Type C and Type D. The current flow at which they trip depends upon the level of overcurrent and is usually determined by a pre-setting of the magnetic device within the circuit breaker. Type B offers the best protection and is normally found in domestic settings, Type C allows for high starting currents common with transformers, and Type D is only used for very specific situations and so should never be installed on standard circuits.

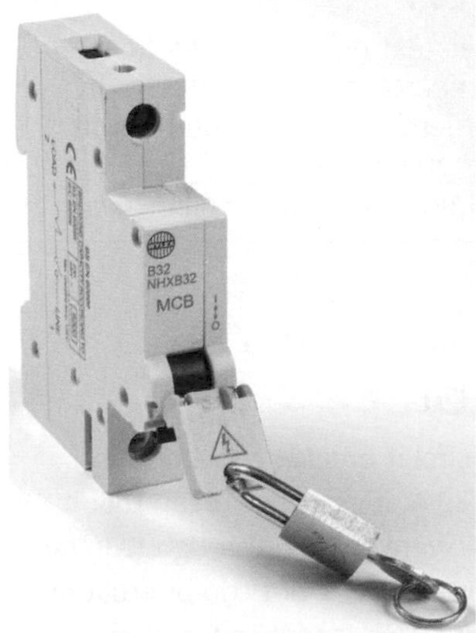

▲ Figure 11.20 A circuit breaker with a lock to secure it in the off position

Residual current devices (RCDs)

An RCD on its own is technically not a circuit protective device as it does not offer overload or **short circuit** protection. It works by monitoring the line and neutral conductors of a circuit. If the current in the two conductors is different, it means that some must be leaking to earth. If this leakage is more than the residual current rating of the device, typically 30 mA, the device will trip the circuit. RCDs are very sensitive and provide very good levels of protection against electric shock.

KEY TERM

Short circuit: a fault between line and neutral.

ACTIVITY

Why are the most common RCD ratings set at 30 mA? Think about shock current.

Answer

Because 40 mA can cause death, having a trip setting less than this saves lives.

RCBOs

An RCBO is a residual current device (RCD) combined with a circuit breaker. This means that it offers the characteristics of an RCD along with the overload and short circuit protection of a circuit breaker.

HEALTH AND SAFETY

RCDs and RCBOs have test buttons on the front of them. It is vitally important these are tested by pushing the button at least every six months. This ensures that the mechanism is kept free and stops it sticking so that it will be reliable when needed.

Fault currents causing disconnection

Just because a fuse or circuit breaker has a nominal rating of, for example, 32 A, it does not mean that it will trip or erupt if 33 A were to flow through it. In fact, most fuses or circuit breakers will not trip or erupt until at least 1.45 times their rating is reached, and, even then, it may take minutes or hours to break the circuit.

To achieve instantaneous fault protection, much higher currents are needed. For example, a 32 A rated B type circuit breaker must have a fault current of at least 160 A in order to trip instantly. Anything less than this will take many seconds, or possibly minutes, to trip it.

For this reason, it is important for earthing to have as low a resistance as possible to cause a high fault current.

HEALTH AND SAFETY

Always remember Ohm's law ($V = IR$) and that the voltage of most electrical systems is fixed at 230 V. This means that the lower the resistance, the higher the fault current and the quicker a circuit will be disconnected to remove the danger. Therefore it is important to have an earthing system with as low a resistance as possible.

INDUSTRY TIP

Earthing is not the same as bonding. Earthing is needed all the time, but bonding may not be needed. If you are unsure whether or not something needs bonding, ask a qualified electrician. If you bond something that is suitably insulated from earth, you may make it a potential risk when it was actually perfectly safe before.

Even if the circuit is protected by an RCD or an RCBO, it is still important to have paths of low resistance to earth even though these devices trip at low fault currents.

7 WORKING ON ELECTRICAL SYSTEMS

When working on electrical systems, it is vitally important to follow safe work practices. Remember that you cannot see or smell electricity, so you need to make sure it is not present before you work on electrical systems or equipment.

Safe isolation procedure

Isolation can be very complex due to the differing industrial, commercial and domestic working environments, some of which require experience and knowledge of the system processes. This section deals with a basic practical procedure for isolation and for the securing of isolation. It also looks at the reasons for safe isolation and the potential risks involved during the isolation process.

How to undertake a basic practical procedure for isolation

First, prepare all of the equipment required for this task. You will need:

- a voltage indicator which has been manufactured and maintained in accordance with Health and Safety Executive (HSE) **Guidance Note GS38**
- a proving unit compatible with the voltage indicator
- a lock and/or multi-lock system (there are many types of lock available)
- warning notices which identify that work is being carried out.

KEY TERM

Guidance Note GS38: a guidance publication from the Health and Safety Executive (HSE) relating to the safety of test equipment, leads and probes used on electrical circuits.

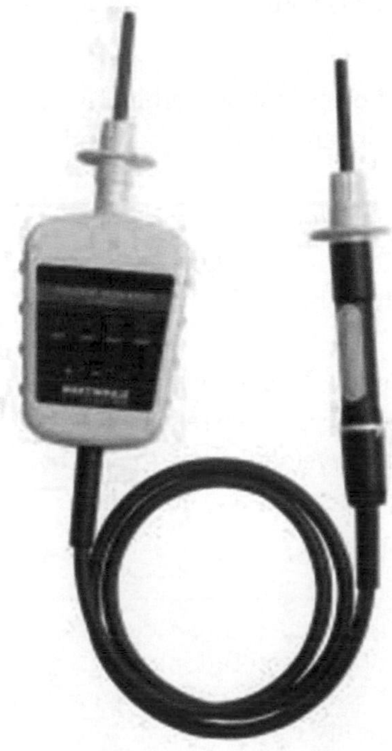

▲ Figure 11.21 Voltage indicator

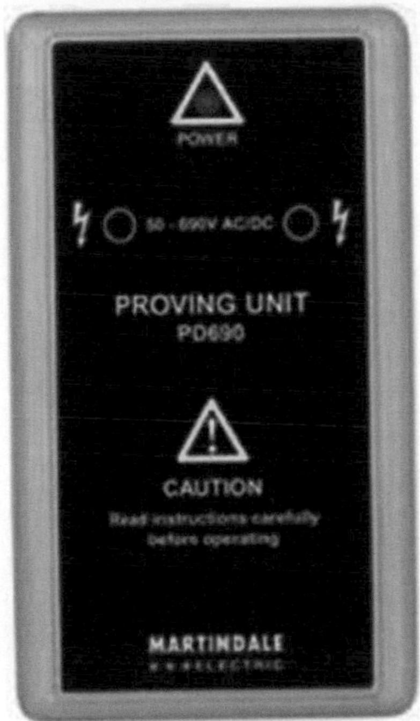

▲ Figure 11.22 Proving unit

▲ Figure 11.23 Lock out devices for circuit breakers

Isolating at the correct point is vital. If you are working on a boiler control system or a hot water cylinder immersion heater and these circuits are clearly marked and identified, isolate at the consumer unit using the identified circuit breaker according to the procedure below.

If there is a clearly identified fused connection unit where the fuse can be withdrawn, then do so at this point, but, if you are making, inspecting or disconnecting anything in that connection unit, it will still be partially live, so the circuit should be isolated using the circuit protective device.

1. **Identify** – identify the equipment or circuit to be worked on and point(s) of isolation.
2. **Isolate** – switch off, isolate and lock off (secure) the equipment or circuit in an appropriate manner. Retain the key and display caution signs with details of the work being carried out. If using a fused connection unit, keep the fuse and carrier in your pocket.
3. **Check** – check the condition of the voltage indicator leads and probes. Confirm that the voltage indicator is functioning correctly by using a proving unit.
4. **Test** – using the voltage indicator, test all the equipment you are going to be working on by testing:
   - L – N
   - L – E
   - N – E

   (L = line, N = neutral, E = earth)
5. **Prove** – using a voltage indicator and proving unit, prove that the voltage indicator is functioning correctly.
6. **Confirm** – confirm that the isolation is secure, and the correct equipment has been isolated. This can be achieved by operating functional switching for the isolated circuit(s). It is good practice to be sure.
7. The relevant work can now be carried out.

▲ Figure 11.24 Switched fused connection unit with indicator

When the work is completed, you must ensure that all electrical barriers and enclosures are in place and that it is safe to switch on the isolated circuit.

- Remove the locking device and danger/warning signs.
- Reinstate the supply.
- Carry out system checks to ensure that the equipment is working correctly.

Always bear in mind, before isolating anything, the implications that this loss of electrical supply may bring.

Considerations include:

- What else will the client lose if a circuit or part of a circuit is isolated?
- Will you need to provide a temporary supply for essential equipment?
- Will you need to timetable the isolation for a time when the loss of services will have minimum impact?

Electrical faults

One of the most common reasons for working on existing electrical systems is that a fault occurs. Fault diagnosis must always be undertaken by competent, skilled persons as faults can create a great deal of danger.

As well as a major risk of electric shock, other consequences exist if faults are not diagnosed and rectified. These include:

- Fire: arcing and sparking from faulty components or, more frequently, from loose connections can cause a fire, leading to a risk to life and damage to property.
- Component failure: acting on early warning signs of faults can lead to less expensive repairs, whereas if faults are allowed to develop further, they could lead to expensive replacements and **downtime**.

KEY TERM

Downtime: the time during which a particular process or action is stopped due to faults or delays, for example the time that a building is without heat due to a faulty component in a boiler and the time needed to obtain the part and replace it.

An electrical fault may be beyond the competency of a person trained to work on mechanical systems. As a result, should faults occur it is very important for the correct person to act on the fault by undertaking the necessary diagnosis and rectification themselves, or report it and authorise those who can. This includes:

- homeowners of private houses, who can appoint an electrician to undertake the work
- tenants of rented accommodation, who can inform their landlord (should you be contracted by the landlord, you can inform them directly)
- the supervisor of a construction site (or, if you are a subcontractor, the site agent or site manager).

8 INSTALLING WIRING SYSTEMS

There are many types of wiring systems used. A wiring system could be a cable on its own or a means of containing or managing the cable. You may encounter the following terms.

- **Wiring system** – this is the term used to describe the type of cable used and the method of supporting it. For example, single-core cable in conduit is a wiring system but, equally, so is a twin and cpc cable clipped to a wall.
- **Support system** – this is the method of supporting a cable and could include systems such as a cable tray, basket or simply clips or cleats.
- **Cable management system** or **cable containment system** – this is a method of supporting and protecting cables, usually by enclosing the cables in conduit, trunking or ducting, for example.

In this chapter, we will only be covering the installation of cables that are commonly used in wiring systems in domestic environments for the supply and control of electro-mechanical systems.

Terminating flexible cables

Tools required

- Ringing tool (Method 1)
- Stripping knife (Method 2)

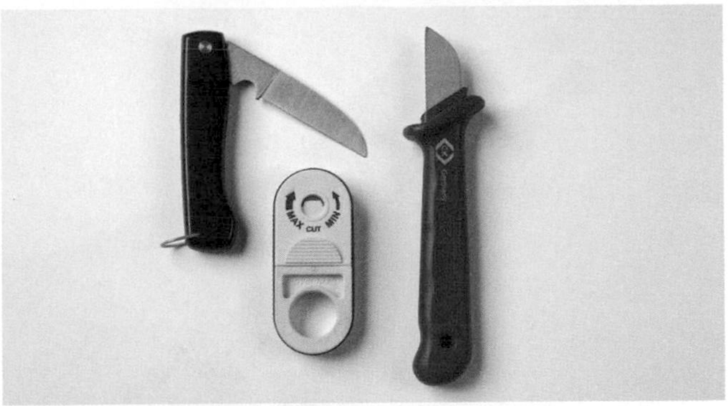

▲ Figure 11.25 Ringing tool (Method 1); stripping knife (Method 2)

Safety considerations

There is a risk of cutting your hands when using knives. Wearing gloves and eye protection is recommended.

Method

Before the flex can be terminated, the outer sheath must be removed. Two methods of removing the outer sheath are outlined here.

Method 1

The outer sheath can be removed with the use of a ringing tool. These come in various shapes and forms, the most basic of which is shown in Figure 11.26. This tool slides over the end of the cable to the required stripping position and is then rotated around the cable, cutting it slightly.

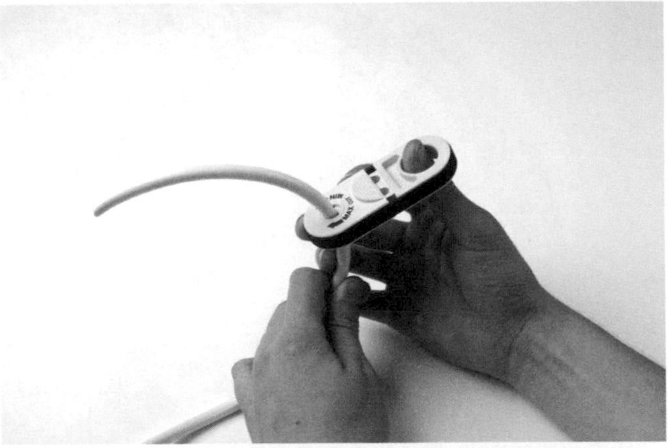

▲ Figure 11.26 Flexible cable sheath being removed by a ringing tool

The ringing tool is removed, the cable is bent to finish the cut and the sheath is then slid off.

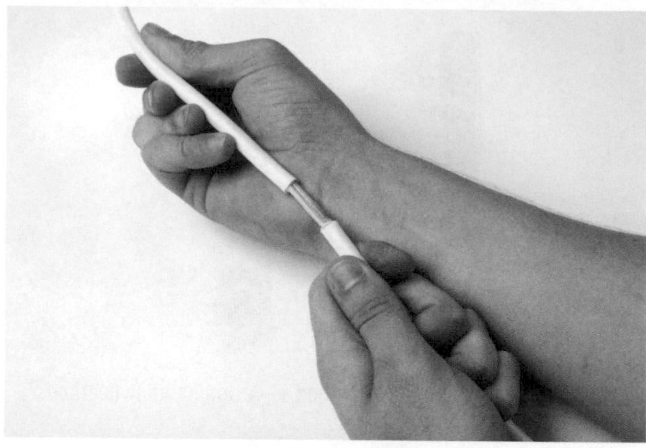

▲ Figure 11.27 Outer sheath being removed

Method 2

There may be times when a ringing tool is not available, so it is important to know how to remove the outer sheath without this tool. Thermoplastic has a tendency to split when it is damaged and pressure is applied, and this property can be used to help strip the sheath.

Bend the cable into a tight bend at the point where the sheath is to be removed. Using a sharp knife, gently score the top of the bend. The thermoplastic will split open like a little mouth (Figure 11.28).

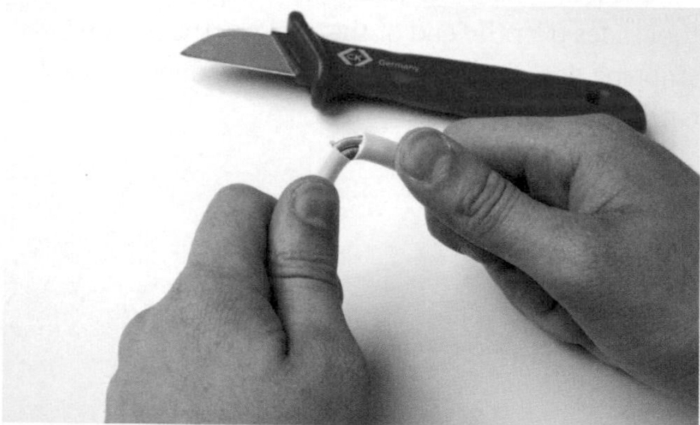

▲ Figure 11.28 Bending the cable to cause the scored sheath to split

Plugs

A plug is a simple device that allows you to safely connect or disconnect an appliance to and from an AC socket. Poorly wired plugs are a common cause of electrical faults. Here is a step-by-step guide to wiring a plug correctly and safely.

1. You will need to strip off about 4–5 cm of the outer cable sleeve. Slit the outer sleeve of the cable lengthways using an electrical knife or flexible cable stripping tool, being careful not to cut into the coloured wires or yourself. Peel the outer sleeve away and cut it off using cable cutters or wire cutters.
2. Separate the wires and cut them to the correct length using wire cutters. Measure the length they need to be against the plug, matching them up to the correct terminal points.
3. Cut off about 5 mm of the insulation using wire strippers or wire cutters. Be careful not to cut into the individual strands of wire. Once you have done this, twist the ends of the wires, so that you have a hard cable to work with, rather than individual strands.
4. Now connect each wire to the correct terminal. To start, slacken the screw on top of the terminal and push the bare wire into the hole created. The correct tool for this type of terminal is an electrician's screwdriver. All electrician's tools have insulation to protect the user from an electric shock. Then retighten the screw down on to the bare cable and terminal body. Make sure the terminals are tight and there are no bare wires or loose strands of wire showing or overhanging the terminal, as a loose wire could cause a short circuit. The terminals are normally stamped with the following letters:
   - L = line (brown)
   - N = neutral (blue)
   - E = earth (yellow and green).

   You may also notice the earth symbol, shown in Figure 11.30. If not, the top terminal is usually earth, the terminal attached to the fuse is line and the last terminal is neutral. As a tip, connect the neutral first.
5. Once the wires are connected, tighten the cord clamp over the cable. Make sure the cord clamp is gripping only the outer sleeve of the cable, and not the coloured wires.
6. Before screwing on the top of the plug, check that you have used the correct size of fuse for the appliance. There are three standard fuse ratings: 3 A, 5 A and 13 A.
7. Before screwing the plug top back on, check that the wires are fitted correctly in the channels provided and that they will not become crushed. Most importantly, recheck the wiring you have completed and make sure there are no conductors or copper showing.

All the tools you use for cutting and removing sheathing from the cable must be specifically made for the purpose. These are known as insulating cutters (or cable cutters) and insulating strippers. You should not use a Stanley knife to carry out this procedure. As with the electrician's screwdriver, these tools are designed to prevent the user getting an electric shock if the system accidentally goes live during the work.

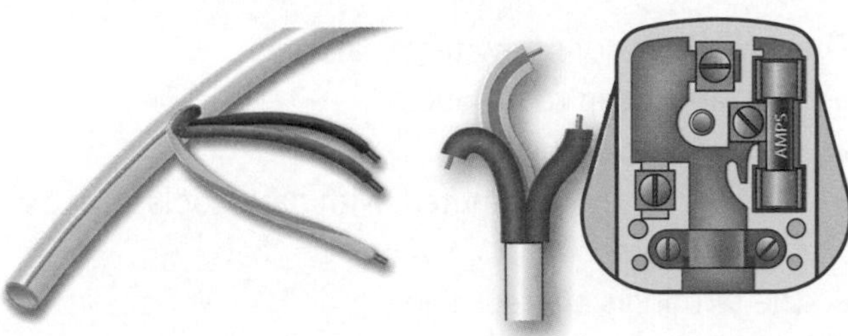

▲ Figure 11.29 Wiring a three-pin plug

▲ Figure 11.30 The earth symbol

Thermoplastic (PVC) cables

Thermoplastic cables are commonly referred to as PVC (polyvinyl chloride) cables and they come in various shapes, sizes and forms, including:

- single-core cable
- twin and cpc flat profile cable
- three-core and cpc flat profile cable.

In domestic installations, the most common cables are the twin and cpc, and the three-core and cpc flat profile cables.

Terminating PVC/PVC flat profile cables

Tools required

Dependent on method used:

- electrician's knife
- side cutters
- pliers.

Safety considerations

- Cuts to hands from using a knife.
- Injury caused by cutters or pliers slipping.

Wearing gloves and eye protection is recommended.

Method

Before the conductors can be connected, the outer sheath of the cable must be removed.

First, identify how much of the sheath should be removed from the cable. The purpose of the sheath is to provide some mechanical protection for the insulation on the conductors. Too much sheath, however, takes up space within the accessory and will put excess strain on the conductors. The sheath should, therefore, be stripped back almost to where the cable enters the accessory, leaving only 10–15 mm, a thumb's width, of sheath within the accessory.

Care must be taken to avoid damage occurring to either the conductors or the insulation around the conductors when stripping the sheath.

Having decided on the length of sheath required, and with the cable in place, score the sheath at the point from which it is to be removed. This can be performed with an electrician's knife. Care should be taken not to cut into the cable.

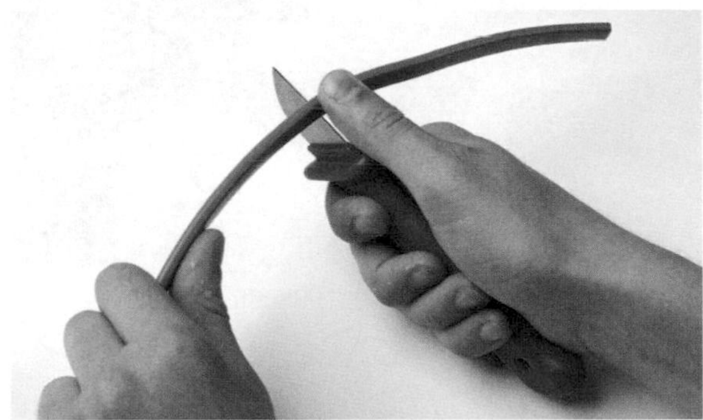

▲ Figure 11.31 Scoring the outer sheath

Snip down the end of the cable with the side cutters, then use a sharp electrician's knife to slice down the cable. To do this, run the blade along the protective conductor. Note that this can damage the protective conductor or, even worse, the insulation of one of the live conductors if it is not done carefully.

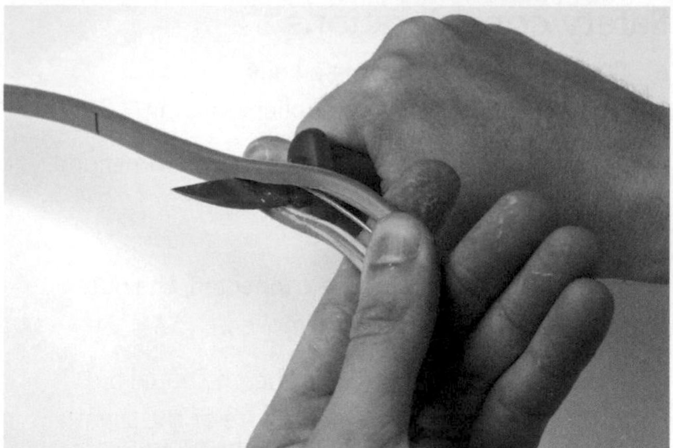

▲ Figure 11.32 Running an electrician's knife along the cpc

Termination

Now the inner wires can be stripped, using a cable stripper tool, to the appropriate length for the terminal. Make sure that the copper conductors do not extend outside of the terminal but, equally important, make sure that the insulation is not clamped inside the terminal.

▲ Figure 11.33 Cable stripper tool

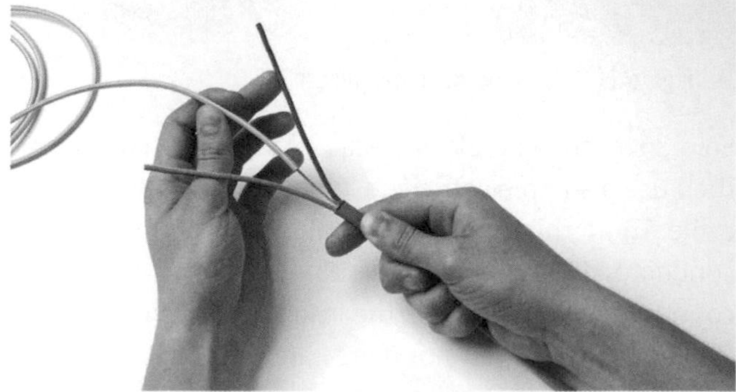

▲ Figure 11.34 Applying sleeving to bare cpc ready for final termination

For the bare cpc, make sure that the green and yellow sleeving is applied correctly to the bare copper cpc. The sleeving must cover all of the bare cpc, apart from a small amount at the end, which should be just long enough to go into the terminal.

The most common type of termination, used to connect the cable conductors to the equipment, is a screw type terminal.

The advantages of screw terminals are:

- they are cheap to produce
- they are reliable
- they are easily terminated, with basic tools such as screwdrivers
- the terminals are reusable.

The disadvantages are that:

- over-tightening may result in damage to the terminal or the conductor
- under-tightening may result in overheating and arcing
- terminals can become loose, due to movement of the conductor in use or due to mechanical vibration
- terminations need to be accessible for inspection.

Simple screw terminals can be made using cable cutters and screwdrivers.

Cable cutters

A variety of different tools can be used to cut cables, the most common being side cutters. Side cutters are one of the most important tools that electricians have in their tool kit. They are used for cutting cables to length, cutting sleeving and cutting nylon tie-wraps, for example. They work on a compression-force basis and are shaped so that the cutting point is along one side.

Screwdrivers

An electrician will use a selection of different sized screwdrivers. They will all be of approved standards.

The most common screwdrivers used are:

- terminal (3–3.5 mm flat head)
- large flat (4–5 mm flat head)
- Pozidriv (PZ2 cross head).

Test your knowledge

1 What is the voltage for a circuit that has a resistance of 12 Ω and a current of 4 A?

a 3 V

b 12 V

c 36 V

d 48 V

2 What is the orbiting part of an atom called?

a Proton

b Electron

c Neutron

d Nucleus

3 What colour is the insulation around the conductor designated as the neutral?

a Brown

b Green

c Blue

d Yellow

4 What effect does a change in temperature have on the resistance of a conductor?

a Resistance increases when temperature increases

b Resistance decreases when temperature increases

c Resistance is not affected by temperature change

d Resistance only changes at absolute zero temperature

5 Which of the following is **not** a standard conductor cross-sectional area?

a 1.0 mm²

b 2.5 mm²

c 4.5 mm²

d 6.0 mm²

6 What is the total permitted percentage voltage drop for a power circuit in an electrical installation?

a 25%

b 3%

c 4%

d 5%

7 What is the total circuit current (*I*) for this circuit?

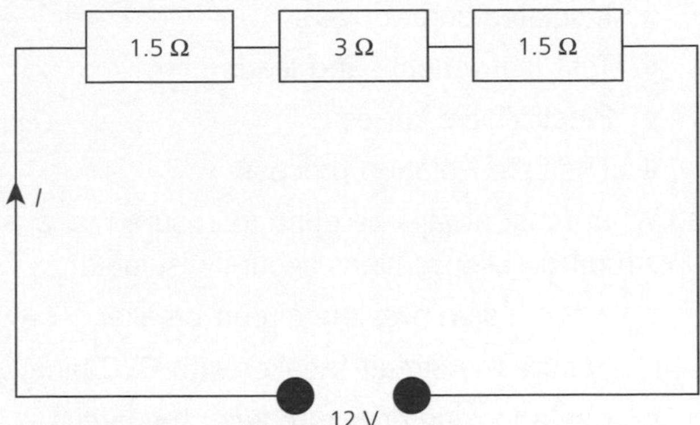

a 2 A

b 6 A

c 12 A

d 72 A

8 What value of voltage will overcome skin resistance, causing an electric shock, for an average person in average conditions?

a 12 V AC

b 25 V AC

c 40 V AC

d 50 V AC

9 What is the purpose of basic protection in an electrical circuit?

a It prevents faults from happening

b It prevents contact with live parts

c It decreases disconnection times

d It increases disconnection times

10 Which earthing arrangement relies on the general mass of earth as the earth return path?

a TT

b TN-C

c TN-S

d TN-C-S

11 What is the BS number for a plug fuse?

a BS 88

b BS 3036

c BS 1362

d BS 4343

12 What does HSE guidance GS38 specifically cover?

a Insulated screwdrivers

b Test instruments and leads

c Electrician's knives

d The safe isolation process

13 What must always be done to ensure that a circuit breaker remains securely isolated?

a Place a sign over the circuit breaker

b Secure the circuit breaker with PVC tape

c Use a locking device to keep the circuit breaker in the off position

d Use a locking device to keep the circuit breaker in the on position

14 What is a short circuit fault?

a A fault between line and earth in a circuit

b An overload caused by too much equipment

c A broken conductor leaving a circuit open

d A fault between line and neutral in a circuit

15 What tool is the most appropriate to remove the insulation from a conductor?

a Side cutters

b Wire stripper

c Electrician's knife

d Rotary stripping tool

16 Describe the relationship between atoms and current flow.

17 Explain why earthing needs to have a low resistance.

18 Describe the term voltage drop.

19 Calculate the voltage at the terminals of a 1200 W item of electrical equipment. The circuit is supplied at 230 V. The resistance of the circuit line conductor is 0.24 Ω and the circuit neutral is 0.22 Ω.

20 Calculate the total circuit current (*I*) for this circuit.

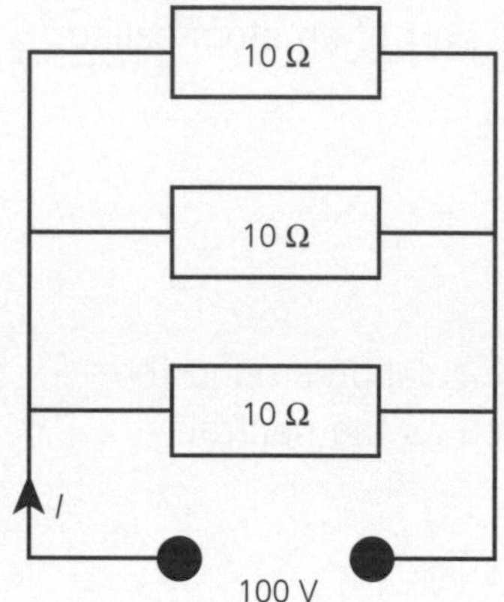

21 Explain the difference between earthing and bonding.

22 Describe why the safe isolation procedure requires the voltage tester to be used on a known supply such as a proving unit.

23 Explain how to ensure that a switched fused connection unit remains securely dead.

24 List three types of cable support.

25 Describe, in detail, the safe isolation process.

Answers can be found online at www.hoddereducation.co.uk/construction

Practice synoptic assessments

Practice assessment 1

You have been called to a customer who is looking to upgrade their central heating and hot water system. The property is a 1950s build which has a gravity hot water and single pipe central heating system installed. The house is semi-detached. The customer has only just moved into the house and they would like to upgrade the systems before they decorate. Downstairs the property has a lounge, kitchen, hallway and toilet. Upstairs the property has three bedrooms and a bathroom. The basic layout of the property is shown in the diagram at the bottom of the page.

The customer has asked for your advice on what systems should be installed and what controls are required to bring the hot water and central heating up to current standards.

1 Recommend a hot water system and central heating system that would suit the customer's needs.
2 Justify your choice by comparing with other systems.
3 Outline the potential disruptions that will occur while installation takes place.
4 Outline any health and safety factors that need to be considered.
5 List the control components that are required for both the hot water and central heating systems.
6 The customer would also like to know the floor area of their kitchen–diner.

(1) Ground floor

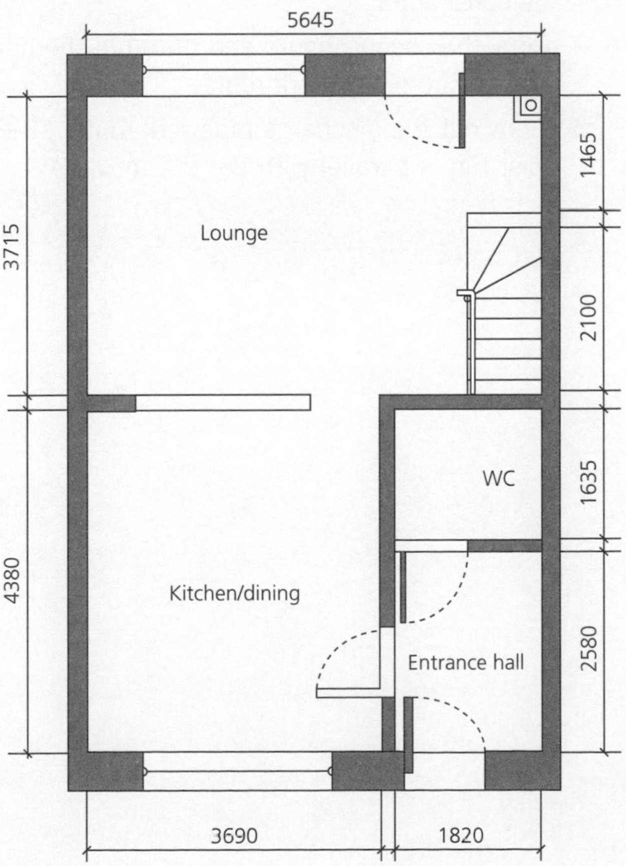

(2) First floor

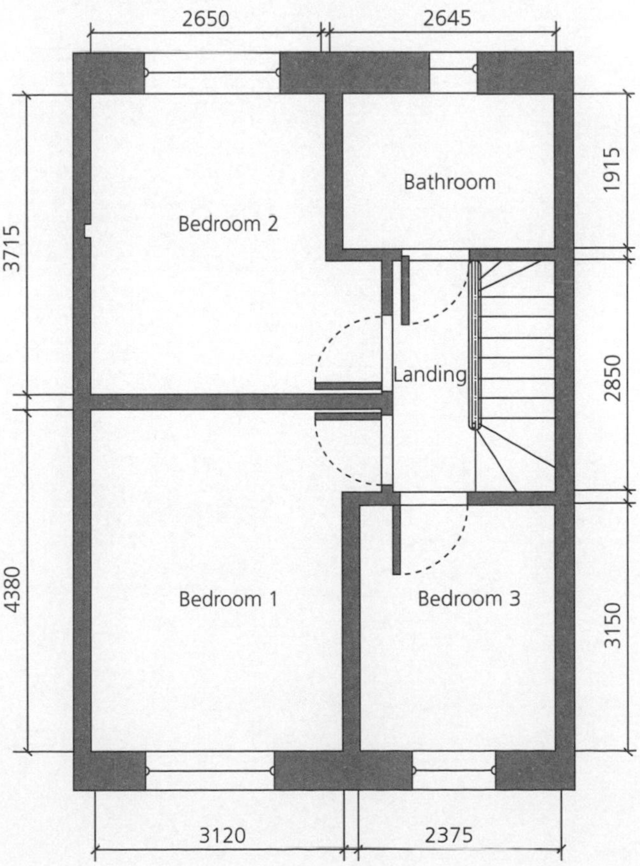

Practice assessment 2

Your company has asked you to visit a customer who is looking to change their bathroom suite. The room is not particularly large and only just fits the existing appliances. At present, a dated bathroom is installed and the customer would like to upgrade to a modern installation that includes a shower.

The first diagram shows the existing layout and dimensions.

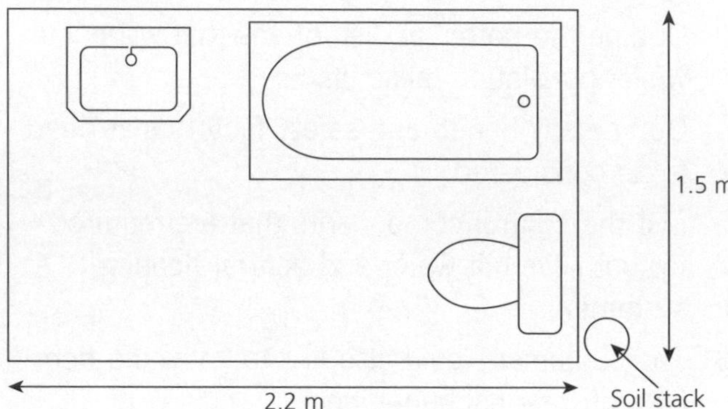

The second diagram shows the empty bathroom, the hot and cold supply and the position of the soil stack.

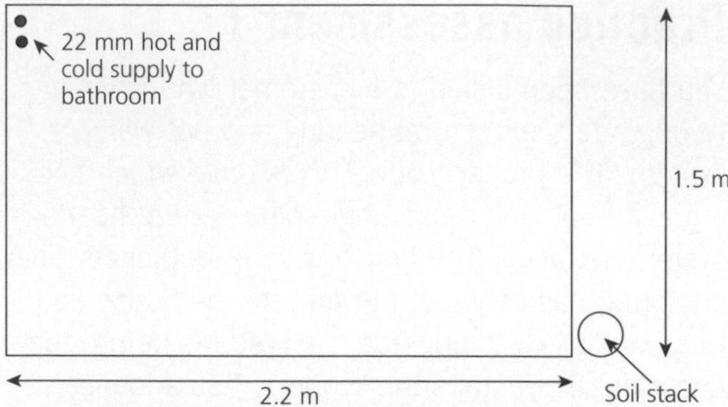

You will need to:

- take out the existing bath, basin and WC
- install a new bath, basin, shower and WC.

1 Offer the customer a couple of options for appliances.
2 State any potential disruptions that may occur during the installation.
3 List any health and safety factors that need to be considered.
4 Copy the second figure and draw out the new hot and cold pipework route.
5 Draw out the discharge pipework route. (The joist run is parallel with the 2.2 m wall.)

Glossary

Acceleration a measure of the rate at which an amount of matter increases its velocity. It is measured in a change of velocity over a period of time and, as such, is directly proportional to force. It will increase and decrease linearly with an increase or decrease in force if the mass remains constant. It is measured in metres per second squared (m/s^2).

Acceleration due to gravity the rate of change of velocity of an object due to the gravitational pull of the Earth. If gravity is the only force acting on an object, then the object will accelerate at a rate of 9.81 m/s^2 downwards towards the ground.

Accent the way in which people pronounce their words.

Accident an unexpected or unplanned event that could result in personal injury, damage and, occasionally, death. When an accident occurs, there are always reasons for it and if there's a reason, then there is usually blame.

Actual capacity (of a cistern) the maximum volume it could hold when filled to its overflowing level.

Acute injury occurs when manual handling or lifting causes immediate pain and injury.

Adhesion the way in which water molecules 'stick' to other molecules they come into contact with.

Advisory, Conciliation and Arbitration Service (ACAS) an organisation that provides free and impartial information and advice to employers and employees on all aspects of workplace relations and employment law.

Advisory recommended but not enforced.

Aesthetically pleasing beautiful in appearance, good-looking, in keeping with the rest of the surroundings.

Air infiltration a process where air can get into a system and cause air locks and corrosion.

Amp (and milliamp) unit of electrical current, the measurement of ampere.

Annealing a process that involves heating the copper to a cherry-red colour and then quenching it in water. This softens the copper tube so that the copper can be worked without fracturing, rippling or deforming.

Anodising coating one metal with another by electrolysis to form a protective barrier against corrosion.

Asbestos a fibrous silicate material highly resistant to heat.

Atom a fundamental piece of matter made up of three kinds of particles called subatomic particles – protons, neutrons and electrons.

Average conditions somewhere with a comfortable indoor temperature of about 20°C and a low humidity.

Benchmarking this is now a compulsory requirement to ensure that systems and appliances are installed in accordance with the regulations and the manufacturer's instructions. It also safeguards any guarantee against bad workmanship.

Bernoulli's principle when a pipe reduces in size, the pressure of the water will drop but the velocity of the water increases. When the pipe increases back to its original size, then the velocity will decrease and the pressure will increase almost to its original pressure.

Blowing this describes the outside surface of a brick being damaged when drilling a hole from the inside.

Boiler cycling the process of the constant firing up and shutting down as the system water cools slightly wastes a lot of fuel energy. Applied to when a heating system has reached temperature, the boiler shuts down on the boiler thermostat. A few minutes later the boiler will fire up again to top up the temperature as the system loses heat and, after a few seconds, shuts down again.

Boiler interlock this is not a physical item, but the way in which a modern system is wired up to prevent the boiler from firing unless there is a demand for heat.

BSP British Standard Pipe.

BSPT British Standard Pipe Thread; the type of thread used on screwed low carbon steel pipes and fittings.

Calorific value the amount of energy stored in the gas in its uncombusted state. It is the amount of energy released when the gas is combusted. It is measured in megajoules (MJ) per cubic metre or MJ/m^3.

Capacity the ability of a cable to carry current. This is directly linked to the cable csa.

Centre to centre measuring from the centre line of one pipe to the centre line of another, so that all the tube centres are uniform. This ensures that the pipework will look perfectly parallel because all of the tubes will be at equal distance from one another.

Chamfer to take off a sharp edge at an angle. If we chamfer a pipe end, we are taking the sharp, square edge off the pipe.

Chronic injury type of injury that can take weeks, months or even years to develop.

Code of practice similar to a British Standard, this is a set of rules that explains how people should behave in their chosen profession.

Cohesion the way in which the water molecules 'stick' to one another to form a mass rather than staying individual. This is because water molecules are attracted to other water molecules.

Colour coding the colour coding of electric cable insulation is
Brown – line conductor (L)
Blue – neutral conductor (N)
Green/yellow – earth (E) or circuit protective conductor (cpc)

Combination ('combi') boiler a boiler that provides central heating and instantaneous hot water.

Competent having the necessary ability, knowledge or skill (trained, tested and received a certificate).

Comply act in accordance with; meet the standards of.

Compression the process of water hitting a bend at forces that cause a shock wave of air upwards.

Conciliation an alternative dispute resolution process whereby the parties to a dispute agree to use the services of a conciliator, who then meets with the parties separately in an attempt to resolve their differences. Collective conciliation is when a group of employees is involved, and individual conciliation is when there is only one employee involved in the dispute.

Condensate the water vapour that is present in the CO_2 emissions resulting from burning gas. As the flue gases cool down, the water vapour condenses back into water droplets that are collected in the boiler and discharged via the condense pipework.

Corrosion any process involving the deterioration or degradation of metal components, where the metal's molecular structure breaks down irreparably.

Crimping the process of pressing the fittings into a copper pipe using a 'press fit' or crimping tool.

CSCS card this stands for Construction Skills Certification Scheme card. Its purpose is to confirm that people who work in the built environment have the necessary competence, and identifies their qualifications. For example, a trainee plumber would carry a small plastic ID craft or operative card that identifies them as a person enrolled on an NVQ programme but not yet qualified.

Delegation sharing or transfer of authority and responsibility, from an employer or supervisor to an employee.

Derived units combinations of the seven base units by a system of multiplication and division calculations. There are 21 derived units of measurement, some of which have special names and symbols.

Deviate change; do differently from the original plan.

Dew point the temperature at which the moisture within a gas is released to form water droplets. When a gas reaches its dew point, the temperature has been cooled to the point where the gas can no longer hold the water and it is released in the form of 'dew', or water droplets.

Dialect a combination of the way people pronounce words, the vocabulary they use and the grammatical structures they use.

Diverted neutral current a high-risk situation in which a problem in the electricity supply causes neutral current to try and find alternative routes back to the local electrical substation transformer.

Downtime the time during which a particular process or action is stopped due to faults or delays, for example the time that a building is without heat due to a faulty component in a boiler and the time needed to obtain the part and replace it.

Dressing the term used by plumbers to describe the preparation of the appliances ready for installation.

Duty holder a person who controls, reduces or eliminates health and safety risks that may arise during the construction of a building or during future maintenance. They must also provide information for the health and safety file.

Duty of care in British law, the moral and legal obligation imposed on an organisation or individual, which necessitates that a standard of reasonable care is adhered to. If the standard of care is not met, then the acts are considered to be negligent and damages may be claimed for in a court of law.

Electrolyte a fluid that allows the passage of electrical current, such as water. The more impurities (such as salts and minerals) there are in the fluid, the more effective it is as an electrolyte.

Equipotential bonding a system where all metal fixtures in a domestic property, such as hot and cold water pipes, central heating pipes, gas pipes, radiators, stainless-steel sinks, pressed-steel enamelled washbasins and steel and cast iron baths, are connected together through earth bonding so that they are at the same potential voltage everywhere.

Erroneous wrong; incorrect.

Exposed conductive part a metallic or conductive material that forms part of the electrical circuit and therefore is likely to become live if an electrical fault occurs. Examples include a metal consumer unit, a metal central-heating pump or a metal-cased boiler.

Extraneous conductive parts conductive metal parts that are not part of the electrical system but provide a direct path to earth, for example steel framed buildings with steel structural parts embedded into the ground or metal gas pipes coming out of the ground.

Fire stop a barrier is placed where the pipe passes through a floor, ceiling or wall to prevent the spread of fire and smoke.

Flashback where the flame burns in the torch body, accompanied by a high-pitched whistling sound. It will occur when flame speed exceeds gas flow rate so that the flame can pass back through the mixing chamber into the hoses. Most likely causes are incorrect gas pressures giving too low a gas velocity, hose leaks or loose connections.

Floc a collection of loosely bound particles or materials. These are bound together by the coagulation process for easy removal from the water.

Foot a ladder stand with one foot on the bottom rung, the other firmly on the ground.

Forced draught the use of a purpose-designed fan to create a positive updraught by forcing the products of combustion up the flue.

Fully pumped system a heating system that uses pumped circulation to both heating and hot water circuits.

Gravity circulation circulation that occurs because heat rises through the water. No pump is required.

Grey water waste water from baths, showers, washing machines, dishwashers and sinks.

Guidance Note GS38 a guidance publication from the Health and Safety Executive (HSE) relating to the safety of test equipment, leads and probes used on electrical circuits.

Hazard a danger; something that can cause harm.

Health and Safety Executive (HSE) the government body in the UK responsible for the encouragement, regulation and enforcement of workplace health, safety and welfare regulations and government legislation.

Heat exchanger a device or vessel that allows heat to be transferred from one water system to another without the two water systems being allowed to come into contact with each other. The transfer of heat between the two systems takes place via conduction (see Chapter 3, Scientific principles).

Immersion heater an electrical element that sits in a body of water, just like in a kettle. When switched on, the electrical current causes the electrical element to heat up, which in turn heats up the water. Most immersion heater elements are rated at 3 kW but cylinders can have 1, 2, 3 or 4 elements. All immersion heaters must comply with BS EN 60335–2–73 and have a resettable double thermostat (RDT) as standard. This enables problems with overheating to be recognised quickly.

In situ in situ, in plumbing terms, simply means pipework or appliances that are already in place. They are already 'in situation', hence the term 'in situ'.

IP2X and IP4X part of the international coding system for the level of Ingress Protection for enclosures. This is the level of protection against the ingress of foreign bodies, such as fingers (first number) and the level of protection against the ingress of water and moisture (second number).

Legislation a law or group of laws that have come into force; health and safety legislation for the plumbing industry includes the Health & Safety at Work Act and the Electricity at Work Regulations.

Level when pipework is perfectly horizontal.

Liaise establish a co-operative working relationship.

Main earthing terminal (MET) every electrical system in a building will have a main earthing terminal (MET) which is either in the main consumer unit housing the fuses/circuit breakers or a separate metal block close to the electricity meter.

Maintenance preserving the working condition of appliances and services.

Mandatory required by law; compulsory.

Mechanical suction suction that is created by an electrical/mechanical pump installed onto a tanker truck.

Method statement the record of how management wants the job to be done. Its main purpose is to guide site work and it must always be available on-site as a live document with an aim to prevent accidents or dangerous situations from occurring.

Molecule the smallest particle of a specific element or compound that retains the chemical properties of that element or compound.

Multi-disciplinary approach using skills from other professions or trades to overcome problems outside the normal scope of your skill set, trade or profession to reach satisfactory solutions, conclusions or outcomes.

Multi-storey tall building that requires boosting or pumping of the water supply pressure given its height.

Naphtha a waxy oil deposit that is present in natural gas in its unrefined state. It is removed and later reused in other products such as cosmetics.

Noggin a term often used on-site to describe a piece of wood that supports or braces timber joists or timber-studded walls. They are particularly common in timber floors as a way of keeping the joists rigid and at specific centres, but they can also be used as supports for appliances such as wash hand basins and radiators that are being fixed to plasterboard.

Nominal capacity (of a cistern) the total volume it could hold when filled to the top of the cistern.

Overheads costs that include such things as site offices and staff salaries.

Overload where more is drawn into a circuit than the circuit is designed for. This is usually caused by misuse such as plugging in too many appliances.

Parasitic circulation circulation that occurs within the same pipe; often called one pipe circulation. It generally occurs in open vent pipes that rise vertically from the open vented hot water storage cylinder. The hotter middle water rises up the vent pipe, and the cooler water, towards the wall of the pipe, falls back to the cylinder. It can be a major source of heat loss from hot water storage cylinders.

Passive when a metal becomes passive, it means that an oxide film has formed that prevents further attack on the metal.

Plumb when pipework is perfectly vertical.

Portable appliance test (PAT test) the process of checking electrical appliances and equipment to ensure they are safe to use.

Potable drinkable, from the French word 'potable', pronounced 'poe-table'.

Primary and secondary water the primary water is the water that is in the boiler, central heating system and the heat exchanger of an indirect-type hot water storage cylinder/vessel. It is called the primary water because it is heated by the primary source of heat and hot water in the dwelling, namely the boiler. The pipes that connect the boiler to the heat exchanger are called the primary flow and the primary return. The secondary water is the stored water in the cylinder itself that is delivered to the hot water outlets and taps. The primary water heats the secondary water indirectly via the heat exchanger.

Prohibit prevent or forbid by law.

Qualitative method divides risks into categories such as low, medium and high.

Quantitative approach ranking a risk with a number.

Rectification putting something right, correcting.

Regulatory body an organisation set up by the government to monitor, control and guide various sectors within industry.

Resistivity the measure of the resistance of a 1 m^3 block of a material at 20°C; it is measured in ohm-metres (Ω m).

Rippling an unwanted, wavy pattern made on the inside face of a machine bend when the bending arm roller is not tight enough.

Risk calculation formula this is a method of using a formula of multiplying likelihood by consequences to provide a number that quantifies the level of risk for a particular job.

Rosin a natural solid, resin-type material obtained from pine trees, which, when heated, forms acidic particles that can irritate the breathing. This could lead to occupational asthma.

Schmutzdecke 'schmutzdecke' comes from the German word meaning 'dirt cover'.

Sealed heating systems heating systems that are sealed from the atmosphere and operate under pressure. They do not contain a feed and expansion cistern. Instead, they have an expansion vessel to take up water expansion and a filling loop to fill the system from the cold water main.

Semi-gravity system a central heating system that has pumped heating circulation but gravity hot water circulation.

Sheeting out sheeting out a tower scaffold means covering the outside of the scaffold with tarpaulins. This can be extremely dangerous as the tarpaulins act like the sails on a ship and could easily blow the scaffold over.

Short circuit a fault between line and neutral.

Single feed, self-venting indirect cylinder often referred to as the 'Primatic' cylinder, which is a trade name of IMI Ltd. Another version of this type of cylinder was also available and may be found in some existing installations. It was known as the 'Aeromatic'. It is slightly different from the Primatic because it has an air release valve on the side of the cylinder near the heat exchanger to bleed air from the heat exchanger.

Soakaway a pit, usually 1 m × 1 m × 1 m, dug into the ground and filled with gravel, into which the rainwater pipe discharges. It allows rainwater to soak naturally away to the water table. A soakaway should be situated at least 5 m away from the property.

Spigot another name for the plain end of a pipe. If the fitting we buy has a plain pipe end, we call this a spigot end.

Stratification in a hot water storage cylinder, water forms in layers of temperature from the top of the cylinder where the water is at its hottest, to the base where it is at its coolest. Stratification is necessary if the cylinder is to perform to its maximum efficiency and manufacturers will purposely design storage vessels and cylinders with stratification in mind. Designers will generally design:

- a vessel that is cylindrical in shape
- a vessel that is designed to be installed upright rather than horizontal
- a vessel with the cold feed entering the cylinder horizontally.

Sweating the term sweating a joint refers to the process of fluxing, heating and soldering the joint to create a watertight connection.

Temper the temper of a metal refers to how hard or soft it is.

Tender to submit a price or quotation for a job or contract.

Terminal the terminal of a flue system is the last section of the flue before the flue gases evacuate to the atmosphere. Different boilers and fuels require different terminals.

Throat the inside face.

Throating a slight indentation that the bending machine makes when the bend is formed.

Toolbox talk a toolbox talk is an informal meeting to deal with matters of health and safety in the workplace and safe working practices. They are normally short meetings conducted on-site before the commencement of the day's work activities. Toolbox talks are an effective way of refreshing operatives' knowledge and communicating the company's health and safety culture.

Turbidity the cloudiness or haziness of water caused by particles that are usually invisible to the naked eye. Turbidity is a key test of water quality.

Velocity the measurement of the rate at which an object changes its position. In order to measure it, we need to know both the speed of the object and the direction in which it is travelling. It is measured in metres per second (m/s).

Water hammer caused by a rapid opening and closing of the float-operated valve. As the water nears the water level in the cistern, the ball valve can begin to bounce quickly up and down and from side to side. This causes the noise to travel down the pipework, resulting in reverberation or a whining noise. It can also be caused by a faulty washer or diaphragm.

Wavering out the process of water in traps in appliances moving with a wave-like motion because of pressure fluctuations due to exposure to winds.

Wheel and axle a mechanical device used to wind up weight; includes a grooved wheel, turned by a cord/chain, and a rigid axle.

Work programme a very detailed document used on projects to record and assess activity against expected time to complete the project. For example, it might highlight that poor quality of work and low safety standards could apply to someone completing work ahead of schedule. It could also demonstrate that very slow progress on a job would impact on labour costs. The competence of the plumber is very important and their performance must be assessed carefully.

Zinc chloride a corrosive substance that can cause skin irritation, burns and eye damage if it gets in the eye.

Zoning a process where living spaces and sleeping spaces are individually controlled via independent time clocks, room thermostats and motorised zone valves.

Index

Note: page numbers in **bold** indicate location of key term definitions.

A

ABS 132
  soil/vent pipes 100–1, 501
AC (alternating current) 550–1
ACAS **180**
acceleration **151**
  due to gravity **151**
accent **178**
accidents **16**
  preventing 16–26
  recording and reporting 40
  responding to 33–40
  tools involved in 76–7
accumulators 226–7, 314, 315–16
acetylene 47, 48
  oxyacetylene equipment 48–50
actual capacity (cisterns) **234**
acute injury **30**
adhesion **141**
advisory (guidance) **15**
aeration of water 211, 373
aesthetically pleasing **109**
air admittance valves (AAVs) 465–6
air gaps, backflow prevention 243–9
air infiltration **399**
air release valves, radiators 390, 423
air source heat pumps 419
airlocks 271–2, 355
aluminium guttering systems 440
amp (and milliamp) **38**
annealing 75, **76**, 85–6
anodising 138, **139**
anti-freeze solution (glycol) 139, 141
anti-gravity valves 394
appointed person 34
aquifers 203
Archimedes' screw 157, 536
architects 166
asbestos **24–6**, 515
  regulation 9
atoms **543–4**
automatic air valves 393
automatic bypass valves 393
automatic disconnection of supply (ADS) 552
average conditions **551**

B

back siphonage 244
  air gap prevention devices 243–9
  DC pipe interrupter 249–50
  HC diverter with automatic return 255
  risks of 240–1
backflow prevention 240–55
  air gaps as method of 243–9
  Document G Guidance 291
  mechanical devices for 250–5
  non-mechanical 249–50
  point-of-use protection 242
  whole site protection 241–2
  zoned protection 242
backflow risks, appliances 241, 348
balanced forces in equilibrium 159–60
balancing
  central heating 393
  secondary circulation 352
base units 128
bathrooms
  layout 484–5
  pipework installation 269–70
  refurbishment 508–9
baths 476–7
  backflow risks 348
  dressing 504, 505
  installing and testing 506
  removal of cast iron 514
  tap hole and waste 477–8
  traps 487
  whirlpool/air spa 478
benchmarking **121**
bending
  copper tube 81–6
  low carbon steel pipe 93–5
  polybutylene pipe 98–9
Bernoulli effect/principle 154, **317**
bidets 475–6
  backflow risk 348
  connections to 269
biomass 419, 528, 529
  delivery requirements 537
  restrictions on use of 538–9
  storage requirements 535–6
black water 208
bladder (bag) type
  accumulators 226–7
  expansion vessels 329–30
blockages
  in guttering 456
  sanitary systems 517–18
blowing **119**
boiler cycling 365, **366**
boiler interlock **369**, 401, 406
boilers
  'combi' 317–18, 363, 382
  condensing boiler pipework 497–8
  flue systems for 385–7
  gas burning 380–4
  low loss headers 396, 397–8
  management systems 405
  oil-fired 384–5
  replacement standards 402
  solid fuel appliances 378–80
boiling point of water 139
  and pressure 327, 328, 337, 374
bonding 552–4
  equipotential **46**
boost mixing valves, shower 262–3
booster/boosting pumps
  break cisterns 237–8
  multi-storey buildings 220–3
  private water supplies 225–6
  replacing 280
  shower 264, 344–5
boreholes 203, 224, 225
bottle traps 487
bottled gases 47–52
Boyle's law 143, 331–2
brass
  compression fittings 97

de-zincification of 136
pipe clips 110
British Standards 176
central heating systems 362, 388, 426
cold water systems 212, 213
fuel systems 526, 531
hot water systems 292, 322, 349
pipe threads 76
rainwater systems 437, 441, 444, 446
sanitation systems 484–5, 499
BSP/T (British Standard Pipe/Thread) **76**
buffer tanks 398
building control officer 167–8
Building Regulations 12–13
central heating systems 361–2
cold water systems 213
Compliance certificates 121, 354
hot water systems 285–6, 353
rainwater systems 442, 446
sanitation systems 498
building services industry
information sources 172–6
legislation specific to 13
role of construction team 164–72
site responsibilities 181–4
burns, treating minor 35

C

C-plan plus (two-pipe) semi-gravity system 368
C-plan (two-pipe) system 366–7
cables, electric 544–5, 561–4
calorific value (CV) **525**
cancellation rights, contracts 174
capacity (of a cable) **545**
capillary action, 'S' traps 492
capillary attraction 140
capillary fittings 87–8
cast iron
boilers 381–2
guttering 439–40, 454, 455–6
health and safety 515
waste pipes 501
CDM Regulations (2015) 10–11, 166
Celsius scale 144
central heating systems 361–435
alternative designs 376–8
appliances and boilers 378–85
decommissioning 431–2
district heating 423–4
electrical controls 400–5
filling and venting 422–3
flue systems 385–7
fuel selection 423
'green' technologies 417–22
heat emitters 387–92
information sources 361–2
installing and testing 424–7
layouts and types 363–76
maintenance 427–31
mechanical components 392–400
system design and control 405–7
underfloor 408–13, 414–17
zoning 413–14
centralised hot water systems 294
'combi' boilers 317–18
instantaneous multi-point 317
open vented systems 295–308
solar thermal 319–20
thermal stores 318–19
unvented systems 308–16
centre to centre **107**
ceramic 132, 468
disc taps 261–2, 278
cesspits 496
chamfer **101**
Charles's law, gases 143
chases in walls 107
chemicals 21–6
CHP (combined heat and power) 529
chronic injury **30**
circuit breakers (CBs) 557
lock out devices 45, 559
circuits, electrical 546–60
circulating pumps 372–3, 412
replacing 427–8
cisterns 230
break 237–8
float operated valves 231–2, 258–60
installation requirements 230–2
large-scale 233–5
materials for 233
multi-storey buildings 220–3
multiple installations 235–7
overflow/warning pipe 233
stagnation prevention 232
water regulations 229–30
clerk of works (CoW) 167
clients 165–6, 181–2
coal 527, 532–3
codes of practice 15, 176, 484, **500**
cohesion **141**
coke 527
cold water systems 201–84
backflow protection 240–56
commissioning 272–3
decommissioning 280–1
frost protection 238–9
information sources 212–14
installation 256–73
maintenance 274–80
multi-storey dwellings 220–3
private water supplies 223–9
regulations 240
sources of water 202–3
storage cisterns 229–38
supply to dwellings 206–8, 214–17
testing 271–2
water distribution 211–12
water treatment process 208–11
colour coding
electric cable insulation **544**
of pipework 256
combination ('combi') boilers 317–18, **363**, 382
combustion 52
and the need for flues 385–7
communication 177–9
with the client 181
effects of poor 180–1
competent persons **6**
Compliance certificates 121, 354, 513
comply **3**
composite valves 333

compression **463**
and loss of tap/trap seal 491
compression fittings/joints
brass 97
copper tube 88–9
low carbon steel 96
polybutylene pipe 99
polyethylene 97
conciliation **180**
condensate **497**
condensing boilers 381, 383
low loss headers 396, 398
termination of condensate 497–8
conduction 147–8
conductivity 134
conductors 544–5
construction site staff 164–72
controls
central heating 400–5
hot water systems 326–33
space heating zones 413–14
convection 148–9
cordless tools 44
corrosion (of metals) **130**, 136–7
protection 138–9, 307, 399–400
COSHH Regulations (2002) 4–5
CPR (cardiopulmonary resuscitation) 38–9
craft operatives 170–1
cranes 33
crimping **79**
cross-connection 255–6
cross-flow 463–4
CSCS card **181**
current 542 *see also* electricity
customers 181–2
communicating with 181–2, 280–1
handover to 121, 174, 354, 514
cuts, treating minor 35
D
DB pipe interrupter 252–3
DC (direct current) 551
DC pipe interrupter 249–50
de-zincification of brass 136
dead leg 322
decommissioning 122, 515–16
central heating systems 431–2
cold water systems 280–1
hot water systems 354–5
rainwater systems 456
sanitary systems 514–15
delegation **186**
derived units **127**, 128
deviate **505**
dew point **383**
dialect **178**
discharge pipework
Document G Guidance 288
sanitary systems 460–6
unvented hot water systems 333–7
district heating 387, 423–4, 529
diverted neutral currents 553, **554**
documents, on-site 172–4
downtime **560**
drainage systems
above-ground 498–9
below-ground 492–6
heath hazards 516
dressing (an appliance) 390, **504**, 505
drinking water filters 266
duty of care **229**
duty holders **7**, 11
dynamic water pressure 153, 315
E
ear defenders 30
earthing 552
arrangements 554–6
electric shock 41
dealing with 38
protection against 550–4
electric storage heaters 423
electrical conductivity 134
electricity
circuits 546–9
conductors and insulators 544–5
health and safety 40–6
isolation of 45–6, 280, 558–60
measuring 549–50
principles of 542–4
protective devices 556–8
regulations 6, 12, 13
systems of supply 554–6
wiring 561–4
electrolyte **136**
electrolytic corrosion 399
hot water storage cylinders 307
electrolytic water conditioner 265
electron flow 543–4
energy 144, 149–50
engineers, specialist 166–7
environmental technologies 417–22
equilibrium 159–60
equipment *see also* tools
regulations 3–4
equipotential bonding **46**
erroneous **213**
Essex flange 324
evaporation 491
excavations, safe working 63–4
expansion
and contraction 148, 342–3, 424, 511
joints 399
vessels 329–32, 394, 422
explosion risks 537–8
exposed conductive parts **552**
extraneous conductive parts 553, **554**
eye injuries/protection 28, 36
F
fan-assisted room sealed boilers 386–7
fan convectors 391
filling
central heating systems 426–7
of closed circuits, guidance 291
initial system fill 350
loop, sealed system 396, 422, 423, 427
filtration of water 209–11, 266
fire safety 52–3
fire stop **119**
fireclay 132, 468
first aid
provision 33–4
regulations 9
fixes, installation 114–15, 186–7

flashback **49**
float operated valves 231–2, 258–60
repairing/replacing 279–80
float switches 227–8
floc **210**
floorboards, lifting 104–5
flow rate 151, 154–5
central heating systems 398
checking hot water 351
guttering 444
flue systems 385–7
fluid categories 204–6
flushing
central heating systems 429
hot water systems 350–1
urinals 480–2
WCs 469–72
fluxes 23, 87–8
force 151–2, 155–7, 158, 159–60
forced draught flues **386**, 387
frost protection 238–9, 404
fuel oil 141, 526
storage of 533–4
fuels for domestic use 524–30
installation regulations 531–2
reasons for choosing 530–1
storage requirements 532–9
full-way gate valves 257
fully pumped system **363**
fume exposure 36–7
functional controls 328–9
fuses 556–7
G
galvanic corrosion 136
Gantt charts 172–3
gas central heating boilers 380–4
gas fuels
LPG 525–6, 534–5
natural gas 142, 423, 524–5
gases 142–3
bottled 47–50
refrigerants 141
safety regulations 13, 291
specific gravity 130
glycol 141
gravity
centre of 158
drainage systems 498–9
water distribution by 211
gravity circulation **363**
convection 148–9
semi-gravity heating 364–8
grey water 206–7, **518**
harvesting 207–8, 421–2, 518–20
ground source heat pumps 418–19
underfloor heating 408, 411
guardrails 61–2
Guidance Note GS38 **559**
guttering systems 437–40
size and type 440–5
thermal expansion 445
H
hand tools 69–77
handover to customer 121, 174, 354, 514
hardness
of materials 134
water 140, 265
hazards **1**
COSHH regulations (2002) 4–5
drainage systems 516
electrical 40–6
identifying 16–26, 190
protective equipment 26–30, 191
head (static water pressure) 152–3
health and safety
accidents 16–26, 33–40
cold water systems 233, 274
confined spaces 64–5
electricity 40–6, 544, 550, 551, 554, 558
excavations and trenches 63–4
hand and power tools 76–7, 79
hazardous substances 16–26
heat-producing tools 47–53
injuries, dealing with 35–9
inspectors 171
ladders 54–8
Legionella risk 353
legislation and regulations 1–16
manual handling 30–2
mechanical lifting aids 32–3
personal protective equipment (PPE) 26–30
rainwater systems 446–7, 448, 454, 455, 456
risk assessments 17–18, 189–95
sanitary systems 475, 514–15, 516, 517
scaffolds and platforms 58–63
Health and Safety Executive (HSE) **10**
Health and Safety at Work etc. Act (1974) 2–3
hearing protection 30
heat emitters 387–92
heat exchangers **299**
cast iron, in boilers 381–2
hot water storage cylinders 298–9, 301–2
solar thermal panels 320
thermal stores 318–19
heat pumps 418–19, 528
heat transfer methods 147–9
height, working at 446–7
access equipment 54–63
regulations (2005) 7
hot water systems 285–360
centralised 295–320
choice of 293–4
components and controls 326–37
decommissioning 354–5
factors affecting 292–3
fault finding 355–6
installation 341–9
layouts and types of 247, 294
localised 320–2
regulations 285–90
safety features 337–40
secondary circulation 322–6, 352
testing and commissioning 349–54
hydraulic bending machine 78, 93
hydraulic flush control valve 482
hydrogen boilers 384

I

IET Wiring Regulations 13, 291
immersion heaters **297–8**, 341–2, 355
incidents, reporting 5–6
induced siphonage 490
industry standards 291–2
infrared thermometers 145, 351
inhibitors
  corrosion 399–400
  limescale 265
injuries **30**
  dealing with 35–9
  reporting of 5–6
inspectors 15–16, 171–2
insulation
  electric cables 544–5
  pipework 238–9, 325, 342
  storage cylinders 305
integral solder rings 87
interconnected cisterns 235–7
intumescent collars 119
IP2X and IP4X **551**
  iron 130 *see also* cast iron
  threaded fittings 95
  isolation (of services) 432 *see also* decommissioning
  cold water supply 272, 274
  electricity 45–6, 280, 558–60
  valves 257–8

J

job specifications 172, 185–6
jointing methods
  copper tubes 86–91
  guttering 437–40, 452
  low carbon steel pipe 95–6
  plastic pipes 97–8, 100–1
  sanitary appliances 209
joists
  and lifting floorboards 104–5
  notching and drilling 105–7
  polybutylene cabling 115
  suspended timber floors 267–8, 416

K

Kelvin scale 145
kerosene 141, 526, 533–4
kinetic lifting 31

L

ladders 54–8, 447
  foot a ladder **57**
large-scale cisterns 233–5
lead 131
  leadlocks 102
  regulation 8–9
  working with 22–3, 270, 515
leakages
  acetylene cylinder 50
  cold water systems 272
  gutters, repairing 453–4
  hot water systems 352
  Legionella risk, hot water systems 353
legislation **1**, 176
  accident reporting 5–6, 40
  building services-specific 13
  construction-specific 10–13
  first aid provision 33–4
  health and safety 1–10
  smoke control 530
levers 155–6
lifting aids, mechanical 32–3
lifting techniques 31–2
limescale inhibitors 265
liquids, properties of 139–42
local authority 167–8
  notifying of works carried out 353, 512–13
localised hot water systems 320–2
low carbon fuels 528–30
low carbon steel pipes 92–6, 111, 120, 425
low loss headers 396–8
LPG (liquid petroleum gas) 50–1, 142, 525–6
  storage 51–2, 534–5
lubricants 141–2

M

macerators 482
main earthing terminal (MET) 553, **554**
main protective bonding conductor 553
maintenance **122**
  central heating systems 427–31
  cold water systems 274–80
  guttering 453–6
  sanitation systems 516–18
  work programme 185
'making good' 123
mandatory (guidance) **15**
  safety sign 20
manifolds 411–12, 414
  microbore 376, 377, 425
manual handling 30–1
  regulations 8
manufacturers' instructions 176
  cold water systems 213–14, 274
  hot water systems 292
materials
  delivery 186
  guttering 437–40, 452
  pipe insulation 239
  pipework 425
  properties of 129–43
  sanitary appliances 467–8
  sanitary pipework 500–1, 514–15
  storage of 108–9
matter, states of 146
mCHP (micro-combined heat and power) 419–20
measurement units 127–9
mechanical advantage 155, 156, 158
mechanical principles 155–60
mechanical suction **496**
metals 130–1
  corrosion 135–7, 138–9
  properties 132–4
method statement 18, **195**, 196
microbore system 376–7
moments of a force (torque) 158
momentum 492
motion: action and reaction 158–9
motorised valves 404
multi-disciplinary approach **167**
multi-section ladders 55–6
multi-storey buildings **220**
  cold water systems in 220–3
  large-scale cisterns 233–7

multiple installations
boilers 397
cisterns 235–7
self-sealing traps 489
waste appliances 466, 490
MuPVC soil/vent pipes 100–1, 501

N

naphtha **525**
natural draught flues 386, 387
natural gas 142, 524–5
noggins 105, **113**
noise issues
cold water systems 271
hot water systems 355–6
nominal capacity (of a cistern) **234**
non-condensing boilers 381, 382
notification procedure 353, 512–13

O

Ohm's law 543, 546–7, 548, 558
oil 423, 526, 532
storage of 533–4
oil-fired boilers 295, 317, 384–5
one-pipe system, central heating 365–6
open vented central heating systems 364–74
open vented hot water storage systems 295–6
cylindered 303–7
direct 295–8
faults with 355–6
indirect 298–303
initial system fill 350
pipe sizes 307–8
secondary circulation installations 323–4
vent pipes 337
vs unvented 316
outriggers 60
over sink water heaters 321, 322
overheads **168**
overload (of a circuit) **545**
protection against 558
oxidation of metals 135
oxidative degradation of plastics 138
oxyacetylene equipment 48–50
oxygen, bottled liquid 47

P

parasitic circulation **295–6**
passive (of a metal) **135**
PAT test **6**
peat 528
performance testing 510–11
personal protective equipment (PPE) 26–30
regulations (1992) 4
pipe gain 85
pipe interrupters 249–50, 252–3
pipework
bending and jointing 80–102
cold water systems 266–70
colour coding 256
installation techniques 114–23
sanitary 460–7
plastics 131–2
degradation of 138
sanitation pipes 100–2
plugs, wiring 562–3
policies and procedures 173, 174–5
polybutylene (PB-1) pipe 98–9, 132
clippings/supports 111–12
expansion 342–3
installing 115, 425–6
life expectancy 266–7
testing 120, 271
polyethylene (PE) pipes 97–8, 132
polypropylene (PP) 102, 132
cisterns 233
push-fit waste pipes 102, 111, 501
portable appliance test (PAT test) **6**
potable **204**
power 149–50, 549
power flushing 429
power tools 77–9
PPE regulations 4
press-fit fittings, copper tube 90–1
pressure
gases 142, 143
water 152–3, 315
pressure filters 210–11
pressure gauge 396
pressure reducing valves 315, 329
pressure relief valves 332, 395–6
pressure testing 426
primary water **298**
private water supply (PWS)
components used 225–9
regulations (2016) 213
system layouts 223–5
professional authority, limits to 175
programmers, heating control 403
progress of work, monitoring 189
prohibition **9**
propane 47, 51, 52, 525–6
proving units 559
pulleys 156–7
pumps *see also* booster/boosting pumps
circulating 372–3, 412
heat 418–19, 528
private water supply 225–6
shower 264, 344–5
push-fit fittings/joints
benefits and drawbacks 501
for copper tube 89–90
plastic 97–8
polybutylene pipe 99
polypropylene waste pipe 102
soil/vent pipes 101
PVC cables 563–4
PVCu 100
clipping distances 111
gutter installation 448–52
gutter leakage 453–4
guttering systems 437–9
jointing methods 100–1
replacing cast iron gutters with 455–6
soil pipe fittings 101–2
thermal expansion 445

Q

qualitative method **193**
quantitative approach **192**
'quick recovery' cylinders 304–5

R

radiation 149
radiators 388–92
faults, rectifying 430–1
positioning 426
replacing 428

valves 392–4, 428–9
versus underfloor heating 408
rainfall intensity 440–1
rainwater cycle 201–2
rainwater harvesting 207–8, 420–1, 518–20
rainwater systems 436–59
decommissioning 456
installation 446–53
layouts 436–45
maintenance 453–6
testing 457
recovery position 39
rectification **115**
recycling of water 207–8, 421–2, 518–20
refrigerants 141
regulatory bodies **531–2**
relative density/specific gravity 129–30, 139
renewable energy
hot water cylinders 301–2
low-carbon fuels 528–30
technologies 417–22
residual current device (RCD) 558
resistance, electric 542–3
in series and parallel 546–9
resistivity **544**
respirators 28–9
reversed return system 377–8
rippling 75, **76**
risk assessments 17–18, 189–95
risk calculation formula **192**
roof area 441–2
room sealed (balanced) flues 386–7
rosin **23**
rust formation 135–6

S

S-plan heating system 370–1, 406
S-plan plus heating system 374–5
safety issues *see also* health and safety
fuel storage 537–8
hand tools 70, 71, 76–7
hot water temperature 327–8, 337
open vent pipe 374
power tools 79
safety signs 19–20
regulations 8
sand filters 209–10
sanitation systems 460–523
appliances 467–98
grey water recycling 518–20
installation of 498–516
pipework 460–7
servicing/maintenance 516–18
scaffolds 58–61, 446
schmutzdecke **210**
scientific principles 127–50
electricity 542–4
energy, heat and power 144–50
force and pressure 151–5
mechanical principles 155–60
properties of materials 129–44
SI measurement units 127–9
screwdrivers 69–70, 564
sealed heating systems **374–6**
components of 376
expansion in 424
filling 396, 427
secondary circulation 322–6, 352
secondary water **298**, 299
sedimentation 209
semi-gravity heating systems **364–8**
and anti-gravity valves 394
septic tanks 196, 496
service valves 115, 236–7
sheeting out **59**
short circuit **558**
shower mixer valves 262–4
installation of 343–7
maintenance of 355
showers
bases/trays 478–9
cubicles/enclosures 479
pumps 264
single check (non-return) valve 329
single feed, self-venting indirect cylinder **302–3**
sink waste disposal units 484
sinks 479–80
siphonage 489–90
siphonic action 153–4
for WC flushing 471, 472
siphonic type WC pan 469–70
skirting heating 392
sludge
corrosion inhibitor 399–400
power flushing 429, 431
problems of 399
and water treatment 209, 210
snagging 115
soakaway **495**
soil pipe/stack 460
connection to the drain 503
jointing methods and fittings 100–2
pipework systems 461–6
unblocking 517–18
waste pipe connections to 502
solar photovoltaic system 418
solar thermal systems 319–20, 417–18, 528
soldering
capillary fittings 87–8
equipment 75
fire precautions 52, 118
risk calculation 192–3
solid fuel 423, 526–8, 532
appliances 378–80
solids
applications of 130–2
corrosion 135–7, 138–9
degradation (plastics) 138
properties of 132–4
solvent weld 501
joints 100
waste pipe fittings 102
solvents, working with 23–4
soundness testing
central heating pipework 425
hot water systems 350
rainwater systems 457
sanitation systems 510
specific gravity 129
of gases 130, 525, 526
of water 129, 139
specific heat capacity 139, 144, 149–50
spigot **101**

spring bending, copper tube 81, 85–6
stagnation prevention, cisterns 232
static water pressure (head) 152–3, 315
steam 140, 142, 146–7
sterilisation, water treatment 211
stop valves 215–17, 257, 258, 272
stratification **343**, 344
strength of a material 132–3
stub stack system 465
supervisors 168–9, 175, 182–4
supplementary bonding 553
surveyors 166, 171
sweating a joint **23**
symbols, plumbing 122–3
T
taps 260–2
  backflow prevention 248–9, 254–5
  holes for 474, 477
  installing 504
  maintenance tasks 275–9
  stop taps 216–17, 257
temper (of a metal) **81**
temperature 144–5
  checking hot water 351
  of gases 143
  measuring 145–6
  sensors 229
temperature relief valves 337
  Document G Guidance 288, 290
temporary continuity bonding 46, 432, 553
tender **187**
terminal (of a flue system) **386**
termination
  condensing boiler pipework 497–8
  discharge pipework 335–7
  electric cables 561–2, 563–4
testing procedures 119–20
  central heating systems 425–6
  hot water systems 349, 350
  PAT testing 43
  rainwater systems 453, 457
  sanitation systems 510–11
thermal conductivity 134, 148, 239
thermal degradation, plastics 138
thermal radiation 149
thermal stores 295, 318–19
thermistors 145
thermocouples 145
thermometers 145–6
thermoplastic cables 563–4
thermoplastics 131
thermostatic mixing valves (TMVs) 337, 339–40
  group installations 338–9
  showers 263–4
  single installations 338
  underfloor heating 412
  uses of 342
thermostatic radiator valves (TRVs) 392
thermostats
  checking 355, 356
  frost and pipe 404
  high-limit 341, 370
  immersion heaters 298, 341–2
  room 403
  safety controls, hot water 327–8
  thermo-mechanical 393–4
throat **84**
throating 75, **76**, 84
time allocation, work tasks 186–7
time clocks 324–5, 403
toe boards 61–2
toolbox talk **189**
tools 69–80
  regulations 3
  storing 108–9
  wiring jobs 561, 563, 564
torque 158
towel warmers 391, 391–2
trace heating 326
trades, construction site 169–71
transducers 228
traps 485–9
  cleaning out 516–17
  reasons for loss of seal 489–92
  testing seal depth 510–11
trenches, working in 63–4
tubular scaffolds 60–1
tubular traps 486–7
tundish, discharge pipework 332–3
turbidity **209**
U
unconscious people 37–8
under-sink heaters 311
underfloor heating systems 408–13
unit conversion tables 128
unvented hot water storage systems 308–11
  accumulators 314, 315–16
  commissioning 349–50
  cylinders 311–14
  discharge pipework 333–7
  functional controls 328–33
  pipework arrangement 311
  secondary circulation installation 323
  shower mixing valves 346–7
  vs open vented 316
unwholesome water 206–8
urinals 480–2
UV (ultraviolet) degradation 138
V
variation orders 187–8
velocity **151**, 154
vent pipes
  hot water systems 295, 307–8
  open 337, 372, 374
  sanitary systems 460–6
venturi mixing shower valve 262–3
voltage (V) 542, 543
  colour coding for site 41–2
  drop in 545, 547
  indicators 45–6, 559, 560
volume of a gas 143
W
W-plan heating system 370, 372, 376, 404
warm air systems 423
washbasins 472–5
  backflow risks 348
  installation of 506–7
  mixing valves 338–9
  traps 485, 487, 489
  waste pipes 462
waste arrangements

baths 477–8
shower trays 479
wash basins 474–5
waste disposal units 483, 484
waste pipes 462–7, 495, 501
connecting to soil stack 502
unblocking 518
waste water lifters 483
Water Act (2003) 206
water conditioners 265
water distribution 211–12
pipework to dwelling 214–17
water filters, drinking water 266
water hammer **259**
water hardness/softness 140
water inspectors 171–2
water meters, fitting 216
water pressure 152–3, 315
checking 351
low due to airlocks 271–2
and poor mains supply 314–15
transducers 228
water properties 139–41
fluid categories 204–6
water softeners 265–6
water sources 202–3
Water Supply (Water Fittings) Regulations (1999) 13, 240
backflow risk 242–3
cold water systems 212–13, 217, 229–30
frost protection 238–9
hot water supply 286–91
water treatment 208–11
scale reduction 264–6
wavering out **491**
WCs 469–72
blocked 518
cistern assembly 505
installation of 507–9
macerators 482
weather compensation controls 404–5
wells, water from 203
wet central heating 363–76
see also radiators
boilers for 383–4
minimum controls 401
wheel and axle **156**
wholesome water 204
wiring
IET regulations 13, 291
plugs 562–3
systems, installing 561–4
work permits 18
work programmes **184–9**
working drawings 122–3, 172
working platforms 62–3
working policies/procedures 174–5
workplace conflicts 179–80

Y

Y-plan heating system 369–70, 372, 375, 376

Z

zinc 131, 136
zinc chloride **23**
zone valves
C-plan plus) 368
S-Plan 370–1, 406
S-plan plus 374–5
zoning **370**, 401, 413–14